Periodic Table of the Elements

H 1																	He 2
Li 3	Be 4											B 5	C 6	N 7	O 8	F 9	Ne 10
Na 11	Mg 12											Al 13	Si 14	P 15	S 16	Cl 17	Ar 18
K 19	Ca 20	Sc 21	Ti 22	V 23	Cr 24	Mn 25	Fe 26	Co 27	Ni 28	Cu 29	Zn 30	Ga 31	Ge 32	As 33	Se 34	Br 35	Kr 36
Rb 37	Sr 38	Y 39	Zr 40	Nb 41	Mo 42	Tc 43	Ru 44	Rh 45	Pd 46	Ag 47	Cd 48	In 49	Sn 50	Sb 51	Te 52	I 53	Xe 54
Cs 55	Ba 56	*	Hf 72	Ta 73	W 74	Re 75	Os 76	Ir 77	Pt 78	Au 79	Hg 80	Tl 81	Pb 82	Bi 83	Po 84	At 85	Rn 86
Fr 87	Ra 88	#	Rf 104	Db 105	Sg 106	Bh 107	Hs 108	Mt 109	Uun 110	Uuu 111	Uub 112		Uuq 114				

* Lanthamides	La 57	Ce 58	Pr 59	Nd 60	Pm 61	Sm 62	Eu 63	Gd 64	Tb 65	Dy 66	Ho 67	Er 68	Tm 69	
# Actinides	Ac 89	Th 90	Pa 91	U 92	Np 93	Pu 94	Am 95	Cm 96	Bk 97	Cf 98	Es 99	Fm 100	Md 101	

Chemistry for Environmental Engineering and Science

The McGraw-Hill Series in Civil and Environmental Engineering

Engineering Economy
Blank and Tarquin: *Engineering Economy*
Humphreys: *Jelen's Cost and Optimization Engineering*
Riggs, Bedworth, Randhawa: *Engineering Economics*
Steiner: *Engineering Economic Principles*

Engineering Math and Statistics
Bannerjee: *The Boundary Element Methods in Engineering*
Barnes: *Statistical Analysis for Engineers and Scientists: A Computer-Based Approach (IBM)*
Ledbed: *Formulas for Structural Dynamics*
Milton and Arnold: *Introduction to Probability and Statistics: Principles and Applications for Engineering and the Computing Sciences*
Reddy: *Introduction to the Finite Element Method*
Rosenkrantz: *Introduction to Probability and Statistics for Scientists and Engineers*
Zienkiewicz and Taylor: *The Finite Element Method: Basic Concepts and Linear Applications*

Fluid Mechanics
Çengel and Turner: *Fundamentals of Thermal-Fluid Sciences*
Finnemore and Franzini: *Fluid Mechanics with Engineering Applications*
Streeter, Bedford, Wylie: *Fluid Mechanics*
White: *Fluid Mechanics*

Geotechnical Engineering
Atkinson: *Introduction to the Mechanics of Soils and Foundations*
Bowles: *Foundation Analysis and Design*
Bowles: *Engineering Properties of Soils and Their Measurement*

Numerical Methods
Chapra and Canale: *Numerical Methods for Engineers*
Heath: *Scientific Computing: An Introductory Survey*

Structures
Gaylord and Stallmeyer: *Design of Steel Structures*
Laursen: *Structural Analysis*
Leet and Bernal: *Reinforced Concrete Design*
Leet and Uang: *Fundamentals of Structural Analysis*
Leonard: *Tension Structures: Behavior and Analysis*
Lin and Cai: *Probabilistic Structural Dynamics: Advanced Theory and Applications*

Nilson: *Design of Concrete Structures*
Nowak and Collins: *Reliability of Structures*
Taly: *Design of Modern Highway Bridges*
Taly: *Reinforced Masonry Structure Design*

Surveying
Anderson and Mikhail: *Surveying: Theory and Practice*
Wolf and DeWitt: *Elements of Photogrammetry (with Applications in GIS)*

Statics, Dynamics, and Mechanics of Materials
Barber: *Intermediate Mechanics of Materials*
Beer and Johnston: *Vector Mechanics for Engineers: Statics*
Beer and Johnston: *Vector Mechanics for Engineers: Dynamics*
Beer and Johnston: *Vector Mechanics for Engineers: Statics and Dynamics*
Beer and Johnston: *Mechanics of Materials*
Young: *Roark's Formulas for Stress and Strain*

Construction Engineering and Project Management
Raymond E. Levitt, Stanford University, Consulting Editor

Barrie and Paulson: *Professional Construction Management*
Bockrath: *Contracts and the Legal Environment for Engineers and Architects*
Callahan, Quackenbush, Rowlings: *Construction Project Scheduling*
Griffis and Farr: *Construction Planning for Engineers*
Hinze: *Construction Contracts*
Oberlender: *Project Management for Engineering and Construction*
Peurifoy, Ledbetter, Schexnayder: *Construction Planning, Equipment, and Methods*
Peurifoy and Oberlender: *Estimating Construction Costs*

Transportation Engineering
Edward K. Morlok, University of Pennsylvania, Consulting Editor

Banks: *Introduction to Transportation Engineering*
Horonjeff and McKelvey: *Planning and Design of Airports*
Kanafani: *Transportation Demand Analysis*
Meyer and Miller: *Urban Transportation Planning*
Wells: *Airport Planning and Management*

Water Resources and Environmental Engineering
George Tchobanoglous, University of California, Davis, Consulting Editor

Bailey and Ollis: *Biochemical Engineering Fundamentals*
Benjamin: *Water Chemistry*
Bishop: *Pollution Prevention: Fundamentals and Practice*
Canter: *Environmental Impact Assessment*
Chanlett: *Environmental Protection*
Chapra: *Surface Water Quality Modeling*
Chow, Maidment, Mays: *Applied Hydrology*
Crites and Tchobanoglous: *Small and Decentralized Wastewater Management Systems*
Davis and Cornwell: *Introduction to Environmental Engineering*
deNevers: *Air Pollution Control Engineering*
Eckenfelder: *Industrial Water Pollution Control*
Eweis, Ergas, Chang, Schroeder: *Bioremediation Principles*
Freeman: *Hazardous Waste Minimization*
LaGrega, Buckingham, Evans: *Hazardous Waste Management*
Linsley, Franzini, Freyberg, Tchobanoglous: *Water Resources Engineering*
McGhee: *Water Supply and Sewage*
Metcalf & Eddy, Inc.: *Wastewater Engineering: Collection and Pumping of Wastewater*
Metcalf & Eddy, Inc.: *Wastewater Engineering: Treatment and Reuse*
Peavy, Rowe, Tchobanoglous: *Environmental Engineering*
Rittmann and McCarty: *Environmental Biotechnology: Principles and Applications*
Rubin: *Introduction to Engineering and the Environment*
Sawyer, McCarty, Parkin: *Chemistry for Environmental Engineering and Science*
Sturm: *Open Channel Hydraulics*
Tchobanoglous, Theisen, Vigil: *Integrated Solid Waste Management: Engineering Principles and Management Issues*
Wentz: *Safety, Health, and Environmental Protection*

Others Titles of Interest
Budynas: *Advanced Strength and Applied Stress Analysis*
Dally and Riley: *Experimental Stress Analysis*
Ugural: *Stresses in Plates and Shells*

Chemistry for Environmental Engineering and Science

Fifth Edition

Clair N. Sawyer

Late Professor of Sanitary Chemistry
Massachusetts Institute of Technology

Perry L. McCarty

Silas H. Palmer Professor Emeritus of Civil and Environmental Engineering
Stanford University

Gene F. Parkin

Professor of Civil and Environmental Engineering
University of Iowa

Boston   Burr Ridge, IL   Dubuque, IA   Madison, WI   New York   San Francisco   St. Louis
Bangkok   Bogotá   Caracas   Kuala Lumpur   Lisbon   London   Madrid   Mexico City
Milan   Montreal   New Delhi   Santiago   Seoul   Singapore   Sydney   Taipei   Toronto

McGraw-Hill Higher Education

A Division of The McGraw-Hill Companies

CHEMISTRY FOR ENVIRONMENTAL ENGINEERING AND SCIENCE
FIFTH EDITION

Published by McGraw-Hill, a business unit of The McGraw-Hill Companies, Inc., 1221 Avenue of the Americas, New York, NY 10020. Copyright © 2003 by The McGraw-Hill Companies, Inc. All rights reserved. Previously published under the titles of *Chemistry for Environmental Engineering.* Copyright © 1994, 1978 by The McGraw-Hill Companies, Inc. All rights reserved. *Chemistry for Sanitary Engineers.* Copyright © 1967, 1960 by McGraw-Hill, Inc. All rights reserved. No part of this publication may be reproduced or distributed in any form or by any means, or stored in a database or retrieval system, without the prior written consent of The McGraw-Hill Companies, Inc., including, but not limited to, in any network or other electronic storage or transmission, or broadcast for distance learning.

Some ancillaries, including electronic and print components, may not be available to customers outside the United States.

This book is printed on acid-free paper.

International   2 3 4 5 6 7 8 9 0 QFR/QFR 0 9 8 7 6 5 4
Domestic   7 8 9 0 QFR/QFR 1 5 4 3

ISBN   978-0-07-248066-5
MHID   0-07-248066-1
ISBN   978-0-07-119888-2 (ISE)
MHID   0-07-119888-1 (ISE)

Publisher: *Elizabeth A. Jones*
Sponsoring editor: *Suzanne Jeans*
Developmental editor: *Kate Scheinman*
Marketing manager: *Sarah Martin*
Senior project manager: *Jayne Klein*
Production supervisor: *Sherry L. Kane*
Media project manager: *Jodi K. Banowetz*
Senior media technology producer: *Phillip Meek*

Designer: *K. Wayne Harms*
Cover designer: *Scan Communications Group, Inc.*
Cover image: *Corbis*
Lead photo research coordinator: *Carrie K. Burger*
Photo research: *David Tietz*
Compositor: *UG / GGS Information Services, Inc.*
Typeface: *10/12 Times Roman*
Printer: *Quad/Graphics Fairfield, PA*

Library of Congress Cataloging-in-Publication Data

Sawyer, Clair N.
    Chemistry for environmental engineering and science / Clair N. Sawyer, Perry L. McCarty, Gene F. Parkin.
—5th ed.
            p.   cm.—(The McGraw-Hill series in civil and environmental engineering)
    Includes index.
    ISBN 0-07-248066-1 (acid-free paper)—ISBN 0-07-119888-1 (ISE)
    1. Environmental chemistry. 2. Environmental chemistry—Laboratory manuals.
3. Sanitary engineering. 4. Sanitary engineering—Laboratory manuals. I. McCarty, Perry L. II. Parkin, Gene F.
III. Title. IV. Series.

TD193 .S28 2003
628′.01′54—dc21

                                                                                                2002025494
                                                                                                    CIP

INTERNATIONAL EDITION ISBN 0-07-119888-1
Copyright © 2003. Exclusive rights by The McGraw-Hill Companies, Inc., for manufacture and export. This book cannot be re-exported from the country to which it is sold by McGraw-Hill. The International Edition is not available in North America.

www.mhhe.com

TO OUR FAMILIES Martha, Annette, Kyle, and Eric who sacrificed much for this current effort.

CONTENTS

PREFACE

Education in environmental engineering and science has historically been conducted at the graduate level, and up to the present time has drawn mainly on students with a civil engineering background. In general, education in civil engineering does not prepare a student well in chemistry and biology. Since a knowledge of these sciences is vital to the environmental engineer, the graduate program must be designed to correct this deficiency. In recent years, students from other engineering disciplines and from the natural sciences have been attracted to this field. Some have a deficiency in chemistry and biology similar to that of the civil engineer and need exposure to general concepts of importance.

A current trend in the United States is the introduction of an undergraduate environmental engineering option or degree program within civil engineering departments. These students also require an introduction to important concepts in chemistry and biology.

This book is written to serve as a textbook for a first course in chemistry for environmental engineering and science students with one year of college-level chemistry. Environmental professionals need a wide background in chemistry, and in recognition of this need, *Chemistry for Environmental Engineering and Science* summarizes important aspects from various areas of chemistry. This treatment should help orient the students, aid them in choosing areas for advanced study, and help them develop a better "feel" for what they should expect to gain from further study.

The purpose of this book is twofold: It (1) brings into focus those aspects of chemistry that are particularly valuable for solving environmental problems, and (2) it lays a groundwork of understanding in the area of specialized quantitative analysis, commonly referred to as water and wastewater analysis, that will serve the student as a basis in all the common phases of environmental engineering practice and research.

Substantial changes continue to occur in the emphasis of courses for environmental engineers and scientists. The trend is toward a more fundamental understanding of the chemical phenomena causing changes in the quality of surface and groundwaters, of waters and wastewaters undergoing treatment, and of air. This fundamental understanding of chemistry is absolutely critical as environmental professionals attempt to solve complex problems such as hazardous waste pollution, air pollution from emission of toxic compounds, radioactive waste disposal, ozone depletion, and global climate change.

Chemistry for Environmental Engineering and Science is organized into two parts. Part One is concerned solely with fundamentals of chemistry needed by environmental engineers and scientists. It includes chapters on general chemistry, physical chemistry, equilibrium chemistry, organic chemistry, biochemistry, colloid chemistry, and nuclear chemistry. Each emphasizes environmental applications. In this new edition, the chapters on general and physical chemistry have been updated, and new homework problems have been added. The chapter on equilibrium chemistry has been revised, with many new example and homework problems. The chapter on organic chemistry includes an added emphasis on organic compounds of environmental significance (e.g., chlorinated solvents). Sections are included on the behavior (fate) of organic compounds in the environment and in engineered systems and on the use of structure-activity relationships. The chapter on biochemistry has been updated. We feel that these revisions make the text even more suitable for lecture courses on environmental chemistry principles.

Part Two is concerned with analytical measurements. A new chapter has been added on statistical analysis of analytical data. All analytical procedures are subject to errors. There is a critical need for students to learn how to evaluate the uncertainties such errors present. This chapter discusses basic methods for evaluating and reporting uncertainties in measurements that are essential for analytical chemists, regulatory agencies, and environmental professionals who use analytical data to make important decisions.

The next several chapters contain general information on quantitative, qualitative, and instrumental methods of analysis, useful as background material for the subsequent chapters concerned with water and wastewater analyses of particular interest to environmental engineers and scientists. These chapters are written to stress the basic chemistry of each analysis and show their environmental significance. They should be particularly useful when used with "Standard Methods for the Examination of Water and Wastewater," published jointly by the American Public Health Association, American Water Works Association, and Water Environment Federation, and giving the details for carrying out each analytical determination. The final chapter stresses trace contaminants, many of which are determined analytically with instrumental procedures discussed in earlier chapters. A listing of U.S. Environmental Protection Agency drinking water standards and World Health Organization drinking water quality guidelines for various trace contaminants are also contained in this chapter. Part Two is considered to be most useful as lecture material to accompany a laboratory course in environmental chemistry. Revisions have been made in other chapters to reflect the many changes in "Standard Methods" that have occurred since the last edition of this text.

Problems are included at the end of most chapters to stress fundamentals and increase the usefulness of this book as a classroom text. Example problems throughout the text help increase the students' understanding of the principles outlined. In Part One of the book, where the emphasis is on chemical fundamentals, answers are included after many homework problems, allowing students to evaluate independently their understanding of the principles emphasized.

To meet textbook requirements, brevity has been an important consideration throughout. For those who believe that we have been too brief, we can only beg their indulgence and recommend that they seek further information in standard references on the subject. Important references are listed at the end of each chapter.

It is inevitable that we have made errors in producing this textbook. For this we apologize. Hopefully they are not so numerous that they impede the student's ability to learn the material. Fortunately, for this new edition, McGraw-Hill is providing a website where we can list errata that can be readily downloaded with no charge to students and faculty. A solution manual for text problems can also be obtained at this website, but by faculty members only. We hope also to use this website to post more example problems and their solutions. There was a request for such by reviewers, and this is one way that we can provide additional material without expanding the number of pages and costs for the text. The website for this textbook can be found at http://www.mhhe.com/sawyer. We would appreciate hearing from students and faculty when errors are found so that we can enter them in a timely manner on the website. Our e-mail addresses are included in the errata section of the website.

Special thanks are due colleagues at the University of Iowa—Michelle Scherer for specific suggestions to improve the text and generous help with new homework problems, and Pedro Alvarez, Keri Hornbuckle, Craig Just, Jerry Schnoor, and Richard Valentine for helpful discussions. Thanks also to Mark Benjamin of the University of Washington for e-mail discussions of activity corrections and other weighty matters. Finally, we wish to express our gratitude to William Burgos, Pennsylvania State University; Cindy Lee, Clemson University; Howard Liljestrand, University of Texas, Austin; John Pardue, Louisiana State University; and Andrew Randall, University of Central Florida, as well as the anonymous reviewers, all of whom were selected by the publisher to provide comments about the textbook and to provide recommendations for change. We appreciate the many thoughtful and detailed comments that were offered and used them extensively in preparing this revision. We hope that the reviewers and other faculty find the changes to be beneficial to them and to their students.

Perry L. McCarty
Gene F. Parkin

ABOUT THE AUTHORS

The late **Clair N. Sawyer** was active in the field of sanitary chemistry for over 30 years. He received a Ph.D. from the University of Wisconsin. As Professor of Sanitary Chemistry at the Massachusetts Institute of Technology, he taught and directed research until 1958. He then was appointed Vice President and Director of Research at Metcalf and Eddy, Inc., and served as consultant on numerous water and wastewater treatment projects in the United States and many foreign countries. After retiring, he served as an environmental consultant for several years. He passed away in 1992. He was the originator and sole author of the first edition, which was published in 1960.

Perry L. McCarty is the Silas H. Palmer Professor Emeritus of civil and environmental engineering at Stanford University. He received a B.S. degree in civil engineering from Wayne State University and S.M. and Sc.D. degrees in sanitary engineering from the Massachusetts Institute of Technology, where he taught for four years. In 1962 he joined the faculty at Stanford University. His research has been directed towards the application of biological processes for the solution of environmental problems. He is an honorary member of the American Water Works Association and the Water Environment Federation, and Fellow in the American Academy of Arts and Sciences, the American Association for the Advancement of Science, and the American Academy of Microbiology. He was elected to the National Academy of Engineering in 1977. He received the Tyler Prize for environmental achievement in 1992 and the Clarke Prize for outstanding achievement in water science and technology in 1997.

Gene F. Parkin is a Professor of Civil and Environmental Engineering at the University of Iowa, and Director of the Center for Health Effects of Environmental Contamination. He received a B.S. degree in civil engineering and an M.S. degree in sanitary engineering from the University of Iowa and a Ph.D. degree in environmental engineering from Stanford University. He taught at Drexel University for eight years before joining the faculty at the University of Iowa in 1986. His teaching interests have been in biological treatment processes and environmental chemistry. His research has been directed toward anaerobic biological processes and bioremediation of waters contaminated with organic chemicals. He has received the J. James R. Croes Medal from the American Society of Civil Engineers and the Harrison Prescott Eddy Medal from the Water Environment Federation. In 1989 he received the Hancher-Finkbine Medallion from the University of Iowa for outstanding teaching and leadership and in 1999 he received a state of Iowa Board of Regents Award for Faculty Excellence.

Fundamentals of Chemistry for Environmental Engineering and Science

1

Introduction

Environmental engineers and scientists have evolved as professionals over the past century to address human health and environmental problems resulting from industrial expansion, resource utilization, and the concentration of growing populations in cities. Such trends over the past two centuries have resulted in the biological and chemical contamination of water supplies, pollution of rivers, and severe fouling of air and land. To address these problems, environmental engineers have worked closely with scientists to learn new chemical, physical, and biological principles that can be applied to help solve these difficult human health and ecological problems.

For many years attention was devoted largely to the development of safe water supplies and the sanitary disposal of human wastes. Because of the success in controlling the spread of enteric diseases through the application of scientific and engineering principles, a new concept of the potentialities of preventive medicine was born. Expanding populations with resultant increased industrial operations, power production, and use of motor-driven vehicles, plus new industries based upon new technology have intensified old problems and created new ones in the fields of water supply, waste disposal, air pollution, and global environmental change. Many of these offer major challenges to environmental engineers and scientists.

Over the years, intensification of old problems and the introduction of new ones have led to basic changes in the philosophy of environmental protection. Originally the major objectives were to produce hygienically safe water supplies and to dispose of wastes in a manner that would prevent the development of nuisance conditions. Many other factors concerned with aesthetics, economics, recreation, and other elements of better living are important considerations and have become part of the responsibilities of the modern environmental engineer and scientist. If one were to develop a list of the most important environmental problems of today, it would include, but not be limited to, water and wastewater treatment, surface water and groundwater contamination, hazardous waste management, radioactive waste management, acid rain, air toxics emission, ozone depletion, and global climate change. Understanding these problems and development of processes to minimize or eliminate them requires a fundamental understanding of chemistry. ■

1.1 | WATER

Water is one of the materials required to sustain life and has long been suspected of being the source of much human illness. It was not until approximately 150 years ago that definite proof of disease transmission through water was established. For many years following, the major consideration was to produce adequate supplies that were hygienically safe. However, source waters (surface water and ground-water) have become increasingly contaminated due to increased industrial and agricultural activity. The public has been more exacting in its demands as time has passed, and today water engineers are expected to produce finished waters that are free of color, turbidity, taste, odor, nitrate, harmful metal ions, and a wide variety of organic chemicals such as pesticides and chlorinated solvents. Health problems associated with some of these chemicals include cancer, birth defects, central nervous system disorders, disruption of the endocrine system, and heart disease. At the present time, more than 85 specific chemicals are listed in the U.S. Environmental Protection Agency's drinking water standards, and the World Health Organization lists over 100 specific chemicals in its guidelines for drinking water quality. In addition, the public desires water that is low in hardness and total solids, noncorrosive, and non-scale-forming. To provide such water, chemists, biologists, and engineers must combine their efforts and talents. Chemists, through their knowledge of colloidal, physical, and organic chemistry, are especially helpful in solving problems related to the removal of color, turbidity, hardness, harmful metal ions, and organic compounds, and to the control of corrosion and scaling. The biologist is often of great help in taste and odor problems that derive from aquatic growths. In a true sense, therefore, all who cooperate in the effort regardless of discipline are environmental engineers.

As populations increase, the demand for water grows accordingly and at a much more rapid rate if the population growth is to be accompanied by improved living standards. The combination of these two factors is placing greater and greater stress on finding adequate supplies. In many cases inferior-quality, and often polluted, water supplies must be utilized to meet the demand. It is to be expected that this condition will continue and grow more complicated as long as population and industrial growth occurs. In many situations in water-short areas, purposeful recycling of treated wastewaters will be required in some degree to avoid serious curtailing of per-capita usage and industrial development. The ingenuity of scientists and engineers is being taxed to the limit to meet this need.

The problems faced by the water-supply community in developed countries are significantly different from those faced by underdeveloped countries. For example, in the United States many of the drinking water standards for organic chemicals are based on the desire to minimize the risk of developing cancer from drinking water containing suspected carcinogens. The level of acceptable risk is currently considered to be one-in-ten-thousand to one-in-a-million. That is, if one drinks water containing the chemical of interest at the level of the drinking water standard over a 70-year lifetime, the risk of developing cancer is increased by 10^{-4} to 10^{-6}. Removing these compounds to these levels is a significant challenge. In the underdeveloped world, however, millions of children under the age of 5 die each year due to water-borne diseases. Thus, the goal of the water-supply engineer in this situation is significantly different.

1.2 | WASTEWATER AND WATER POLLUTION CONTROL

The disposal of human wastes has always constituted a serious problem. With the development of urban areas, it became necessary, from public health and aesthetic considerations, to provide drainage or sewer systems to carry such wastes away from the area. The normal repository was usually the nearest watercourse. It soon became apparent that rivers and other receiving bodies of water have a limited ability to handle waste materials without creating nuisance conditions. This led to the development of purification or treatment facilities in which chemists, biologists, and engineers have played important roles. The chemist in particular has been responsible for the development of test methods for evaluating the effectiveness of treatment processes and providing a knowledge of the biochemical and physicochemical changes involved. Great strides have been made in the art and science of waste treatment in the past few decades. These have been made possible by the fundamental knowledge of wastewater treatment established by scientists with a wide variety of training. It has been the responsibility of the engineers to synthesize this basic knowledge into practical systems of wastewater treatment that are effective and economical.

It has long been known that all natural bodies of water have the ability to oxidize organic matter without the development of nuisance conditions, provided that the organic and nitrogen (primarily ammonia) loading is kept within the limits of the oxygen resources of the water. It is also known that certain levels of dissolved oxygen must be maintained at all times if certain forms of aquatic life are to be preserved. A great deal of research has been conducted to establish these limits. Such surveys require the combined efforts of biologists, chemists, and engineers if their full value is to be realized. In the past, streams have been classified into the following four broad categories: (1) those to be used for the transportation of wastes without regard to aquatic growths but maintained to avoid the development of nuisance conditions, (2) those in which the pollutional load will be restricted to allow fish to flourish, (3) those to be used for recreational purposes, and (4) those that are used for water supplies.

The desire in the United States to obtain maximum benefit from all streams and rivers and the inability to readily measure contaminants beyond suspended solids and oxygen consuming potential led in the 1970s away from the stream classification approach to requirements of highest practical degree of treatment for all wastewaters. Effluent quality or effluent standards thus superseded stream standards. However, it has became apparent in more recent years that this approach has been inadequate in many instances to be sufficiently protective of surface water quality as pollution often results from a variety of sources besides wastewater point discharges and from a broad range of chemicals. Nonpoint sources including agricultural runoff and air pollutant fallout can prevent even the best efforts at point source control to mitigate some contamination problems of concern. To address such problems, a program of total daily maximum load (TDML) may be imposed, which is a comprehensive mechanism for pollution prevention that considers both point and non-point source contributions and their optimum control for meeting water quality standards. This newer approach places a higher requirement on environmental engineers and scientists to understand contaminant fate and effects, as well as to acquire expertise in pollution control strategies beyond the conventional handling of effluent waste streams.

Historically, the major concern with regard to pollution of surface waters was with their oxygen resources as described. However, in recent decades, an increasing concern is the pollution of surface waters and groundwaters with other pollutants of primarily industrial or agricultural origin. During the past half century, many new chemicals have been produced for agricultural purposes. Some of them are used for weed control, others for pest control. There has also been a dramatic increase in the application of nitrogen fertilizers. Residues of these materials are often carried to watercourses during periods of heavy rainfall and have had serious effects upon the biota of streams. A great deal of research by chemists and biologists has demonstrated which of the materials have been most damaging to the environment, and many products have been outlawed. Continuing studies will be needed, but hopefully new products will be more environmentally friendly and will be kept from general use until proven equal or even less harmful than those in current use. The wide variety of organic chemicals and heavy metals produced and used by industry has also been shown to contaminate surface waters and groundwaters. These compounds are of public health concern, and they also may have an adverse impact on desirable aquatic life. Many municipal wastewater treatment plants are required to remove such chemicals prior to discharge to receiving waters. A sound knowledge of chemistry is required to understand the environmental fate of these chemicals and to develop methods for their removal.

1.3 | INDUSTRIAL AND HAZARDOUS WASTES

A most challenging field in environmental engineering practice is the treatment and disposal of industrial and hazardous wastes. Because of the great variety of wastes produced from established industries and the introduction of wastes from new processes, a knowledge of chemistry is essential to a solution of most of the problems. Some may be solved with a knowledge of inorganic chemistry; others may require a knowledge of organic, physical, or colloidal chemistry, biochemistry, or even radiochemistry. It is to be expected that, as further technological advances are made and industrial wastes of even greater variety appear, chemistry will serve as the basis for the development and selection of treatment methods.

The problems associated with managing hazardous wastes are particularly complex. Over 100 million tons of hazardous wastes are generated each year in the United States. The U.S. Environmental Protection Agency has placed well over 1200 sites that are contaminated with hazardous chemicals on the National Priority or Superfund list because of their potential threat to human health and the environment, and most likely will add many more in the near future. There is other widespread contamination of soils and groundwater requiring cleanup that has resulted from industrial operations, leaking underground storage tanks, and federal activities, especially those within the Department of Defense and the Department of Energy. The management of the newly generated hazardous wastes and the cleanup of past contamination requires the combined efforts of many scientists and engineers. Another significant aspect of this problem is analytical chemistry: sampling, separation, and quantification of the myriad of chemicals present in industrial wastes, leachates, and contaminated surface waters and aquifers is most challenging.

It should be emphasized that many of the industrial and hazardous wastes problems of the future will be solved by minimizing the quantities of these materials produced and used through product substitution, waste recovery and recycling, and waste minimization. It is a generally accepted axiom that it is much more cost effective to prevent pollution rather than clean it up.

1.4 | AIR POLLUTION AND GLOBAL ENVIRONMENTAL CHANGE

Although the problems of water supply and liquid-waste disposal are of major importance to urban populations, their solution alone does not ensure a completely satisfactory environment. Pollution of the atmosphere increases in almost direct ratio to the population density and is largely related to the products of combustion from heating plants, incinerators, and automobiles, plus gases, fumes, and smokes arising from industrial processes. The intensity of most air pollution problems is usually related to the amount of particulate matter emitted into the atmosphere and to the atmospheric conditions that exist. In general, visible particulate matter can be controlled by adequate regulations. The most serious situations develop where local conditions favor atmospheric inversions and the products of combustion and of industrial processing are contained within a confined air mass. A notable example is the situation in Los Angeles, where inversions occur frequently; they also occur, though less often, in several other metropolitan areas.

In cases where atmospheric inversions occur over metropolitan areas under cloudless skies, a haze commonly called "smog" is produced in the atmosphere. Under such conditions the atmosphere is often highly irritating to the eyes and to the respiratory tract and is far too intense to be accounted for by the materials emitted to the atmosphere from the separate sources. Research on this problem has been extensive. Many theories were advanced as to the cause, but the consensus is that photochemical action between oxides of nitrogen and unsaturated hydrocarbons from automobile exhaust gases combine to form several products of health concern such as ozone, formaldehyde, and organic compounds of nitrogen. These substances can condense on particulate matter in the atmosphere to form a fog. A knowledge of chemistry has played an important role in finding the cause of this enigma.

Air can become contaminated with pollutants from motor vehicles, factories, power plants, and many other sources. Such pollutants can cause cancer or other serious health effects, such as reproductive or birth defects, damage to the immune system, and respiratory problems, or they may cause adverse effects to the environment. The 1990 Clean Air Act amendments list 188 toxic air pollutants that the U.S. Environmental Protection Agency is required to regulate. These include particulate matter; volatile organic compounds such as benzene and toluene; halogen compounds such as tetrachloroethene, dichloromethane, and dioxin; heavy metals such as cadmium, mercury, chromium, and lead; and other hazardous compounds such as asbestos. Indoor air pollutants in the home are also of concern because this closed environment is where people generally spend most of their time. Indoor air pollution can result from combustion sources such as oil, gas, kerosene, coal, or tobacco products. Building materials and furnishings may include toxic chemicals that may be slowly released to the air such as asbestos and radon, and products used for household cleaning and maintenance, personal care, or hobbies often contain volatile chemicals that can cause harm.

Air pollution of quite another type was of major concern a few decades ago. This resulted from radioactive materials that gained entrance to the atmosphere through nuclear explosions. The nuclides that were dispersed and settled as "fallout" varied greatly in their effect upon living plants and animals. Although limitations on atmospheric nuclear testing have greatly lessened this problem, other uses of nuclear energy have raised new fears over release of radioactive materials to the atmosphere. These fears have resulted from the development and installation of nuclear power plants. Experience has indicated these particular fears have little basis with properly designed and operated plants, the accidents at Chernobyl and Three-Mile Island notwithstanding. The major threat to the environment remains the transportation of "nuclear ash," separation, and safe disposal of the waste radioactive materials. This constitutes major challenges with the ever-diminishing supplies of fossil fuels and other problems associated with fossil fuel combustion, nuclear power represents a potential alternative energy source (along with renewable sources such as solar energy, wind, and biomass). A current concern with airborne radioactivity involves the presence of high levels of radon in homes built on soils with radon-containing minerals.

It has become increasingly apparent in recent years that pollution problems are becoming more global in nature. That is, human activities in one region have a significant impact on the environmental quality many miles removed from that region. This was true of the radioactive fallout just described. Current concerns are with stratospheric ozone depletion and global warming. A knowledge of chemical principles is once again playing a role in helping us to understand how these problems developed and how they might be solved. It has been shown that a complex photochemical reaction involving chlorofluorocarbons, emitted at the earth's surface and transported to the stratosphere, is destroying the stratospheric ozone layer. This layer is important in protecting the earth's surface from cancer-causing ultraviolet radiation. Because of this discovery, chlorofluorocarbons are being banned from use globally and replacement compounds are being sought.

Some gases such as carbon dioxide, water vapor, methane, chlorofluorocarbons, oxides of nitrogen, and ozone absorb thermal radiation near the earth's surface and are called "greenhouse gases." The concentrations of these gases are increasing in the earth's atmosphere due to human activities. Since they do trap heat trying to escape from the earth, they have the potential for warming the earth, a process sometimes called global warming. Although there is consensus that the concentrations of these gases are increasing, there is controversy over the extent to which this will lead to a general increase in temperature. Much remains to be learned here, and increased knowledge of environmental chemistry will contribute much to better understanding of this problem.

1.5 | SUMMARY

From the discussions presented it should be apparent that the solution of many environmental problems has required the concerted efforts of scientists and engineers and that chemists, in many instances, have played an indispensable role. It is to be expected that problems arising in the future will be fully as complex as those of the past and that chemistry will continue to be an important factor. Engineers with sound chemical training should find that their knowledge is a great aid and advantage in conquering unsolved problems and that liaison with scientists working on the same or allied problems will be facilitated. The chapters following are dedicated to that purpose.

2

Basic Concepts from General Chemistry

The factual information and basic concepts taught in introductory chemistry vary considerably, depending upon the institution and the interests of the students. In many schools, engineers are given a considerably different course from that given to science majors. Because of these differences and because certain fundamental information is essential for environmental engineering, a review of certain phases of general chemistry is warranted.

2.1 | ELEMENTS, SYMBOLS, ATOMIC WEIGHTS, GRAM ATOMIC WEIGHTS

Remembering the names of the common elements poses no particular problem to the average student. However, the proper symbol does not always come to mind. This is mainly because many of the symbols are derived from Latin, Greek, or German names of the elements, and sometimes because of a similarity of names which makes a multiple choice of symbols possible. This similarity is well illustrated by the symbols for magnesium, Mg, and manganese, Mn, which are commonly confused.

To remember the symbols for magnesium, manganese, and those derived from Latin or other foreign names, one must rely entirely upon memory or association with the uncommon name. A list of the elements whose symbols are derived from Latin, Greek, or German names is given in Table 2.1.

Atomic weights of the elements refer to the relative weights of the atoms as compared with some standard. In 1961 the ^{12}C isotope of carbon was adopted as the atomic weight standard with a value of exactly 12. According to this standard, the atomic weight of oxygen is 15.9994, or 16 for all practical purposes.

It is not necessary to remember the atomic weights of the elements, because tables giving these values are readily available. It will save time, however, to remember the weight of the more commonly used elements such as hydrogen, oxygen, carbon, calcium, magnesium, sodium, sulfur, aluminum, chlorine, and a few others. It

Table 2.1 | List of elements with symbols derived from
Latin, Greek, or German names

Element	Name from which symbol is derived	Symbol
Antimony	Stibium, L.	Sb
Copper	Cuprum, L.	Cu
Gold	Aurium, L.	Au
Iron	Ferrum, L.	Fe
Lead	Plumbum, L.	Pb
Mercury	Hydrargyrum, Gr.	Hg
Potassium	Kalium, L.	K
Silver	Argentum, L.	Ag
Sodium	Natrium, L.	Na
Tin	Stannum, L.	Sn
Tungsten	Wolfram, G.	W

is usually sufficient for all practical purposes to round off the atomic weights at three significant figures: thus the atomic weight for aluminum is called 27.0, chlorine 35.5, gold 197, iodine 127, and so on.

In general, elements do not have atomic weights that are whole numbers because they consist of a mixture of isotopes. Chlorine is a good example. Its atomic weight of 35.45 is due to the fact that it consists of two isotopes with atomic weights of 35 and 37. Cadmium contains eight isotopes with atomic weights ranging from 110 to 116.

The *gram atomic weight* of an element refers to a quantity of the element in grams corresponding to the atomic weight. It has principal significance in the solution of problems involving weight relationships.

2.2 | COMPOUNDS, FORMULAS, FORMULA WEIGHTS, GRAM MOLECULAR WEIGHTS, MOLE, EQUIVALENT WEIGHTS, EQUIVALENTS

Although the concept of chemical compounds is readily established, association of the proper and correct formula for each compound does not always follow. This difficulty is sometimes due to faulty use of symbols, but much more often to a lack of knowledge regarding valence. The subject of valence will be discussed in Sec. 2.4. If strict attention is paid to correct symbols and valences, errors in writing formulas will be eliminated.

Calculation of formula weights poses no real problem except when rather complex formulas are involved. Most difficulties in this regard can be overcome by writing structural formulas and applying some effort in the form of practice. The importance of correct formula weights as the basis for engineering calculations should be emphasized.

The term *gram molecular weight* (GMW, the short-hand symbol being MW), or *formula weight* (FW), refers to the molecular weight in grams of any particular compound. This is also referred to as a *mole*. Its chief significance is in the preparation of *molar* or *molal* solutions. A molar solution consists of 1 formula weight dissolved in enough water to make 1 liter of solution, whereas a molal solution consists of 1 formula weight dissolved in 1 kilogram (kg) of water, the resulting solution having a volume slightly in excess of 1 liter.

The concepts of *equivalents* and *normality* are very useful. These concepts are discussed in more detail in Chap. 11 and are introduced here. The term *equivalent weight* (EW) can be defined as follows:

$$EW = \frac{FW}{Z}$$

where Z = (1) the absolute value of the ion charge, (2) the number of H^+ or OH^- ions a species can react with or yield in an acid-base reaction, or (3) the absolute value of the change in valence occurring in an oxidation-reduction reaction. One equivalent is then defined as 1 mol of a compound divided by its EW. The usefulness of this concept is demonstrated by Example 2.1.

EXAMPLE 2.1

(a)   What is the equivalent weight of the calcium ion (Ca^{2+})?

$$EW = FW/Z = 40 \text{ g per mole}/2 = 20 \text{ g per equivalent}$$

(b)   What is the equivalent weight of calcium carbonate ($CaCO_3$)?

One way of solving this problem is to consider the following reaction:

$$CaCO_3 + 2H^+ = Ca^{2+} + H_2CO_3$$
$$EW = FW/Z = (40 + 12 + (3 \times 16))/2$$
$$= 100 \text{ g per mole}/2 = 50 \text{ g per equivalent}$$

Another way is to realize that $CaCO_3$ is made up of Ca^{2+} and CO_3^{2-}, the absolute value of the ion charge of each being 2. Thus,

$$EW = 100/2 = 50 \text{ g per equivalent}$$

(c)   What concentration is 40 mg/L of Ca^{2+} when expressed as $CaCO_3$ (hardness of water is often expressed as mg/L of $CaCO_3$)?

A major use of the concept of equivalents is that one equivalent of an ion or molecule is "chemically" equivalent to one equivalent of a different ion or molecule. Thus, if concentrations are expressed in terms of equivalents per liter (eq/L), they can be added, subtracted, or converted easily.

$$\text{eq/L } Ca^{2+} = (40 \text{ mg/L}) \times (1 \text{ eq}/20 \text{ g}) \times (1 \text{ g}/1000 \text{ mg}) = 0.002 \text{ eq/L of } Ca^{2+}$$
$$\text{mg/L as } CaCO_3 = (0.002 \text{ eq/L}) \times (50 \text{ g/eq}) \times (1000 \text{ mg/g})$$
$$= 100 \text{ mg/L as } CaCO_3$$

The student should note that concentrations can also be expressed as milliequivalents per liter (meq/L) by multiplying eq/L by 1000 (or by expressing EW in mg).

2.3 | AVOGADRO'S NUMBER

A significant fact is that by definition a mole contains the same number of molecules, whatever the substance. This number is called Avogadro's number and is approximately equal to 6.02×10^{23}. Avogadro's number can be expressed as atoms per mole, molecules per mole, ions per mole, electrons per mole, or particles per mole, depending on the context.

$$6.02 \times 10^{23} \text{ O atoms} = 16.0 \text{ g O}$$
$$6.02 \times 10^{23} \text{ H atoms} = 1.01 \text{ g H}$$
$$6.02 \times 10^{23} \text{ H}_2\text{O molecules} = 18.0 \text{ g H}_2\text{O}$$
$$6.02 \times 10^{23} \text{ OH}^- \text{ ions} = 17.0 \text{ g OH}^-$$

The enormous size of this number is incomprehensible. Some concept of its magnitude may be gained from a consideration that the life span of the average U.S. citizen is of the order of 2.2×10^9 seconds (s) and that a person would have to live about 3×10^{14} lives to count to Avogadro's number.

2.4 | VALENCY, OXIDATION STATE, AND BONDING

A knowledge of valency and bonding theory serves as the key to correct formulas. In general, the writing of formulas with elements and radicals that have a fixed valance (or *oxidation state*) is easy, if a knowledge of electrostatics is applied. The real difficulty stems from elements that can assume several oxidation states, from which a variety of ions, molecules, and radicals can result, and a lack of knowledge of nomenclature, which is not always consistent.

Molecules, some ions, and radicals consist of two or more atoms bonded together in some definite manner. In general, the bonds may be *ionic* or *covalent*. An *ionic bond* is formed by the transfer of electrons from one atom to the other. One atom then takes on a positive charge (the *cation*) and the other a negative charge (the *anion*). The *ion pair* that results is held together loosely by electrostatic attraction. In other cases, electrons are not transferred, but are shared between atoms. In elementary molecules with identical atoms, such as Cl_2, N_2, and O_2, the electrons are shared equally to form a *covalent bond*. On the other hand, in heteronuclear molecules which consist of unlike atoms, the electrons forming the bond are shared unequally. For this case the bonding is termed *polar covalent*.

The valency or oxidation number of an atom is determined by the number of electrons that it can take on, give up, or share with other atoms. According to valency theory, most atoms consist of neutrons, protons $(+)$, and electrons $(-)$. The neutrons and protons are contained within the nucleus, and a number of electrons, corresponding to the number of protons (atomic number) in the nucleus, are arranged in orderly shells outside. The outer shell contains the valence electrons. If electrons are lost, the atom becomes a positively charged ion, and if electrons are gained, the atom becomes a negatively charged ion. Except for inert elements (such as argon) that already have complete shells, atoms tend to gain or lose electrons so as to assume or approach complete shells. To do this, they must team up with another atom in some

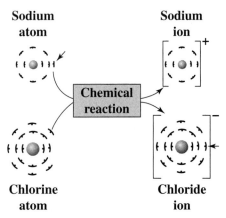

Sodium
atom

Sodium
ion

Chemical
reaction

Chlorine
atom

Chloride
ion

Figure 2.1 |
Electron transfer during a chemical reaction,
producing a sodium ion with an oxidation
state of $+1$ and a chloride ion with an
oxidation state of -1.

manner. In the formation of ions, atoms of two elements undergo reduction and oxidation: one gains electrons and the other loses electrons. In the exchange, the metal or metal-like element loses electrons to gain or approach a stable condition with no electrons in its outer shell. The nonmetal steals electrons from the metal to complete it outer shell to eight electrons, a stable configuration. This exchange is normally accomplished by the release of a great deal of energy. This simple type of reaction is well illustrated by the one between sodium and chlorine, as shown in Fig. 2.1.

The chlorine atom also serves as an example of polar covalent bonding in its various possible combinations with oxygen. The chlorine atom contains several electrons in its outer shell. Oxygen has six electrons in its outer shell and needs two more to complete the shell. These it can obtain in various ways by sharing electrons with the chlorine atom, forming various molecular species which may or may not be charged as illustrated in Fig. 2.2. The electrons contained in the outer shells are represented by dots for simplicity. With oxygen, chlorine tends to share one, three, four, five, or seven of its electrons, to form Cl_2O, ClO_2^-, ClO_2, ClO_3^-, and Cl_2O_7. The oxides from which ClO_2^- and ClO_3^- are derived have never been isolated. However, compounds of chlorine with oxidation numbers of $+1$, $+3$, $+4$, $+5$, and $+7$ are well defined. Sulfur, nitrogen, and the halogens are nonmetals that are capable of exhibiting a wide range of oxidation numbers because of their ability to take on or share electrons to complete the outer shell to eight or to give up one or more electrons to reach a stable configuration. Manganese, chromium, copper, and iron are examples of metals that can obtain several oxidation states by yielding or sharing one or more electrons. Manganese is an extreme case in that it can yield or share two, three, four, six, or seven electrons. The oxidation numbers and valences of some important elements are given in Table 2.2. Such information is useful in balancing oxidation-reduction reactions (Sec. 2.7).

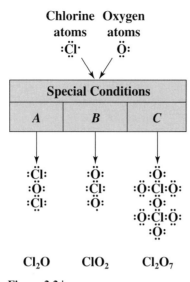

Figure 2.2 |
Multiple oxidation states of chlorine
due to sharing of electrons. Chlorine
forms similar compounds in all states
of oxidation from +1 to +7, except
for +6.

Table 2.2 | Oxidation numbers (valences) of some important elements

	Element	Most common oxidation numbers (valences)
Fixed		
	O	−2, 0
	H	+1, 0
	Ca, Mg	+2, 0
	K, Na	+1, 0
Variable		
	N	+5, +4, +3, +2, +1, 0, −3
	Cl	+7, +5, +4, +3, +1, 0, −1
	Mn	+7, +6, +4, +3, +2, 0
	S	+6, +4, +2, 0, −2
	Fe	+3, +2, 0
	Cu	+2, +1, 0
	Cr	+6, +3, 0
	C	+4, +3, +2, +1, 0, −1, −2, −3, −4

2.5 | NOMENCLATURE

There are only a very few hard-and-fast rules concerning nomenclature of inorganic compounds (the naming of organic compounds is covered in Chap. 5). One concerns binary compounds (those formed from two elements); they all have the ending *-ide*. For example, anhydrous HCl is hydrogen chloride. Most nomenclature problems arise from the acids containing oxygen (*oxoacids* and their associated *oxoanions*, which are also called *oxyacids* and *oxyanions*). In general, the nomenclature is related to the oxidation state of the element that characterizes the acid. The acids having the highest oxidation state are usually called *-ic,* e.g., sulfuric, phosphoric, and chromic. They give rise to *-ate* salts. The acids that exist in the next lowest oxidation state are called *-ous,* e.g., sulfurous, phosphorous, and chromous. They give rise to *-ite* salts. If acids of a lower oxidation state exist, they are called *hypo · · · ous,* e.g., hypochlorous or hypophosphorous, and their salts are called *hypo · · · ites.*

Occasionally, as with the oxy acids of the halogens, more than three acids are known. In such cases the acid in the highest oxidation state is given the prefix *per,* e.g., perchloric or periodic, and their salts are called *per · · · ates.* The acid derived from manganese in an oxidation state of 7 ($HMnO_4$) is known as permanganic acid, and its salts are the familiar permanganates. All *per · · · acids* contain an element that has an oxidation state of 7, which appears to be the reason that $HMnO_4$ is given a prefix *per.* The only other well-defined acid of manganese is manganic, in which the manganese has an oxidation state of 6. A summary of this nomenclature is given in Table 2.3.

Table 2.3 | Nomenclature of oxygen-bearing acids and their salts

Name of acid	Formula	Name of salt
On the basis of oxidation state		
Sulfurous	H_2SO_3	Sulfite
Sulfuric	H_2SO_4	Sulfate
Nitrous	HNO_2	Nitrite
Nitric	HNO_3	Nitrate
Hypochlorous	$HOCl$	Hypochlorite
Chlorous	$HOCl_2$	Chlorite
Chloric	$HOCl_3$	Chlorate
Perchloric	$HOCl_4$	Perchlorate
Permanganic	$HMnO_4$	Permanganate
Phosphorous	H_3PO_3	Phosphite
Phosphoric	H_3PO_4	Phosphate
Chromic	H_2CrO_4	Chromate
On the basis of hydration		
Orthophosphoric	H_3PO_4	Orthophosphate
Metaphosphoric	HPO_3	Metaphosphate
Pyrophosphoric	$H_4P_2O_7$	Pyrophosphate
Pyrochromic	$H_2Cr_2O_7$	Dichromate

In the past, acids were also named in terms of their degree of hydration: ortho, meta, and pyro. The variety of phosphate acids and salts and dichromate are the most important examples still in common use (Table 2.3). The ortho acids consist of the highest hydrated form of the acid anhydride, e.g., sulfuric (H_2SO_4), phosphoric (H_3PO_4), phosphorous (H_3PO_3), and chromic (H_2CrO_4). The meta acids are derived from the ortho acids by removal of one molecule of water from each molecule of acid as follows:

$$H_3PO_4 \xrightarrow{\Delta} \underset{\text{Meta}}{HPO_3} + H_2O \uparrow$$
$$\underset{\text{Ortho}}{}$$

where Δ = heat and $\uparrow$ indicates that water escapes. The ortho acids give rise to ortho salts and the meta acids to meta salts. The pyro acids may be derived, theoretically, from ortho acids by removal of one molecule of water from two molecules of acid, as follows:

$$2H_3PO_4 \xrightarrow{\Delta} H_4P_2O_7 + H_2O \uparrow$$
$$2H_2CrO_4 \xrightarrow{\Delta} \underset{\text{Pyro}}{H_2Cr_2O_7} + H_2O \uparrow$$
$$\underset{\text{Ortho}}{}$$

Free pyro acids are not known, but well-defined salts are common. The pyro salts of chromic acids are commonly called dichromate, an example of deviation from the general rule.

2.6 | CHEMICAL EQUATIONS: WEIGHT RELATIONSHIPS AND CONSERVATION OF MASS AND CHARGE

A fundamental rule that must be observed at all times is that expressions of chemical reactions become equations only when they are balanced. Mass must be conserved; that is, the total number of each kind of atom must be the same on both sides of the equation. Also, the sum of the charge on one side of the equation must equal that on the other. In order to balance a chemical equation, it is essential that it represent a reaction in true manner, and all formulas used must be correct. Unless these conditions are complied with, weight relationships are meaningless. Weight relationships serve as the basis for the sizing of chemical feeding equipment, necessary storage space for chemicals, structural design, and cost estimates in engineering considerations. Their importance should not need further emphasis.

EXAMPLE 2.2

$$NaOH + HCl \rightarrow NaCl + H_2O$$

Weight relationship	(23+16+1)	(1+35.5)	(23+35.5)	(2×1)+16
	40	36.5	58.5	18

Thus, 40 g of NaOH combines with 36.5 g of HCl to form 58.5 g of NaCl and 18 g of H_2O.

EXAMPLE 2.3

$$2HCl + Na_2CO_3 \rightarrow H_2O + 2NaCl + CO_2$$

Weight relationship

| $2(1+35.5)$ | $(2\times23)+12+(3\times16)$ | $(2\times1)+16$ | $2(23+35.5)$ | $12+(2\times16)$ |
| 73 | 106 | 18 | 117 | 44 |

In this case 2 mol or 73 g of HCl combine with 1 mol or 106 g of $NaCO_3$ to form 1 mol or 18 g of H_2O, 2 mol or 117 g of NaCl, and 1 mol or 44 g of CO_2.

2.7 | OXIDATION-REDUCTION EQUATIONS

Concepts of oxidation and reduction are based upon the idea of atomic structure and electron transfer as described in Sec. 2.4. An atom, molecule, or ion is said to undergo *oxidation* when it *loses* an electron, and to undergo *reduction* when it *gains* an electron. With reference to Fig. 2.1, when sodium reacts with chlorine to form sodium chloride, the sodium atom loses an electron and becomes oxidized to the sodium ion, Na^+. Chloride gains an electron and is reduced to the anion, Cl^-.

When oxidation-reduction reactions occur between atoms to form molecules or ions with polar covalent bonds, certain assumptions are required in order to maintain a consistent concept. A good illustration is the reaction that occurs when hydrogen burns in oxygen.

$$2H_2 + O_2 \rightarrow 2(H^+ - O^{2-} - H^+) \tag{2.1}$$

or

$$2H{:}H + {:}\ddot{O}{:}\ddot{O}{:} \rightarrow 2{:}\overset{\overset{\textstyle H}{}}{\underset{..}{O}}{:}H$$

H_2 and O_2 are *homonuclear covalent molecules.* We adopt the convention that the electrons are shared equally by the homonuclear cores: no atom gains or loses electrons in the formation of the molecule from its atoms: thus, the *oxidation number* (sometimes called oxidation state or valence) is zero. Water is a *heteronuclear polar covalent molecule.* In H_2O, the electrons are shared unequally by hydrogen and oxygen; the oxygen atom tends to have a greater holding power on the electrons and is said to be more *electronegative* than hydrogen. This leads to a polar covalent bond in which the oxygen part of the molecule tends to take on a negative charge and the hydrogen a positive charge. To calculate the oxidation number, we adopt the convention that the more electronegative element effectively acquires complete control of the shared electrons. This is equivalent to exaggerating the polar covalent bond into an ionic bond. In the formation of the water molecule, each hydrogen atom takes on an oxidation number or valence of $+1$ (becomes oxidized), and the oxygen atom takes on an oxidation number of -2 (becomes reduced). Hydrogen and oxygen, when part of essentially all heteronuclear molecules and ions of interest in environmental engineering and science, take on these oxidation numbers.

Sometimes there is difficulty in naming compounds containing elements that can exhibit more than one oxidation number. A scheme now frequently used is one recommended by the International Union of Pure and Applied Chemistry (IUPAC). Here, the oxidation number of the element in the positive oxidation state is indicated by a roman numeral in parentheses following the name of the element: thus, $FeCl_2$ is iron(II) chloride; $FeCl_3$ is iron(III) chloride; and Cl_2O_7 is chlorine(VII) oxide. Also, when speaking of an element one can refer to Fe(III), which means iron with a $+3$ oxidation state, without specifically considering whether the iron is present as the ion Fe^{3+} or whether it occurs within a heteronuclear ion or molecule. There are times when use of this nomenclature can help greatly to avoid confusion.

With these concepts of oxidation and reduction, general definitions of oxidizing agents and reducing agents can be derived. An *oxidizing agent* is any substance that can add electrons, e.g.,

O(0), Cl(0), Fe(III), Cr(VI), Mn(IV), Mn(VII), N(V), N(III), S(0), S(IV), S(VI)

A *reducing agent* is any substance that can give up electrons, e.g.,

H(0), Fe(0), Mg(0), Fe(II), Cr(II), Mn(IV), N(III), Cl($-$I), S(0), S($-$II), S(IV)

It will be noted that Mn(IV), N(III), S(0), and S(IV) appear in both of the given series. Any element in an intermediate state of oxidation can serve as a reducing or as an oxidizing agent under proper conditions.

It is a fundamental rule that oxidation cannot occur without reduction, and the gain of electrons by the oxidizing agent must equal the loss of electrons by the reducing agent. Balancing oxidation-reduction reactions involves conservation of charge: the number of electrons gained must equal the number of electrons lost.

Simple Oxidation-Reduction Reactions

$$H_2^\circ + Cl_2^\circ \rightarrow 2H^+Cl^- \tag{2.2}$$

$$4Fe^\circ + 3O_2^\circ \rightarrow 2Fe_2^{3+}O_3^{2-} \tag{2.3}$$

$$Mg^\circ + H_2^+SO_4^{2-} \rightarrow Mg^{2+}SO_4^{2-} + H_2^\circ \tag{2.4}$$

$$2Fe^{2+} + Cl_2^\circ \rightarrow 2Fe^{3+} + 2Cl^- \tag{2.5}$$

$$2I^- + Cl_2^\circ \rightarrow I_2^\circ + 2Cl^- \tag{2.6}$$

In each of Eqs. (2.2) to (2.6), the oxidizing agent gains the same number of electrons as are lost by the reducing agent and the superscript $^\circ$ (e.g., H_2°) refers to an oxidation state of 0.

Complex Oxidation-Reduction Reactions

Many oxidation-reduction reactions require the presence of a third compound, usually an acid or water, to progress. It is a rule that when the oxidizing agent is a compound containing oxygen, such as $KMnO_4$ or $K_2Cr_2O_7$, one of the products is water.

The balancing of complex oxidation-reduction equations is simplified if the following three steps are followed:

1. Write the skeleton equation. This may be in either the molecular or ionic form but must be a *true* representation of the reaction that occurs.
2. Balance the equation with respect to oxidation number change; that is, balance the gain and loss of electrons.
3. Complete the equation in the usual manner.

A few illustrations will serve to show how the scheme is applied.

EXAMPLE 2.4

Step 1

$$KMnO_4 + FeSO_4 + H_2SO_4 \rightarrow Fe_2(SO_4)_3 + K_2SO_4 + MnSO_4 + H_2O \qquad (a)$$

or

$$MnO_4^- + Fe^{2+} + H^+ \rightarrow Fe^{3+} + Mn^{2+} + H_2O \qquad (b)$$

Equation (*b*) is the ionic form of Eq. (*a*), if one recognizes that the potassium and sulfate ions do not enter into the reaction (they are not oxidized or reduced). Note that neither equation is balanced at this stage of the example.

Step 2

$$\overset{+5 \times 2 = 10e \text{ gain}}{2KMn^{7+}O_4 + 10Fe^{2+}SO_4 + H_2SO_4 \rightarrow 5Fe_2^{3+}(SO_4)_3 + K_2SO_4 + 2Mn^{2+}SO_4 + H_2O}$$

$$-1 \times 10 = 10e \text{ loss}$$

or

$$\overset{+5e \text{ gain}}{MnO_4^- + 5Fe^{2+} + H^+ \rightarrow 5Fe^{3+} + Mn^{2+} + H_2O}$$

$$-5e \text{ loss}$$

In this step a total of 10 electrons are involved in the molecular equation because there are two atoms of iron in each molecule of ferric sulfate. The least common multiple of 2 and 5 is 10.

Step 3

$$2KMnO_4 + 10FeSO_4 + 8H_2SO_4 \rightarrow 5Fe_2(SO_4)_3 + K_2SO_4 + 2MnSO_4 + 8H_2O \quad (2.7)$$

or

$$MnO_4^- + 5Fe^{2+} + 8H^+ \rightarrow 5Fe^{3+} + Mn^{2+} + 4H_2O \qquad (2.8)$$

EXAMPLE 2.5

Step 1

$$K_2Cr_2O_7 + KI + H_2SO_4 \rightarrow Cr_2(SO_4)_3 + K_2SO_4 + I_2 + H_2O$$

or

$$Cr_2O_7^{2-} + I^- + H^+ \rightarrow Cr^{3+} + I_2 + H_2O$$

Step 2

$$+ 3 \times 2 = 6e \text{ gain}$$

$$K_2Cr_2^{6+}O_7 + 6KI^- + H_2SO_4 \rightarrow Cr_2^{3+}(SO_4)_3 + K_2SO_4 + 3I_2^\circ + H_2O$$

$$-1 \times 6 = 6e \text{ loss}$$

$$+ 3 \times 2 = 6e \text{ gain}$$

$$Cr_2O_7^{2-} + 6I^- + H^+ \rightarrow 2Cr^{3+} + 3I_2^\circ + H_2O$$

$$-1 \times 6 = 6e \text{ loss}$$

Step 3

$$K_2Cr_2O_7 + 6KI + 7H_2SO_4 \rightarrow Cr_2(SO_4)_3 + 4K_2SO_4 + 3I_2 + 7H_2O \qquad (2.9)$$

or
$$Cr_2O_7^{2-} + 6I^- + 14H^+ \rightarrow 2Cr^{3+} + 3I_2 + 7H_2O \qquad (2.10)$$

An even more complicated oxidation-reduction equation is involved when one element in a high oxidation state oxidizes the same element in a lower oxidation state, with all the particular elements from both the oxidizing and reducing agents appearing in the same final oxidation state. The action of potassium bi-iodate with potassium iodide is an excellent example since it is the basis of a reaction commonly used to release iodine for the standardization of sodium thiosulfate solutions.

Step 1

$$KH(IO_3)_2 + KI + H_2SO_4 \rightarrow I_2 + I_2 + K_2SO_4 + H_2O$$

Since free iodine is formed from both the oxidizing and reducing agents, it is desirable to repeat I_2 in the equation.

Step 2

$$+5 \times 2 = 10e \text{ gain}$$

$$KH(I^{5+}O_3)_2 + 10KI^- + H_2SO_4 \rightarrow I_2^\circ + 5I_2^\circ + K_2SO_4 + H_2O$$

$$-1 \times 10 = 10e \text{ loss}$$

Step 3

$$2KH(IO_3)_2 + 20KI + 11H_2SO_4 \rightarrow 12I_2 + 11K_2SO_4 + 12H_2O \qquad (2.11)$$

In order to balance the equation in this step, it is necessary to multiply by 2 the parts that are balanced in step 2 because of the limitations imposed by potassium. Ionically this equation results in

$$IO_3^- + 5I^- + 6H^+ \rightarrow 3I_2 + 3H_2O \qquad (2.12)$$

Use of Half Reactions

Another procedure that can simplify the development of complex oxidation-reduction reactions, including those involving organic compounds, involves the use

Table 2.4 | Half reactions

Reaction number	Element reduced	Half reaction	ΔG^0, kJ/mol	E^0, volts	pE^0
1	C	$\frac{1}{4}CO_2(g) + \frac{7}{8}H^+ + e^- = \frac{1}{8}CH_3COO^- + \frac{1}{4}H_2O$	-6.88	0.071	1.20
2	C	$\frac{1}{4}CO_2(g) + H^+ + e^- = \frac{1}{24}C_6H_{12}O_6 + \frac{1}{4}H_2O$	1.08	-0.011	-0.19
3	Cl	$\frac{1}{2}Cl_2(g) + e^- = Cl^-$	-131.30	1.361	23.00
4	Cl	$\frac{1}{2}Cl_2(aq) + e^- = Cl^-$	-134.75	1.397	23.60
5	Cl	$\frac{1}{2}ClO^- + H^+ + e^- = \frac{1}{2}Cl^- + \frac{1}{2}H_2O$	-165.84	1.719	29.05
6	Cl	$\frac{1}{8}ClO_4^- + H^+ + e^- = \frac{1}{8}Cl^- + \frac{1}{2}H_2O$	-133.93	1.388	23.46
7	Cr	$\frac{1}{6}Cr_2O_7^{2-} + \frac{7}{3}H^+ + e^- = \frac{1}{3}Cr^{3+} + \frac{7}{6}H_2O$	-131.71	1.365	23.07
8	Cu	$\frac{1}{2}Cu^{2+} + e^- = \frac{1}{2}Cu(s)$	-32.75	0.339	5.74
9	Fe	$\frac{1}{2}Fe^{2+} + e^- = \frac{1}{2}Fe(s)$	39.44	-0.409	-6.91
10	Fe	$Fe^{3+} + e^- = Fe^{2+}$	-74.27	0.770	13.01
11	Fe	$\frac{1}{3}Fe^{3+} + e^- = \frac{1}{3}Fe(s)$	1.53	-0.016	-0.27
12	H	$H^+ + e^- = \frac{1}{2}H_2(g)$	0.00	0.000	0.00
13	Hg	$\frac{1}{2}Hg^{2+} + e^- = \frac{1}{2}Hg(l)$	-76.80	0.796	13.45
14	I	$\frac{1}{2}I_2(s) + e^- = I^-$	-51.59	0.535	9.04
15	I	$\frac{1}{5}IO_3^- + \frac{6}{5}H^- + e^- = \frac{1}{10}I_2(s) + \frac{3}{5}H_2O$	-116.71	1.210	20.44
16	Mn	$\frac{1}{2}MnO_2(s) + 2H^+ + e^- = \frac{1}{2}Mn^{2+} + H_2O$	-118.63	1.230	20.78
17	Mn	$\frac{1}{5}MnO_4^- + \frac{8}{5}H^+ + e^- = \frac{1}{5}Mn^{2+} + \frac{4}{5}H_2O$	-143.90	1.491	25.21
18	Mn	$\frac{1}{3}MnO_4^- + \frac{4}{3}H^+ + e^- = \frac{1}{3}MnO_2 + \frac{2}{3}H_2O$	-162.00	1.679	28.38
19	N	$\frac{1}{6}NO_2^- + \frac{4}{3}H^+ + e^- = \frac{1}{6}NH_4^+ + \frac{1}{3}H_2O$	-86.09	0.892	15.08
20	N	$\frac{1}{8}NO_3^- + \frac{5}{4}H^+ + e^- = \frac{1}{8}NH_4^+ + \frac{3}{8}H_2O$	-84.95	0.880	14.88
21	N	$\frac{1}{3}NO_2^- + \frac{4}{3}H^+ + e^- = \frac{1}{6}N_2(g) + \frac{2}{3}H_2O$	-145.72	1.510	25.52
22	N	$\frac{1}{5}NO_3^- + \frac{6}{5}H^+ + e^- = \frac{1}{10}N_2(g) + \frac{3}{5}H_2O$	-120.04	1.244	21.03
23	O	$\frac{1}{4}O_2(g) + H^+ + e^- = \frac{1}{2}H_2O$	-118.59	1.229	20.77
24	S	$\frac{1}{6}SO_4^{2-} + \frac{4}{3}H^+ + e^- = \frac{1}{6}S(s) + \frac{2}{3}H_2O$	-34.02	0.353	5.96
25	S	$\frac{1}{8}SO_4^{2-} + \frac{5}{4}H^+ + e^- = \frac{1}{8}H_2S(aq) + \frac{1}{2}H_2O$	-29.00	0.301	5.08
26	S	$\frac{1}{4}SO_4^{2-} + \frac{5}{4}H^+ + e^- = \frac{1}{8}S_2O_3^{2-} + \frac{5}{8}H_2O$	-26.26	0.272	4.60
27	S	$\frac{1}{2}SO_4^{2-} + H^+ + e^- = \frac{1}{2}SO_3^{2-} + \frac{1}{2}H_2O$	10.43	-0.108	-1.83
28	Zn	$\frac{1}{2}Zn^{2+} + e^- = \frac{1}{2}Zn(s)$	73.55	-0.762	-12.88

Note: $E^0 = \dfrac{-\Delta G^0}{96.485 \text{ kJ/volt-eq}}$ and $pE^0 = 16.9E^0$

of half reactions. A series of half reactions is shown in Table 2.4. The generally accepted convention is to write half reactions as reductions. Half reactions are balanced oxidation-reduction reactions for a single element. They are not complete reactions because electrons are shown as one of the reactants. Free electrons cannot occur in solution. A complete reaction is made by adding one half reaction to the reverse of another. For example, Eq. (2.12) can be developed by adding the reverse of reaction 14 to reaction 15 from Table 2.4.

Reaction 15	$\frac{1}{5}IO_3^- + \frac{6}{5}H^+ + e^- = \frac{1}{10}I_2 + \frac{3}{5}H_2O$
Reverse of reaction 14	$I^- = \frac{1}{2}I_2 + e^-$
Sum	$\frac{1}{5}IO_3^- + I^- + \frac{6}{5}H^+ = \frac{3}{5}I_2 + \frac{3}{5}H_2O$
$\times 5$ to give	
Eq. (2.12)	$IO_3^- + 5I^- + 6H^+ = 3I_2 + 3H_2O$

The significance of the value for ΔG^0, E^0, and pE^0 listed in Table 2.4 will be discussed later in Sec. 4.10. A great number of oxidation-reduction reactions of interest in water chemistry can be produced through combinations of the half reactions listed. Additional half reactions can be readily developed as follows, using for illustration I_2 as the reduced species and IO_3^- as the oxidized species for iodine.

Step 1. Begin the half reaction with the species containing the more oxidized form of the element on the left and the more reduced form on the right, and balance for the element.

$$2IO_3^- = I_2$$

Step 2. Add a sufficient number of moles of water to either side of the equation to produce an oxygen balance.

$$2IO_3^- = I_2 + 6H_2O$$

Step 3. Add sufficient H^+ to either side of the reaction to produce a hydrogen balance.

$$2IO_3^- + 12H^+ = I_2 + 6H_2O$$

Step 4. Add electrons to the left side of the reaction to make a charge balance. This results in a balanced half reaction.

$$2IO_3^- + 12H^+ + 10e^- = I_2 + 6H_2O$$

Step 5. Divide the equation by the number of electrons indicated by step 4 to normalize the reaction to one electron equivalent.

$$\frac{1}{5}IO_3^- + \frac{6}{5}H^+ + e^- = \frac{1}{10}I_2 + \frac{3}{5}H_2O$$

2.8 | METALS AND NONMETALS

The division between metals and nonmetals is not as distinct as we often desire it to be. In general, those elements that easily lose electrons to form positive ions are called metals. In the free state, metals usually conduct electric current readily. Elements that hold electrons firmly and tend to gain electrons to form negative ions are called nonmetals. Two tests are commonly applied in making a decision: (1) Most metals will form the cation portion of salts having oxygen-bearing anions, such as nitrate and sulfate, and nonmetals do not. (2) Metals form at least one oxide with reasonably strong basic characteristics. The latter is not of general application because of the limited solubility of some oxides and hydroxides.

2.9 | THE GAS LAWS

The gas laws, particularly their influence on the solution or removal of gases from liquids, are of particular significance to the environmental engineer and scientist.

Boyle's Law

Boyle's law states: *The volume of a gas varies inversely with its pressure at constant temperature.* This law is so simple and usually so well understood that further elaboration seems unnecessary. Its principal application is in converting observations of gas volume from field conditions to some standard condition. This is particularly significant at high altitudes, such as at Denver and Salt Lake City.

Charles' Law

Charles' law states: *The volume of gas at constant pressure varies in direct proportion to the absolute temperature.* Interpretation of this law poses no problems, provided that the absolute-temperature scale is used. Charles' law finds its greatest use in the calculation of pressures in fixed-volume containers with variable temperature. In conjunction with Boyle's law, it serves as the basis for sizing gas holders.

Generalized Gas Law

For a given quantity of a gas, Boyle's law and Charles' law can be combined in the form

$$PV = \beta T \tag{2.13}$$

where β = constant proportional to weight of gas
  P = pressure of gas
  V = volume of gas
  T = absolute temperature of gas

It has been shown that the constant β is a function of the number of moles of gas present, and that a more universal, idealized gas law (termed the *ideal gas law*) which is quite general for any gas can be expressed as

$$PV = nRT \qquad (2.14)$$

where n equals the number of moles of gas in the particular sample and R is a universal constant for all gases. The numerical value for R depends on the units chosen for the measurement of P, V, and T. A useful way to evaluate R is to remember that 1 mol of an ideal gas at 1 atm pressure occupies a volume of 22.414 liters at 273 kelvins (K). From this, R can be evaluated to be 0.082 liter atmosphere per mole per kelvin (0.082 L-atm/mol-K).

EXAMPLE 2.6

What tank volume is required to hold 10,000 kg of methane gas (CH_4) at 25 degrees Celsius (°C) and 2 atm pressure?

The molecular weight of CH_4 gas is $12 + 4(1) = 16$ g. The number of moles in 10,000 kg is $10,000,000/16 = 625,000$ mol. According to the general gas law,

$$V = \frac{nRT}{P} = \frac{625,000(0.082)(273 + 25)}{2} = 7.64 \times 10^6 \text{ liters}$$

Thus, a tank with a volume of 7.64×10^6 liters or 2.7×10^5 cubic feet (ft^3) would be required.

Dalton's Law of Partial Pressures

This law has been presented in a number of ways, but in essence it may be stated as follows: *In a mixture of gases, such as air, each gas exerts pressure independently of the others. The partial pressure of each gas is proportional to the amount (percent by volume) of that gas in the mixture, or in other words, it is equal to the pressure that gas would exert if it were the sole occupant of the volume available to the mixture.* The basic concept of this law, in combination with Henry's law, serves in many engineering considerations and calculations.

Henry's Law

Henry's law states: *The mass of any gas that will dissolve in a given volume of a liquid, at constant temperature, is directly proportional to the pressure that the gas exerts above the liquid.* In equation form,

$$K_H = \frac{P_{gas}}{C_{equil}} \qquad (2.15)$$

where C_{equil} = concentration of gas dissolved in liquid at equilibrium

           P_{gas} = partial pressure of gas above liquid

           K_H = Henry's law constant for gas at given temperature

Henry's law is undoubtedly the most important of all the gas laws in problems involving liquids. With a firm knowledge of Dalton's and Henry's laws, one should be capable of coping with all problems involving gas transfer into and out of liquids. As an example, the Henry's law constant, K_H, for oxygen in water at 20°C is 0.73 atm-m^3/mol. Since air contains 21 percent by volume of oxygen, the partial pressure of oxygen in air according to Dalton's law would be 0.21 atm when the total air pressure is 1 atm. Therefore, the equilibrium concentration of oxygen in water at 20°C and in the presence of 1 atm of air would be 0.21/0.73 = 0.288 mol/m^3 or 0.288(32,000)/1000 = 9.2 mg/L.

In the environmental engineering field, many of the problems related to the transfer of gases into liquids involve addition of oxygen by aeration to maintain aerobic conditions. The removal of gases from liquids is also accomplished by aeration devices of one sort or another. Usually the processes involve gas transfer at or near atmospheric pressure from air bubbles passing through a liquid, liquid drops falling through air, or thin films of liquid flowing over surfaces exposed to the air. Although Henry's law is an equilibrium law and is not directly concerned with the kinetics of gas transfer, it serves to indicate how far a liquid-gas system is from equilibrium, which in turn is a factor in the rate of gas transfer. Thus, the rate of dissolution of oxygen is proportional to the difference between the equilibrium concentration as given by Henry's law and the actual concentration in the liquid:

$$\frac{dC}{dt} \propto (C_{\text{equil}} - C_{\text{actual}})$$

This concept serves as the basis for calculations in aerobic methods of waste treatment, such as the activated sludge process, and in the evaluation of the reaeration capacity of lakes and streams.

The removal of undesirable gases, such as carbon dioxide, hydrogen sulfide, hydrogen cyanide, and a wide variety of volatile organic compounds, such as trichloroethene and carbon tetrachloride, from liquids is also commonly accomplished by some form of aeration. The general principles involved are the same as in the transfer of gases into the liquid. However, in this case the normal partial pressure of the gas in air is very low or zero, so based on Henry's law, C_{equil} is also low and much less than C_{actual}. Thus, the rate of transfer given by the above equation is negative, and the gas leaves rather than enters the solution. The same principles apply to the removal of any volatile substance dissolved in water as long as it exerts a significant vapor pressure at the temperature involved. Examples and discussion are given in Sec. 5.34.

Graham's Law

Graham's law is concerned with the diffusion of gases, and it states: *The rates of effusion (escape of a gas through a tiny hole) of gases are inversely proportional to the square roots of their molecular masses.* This law can be illustrated by a comparison of the rates of diffusion of hydrogen, oxygen, chlorine, and bromine, which have molecular weights of approximately 2, 32, 71.5, and 160, respectively. On the basis of Graham's law, oxygen diffuses about one-fourth, chlorine about one-sixth, and bromine about one-ninth as fast as hydrogen. This law finds its greatest application in the field of industrial hygiene and air pollution control. Molecular mass is also important in the rate of gas transfer into (e.g., oxygen for aeration processes) and out of (e.g., removal of contaminant gases from water) aqueous solution.

Gay-Lussac's Law of Combining Volumes

Gay-Lussac's law is basic to an understanding of gas analysis. The law states: *The volumes of all gases that react and that are produced during the course of a reaction are related, numerically, to one another as a group of small, whole numbers.* This law may be illustrated as follows:

$$C + O_2 \rightarrow CO_2 \qquad (2.16)$$
$$\underset{1 \text{ vol}}{} \quad \underset{1 \text{ vol}}{}$$

One volume of oxygen combines with carbon (a solid) to yield 1 volume of carbon dioxide, or

$$CH_4 + 2O_2 \rightarrow CO_2 + 2H_2O \qquad (2.17)$$
$$\underset{1 \text{ vol}}{} \quad \underset{2 \text{ vol}}{} \quad \underset{1 \text{ vol}}{} \quad \underset{0 \text{ vol}}{}$$

Two volumes of oxygen combine with 1 volume of methane to form 1 volume of carbon dioxide. If the temperature of the system is held above 100°C, 2 volumes of water vapor will result. Usually the temperature of the system is brought back to room temperature, the water vapor condenses, and the volume of water is considered zero because it is segregated from the gaseous phase and does not interfere with measurement of the volume of gaseous products.

2.10 | SOLUTIONS

The concepts of unsaturated, saturated, and supersaturated solutions are usually firmly entrenched in the minds of those who have studied general science or chemistry. The terms *molar* and *molal,* as described in Sec. 2.2, are not so well understood. Molal solutions are normally used when the physical properties of solutions, such as vapor pressure, freezing point, and boiling point, are involved. Molar concentrations are generally of interest for equilibrium calculations of various kinds. *Normal* solutions are commonly used for making analytical measurements and are described in Sec. 11.4.

Vapor Pressure

The presence of a nonvolatile solute in a liquid always lowers the vapor pressure of the solution. Thus, when sugar, sodium chloride, or a similar substance is dissolved in water, the vapor pressure is decreased. This phenomenon is believed to be due to a physical blocking effect at the surface of the liquid where particles (ions or molecules) of the solute happen to be.

Raoult's Law

Raoult's law states: *The vapor (partial) pressure (P$_i$) of a compound in a solution mixture of compounds is equal to the mole fraction of that compound (X$_i$) in the mixture multiplied by its vapor pressure in pure form (P°):*

$$P_i = X_i P° \tag{2.18}$$

Values of $P°$ for many compounds can be found in standard reference books such as the *Handbook of Chemistry and Physics*.[1] Mole fraction is a measure of the concentration of species i in the mixture. It is a dimensionless value that equals the ratio of the number of moles of species i (n_i) present divided by the total number of moles of all species present (n_j):

$$X_i = \frac{n_i}{\sum\limits_{j} n_j} \tag{2.19}$$

Raoult's law is similar to Henry's law in that it relates the vapor pressure above a solution to the concentration in the solution. For ideal solutions, they are equivalent expressions. In practice and with the usual nonideal solutions, however, Henry's law is generally used when the concentration of species i in the mixture is low (for example, in dilute aqueous solution where the mole fraction of water approaches 1.0) while Raoult's law is generally applied when the concentration of species i in the mixture is high (for example, petroleum mixtures and mixtures of hazardous chemicals, i.e., chlorinated solvents). Deviations from ideal behavior are sometimes corrected through use of activity coefficients and fugacities as discussed in Sec. 2.12.

EXAMPLE 2.7

Benzene, toluene, ethylbenzene, and xylene (BTEX) are common constituents of gasoline. Vapor pressures of the pure liquids are, respectively, 0.126, 0.0380, 0.0126, and 0.0117 atm at 25°C.[2] Assuming an equimolar mixture of these liquids obeys Raoult's law, calculate the vapor pressure exerted by each chemical and the total vapor pressure exerted by the mixture.

According to Raoult's law: $P_i = X_i P°$. In this example, $X_i = 0.25$ for each chemical.

$$P_{\text{benzene}} = 0.25(0.126) = 0.0315 \text{ atm}$$
$$P_{\text{toluene}} = 0.25(0.0380) = 0.0095 \text{ atm}$$
$$P_{\text{ethylbenzene}} = 0.25(0.0126) = 0.0032 \text{ atm}$$
$$P_{\text{xylene}} = 0.25(0.0117) = 0.0029 \text{ atm}$$
$$P_{\text{total}} = 0.0315 + 0.0095 + 0.0032 + 0.0029 = 0.0471 \text{ atm}$$

---

[1]D. R. Lide (ed.): "Handbook of Chemistry and Physics," 82nd ed.; CRC Press, Boca Raton, FL, 2001.

[2]R. P. Schwarzenbach, P. M. Gschwend, and D. M. Imboden, "Environmental Organic Chemistry," Wiley, New York, 1993.

If the mixture of BTEX given in Example 2.7 is in equilibrium with water, use Raoult's law and Henry's law to estimate the solubility of benzene in water. K_H for benzene at 25°C is 0.0055 atm-m^3/mol.[3]

EXAMPLE 2.8

From Example 2.7, $P_{benzene}$ = 0.0315 atm. From Henry's law, Eq. (2.15)

$$C_{equil} = P_{gas}/K_H$$

Thus,

$$C_{equil} = 0.0315/0.0055 = 5.73 \text{ mol/m}^3 \quad \text{or} \quad 0.00573 \text{ M}$$

The FW of benzene is 78, so this represents an aqueous concentration of 447 mg/L.

Another use of Raoult's law is in understanding the effect that solutes have upon the freezing and boiling points of water and other solvents. Application of Raoult's law has shown that molal solutions of nonelectrolytes in water, such as sugar containing 6.02 × 10^{23} (Avogadro's number) molecules or particles, have their vapor pressures decreased to the same degree, and the boiling point is raised 0.52°C while the freezing point is depressed 1.86°C. A molal solution of an electrolyte such as NaCl, which yields two ions, produces nearly twice as great an effect because, after solution and ionization occur, the molal solution contains nearly two times Avogadro's number of particles.

2.11 | EQUILIBRIUM AND LE CHATELIER'S PRINCIPLE

Nearly all chemical reactions are reversible in some degree. When a reaction proceeds to a point where the combination of reactants to form products is just balanced by the reverse reaction of products combining to form reactants, then the reaction is said to have reached *equilibrium*. The concentration of the reactants and of the products is important in determining the final state of equilibrium. For a system in equilibrium as expressed by the classical equation

$$A + B \rightleftharpoons C + D \qquad (2.20)$$

an increase in either A or B will shift the equilibrium further to the right. Conversely, an increase of either C or D will shift the equilibrium to the left. The shifting of an equilibrium in response to a change of concentration is an example of the well-known principle of Le Chatelier, which states: *A reaction, at equilibrium, will adjust itself in such a way as to relieve any force, or stress, that disturbs the equilibrium.*

A chemical reaction in true equilibrium can be expressed as

$$\frac{[C][D]}{[A][B]} = K \qquad (2.21)$$

---

[3]Ibid.

where K is constant for a given temperature and is called the *equilibrium constant,* and the designation [ ] *signifies the concentration* of the reacting substances. It is easily seen that in this form, called the *equilibrium relationship,* a change in concentration of any of the four substances will change the concentrations of all the others. This expression is widely used, and a clear understanding of its implications is necessary in all phases of science and applied sciences, such as environmental engineering. For the general reaction,

$$a\text{A} + b\text{B} + \cdots \rightleftharpoons c\text{C} + d\text{D} + \cdots$$

in which a, b, c, and d are the number of molecules of the respective substances entering into the reaction, the equilibrium relationship is

$$\frac{[\text{C}]^c[\text{D}]^d \cdots}{[\text{A}]^a[\text{B}]^b \cdots} = K \tag{2.22}$$

This expression is sometimes called the *law of mass action.* If the reactants and products of a reaction are dissolved in a solvent such as water, then concentrations in the equilibrium relationship are ordinarily expressed in moles per liter. Methods of expressing concentrations for solids, gases, and solvents are discussed in Sec. 2.12.

2.12 | ACTIVITY AND ACTIVITY COEFFICIENTS

Equilibrium relationships are frequently applied to equilibria involving salts, acids, bases, gases, and other molecules in aqueous solution. As solutions of these materials become more concentrated, their quantitative effect on equilibria becomes progressively less than that calculated solely on the basis of changes in concentration. Thus, the effective concentration, or *activity,* of ions and molecules is changed from that of the actual concentration. When activity does not equal concentration, the species is behaving *nonideally.* This has been explained partially as resulting from forces of attraction between positive and negative ions and is related to the thermodynamics of the reaction. *Ideal* behavior occurs when activity equals concentration.

Therefore, activities or effective concentrations, rather than actual concentrations, should be used in equilibrium relationships for accurate results.

$$\frac{\{\text{C}\}^c\{\text{D}\}^d}{\{\text{A}\}^a\{\text{B}\}^b} = K \tag{2.23}$$

where the braces { } distinguish activity from concentration.

We need some general rules for expressing activity in these relationships. In many texts, activity and concentration are related as follows:

$$\gamma_i = \frac{\{i\}}{[i]} \tag{2.24}$$

where γ_i = activity coefficient for species i
$\{i\}$ = activity of species i
$[i]$ = concentration of species i

To be strictly correct, the concentration of species i needs to be reported relative to a *reference state* for species i as follows:

$$\gamma_i = \frac{\{i\}}{[i]/[i]_o} \qquad (2.25)$$

where $[i]_o$ is the concentration of species i in the reference state.

So, activity $\{i\}$ is actually unitless. The reference state for ions and molecules dissolved in water is typically 1.0 M. The reference state for solvents is the pure solvent with a mole fraction = 1.0. The reference state for solids is the pure solid with a mole fraction = 1.0. The reference state for gases is pure gas (i.e., 1 atm pressure). Thus, in most cases, $[i]_o = 1.0$, and Eq. (2.25) reduces to Eq. (2.24). We will use Eq. (2.24) in this book.

Unfortunately, the activity coefficient is not usually an easy number to determine with precision. Although numerical calculations from equilibrium relationships may be in error if actual concentrations are used in place of activities, the error is not very great for dilute aqueous solutions. Also, a high degree of precision is seldom required in equilibrium analytical computations. For this reason, activity coefficients will in general be assumed to be unity for equilibrium calculations in this book. A more detailed discussion of this subject, together with methods for estimating the activity coefficient, is given in Sec. 4.3.

It is up to the individual to decide whether to use concentrations or activities in the equilibrium relationships. This will of course depend on the nature and concentration of ions and molecules making up the solution of interest. However, certain conventions for expressing concentrations or activities of solvents, solutes, solids, and gases need to be understood, especially if published values for equilibrium constants are to be used. The methods of expression are as follows.

1. For ions and molecules in solution:

$$\{i\} = \gamma_i [i]$$

where $\{i\}$ = activity of species i

$[i]$ = concentration of species i, mol/L

γ_i = activity coefficient for species i

In dilute solutions, γ_i approaches 1.0 and $\{i\} = [i]$.

2. For a solvent in solution or for mixtures of liquids:

$$\{i\} = \gamma_i X_i$$

where X_i is the mole fraction of species i (defined in Sec. 2.10). In aqueous solutions, the solvent is water. So, $X_{H_2O} \approx 1$ (this really means that $[H_2O]/[H_2O]_o = 1$). Here, $\gamma_{H_2O} = 1$ for the standard reference state. Therefore, $\{H_2O\} = 1.0$. This is somewhat confusing to students. However, by convention, we typically assume that $\{H_2O\} = [H_2O] = 1.0$. Thus, water is left out of an equilibrium constant expression for dilute aqueous solutions. It should be noted that for seawater (contains lots of dissolved salts), $\gamma_{seawater} \approx 0.98$, which can have a significant effect on equilibrium calculations.

3. For pure solids or pure liquids in equilibrium with aqueous solution:

$$\{i\} = [i] = 1.0$$

As with water, the concentration or activity of pure solids or pure liquids do not need to be included in equilibrium constant expressions.

4. For gases in equilibrium with aqueous solution:

$$\{i\} = \gamma_i P_i$$

where P_i is the partial pressure of gas i in atmospheres (or bars). When total pressure decreases, γ_{gas} approaches 1.0. When reactions take place at atmospheric pressure, $\{i\} \approx P_i$. Air is about 21 percent oxygen, so $P_{O_2} = 0.21$ atm $= \{O_2\} = [O_2]$.

With gases, the concept of *fugacity*, which has units of partial pressure, has also been used. Morel and Hering[4] describe fugacity as being to gases as activity is to ions in solution. That is, fugacity accounts for the nonideal behavior of gases. Mackay[5] has described how fugacity might be used in environmental engineering and science. Schwarzenbach et al.[6] use fugacity as a measure of the "escaping tendency" from a phase.

2.13 | VARIATIONS OF THE EQUILIBRIUM RELATIONSHIP

Equations (2.22) and (2.23) are general forms of the equilibrium relationship. They are useful in helping to understand the various ways in which substances may be distributed in aqueous solution and methods for their control. *Homogeneous chemical equilibria* are characterized by all reactants and products of the reaction occurring in the same physical state or phase, such as reactions between gases or between materials dissolved in water. Examples of homogeneous equilibria in water are the ionization of weak acids and bases, and complex formation.

Heterogeneous chemical equilibria are characterized by substances occurring in two or more physical phases. Examples are equilibria for the solubility of a gas in a liquid, the solubility of solids in water, the distribution of a material between two different solvents, the equilibrium of a substance between its liquid phase and gaseous phase, or between its liquid phase and solid phase. Examples of homogeneous and heterogeneous equilibria of particular concern to environmental engineering and science are given here. These are considered in greater detail in Chaps. 3 and 4.

[4]F. M. M. Morel and J. G. Hering, "Principles and Applications of Aquatic Chemistry," Wiley, New York, 1993.

[5]D. Mackay, Finding Fugacity Feasible, *Env. Sci. & Tech.,* **13**: 1218–1223 (1979).

[6]R. P. Schwarzenbach, P. M. Gschwend, and D. M. Imboden, "Environmental Organic Chemistry."

Ionization

The theory of ionization stems from a doctoral dissertation completed by Svante Arrhenius in 1887. According to the original theory of Arrhenius, all acids, bases, and salts dissociate into ions when placed in solution in water. He noted that equivalent[7] solutions of different compounds often varied greatly in conductivity. This phenomenon was attributed by Arrhenius to a difference in degree of dissociation or ionization and serves today to explain many of the observed phenomena in aqueous solution.

Ion Product of Water

One of the most important equilibria of concern in dealing with aqueous solutions is the dissociation of water into a hydrogen ion, or proton, and hydroxyl ion.

$$H_2O \rightleftharpoons H^+ + OH^- \tag{2.26}$$

A proton is a very small particle and as such would have an extremely large charge-to-volume ratio. As a result, it will attach itself to almost anything that does not have a large positive charge. In aqueous solution, it readily becomes attached to water molecules, so that the following equation is a more correct description of water dissociation than Eq. (2.26):

$$2H_2O \rightleftharpoons H_3O^+ + OH^- \tag{2.27}$$

where H_3O^+ is called the *hydronium ion*. The hydronium ion can also react with water to form other hydrated species, so even Eq. (2.27) is not a completely accurate description for the dissociation of water. For many practical purposes, the simple dissociation indicated by Eq. (2.26) can be assumed. This leads to the ion product for water, which can be written as follows (using molar concentration instead of activity):

$$\frac{[H^+][OH^-]}{[H_2O]} = K_W \tag{2.28}$$

However, by convention $[H_2O]$ is taken to equal 1, so that the ion product for water in its simplest form is

$$[H^+][OH^-] = K_W = 10^{-14} \qquad \text{at } 25°C \tag{2.29}$$

In satisfying this equilibrium, the numerical values of $[H^+]$ and $[OH^-]$ include *all* the H^+ and OH^- ions present, regardless of whether these ions are produced by the water alone or are contributed by other constituents in the water.

Ionization of Acids and Bases

The classical definition of an acid is a compound that yields a proton (H^+) upon addition to water. A base yields a hydroxide ion (OH^-) upon addition to water. Arrhenius' theory of ionization can help explain the variation in strength of acids and bases. All strong acids and bases are considered to approach 100 percent ionization

---

[7]See Sec. 11.4 for definition.

in dilute solutions; that is, they completely ionize and dissociate. For example, consider the case of a strong acid HA:

$$HA \rightleftharpoons H^+ + A^-$$

$$\frac{[H^+][A^-]}{[HA]} = K_A$$

For a strong acid, K_A is very large since there will be almost no undissociated acid (HA) in solution.

The weak acids and bases, however, are so poorly ionized that in most cases it is impractical to express the degree of ionization as a percentage. The equilibrium relationship, however, can be used as will now be illustrated.

For a typical monoprotic (yields one proton) acid (acetic acid),

$$HAc \rightleftharpoons H^+ + Ac^-$$

$$\frac{[H^+][Ac^-]}{[HAc]} = K_A = 1.75 \times 10^{-5} \quad \text{at } 25°C \quad (2.30)$$

where Ac^- is used to designate the acetate ion.

For a typical diprotic (may yield two protons) acid (carbonic acid),

$$H_2CO_3 \rightleftharpoons H^+ + HCO_3^-$$

$$\frac{[H^+][HCO_3^-]}{[H_2CO_3]} = K_{A1} = 4.45 \times 10^{-7} \quad \text{at } 25°C \quad (2.31)$$

$$HCO_3^- \rightleftharpoons H^+ + CO_3^{2-}$$

$$\frac{[H^+][CO_3^{2-}]}{[HCO_3^-]} = K_{A2} = 4.69 \times 10^{-11} \quad \text{at } 25°C \quad (2.32)$$

For a typical base (ammonia, NH_3),

$$NH_3 + H_2O \rightleftharpoons NH_4^+ + OH^-$$

$$\frac{[NH_4^+][OH^-]}{[NH_3]} = K_B = 1.75 \times 10^{-5} \quad \text{at } 25°C \quad (2.33)$$

The concentration of water was not included in Eq. (2.33) for the reason discussed in Sec. 2.12. Tables giving ionization constants of weak acids, bases, and salts may be found in the usual handbooks and many textbooks of quantitative analysis or physical chemistry.

EXAMPLE 2.9

If 3 g of acetic acid (HAc = CH_3COOH) is added to enough distilled water to make 1 liter of solution, what will be the acetate ion concentration?

The molar concentration of the solution is $\frac{3}{60}$ or 0.05 M. If x moles of the acetic acid ionize to form H^+ and Ac^- ions, then

$$[HAc] = (0.05 - x) \quad \text{mol/L}$$

and
$$[H^+] = [Ac^-] = x \quad \text{mol/L}$$

$$K_A = \frac{[H^+][Ac^-]}{[HAc]} = \frac{(x)(x)}{0.05 - x} = 1.75 \times 10^{-5}$$

Solving, we obtain $x = [Ac^-] = [H^+] = 9.27 \times 10^{-4}$ mol/L. Thus, 0.05 M acetic acid is $9.27(10^{-4})(100)/0.05$ or 1.85 percent ionized. The pH, defined here as $-\log [H^+]$, equals 3.03. Much more detail on acid-base reactions is given in Sec. 4.5.

Complex Ions

Complex ions consist of one or more central ions (usually metals) that are associated with one or more ions or molecules (called *ligands*) which act to stabilize the central ion and keep it in solution. For example, if Hg^{2+} and Cl^- are present in water, they will combine to form the undissociated but soluble species $HgCl_2(aq)$, where (aq) is used to designate that the species is in solution. Chloride can also combine with mercury in other proportions to form a variety of complexes. Equilibrium relationships can be developed for the various mercury-chloride species from the following reactions:

$$Hg^{2+} + Cl^- \rightleftharpoons HgCl^+ \tag{2.34}$$

$$HgCl^+ + Cl^- \rightleftharpoons HgCl_2 \tag{2.35}$$

$$HgCl_2 + Cl^- \rightleftharpoons HgCl_3^- \tag{2.36}$$

$$HgCl_3^- + Cl^- \rightleftharpoons HgCl_4^{2-} \tag{2.37}$$

Equilibrium relationships associated with the reactions of Eqs. (2.34) to (2.37) are as follows (using molar concentration instead of activity):

$$\frac{[HgCl^+]}{[Hg^{2+}][Cl^-]} = K_1 = 10^{6.72} = 5.25 \times 10^6 \tag{2.38}$$

$$\frac{[HgCl_2]}{[HgCl^+][Cl^-]} = K_2 = 10^{6.51} = 3.24 \times 10^6 \tag{2.39}$$

$$\frac{[HgCl_3^-]}{[HgCl_2][Cl^-]} = K_3 = 10 \tag{2.40}$$

$$\frac{[HgCl_4^{2-}]}{[HgCl_3^-][Cl^-]} = K_4 = 10^{0.97} = 9.33 \tag{2.41}$$

The usual convention is to write complex reactions as indicated in Eqs. (2.34) to (2.37), as formation rather than as dissociation reactions. The equilibrium constant is then called a *formation* or *stablity constant*. The subscript on the stability constant indicates the number of chloride ions in the complex formed by the reaction.

It is sometimes convenient to consider overall reactions for the formation of complexes, and these can be developed by combination of the stepwise reactions indicated above (Eqs. (2.34) to (2.37)).

$$Hg^{2+} + Cl^- \rightleftharpoons HgCl^+ \qquad \frac{[HgCl^+]}{[Hg^{2+}][Cl^-]} = \beta_1 \qquad (2.42)$$

$$Hg^{2+} + 2Cl^- \rightleftharpoons HgCl_2 \qquad \frac{[HgCl_2]}{[Hg^{2+}][Cl^-]^2} = \beta_2 \qquad (2.43)$$

$$Hg^{2-} + 3Cl^- \rightleftharpoons HgCl_3^- \qquad \frac{[HgCl_3^-]}{[Hg^{2+}][Cl^-]^3} = \beta_3 \qquad (2.44)$$

$$Hg^{2+} + 4Cl^- \rightleftharpoons HgCl_4^{2-} \qquad \frac{[HgCl_4^{2-}]}{[Hg^{2+}][Cl^-]^4} = \beta_4 \qquad (2.45)$$

Here the subscript on the equilibrium constant has the same meaning as in Eqs. (2.38) to (2.41). It can easily be shown that $\beta_1 = K_1, \beta_2 = K_1 K_2, \beta_3 = K_1 K_2 K_3$, and $\beta_4 = K_1 K_2 K_3 K_4$. Sometimes the reciprocal of the overall formation constant of the highest complex is given as the *instability constant* for that reaction. Thus,

$$HgCl_4^{2-} \rightleftharpoons Hg^{2+} + 4Cl^- \qquad \frac{[Hg^{2+}][Cl^-]^4}{[HgCl_4^{2-}]} = K_{inst} \qquad (2.46)$$

EXAMPLE 2.10

The chloride concentration in a typical freshwater stream is 10^{-3} M. If the $HgCl_2(aq)$ concentration is 10^{-8} M (about the accepted limit for Hg in drinking water), what will be the concentrations of Hg^{2+}, $HgCl^+$, $HgCl_3^-$, and $HgCl_4^{2-}$?

From Eq. (2.39)

$$[HgCl^+] = \frac{[HgCl_2]}{K_2[Cl^-]} = \frac{(10^{-8})}{10^{6.51}(10^{-3})} = 3.1 \times 10^{-12} M$$

From Eq. (2.38)

$$[Hg^{2+}] = \frac{[HgCl^+]}{K_1[Cl^-]} = \frac{3.1(10^{-12})}{10^{6.72}(10^{-3})} = 5.9 \times 10^{-16} M$$

From Eq. (2.40)

$$[HgCl_3^-] = K_3[HgCl_2][Cl^-] = 10(10^{-8})(10^{-3}) = 10^{-10} M$$

From Eq. (2.41)

$$[HgCl_4^{2-}] = K_4[HgCl_3^-][Cl^-] = 9.33(10^{-10})(10^{-3}) = 9.3 \times 10^{-13} M$$

These calculations indicate that most of the mercury in natural waters occurs as the neutral $HgCl_2(aq)$ species. Little of the mercury occurs in ionized forms. The sorptive behavior of metals is significantly affected by complexation, pH, and charge, as well as the nature of the adsorbent (solid). Additional details are given in Secs. 3.12 and 4.9.

Other soluble molecules or ions can act as ligands to form complexes with metals such as mercury. Among the ligands are H^+, OH^-, CO_3^{2-}, NH_3, F^-, CN^-, $S_2O_3^{2-}$, and many other inorganic and organic species. NH_3 complexes with metals are common. For example, with silver

$$Ag^+ + 2NH_3 \rightleftharpoons Ag(NH_3)_2^+$$

$$\frac{[Ag(NH_3)_2^+]}{[Ag^+][NH_3]^2} = \beta_2 = 1.74 \times 10^7 \tag{2.47}$$

All such ions are readily destroyed by creating conditions, physically or chemically, that will remove one of the dissociation products.

The silver-ammonia complex ion can be destroyed by adding a source of hydrogen ions. In this case destruction is caused by the formation of a more stable complex ion, NH_4^+. The ammonium ion (NH_4^+) exists in equilibrium, written in the formation direction, as follows:

$$NH_3 + H^+ \rightleftharpoons NH_4^+$$

$$\frac{[NH_4^+]}{[NH_3][H^+]} = K = 1.8 \times 10^9 \tag{2.48}$$

Addition of a strong base such as sodium hydroxide will decrease the $[H^+]$ concentration through the formation of poorly ionized water, and the equilibrium for the silver-ammonia complex will be shifted far to the left but not completely. The equilibrium may be completely destroyed by boiling the solution to expel ammonia. This is the basis of the separation and determination of ammonia nitrogen by the distillation technique.

Solubility Product

A fundamental concept is that all solids, no matter how insoluble, are soluble to some degree. For example, silver chloride and barium sulfate are considered to be very insoluble. However, in contact with water they do dissolve, slightly, and form the following equilibria:[8]

$$AgCl(s) \rightleftharpoons Ag^+ + Cl^- \tag{2.49}$$

$$BaSO_4(s) \rightleftharpoons Ba^{2+} + SO_4^{2-} \tag{2.50}$$

$$Cu(OH)_2(s) \rightleftharpoons Cu^{2+} + 2OH^- \tag{2.51}$$

Crystals of compounds consist of ions arranged in an orderly manner. Thus, when crystals of a compound are placed in water, the ions at the surface migrate

---

[8]The (s) represents solid or precipitated material.

into the water and will continue to do so until the salt is completely dissolved or a condition of saturation is attained. With so-called insoluble substances, the saturation value is very small and is reached quickly. In silver chloride, barium sulfate, and other insoluble compounds, the ionic concentrations that can exist in equilibrium with the solid of crystalline material are very small.

The equilibrium that exists between crystals of a compound in the solid state and its ions in solution is amenable to consideration under the equilibrium relationship and can be treated mathematically as though the equilibrium were homogeneous in nature. For example, consider silver chloride at equilibrium, as shown in Eq. (2.49) and using molar concentrations instead of activities:

$$\frac{[Ag^+][Cl^-]}{[AgCl(s)]} = K \tag{2.52}$$

but $[AgCl(s)]$ represents the silver chloride that exists in the solid state. Thus, in Eq. (2.52), $[AgCl(s)]$ is taken to equal 1; so

$$[Ag^+][Cl^-] = K_{sp} \tag{2.53}$$

where the constant K_{sp} is called the *solubility-product* constant.

For more complex substances, such as tricalcium phosphate, that ionize as follows:

$$Ca_3(PO_4)_3(s) \rightleftharpoons 3Ca^{2+} + 2PO_4^{3-} \tag{2.54}$$

the solubility-product expression in accordance with Eq. (2.53) is

$$[Ca^{2+}]^3[PO_4^{3-}]^2 = K_{sp} \tag{2.55}$$

Solubility products for nearly all insoluble substances may be obtained by reference to qualitative and quantitative textbooks or chemical handbooks. Typical solubility-product constants of interest in water chemistry are listed in Table 2.5.

A prediction of relative solubilities of compounds cannot be made by a simple comparison of solubility-product values because of the squares and cubes that enter into the calculation when more than two ions are derived from one molecule, as shown in Eqs. (2.54) and (2.55). Barium sulfate, which yields two ions, and calcium fluoride, which yields three ions, may be used to illustrate the point. If we let S represent the solubility of the metal, then for $BaSO_4$

$$[Ba^{2+}] = S = [SO_4^{2-}]$$

and
$$[Ba^{2+}][SO_4^{2-}] = S^2 = K_{sp} = 1 \times 10^{-10} \tag{2.56}$$

Table 2.5 | Typical solubility-product constants

Equilibrium equation	K_{sp} at 25°C	Significance in environmental engineering and science
$MgCO_3(s) \rightleftharpoons Mg^{2+} + CO_3^{2-}$	4×10^{-5}	Hardness removal, scaling
$Mg(OH)_2(s) \rightleftharpoons Mg^{2+} + 2OH^-$	9×10^{-12}	Hardness removal, scaling
$CaCO_3(s) \rightleftharpoons Ca^{2+} + CO_3^{2-}$	5×10^{-9}	Hardness removal, scaling
$Ca(OH)_2(s) \rightleftharpoons Ca^{2+} + 2OH^-$	5×10^{-9}	Hardness removal
$CaSO_4(s) \rightleftharpoons Ca^{2+} + SO_4^{2-}$	2×10^{-5}	Flue gas desulfurization
$AgCl(s) \rightleftharpoons Ag^+ + Cl^-$	3×10^{-10}	Chloride analysis, heavy metal removal
$Ag_2CO_3(s) \rightleftharpoons 2Ag^+ + CO_3^{2-}$	8.5×10^{-12}	Heavy metal removal and fate
$AgOH(s) \rightleftharpoons Ag^+ + OH^-$	2×10^{-8}	Heavy metal removal
$Ag_2SO_4(s) \rightleftharpoons 2Ag^+ + SO_4^{2-}$	1.6×10^{-5}	Heavy metal removal
$Ag_2S(s) \rightleftharpoons 2Ag^+ + S^{2-}$	1×10^{-49}	Heavy metal removal and fate
$CdCO_3(s) \rightleftharpoons Cd^{2+} + CO_3^{2-}$	1.8×10^{-14}	Heavy metal removal and fate
$Cd(OH)_2(s) \rightleftharpoons Cd^{2+} + 2OH^-$	2×10^{-14}	Heavy metal removal
$CdS(s) \rightleftharpoons Cd^{2+} + S^{2-}$	1.4×10^{-29}	Heavy metal removal and fate
$Cr(OH)_3(s) \rightleftharpoons Cr^{3+} + 3OH^-$	6×10^{-31}	Heavy metal removal
$CuCO_3(s) \rightleftharpoons Cu^{2+} + CO_3^{2-}$	2.3×10^{-10}	Heavy metal removal and fate
$Cu(OH)_2(s) \rightleftharpoons Cu^{2+} + 2OH^-$	2×10^{-19}	Heavy metal removal
$CuS(s) \rightleftharpoons Cu^{2+} + S^{2-}$	1×10^{-36}	Heavy metal removal and fate
$NiCO_3(s) \rightleftharpoons Ni^{2+} + CO_3^{2-}$	1.5×10^{-7}	Heavy metal removal and fate
$Ni(OH)_2(s) \rightleftharpoons Ni^{2+} + 2OH^-$	2×10^{-16}	Heavy metal removal
$NiS(s) \rightleftharpoons Ni^{2+} + S^{2-}$	1.4×10^{-24}	Heavy metal removal and fate
$PbCO_3(s) \rightleftharpoons Pb^{2+} + CO_3^{2-}$	7.4×10^{-14}	Heavy metal removal and fate
$Pb(OH)_2(s) \rightleftharpoons Pb^{2+} + 2OH^-$	2.5×10^{-16}	Heavy metal removal
$PbS(s) \rightleftharpoons Pb^{2+} + S^{2-}$	1×10^{-28}	Heavy metal removal and fate
$PbSO_4(s) \rightleftharpoons Pb^{2+} + SO_4^{2-}$	1.6×10^{-8}	Heavy metal removal
$ZnCO_3(H_2O)\,(s) \rightleftharpoons Zn^{2+} + CO_3^{2-} + H_2O$	5.5×10^{-11}	Heavy metal removal and fate
$Zn(OH)_2(s) \rightleftharpoons Zn^{2+} + 2OH^-$	8×10^{-18}	Heavy metal removal
$ZnS(s) \rightleftharpoons Zn^{2+} + S^{2-}$	1×10^{-22}	Heavy metal removal and fate
$Al(OH)_3(s) \rightleftharpoons Al^{3+} + 3OH^-$	1×10^{-32}	Coagulation
$Fe(OH)_3(s) \rightleftharpoons Fe^{3+} + 3OH^-$	6×10^{-38}	Coagulation, iron removal, corrosion

(Cont.)

Table 2.5 | Typical solubility-product constants (*Cont.*)

Equilibrium equation	K_{sp} at 25°C	Significance in environmental engineering and science
$Fe(OH)_2(s) \rightleftharpoons Fe^{2+} + 2OH^-$	5×10^{-15}	Coagulation, iron removal, corrosion
$FeCO_3(s) \rightleftharpoons Fe^{2+} + CO_3^{2-}$	2.8×10^{-11}	Iron removal and fate
$FeS(s) \rightleftharpoons Fe^{2+} + S^{2-}$	5×10^{-18}	Iron removal and fate; abiotic reductant
$Mn(OH)_3(s) \rightleftharpoons Mn^{3+} + 3OH^-$	1×10^{-36}	Manganese removal
$Mn(OH)_2(s) \rightleftharpoons Mn^{2+} + 2OH^-$	5×10^{-15}	Manganese removal
$AlPO_4(s) \rightleftharpoons Al^{3+} + PO_4^{3-}$	3.2×10^{-23}	Phosphate removal
$Ca_3(PO_4)_2(s) \rightleftharpoons 3Ca^{2+} + 2PO_4^{3-}$	1×10^{-27}	Phosphate removal
$CaHPO_4(s) \rightleftharpoons Ca^{2+} + HPO_4^{2-}$	3×10^{-7}	Phosphate removal
$Ca_5(PO_4)_3OH(s) \rightleftharpoons 5Ca^{2+} + 3PO_4^{3-} + OH^-$	8×10^{-55}	Phosphate removal
$FePO_4(s) \rightleftharpoons Fe^{3+} + PO_4^{3-}$	1.3×10^{-22}	Phosphate removal
$Fe_3(PO_4)_2(s) \rightleftharpoons 3Fe^{2+} + 2PO_4^{3-}$	1×10^{-36}	Phosphate removal
$CaF_2(s) \rightleftharpoons Ca^{2+} + 2F^-$	3×10^{-11}	Fluoridation
$BaSO_4(s) \rightleftharpoons Ba^{2+} + SO_4^{2-}$	1×10^{-10}	Sulfate analysis

Thus, $S = 1 \times 10^{-5}$ and the solubility of $BaSO_4(s)$ is 1×10^{-5} M. Similarly, for $CaF_2(s)$

$$[Ca^{2+}] = S$$
$$[F^-] = 2S$$

and
$$[Ca^{2+}][F^-]^2 = (S)(2S)^2 = K_{sp} = 3 \times 10^{-11} \tag{2.57}$$

Thus, $S = 1.96 \times 10^{-4}$ and the solubility of $CaF_2(s)$ is 1.96×10^{-4} M.

It will be noted that calcium fluoride is about 20 times more soluble than barium sulfate. From this it is obvious that the most soluble material $CaF_2(s)$ has the smallest solubility product because of the squaring of the fluoride concentration. The case of compounds that yield more than three ions is even more exaggerated.

There are two corollary statements related to the solubility-product principle, an understanding of which is basic to explaining the phenomena of precipitation and solution of precipitates. They may be expressed as follows:

1. In an unsaturated solution, the product of the molar concentration of the ions is less than the solubility-product constant, or for a species AB, $[A^+][B^-] < K_{sp}$.

2. In a supersaturated solution, the product of the molar concentration of the ions is greater than the solubility-product constant, or $[A^+][B^-] > K_{sp}$. In the former case, if undissolved AB is present, it will dissolve to the extent that $[A^+][B^-] = K_{sp}$, and a saturated solution results. In the second case, nothing

will happen until such time as crystals of AB are introduced into the solution or internal forces allow formation of crystal nuclei; then precipitation will occur until the ionic concentrations are reduced equal to those of a saturated solution.

Common Ion Effect

The advantage of relating solubilities to the equilibrium relationship is that this allows mathematical treatment of the equilibrium and prediction of the effect of adding a common ion to a solution containing a slightly soluble salt. For example, consider a solution that has been saturated with barium sulfate. As indicated in Eq. (2.56), both $[Ba^{2+}]$ and $[SO_4^{2-}]$ would equal 1×10^{-5} M. Now, if the barium ion concentration should be increased by addition from an outside source, such as $BaCl_2$, the concentration of sulfate ion must decrease and the amount of precipitated $BaSO_4(s)$ must increase in order for K_{sp} to remain the same. To illustrate, assume that 10×10^{-5} mol/L of $BaCl_2$ is added to the solution saturated with barium sulfate. This will result in the formation of an additional x moles of precipitated $BaSO_4$. The following changes in $[Ba^{2+}]$ and $[SO_4^{2-}]$ must then take place:

$$\underset{\substack{(1\times10^{-5} + \\ 10\times10^{-5} - x)}}{Ba^{2+}} + \underset{(1\times10^{-5} - x)}{SO_4^{2-}} \rightarrow \underset{x}{BaSO_4(s)}$$

According to the solubility-product principle,

$$(11.0 \times 10^{-5} - x)(1.0 \times 10^{-5} - x) = K_{sp} = 1 \times 10^{-10}$$

By solving for x, it is found that an additional 0.90×10^{-5} mol/L of precipitated $BaSO_4$ is formed and that the new equilibrium concentrations of barium and sulfate ions are

$$[Ba^{2+}] = (11.0 \times 10^{-5}) - (0.90 \times 10^{-5}) = 10.1 \times 10^{-5}$$

$$[SO_4^{2-}] = (1.0 \times 10^{-5}) - (0.90 \times 10^{-5}) = 0.10 \times 10^{-5}$$

These calculations indicate that the $[SO_4^{2-}]$ is reduced considerably. This example is an application of the *common ion effect,* which is used extensively in environmental engineering practice as well as in qualitative and quantitative analysis to accomplish essentially complete precipitation of desired ions. For example, hardness (Ca^{2+} and Mg^{2+}) is removed by adding CO_3^{2-} to precipitate $CaCO_3(s)$ and OH^- to precipitate $Mg(OH)_2(s)$.

Diverse Ion Effect

The *diverse ion effect* describes the adverse effect that unrelated ions often have upon the solubility of some relatively insoluble substances. Such ions, theoretically, play no part in the chemical equilibrium involved but often increase the solubility of desired precipitates to such an extent that undesirable results are obtained. Nitrate ion has such an effect on silver chloride. Because of the diverse ion effect, it is general practice to keep the concentration of extraneous ions as low as possible during qualitative and quantitative work.

The diverse ion effect can be explained by considering activity corrections to the solubility product. Using $CaCO_3(s)$ as an example:

$$K_{sp} = \{Ca^{2+}\}\{CO_3^{2-}\} = \gamma[Ca^{2+}]\gamma[CO_3^{2-}]$$

Since increasing ionic strength (increasing the concentration of unrelated ions) causes γ to decrease to less than 1.0 (see Sec. 4.3), this increases the molar concentration of Ca^{2+} and CO_3^{2-} in solution at equilibrium.

Sometimes a common ion may serve very well up to certain concentrations, but when used at higher concentrations, it appears to have a diverse ion effect. In this case, the usual explanation is that complex ion formation is taking place, as described previously. Hydrochloric acid acts in such a manner when it is used as the agent to precipitate silver ion. When hydrochloric acid is added in excess, the soluble $AgCl_2^-$ complex is formed:

$$AgCl(s) + Cl^- \rightarrow AgCl_2^- \tag{2.58}$$

Therefore, the amount of hydrochloric acid used should be carefully controlled.

2.14 | WAYS OF SHIFTING CHEMICAL EQUILIBRIA

Environmental engineers deal routinely with materials that are in either homogeneous or heterogeneous equilibrium. They, like analytical chemists, must be able to apply stresses to their systems in accordance with Le Chatelier's principle to bring about desired changes. Many of the stresses that they apply are exactly the same in character as those used by chemists. Therefore, it is important to consider the ways in which equilibria can be shifted to bring about essentially complete reactions. Five methods are commonly employed.

Formation of Insoluble Substances

All precipitation reactions are examples of this method of equilibrium shift. In this case a knowledge of the solubility-product principle, solubility-product constants, and the common ion effect is brought into service. The removal of metal ions from industrial wastes, such as copper and brass wastes, by precipitation with calcium hydroxide, and the softening of hard waters by lime–soda ash treatment are excellent examples of how this method of shifting a chemical equilibrium is applied to gain an objective. Equations associated with these precipitation reactions are listed in Table 2.5.

Formation of a Weakly Ionized Compound

Certain systems that are in equilibrium can be destroyed by adding a reagent that will combine with one of the ions to form a poorly ionized compound. The neutralization of acid and of caustic wastes is based upon such information, since the reaction involved is between hydrogen ions and hydroxyl ions to form poorly ionized water:

$$H^+(\text{acid waste}) + NaOH \rightarrow H_2O + Na^+ \tag{2.59}$$

Similar reactions are frequently used to bring the pH of industrial wastes into a favorable range for subsequent biological treatment.

This method of equilibrium shifting routinely is used to dissolve precipitates of the metallic hydroxides, such as ferric and aluminum hydroxide:

$$Fe(OH)_3(s) + 3H^+ \rightarrow Fe^{3+} + 3H_2O \tag{2.60}$$

$$Al(OH)_3(s) + 3H^+ \rightarrow Al^{3+} + 3H_2O \tag{2.61}$$

Formation of Complex Ion

Complex-ion formation can be used to dissolve insoluble salts and hydroxides. Silver chloride, for example, dissolves readily in ammonium hydroxide solution. This occurs because silver ion combines with molecular NH_3 contained in the ammonium hydroxide to form $[Ag(NH_3)_2]^+$; as a result the solution becomes unsaturated with respect to silver and chloride ions and solid silver chloride passes into solution in an attempt to form a saturated solution. The net effect is as follows:

$$AgCl(s) + 2NH_3 \rightarrow [Ag(NH_3)_2]^+ + Cl^- \tag{2.62}$$

If enough ammonium hydroxide is present, all silver chloride passes into solution. The equilibrium relationship for Eq. (2.62) may be written as

$$\frac{[Ag(NH_3)_2^+][Cl^-]}{[NH_3]^2} = K \tag{2.63}$$

A value for the equilibrium constant K may be determined from solubility-product data for silver chloride and from the equilibrium constant for the ammonia-silver complex. Illustrations of such computations are given in Section 4.9.

Zinc and copper hydroxide dissolve in ammonium hydroxide for the same reason given in the preceding paragraph. The complex ions formed are $[Zn(NH_3)_4]^{2+}$ and $[Cu(NH_3)_4]^{2+}$. These reactions illustrate why ammonium hydroxide would not be a good reagent for precipitating copper and zinc ions from a brass-mill waste.

Industrial wastes containing sodium cyanide are particularly toxic to fish, even though the concentration of cyanide ion may be reduced to very low levels by dilution with river water. The destruction or inactivation of cyanide ion may be accomplished by complex-ion formation. Formerly, before development of more efficient methods, cyanide ion was reduced to low levels by treatment with ferrous sulfate. In the reaction, Fe^{2+} combined with CN^- to give the complex ion $Fe(CN)_6^{4-}$, which could be precipitated as Prussian blue by subsequent oxidation of excess Fe^{2+} to Fe^{3+}, in the presence of potassium ions,

$$Fe^{3+} + K^+ + [Fe(CN)_6]^{4-} \rightarrow KFe[Fe(CN)_6](s) \tag{2.64}$$

Formation of a Gaseous Product

In reactions involving the formation of a gaseous product, the reactions go to practical completion because the gas escapes from the sphere of the reaction. This method of forcing a reaction to completion is used when dissolving metallic sulfides, such as ferrous sulfide, in hydrochloric acid:[9]

$$FeS(s) + 2H^+ \rightarrow H_2S \uparrow + Fe^{2+} \tag{2.65}$$

This does not necessarily mean that all metallic sulfides are soluble in hydrochloric acid, for the sulfides of copper and mercury are not. This can be explained by considering that copper and mercury sulfides are so insoluble that the sulfide ion liberated by them at equilibrium is of such small concentration that not enough un-ionized hydrogen sulfide (H_2S) is formed, in the presence of concentrated hydrochloric acid, to be released as a gas. Consequently, the sulfides do not dissolve.

In industrial waste treatment, cyanides were formerly removed from aqueous solution by treatment with sulfuric acid:

$$2CN^- + 2H^+ + SO_4^{2-} \rightarrow 2HCN \uparrow + SO_4^{2-} \tag{2.66}$$

The hydrogen cyanide released as a gas was diluted with large volumes of air and forced up through tall stacks to get dispersion and hopefully to avoid serious atmospheric-pollution problems. More modern methods accomplish destruction of cyanide ion by oxidation or other techniques.

Oxidation and Reduction

A very sure method of sending reactions to completion is oxidation and reduction. In this way one or more of the ions involved in the equilibrium reaction can be destroyed, and the reaction will proceed to completion. A classic example is the destruction of cyanide ion by chlorination according to the following equation:

$$2CN^- + 5Cl_2 + 8OH^- \rightarrow 10Cl^- + 2CO_2 + N_2 \uparrow + 4H_2O \tag{2.67}$$

From the discussion already presented, this reaction would go to completion for two reasons.

1. Cyanide ion is oxidized to carbon dioxide and nitrogen.

2. The nitrogen escapes as a gas.

Since formation of nitrogen is dependent upon oxidation of cyanide ion, the reaction is considered complete whether the nitrogen escapes or not.

---

[9]The arrow represents production and release of gas from solution.

2.15 | AMPHOTERIC HYDROXIDES

The oxides or hydroxides of metals are basic in character and react with acids to form salts. The insoluble metallic hydroxides, such as ferric hydroxide, dissolve readily in acids to form salts but are insoluble in solutions of bases. Likewise the oxides of nonmetals are acidic in character, and insoluble forms are soluble in bases but not in acids. These characteristics serve as one basis of differentiating between metals and nonmetals.

The hydroxides of aluminum, zinc, chromium, and a few other elements are soluble in both acids and bases. They are known as *amphoteric* hydroxides, and advantage is often taken of this fact to accomplish separations in qualitative analysis and in chemical processing.

An insoluble metallic hydroxide, of course, exists in equilibrium with its ions. For example, ferric hydroxide dissolves to a limited extent to produce Fe^{3+} and OH^- ions:

$$Fe(OH)_3(s) \rightleftharpoons Fe^{3+} + 3OH^- \qquad (2.68)$$

and at saturation

$$[Fe^{3+}][OH^-]^3 = K_{sp} \qquad (2.69)$$

When a strong acid is added, the OH^- ions combine with the H^+ ions of the acid to form poorly ionized water, and the $[OH^-]$ decreases so that

$$[Fe^{3+}][OH^-]^3 < K_{sp} \qquad (2.70)$$

Therefore, ferric hydroxide dissolves in an attempt to establish conditions represented in Eq. (2.69). If enough acid is added, eventually all the ferric hydroxide will dissolve. A similar situation holds for insoluble nonmetallic oxides that are soluble in bases, except that the insoluble oxide is in equilibrium with H^+ and some acid radical.

An amphoteric hydroxide forms complexes with hydroxide. The complexes formed under both acidic and basic conditions are charged ions and soluble in water. Those formed under conditions between these extremes are neutral in charge and insoluble in water. Aluminum hydroxide will serve as an example. Stepwise reactions for Al^{3+} and OH^- are as follows:

$$Al^{3+} + OH^- \rightleftharpoons Al(OH)^{2+} \qquad (2.71)$$
$$Al(OH)^{2+} + OH^- \rightleftharpoons Al(OH)_2^+ \qquad (2.72)$$
$$Al(OH)_2^+ + OH^- \rightleftharpoons Al(OH)_3\,(aq) \qquad (2.73)$$
$$Al(OH)_3 + OH^- \rightleftharpoons Al(OH)_4^- \qquad (2.74)$$

Under strongly acidic conditions, the OH^- concentration is low and the species present will be the positively charged ions, Al^{3+} and $Al(OH)^{2+}$. If a strong base is gradually added, the OH^- concentration will increase and these ions will begin to add to the aluminum complexes, reducing the charges until the neutral and insoluble species $Al(OH)_3(s)$ is formed, which will then precipitate from solution. As base continues to be added and the OH^- concentration in solution increases further, the negatively charged and soluble $Al(OH)_4^-$ ion is formed, and the precipitate dissolves.

Actually, metal ions such as Al^{3+} and Fe^{3+} do not occur as such in solution, just as H^+ does not. Because of their strong positive charges, they readily become hydrated to form species such as $[Fe(H_2O)_6]^{3+}$ and $[Al(H_2O)_6]^{3+}$. However, the symbols Fe^{3+} and Al^{3+} will be used in this book to represent these species in order to reduce the complexity of the equations.

The amphoteric property of aluminum hydroxide is a factor limiting its use as a coagulant in water purification and industrial waste treatment. The amphoteric properties of zinc and chromium hydroxides are important considerations in treating industrial wastes containing Zn^{2+} and Cr^{3+}.

PROBLEMS

2.1 Calculate the formula weight and equivalent weight of (*a*) $MgCO_3$, (*b*) $NaNO_3$, (*c*) CO_2, and (*d*) K_2HPO_4. You may have to assume a reaction to determine Z.
Answer: (*a*) 84.3, 42.15; (*b*) 85, 85; (*c*) 44, 22; (*d*) 174.2, 87.1

2.2 Calculate the formula weight and equivalent weight of (*a*) $BaSO_4$, (*b*) Na_2CO_3, (*c*) H_2SO_4, and (*d*) $Mg(OH)_2$.

2.3 What is the molar concentration of a solution containing 10 g/L of (*a*) $NaOH$, (*b*) Na_2SO_4, (*c*) $K_2Cr_2O_7$, and (*d*) KCl?
Answer: (*a*) 0.25 M; (*b*) 0.0704 M; (*c*) 0.034 M; (*d*) 0.134 M

2.4 (*a*) Calculate the weight of $KMnO_4$ contained in 2 liters of a 0.15 M solution.

　　　(*b*) Calculate the weight of $KMnO_4$ contained in 2 liters of a 0.15 N solution.

2.5 A water sample contains 44 mg/L of calcium ion and 19 mg/L of magnesium ion. What is the hardness expressed as mg/L of $CaCO_3$? Note that hardness is the sum of the multivalent cations.
Answer: 188 mg/L

2.6 A water sample contains 118 mg/L of bicarbonate ion, 19 mg/L of carbonate ion, and has a pH of 9.5. What is the alkalinity of this water expressed as mg/L of $CaCO_3$? Note that alkalinity (Alk) is sometimes defined by the following expression: eq/L of $Alk = [HCO_3^-] + 2[CO_3^{2-}] + [OH^-] - [H^+]$ where [ ] represents concentration in mol/L.

2.7 Balance the following equations:

(*a*) $CaCl_2 + Na_2CO_3 \rightarrow CaCO_3 + NaCl$

(*b*) $Ca_3(PO_4)_2 + H_3PO_4 \rightarrow Ca(H_2PO_4)_2$

(*c*) $MnO_2 + NaCl + H_2SO_4 \rightarrow MnSO_4 + H_2O + Cl_2 + Na_2SO_4$

(*d* $Ca(H_2PO_4)_2 + NaHCO_3 \rightarrow CaHPO_4 + Na_2HPO_4 + H_2O + CO_2$

2.8 Balance the following equations:

(*a*) $FeS + HCl \rightarrow FeCl_2 + H_2S$

(*b*) $Cl_2 + KOH \rightarrow KCl + KClO_3 + H_2O$

(*c*) $FeSO_4 + K_2Cr_2O_7 + H_2SO_4 \rightarrow Fe_2(SO_4)_3 + Cr_2(SO_4)_3 + K_2SO_4 + H_2O$

(*d*) $Al_2(SO_4)_3 \cdot 14 H_2O + Ca(HCO_3)_2 \rightarrow Al(OH)_3 + CaSO_4 + H_2O + CO_2$

2.9 Balance the following equations:

(a) $Fe(OH)_2 + H_2O + O_2 \rightarrow Fe(OH)_3$

(b) $KI + HNO_2 + H_2SO_4 \rightarrow NO + I_2 + H_2O + K_2SO_4$

(c) $H_2C_2O_4 + KMnO_4 + H_2SO_4 \rightarrow CO_2 + MnSO_4 + K_2SO_4 + H_2O$

(d) $SO_3^{2-} + Fe^{3+} + H_2O \rightarrow SO_4^{2-} + Fe^{2+} + H^+$

2.10 Balance the following equations:

(a) $HClO \rightarrow HClO_3 + HCl$

(b) $NO_2^- + MnO_4^- + H^+ \rightarrow NO_3^- + Mn^{2+} + H_2O$

(c) $Cl^- + NO_3^- + H^+ \rightarrow Cl_2 + NO + H_2O$

(d) $I_2 + IO_3^- + H^+ + Cl^- \rightarrow ICl_2^- + H_2O$

2.11 Using half reactions, write complete balanced oxidation-reduction equations for the following:

(a) Oxidation of I^- to I_2 and reduction of MnO_2 to Mn^{2+}

(b) Oxidation of $S_2O_3^{2-}$ to SO_4^{2-} and reduction of Cl_2 to Cl^-

(c) Oxidation of NH_4^+ to NO_3^- and reduction of O_2 to H_2O

(d) Oxidation of CH_3COO^- to CO_2 and reduction of $Cr_2O_7^{2-}$ to Cr^{3+}

(e) Oxidation of $C_6H_{12}O_6$ to CO_2 and reduction of NO_3^- to N_2

Answers:

(a) $2I^- + MnO_2 + 4H^+ \rightarrow I_2 + Mn^{2+} + 2H_2O$

(b) $S_2O_3^{2-} + 4Cl_2 + 5H_2O \rightarrow 2SO_4^{2-} + 8Cl^- + 10H^+$

(c) $NH_4^+ + 2O_2 \rightarrow NO_3^- + H_2O + 2H^+$

(d) $3CH_3COO^- + 4Cr_2O_7^{2-} + 35H^+ \rightarrow 6CO_2 + 8Cr^{3+} + 22H_2O$

(e) $5C_6H_{12}O_6 + 24NO_3^- + 24H^+ \rightarrow 30CO_2 + 12N_2 + 42H_2O$

2.12 Using half reactions, write complete balanced oxidation-reduction equations for the following:

(a) Oxidation of Mn^{2+} to MnO_2 and reduction of O_2 to H_2O

(b) Oxidation of $S_2O_3^{2-}$ to SO_4^{2-} and reduction of I_2 to I^-

(c) Oxidation of NH_4^+ to NO_2^- and reduction of O_2 to H_2O

(d) Oxidation of $C_6H_{12}O_6$ to CO_2 and reduction of $Cr_2O_7^{2-}$ to Cr^{3+}

(e) Oxidation of CH_3COO^- to CO_2 and reduction of SO_4^{2-} to H_2S

2.13 Construct half reactions for the following reductions:

(a) SO_4^{2-} to S

(b) NO_3^- to NO_2^-

(c) CH_3COO^- to $CH_3CH_2CH_2COO^-$

Answers:

(a) $\frac{1}{6}SO_4^{2-} + \frac{4}{3}H^+ + e^- = \frac{1}{6}S + \frac{2}{3}H_2O$

(b) $\frac{1}{2}NO_3^- + H^+ + e^- = \frac{1}{2}NO_2^- + \frac{1}{2}H_2O$

(c) $\frac{1}{2}CH_3COO^- + \frac{5}{4}H^+ + e^- = \frac{1}{4}CH_3CH_2CH_2COO^- + \frac{1}{2}H_2O$

2.14 Construct half reactions for the following reductions:

(a) CO_2 to CH_4

(b) S_2 to H_2S

(c) $CH_3COO^- + CO_2$ to $CH_3CH_2COO^-$

2.15 Develop the appropriate half reactions, and from these construct the complete oxidation-reduction equation for oxidation of H_2S to S and reduction of Fe^{3+} to Fe^{2+}.

Answer: $H_2S + 2Fe^{3+} \rightarrow S + 2Fe^{2+} + 2H^+$

2.16 Develop the appropriate half reactions, and from these construct the complete oxidation-reduction equation for oxidation of CH_3CH_2OH to CO_2 and reduction of NO_3^- to NO_2^-.

2.17 How many moles of H_2SO_4 are required to form 65 g of $CaSO_4$ from $CaCO_3$?

Answer: 0.478

2.18 How many grams of iodine (I_2) are formed from the oxidation of an excess of KI by 6 g of $K_2Cr_2O_7$, under acid conditions [see Eq. (2.9)]?

2.19 Calculate the volume in cubic feet occupied by 120 lb of carbon dioxide at 1.5 atm and 40°C.

Answer: 750 ft^3

2.20 Determine the weight in grams of oxygen contained in a 10-liter volume under a pressure of 5 atm and at a temperature of 0°C.

2.21 What volume of oxygen at 25°C and 0.21 atm is required for combustion of 25 g of methane gas?

Answer: 364 liters

2.22 If 6 g of ethane gas (CH_3CH_3) is burned in oxygen, (*a*) how many moles of water are formed; (*b*) how many moles of carbon dioxide are formed; (*c*) what is the volume in liters of carbon dioxide formed at 1 atm pressure and 20°C?

2.23 A gas mixture at 25°C and 1 atm contains 100 mg/L of H_2S gas. What is the partial pressure exerted by this gas?

Answer: 0.072 atm

2.24 A 30-liter volume of gas at 25°C contains 12 g of methane, 1 g of nitrogen, and 15 g of carbon dioxide. Calculate (*a*) the moles of each gas present, (*b*) the partial pressure exerted by each gas, (*c*) the total pressure exerted by the mixture, and (*d*) the percentage by volume of each gas in the mixture.

2.25 Five liters of water are equilibrated with a gas mixture containing carbon dioxide at a partial pressure of 0.3 atm. If the Henry's law constant for carbon dioxide solubility is 31.6 atm-L/mol, at 25°C how many grams of carbon dioxide are dissolved in the water?

Answer: 2.09 g

2.26 What is the concentration of oxygen dissolved in water at 20°C in equilibrium with a gas mixture at 0.81 atm and containing 21 percent by volume of oxygen?

2.27 A non-aqueous-phase liquid (NAPL) mixture contains 100 kg of tetrachloroethene [C_2Cl_4; also called perchloroethene (PCE)], 100 kg of benzene (C_6H_6), 100 kg of toluene (C_7H_8), 80 kg of ethylbenzene (C_8H_{10}), and 60 kg of xylene (C_8H_{10}). Calculate the mole fraction of each compound in the NAPL.

2.28 Consider an NAPL mixture of tetrachloroethene (PCE) and fuel containing BTEX (benzene, toluene, ethylbenzene, and xylene) with the following mole fractions: PCE = 0.10, B = 0.15, T = 0.15, E = 0.08, and X = 0.10. The vapor pressures (in atmospheres at 25°C) of the pure compounds[10] are PCE = 0.0251 B = 0.126, T = 0.0380, E = 0.0126, and X = 0.0117, and Henry's constant K_H (in atm-L/mol at 25°C) for PCE is 26.9.

---

[10]Schwarzenbach, Gschwend, and Imboden, "Environmental Organic Chemistry," Wiley, New York, 1993.

(a) Assuming Raoult's and Henry's laws apply, estimate the aqueous concentration of PCE in water that is in equilibrium with this NAPL mixture.

(b) Estimate the reduction in aqueous PCE concentration (solubility) due to the presence of BTEX. *Hint:* First estimate the solubility when PCE is the sole NAPL (mole fraction = 1.0).

2.29 The aqueous concentration of trichloroethene (TCE) is measured to be 20 mg/L. The water ($T = 25°C$) is in contact with a dense non-aqueous-phase liquid (DNAPL) containing TCE. Assuming Henry's and Raoult's laws hold, estimate the mole fraction of TCE in the DNAPL. At 25°C, Henry's constant for TCE is 11.6 atm-L/mol and the vapor pressure of pure TCE is 0.0977 atm.[11]

2.30 Calculate the percent ionization and hydrogen ion concentration at 25°C in a solution containing (a) 0.10 M H_2CO_3, and (b) 0.01 M H_2CO_3.
        Answer: (a) 0.211 percent, 2.11×10^{-4} mol/L; (b) 0.67 percent, 6.67×10^{-5} mol/L

2.31 Calculate the percent ionization and hydrogen ion concentration in a solution containing 0.05 M hypochlorous acid (HOCl), which has an ionization constant at 25°C of 2.9×10^{-8}.

2.32 (a) Write reactions and equilibrium relationships for the first four complexes formed between cadmium(II) and chloride.

(b) Write the equilibrium relationship for the instability constant for $CdCl_4^{2-}$.

2.33 (a) Write reactions and equilibrium relationships for the first five complexes formed between copper(II) and NH_3. A sixth complex is theoretically possible, but its formation has not yet been detected.

(b) Write the overall reaction and equilibrium relationship for the fourth complex between copper(II) and NH_3.

2.34 Equilibrium constants for the complexes between cadmium(II) and chloride are $K_1 = 100$, $K_2 = 4.0$, $K_3 = 0.63$, and $K_4 = 0.20$. Calculate the molar concentration of each of the first four cadmium chloride complexes in a water sample if $Cd^{2+} = 10^{-8}$ M and $Cl^- = 10^{-3}$ M. Identify the most prevalent cadmium species.
        Answer: 1×10^{-9}, 4×10^{-12}, 2.52×10^{-15}, 5.04×10^{-19}, Cd^{2+}

2.35 Do Prob. 2.34, but assume that the chloride concentration is similar to that of seawater or about 0.5 M.

2.36 Cadmium (Cd) is reported to be much more toxic as the free metal (Cd^{2+}) than when complexed with ligands. The total soluble Cd concentration ($C_{T,Cd} = [Cd^{2+}]$ + the sum of all soluble Cd complexes) in a water is measured to be 10^{-4} M. In order to ensure that the toxicity of Cd is small, it is desirable to have Cd^{2+} be less than 10^{-7} M. An environmental engineer suggests adding 0.5 M NaCl to complex the Cd^{2+}. Ignore activity corrections and the formation of any solids. Consider only the complexes of chloride with cadmium—there are 4 ($\log \beta_1 = 2.0$, $\log \beta_2 = 2.6$, $\log \beta_3 = 2.4$, and $\log \beta_4 = 1.7$).

(a) Will this help keep $[Cd^{2+}]$ below 10^{-7} M?

(b) What is $[Cd^{2+}]$?

---

[11]Ibid.

2.37 Calculate the concentration in (a) moles per liter, (b) milligrams per liter, and (c) number of ions per liter, for the sulfate ion in a saturated solution of barium sulfate to which barium chloride is added until $[Ba^{2+}] = 0.0001$ M.
   Answer: (a) 10^{-6} mol/L; (b) 0.096 mg/L; (c) 6.02×10^{17} ions/L

2.38 Calculate the concentration in (a) moles per liter, (b) milligrams per liter, and (c) number of ions per liter, for the chloride ion in a saturated solution of silver chloride to which silver nitrate is added until $[Ag^+] = 0.0001$ M.

2.39 (a) Write the expression for the solubility-product constant of (1) AgCl, (2) CuS, (3) $MgNH_4PO_4$, (4) $Au(OH)_3$, (5) Ag_2CrO_4, and (6) $BaCO_3$. (b) What ions can be added to solutions containing the compounds listed in (a) which, in each case, will lower the concentration of the cation? Explain why.

2.40 Why does solubility often increase as still larger quantities of a common ion are added?

2.41 Tell what is meant by (a) common ion effect, (b) complex ion, (c) solubility-product constant, (d) amphoteric hydroxide, (e) diverse ion effect, (f) heterogeneous equilibrium, (g) homogeneous equilibrium, (h) driving reaction to completion, and (i) saturated solution.

2.42 From each of the following values of water solubility, evaluate the corresponding solubility-product constant: (a) $Mg_3(PO_4)_2$, 6.1×10^{-5} mol/L, (b) FeS, 6.3×10^{-9} mol/L, (c) $Zn_3(PO_4)_2$, 1.6×10^{-7} mol/L, and (d) CuF_2, 7.4×10^{-3} mol/L.

2.43 Which is the better chemical for removing calcium ions from solution, sodium hydroxide or sodium carbonate? Why?

2.44 Which is the better chemical for removing magnesium ions from solution, sodium hydroxide or sodium carbonate? Why?

2.45 Which of the following is the better chemical for removing calcium ions from water? Make sure you show the calculations necessary to justify your answer. For this problem, ignore calcium complexes.
   (a) NaOH
   (b) K_2CO_3
   (c) K_2SO_4
   (d) KF

2.46 Given the reaction $NH_3 + H_2O \rightleftharpoons NH_4^+ + OH^-$, how many ways can you think of for forcing the reaction (a) to the right? (b) to the left? (c) Under what conditions might it go to completion in either direction?

2.47 Determine how many milligrams per liter of magnesium ion will dissolve in water which is (a) 10^{-5} M in OH^-, (b) 10^{-3} M in OH^-.

2.48 A metal-plating waste contains 20 mg/L Cu^{2+}, and it is desired to add $Ca(OH)_2$ to precipitate all but 0.5 mg/L of the copper. To what concentration in moles per liter must the hydroxide concentration be raised to accomplish this?
   Answer: 1.6×10^{-7} mol/L

2.49 A waste contains 50 mg/L of Zn^{2+}. How high must the pH be raised to precipitate all but 1 mg/L of the zinc? What adverse effect might occur if the pH were raised too high?

2.50 Calculate the pH required to decrease the iron concentration in a water supply to 0.03 mg/L if (a) the iron is in the Fe^{2+} form, (b) the iron is in the Fe^{3+} form.
   Answer: (a) 10.0; (b) 3.7

2.51 Calculate the pH required to decrease the manganese concentration in a water supply to 0.01 mg/L if (a) the manganese is in the Mn^{2+} form, (b) the manganese is in the Mn^{3+} form.

2.52 It is desired to fluoridate a water supply by adding sufficient sodium fluoride to increase the fluoride concentration to 1 mg/L. Will this fluoride concentration be soluble in a water supply containing 200 mg/L of calcium? Show your computations.

2.53 At what Ca^{2+} concentration in milligrams per liter will precipitation of $Ca_3(PO_4)_2$ occur in a solution containing 10^{-6} M PO_4^{3-}?

2.54 If the CO_3^{2-} concentration in a water sample is 100 mg/L, what is the maximum solubility in milligrams per liter of (a) Ca^{2+}? (b) Mg^{2+}?
        Answer: (a) 0.12 mg/L; (b) 583 mg/L

2.55 (a) If excess AgCl is mixed in distilled water, what will be the concentration of silver in solution in milligrams per liter at equilibrium? (b) If ammonium hydroxide is added to the solution in (a) so that the resultant NH_3 concentration is 0.01 mol/L, what will the total concentration of silver in solution then be? Assume K in Eq. (2.63) is equal to 5×10^{-3}.

REFERENCES

Benjamin, M. M.: "Water Chemistry," McGraw-Hill, New York, 2002.

Chang, R.: "Chemistry," 7th ed., McGraw-Hill, New York, 2002.

Dean, J. A.: "Lange's Handbook of Chemistry," 15th ed., McGraw-Hill, New York, 1999.

Kotz, J. C., and K. S. Purcell: "Chemistry and Chemical Reactivity," 2nd ed., Saunders, Philadelphia, 1991.

Levine, I. N.: "Physical Chemistry," 5th ed., McGraw-Hill, New York, 2002.

Lide, D. R. (ed.): "Handbook of Chemistry and Physics," 82nd ed., CRC Press, Boca Raton, FL, 2001.

Mahan, B. H., and R. J. Meyers: "University Chemistry," 4th ed., Benjamin/Cummings, Menlo Park, CA, 1987.

Oxtoby, D. W., W. A. Freeman, and T. F. Block: "Chemistry: Science of Change," Saunders, Philadelphia, 1998.

Pankow, J. F.: "Aquatic Chemistry Concepts," Lewis, Chelsea, MI, 1991.

Pauling, L.: "General Chemistry," 3rd ed., Freeman, San Francisco, 1970.

3

Basic Concepts from Physical Chemistry

3.1 | INTRODUCTION

That portion of science dealing with laws or generalizations related to chemical phenomena is called *physical chemistry*. A sound knowledge of the physics is fully as important in the study of physical chemistry as a good grounding in chemistry. Some physical chemistry has always been taught in general, qualitative, and quantitative chemistry without being described as such. The trend has been to include more and more physical chemistry in such courses, and even in high-school chemistry, as those in the teaching profession have realized the need and gained stimulation for teaching "why" as well as "what" chemical reactions occur.

The subject of valency and bonding, oxidation-reduction reactions, the gas laws, Raoult's law, equilibrium relationships, the theory of ionization, the solubility-product principle, and other discussions pertaining to heterogeneous and homogeneous equilibria are all examples of topics in physical chemistry that have been developed previously. Some of them require further amplification, and a great many additional concepts need to be developed.

3.2 | THERMODYNAMICS

Thermodynamics is the study of energy changes accompanying physical and chemical processes. The energy changes associated with chemical reactions are of considerable importance and will be briefly discussed to indicate the relationships of most interest in environmental engineering and science. First, it is necessary to review the relationship between heat and work.

Heat and Work

Heat and work are related forms of energy. Heat energy can be converted into work, and work can be converted into heat energy. Steam engines and frictional losses are examples of each conversion.

Heat is that form of energy which passes from one body to another solely as a result of a difference in temperature. On the molecular scale it is known that the temperature of a substance is related to the average translational energy of the molecules, and that flow of heat results from transfer of this molecular energy. The basic unit of heat is the calorie, the heat required to raise the temperature of one gram of water one degree Celsius. In engineering it is common to measure heat in British thermal units (Btu), which is the heat required to raise one pound of water one degree Fahrenheit. One Btu is equivalent to 252 calories (cal) or 1054 joules (J).

The *specific heat* of a substance is the heat required to raise one gram of the material one degree Celsius, or

$$C = \frac{q}{M\,\Delta T} \tag{3.1}$$

where
C = specific heat
q = heat added, cal or J
M = weight of material, g
ΔT = rise in temperature of material, °C

For water the specific heat is 1.000 cal or 4.184 J per gram-degree Celsius only at 15°C, but varies from this value by less than 1 percent over the entire range from 0 to 100°C. Therefore, the assumption of a constant specific heat over a limited range in temperature is frequently a good one. Two values for specific heat are frequently reported: C_v is the specific heat at constant volume, and C_p is the specific heat at constant pressure. For liquids and solids there is little difference between these two values. For gases, however, C_p is greater than C_v because of the extra heat energy required to expand a gas against a constant pressure.

The *heat of fusion* is the heat required to melt a substance at its normal melting temperature. The *heat of vaporization* is the heat required to evaporate the substance at its normal boiling point. The values for water are 333 and 2258 J per gram, respectively. Environmental engineers and scientists encounter a wide variety of heating, evaporation, drying, and incineration problems involving a knowledge of specific heat, heat of fusion, and heat of vaporization. In addition, many industrial processes involve evaporation or drying in which cooling water is needed. In essence, the heat of vaporization is transferred to the coolant during condensation, and the warmed water may become a thermal pollution problem in terms of the receiving body of water. This may become a matter of concern when water use by industry and the public increases.

Work in chemical systems usually involves work of expansion. The system may either do work on its surroundings or have work done on itself. This depends upon whether the volume of the system is expanding or contracting. Work (*dw*) is normally measured in terms of force times distance, which for a closed system is equivalent to pressure (*P*) times the change in volume (*dV*):

$$dw = P\,dV \tag{3.2}$$

Work is measured in foot-pounds or in joules (J). Since heat and work are both forms of energy, they can be equated. Thus, 1 cal of heat is equivalent to 4.184 J. In other units, 1 Btu of heat is equivalent to 778 ft-lb.

Energy

The first law of thermodynamics states that energy can be neither created nor destroyed. In chemical systems the energy involved is most easily handled in terms of three quantities: the work that is performed, the heat that flows, and the energy stored in the system. The chemical system may range from the contents of a laboratory beaker to a full-scale, activated-sludge plant. According to the conservation-of-energy law, any heat or work that flows into or out of the system must result in a change in the total energy stored in the system. In equation form,

$$\Delta E = q - w \tag{3.3}$$

where ΔE = change in *internal* energy of system
   q = heat flowing *into* system
   w = work done *by* system

It is important to note that if heat is absorbed by the system, q has a positive value. If the system gives off heat, q has a negative value. Also, if the system does work on the surroundings, w has a positive value, but if the surroundings do work on the system, w has a negative value.

In chemical systems, work performed is usually expansion work as given by Eq. (3.2). When the system expands in volume, it does work on its surroundings and w is positive. When it contracts in volume, work is done on the system and w is negative. If the volume of the system remains constant, then no expansion work can be done and w equals zero. For this case,

$$\Delta E = q_v \qquad (V \text{ constant}) \tag{3.4}$$

where q_v is the heat absorbed in a constant-volume system. Thus, in a constant-volume system, the change in internal energy is just equal to the heat absorbed. Although this is convenient for constant-volume systems, most chemical systems of interest to environmental engineers and scientists are open to the atmosphere and so operate under constant pressure, rather than constant volume. For such systems, the concept of enthalpy has been developed.

Enthalpy

The enthalpy H of a system is defined as follows:

$$H = E + PV \tag{3.5}$$

where E = internal energy of system
   P = pressure on system
   V = volume of system

Consider a constant-pressure system in which some chemical change has taken place, resulting in a change in internal energy. The heat absorbed at constant pressure is q_p, and the work done by the system is given by an integration of Eq. (3.2). For constant temperature and pressure this integration gives $w = P(V_2 - V_1)$, where V_1 is the initial volume of the system and V_2 is then final volume after the change. The change in internal energy of such a system then becomes

$$\Delta E = E_2 - E_1 = q_p - w = q_p - P(V_2 - V_1)$$

Rearranging, we obtain

$$(E_2 + PV_2) - (E_1 + PV_1) = q_p$$

The terms in parentheses are just equal to the final and initial enthalpy of the system, and thus

$$H_2 - H_1 = q_p$$

or
$$\Delta H = q_p \qquad (T \text{ and } P \text{ constant}) \qquad (3.6)$$

Thus, *the quantity of heat absorbed by a system at constant temperature and pressure is equal to the change in system enthalpy.* Chemical changes that are accompanied by the absorption of heat, making ΔH positive, are called *endothermic* reactions. Those accompanied by the evolution of heat, making ΔH negative, are called *exothermic.*

The total enthalpy (H) of a system would be difficult to measure. We are normally interested, however, only in the *change* in enthalpy and not in its absolute value. By developing a standard basis for comparison, it is possible to calculate the change in enthalpy or the heat of a given reaction from tabulated measurements from quite different reactions.

A convenient standard state for a substance may be taken as the stable state of the compound at 25°C and 1 atm pressure. For example, under these conditions, the standard state for oxygen is a gas, for mercury a liquid, and for sulfur it is rhombic crystals. By convention, the enthalpies of the *chemical elements* in this standard state are set equal to zero. The standard enthalpy of any *compound* is then the heat of the reaction by which it is formed from its elements; reactants and products all being in the standard state at 25°C and 1 atm.

For example,

$$H_2(g) + \tfrac{1}{2}O_2(g) \rightarrow H_2O(l) \qquad \Delta H_f^0 = -286{,}000 \text{ J}$$
$$C(s) + O_2(g) \rightarrow CO_2(g) \qquad \Delta H_f^0 = -394{,}000 \text{ J}$$

The symbols in parentheses after each element or compound indicate the standard state of the elements or compounds. The superscript zero and subscript f on the enthalpy indicates a standard heat of formation with reactants and products at 1 atm and 25°C (298 K). Standard enthalpies for many compounds of interest are listed in Table 3.1. Many other values can be found in chemical handbooks such as those listed at the end of the chapter. An expanded listing is also given in App. A.

Standard enthalpy values can be used to determine the heat given off by a variety of reactions. In making such calculations it is very important to note the state

Table 3.1 | Standard Free Energies and Enthalpies of Formation at 25°C

Sub-stance	State*	ΔG_f^0, kJ/mol	ΔH_f^0, kJ/mol	Sub-stance	State*	ΔG_f^0, kJ/mol	ΔH_f^0, kJ/mol
Ca^{2+}	aq	−553.6	−542.8	H_2O	l	−237.18	−285.80
$CaCO_3$	s	−1128.76	−1206.87	H_2O	g	−228.60	−241.8
CaF_2	s	−1175.6	−1228.0	HS^-	aq	12.05	−17.6
$Ca(OH)_2$	s	−896.76	−986.6	H_2S	g	−33.4	−20.6
$CaSO_4$	s	−1322.0	−1434.5	H_2S	aq	−27.87	−39.3
CH_4	g	−50.79	−74.6	H_2SO_4	l	−690.0	−814
CH_3CH_3	g	−32.89	−84.67	Mg^{2+}	aq	−454.8	−466.9
CH_3COOH	aq	−399.6	−488.4	$Mg(OH)_2$	s	−833.5	−924.5
CH_3COO^-	aq	−369.41	−486.0	Na^+	aq	−261.9	−240.1
CH_3CH_2OH	l	−174.8	−277.7	NH_3	g	−16.4	−45.9
$C_6H_{12}O_6$	aq	−917.22		NH_3	aq	−26.65	−80.83
Cl_2	g	0	0	NH_4^+	aq	−79.37	−132.5
Cl_2	aq	6.9	−23.4	NO_2^-	aq	−37.2	−104.6
Cl^-	aq	−131.3	−167.20	NO_3^-	aq	−111.34	−207.4
CO_2	g	−394.38	−393.51	O_2	g	0	0
CO_2	aq	−386.23	−412.92	O_2	aq	16.32	−11.71
CO_3^{2-}	aq	−527.80	−677.10	OH^-	aq	−157.2	−230.0
F^-	aq	−278.8	−332.6	S^{2-}	aq	79.5	30.1
Fe^{2+}	aq	−78.87	−89	SO_4^{2-}	aq	−744.63	−909.3
Fe^{3+}	aq	−4.6	−48.5	Zn^{2+}	aq	−147.1	−153.9
HCO_3^-	aq	−586.85	−692.00	ZnS	s	−201.3	−206.0
H_2CO_3	aq	−623	−699				

* aq = aqueous, s = solid, l = liquid
Note: 1 kcal = 4.184 kJ

in which the products and reactants exist, as this can make a significant difference in the heat of reaction. The procedure used is to write a balanced equation for the reaction. The heat of the reaction is then equal to the sum of the standard enthalpies of the products minus the sum of the standard enthalpies of the reactants. It should be noted that standard enthalpies of formation are given in terms of kilojoules per mole, and these values must be multiplied by the number of moles entering into the reaction.

EXAMPLE 3.1

Calculate both the net and gross heat of combustion of methane gas.

The *gross heat* is the heat released if the water vapor formed upon combustion is condensed to form liquid water.

$$CH_4(g) + 2O_2(g) \rightarrow CO_2(g) + 2H_2O(l)$$
$$-74.6 \qquad 0 \qquad -393.51 \quad 2(-285.80)$$
$$\Delta H^0 \text{ for combustion} = (-393.51) + 2(-285.80) - (-74.6)$$
$$= -890.51 \text{ kJ/mol of methane}$$

The *net heat* is the heat released if the water remains as a vapor or gas. This is the value of usual interest.

$$CH_4(g) + 2O_2(g) \rightarrow CO_2(g) + 2H_2O(g)$$
$$ -74.6 \quad\quad 0 \quad\quad -393.51 \quad 2(-241.8)$$
$$\Delta H^0 \text{ for combustion} = (-393.51) + 2(-241.8) - (-74.6)$$
$$= -802.51 \text{ kJ/mol of methane}$$

Since the values of ΔH^0 for the combustion are negative, heat is given off by the combustion of methane gas.

EXAMPLE 3.2

Calculate the approximate rise in solution temperature if 1 liter of 1 N H_2SO_4 is mixed with 1 liter of 1 N NaOH.

Sulfuric acid is a strong acid and sodium hydroxide is a strong base and so they are completely ionized in solution, as is the Na_2SO_4 which is formed when the given solutions are mixed. Therefore, the enthalpies of the aqueous solutions are equal to the sum of the enthalpies for the individual ions.

$$\Delta H^0_{(H_2SO_4)}(aq) = 2\Delta H^0_{(H^+)} + \Delta H^0_{(SO_4^{2-})} = 2(0) + (-909.3) = -909.3$$
$$\Delta H^0_{(NaOH)}(aq) = \Delta H^0_{(Na^+)} + \Delta H^0_{(OH^-)} = (-240.1) + (-230.0) = -470.1$$
$$\Delta H^0_{(Na_2SO_4)}(aq) = 2\Delta H^0_{(Na^+)} + \Delta H^0_{(SO_4^{2-})} = 2(-240.1) + (-909.3) = -1389.5$$

The neutralization reaction that occurs is

$$\tfrac{1}{2}H_2SO_4(aq) + NaOH(aq) \rightarrow H_2O(l) + \tfrac{1}{2}Na_2SO_4(aq)$$
$$\Delta H_f^0 \ \tfrac{1}{2}(-909.3) \quad\quad -470.1 \quad\quad -285.80 \quad \tfrac{1}{2}(-1389.5)$$

$$\Delta H^0 \text{ for neutralization} = (-285.80) + \tfrac{1}{2}(-1389.5) - \tfrac{1}{2}(-909.3) - (-470.1)$$
$$= -55.80 \text{ kJ/mol} = -13.34 \text{ kcal/mol}$$

The final volume of the mixed solution is 2 liters, which would weigh about 2000 g. Since this solution is mainly water, the specific heat would be 1 cal/g-°C, and, from Eq. (3.1),

$$T = \frac{13,340}{2000 \times 1} = 6.67°C$$

Thus, if the initial temperatures of the solutions were 25°C, the temperature would rise to a value somewhat over 31°C.

A knowledge of heats of reaction, as well as heats of fusion, vaporization, and specific heats, is used for incineration, combustion, wet-air oxidation, heating of digesters, chemical handling, and thermal pollution studies, among others.

Entropy

A large part of chemistry is concerned, in one way or another, with the state of equilibrium and the tendency of systems to move spontaneously in the direction of the equilibrium state. The concept of *entropy* was developed from the search for a thermodynamic function that would serve as a general criterion of spontaneity for physical and chemical changes. The concept of entropy is based on the second law of thermodynamics, which in essence states that all systems tend to approach a state of equilibrium. The significance of the equilibrium state is realized from the fact that work can be obtained from a system only when the system is not already at equilibrium. If a system is at equilibrium, no process tends to occur spontaneously, and no chemical or physical changes are brought about.

The chemist's interest in entropy is related to the use of this concept to indicate something about the position of equilibrium in a chemical process. Entropy is defined by the following differential equation:

$$dS = \frac{dq_{rev}}{T} \tag{3.7}$$

where S is the entropy of the system, and T is the absolute temperature. The quantity q_{rev} is the amount of heat that the system absorbs if a chemical change is brought about in an infinitely slow, reversible manner. As with enthalpy, it is the change in entropy in a system that is of usual interest, and this is evaluated as follows:

$$\Delta S = S_2 - S_1 = \int_1^2 \frac{dq_{rev}}{T} \tag{3.8}$$

On the basis of the third law of thermodynamics, the entropy of a substance at 0 K is zero. Because of this, the absolute entropy of elements and compounds at some standard state can be determined by integration of Eq. (3.8), using the initial state to represent the equilibrium state of 0 K.

The significance of entropy is that when a spontaneous change occurs in a system, it will always be found that if the total entropy change for everything involved is calculated, a positive value is obtained. Thus, *all spontaneous changes in an isolated system occur with an increase of entropy.* If one wanted to determine whether a chemical or physical change from one state *a* to another state *b* could occur in a system, a calculation of entropy change would give the desired information. If ΔS for the whole system were positive, the change could occur spontaneously; if ΔS were negative, the change would tend to occur in the reverse direction, i.e., from *b* to *a*. However, if ΔS were zero, the system would be at equilibrium, and the change could not take place spontaneously in either direction.

On the molecular scale, entropy has a statistical basis. Systems tend to move from a highly ordered state to a more random state. The more highly probable or random a system becomes, the higher will be its entropy. For example, if two different ideal gases are placed together in a closed container, the gas molecules will not remain isolated, but will become randomly mixed. This spontaneous process occurs without a change in internal energy in the container, but with an increase in entropy.

Free Energy

In the original concepts of thermodynamics it was incorrectly assumed that energy given out by a reaction was a measure of the driving force. It is now seen that in an isolated system, where energy cannot be gained or lost, the entropy change is the driving force. In more general systems, such as those used in environmental engineering practice, both energy and entropy factors must be considered in order to determine what processes will occur spontaneously. For this purpose the concept of *free energy* has been developed. Free energy (G) is defined as

$$G = H - TS \tag{3.9}$$

where H = enthalpy, J

T = absolute temperature, K (K = °C + 273.15)

S = entropy, J/K

At constant temperature and pressure, the change in free energy for a given reaction is

$$\Delta G = \Delta H - T\,\Delta S \qquad (T \text{ and } P \text{ constant}) \tag{3.10}$$

From Eq. (3.5) and for P = constant,

$$\Delta H = H_2 - H_1 = (E_2 + P_2 V_2) - (E_1 + P_1 V_1) = \Delta E + P\,\Delta V$$

Combining with Eq. (3.3) for E, we obtain $\Delta H = q - w + P\,\Delta V$.

From Eq. (3.8) at constant, T, we see that $T\,\Delta S$ is equal to q_{rev}. If the system change is brought about very slowly so that energy losses are at a minimum, then q becomes q_{rev}, and w becomes w_{max}, the maximum quantity of work that could be obtained from the change.

Therefore,

$$\Delta G = q_{rev} - w_{max} + P\,\Delta V - q_{rev}$$

and

$$-\Delta G = w_{max} - P\,\Delta V$$

$P\,\Delta V$ gives the portion of the work that must be "wasted" in expanding the system against the confining pressure. Therefore, $-\Delta G$ gives the difference between the maximum work and the wasted work; in other words, it gives the *useful* work available from the system change:

$$-\Delta G = w_{useful} \qquad (T \text{ and } P \text{ constant}) \tag{3.11}$$

In principle, any process that tends to proceed spontaneously can be made to do useful work. Since the free-energy change measures the useful work that might be obtained from a constant-pressure process, it is a measure of the spontaneity of the process.

Consider a change from a to b in a constant-pressure system. If ΔG for such a change is negative, then the process can proceed; if ΔG is positive, the process can proceed, but in the reverse direction, i.e., from b to a; if ΔG is zero, the system is in equilibrium, and the process cannot proceed in either direction. This is a particularly significant relationship, and for the chemist it is one of the most important in thermodynamics.

In order for the concept of free energy to be useful, a reference point for determining free-energy changes must be available. As in the case of enthalpy, a zero value is assigned to free energies of the stable form of the elements at 25°C and 1 atm pressure. In addition, the hydrogen ion at unit activity (approximately one normal solution) is assigned a standard free energy of zero.

The standard free energy of a compound (ΔG_f^0) is the free energy of formation of that compound from its elements, considering reactants and products all to be in the standard state at 25°C and 1 atm (that is, at unit activity). A list of standard free energies of formation of various compounds is given in Table 3.1 and App. A.

Free-energy changes accompanying a chemical reaction can now be calculated, and the direction in which the reaction will proceed can be determined in a qualitative manner from the sign of the free-energy change. It is apparent from a knowledge of equilibrium reactions that a reaction will proceed in a given direction only until the system reaches a state of equilibrium. It can be shown that free energies can also be used to determine the equilibrium state to which the reaction carries the system, as well as the reaction direction. Without going into the details, it should also be apparent that the direction of a reaction is dependent upon the concentration of reactants and products, and so this must be considered in free-energy calculations. Consider the following reaction:

$$a\mathrm{A} + b\mathrm{B} \rightleftharpoons c\mathrm{C} + d\mathrm{D}$$

The free energy of this reaction, considering the concentration of the various reactants and products, is given by the following equation:

$$\Delta G = \Delta G^0 + RT \ln \frac{\{\mathrm{C}\}^c\{\mathrm{D}\}^d}{\{\mathrm{A}\}^a\{\mathrm{B}\}^b} \tag{3.12}$$

where ΔG = reaction free-energy change,
ΔG^0 = standard free-energy change,
R = universal gas constant = 8.314 J/K-mol = 1.99 cal/K-mol
T = absolute temperature, K

The terms in the braces are the activities of the various reactants and products. The convention outlined in Sec. 2.12 for expressing activities must be followed. The term $\{\mathrm{C}\}^c\{\mathrm{D}\}^d/\{\mathrm{A}\}^a\{\mathrm{B}\}^b$ is called the reaction quotient Q.

The development of Eq. (3.12) is beyond the scope of this book but is given in most books on physical chemistry. This equation is important, as it allows the prediction of reaction direction for any activity (or concentration, approximately) of products and reactants of interest. If the free-energy change (ΔG) is negative, the reaction will proceed to the right as written. If ΔG is positive, the reaction will proceed in the reverse direction. If a reaction is allowed to proceed to a state of equilibrium, it will reach a position for which no further driving force is operative. At this point ΔG will be zero, and thus

$$\Delta G^0 = -RT \ln \left(\frac{\{\mathrm{C}\}^c\{\mathrm{D}\}^d}{\{\mathrm{A}\}^a\{\mathrm{B}\}^b} \right)_{\text{equilibrium}} \tag{3.13}$$

The subscript "equilibrium" is to indicate that this equation is true only when the system is at equilibrium. It should now be noted that at equilibrium the ratio of product concentrations to reactant concentrations is equal to the equilibrium constant:

$$\frac{\{\mathrm{C}\}^c\{\mathrm{D}\}^d}{\{\mathrm{A}\}^a\{\mathrm{B}\}^b} = K$$

Thus, Eq. (3.13) can be written in the following form:

$$\Delta G^0 = -RT \ln K \tag{3.14}$$

This equation is one of the most important results of thermodynamics. It shows the relationship between the equilibrium constant of a reaction and its thermochemical properties. By use of this equation, the equilibrium constant for a reaction can be calculated from thermochemical properties of the reactants and products which may have been determined from entirely different reactions.

Comparison of Q with K indicates whether the reaction is proceeding to the right or the left, or if the reaction is at equilibrium. If Q is less than K, the reaction will proceed from left to right; if Q is greater than K, the reaction will proceed from right to left; if Q equals K, the reaction is at equilibrium.

Determine the first ionization constant for carbonic acid at 25°C from free-energy considerations.

EXAMPLE 3.3

The equation and free energies for the first ionization are

$$H_2CO_3(aq) = H^+(aq) + HCO_3^-(aq)$$
$$-623 \qquad 0 \qquad -586.85$$

ΔG^0 for reaction $= (-586.85) + (0) - (-623) = 36.15$ kJ

$\ln K_1 = -\Delta G^0/RT = -36{,}150/8.314(298) = -14.59$
Therefore, $K_1 = 4.61 \times 10^{-7}$

So
$$\frac{[H^+][HCO_3^-]}{[H_2CO_3]} = 4.61 \times 10^{-7}$$

Determine the solubility constant for carbon dioxide gas in water at 25°C from free-energy considerations.

EXAMPLE 3.4

The equation and free energies of interest are

$$CO_2(aq) = CO_2(g)$$
$$-386.23 \qquad -394.38$$

ΔG^0 for reaction $= (-394.38) - (-386.23) = -8.15$ kJ

$\ln K_{sol} = -(-8{,}150)/8.314(298) = 3.29$

Therefore, $K_{sol} = 26.8$ atm-L/mol

So
$$\frac{CO_2(g) \text{ in atm}}{CO_2(aq) \text{ in mol/L}} = 26.8 \text{ atm-L/mol}$$

Thus, the Henry's law constant discussed in Sec. 2.9 can be determined from free-energy calculations.

| EXAMPLE 3.5 | Calculate the solubility product for calcium fluoride at 25°C from free-energy considerations. |

The equation and free energies for this solubility are

$$CaF_2(s) = Ca^{2+}(aq) + 2F^-(aq)$$
$$-1175.6 \qquad -553.6 \qquad 2(-278.8)$$
$$\Delta G^0 \text{ for reaction} = (-553.6) + 2(-278.8) - (-1175.6) = 64.4 \text{ kJ}$$

$$\ln K_{sp} = -64{,}400/8.314(298) = -25.99$$

Therefore, $\qquad K_{sp} = 5.14 \times 10^{-12}$

And so, on the basis of these calculations,

$$[Ca^{2+}][F^-]^2 = 5.14 \times 10^{-12}$$

| EXAMPLE 3.6 | If a water sample has a Ca^{2+} concentration of 20 mg/L and a F^- concentration of 0.1 mg/L, will $CaF_2(s)$ dissolve or precipitate, or is the water in equilibrium with respect to $CaF_2(s)$? Ignore activity corrections. |

$$Q = [Ca^{2+}][F^-]^2$$
$$[Ca^{2+}] = (20 \text{ mg/L})/(40{,}000 \text{ mg/mol}) = 5.0 \times 10^{-4} \text{ M}$$
$$[F^-] = (0.1 \text{ mg/L})/(19{,}000 \text{ mg/mol}) = 5.3 \times 10^{-6} \text{ M}$$
$$Q = (5.0 \times 10^{-4} \text{ M})(5.3 \times 10^{-6} \text{ M})^2 = 1.4 \times 10^{-14}$$

Since Q is less than K_{sp} ($1.4 \times 10^{-14} < 5.14 \times 10^{-12}$), the reaction is proceeding from left to right and $CaF_2(s)$ will dissolve further.

Another way to approach the problem is to realize from Eqs. (3.12) and (3.14) that,

$$\Delta G = RT \ln(Q/K)$$

In this case, Q is less than K, making ΔG negative. Thus, the reaction proceeds from left to right as written.

Temperature Dependence of Equilibrium Constant

From a consideration of the relationships between free energy and the equilibrium constant and between free energy and enthalpy, a thermodynamic basis for predicting the change in equilibrium constant with temperature can be obtained. In differential form, this relationship is

$$\frac{d \ln K}{dT} = \frac{\Delta H^0}{RT^2} \tag{3.15}$$

This equation indicates that for exothermic reactions the equilibrium constant decreases with increasing temperature, while for endothermic reactions it increases.

Over the rather limited temperature range of environmental interest, ΔH^0 is normally constant enough that the integrated form can be used:

$$\ln\frac{K_{(T_2)}}{K_{(T_1)}} = -\frac{\Delta H^0}{R}\left(\frac{T_1 - T_2}{T_1 T_2}\right) \tag{3.16}$$

Thus, if the equilibrium constant for a reaction is known for one temperature, the value for another temperature can be calculated from standard enthalpy values for the products and reactants of the reaction.

Calculate the ionization constant K_1 at 10°C for carbonic acid, assuming $K_1 = 4.61 \times 10^{-7}$ at 25°C as calculated in Example 3.3. The equation and enthalpies of formation for this reaction are

$$H_2CO_3(aq) = H^+(aq) + HCO_3^-(aq)$$
$$\Delta H_f^0 \qquad -699 \qquad 0 \qquad -692.00$$
$$\Delta H^0 \text{ for reaction} = (-692.00) - (-699) = 7 \text{ kJ/mol}$$

Using Eq. (3.16), we obtain

$$\ln\frac{K_{10}}{K_{25}} = \frac{-7,000}{8.314}\left[\frac{298 - 283}{(298)(283)}\right] = -0.150$$

Therefore, $K_{10} = e^{-0.150}(4.61 \times 10^{-7}) = 3.97 \times 10^{-7}$.

EXAMPLE 3.7

3.3 | VAPOR PRESSURE OF LIQUIDS

According to the *kinetic-molecular theory,* liquids as well as gases are in constant agitation, and molecules are constantly flying from the surface of the liquid into the atmosphere above. In open systems most of these particles never return, and the liquid is said to be undergoing evaporation. In a closed system, however, particles return to the liquid phase in proportion to their concentration in the gaseous phase. Eventually the rate of return equals the rate of flight, and a condition of equilibrium is established. The vapor is then said to be saturated. The pressure exerted by the vapor under these conditions is known as the *vapor pressure.* The vapor pressure of all liquids increases with temperature. The vapor-pressure values of water and a few organic liquids are given in Table 3.2. It will be noted that vapor pressures do not rise in a regular manner. For rough approximations, the vapor pressure may be considered to increase about 1.5 times for each 10°C rise in temperature.

The Rankine formula, or one of its modifications, is commonly used to calculate vapor pressures,

$$\log p = \frac{A}{T} + B \log T + C \tag{3.17}$$

where
$$p = \text{vapor pressure}$$
$$T = \text{temperature}$$
$$A, B, C = \text{constants}$$

Table 3.2 | Vapor Pressure of Liquids, atm

Temp., °C	Water	Ethyl Alcohol	*n*-Hexane	Benzene
0	0.006	0.016	0.060	0.035
10	0.012	0.031	0.099	0.060
20	0.023	0.058	0.158	0.098
30	0.042	0.104	0.244	0.156
40	0.073	0.178	0.364	0.238
50	0.122	0.292	0.528	0.354
60	0.197	0.464	0.745	0.511
70	0.308	0.714	1.036	0.720
80	0.467	1.068	1.397	0.968
90	0.692	1.562	1.851	1.337
100	1.000	2.228	2.416	1.768

Vapor-pressure data for many of the compounds of environmental interest can be found in the "International Critical Tables of Numerical Data, Physics, Chemistry and Technology" or standard handbooks of chemistry. Appropriate formulas and constants are usually given to allow calculation of vapor-pressure values for any temperature.

When the vapor pressure of a liquid becomes equal to the pressure of the atmosphere above it, the liquid is said to have reached its *boiling point*. Violent agitation of the liquid occurs under these conditions as a result of the transformation of liquid to gas at the source of heat, migration of the bubbles of vapor through the liquid, and their escape from the liquid surface.

Liquids with appreciable vapor pressure may be caused to boil over a wide range of temperatures by decreasing or increasing the pressure. Water boils at room temperature if the pressure above it is reduced to about 0.028 atm. On the other hand, the boiling point of water in a steam boiler operating at 13.6 atm is 194.4°C. Application of this principle can be employed in the wet oxidation process for combustion of organic sludges and other organic wastes. In this process part of the organic matter is chemically oxidized in an aqueous phase by dissolved oxygen in a specially designed reactor in which the water temperature is elevated to between 250 and 300°C. To maintain a temperature in this range without boiling the water requires pressures between 40 and 85 atm, respectively. The oxidizability of the waste organics, as well as the oxidation rate, increases markedly with temperature. However, the maximum temperature that can be used in such a reactor is set by the *critical temperature* above which water can no longer exist in a liquid phase, regardless of pressure. This temperature is 374°C. The *critical pressure* that suffices to keep water in liquid form just below this critical temperature is 218 atm.

3.4 | SURFACE TENSION

According to the kinetic-molecular theory, molecules of a liquid attract each other. At the surface, the molecules are subjected to an unbalanced force, since the molecules in the gaseous phase are so widely dispersed. As a result, the molecules at the *surface* are under *tension* and form a thin skinlike layer that adjusts itself to give a minimum surface area. This property of surface tension causes liquid droplets to assume a spherical shape; water to rise in a capillary tube; and liquids, such as water, to move through porous materials that they are capable of wetting. The movement of water through soils is an excellent example.

Surface tension may be most accurately determined by measuring the height to which a liquid will rise in a capillary tube. Most liquids, like water, wet the walls of a glass tube and the liquid adhering to the walls pulls liquid up into the tube to decrease the total surface area in relation to its surface tension. This is the basis of capillary action so important in supplying water and nutrients to plant and animal tissues. Under static conditions, such as occur in a glass tube used to measure surface tension, the opposing forces are equal. The downward force may be expressed as $\pi r^2 h\rho g$, and the upward force as $2\pi r\gamma \cos \theta$.

$$\pi r^2 h\rho g = 2\pi r\gamma \cos \theta \tag{3.18}$$

or

$$\gamma = \frac{h\rho g r}{2 \cos \theta} \tag{3.19}$$

where
γ = surface tension, N/m
ρ = density of liquid, (kg/m^3)
g = acceleration due to gravity, m/s^2
h = height, m
r = radius, m
θ = angle of contact that liquid makes with wall of capillary tube.

For water and for many other liquids, θ is so small that $\cos \theta$ may be considered equal to 1. Then

$$\gamma = \frac{1}{2} h\rho g r \quad \text{or} \quad h = \frac{2\gamma}{\rho g}\frac{1}{r} \tag{3.20}$$

and the relationship between the height to which a liquid will rise in a capillary and its radius is readily apparent. The capillaries in the giant sequoia trees, which reach a height of over 100 m, must be extremely small.

Capillary action is also important for liquids that do not wet a surface with which they are in contact. Examples are mercury in contact with glass, or oil and chlorinated solvents in contact with sand. Mercury tends to pull away and forms droplets on a glass surface rather than spreading over it. In a capillary tube, mercury forms a convex, rather than a concave angle with the glass walls, and tends to be pulled downward rather than moving upward. When soil and groundwater become contaminated with petroleum hydrocarbons or chlorinated solvents, these non-aqueous-phase liquids (NAPL) tend to pull themselves together into droplets within the soil pores, rather than spreading like water, making them extremely difficult to remove. Here, the capillary forces involved create very challenging problems for the engineer trying to clean contaminated soils and groundwater.

Poiseuille Law

The behavior of liquids when flowing through capillary tubes, in relation to their viscosity, was studied by G. Hagen and J. L. Poiseuille. They summarized their findings in the equation

$$\mu = \frac{\pi P r^4}{8Vl}t \tag{3.21}$$

where V = volume of liquid

  μ = viscosity of liquid

  l = length of capillary tube

  r = radius of capillary tube

  t = time

  P = pressure of liquid in capillary tube

Environmental engineers and scientists are often confronted with problems involving the flow of liquids through capillaries. A notable example is in the filtration of sludge in which the void areas are considered as tortuous capillaries. Since water of a uniform viscosity is the liquid to be removed, and the volume or rate of movement is of interest, the *Hagen-Poiseuille equation* is important and is usually written as

$$Q = \frac{\pi P r^4}{8\mu l} \tag{3.22}$$

where P, r, μ, and l are as defined for Eq. (3.21). The value Q equals V/t, the flow rate through the capillary. From Eq. (3.22), the importance of the diameter of the capillaries is immediately apparent as being the principal factor in determining the pressure differential needed to maintain a constant liquid volume flow. A knowledge of the Hagen-Poiseuille law is helpful in explaining how filter aids and chemical-conditioning agents such as lime are beneficial in filtration operations and as a basic concept for planning research on filtration or related problems.

3.5 | BINARY MIXTURES

Binary mixtures of miscible liquids such as water and ethanol are of interest because of the differences in vapor pressure that they exhibit and the influence that vapor pressure has upon their separation by distillation. All mixtures fall into one of three classes, and their properties are considerably different.

Class I

Class I includes all mixtures with total vapor pressure, regardless of the composition of the mixture, that is always less than that of the most volatile component and always more than that of the least volatile component; consequently, the boiling point of Class I mixtures is always between those of the two components. The com-

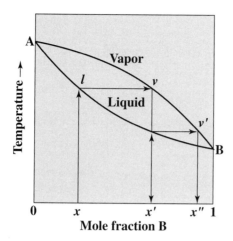

Figure 3.1
Composition of liquid and vapor phases
during distillation of Class I binary mixtures.

position of the vapor is always richer in the more volatile component than the liquid
from which it distills. Such mixtures are amenable to essentially complete separation by means of fractional distillation.

A diagram showing the composition of the liquid and vapor phases, and how
separation of the two components of Class I mixtures can be accomplished by fractional distillation, is given in Fig. 3.1. If a mixture of A and B having the composition represented by x is heated to its boiling point, the liquid will have a temperature
corresponding to l, and the vapor produced will have a composition corresponding
to v on the vapor curve. If the vapor at v is condensed, a liquid corresponding to x',
much richer in B, is obtained. Redistillation of the mixture x' results in a vapor with
a composition x''. Through successive condensations and evaporations, normally accomplished by a fractionating column, a distillate of essentially pure B can be obtained, and A will remain as a relatively pure residue in the still. A wide variety of
compounds form Class I binary mixtures.

Class II

Class II binary mixtures include those that at certain mole ratios have vapor pressures less than either of the components and, consequently, at these ratios have boiling points that are greater than that of either of the components. Upon distillation of
such mixtures, one or the other of the components may be fractionated into relatively pure form until the liquid mixture reaches a composition of minimum vapor
pressure or maximum boiling point. From that point on, a constant boiling mixture
is obtained, the compositions of vapor and liquid are identical, and further separation is impossible by this means.

A diagram showing the composition of the liquid and vapor phases of a Class II
binary mixture is given in Fig. 3.2. If a mixture corresponding to x' is distilled, the

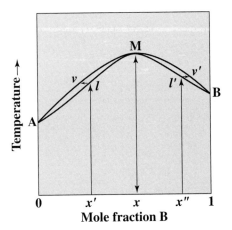

Figure 3.2
Composition of liquid and vapor phases
during distillation of a Class II binary mixture.

vapor formed is rich in component A and may be recovered in part in relatively pure form by fractionation. However, as A is removed from the liquid phase, the liquid phase grows richer in B until it equals the composition shown by x. At this point the compositions of the vapor and liquid phases are identical, and a constant boiling mixture that cannot be fractionated results. Likewise, if a mixture with a composition represented by x'' is distilled, a distillate of B can be obtained, the liquid remaining will approach a composition equal to x, and a constant boiling mixture will result. Hydrochloric, hydrobromic, hydroiodic, hydrofluoric, nitric, and formic acid in aqueous solution are all binary mixtures of Class II. The constant boiling mixture of hydrochloric acid at 1 atm pressure contains 20.2 percent HCl and is often used as a primary standard in quantitative analysis.

Class III

Class III binary mixtures include those that at certain mole ratios have vapor pressure greater than that of either of the components, and therefore the boiling points at such mole ratios are lower than that of either component. Upon distillation of Class III mixtures, the results are opposite to those obtained with Class II mixtures. A distillate is obtained that contains both components in a constant ratio, and the residue remaining in the flask consists of one or the other component in pure form.

A diagram showing the composition of the liquid and vapor phases of a Class III binary mixture is given in Fig. 3.3. A mixture corresponding to x' will produce a vapor with composition v. Fractionation of this vapor will produce a distillate with composition x, and the liquid phase will grow richer in component A. Distillation of a mixture corresponding to x'' will produce a vapor with a composition v'. Fractionation will yield a distillate with composition x, and the liquid will become richer in component B. Eventually either pure A or pure B will remain in the liquid phase.

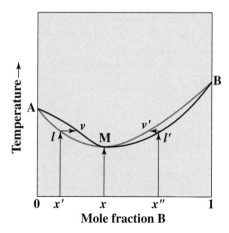

Figure 3.3
Composition of liquid and vapor phases
during distillation of a Class III binary mixture.

Ethanol (ethyl alcohol) and water form a binary mixture of this class. The distillate, regardless of the composition of the original mixture, will always contain 95.6 percent of alcohol at 1 atm pressure as long as both components are present in the liquid phase.

3.6 | SOLUTIONS OF SOLIDS IN LIQUIDS

The amount of a solid that will dissolve in a liquid is a function of the temperature, the nature of the solvent, and the nature of the solute (solid). The broad concepts of unsaturated, saturated, and supersaturated solutions are generally treated quite adequately in courses in general science and chemistry. However, the significance of crystal or particle size and the influence of temperature on solubility are not always adequately discussed.

Significance of Particle Size

The solubility of solids has been shown to increase as particle size diminishes. For example, coarse granular $CaSO_4(s)$ dissolves to the extent of 2.08 g/L at 25°C, whereas finely divided $CaSO_4(s)$ dissolves to the extent of 2.54 g/L. This phenomenon is considered to be due to the increased ratio of surface area to mass and to an increase in vapor pressure of the solid as particle size decreases. Particles of colloidal size are considered to have the greatest solubility because of their submicroscopic size.

In quantitative analysis, advantage is often taken of the fact that solubility varies with particle size. In gravimetric analysis, such as in the determination of sulfate by precipitation as $BaSO_4(s)$, the precipitate first formed is highly colloidal in nature. However, if some care is used in the precipitation procedure, a few crystals of larger size will be formed. If the precipitated material is allowed to stand for a period of time before filtration, the colloidal-size particles will tend to dissolve and then precipitate out on the large crystals present. The rate of transfer is a function of the number of crystals present, the differential in solubilities, and the temperature. The difference in solubility is known to increase with temperature, as well as the rate of exchange; consequently, "digestion" of precipitates is normally done at temperatures near the boiling point of the solvent.

Temperature Relationships

In general, the solubility of solids in liquids increases as the temperature increases. There are a number of exceptions, however. The influence of temperature on solubility depends mainly upon the total heat effects of the solution. If the heat of solution is endothermic, the solubility increases with an increase in temperature; if the heat of solution is exothermic, the solubility decreases with an increase in temperature; and if there is little thermal change, the solubility is influenced very little by change of temperature. These considerations are all in accord with Le Chatelier's principle and the thermodynamic principles discussed in Sec. 3.2.

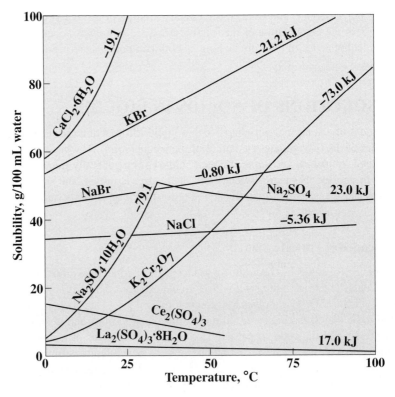

Figure 3.4
Relationship between solubility in water and heat of solution.

Figure 3.4 shows solubility curves for a number of solids in water and illustrates the relationship to heat of solution. The solubility curves for some solids, such as sodium sulfate, show abrupt changes because of a change in molecular composition and heat of solution.

3.7 | MEMBRANE PROCESSES: OSMOSIS AND DIALYSIS

Membrane processes are being used increasingly to remove particulates and solutes from waters, wastewaters, and gas. More stringent environmental regulations coupled with advances in membrane technology and more favorable economics are primarily responsible for this increased use. In general, membrane processes are classified by the size of the pollutants being removed. The following table gives a general classification of the common membrane processes (1 μm = 1 micron = 10^{-6} m).

Membrane process	Applicable particle or solute size range, μm
Macrofiltration	>10–100
Microfiltration	0.1–10
Ultrafiltration	0.005–0.1
Nanofiltration	0.001–0.005
Reverse osmosis	<0.001
Dialysis	<0.001

Reverse osmosis and dialysis can remove dissolved ions from solution. In the rest of this section, the principles of osmosis and dialysis are discussed in some detail.

Osmosis

Osmosis is the movement of a solvent through a membrane that is impermeable to a solute. The direction of flow is from the more dilute to the more concentrated solution. For example, if a salt solution is separated from water by means of a semipermeable membrane, as shown in Fig. 3.5, water will pass through the membrane in both directions, but it will pass more rapidly in the direction of the salt solution. As a result, a difference in hydrostatic pressure develops. The tendency for the solvent to flow can be opposed by applying pressure to the salt solution. The excess pressure that must be applied to the solution to produce equilibrium is known as the *osmotic pressure* and is denoted by π.

The net flow of solvent across a membrane results in response to a driving force which can be estimated by the difference in vapor pressure of the solvent on either side of the membrane. The transfer of solvent across the membrane from the less concentrated to the more concentrated solution will continue until the effect of hydrostatic pressure overcomes the driving force of the vapor pressure differential. For an incompressible solvent, the osmotic pressure at equilibrium π (expressed in atmospheres) can be estimated from the following:

$$\pi = \frac{RT}{V_A} \ln \frac{P_A^0}{P_A} \tag{3.23}$$

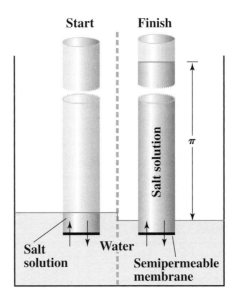

Figure 3.5
The process of osmosis and the development of osmotic pressure.

where $R = 0.08206$ L atm/mol $\cdot K$

T = temperature, K

P_A^0 = vapor pressure of solvent in dilute solution

P_A = vapor pressure of solvent in concentrated solution

V_A = volume per mole of solvent ($= 0.018$ L/mol for water)

Raoult's law, discussed in Sec. 2.10, indicates that for dilute solutions the reduction in vapor pressure of a solvent is directly proportional to the mole fraction of other species in solution. From this fact, the osmotic pressure can be related to the molar concentration of dissolved substances c in the concentrated solution:

$$\pi = cRT \qquad (3.24)$$

This equation is valid in a strict sense only for dilute solutions in which Raoult's law holds true.

An application of osmotic pressure principles is in the demineralization of salt-laden (brackish) water by the *reverse osmosis* process. Reverse osmosis is also used to remove specific contaminants such as nitrate and radium from waters. As the name implies, this process is the reverse of osmosis, and water is caused to flow in a reverse manner through a semipermeable membrane from brackish to dilute fresh water. This is accomplished by exerting a pressure on the brackish water in excess of the osmotic pressure. The semipermeable membrane acts like a filter to retain the ions and particles in solution on the brackish water side, while permitting water alone to pass through the membrane. Theoretically, the process will work if a pressure just in excess of the osmotic pressure is used. In practice, however, a considerably higher pressure is necessary to obtain an appreciable flow of water through the membrane. Also, as fresh water passes through the membrane, the concentration of salts in the brackish water remaining increases, creating a greater osmotic pressure differential. The theoretical minimum energy required to remove salts from water in such a process is equal to the osmotic pressure multiplied by the volume of water being demineralized.

EXAMPLE 3.8

The molar concentration of the major ions in a brackish groundwater supply are as follows: Na^+, 0.02; Mg^{2+}, 0.015; Ca^{2+}, 0.01; K^+, 0.001; Cl^-, 0.025; HCO_3^-, 0.001; NO_3^-, 0.02; and SO_4^{2-}, 0.012.

(a) What would be the approximate osmotic pressure difference across a semipermeable membrane that had brackish water on one side and mineral-free water on the other, assuming the temperature is 25°C?

The molar concentration of particles in the brackish water is

$$c = 0.02 + 0.015 + 0.01 + 0.001 + 0.025 + 0.001 + 0.002 + 0.012$$
$$= 0.086 \text{ M}$$

From Eq. (3.24)

$$\pi = cRT = \frac{0.086 \text{ mol}}{\text{liter}} \times \frac{0.08206 \text{ L}-\text{atm}}{\text{K}-\text{mol}} \times (273 + 25) \text{ K}$$
$$= 2.10 \text{ atm or } 30.3 \text{ pounds per square inch (psi)}$$

(b) If in part (a), a yield of 75 percent fresh water were desired, what minimum pressure would be required to balance the osmotic pressure difference that will develop?

For a 75 percent yield, the salts originally present in four volumes of brackish water would be concentrated in one volume of brackish water left behind the membrane after three volumes of fresh water have passed through the membrane. Thus, the molar concentration of salt in the brackish water would be four times that of the original brackish water or 0.344 M. Then,

$$\pi = 0.344 \times 0.08206 \times 298 = 8.41 \text{ atm or } 124 \text{ psi}$$

At this point the pressure required to push the fresh water through the membrane would be in excess of 124 psi.

Dialysis

By choice of a membrane of a particular permeability, which is wetted by the solvent, it is possible to cause ions to pass through the membrane while large molecules of organic substances or colloidal particles are unable to pass. Thus, a separation of solutes can be accomplished. This process is termed *dialysis*.

Dialysis is used extensively to remove electrolytes from colloidal suspensions to render the latter more stable. Dialysis is used to recover sodium hydroxide from certain industrial wastes that have become contaminated with organic substances, as

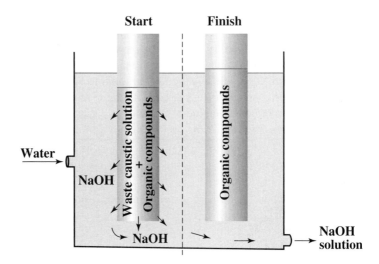

Figure 3.6
A simple dialysis cell for recovery of sodium hydroxide from an industrial waste.

shown in Fig. 3.6. In the process, the waste material is placed in cells with permeable membranes, and the cells are surrounded with water. The sodium and hydroxide ions pass through the cell wall into the surrounding water and some water may pass into the cell. The NaOH solution is evaporated to recover the sodium hydroxide, and the organic waste remaining in the cells is disposed of separately. Waste caustic solutions must be quite concentrated before recovery by dialysis can be justified economically. Mercerizing wastes of the cotton textile industry are an example.

In another application, the dialysis principle can be used for demineralization of brackish water. In this case, the brackish water is placed both inside and outside the cell. Electrodes are placed in the water outside the cell, and when a current is applied, the ions within the cell are caused to flow through the semipermeable membrane and to concentrate in the water outside. The cations flow toward the cathode and the anions flow toward the anode, as discussed in Sec. 3.10. By this method, the water within the cell is demineralized. This process, termed *electrodialysis,* uses electrical energy to cause the flow of ions against a concentration gradient. In practice, a large number of thin, continuous-flow cells are used to make the process efficient for large-scale usage. Electrodialysis is also used to remove specific pollutants such as heavy metals, nitrate, hardness (Ca^{2+} and Mg^{2+}), and radium from contaminated waters.

3.8 | PRINCIPLES OF SOLVENT EXTRACTION

Industrial and hazardous wastes often contain valuable constituents that can be recovered most effectively and economically by means of extraction with an immiscible solvent [commonly called NAPL (non-aqueous-phase liquid)], such as petroleum ether, diethyl ether, benzene, hexane, dichloromethane, or some other organic solvent. Also, many methods for water analysis involve extraction of a constituent or complex from the water sample as one step in the determination. This is true of some procedures for measurement of surface active agents, organic compounds, and various heavy metals. Because of the importance of this operation in environmental engineering practice, a discussion of the principles involved is merited.

When an aqueous solution is intimately mixed with an immiscible solvent, the solutes contained in the water distribute themselves in relation to their solubilities in the two solvents. For low-to-moderate concentrations of solute, the ratio of distribution is always the same:

$$\frac{C_{solvent}}{C_{water}} = \frac{C_s}{C_w} = K \tag{3.25}$$

The equilibrium constant K, or the ratio of distribution, is known as the *distribution coefficient*. In actual practice the immiscible solvent is selected for its ability to dissolve the desired material, and the values for K are normally greater than 1. A distribution coefficient of particular interest is the octanol-water partition coefficient (K_{ow}). Use of K_{ow} is described in Sec. 5.34.

If the volume of solvent used is equal to the volume of the sample being extracted, the mathematics involved are rather simple. For a system with a distribution coefficient of 9, 90 percent of the material would be extracted in the first step, and 90 percent of the material remaining in each successive step. After three extractions with fresh solvent, 99.9 percent of the material would be removed.

In actual practice it is seldom feasible to use a volume of solvent equal to the waste volume, and calculations become somewhat involved. The question in industrial waste treatment that usually requires answering is this: How much remains in the aqueous phase after n extractions? The expression defining the distribution coefficient may be written in terms of the amounts of the substance extracted and the volumes of the liquids involved,

$$K = \frac{C_s}{C_w} = \frac{(W_0 - W_1)/V_s}{W_1/V_w} \tag{3.26}$$

where W_0 = weight of substance originally present in aqueous phase
   W_1 = weight remaining in water after one extraction
   V_s = volume of solvent
   V_w = volume of water

Simplifying, we obtain

$$W_1 = W_0 \frac{V_w}{KV_s + V_w} \tag{3.27}$$

In the second step of the extraction,

$$W_2 = W_1 \frac{V_w}{KV_s + V_w} \tag{3.28}$$

or, in terms of the original sample,

$$W_2 = W_0 \frac{V_w}{KV_s + V_w} \frac{V_w}{KV_s + V_W}$$

$$= W_0 \left(\frac{V_w}{KV_s + V_w} \right)^2 \tag{3.29}$$

and after n extractions the weight of substance remaining in the water is

$$W_n = W_0 \left(\frac{V_w}{KV_s + V_w} \right)^n \tag{3.30}$$

Equation (3.30) has general application and may be used to calculate the volume of a solvent needed to reduce the concentration of a material in the aqueous phase to definite levels with a fixed number of extractions, or the number of extractions needed with a fixed volume of a solvent, provided that the distribution coefficient is known.

A related phenomenon of significance to environmental engineers is the "cosolvent" effect. An important example is the use of ethanol as a fuel oxygenate in gasoline formulations. The purpose of adding ethanol to gasoline is to increase the oxygen content as mandated by the 1990 Clean Air Act amendments. The presence of ethanol in gasoline may cause unintended problems.[1] The presence of ethanol, which is miscible with water, as a cosolvent will increase the water solubility of gasoline components such as benzene, toluene, ethylbenzene, and xylene (BTEX). Second, if the ethanol-gasoline mixture (ethanol and gasoline are miscible) becomes contaminated with enough water, the gasoline will separate into two phases: a gasoline-rich phase that will float on top of an ethanol-water phase. This behavior prevents distribution of ethanol-gasoline formulations by pipeline.

Another, more complex example is cosolvent flushing of soils contaminated with NAPL [they may be less dense than water (LNAPL) or more dense than water (DNAPL)]. Water-miscible cosolvents such as ethanol may be added to increase the solubility of the NAPL and enhance its extraction and recovery.[2] Surfactants have also been employed.[3] In this application, the effects of properties such as surface tension (capillary forces), density, and viscosity, among others, are also important.

3.9 | ELECTROCHEMISTRY

Electrochemistry is concerned with the relationships between electrical and chemical phenomena. A knowledge of electrochemistry has several applications in environmental engineering. It is germane to an understanding of corrosion as well as

---

[1]S. E. Powers, D. Rice, B. Dooher, and P. J. J. Alvarez, Will Ethanol-Blended Gasoline Affect Groundwater Quality, *Env. Sci. Tech.,* **30:** 24A–30A (2001).

[2]J. W. Jawitz, R. K. Sillan, M. D. Annable, P. S. C. Rao, and K. Warner, In-Situ Alcohol Flushing of a DNAPL Source Zone at a Dry Cleaner Site, *Env. Sci. Tech.,* **34:** 3722–3729 (2000).

[3]K. D. Pennel, G. A. Pope, and L. M. Abriola, Influence of Viscous and Buoyancy Forces on the Mobilization of Residual Tetrachloroethylene during Surfactant Flushing, *Env. Sci. Tech.,* **30:** 1328–1335 (1996).

to a study of solutions of electrolytes and the phenomena occurring at electrodes immersed in such solutions. Many analytical procedures of interest are based on electrochemical measurements. Also, automatic continuous stream monitors, which have wide application, use electrochemical methods to translate chemical characteristics into electrical impulses that can be recorded. Electrochemical principles are also useful in understanding oxidation-reduction reactions (Sec. 4.10). A brief review of the more fundamental concepts of electrochemistry will be presented. More detailed information can be obtained from standard texts on physical chemistry.

Current Flow in Solution

An electric current can flow through a solution of an electrolyte as well as through metallic conductors. However, there are some basic differences that are of importance and are summarized here:

Characteristics of current flow through a metal
1. Chemical properties of metal are not altered.
2. Current is carried by electrons
3. Increased temperature increases resistance.

Characteristics of current flow through a solution
1. Chemical change occurs in the solution.
2. Current is carried by ions.
3. Increased temperature decreases resistance
4. Resistance is normally greater than with metals.

A significant feature of current flow through a solution is that the current is carried by ions which move toward electrodes immersed in the solution. Also, a chemical change takes place in the solution at the electrodes, and this alters the chemical properties of the solution. These two phenomena, current flow or conductivity and chemical change at electrodes, will first be considered separately.

Conductivity

The conductivity of a solution is a measure of its ability to carry an electric current and varies both with the number and type of ions the solution contains. Conductivity can be measured in a *conductivity cell* connected to a Wheatstone bridge circuit as shown in Fig. 3.7. Such an arrangement allows measurement of the electrical resistance provided by the cell. The measurement consists of altering the variable resistance until no current flows through the detecting circuit containing the meter A. Modern instruments can do this automatically to give a direct readout. When this state of balance is achieved, the potential at D must be the same as that at E, and the resistance offered by the solution is determined by the relationship

$$X = R_3 \frac{R_1}{R_2} \tag{3.31}$$

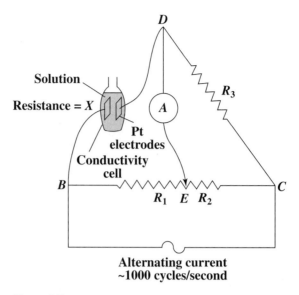

Figure 3.7
Conductivity-measuring apparatus.

Special care must be taken if this measured resistance is to be meaningful. If direct current is used, the apparent resistance changes with time, because of a *polarization* effect at the electrodes. This unwanted effect can be overcome by rapidly changing the direction of current flow. This is done by using an alternating current of several thousand cycles per second. In addition, a more reproducible state of balance is normally obtained when the platinum electrodes of the cell are coated with platinum black. For continuous monitoring and field studies, electrodes made of durable common metals such as stainless steel are often used.

When these precautions are taken, it can be shown that the conductivity cell filled with an electrolytic solution obeys Ohm's law:

$$E = IR \tag{3.32}$$

where E = electromotive force, volts

        I = current, Amperes

       R = resistance of cell contents, ohm

The resistance depends upon the dimensions of the conductor:

$$R = \rho \frac{l}{A} \tag{3.33}$$

where l is the length and A is the cross-sectional area of the conductor. The value ρ (ohm-cm) is called the *specific resistance* of the conductor. Our interest is normally in the *specific conductance* of a solution rather than in its specific resistance. These quantities are reciprocally related as follows:

$$\kappa = \frac{1}{\rho} \tag{3.34}$$

where κ is the specific conductance and has units of 1/ohm-cm, one unit of which is called a siemen (S). The specific conductance can be thought of as the conductance afforded by 1 cm^3 of a solution of electrolyte.

In practice, a conductivity cell is calibrated by determining the resistance R_s of a standard solution, and from this the cell constant C is determined:

$$C = \kappa_s R_s \tag{3.35}$$

Normally, 0.0100 N KCl is used as a standard solution for this calibration and has a specific conductance κ_s of 0.0014118 S at 25°C, or in more convenient units, 1411.8 μS. The specific conductance of an unknown sample can be determined by measuring its resistance R in the cell and then using the following relationship:

$$\text{Specific conductance} = \frac{C}{R} \tag{3.36}$$

Specific conductance has a marked temperature dependence, and caution must be taken to measure the resistance of the standard and the unknown at the same temperature.

Specific conductance measurements are frequently used in water analysis to obtain a rapid estimate of the dissolved solids content of a water sample. If a flow-through cell is used and water from a river or waste stream is pumped through the cell, a continuous recording of specific conductance can be obtained. The dissolved solids content can be approximated by multiplying the specific conductance in microsiemens by an empirical factor varying from about 0.55 to 0.9. The proper factor to use depends upon the ionic components in the solution, as will be indicated by the introduction of a new parameter, the *equivalent conductance* Λ, which is defined as follows:

$$\Lambda = \frac{1000}{N} \kappa \tag{3.37}$$

where N is the normality of the salt solution. For an ideal ionic solution, κ should vary directly with N, and thus Λ should remain constant with varying solution normality. However, because of deviation from ideal behavior, Λ decreases somewhat as the salt concentration increases.

Current is carried by both anions and cations of a salt, but to a different degree. The equivalent conductance of a salt is thus the sum of the equivalent ionic conductances of the cation λ_0^+ and the anion λ_0^-:

$$\Lambda_0 = \lambda_0^+ + \lambda_0^- \tag{3.38}$$

The zero subscript is used to indicate equivalent conductance at infinite dilution, where the deviation from ideal behavior is at a minimum. Several values for equivalent ionic conductance are shown in Table 3.3. It should be noted that these values are strongly temperature-dependent. It is apparent that the equivalent ionic conductances are in general of the same order of magnitude, with the exception of the hydrogen ion and the hydroxyl ion. The latter two are more mobile than the others in aqueous solution and so can carry a larger portion of the current. This fact should be considered when estimating the dissolved solids concentration from conductivity

Table 3.3 | Equivalent ionic conductance at infinite dilution at 25°C in S-cm³/eq

Cation	λ_0^+	Anion	λ_0^-
H^+	350.1	OH^-	199.2
Na^+	50.1	HCO_3^-	44.5
K^+	73.5	Cl^-	76.3
NH_4^+	73.7	NO_3^-	71.4
$\frac{1}{2}Ca^{2-}$	59.5	CH_3COO^-	41.0
$\frac{1}{2}Mg^{2-}$	53.1	$\frac{1}{2}SO_4^{2-}$	80.0
$\frac{1}{2}Ba^{2+}$	45	F^-	55.4

Source: J. A. Dean, "Lange's Handbook of Chemistry," 15th ed., McGraw-Hill, Inc., New York, 1999.

measurements of solutions with either a high or a low pH. If the approximate chemical composition of a water solution is known, the equivalent ionic conductance values will allow a better choice of the appropriate factor for conversion from conductance to dissolved solids concentration.

Another important point is that only ions can carry a current. Thus, the unionized species of weak acids or bases will not carry a current, although they are a portion of the total dissolved solids in a water sample. Also, uncharged soluble organic materials, such as ethanol and glucose, cannot carry a current and so are not measured by conductance.

EXAMPLE 3.9

The specific conductance κ of a sodium chloride solution at 25°C is 125×10^{-6} S. What is the approximate concentration of sodium chloride in mg/L?

The equivalent conductance for sodium chloride, if we use Eq. (3.38) and the values from Table 3.3, is

$$\Lambda_0 = \lambda_0^+ + \lambda_0^- = 50.1 + 76.3 = 126.4 \text{ S-cm}^3/\text{eq}$$

The approximate normality of the solution, if we use Eq. (3.37), is

$$N \cong 1000(\kappa)/\Lambda_0 = 1000(125 \times 10^{-6})/126.4 = 0.99 \times 10^{-3} \text{ eq/L}$$

The equivalent weight of NaCl is 58.5, and so

$$\text{NaCl concentration} \cong (0.99 \times 10^{-3})(58.5 \times 10^3) = 58 \text{ mg/L}$$

Current and Chemical Change

When electrodes are introduced into a water solution in such a way as to allow a direct current to flow through the solution, a chemical change will take place at the electrodes. The nature of the chemical change depends upon the composition of the solution, the nature of the electrodes, and the magnitude of the imposed electromotive force.

Consider first the chemical changes occurring when platinum electrodes are introduced into a solution of HCl as indicated in Fig. 3.8. When a voltage of about 1.3

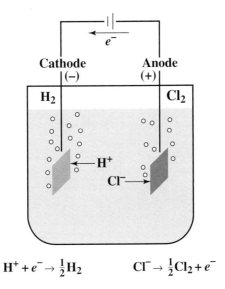

$$H^+ + e^- \rightarrow \tfrac{1}{2}H_2 \qquad\qquad Cl^- \rightarrow \tfrac{1}{2}Cl_2 + e^-$$

Figure 3.8
Electrolysis of a hydrochloric acid solution.

volts is applied, it is found that H_2 is evolved at the cathode and Cl_2 at the anode. The current movement is as follows: Electrons flow through the external metallic conductor in the direction shown, as a result of the driving force of the battery. Such a flow maintains the negative charge at the cathode and the positive charge at the anode. The flow of current through the solution is maintained by movement of the cations (H^+) to the negatively charged cathode and anions (Cl^-) to the positively charged anode.

When an H^+ ion reaches the cathode, it picks up an electron and is reduced to H_2 gas, according to the half reaction

$$H^+ + e^- \rightarrow \tfrac{1}{2}H_2(g) \tag{3.39}$$

When the Cl^- ion reaches the anode, it gives up an electron and is oxidized to Cl_2 gas by the following half reaction:

$$Cl^- \rightarrow \tfrac{1}{2}Cl_2(g) + e^- \tag{3.40}$$

The electrons released by the Cl^- ions are "pumped" through the external circuit by the driving force of the battery to be picked up by the H^+ ions at the cathode. Thus, the battery acts as a driving force to keep the current flowing and the reaction going. As indicated above, *reduction* takes place at the cathode and *oxidation* at the anode. The overall chemical change which takes place in the solution is as follows:

Reduction at cathode	$H^+ + e^- \rightarrow \tfrac{1}{2}H_2(g)$
Oxidation at anode	$Cl^- \rightarrow \tfrac{1}{2}Cl_2(g) + e^-$
Net change	$H^+ + Cl^- \rightarrow \tfrac{1}{2}H_2(g) + \tfrac{1}{2}Cl_2(g)$

The flow of electrons in the external circuit is necessary to bring about the chemical change. It is apparent that in order to bring about an equivalent of chemical change at an electrode, an Avogadro's number of electrons must flow through the external circuit. This quantity of electrons is called the faraday, (F). The rate of flow of electrons gives the current I, which is normally measured in amperes. One faraday is equivalent to an ampere of current flowing for 96,485 seconds. An ampere is also defined as a coulomb per second, so that a faraday is equivalent to 96,485 coulombs.

EXAMPLE 3.10

If a current is passed through a sodium chloride solution, hydrogen gas is evolved at the cathode and chlorine gas at the anode according to the following equations:

Cathode	$H_2O + e^- \rightarrow OH^- + \frac{1}{2}H_2$
Anode	$Cl^- \rightarrow \frac{1}{2}Cl_2 + e^-$
Net	$H_2O + Cl^- \rightarrow OH^- + \frac{1}{2}H_2 + \frac{1}{2}Cl_2$

How many grams of chlorine are produced if a 0.2-A current is passed through the solution for 24 h?

The amount of current flowing in this time period is

$$(0.2 \text{ coulomb/s})(24 \times 60 \times 60 \text{ s}) = 17,300 \text{ coulombs}$$

Thus, the equivalents of chemical change taking place is 17,300/96,485 or 0.179 equivalent. Since the equivalent weight of chlorine is 35.5,

$$\text{Chlorine formed} = 0.179(35.5) = 6.35 \text{ g}$$

Electrochemical Cell

As indicated previously, the nature of a chemical change occurring at an electrode is partially dependent upon the type of electrode used. A single electrode dipping into a solution is said to constitute a *half-cell;* the combination of two half-cells as indicated in Fig. 3.8 is a typical electrochemical cell. Some half-cell systems are used for analysis of various constituents and properties in water. These are discussed in detail in Chap. 12 of this book.

When two half-cells are connected so that ions can pass between them, an electrochemical or galvanic cell is obtained. The electrochemical cell shown in Fig. 3.8 is quite simple, as only one solution is involved. However, in most electrochemical analyses the solutions associated with each half-cell are different, and they must be kept from mixing. In such a case, some type of salt bridge is used which allows passage of ions while keeping interdiffusion of the solutions to a minimum (see Fig. 3.9).

If the two electrodes of an electrochemical cell are connected through a metallic conductor, electrons will flow through the external circuit, and a chemical change will begin to take place in the solutions. If a voltmeter is connected across the half-cells as indicated in Fig. 3.9, it will be found that electromotive force (emf)

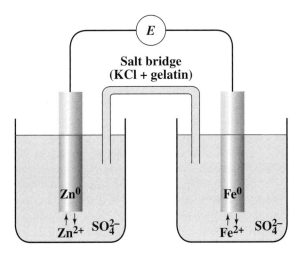

Figure 3.9
An electrochemical cell.

is being generated by the cell. This emf is a measure of the driving force of the chemical reaction that is occurring in the half-cell solutions. Thus, it gives a measure of the chemical potential or free energy of the reaction. From this fact, a relationship between electrochemical potential and chemical free energy can be found. Electric energy is measured in terms of the *joule,* which is the energy generated by the flow of 1 A in 1 second against an emf of 1 volt. Electric energy is given by the product, *EIt* and has units of the volt coulomb or joule.

The electric energy expended in bringing about 1 mol of chemical change is zEF, where z is the number of electron-equivalents per mole, F is the faraday or coulombs per equivalent, and E is the emf of the cell in volts. By convention, if the reaction proceeds, E is positive, so the relation between free energy and electrical energy is

$$\Delta G = -zFE \tag{3.41}$$

Consider the following chemical reaction:

$$a\text{A} + b\text{B} \rightleftharpoons c\text{C} + d\text{D}$$

If we substitute the relationship for E from Eq. (3.41) into the free-energy equation (3.12), the following relationship between cell emf and concentration of reactants and products results:

$$E = E^0 - \frac{RT}{zF} \ln \frac{\{\text{C}\}^c \{\text{D}\}^d}{\{\text{A}\}^a \{\text{B}\}^b} \tag{3.42}$$

Here the value of the gas constant R, in electrical units, is 8.314 J/K-mol. This important equation indicates the relationship between the standard electrode potential of a cell and the activities of the products and reactants. It is sometimes called the Nernst equation. For a temperature of 25°C, and converting ln to log, Eq. (3.42) becomes:

$$E = E^0 - \frac{0.059}{z} \log \frac{\{C\}^c \{D\}^d}{\{A\}^a \{B\}^b} \tag{3.43}$$

When the activities of products and reactants are unity, the logarithmic term is unity, and $E = E^0$ (that is, the emf for the standard state).

The emf of a cell can be calculated from tabulated values just like free-energy and enthalpy values. Such tables list the standard potentials of various half-cells with respect to the standard hydrogen electrode, which is assigned by convention the value $E^0 = 0$. By taking the difference between the standard potentials of two half-cells, the potential of the whole cell can be determined. Standard potentials for various half-cells of interest are listed in Table 3.4. The E^0 values listed in Table 3.4 are for a reaction written for 1 mol of e^- change, for example:

$$\tfrac{1}{2}Fe^{2+} + e^- = \tfrac{1}{2}Fe(s)$$

If an electrochemical cell has reached a state of equilibrium, no current can flow, and the emf of the cell is zero. For this case a relationship between the standard-cell potential and the equilibrium constant for the reaction can be obtained by using Eqs. (3.14) and (3.41):

$$E^0 = -\frac{\Delta G^0}{zF} = \frac{RT}{zF} \ln K \tag{3.44}$$

Table 3.4 | Standard electrode potentials in water at 25°C

Half-cell reaction	E^0, volts
$O_2(g) + 4H^+ + 4e^- \rightarrow 2H_2O$	1.229
$Ag^+ + e^- \rightarrow Ag(s)$	0.799
$Fe^{3+} + e^- \rightarrow Fe^{2+}$	0.771
$Ag_2CrO_4(s) + 2e^- \rightarrow 2\,Ag(s) + CrO_4^{2-}$	0.446
$Cu^{2+} + 2e^- \rightarrow Cu(s)$	0.337
$AgCl(s) + e^- \rightarrow Ag(s) + Cl^-$	0.222
$S(s) + 2H^+ + 2e^- \rightarrow H_2S(g)$	0.141
$2H^+ + 2e^- \rightarrow H_2(g)$	0.000
$Pb^{2+} + 2e^- \rightarrow Pb(s)$	−0.126
$Sn^{2+} + 2e^- \rightarrow Sn(s)$	−0.136
$Fe^{2+} + 2e^- \rightarrow Fe(s)$	−0.441
$S(s) + 2e^- \rightarrow S^{2-}$	−0.48
$Zn^{2+} + 2e^- \rightarrow Zn(s)$	−0.763
$Zn(OH)_2(s) + 2e^- \rightarrow Zn(s) + 2OH^-$	−1.245
$ZnS(s) + 2e^- \rightarrow Zn(s) + S^{2-}$	−1.44
$Al^{3+} + 3e^- \rightarrow Al(s)$	−1.66
$Mg^{2+} + 2e^- \rightarrow Mg(s)$	−2.37
$Mg(OH)_2(s) + 2e^- \rightarrow Mg(s) + 2OH^-$	−2.69

Source: W. M. Latimer, "Oxidation Potentials," 2nd ed., Prentice-Hall, Englewood Cliffs, NJ, 1952.

Alternatively, from Eq. (3.43):

$$E^0 = \frac{0.059}{z} \log K \quad \text{or} \quad \log K = 16.9zE^0 \qquad (3.45)$$

EXAMPLE 3.11

Determine the solubility product constant at 25°C for silver chloride, using standard electrode potentials.

The equilibrium of interest is

$$AgCl(s) \rightleftharpoons Ag^+ + Cl^-$$

and

$$[Ag^+][Cl^-] = K_{sp}$$

A cell consisting of the following two half-cells from Table 3.4 will produce the overall reaction of interest:

		E^0, volts
$Ag(s) \rightarrow Ag^+ + e^-$		-0.799
$AgCl(s) + e^- \rightarrow Ag(s) + Cl^-$		0.222
Net	$AgCl(s) \rightarrow Ag^+ + Cl^-$	-0.577

Thus, E^0 for the net reaction as written is -0.577 volt, and K_{sp} can be determined by using Eq. (3.44):

$$E^0 = \frac{RT}{zF} \ln K_{sp}$$

$$-0.577 = \frac{8.314(273 + 25)}{(1)(96,485)} \ln K_{sp}$$

$$\ln K_{sp} = -22.5 \quad \text{and} \quad K_{sp} = 1.7 \times 10^{-10}$$

EXAMPLE 3.12

Can $Cl_2(g)$ be used to oxidize NH_4^+ to NO_3^-?

The half reactions listed in Table 2.4 can be used to construct a balanced oxidation-reduction reaction and an E^0 can be calculated for the overall reaction. For this particular problem, half reaction 20 in Table 2.4 can be reversed (with a corresponding sign change for E^0) and added to half reaction 3 to produce the overall oxidation-reduction reaction:

	E^0, volts
$\frac{1}{8}NH_4^+ + \frac{3}{8}H_2O = \frac{1}{8}NO_3^- + \frac{5}{4}H^+ + e^-$	-0.882
$\frac{1}{2}Cl_2(g) + e^- = Cl^-$	1.361
Net $\quad \frac{1}{8}NH_4^+ + \frac{1}{2}Cl_2(g) + \frac{3}{8}H_2O = \frac{1}{8}NO_3^- + Cl^- + \frac{5}{4}H^+$	0.479

Since E^0 is positive, yes, the reaction can theoretically proceed as written.

Additional discussion of these concepts is given in Sec. 4.10 on oxidation-reduction reactions.

Galvanic Protection

If a zinc metal electrode is connected to an iron metal electrode by means of a conducting salt bridge and an external metallic conductor as indicated in Fig. 3.9, a cell will result. The cell reaction and standard potential of the cell will be

		E^0
Zn electrode	$Zn(s) \rightarrow Zn^{2+} + 2e^-$	0.763
Fe electrode	$Fe^{2+} + 2e^- \rightarrow Fe(s)$	-0.441
Net	$Zn(s) + Fe^{2+} \rightarrow Zn^{2+} + Fe(s)$	0.322

Since the potential is positive, the reaction can proceed as written when products and reactants are near unit activity. Zinc ions will tend to pass into solution, and iron ions will tend to plate out on the iron electrode. Thus, the iron is kept from passing into solution, while the zinc acts in a *sacrificial* manner. As electrochemical cells are sometimes called galvanic cells, this method of corrosion prevention is called *galvanic protection*. It is the basic principle involved in the protection of iron by galvanizing with zinc.

   From this consideration, it is apparent that when two metals are in electrical contact, the metal with the greater single-electrode potential will sacrifice itself to protect the other. Since the protected electrode assumes a negative charge, it is the cathode, and this can be called *cathodic protection*. From these considerations, the engineer can explain why discontinuous coatings of tin aggravate the rusting of iron. These principles are also the basis of regulations prohibiting the joining of copper and iron pipe without the use of insulating connectors.

   If a battery is placed in the external circuit connecting two half-cells, either electrode can be made to be the cathode and thus be protected. Hence, electric energy can be made to counterbalance the chemical energy of the cell, and so reverse the reaction. Such electrochemical principles are widely used for the cathodic protection of steel pipelines, tanks, and structures by means of sacrificial anodes or artificially impressed negative potentials.

3.10 | CHEMICAL KINETICS

Chemical kinetics is concerned with the speed or velocity of reactions. Many reactions have rates that at a given temperature are proportional to the concentration of one, two, or more of the reactants raised to a small integral power. For example, if a reaction is considered in which A, B, and C are possible reactants, then the rate equations that express the concentration dependence of the reaction rate may take one of the following forms, among others:

$$\text{Rate} = kC_a \qquad\qquad\qquad\qquad\qquad \text{first order}$$

$$\text{Rate} = kC_a^2 \quad \text{or} \quad kC_aC_b \qquad\qquad\quad \text{second order}$$

$$\text{Rate} = kC_a^3 \quad \text{or} \quad kC_a^2C_b \quad \text{or} \quad kC_aC_bC_c \quad \text{third order}$$

where C_a, C_b, and C_c represent the concentrations of reactants A, B, and C, respectively. Reactions that proceed according to such simple expressions are said to be reactions of the first, second, or third order as indicated, with the *order* of the reaction being defined as the sum of the exponents of the concentration terms in the reaction equations. Not all reactions have such simple rate equations. Some involve concentrations raised to a fractional power, while others consist of more complex algebraic expressions. Attention must be paid to the units of the reaction rate constant k. They are different for the different reaction orders and depend on the concentration units used.

Environmental engineers and scientists deal with many reactions that proceed slowly and require rate expressions so that the reactions can be dealt with on a practical basis. This is true for biotransformation reactions, microbial growth and decay, aeration, radioactive decay, disinfection, chemical hydrolysis, and oxidation and reduction. In general, first-order reactions are the most common, although reactions of other orders or of a more complex nature are sometimes involved. Additional detail is given in Sec. 5.34.

Zero-Order Reactions

A *zero-order* reaction is one in which the rate of reaction is independent of concentration:

$$-\frac{dC}{dt} = k \tag{3.46}$$

where C is the concentration of reactant and k is the rate constant in units of concentration/time. Many biologically induced reactions, particularly those involving growth on simple, soluble substrates, appear to occur in a linear manner over fairly large ranges of concentrations. That is, a plot of concentration versus time yields a straight line. These are classified as zero-order reactions. Examples of environmental interest include the oxidation of ammonia to nitrite and the oxidation of glucose by aerobic bacteria. All such biological reactions, however, become slower as the substrate concentration approaches zero, as will be discussed under Enzyme Reactions.

First-Order Reactions

The decomposition of a radioactive element is the simplest example of a true *first-order* reaction. In such a reaction the rate of decomposition is directly proportional to the amount of undecayed material and may be expressed mathematically as

$$-\frac{dC}{dt} = kC \tag{3.47}$$

where the minus sign indicates a loss of material with time, C is its concentration, and k is the rate constant for the reaction and has units of reciprocal time.

If the initial concentration, at time $t = 0$, is C_0, and if at some later time t the concentration has fallen to C, the integration of Eq. (3.47) gives

$$-\int_{C_0}^{C} \frac{dC}{C} = k \int_{0}^{t} dt$$

and
$$-\ln \frac{C}{C_0} = kt \quad \text{or} \quad \ln \frac{C}{C_0} = -kt \tag{3.48}$$

or
$$C = C_0 \, e^{-kt} \tag{3.49}$$

Converting to $\log_{10}$, we see that Eqs. (3.48) and (3.49) become

$$\log_{10} \frac{C}{C_0} = -\frac{kt}{2.303} = -k't \quad \text{and} \quad C = C_0 10^{-k't} \tag{3.50}$$

As can be seen from Eq. (3.48), a plot of $\ln(C/C_0)$ versus t will yield a straight line. This is a common way of proving whether or not a reaction is first-order. The rate constant k can be determined directly from the slope of this line ($k = -$slope, since the rate constant cannot be negative). Similarly, a plot of $\log_{10}(C/C_0)$ versus t will yield a straight line. The rate constant k can be evaluated from this plot by multiplying the slope of the plotted line by -2.303.

For radioactive substances it is customary to express decomposition rates in terms of *half-life*, or the time required for the amount of substance to decrease to half its initial value. For a first-order reaction, the half-life, denoted by $t_{1/2}$, can be found from Eq. (3.48) by inserting the requirement that at $t = t_{1/2}$ the concentration of $C = \frac{1}{2}C_0$. This gives

$$t_{1/2} = \frac{\ln 2}{k} = \frac{0.693}{k} \tag{3.51}$$

There is a growing tendency to use the concept of half-life for a variety of environmental phenomena (see Sec. 5.34).

Environmental engineers also find application for the concepts of first-order reactions in areas that do not involve decomposition reactions. For example, the dissolution of gases into water and the removal of gases from water, under a given set of conditions, can be described by a first-order reaction (Sec. 2.9). In other cases, reactions that may in fact not be true first-order reactions can be approximated as such. The rate of death of microorganisms by disinfection is frequently considered to be a first-order reaction and dependent upon the concentration of live microorganisms remaining. Also, the decomposition of organic matter by bacteria in the biochemical oxygen demand (BOD) test is normally considered to be a first-order reaction dependent only upon the concentration of organic matter remaining. In actual fact, this is a very complex reaction, and the limitations of a first-order assumption should be well understood to prevent misinterpretation of BOD data. The kinetics of the BOD test are discussed in more detail in Sec. 23.2.

EXAMPLE 3.13

Strontium 90 (^{90}Sr) is a radioactive nuclide of public health significance and has a half-life of 29 years. How long would a given amount of ^{90}Sr need to be stored to obtain a 99.9 percent reduction in quantity?

From Eq. (3.51), we have $k = 0.693/t_{1/2} = 0.693/29$ yr $= 2.39 \times 10^{-2}$ yr^{-1}. Time required is that for C to be reduced 99.9 percent to $0.001 C_0$. Thus, from Eq. (3.48),

$$-kt = \ln(0.001 C_0/C_0) = \ln 0.001$$

and
$$t = -\ln 0.001/2.39 \times 10^{-2} = 289 \text{ yr}$$

EXAMPLE 3.14

The following are data from an experiment to assess the disinfection of a water supply with a given dose of chlorine. Assuming first-order kinetics, determine the rate constant.

Time, min	Percent coliform bacteria remaining	C/C_0	$\ln(C/C_0)$
0	100	1.0	0.000
10	70	0.70	−0.357
20	21	0.21	−1.561
30	6.3	0.063	−2.765
60	0.6	0.006	−5.116

From Eq. (3.48), a plot of $\ln(C/C_0)$ versus time should yield a straight line with a slope of $-k$. A plot of the data is given below and includes the equation of the line developed from a linear regression of the data. The student should note that r^2, termed the coefficient of determination, is a measure of the goodness of fit of the data with the rate expression used. The larger the r^2 (with a maximum of 1.000), the better the fit. The equation of the regression line and values for r^2 can be calculated using methods described in Chap. 10, or can be determined using hand-held calculators and computer graphics software.

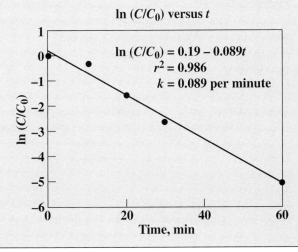

ln (C/C_0) versus t

$$\ln (C/C_0) = 0.19 - 0.089t$$
$$r^2 = 0.986$$
$$k = 0.089 \text{ per minute}$$

Gas-Liquid Mass-Transfer Kinetics.   The rate of gas transfer into or from aqueous solution can be described using the first-order relationship given in Sec. 2.9:

$$\frac{dC}{dt} = K_L a(C_{equil} - C) \tag{3.52}$$

where C = concentration in water (mass/volume)

       C_{equil} = concentration in water (mass/volume) that would be in equilibrium with concentration in gas phase (Henry's law)

       $K_L a$ = first-order, overall mass-transfer rate coefficient (time^{-1})

Mass transfer (flux) into or from aqueous solution can be thought of as resistance in series: transfer from the bulk gas phase across a stagnant gas film to the gas-water interface, followed by transfer across a stagnant liquid film to the bulk water phase ("two-film" theory). Reaction rates can be limited by flux across the gas film or the liquid film. This is best understood using the following relationship:

$$\frac{1}{K_L a} = \frac{RT}{K_H k_g a} + \frac{1}{k_w a} \tag{3.53}$$

where R = 0.08206 L-atm/mol-K

       T = temperature, K

       a = interfacial area to volume of water, m^{-1}

       K_H = Henry's constant, L-atm/mol

       k_g = gas-phase mass-transfer rate coefficient, m/s

       k_w = water-phase mass-transfer rate coefficient, m/s

For compounds with large K_H, the second term will dominate the expression and water-film mass transfer will control overall rates. For compounds with small K_H, the first term dominates and gas-film mass transfer controls rates. A practical implication of such considerations is whether mixing the liquid or gas will enhance mass-transfer rates. In addition to the two-film model, surface renewal and penetration models have also been used to describe gas-water mass transfer.[4]

Second-Order Reactions

A *second-order* reaction is one in which the rate of the reaction is proportional to the square of the concentration of one of the reactants or to the product of the concentrations of two different reactants. Thus, if the overall second-order reaction were of the form

$$A + B \rightarrow products$$

---

[4]W. J. Weber, Jr., and F. A. DiGiano, "Process Dynamics in Environmental Systems," John Wiley & Sons, New York, 1996.

then the rate law for this situation might be

$$-\frac{dC_a}{dt} = k_a C_a^2 \tag{3.54}$$

or

$$-\frac{dC_a}{dt} = k_a C_a C_b \tag{3.55}$$

where C_a and C_b are the concentrations of A and B, respectively. The decrease in B could be formulated in a similar manner:

$$-\frac{dC_b}{dt} = k_b C_b^2 \tag{3.56}$$

or

$$-\frac{dC_b}{dt} = k_b C_a C_b \tag{3.57}$$

The integrated forms of Eqs. (3.54) and (3.56) are

$$1/A = 1/A_0 + k_a t \tag{3.58}$$

$$1/B = 1/B_0 + k_b t \tag{3.59}$$

Thus, if a plot of 1/A or 1/B versus t gives a straight line, a second-order reaction is implicated. The interested student can find the integrated forms of Eqs. (3.55) and (3.57) in most textbooks on physical chemistry.

 An important use of a second-order reaction is in describing cometabolic biotransformation of some halogenated organic compounds or biological transformations in general where the concentration of the compound transformed is very low (< 1 mg/L). In this case, the rate expression is

$$-dC/dt = kCX \tag{3.60}$$

where C is the concentration of the organic compound transformed and X is the concentration of bacteria. This reaction is of the type described by Eq. (3.55). During such biotransformations, X might also be considered constant and the rate expression becomes

$$-dC/dt = k'C \tag{3.61}$$

where k' is equal to kX, and the reaction in this form is typically termed a *pseudo-first-order* reaction with k' being the pseudo-first-order rate constant. Of course, the reaction is first-order with respect to C regardless of whether X is constant. This type of rate expression is used to describe other reactions of importance to environmental engineers and scientists.

Consecutive Reactions

Consecutive reactions are complex reactions of great environmental importance, and so equations describing the kinetics of such reactions are of real interest. In consecutive reactions, the products of one reaction become the reactants of a following reaction:

$$A \xrightarrow{k_1} B \xrightarrow{k_2} C \tag{3.62}$$

Here reactant A is converted to product B at a rate determined by rate constant k_1. Product B in turn becomes the reactant for the second step and is converted to product C as determined by rate constant k_2. If the rates of each of the consecutive reactions are considered to be first-order, then the differential equations which describe the rates of decomposition and formation of the reactants and products are as follows:

$$-dC_a/dt = k_1 C_a \tag{3.63}$$

$$dC_b/dt = k_1 C_a - k_2 C_b \tag{3.64}$$

and
$$dC_c/dt = k_2 C_b \tag{3.65}$$

If at $t = 0$ we have $C_a = C_a^0$, $C_b = C_b^0$, and $C_c = C_c^0$, then a solution for the concentration of each constituent at some time t is as follows:

$$C_a = C_a^0 e^{-k_1 t} \tag{3.66}$$

$$C_b = \frac{k_1 C_a^0}{k_2 - k_1}(e^{-k_1 t} - e^{-k_2 t}) + C_b^0 e^{-k_2 t} \tag{3.67}$$

$$C_c = C_a^0\left(1 - \frac{k_2 e^{-k_1 t} - k_1 e^{-k_2 t}}{k_2 - k_1}\right) + C_b^0(1 - e^{-k_2 t}) + C_c^0 \tag{3.68}$$

Equation (3.67) is widely used to describe the oxygen deficit in a stream caused by organic pollution. In this case C_b can be considered the oxygen deficit being created in the first step by the biological oxidation of organic matter with concentration C_a. At the same time, the oxygen deficit is being decreased by atmospheric reaeration [see Eq. (3.52)] to give the second step in the consecutive reactions. Equation (3.67) when used for this case is the well-known Streeter-Phelps equation.

A consecutive reaction can also be used to describe the bacterial nitrification of ammonia. Here ammonia is oxidized by *Nitrosomonas* bacteria to nitrite, which is then oxidized in the second step by *Nitrobacter* bacteria to nitrate as indicated by the following sequence:

$$NH_3 \xrightarrow[\text{\textit{Nitrosomonas}}]{O_2} NO_2^- \xrightarrow[\text{\textit{Nitrobacter}}]{O_2} NO_3^- \tag{3.69}$$

The buildup and decay of the various forms of nitrogen in this consecutive reaction are sometimes assumed for simplicity to follow first-order kinetics. The changes in nitrogen forms which would occur with this assumption are illustrated in Fig. 3.10. The concentrations of NO_2^- and NO_3^- were set equal to zero when $t = 0$, and k_1 was assumed to equal $2k_2$. In actual fact, the kinetics for nitrification are much more complex, so one should consider the limitations of this assumption before applying these equations in practice. The changes in nitrogen forms indicated in Fig. 3.10 are typical of those frequently noted in trickling filters where ammonia is oxidized or in rivers downstream from an ammonia discharge.

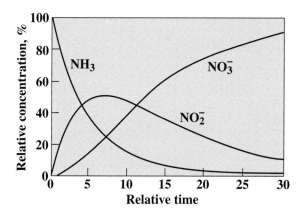

Figure 3.10
Nitrogen changes during nitrification, assuming consecutive first-order reactions.

Consecutive-type kinetics are frequently used to describe the growth and decay of microorganisms in biological treatment processes. They can also describe the consecutive steps in the decomposition of organic matter as it occurs in anaerobic waste treatment. Thus, consecutive-type kinetics are widely applicable in environmental engineering and science practice.

Enzyme Reactions

Another complex kinetics expression to describe the rate of biological waste treatment was first used by Michaelis and Menten[5] to describe enzyme reactions. Since bacterial decomposition involves a series of enzyme-catalyzed steps, the Michaelis-Menten expression can be empirically extended to describe the kinetics of bacterial growth and waste decomposition. The resulting equation is termed the Monod equation. Figure 3.11 indicates the normally observed relationship between substrate or waste concentration, designated as S, and speed of waste utilization per unit mass of enzyme or bacteria, designated as V/E.

The Michaelis-Menten relationship for enzyme reactions in a simplified form assumes the following reaction, where E_f is free enzyme, S is substrate, and E_cS is enzyme-substrate complex:

$$E_f + S \underset{k_{-1}}{\overset{k_1}{\rightleftharpoons}} E_cS \overset{k}{\rightarrow} E_f + \text{products} \quad (3.70)$$

E_cS is formed at rate k_1 when free enzyme and substrate combine. The complex is unstable and decomposes either back to the original free enzyme and substrate at rate k_{-1}, or into free enzyme and reaction products at rate k. The total enzyme

---

[5]Michaelis and Menten, *Biochem. Zeit.,* **49:** 333 (1913).

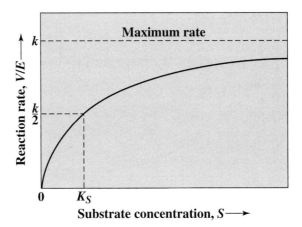

Figure 3.11
The relationship between substrate concentration and
reaction rate for enzyme-type reactions.

concentration in the system, E, remains constant and is equal to $[E_f] + [E_cS]$. On the basis of the Eq. (3.70), the rate of formation of enzyme-substrate complex is

$$d[E_cS]/dt = k_1 [E_f]S - (k_{-1} + k)[E_cS]$$

$$= k_1 S(E - [E_cS]) - (k_{-1} + k)[E_cS] \qquad (3.71)$$

The rate of complex formation is generally much faster than the overall reaction rate, so for the purpose of determining the overall reaction rate, the complex can be considered as being at pseudo-steady-state concentration, that is, $d[E_cS]/dt = 0$. Therefore,

$$k_1S(E - [E_cS]) = (k_{-1} + k)[E_cS]$$

Rearranging, we obtain

$$\frac{S(E - [E_cS])}{[E_cS]} = \frac{(k_{-1} + k)}{k_1} = K_s \qquad (3.72)$$

or

$$[E_cS] = \frac{ES}{K_s + S} \qquad (3.73)$$

The rate of product formation is equal to the overall velocity (rate) of the reaction and is given by $V = k[E_cS]$, and thus from the relationship given by Eq. (3.73), the overall rate as a function of E and S becomes

$$V = \frac{kES}{K_s + S}$$

or

$$\frac{V}{E} = \frac{kS}{K_s + S} \qquad (3.74)$$

The significance of the constants k and K_s is indicated in Fig. 3.11. The constant k gives the maximum rate of the reaction, and K_s is equal to the substrate concentration at which the reaction rate is one-half of maximum. K_s is commonly called the "half-velocity" constant. Two limiting cases for Eq. (3.74) are apparent:

$$V/E \cong k'S \qquad \text{when } S \ll K_s \qquad (k' = k/K_s) \qquad (3.75)$$

and $\qquad\qquad V/E \cong k \qquad \text{when } S \gg K_s \qquad\qquad\qquad (3.76)$

Equation (3.75) indicates that when the substrate concentration is low compared to K_s, the rate of the enzyme reaction is directly proportional to S. Therefore, the reaction can be described as *first-order* with respect to substrate. However, when S is much greater than K_s, the reaction rate is a maximum and independent of the concentration S. The reaction is then said to be *zero-order* with respect to substrate.

Both the continuous Eq. (3.74) and the discontinuous set of Eqs. (3.75) and (3.76) are frequently used to describe biological reaction rates. All are somewhat empirical when used to describe complex biological processes, but they give a sufficiently adequate description of the overall process to yield practical results.

EXAMPLE 3.15

A study was made to evaluate the constants so that the Michaelis-Menten relationship could be used to describe waste utilization by bacteria. It was found that 1 g of bacteria could decompose the waste at a maximum rate of 20 g/day when the waste concentration was high. Also, it was found that this same quantity of bacteria would decompose waste at a rate of 10 g/day when the waste concentration surrounding the bacteria was 15 mg/L. What would be the rate of waste decomposition by 2 g of bacteria if the waste concentration were maintained at 5 mg/L?

The constant k gives the maximum rate of waste utilization, and so for this case it is equal to 20 g/day-g. The constant K_s is equal to the substrate concentration at which the rate is $\frac{1}{2}$ of maximum or 10 g/day-g. Therefore, for this example, $K_s = 15$ mg/L, and from Eq. (3.74),

$$\frac{V}{E} = \frac{20S}{15 + S}$$

and assuming E = weight of bacteria = 2 g, and S = 5 mg/L,

$$V = 2(20)(5)/(15 + 5) = 10 \text{ g/day}$$

While Example 3.15 is interesting, the student should be aware that it is not good practice to attempt to determine reaction rate constants from only one or two measurements.

Temperature Dependence of Reaction Rates

In general, the rates of most chemical and biological reactions increase with temperature. An approximate rule is that the rate of a reaction will about double for each 10°C rise in temperature. In biological reactions, this rule will hold more or less true up to a certain optimum temperature. Above this, the rate decreases, probably owing to destruction of enzymes at the higher temperatures.

The change in rate constant with temperature can be expressed mathematically by the Arrhenius equation,

$$d \ln k/dT = E_a/RT^2 \tag{3.77}$$

where $d \ln k/dT$ = change in natural log of rate constant with temperature

R = universal gas constant

E_a = constant for reaction termed *activation energy*

Integrating between limits gives

$$\ln \frac{k_2}{k_1} = \frac{E_a(T_2 - T_1)}{RT_2T_1} \tag{3.78}$$

where k_2 and k_1 are the rate constants at temperatures T_2 and T_1, respectively. Temperature is expressed in kelvins.

Most processes of concern to environmental engineers and scientists operate over a small temperature range near ambient temperatures. For this case the product T_2T_1 changes very little, and for practical purposes it can be considered constant. Thus, E_a/RT_2T_1 can be considered equal to a constant A, so that an approximate formula for temperature dependence of reaction rates can be used:

$$\ln \frac{k_2}{k_1} = A(T_2 - T_1) \tag{3.79}$$

or

$$k_2 = k_1 \, e^{A(T_2 - T_1)} \tag{3.80}$$

Another common form of the equation describing temperature dependence is

$$k_2 = k_1 \, A^{(T_2 - T_1)} \tag{3.81}$$

Of course, the values for A are different in Eqs. (3.80) and (3.81). The value of A in Eq. (3.81) is equal to e^{E_a/RT_2T_1}. Both Eqs. (3.80) and (3.81) are commonly used to express the effect of temperature on reaction rates. Although A is supposed to be a constant, it sometimes varies significantly even over a limited temperature range. For example, for the BOD reaction rate (see Chap. 23), A [Eq. (3.80)] has been indicated[6] to vary from 0.135 in the temperature range from 4 to 20°C, down to 0.056 in the temperature range from 20 to 30°C. Thus, caution must be exercised in using an A value beyond the temperature range for which it was evaluated.

3.11 | CATALYSIS

Catalysts are compounds that change the rate of a chemical reaction. They may be positive or negative in effect. Regardless of their actual role in the reaction, they are recoverable in their original form at the end of the reaction. It is impor-

---

[6]G. J. Schroepfer, M. L. Robins, and R. H. Susag, The Research Program on the Mississippi River in the Vicinity of Minneapolis and St. Paul, "Advances in Water Pollution Research," vol. I, Pergamon, London, 1964.

tant to remember that catalysts have no influence on the final equilibrium of a reaction. They simply alter the speed with which the equilibrium is attained by changing the activation energy. Positive catalysts have one other property of interest to environmental engineers and scientists. They can initiate and maintain reactions at concentration levels below those at which ordinary reactions would occur.

An example of how catalysts are used is in the control of air pollution. Hydrogen sulfide is catalytically oxidized to sulfur dioxide at concentrations normally incapable of supporting combustion. Catalytic devices are now used to oxidize hydrocarbons and carbon monoxide in the exhaust gases of automobiles, trucks, and buses as one means of controlling smog problems. Enzymes produced by bacteria and other microorganisms are organic catalysts which permit the occurrence at room temperature of a great many reactions of importance such as hydrolysis, oxidation, and reduction of both inorganic and organic pollutants.

3.12 | ADSORPTION

Sorption processes are very important to the fate and transport of contaminants in the environment and for the removal of contaminants in engineered reactors. *Sorption* is most often defined as the concentration or movement of contaminants from one phase to another. *Absorption* involves the partitioning of a contaminant from one phase into another phase. Examples include the dissolution (absorption) of oxygen gas into water and the absorption of the pesticide DDT into the organic solvent hexane. *Adsorption* is the process by which ions or molecules present in one phase tend to condense and concentrate on the surface of another phase. Additional discussion of sorption processes is given in Sec. 4.9, Sec. 5.34, and Chap. 7. Adsorption is discussed in detail in this section.

Adsorption of contaminants present in air or water onto activated carbon is frequently used for purification of the air or water. The material being concentrated is the *adsorbate,* and the adsorbing solid is termed the *adsorbent.* There are three general types of adsorption, *physical, chemical,* and *exchange* adsorption. Physical adsorption is relatively nonspecific and is due to the operation of weak forces of attraction or van der Waals' forces between molecules. Here, the adsorbed molecule is not affixed to a particular site on the solid surface but is free to move about over the surface. In addition, the adsorbed material may condense and form several superimposed layers on the surface of the adsorbent. Physical adsorption is generally quite reversible; i.e., with a decrease in concentration the material is desorbed to the same extent that it was originally adsorbed.

Chemical adsorption (sometimes called chemisorption), on the other hand, is the result of much stronger forces, comparable with those leading to the formation of chemical compounds. Normally the adsorbed material forms a layer over the surface which is only one molecule thick, and the molecules are not considered free to move from one surface site to another. When the surface is covered by the monomolecular layer, the capacity of the adsorbent is essentially exhausted. Also, chemical adsorption is seldom reversible. The adsorbent must generally be heated to higher temperatures to remove the adsorbed materials.

Exchange adsorption is used to describe adsorption characterized by electrical attraction between the adsorbate and the surface. Ion exchange is included in this class. Here, ions of a substance concentrate at the surface as a result of electrostatic attraction to sites of opposite charge on the surface. In general, ions with greater charge, such as trivalent ions, are attracted more strongly toward a site of opposite charge than are molecules with lesser charge, such as monovalent ions. Also, the smaller the size of the ion (hydrated radius), the greater the attraction. Although there are significant differences among the three types of adsorption, there are instances in which it is difficult to assign a given adsorption to a single type.

Since adsorption is a surface phenomenon, the rate and extent of adsorption are functions of the surface area of the solids used. Activated carbon is used extensively for adsorptive purposes because of its tremendous surface area in relation to mass. It is generally made from a wood product or coal by heating to temperatures between 300 and 1000°C in one of a variety of possible gaseous atmospheres such as CO_2, air, or water vapor, and then quickly quenching in air or water. The interior of the wood cells is cleaned out by this procedure, leaving a structure with remarkably small and uniform pores. Surface areas in the range of 1000 m^2 per gram of activated carbon result, with pore sizes in the general range of 10 to 1000 angstroms (Å) in diameter. At a given temperature and pressure, a sample of activated carbon will adsorb a definite quantity of a gas. If the pressure is increased, it will adsorb more; if the pressure is decreased, it will adsorb less. If the quantities of adsorbed gas are plotted against pressure, curves of the sort shown in Fig. 3.12 are obtained.

Adsorption of solutes from solution follow the same general laws as gases. This is illustrated in Fig. 3.13, which shows data for the adsorption of acetic and benzoic acids. The curves are of the same nature as those shown in Fig. 3.12. From

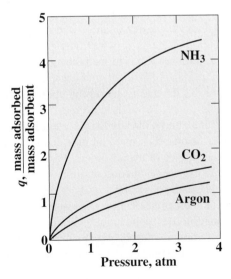

Figure 3.12
Adsorption of gases on charcoal in relation
to pressure at constant temperature.

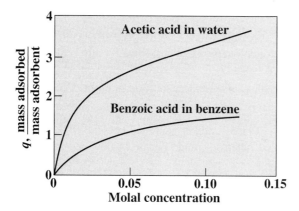

Figure 3.13
Adsorption of solutes on charcoal; temperature and pressure
constant.

these data, it may be concluded that the quantity of substance adsorbed by a given
sample of adsorbent depends upon the nature of the material and its concentration.
Temperature is also a factor which is not demonstrated by the data presented.

Adsorption Isotherms

An *adsorption isotherm* is a quantitative relationship describing the equilibrium between the concentration of adsorbate in solution (mass/volume) and its sorbed concentration (mass adsorbate/mass adsorbent). The term isotherm is used to signify that the
relationship is for a given temperature. Four commonly used isotherms are *linear,
Langmuir, Freundlich,* and *BET.* Which isotherm should be used depends on a variety
of factors such as situation (e.g., engineered reactor versus natural environment); nature, concentration, and number of adsorbates (e.g., hydrophobic versus hydrophilic,
organic compound versus metal, neutral versus charged species, high versus low concentration, single contaminant versus multiple contaminants); type of adsorbent [e.g.,
granular activated carbon (GAC) versus ion exchange resin versus iron oxide minerals
versus aquifer material]; type of fluid (e.g., gas versus water versus organic solvent);
and other environmental factors (e.g., pH, ionic strength). The *linear* isotherm is a limited, special case of the *Freundlich* isotherm. It is often used to describe sorption of organic chemicals in the natural environment and is discussed in more detail in Sec. 5.34.

Langmuir Isotherm   This isotherm assumes that a single adsorbate binds to a single site on the adsorbent and that all surface sites on the adsorbent have the same
affinity for the adsorbate. Surface complexation theory[7] can be used to develop the
Langmuir isotherm:

$$q = q_m \frac{K_{ads}C}{1 + K_{ads}C} \tag{3.82}$$

[7]M. M. Benjamin, "Water Chemistry," McGraw-Hill, New York, 2002.

where q = (some texts use Γ) sorbed concentration (mass adsorbate/mass adsorbent) (sometimes called adsorption density)

q_m = maximum capacity of adsorbent for adsorbate (mass adsorbate/mass adsorbent)

C = aqueous concentration of adsorbate (mass/volume)

K_{ads} = measure of affinity of adsorbate for adsorbent

As C gets larger and larger, adsorption sites become filled and q approaches q_m. Evaluation of the coefficients q_m and K_{ads} can be obtained using the linearized form of Eq. (3.82) as shown in Fig. 3.14:

$$\frac{1}{q} = \frac{1}{q_m K_{ads}}\left(\frac{1}{C}\right) + \frac{1}{q_m} \qquad (3.83)$$

The Langmuir isotherm can be modified to account for competitive adsorption by more than one adsorbate and for adsorbents that have sites with different affinities for a given adsorbate.[8,9]

Freundlich Isotherm   Freundlich studied the adsorption phenomenon extensively and showed that adsorption from solutions can be expressed by the following equation:

$$q = KC^{1/n} \qquad (3.84)$$

The Freundlich isotherm can be derived from the Langmuir isotherm by assuming that there exists a distribution of sites on the adsorbent that have different affinities for different adsorbates with each site behaving according to the Langmuir isotherm.[10] Here, K is a measure of the capacity of the adsorbent (mass adsorbate/mass adsorbent) and n is a measure of how affinity for the adsorbate changes with changes in adsorption density. When $n = 1$, the Freundlich isotherm becomes a *linear* isotherm and indicates that all sites on the adsorbent have equal affinity for the adsorbate(s). Values of $n > 1$ indicate that affinities decrease with increasing adsorption density. Evaluation of the coefficients K and n can be accomplished using the linearized form of Eq. (3.84) (see Fig. 3.15):

$$\log q = \log K + \frac{1}{n}\log C \qquad (3.85)$$

Evaluation of coefficients for both Langmuir and Freundlich isotherms can be done using the same experimental data (see Example 3.16). Typically, different masses of adsorbent are added to a solution containing the adsorbate(s) of interest. These solutions are mixed and allowed to come to equilibrium (times may range from an hour to tens of hours). The concentration of adsorbate(s) remaining is measured. By knowing the initial concentration, q (mass adsorbate removed/mass adsorbent) can

---

[8]Ibid.

[9]Weber and DiGiano, "Process Dynamics in Environmental Systems."

[10]Benjamin, "Water Chemistry."

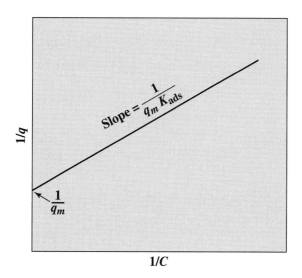

Figure 3.14
Straight-line form of the Langmuir isotherm.

be calculated. Equations (3.83) and (3.85) can then be used to determine values of q_m, K_{ads}, K, and n using techniques such as least-squares linear regression. Statistical analysis of these regressions should allow determination of which isotherm works best for a given situation. Details of such analyses are given in Chap. 10.

BET Isotherm The BET isotherm was developed by Brunauer, Emmett, and Teller as an extension of the Langmuir isotherm to account for multilayer adsorption (adsorption of multiple layers of adsorbate). This model assumes that a number

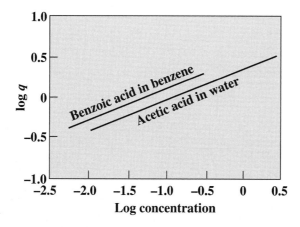

Figure 3.15
Logarithmic plot of adsorption data.

of layers of adsorbate accumulate at the surface and that the Langmuir isotherm applies to each layer. It is somewhat more complex than the Langmuir isotherm and takes the form

$$\frac{q}{q_m} = \frac{bC}{(C_s - C)[1 + (b - 1)C/C_s]} \tag{3.86}$$

The value C_s represents the saturation concentrations for the adsorbate in solution. Of course, when C exceeds C_s, the solute precipitates or condenses from solution as a solid or liquid and concentrates on the surface. The BET equation can be put into the form

$$\frac{C}{q(C_s - C)} = \frac{1}{bq_m} + \frac{b - 1}{bq_m}\left(\frac{C}{C_s}\right) \tag{3.87}$$

With this equation, C_s and b can be obtained from the slope and intercept of the straight line best fitting of the plot of the left side of Eq. (3.87) versus C/C_s. The shape of the BET isotherm and its straight line form are shown in Fig. 3.16.

The adsorption isotherms are equilibrium equations and apply to conditions resulting after the adsorbate-containing phase has been in contact with the adsorbent for sufficient time to reach equilibrium. However, in any practical process for the removal of a contaminant from a gas or liquid, the rate at which the material is adsorbed onto the solid becomes an important consideration. Essentially three steps can be identified in the removal of a contaminant by adsorption. First, it must move from the liquid or gaseous phase through a boundary layer in the fluid to the exterior of the adsorbent. Next, it must pass by diffusion into and through the pores of the adsorbent. Finally, it must become attached to the adsorbent. If the phase containing the adsorbent is quiescent, then diffusion through the boundary layer may be the slowest and rate-determining step. In this case, if the fluid is agitated,

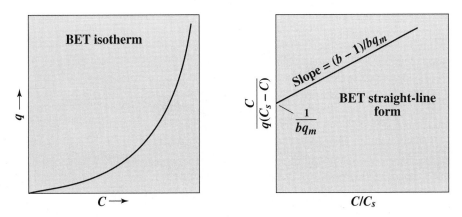

Figure 3.16
Plots of the BET isotherm and its straight-line form.

the thickness of the boundary layer becomes reduced and the rate of adsorption will increase. At increased turbulence, however, a point will be reached where diffusion through the pores becomes the slowest step so that further increased turbulence will not result in greater rates of adsorption. Thus, depending upon the general characteristics of the material being adsorbed and the relative rates of diffusion through the boundary layer and into the pores, increased agitation of the fluid containing the material may or may not increase the rate of adsorption.

For adsorption, say, of a contaminant in water onto activated carbon, both the rate and the extent of adsorption are dependent upon the characteristics of the molecule being adsorbed and of the adsorbent. The extent of adsorption is governed to some extent by the degree of solubility of the substance in water. The less soluble the material, the more likely it is to become adsorbed. With molecules containing both hydrophilic (water liking) and hydrophobic (water disliking) groups, the hydrophobic end of the molecule will tend to become attached to the surface. Next, the relative affinity of the material for the surface is a factor as already discussed in relation to the three general types of adsorption. Finally, the size of the molecule is of significance, as this affects its ability to fit within the pores of the adsorbent, and its rate of diffusion to a surface. Except for exchange adsorption, ions tend to be less readily adsorbed than neutral species. Many organics form negative ions at high pH, positive ions at low pH, and neutral species at intermediate pH ranges. Generally, adsorption is increased at pH ranges where the species at neutral in charge. In addition, pH affects the charge on the surface, altering its ability to adsorb materials. In water or wastewater samples, there are many different materials, each with different adsorption properties. Each competes in some way with the adsorption of the others. For this reason a given material may adsorb to a much less extent in a mixture of materials than if it were the only material in the solution.

One of the most important uses of adsorption in environmental practice has been for the removal of organic materials from waters, wastewaters, and air. Examples include removal of taste- and odor-producing organic materials and other trace organic contaminants such as trihalomethanes, pesticides, and chlorinated organic compounds from drinking waters and air, removal of residual organic contaminants from treated wastewater effluents, and treatment of leachates, industrial wastewaters, and hazardous wastes. Activated carbon, granular or powdered, may be mixed with the water and then removed along with the adsorbed materials by settling or filtration. When large quantities of organic material must be removed, more efficient usage of carbon and a higher quality of water can be obtained by passing the water through a carbon filter bed of large depth. This is a practical application of the principles expressed by the adsorption isotherms and, in some ways, can be compared to countercurrent extraction. Contaminants in air are removed by the passage of air through a bed of activated carbon. Use of such systems to manage a wide variety of contamination problems is now quite common. Development of special activated carbons that can be regenerated efficiently has improved process economics considerably.

EXAMPLE 3.16 | Granular activated carbon (GAC) was tested for its ability to remove soluble organic ni-trogen (SON) from treated wastewater. Different masses of GAC were added to 1 liter of the wastewater (initial SON concentration = 0.9 mg/L) and contacted for 2 h at 20°C and pH = 7.5. Using the data given in the table and linear regression analysis (see Chap. 10), determine whether the Langmuir or Freundlich isotherm best describes the data.

Mass GAC added	C, mg/L of SON
0.2	0.77
0.5	0.65
2.0	0.32
5.0	0.19
10	0.14
20	0.09
50	0.06

mg SON removed = $V(C_O - C) = (1 \text{ liter})(0.9 - C)$

GAC, (g)	C_O, mg/L	C, mg/L	$q = \dfrac{\text{mg SON removed}}{\text{g GAC}}$
0.2	0.90	0.77	0.650
0.5	0.90	0.65	0.500
2.0	0.90	0.32	0.290
5.0	0.90	0.19	0.142
10.0	0.90	0.14	0.076
20.0	0.90	0.09	0.041
50.0	0.90	0.06	0.017

To check the Freundlich isotherm:

$$\log q = \log K + \frac{1}{n} \log C$$

we need to plot $\log q$ versus $\log C$. To check the Langmuir isotherm:

$$\frac{1}{q} = \frac{1}{q_m K_{\text{ads}}}\left(\frac{1}{C}\right) + \frac{1}{q_m}$$

we need to plot $1/q$ versus $1/C$.

q	C	$1/q$	$1/C$	$\log q$	$\log C$
0.650	0.77	1.54	1.30	−0.187	−0.114
0.500	0.65	2.00	1.54	−0.301	−0.187
0.290	0.32	3.45	3.13	−0.538	−0.495
0.142	0.19	7.04	5.26	−0.848	−0.721
0.076	0.14	13.2	7.14	−1.12	−0.854
0.041	0.09	24.4	11.1	−1.39	−1.05
0.017	0.06	58.8	16.7	−1.77	−1.22

Freundlich isotherm plot:

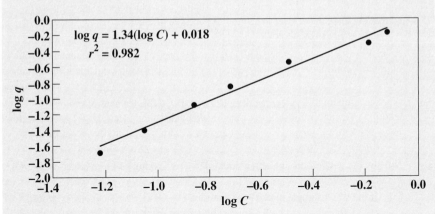

Langmuir isotherm plot:

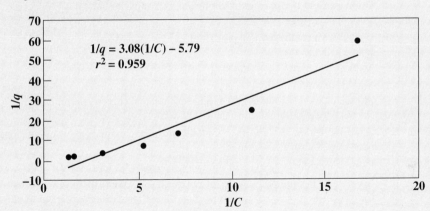

Inspection of the two plots indicates that the Freundlich isotherm works better for these data. Thus,

$$\log\ q = 1.34\ \log\ C + 0.018$$

or

$$\log \frac{q}{C^{1.34}} = 0.018$$

from which,

$$q = 10^{0.018}\ C^{1.34} = 1.04 C^{1.34}$$

$$n = \frac{1}{1.34} = 0.746$$

PROBLEMS

3.1 Determine the *net* heat of combustion of ethane gas from standard enthalpies of formation. Give the answer in kilojoules per mole of ethane.
  Answer: -1428 kJ/mol

3.2 A small quantity of hydrogen gas is sometimes present in the gas from an anaerobic digester. Determine the *net* heat value in kilojoules per mole available from burning this gas.

3.3 A waste containing 1 percent sulfuric acid (10,000 mg/L) is neutralized by the addition of a concentrated lime slurry [made by dissolving $Ca(OH)_2(s)$ in water]. If the water temperature is 15°C before neutralization, what is it after neutralization?
  Answer: 17.7°C

3.4 (*a*) How many kilojoules of heat are required to evaporate 1 liter of water at 1 atm if the initial water temperature is 20°C?

  (*b*) If the water sample in part (*a*) contained 30,000 mg/L of acetic acid, would sufficient heat be liberated by combustion of the acetic acid to satisfy the heat requirements for water evaporation?

3.5 (*a*) Using standard free energies of formation, determine the solubility product for zinc sulfide $(ZnS(s))$ at 25°C.

  (*b*) Which way may the reaction proceed if Zn^{2+} is 10^{-3} M and S^{2-} is 10^{-3} M? Ignore activity corrections.
  Answer: 3.67×10^{-24}, precipitating

3.6 (*a*) Mg^{2+} is one of the major components of hardness. It can be removed from water by precipitation as $Mg(OH)_2(s)$. Using standard free energies of formation, calculate the equilibrium constant K_{sp} for $Mg(OH)_2(s)$ at 25°C.

  (*b*) Is the dissolution of $Mg(OH)_2(s)$ endothermic or exothermic? Show calculations necessary to justify your answer.

3.7 (*a*) Using standard free energies of formation, calculate K_{sp} for $CaCO_3(s)$ at 25°C.

  (*b*) If a water is in equilibrium with $CaCO_3(s)$ at a temperature of 25°C, what is the Ca^{2+} concentration in milligrams per liter if the CO_3^{2-} concentration is 5 mg/L?

  (*c*) Calculate K_{sp} for a temperature of 16°C, which is typical of many groundwaters.

  (*d*) A sample of groundwater with a temperature of 16°C has 100 mg/L of Ca^{2+} and 10 mg/L of CO_3^{2-}. Is this water in equilibrium with $CaCO_3(s)$? If not, is $CaCO_3(s)$ dissolving or precipitating?
  Answer: 4.99×10^{-9}, 2.4 mg/L, 5.88×10^{-9}, no, precipitating

3.8 (*a*) Determine the solubility constant (Henry's law constant) at 25°C for hydrogen sulfide gas in water from standard free energies of formation.

  (*b*) Using standard enthalpies of formation and the solubility constant from part (*a*), estimate the solubility constant for hydrogen sulfide gas at 10°C.

  (*c*) The odor threshold (the concentration at which the average person can detect the odor) for H_2S ("rotten-egg" gas) is reported to be approximately 1 ppm (10^{-6} atm) of H_2S. Assuming equilibrium between this gaseous concentration and water, what is the aqueous concentration of H_2S at 25°C? At 10°C?

3.9 (*a*) Determine the solubility constant (Henry's law constant) at 20°C for carbon dioxide gas in water from standard free energies of formation.

(b) If the dissolved CO_2 concentration in a lake at 20°C is 2.2 mg/L, is the lake in equilibrium with atmospheric CO_2 (partial pressure of $10^{-3.5}$ atm)?
Answer: 23.4 atm-L/mol, volatilizing

3.10 Estimate the ionization constant for acetic acid at 35°C from thermodynamic considerations.

3.11 (a) The following reaction describes the dissociation of ammonium in water:

$$NH_4^+(aq) \rightleftharpoons NH_3(aq) + H^+(aq)$$

From standard enthalpies and free energies of formation, calculate the equilibrium (dissociation) constant at 25°C and 10°C.

(b) It is the un-ionized form of ammonia (NH_3) that is toxic to fish. For a water with a pH of 7.0, what is the ratio of NH_3 to NH_4^+ at 25°C and 10°C?
Answer: 5.74×10^{-10}, 1.90×10^{-10}, 5.74×10^{-3}, 1.90×10^{-3} (NH_3 concentration, relative to NH_4^+, decreases by a factor of approximately 3)

3.12 Estimate the first ionization constant for hydrosulfuric acid at 35°C from thermodynamic considerations.
Answer: 1.33×10^{-7}

3.13 The oxidation of ferrous iron (Fe^{2+}) to ferric iron (Fe^{3+}) by molecular oxygen can be described by the following reaction:

$$4Fe^{2+}(aq) + O_2(g) + 4H^+(aq) = 4Fe^{3+}(aq) + 2H_2O(l)$$

(a) Using free energies of formation, calculate the equilibrium constant for this reaction at 25°C.

(b) For a water in equilibrium with the atmosphere, a pH of 2.0, and a total soluble iron concentration of 1 mg/L, calculate the concentrations of $Fe^{2+}(aq)$ and $Fe^{3+}(aq)$.
Answer: 1.19×10^{31}, 4.51×10^{-11} M, 1.79×10^{-5} M

3.14 Consider the following acid-base reaction:

$$Fe^{3+}(aq) + 3H_2O \rightleftharpoons Fe(OH)_3(s) + 3H^+$$

(a) From thermodynamic considerations calculate the equilibrium constant K at 25°C.

(b) For the value of K you calculated in part (a), if a solution contains 10^{-4} M Fe^{3+} and has a pH of 7.5, will $Fe(OH)_3(s)$ form? Show all calculations necessary to justify your answer.

3.15 Consider the corrosion of elemental iron in anaerobic waters (Fe^0 oxidized to Fe^{2+}) with the reduction of protons (H^+) to form hydrogen gas(H_2):

$$Fe^0 + 2H^+ = Fe^{2+} + H_2(g)$$

(a) From thermodynamic considerations, calculate the equilibrium constant K at 25°C.

(b) For $T = 25$°C, pH = 10, and $[Fe^{2+}] = 10^{-3}$ M, at what partial pressure of H_2 does this reaction become unfavorable?

3.16 Consider the oxidation of ammonium (NH_4^+) to nitrate (NO_3^-) by molecular oxygen:

$$NH_4^+ + 2O_2(g) = NO_3^- + 2H^+ + H_2O$$

From thermodynamic considerations, calculate the equilibrium constant K at 25°C. Is this reaction thermodynamically favorable? Justify your answer.

3.17 (*a*)   Calculate the standard free energy of the reaction for the biological decomposition of 1 mol of acetate under both aerobic and anaerobic conditions:

Aerobic: $CH_3COO^-(aq) + 2O_2(g) \rightarrow HCO_3^-(aq) + H_2O(l) + CO_2(g)$

Anaerobic: $CH_3COO^-(aq) + H_2O(l) \rightarrow HCO_3^-(aq) + CH_4(g)$

(*b*)   For a given quantity of acetate waste, which system would you expect to be capable of supporting the growth of the largest biological population? Why?

3.18 (*a*)   One of the major intermediates in the anaerobic biological degradation of organic matter to methane is propionate ($CH_3CH_2COO^-$). Propionate conversion to methane can be described by the following reaction:

$$4CH_3CH_2COO^-(aq) + 6H_2O(l) = 7CH_4(g) + CO_2(g) + 4HCO_3^-(aq)$$

Assuming that reactants and products are at unit activity, is methane production from propionate thermodynamically favorable?

(*b*)   However, it is known that methane is not formed directly from propionate. Rather, propionate is converted by bacteria to acetate (CH_3COO^-) and hydrogen as follows:

$$CH_3CH_2COO^-(aq) + 2H_2O(l) = CH_3COO^-(aq) + 3H_2(g) + CO_2(g)$$

Show that this reaction is not thermodynamically favorable. In a well-operating anaerobic biological treatment system, propionate is efficiently converted to methane. Offer an explanation how this is possible given that propionate conversion to acetate and hydrogen is not thermodynamically favorable.

Answer: (*a*) Yes, since $\Delta G^0 = -229.9$ kJ has a negative value, (*b*) Reaction not possible under standard conditions since $\Delta G^0 = 71.65$ kJ is positive.

3.19   At 1 atm the boiling temperature for isopropanol (isopropyl alcohol) is 82.5°C and for water is 100°C. A mixture containing 12.5 percent water and 87.9 percent isopropanol by weight has a boiling temperature of 80.4°C, and at this temperature the composition of the vapor is the same as the liquid. If the waste from an industry contained 10,000 mg/L of isopropanol, would it be possible to remove the alcohol from the water by fractional distillation of the waste? Why?

Answer: Yes

3.20   An industrial wastewater contains 10 percent by weight of an organic solvent and has a boiling temperature of 105°C at 1 atm. The vapor is found to be richer in organic solvent than the liquid waste. The boiling temperature of pure solvent is 80°C. Can pure solvent be obtained by fractional distillation? Pure water? Explain why.

3.21 (*a*)   At 1 atm, *n*-butanol (*n*-butyl alcohol) boils at 117.8°C. A binary mixture containing 2.52 mol of water per mole of this alcohol boils at 92.4°C, and with this mixture the composition of the vapor is equal to that of the liquid mixture. Sketch roughly the temperature composition diagram for the liquid-vapor equilibrium for mixtures of water and *n*-butanol at atmospheric pressure.

(*b*)   An industrial waste contains 90 percent water and 10 percent *n*-butanol by weight. What would be the composition of the distillate and the residue from fractional distillation of the waste?

Answer: (*b*) distillate 0.716 mol fraction H_2O, residue pure H_2O

3.22 The boiling temperature for butyric acid at 1 atm is 163.5°C. When wastewaters containing low concentrations of butyric acid are distilled, it is found that the distillate is richer in butyric acid than the wastewater being distilled. In what class would a butyric acid–water binary mixture be placed? Illustrate.

3.23 What approximate osmotic pressure would be created across a semipermeable membrane if water containing 0.01 M Na_2SO_4, 0.02 M $MgCl_2$, and 0.03 M $CaCl_2$ were placed on one side of the membrane and distilled water were on the other?
Answer: 4.4 atm

3.24 (*a*) 20,000 mg/L of NaCl would lower the vapor pressure of water at 100°C by about 8.4 mm Hg. Estimate the osmotic pressure across a semipermeable membrane containing a brackish water with this sodium chloride concentration on one side and distilled water on the other.

(*b*) Calculate the theoretical minimum energy requirement to remove the salt from 1000 gallons of the brackish water. Express energy required in units of liter-atmospheres, foot-pounds, and kilowatt-hours.

3.25 Phenol is approximately 12 times more soluble in 1 volume of isopropyl ether than it is in 1 volume of water. How many extractions are required to reduce the concentration of phenol below 100 mg/L in a waste containing 2000 mg/L of phenol if an ether-to-wastewater ratio of 0.2 by volume is used in each extraction?
Answer: 3

3.26 The solubility of picric acid at 20°C is 9.56 g per 100 g of benzene, and 1.4 g per 100 g of water. If an industrial wastewater contains 5000 mg/L of picric acid, what concentration would remain after 1 extraction with 1 lb of benzene for each 2 lb of water?

3.27 What is the approximate specific conductance at 25°C of a solution containing 100 mg/L of $CaCl_2$ and 75 mg/L of Na_2SO_4?
Answer: 384 μS

3.28 A standard KCl solution (0.01 N), when placed in a conductivity cell at 25°C, was found to produce a resistance of 1000 ohms. A $MgCl_2$ solution was then placed in the cell, and the measured resistance at 25°C was 3000 ohms. Approximately what is the concentration of $MgCl_2$ in milligrams per liter?

3.29 The specific conductance of a $CaCl_2$ solution is 200×10^{-6} S. Estimate the concentration of $CaCl_2$ in milligrams per liter.
Answer: 81.8 mg/L

3.30 Ions can contribute to the conductivity of a solution, but un-ionized molecules cannot. On this basis what is the approximate specific conductance at 25°C of a solution containing 1000 mg/L of acetic acid, if the ionization constant for this acid is 1.75×10^{-5}?

3.31 What weight of silver will pass into solution from a silver anode by the passage of 0.02 A of current through the solution for 24 h?
Answer: 1.93 g

3.32 When an electric current is allowed to pass through a dilute sulfuric acid solution, hydrogen gas is evolved at the cathode and oxygen gas at the anode. What volumes of gases, measured at 1 atm pressure and 0°C, will be obtained at the electrodes when 1 A of current is passed through the solution for a 1-h period?

3.33 On the basis of standard electrochemical potentials, which of the following metals could act in a sacrificial manner to protect iron from corrosion: aluminum, copper, lead, magnesium, silver, tin, and zinc?

Answer: aluminum, magnesium, zinc

3.34 A zinc and an iron bar, connected by a copper wire, were introduced into a solution containing 2000 mg/L of Zn^{2+} ions and 5 mg/L of Fe^{2+} ions. What reaction took place?

3.35 Estimate the solubility-product constant for $Mg(OH)_2(s)$ at 25°C from standard electrode potentials.

Answer: 1.5×10^{-11}

3.36 In a study of the natural die-off of coliform organisms in a stream, it was found that 36 percent of the organisms died within 10 h and 59 percent died within 20 h. If the rate of die-off followed first-order kinetics and were proportional to the number remaining, how long would it take to obtain a 99 percent reduction in coliform organisms?

3.37 A stream flowing with a velocity of 2 mph contains no BOD, but has an oxygen deficit of 6 mg/L. Ten miles downstream this deficit has been reduced to 4 mg/L through absorption from the atmosphere. The stream conditions are uniform throughout its length. Assuming the rate of aeration is proportional to the deficit, what would be the deficit 35 miles downstream from the original point?

Answer: 1.45 mg/L

3.38 The radioactive nuclide P^{32} has a half-life of 14.3 days. How long would a waste containing 10 mg/L of this nuclide have to be stored in order to reduce the concentration to 0.3 mg/L?

3.39 The half-life of atrazine (a herbicide) is estimated to be approximately 14 days. What fraction of the initial atrazine will remain after 100 days?

3.40 Laboratory studies with a groundwater sample indicate that trichloroethylene (TCE) is degraded according to first-order kinetics with a half-life of 150 days. If the current groundwater concentration is 15 μg/L, how long will it take before the TCE concentration is reduced to the drinking water standard of 5 μg/L? Show your work.

3.41 An experiment was conducted to characterize the rate of a chemical reaction. Concentrations (mg/L) were measured at different times (h) and the data were plotted as follows:

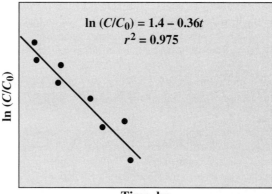

Is this a zero-, first-, or second-order reaction? Why? What is the value of the rate constant (include the proper units)?

3.42 The following represents experimental data collected by measuring the disappearance of a compound (concentration in mg/L) with time.

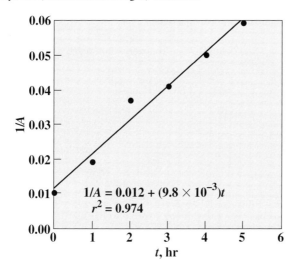

$$1/A = 0.012 + (9.8 \times 10^{-3})t$$
$$r^2 = 0.974$$

(*a*) Is this a zero-order, first-order, or second-order reaction? Why?

(*b*) What is the rate coefficient, reported with the correct units?

3.43 Given the following laboratory data concerning the disappearance of a chemical in water, do the data fit a zero-order, first-order, or second-order rate expression? Provide sufficient justification for your answer, and report the correct rate constant in the correct units.

Time, h	Concentration, mg/L
0	100
0.5	61
1	37
2	14
3	5.0
5	0.67

Answer: first order, $1.0 \ h^{-1}$

3.44 Bacterial decay has been described by a first-order rate expression. If the first-order rate constant for bacterial decay is $0.10 \ day^{-1}$ at 20°C (typical for aerobic, heterotrophic bacteria), and if an A [Eq. (3.81)] of 1.035 is assumed to describe the effect of temperature on the rate constant, what is the first-order rate constant at 30°C? What is the value of E_a?

3.45 Anaerobic biotransformation of chlorinated organics such as chloroform can be modeled using a second-order rate expression. This process is typically called *cometabolism,* and it is typically assumed that the concentration of bacteria remains constant. If the second-order rate coefficient is 0.005 L/mg-d and the concentration of bacteria is 100 mg/L, how many days will it take for an initial concentration of chloroform to be reduced to 0.01 mg/L ($10\mu g/L$, which represents a 99 percent reduction)?

3.46 Biotransformation of chlorinated organic compounds such as 1,1,1-trichloroethane can be modeled using a second-order rate expression. Given the following laboratory data for an experiment measuring the biotransformation of 1,1,1-trichloroethane by 100 mg/L of bacteria, determine the pseudo-first-order rate constant and the second-order rate constant. Plot the data using appropriate graph paper or using an appropriate graphics software package.

Time, h	Concentration, mg/L
0	0.50
2	0.48
5	0.45
10	0.41
24	0.30
48	0.18

Answer: pseudo-first-order $k = 0.0213$ h^{-1}; second-order $k = 2.13 \times 10^{-4}$ L/mg-h

3.47 The transformation of 1,1,1-trichloroethane (1,1,1-TCA) by the mineral iron sulfide (FeS) has been shown to be a second-order reaction, depending on the concentration of 1,1,1-TCA remaining (M) and the surface-area concentration of FeS(m^2/L). The transformation of 1,1,1-TCA by anaerobic bacteria is also second-order, depending on the 1,1,1-TCA concentration (M) and the concentration of bacteria (mg/L). It is hypothesized that when FeS and bacteria are combined in one system, total removal can be described by parallel and independent second-order rate expressions.

(*a*) Write the rate expression describing the total removal of 1,1,1-TCA by the combination of FeS and bacteria.

(*b*) Convert your answer for part (*a*) into a pseudo-first-order expression such that removal is first-order with respect to 1,1,1-TCA only and containing a single rate coefficient. What is the equation for the rate coefficient?

3.48 Differentiate between chemical adsorption, exchange adsorption, and physical adsorption.

3.49 An evaluation of the ability of activated carbon to reduce the odor of a water with a threshold odor of 30 was made, using the Freundlich adsorption isotherm. By plotting the log of odor removed per unit dose of activated carbon versus residual odor, the constants K and n in Eq. (3.84) were found to be 0.5 and 1.0, respectively. What activated carbon dosage in milligrams per liter would be required to reduce the threshold odor to 4 units?

REFERENCES

Adamson, A. W.: "A Textbook of Physical Chemistry," 3rd ed., Academic Press, Orlando, FL, 1986.

Atkins, P. W.: "Physical Chemistry," 4th ed., Oxford Univ. Press, Oxford, Great Britain, 1990.

Benjamin, M. M.: "Water Chemistry," McGraw-Hill, New York, 2002.

Daniels, F., and R. A. Alberty: "Physical Chemistry," 5th ed., Wiley, New York, 1979.

Dean, J. A. (ed.): "Lang's Handbook of Chemistry," 15th ed., McGraw-Hill, New York, 1999.

Levine, I. N.: "Physical Chemistry," 5th ed., McGraw-Hill, New York, 2002.

Lide, D. R. (ed.): "Handbook of Chemistry and Physics," 82nd ed.: CRC Press, Boca Raton, FL, 2001.

Noggle, J. H.: "Physical Chemistry," 2nd ed., Scott Foresman, Boston, 1989.

Snoeyink, V., and D. Jenkins: "Water Chemistry," Wiley, New York, 1980.

Basic Concepts
from Equilibrium Chemistry

4.1 | INTRODUCTION

A knowledge of equilibrium chemistry has become increasingly important for determining quantitatively the relationships between the various constituents in natural and contaminated waters, and for understanding the effect of alterations in the water on the various chemical species present. This is a matter of growing importance because of the use of physical-chemical methods for treating wastewaters and sanitary-landfill leachates and the need to understand the fate and transport of both organic and inorganic compounds in the environment. Thus, a good knowledge of equilibrium chemistry is helpful in understanding the effects of pollutant discharge to the environment as well as in evaluating how best to treat waters and wastewaters in order to rid them of harmful substances. Equilibrium chemistry draws heavily upon the equilibrium relationships discussed in Chap. 2 and the thermodynamics of equilibrium relationships as discussed in Chap. 3. The reader should have a good understanding of these relationships before proceeding with this chapter.

4.2 | LIMITATIONS OF EQUILIBRIUM CALCULATIONS

Before considering equilibria in water systems in detail, it is important to be aware of some of the limitations of equilibrium calculations. Most waters, especially surface waters and wastewaters, undergo dynamic changes resulting from the constant introduction of energy from the sun and materials, both inorganic and organic, from inflowing rivers and wastewater streams. Equilibrium calculations cannot be expected to describe accurately the relationships between all the various species in such a system since equilibrium may not be attained. They can, however, indicate the direction in which reactions would tend to proceed and the extent to which reactions would go if equilibrium were attained. They can indicate whether certain chemical transformations are possible.

Some reactions, such as those between soluble acids and bases, occur very rapidly so that equilibrium may occur within seconds. Others, such as oxidation-reduction reactions, may occur slowly under conditions existing in natural waters so that equilibrium may not be achieved within years or sometimes centuries. On the other hand, many oxidation-reduction reactions which otherwise may tend slowly toward equilibrium may be catalyzed by bacteria which can then capture the energy released for growth. Precipitation reactions may occur readily when an excess of a required chemical constituent is added to water, but the reaction is likely to stop before true equilibrium is reached. This may be due to the initial formation of poorly crystallized or amorphous precipitates, which have greater solubilities. Reconstitution of such amorphous solids into crystalline solids occurs very slowly. Equilibrium calculations can help indicate the direction toward which the chemical nature of an aquatic system will tend to change but cannot indicate the rate at which the change will proceed. Thus, an understanding of the kinetics of change is necessary in a realistic evaluation of the potential for change predicted from equilibrium considerations. Unfortunately, there continues to be a lack of kinetic data for many transformations of interest in environmental engineering and science.

Another limitation is the lack of availability of accurate equilibrium constants for many of the reactions of interest in natural waters. While the equilibrium constant for dissociation of water is known quite accurately ($1.008 \pm 0.001 \times 10^{-14}$ at 20°C), the best-known values for weak acids and bases generally vary in accuracy from ± 0.5 to ± 10 percent. Solubility products are sometimes known with similar accuracy, but several, such as for some sulfides, have values as reported by different investigators that differ by several orders of magnitude. Formation (stability) constants of complex ions also are of widely varying accuracy so that caution is indicated when they are used. Such poor accuracy in values generally results from the inability to measure constituents at low concentrations sometimes associated with equilibria, the uncertainty of whether true equilibria have been attained, and the uncertainty over the actual reactions that occur in solution. For example, most of the earlier determinations of the solubility product for $Al(OH)_3$ (gibbsite) are inaccurate because the solid phase produced by precipitation was sometimes gibbsite, but more often a mixture of gibbsite with other hydrous and hydrated oxides.

These uncertainties would appear to restrict the usefulness of equilibrium calculations severely. However, there are many cases where equilibrium constants are known with sufficient accuracy for most practical environmental purposes and where equilibrium is readily attained. In order to approach a given problem correctly, the accuracy with which predictions can be made and the limitations of the predictions must be well understood.

4.3 | ION ACTIVITY COEFFICIENTS

The fundamental aspects of electrolytic dissociation or ionization were discussed in Sec. 2.13. In Sec. 2.12 it was noted that as solutions of ionized materials become more concentrated, their quantitative effect in equilibrium relationships becomes progressively less than calculated solely from the change in molar concentration. In order to overcome this deficiency, the concept of activity or effective concentration was advanced. The activity of an ion or molecule can be found by multiplying its molar concentration by an activity coefficient γ, as indicated by Eq. (2.24).

For many practical purposes, rough calculations obtained by use of molar concentrations rather than activities are sufficient. Also, in other cases where some of the limitations discussed in Sec. 4.2 apply, the more refined use of activities may not be justified. However, in certain cases it may be practical and desirable to obtain more precise results. For these cases procedures for estimating the activities of ions, which are apt to deviate most from ideality, are desirable.

Lewis and Randall recognized that the activity coefficients for ions in an electrolyte were related to the concentration of charged particles in the solution. They introduced the concept of *ionic strength* as an empirical measure of the interactions among all the ions in a solution that caused deviation from ideal behavior. The ionic strength μ is a characteristic of the solution and is defined as

$$\mu = \tfrac{1}{2}\sum_i C_i Z_i^2 \tag{4.1}$$

where C_i is the molar concentration of the ith ion, Z_i is its charge, and the summation extends over all the ions in the solution. It is important to keep in mind that the ionic strength is a general property of the solution and not a property of any particular ion in the solution. Lewis and Randall found that in dilute solutions the activity coefficient of a given ion was the same in all solutions of the same ionic strength. For fresh natural waters, Langelier[1] indicated that the ionic strength could be approximated by multiplying the milligrams per liter of dissolved solids by 2.5×10^{-5}.

The concept of Lewis and Randall was extended by Debye and Hückel into a general theory that expressed the relationship between the activity coefficient for an ion and the ionic strength of the solution in which it was contained. This theory is applicable only to dilute solutions, those with ionic strengths less than about 0.1. However, excepting seawater, most water solutions of interest in environmental engineering and science are more dilute than this, and thus the theory is applicable. For this case the Güntelberg approximation derived from the Debye-Hückel basic relationship is frequently used:

$$\log \gamma = -0.5\, Z^2 \frac{\sqrt{\mu}}{1 + \sqrt{\mu}} \tag{4.2}$$

Here Z is the charge on the ion for which the activity coefficient is being determined. This theory was developed from the concept that oppositely charged ions attract each other and cause deviations from the behavior that would be produced by

---

[1]W. F. Langelier, The Analytical Control of Anti-Corrosion Water Treatment, *J. Amer. Water Works Assoc.*, **28:** 1500 (1936).

an equal number of uncharged particles. The Davies relationship is reported to work for solutions with μ up to 0.5 M:

$$\log \gamma = -0.5 \, Z^2 \left(\frac{\sqrt{\mu}}{1 + \sqrt{\mu}} - 0.2\mu \right) \tag{4.3}$$

It is generally believed that there is no satisfactory relationship that provides a good estimate for γ for μ greater than 0.5 M.

Calculate the activity coefficients and the activities of each ion in a solution containing 0.01 M $MgCl_2$ and 0.02 M Na_2SO_4. Use both the Güntelberg and Davies relationships.

EXAMPLE 4.1

When the salts are dissolved, the following molar concentrations C of each ion with charge Z result:

Ion	C(M)	Z	CZ^2
Mg^{2+}	0.01	+2	0.04
Na^+	0.04	+1	0.04
Cl^-	0.02	-1	0.02
SO_4^{2-}	0.02	-2	0.08
			$\sum C_i Z_i^2 = 0.18$

From Eq. (4.1),

$$\mu = \frac{1}{2}\sum C_i Z_i^2 = 0.09$$

From Eq. (4.2),

$$\log \gamma = -0.5Z^2 \frac{\sqrt{0.09}}{1 + \sqrt{0.09}} = -0.115Z^2$$

From Eq. (4.3),

$$\log \gamma = -0.5Z^2 \left(\frac{\sqrt{0.09}}{1 + \sqrt{0.09}} - 0.2(0.09) \right) = -0.106Z^2$$

For ions with $Z = 1$, we have $\log \gamma = -0.115$ and $\gamma = 0.77$ using the Güntelberg approximation and $\log \gamma = -0.106$ and $\gamma = 0.78$ using the Davies relationship. For ions with $Z = 2$, we have $\log \gamma = -0.460$ and $\gamma = 0.35$ using the Güntelberg approximation and $\log \gamma = -0.424$ and $\gamma = 0.38$ using the Davies relationship. Therefore, the activity coefficient and activity for each ion are as follows:

Ion	γ	Güntelberg activity $= \gamma C$	γ	Davies activity $= \gamma C$
Mg^{2+}	0.35	0.0035	0.38	0.0038
Na^+	0.77	0.031	0.78	0.031
Cl^-	0.77	0.015	0.78	0.016
SO_4^{2-}	0.35	0.0070	0.38	0.0076

One can see that the effective concentration (activity) of divalent ions is significantly smaller than the analytical concentration!

Values of γ are greater than 1 for many nonelectrolytes (uncharged or neutral species). As the dissolved ion concentration in water increases, it becomes more thermodynamically favorable for most neutral species to leave the water. There are relationships available to make activity corrections for neutral inorganic species [e.g., $NH_3(aq)$, $CO_2(aq)$, $O_2(g)$].[2,3] With neutral organic molecules, the situation is a bit more complicated. Values for γ are typically much greater than 1 owing to the fact that the standard reference state (see Sec. 2.12) assumed is the pure liquid at mole fraction $= 1.0$. Thus, one must express the aqueous concentration as mole fraction. Schwarzenbach et al. describe in detail how to calculate values of γ for neutral organic molecules.[4]

For most applications in environmental engineering and science, γ is close enough to 1.0 to warrant the assumption that activity equals concentration. In this book in general, activity coefficients for ions and nonelectrolytes will be assumed equal to 1.0 unless otherwise noted. However, some example and end-of-chapter problems will involve activity corrections to familiarize students with these concepts.

4.4 | SOLUTION TO EQUILIBRIUM PROBLEMS

In solving equilibrium problems it is helpful to know qualitatively what can be expected to occur. Le Chatelier's principle is helpful for this purpose. It states that a chemical system will respond to change with processes that tend to reduce the effect of the change. For example, if SO_4^{2-} is added to a solution saturated with respect to $CaSO_4(s)$, some of the calcium ions in solution will react with added SO_4^{2-}, thus reducing the effect of the change, i.e., reducing the increase in SO_4^{2-} concentration. Le Chatelier's principle in our present context is a qualitative statement of the fact that the equilibrium constants applicable in a chemical system must be obeyed.

Conceptually, the solution to an equilibrium problem is straightforward. First, for any chemical reaction, the principle of conservation of mass must be obeyed. Conservation of mass means that the mass of each element present must remain the same before, during, and after the reaction. It also means that the quantity of a given chemical entity not destroyed by the reaction must remain constant. For example, when sodium hydroxide is added to an acetic acid solution, the acetic acid present as CH_3COOH is at least partially ionized to form the acetate ion, CH_3COO^-. Conservation of mass requires that the sum of molar concentrations of acetic acid plus acetate, $[CH_3COOH]$ + $[CH_3COO^-]$, remains constant during the sodium hydroxide addition. This sum will be represented by the term C_T, the total concentration in solution. Some texts and available software use TOTX to represent the total concentration of species X in the system.

[2]V. L. Snoeyink and D. Jenkins, "Water Chemistry," Wiley, New York, 1980.

[3]J. F. Pankow, "Aquatic Chemistry Concepts," Lewis, Chelsea, MI, 1991.

[4]R. P. Schwarzenbach, P. M. Gschwend, and O. M. Imboden, "Environmental Organic Chemistry," Wiley, New York, 1993.

Second, electroneutrality must be maintained. This means that all positively charged species in solution must be balanced by equivalent numbers of negatively charged species. For example, in a solution of sodium chloride in water,

$$[Na^+] + [H^+] = [Cl^-] + [OH^-]$$

or, in a solution of calcium hydroxide,

$$2[Ca^{2+}] + [H^+] = [OH^-]$$

The calcium concentration is multiplied by 2 because it is divalent; thus, 1 mol of Ca^{2+} is equivalent to 2 mol of positive charge. To illustrate further, if 1 mol of hydrated lime, $Ca(OH)_2$, were added to 1 liter of water, it would dissociate to yield 1 mol of Ca^{2+} and 2 mol of OH^-. Thus, $[Ca^{2+}] = 1$ and $[OH^-] = 2$. $[H^+]$ in this case would be very small and can be ignored. Using the above equation, we see that electroneutrality is maintained since $2(1) + 0 = 2$.

A third tool, called the proton condition, may also be helpful in solving some acid-base equilibrium problems. The proton condition can be thought of as a mass balance on protons relative to appropriate reference species. That is, the species with an excess of protons must be balanced by the species with a deficiency in protons. Water is always used as a reference species for H^+ and OH^-. Relative to H_2O, H^+ (H_3O^+) has an excess of one proton while OH^- has a deficiency of one proton. For other acid-base species in the solution, the reference species is usually assumed to be the species added. For example, if acetic acid is added to water, CH_3COOH would be the reference species and CH_3COO^- has a deficiency of one proton relative to CH_3COOH. This concept is sometimes difficult for students to grasp. The example problems contained in this chapter will hopefully demonstrate the utility of the proton condition.

Finally, all reactions involved must proceed toward a state of equilibrium at which all appropriate equilibrium relationships are satisfied. Steps involved in the solution of an equilibrium problem involving only the aqueous phase are as follows:

1. Carefully define the equilibrium problem being considered: e.g., What chemical reactions are taking place? What is reacting with what? What are conditions at the outset and at the end of equilibrium?

2. List all constituents of the system that are present at the outset. All systems involving water will of course include H_2O, H^+, and OH^-. The list should also include all other elements, ions, and neutral species that are present initially.

3. For each element initially present, list all the likely forms or species containing it that are likely to be present after equilibrium is attained.

4. Identify the concentrations of each species for each element or entity under initial conditions so that appropriate mass and charge balances can be made.

5. List all appropriate equilibrium relationships between the species of concern, together with associated equilibrium constants. As noted in Chap.3, equilibrium constants can be calculated and adjusted for changes in temperature using thermodynamic considerations.

6. List all mass and charge balance relationships for the system, and for appropriate acid-base problems list the appropriate proton condition. These together with the relationships from step 5 must result in as many independent equations as there are unknown species present.

7. A simultaneous solution of the equations from steps 5 and 6 will give the concentration of each species at equilibrium.

If other phases, such as gaseous or solid phases, are also involved, then equations expressing mass and charge balances between and within each phase must be written and included in the calculations.

EXAMPLE 4.2

A mass equal to 10^{-2} mol of acetic acid is added to sufficient water to make 1 liter of solution at 25°C. What is the equilibrium concentration of all species involved? Ignore activity corrections.

1. The problem is one of ionization of a single weak monoprotic acetic in water. When acetic acid is added to water, it ionizes as follows: $CH_3COOH \rightleftharpoons CH_3COO^- + H^+$.

2. Elements or chemical entities originally present are CH_3COOH, H_2O, H^+, and OH^-.

3. Chemical species likely to be present after equilibrium is attained are

    Acetic acid: CH_3COOH (HAc) and CH_3COO^- (Ac$^-$)
    Water: H_2O, H^+, OH^-

4. Initial conditions are

    $[HAc] = 10^{-2}$ mol/L $= C_T$
    $[H_2O] = 1$ liter of solution (mass change will be negligible during ionization), and
    $[H^+] = [OH^-] = 10^{-7}$ mol/L

5. Equilibrium relationships of interest are (noting that molar concentration equals activity for this problem)

$$[H^+][OH^-] = K_W = 10^{-14} \qquad \text{at 25°C} \qquad (4.4)$$

$$\frac{[H^+][Ac^-]}{[HAc]} = K_A = 1.8 \times 10^{-5} \qquad \text{at 25°C} \qquad (4.5)$$

6. Mass and charge balance realtionships. The acetic acid entity is conserved during ionization such that

$$[HAc] + [Ac^-] = C_T \qquad (4.6)$$

The charge balance between all cations and anions is

$$[H^+] = [OH^-] + [Ac^-] \qquad (4.7)$$

The proton condition, with H_2O and HAc being the reference species, is

$$[H^+] = [OH^-] + [Ac^-]$$

In this case the charge balance and proton condition are identical.

7. The four equations [Eqs. (4.4) through (4.7)] can be solved to determine the concentration of the four unknown species, HAc, Ac$^-$, H$^+$, and OH$^-$. Solution for [Ac$^-$] through a combination of Eqs. (4.4) and (4.7) yields

$$[Ac^-] = [H^+] - K_W/[H^+] \tag{4.8}$$

Next, combination of Eqs. (4.5) and (4.6) gives

$$[H^+][Ac^-] = K_A (C_T - [Ac^-]) \tag{4.9}$$

Finally, a substitution of the value for [Ac$^-$] from Eq. (4.8) in Eq. (4.9) eliminates all variables but [H$^+$]. By rearranging terms, the following final equation results:

$$[H^+]^3 + K_A[H^+]^2 - (K_A C_T + K_W)[H^+] - K_A K_W = 0 \tag{4.10}$$

Equation (4.10) can be solved for [H$^+$], which can then be substituted into Eqs. (4.4) through (4.7) to determine the concentrations of the other species at equilibrium. The results are

$$[H^+] = 4.15 \times 10^{-4}$$
$$[OH^-] = 2.41 \times 10^{-11}$$
$$[HAc] = 9.59 \times 10^{-3}$$
$$[Ac^-] = 4.15 \times 10^{-4}$$

Solution of polynomial equations as in Example 4.2 can be accomplished by trial and error using programmable calculators or spreadsheet software available for use with personal computers. Close approximations can also often be obtained by use of simplifying assumptions, as discussed in Sec. 4.5. Graphical procedures may permit more rapid solutions and have the advantage that they give a better intuitive feel for the phenomena involved. Graphical solutions are demonstrated in Secs. 4.5 to 4.10. Computer methods are also available and, in fact, are generally required for the solution of complex equilibria involving many species and equilibrium relationships. Additional details are presented in each of the remaining sections of this chapter.

4.5 | ACIDS AND BASES

The concept of acids and bases was briefly introduced in Sec. 2.13. Strong acids and bases are considered to be completely ionized in dilute solution, while weak acids and bases are only partially ionized. Acids tend to increase the hydrogen-ion concentration in solution, while bases increase the hydroxide concentration. The product of the activity (or approximately, the molar concentration) of these two ions remains constant at a given temperature and equals K_W. Expression of hydrogen-ion concentration in terms of molar concentration or activity is rather cumbersome, and so the pH method of expression was developed.

The pH of water is a highly important characteristic as it affects equilibria between most chemical species, effectiveness of coagulation, potential of water to be corrosive, suitability of water to support living organisms, and most other quality characteristics of water. The environmental engineer and scientist should thus have a good understanding of factors affecting the pH of natural water.

The pH and p(x) Concept

In 1909 Sørensen proposed to express the hydrogen-ion concentration in terms of its negative logarithm and designated such value as pH^+. His symbol has been superseded by the simple designation pH. The terms may be represented by

$$pH = -\log \{H^+\} \quad \text{or} \quad pH = \log \frac{1}{\{H^+\}} \tag{4.11}$$

With water and in the absence of foreign materials, activity is expressed as molar concentration and $[H^+]$ equals $[OH^-]$ as required by electroneutrality, and the product at 25°C equals K_W or 10^{-14}. These conditions mean that $\{H^+\} = \{OH^-\} = 10^{-7}$, and the pH equals 7, which is considered the "neutral" pH for water. The pH scale is usually represented as ranging from 0 to 14. Values of pH lower than 7 indicate the hydrogen-ion concentration is greater than the hydroxide-ion concentration, and the water is termed acidic. The opposite condition is implied when the pH exceeds 7, and the water is termed basic. It should be noted that when pH is measured using a pH electrode (see Chap. 16), it is the hydrogen-ion activity that is measured and not the molar concentration.

The method of expressing hydrogen-ion activity or concentration as pH is also useful for expressing other small numbers such as the concentration of other ions or ionization constants for solutions of weak acids and bases. For this purpose, the p(x) notation is used, with p(x) defined as

$$p(x) = -\log_{10} x = \log_{10} \frac{1}{x} \tag{4.12}$$

where the quantity x may be the concentration of a given chemical species, an equilibrium constant, or the like. Thus, just as pH is the negative logarithm of the hydrogen-ion activity, pOH signifies the negative logarithm of the hydroxide-ion activity, and pK_W the negative logarithm of the ionization constant for water. From the mass action equation for water,

$$\{H^+\}\{OH^-\} = K_W$$

it follows that

$$-\log \{H^+\} - \log \{OH^-\} = -\log K_W \tag{4.13}$$

and that

$$pH + pOH = pK_W \tag{4.14}$$

Since $K_W = 1 \times 10^{-14}$ at 25°C, it follows that at this temperature $pK_W = 14$.

For weak acids and bases, pK_A is the negative logarithm of the ionization constant for weak acids and pK_B the negative logarithm of the ionization constant for weak bases. The ionization constants and pK_A and pK_B values for several weak acids and bases of environmental interest are listed in Tables 4.1 and 4.2, respec-

Table 4.1 | Typical ionization constants for weak acids at 25°C

Acid	Equilibrium equation	K_A	pK_A	Environmental significance
Acetic	$CH_3COOH \rightleftharpoons H^+ + CH_3COO^-$	1.8×10^{-5}	4.74	Organic wastes
Ammonium	$NH_4^+ \rightleftharpoons H^+ + NH_3$	5.56×10^{-10}	9.26	Nitrification
Boric	$H_3BO_3 \rightleftharpoons H^+ + H_2BO_3^-$	$5.8 \times 10^{-10}(K_{A1})$	9.24	Nitrogen analysis
Carbonic†	$H_2CO_3^* \rightleftharpoons H^+ + HCO_3^-$	$4.3 \times 10^{-7}(K_{A1})$	6.37	Many applications
	$HCO_3^- \rightleftharpoons H^+ + CO_3^{2-}$	$4.7 \times 10^{-11}(K_{A2})$	10.33	
Hypobromous	$HOBr \rightleftharpoons H^+ + OBr^-$	2.3×10^{-9}	8.64	Disinfection
Hydrofluoric	$HF \rightleftharpoons H^+ + F^-$	7.1×10^{-4}	3.15	Fluoridation
Hydrocyanic	$HCN \rightleftharpoons H^+ + CN^-$	4.8×10^{-10}	9.32	Toxicity
Hydrogen sulfide	$H_2S \rightleftharpoons H^+ + HS^-$	$9.1 \times 10^{-8}(K_{A1})$	7.04	Odors, corrosion
	$HS^- \rightleftharpoons H^+ + S^{2-}$	$1.3 \times 10^{-13}(K_{A2})$	12.89	
Hypochlorous	$HOCl \rightleftharpoons H^+ + OCl^-$	2.9×10^{-8}	7.54	Disinfection
Phenol	$C_6H_5OH \rightleftharpoons H^+ + C_6H_5O^-$	1.2×10^{-10}	9.92	Tastes, industrial waste
Phosphoric	$H_3PO_4 \rightleftharpoons H^+ + H_2PO_4^-$	$7.5 \times 10^{-3}(K_{A1})$	2.12	Analytical buffer, plant
	$H_2PO_4^- \rightleftharpoons H^+ + HPO_4^{2-}$	$6.2 \times 10^{-8}(K_{A2})$	7.21	nutrient
	$HPO_4^{2-} \rightleftharpoons H^+ + PO_4^{3-}$	$4.8 \times 10^{-13}(K_{A3})$	12.32	
Propionic	$CH_3CH_2COOH \rightleftharpoons H^+ + CH_3CH_2COO^-$	1.3×10^{-5}	4.89	Organic wastes, anaerobic digestion

†By convention, $[H_2CO_3^*]$ is taken to equal the sum of the actual carbonic acid concentration $[H_2CO_3]$ plus the dissolved carbon dioxide concentration $[CO_2(aq)]$.

tively. It is sometimes useful to note that for a weak acid and its conjugate base (or a weak base and it conjugate acid) the following holds (at 25°C):

$$pK_A + pK_B = 14 \tag{4.15}$$

or

$$K_A K_B = K_W = 10^{-14} \tag{4.16}$$

For example, pK_A for acetic acid from Table 4.1 is 4.74 and pK_B for acetate, its conjugate base, from Table 4.2 is 9.26. Thus, $4.74 + 9.26 = 14$.

Table 4.2 | Typical ionization constants for weak bases and salts of weak acids at 25°C

Substance	Equilibrium equation	K_B	pK_B	Significance in environmental engineering
Acetate	$CH_3COO^- + H_2O \rightleftharpoons CH_3COOH + OH^-$	5.56×10^{-10}	9.26	Organic wastes
Ammonia	$NH_3 + H_2O \rightleftharpoons NH_4^+ + OH^-$	1.8×10^{-5}	4.74	Disinfection, nutrient
Borate	$H_2BO_3^- + H_2O \rightleftharpoons H_3BO_3 + OH^-$	1.72×10^{-5}	4.76	Nitrogen analysis
Carbonate	$CO_3^{2-} + H_2O \rightleftharpoons HCO_3^- + OH^-$	$2.13 \times 10^{-4}(K_{B2})$	3.67	Many applications
	$HCO_3^- + H_2O \rightleftharpoons H_2CO_3^* + OH^-$	$2.33 \times 10^{-8}(K_{B1})$	7.63	
Calcium hydroxide	$CaOH^+ \rightleftharpoons Ca^{2+} + OH^-$	$3.5 \times 10^{-2}(K_{B2})$	1.46	Softening
Magnesium hydroxide	$MgOH^+ \rightleftharpoons Mg^{2+} + OH^-$	$2.6 \times 10^{-3}(K_{B2})$	2.59	Softening

Solving Acid-Base Equilibrium Problems

When solving acid-base equilibrium problems, we will assume that equilibrium occurs very quickly; thus, we will ignore kinetic considerations. Strong acids have large values of K_A compared to weak acids (Table 4.1) and tend to completely dissociate (ionize) when added to water. Common strong acids include perchloric [$HClO_4$, pK_A (25°C) $= -7$], hydrochloric [HCl, pK_A (25°C) $\cong -3$], sulfuric [H_2SO_4, pK_{A1} (25°C) $\cong -3$, pK_{A2} (25°C) $= 1.99$], and nitric [HNO_3, pK_A (25°C) $= -1.3$]. Similarly, strong bases [e.g., hydroxide salts such as NaOH, KOH, Ca(OH)$_2$] can be assumed to completely ionize upon addition to water. The only time one cannot assume complete ionization for strong acids or bases is when the added concentration is near 10^{-7} M. This is illustrated in Example 4.3.

The tools to be used for solving acid-base equilibrium problems are (1) equilibrium relationships, (2) mass balances, (3) charge balance, and when appropriate, (4) the proton condition. The general goal is to identify all unknowns and then generate an equivalent number of equations, that is, to generate n equations in n unknowns. The task is then to solve this set of simultaneous equations. A variety of solution techniques can be used. Spreadsheets can be used, graphical methods can be used, and simplifying assumptions can be used to reduce the number of unknowns. For very complex problems, computer solutions are required. Following is a series of example problems that demonstrate some of these solution techniques.

Exact Solutions   One way of taking advantage of the capabilities of spreadsheet software is to generate a single equation with $\{H^+\}$ (or $[H^+]$) as the only unknown. This can be done for simple problems as was shown in Example 4.2 for the monoprotic (yields one proton) acetic acid. Alternatively, one can use *ionization fractions*, called α values, to make substitutions in the charge balance. This concept is explained below.

Equations for ionization fractions are developed from mass-balance and equilibrium-constant equations. Let's consider the "most complicated" case, a triprotic acid. We will ignore activity corrections and use molar concentration in equilibrium-constant expressions.

$$H_3A \rightleftharpoons H_2A^- + H^+ \qquad K_{A1} = \frac{[H_2A^-][H^+]}{[H_3A]} \qquad (4.17)$$

$$H_2A^- \rightleftharpoons HA^{2-} + H^+ \qquad K_{A2} = \frac{[HA^{2-}][H^+]}{[H_2A^-]} \qquad (4.18)$$

$$HA^{2-} \rightleftharpoons A^{3-} + H^+ \qquad K_{A3} = \frac{[A^{3-}][H^+]}{[HA^{2-}]} \qquad (4.19)$$

$$C_{T,H_3A} = [H_3A] + [H_2A^-] + [HA^{2-}] + [A^{3-}] \qquad (4.20)$$

Values for α are defined in the following equations, where the subscript represents the number of protons "missing" from the given species:

$$\alpha_0 = \frac{[H_3A]}{C_{T,H_3A}} = \frac{\text{concentration of species missing 0 protons}}{\text{total concentration of species in solution}}$$

$$\alpha_1 = \frac{[H_2A^-]}{C_{T,H_3A}} = \frac{\text{concentration of species missing 1 proton}}{\text{total concentration of species in solution}}$$

$$\alpha_2 = \frac{[HA^{2-}]}{C_{T,H_3A}} = \frac{\text{concentration of species missing 2 protons}}{\text{total concentration of species in solution}}$$

$$\alpha_3 = \frac{[A^{3-}]}{C_{T,H_3A}} = \frac{\text{concentration of species missing 3 protons}}{\text{total concentration of species in solution}}$$

Using these equations, we can develop expressions where the only unknown is $[H^+]$, as long as C_{T,H_3A} is known. As an example, let's try to develop an equation for $[H_3A]$ that is only a function of $[H^+]$, C_{T,H_3A}, and the equilibrium constants K_{A1}, K_{A2}, and K_{A3}. We will ignore activity corrections for now. Let's start by using the equilibrium constant equations to solve for all in terms of $[H_3A]$.

From Eq. (4.17): $\quad [H_2A^-] = \dfrac{K_{A1}[H_3A]}{[H^+]}$

From Eq. (4.18): $\quad [HA^{2-}] = \dfrac{K_{A2}[H_2A^-]}{[H^+]} = \dfrac{K_{A2}K_{A1}[H_3A]}{[H^+]^2}$

From Eq. (4.19): $\quad [A^{3-}] = \dfrac{K_{A3}[HA^{2-}]}{[H^+]} = \dfrac{K_{A3}K_{A2}\,K_{A1}[H_3A]}{[H^+]^3}$

From Eq. (4.20):

$$C_{T,H_3A} = [H_3A] + [H_2A^-] + [HA^{2-}] + [A^{3-}]$$

$$= [H_3A] + \frac{K_{A1}[H_3A]}{[H^+]} + \frac{K_{A2}K_{A1}[H_3A]}{[H^+]^2} + \frac{K_{A3}K_{A2}K_{A1}[H_3A]}{[H^+]^3}$$

$$= [H_3A]\left(1 + \frac{K_{A1}}{[H^+]} + \frac{K_{A2}K_{A1}}{[H^+]^2} + \frac{K_{A3}K_{A2}K_{A1}}{[H^+]^3}\right)$$

From the definition for α-values:

$$\alpha_0 = \frac{[H_3A]}{C_{T,H_3A}}$$

Making appropriate substitutions gives

$$\alpha_0 = \frac{1}{1 + \dfrac{K_{A1}}{[H^+]} + \dfrac{K_{A1}\,K_{A2}}{[H^+]^2} + \dfrac{K_{A1}\,K_{A2}K_{A3}}{[H^+]^3}} \tag{4.21}$$

One can develop similar expressions for $[H_2A^-]$, $[HA^{2-}]$, and $[A^{3-}]$:

$$\alpha_1 = \frac{1}{\dfrac{[H^+]}{K_{A1}} + 1 + \dfrac{K_{A2}}{[H^+]} + \dfrac{K_{A2}\,K_{A3}}{[H^+]^2}} \tag{4.22}$$

$$\alpha_2 = \frac{1}{\dfrac{[H^+]^2}{K_{A1}\,K_{A2}} + \dfrac{[H^+]}{K_{A2}} + 1 + \dfrac{K_{A3}}{[H^+]}} \tag{4.23}$$

$$\alpha_3 = \cfrac{1}{\cfrac{[H^+]^3}{K_{A1}\,K_{A2}K_{A3}} + \cfrac{[H^+]^2}{K_{A2}K_{A3}} + \cfrac{[H^+]}{K_{A3}} + 1} \qquad (4.24)$$

The student should note that for a *diprotic* acid, there is no K_{A3} and thus no α_3 (only α_0, α_1, and α_2). For a *monoprotic* acid, there is no K_{A3} or K_{A2} and thus no α_3 or α_2 (only α_0 and α_1). We can now use these equations along with the charge balance and, in some cases the proton condition, to develop a "master equation" with only $[H^+]$ as the unknown. Spreadsheets can then be used to obtain the solution.

EXAMPLE 4.3

Determine the equilibrium pH of a solution made by adding HCl to water to give a concentration of 10^{-2} M at 25°C. Ignore activity corrections.

Since HCl is a strong acid ($K_A = 10^3$), one could assume that it completely ionizes upon addition to water, giving $[H^+] = [Cl^-] = 10^{-2}$ M. As such, pH = 2.0. However, let's set up the problem for an exact solution.

Equlibrium constants of interest are (noting that molar concentration equals activity for this problem):

$$[H^+][OH^-] = K_W = 10^{-14} \qquad \text{at 25°C}$$
$$\frac{[H^+][Cl^-]}{[HCl]} = K_A = 10^3 \qquad \text{at 25°C}$$

The mass balance on hydrochloric acid is

$$[HCl] + [Cl^-] = C_{T,HCl} = 0.01$$

The charge balance is

$$[H^+] = [OH^-] + [Cl^-]$$

From the definition of α_1 for a monoprotic acid, $[Cl^-] = \alpha_1 C_{T,HCl}$. Making substitutions in the charge balance gives

$$[H^+] = \frac{K_W}{[H^+]} + \alpha_1 C_{T,HCl} \qquad (4.25)$$

Since $C_{T,HCl}$ is known and α_1 is only a function of $[H^+]$ and K_A [Eq. (4.22)], we have an equation with one unknown, $[H^+]$. If Eq. (4.22) is substituted into Eq. (4.25), and the resulting equation rearranged, the following is obtained:

$$[H^+]^3 + K_A[H^+]^2 - (K_A\,C_{T,HCl} + K_W)[H^+] - K_A K_W = 0$$

This is Eq. (4.10)! Thus either Eq. (4.10) or Eq. (4.25) is applicable to all solutions made by adding a monoprotic acid to water. How much acid is added (C_T) and the nature of the monprotic acid (K_A) determine the equilibrium pH. In this example, $C_{T,HCl} = 0.01$ M and $K_A = 10^3$. A spreadsheet solution for this problem is given here. "Left side" refers to the left side of Eq. (4.25); "right side" refers to the right side of Eq. (4.25); and the "% error" is calculated using the following: % error = $100(1 - \text{right side/left side})$.

pH	[H$^+$]	α_1	Left side	Right side	% error
1.00	0.100000	0.999900	0.100000	0.009999	90.001000
1.91	0.012303	0.999988	0.012303	0.010000	18.717948
1.92	0.012023	0.999988	0.012023	0.010000	16.824623
1.93	0.011749	0.999988	0.011749	0.010000	14.887196
1.94	0.011482	0.999989	0.011482	0.010000	12.904641
1.95	0.011220	0.999989	0.011220	0.010000	10.875906
1.96	0.010965	0.999989	0.010965	0.010000	8.799916
1.97	0.010715	0.999989	0.010715	0.010000	6.675570
1.98	0.010471	0.999990	0.010471	0.010000	4.501741
1.99	0.010233	0.999990	0.010233	0.010000	2.277278
2.00	**0.010000**	**0.999990**	**0.010000**	**0.010000**	**0.001000**
2.01	0.009772	0.999990	0.009772	0.010000	−2.328299
2.02	0.009550	0.999990	0.009550	0.010000	−4.711855
2.03	0.009333	0.999991	0.009333	0.010000	−7.150931

The solution is pH = 2.0. In general, when high concentrations of strong acids are added to water, the pH equals − log C_T. (Note: The "Solver" routine in a spreadsheet such as Excel© can be used to obtain rapid convergence for trial and error solutions.)

EXAMPLE 4.4

Rework Example 4.3 for an HCl addition of 10^{-8} M. In this case, $C_{T,HCl} = 10^{-8}$ M. Here, we cannot ignore the ionization of water. All the equations developed for Example 4.3 hold here; the only difference is in $C_{T,HCl}$. The equation to be solved is again Eq. (4.25):

$$[H^+] = \frac{K_W}{[H^+]} + \alpha_1 C_{T,HCl}$$

The spreadsheet solution for $C_{T,HCl}$ is given here.

pH	[H$^+$]	α_1	Left side	Right side	% error
6.70	2.00E-07	1.00E+00	2.00E-07	6.01E-08	69.869263
6.80	1.58E-07	1.00E+00	1.58E-07	7.31E-08	53.879710
6.90	1.26E-07	1.00E+00	1.26E-07	8.94E-08	28.960983
7.00	1.00E-07	1.00E+00	1.00E-07	1.10E-07	−10.000000
6.91	1.23E-07	1.00E+00	1.23E-07	9.13E-08	25.802350
6.92	1.20E-07	1.00E+00	1.20E-07	9.32E-08	22.499265
6.93	1.17E-07	1.00E+00	1.17E-07	9.51E-08	19.045024
6.94	1.15E-07	1.00E+00	1.15E-07	9.71E-08	15.432607
6.95	1.12E-07	1.00E+00	1.12E-07	9.91E-08	11.654667
6.96	1.10E-07	1.00D+00	1.10E-07	1.01E-07	7.703514
6.97	1.07E-07	1.00E+00	1.07E-07	1.03E-07	3.571098
6.98	**1.05E-07**	**1.00E+00**	**1.05E-07**	**1.05E-07**	**−0.751010**
6.99	1.02E-07	1.00E+00	1.02E-07	1.08E-07	−5.271631

The solution here is 6.98. If desired, one could reduce the percent error by trying pH values between 6.97 and 6.98.

EXAMPLE 4.5

Determine the equilibrium pH of a solution made by adding enough Na_2HPO_4 to water to give a concentration of 10^{-4} M at 25°C. The Na_2HPO_4 completely dissolves in water. Ignore activity corrections.

The equilibrium relationships of interest are

$$[H^+][OH^-] = K_W = 10^{-14} \qquad \text{at } 25°C$$

$$\frac{[H^+][H_2PO_4^-]}{[H_3PO_4]} = K_{A1} = 7.5 \times 10^{-3} \qquad \text{at } 25°C$$

$$\frac{[H^+][HPO_4^{2-}]}{[H_2PO_4^-]} = K_{A2} = 6.2 \times 10^{-8} \qquad \text{at } 25°C$$

$$\frac{[H^+][PO_4^{3-}]}{[HPO_4^{2-}]} = K_{A3} = 4.8 \times 10^{-13} \qquad \text{at } 25°C$$

The mass balance relationships are

$$C_{T,PO_4} = 10^{-4} = [H_3PO_4] + [H_2PO_4^-] + [HPO_4^{2-}] + [PO_4^{3-}]$$
$$C_{T,Na} = 2 \times 10^{-4} = [Na^+]$$

The charge balance is

$$[Na^+] + [H^+] = [OH^-] + [H_2PO_4^-] + 2[HPO_4^{2-}] + 3[PO_4^{3-}]$$

Using the definitions of α values and Eqs. (4.22) to (4.24), we can make substitutions in the charge balance to give

$$C_{T,Na} + [H^+] = \frac{K_W}{[H^+]} + \alpha_1 C_{T,PO_4} + 2\alpha_2 C_{T,PO_4} + 3\alpha_3 C_{T,PO_4} = \frac{K_W}{[H^+]} + C_{T,PO_4}(\alpha_1 + 2\alpha_2 + 3\alpha_3)$$

Since $C_{T,Na}$ and C_{T,PO_4} are known and α's are only functions of $[H^+]$ and K_A's, we have an equation with one unknown, $[H^+]$. The spreadsheet solution to this problem is given here.

pH	$[H^+]$	α_1	α_2	α_3	Left side	Right side	% error
9.00	1.00E-09	1.59E-02	9.84E-01	4.72E-04	2.00E-04	2.08E-04	-4.2298
8.00	1.00E-08	1.39E-01	8.61E-01	4.13E-05	2.00E-04	1.87E-04	6.4468
8.70	2.00E-09	3.12E-02	9.69E-01	2.33E-04	2.00E-04	2.02E-04	-0.9580
8.60	2.51E-09	3.89E-02	9.61E-01	1.84E-04	2.00E-04	2.00E-04	-0.0520
8.50	3.16E-09	4.85E-02	9.51E-01	1.44E-04	2.00E-04	1.98E-04	0.8393
8.51	3.09E-09	4.75E-02	9.52E-01	1.48E-04	2.00E-04	1.99E-04	0.7497
8.52	3.02E-09	4.64E-02	9.53E-01	1.52E-04	2.00E-04	1.99E.04	0.6602
8.53	2.95E-09	4.54E-02	9.54E-01	1.55E-04	2.00E-04	1.99E-04	0.5710
8.54	2.88E-09	4.44E-02	9.55E-01	1.59E-04	2.00E-04	1.99E-04	0.4819
8.55	2.82E-09	4.35E-02	9.56E-01	1.63E-04	2.00E-04	1.99E-04	0.3929
8.56	2.75E-09	4.25E-02	9.57E-01	1.67E-04	2.00E-04	1.99E-04	0.3040
8.57	2.69E-09	4.16E-02	9.58E-01	1.71E-04	2.00E-04	2.00E-04	0.2151
8.58	2.63E-09	4.07E-02	9.59E-01	1.75E-04	2.00E-04	2.00E-04	0.1261
8.59	**2.57E-09**	**3.98E-02**	**9.60E-01**	**1.79E-04**	**2.00E-04**	**2.00E-04**	**0.0371**
8.60	2.51E-09	3.89E-02	9.61E-01	1.84E-04	2.00E-04	2.00E-04	-0.0520

The solution here is 8.59. If desired, one could reduce the percent error by trying pH values between 8.59 and 8.60.

Approximate Solutions   The use of spreadsheets makes solution of problems such as Examples 4.3 to 4.5 fairly straightforward. However, the student may not develop an intuitive feel for the effect of C_T, strength of the acid (K_A values), or the species added (acid versus conjugate base). Making simplifying assumptions helps develop this "chemical intuition" [and also helps in solving problems when spreadsheets are not available (like during exams!)]. There are a few simple guidelines to follow when making these assumptions. First, compare C_T with 10^{-7}. When C_T is much less than 10^{-7} M, assume pH equals 7.0. If C_T is near 10^{-7} M or higher, one may then look at the form of the compound added. If the acid form is added, it is reasonable to assume that $\{H^+\}$ is greater than $\{OH^-\}$ and that perhaps $\{HA\}$ is greater than $\{A^-\}$. If the base form is added, it is reasonable to assume that $\{OH^-\}$ is greater than $\{H^+\}$ and that $\{A^-\}$ is perhaps greater than $\{HA\}$. Assumptions for strong acids and bases are slightly different than for weak acids and bases. With more complicated problems (e.g., diprotic and triprotic acids), additional assumptions will be required. A useful general strategy is to eliminate enough unknowns to generate a quadratic equation. After the solution is obtained, the assumptions must be checked to see if they are reasonable. Several illustrative examples follow.

Rework Example 4.4 using simplifying assumptions. | **EXAMPLE 4.6**

Equilibrium constants of interest are (noting that molar concentration equals activity for this problem):

$$[H^+][OH^-] = K_W = 10^{-14} \quad \text{at } 25°C$$

$$\frac{[H^+][Cl^-]}{[HCl]} = K_A = 10^3 \quad \text{at } 25°C$$

The mass balance on hydrochloric acid is

$$[HCl] + [Cl^-] = C_{T,HCl} = 10^{-8}$$

The charge balance is

$$[H^+] = [OH^-] + [Cl^-]$$

Since $C_{T,HCl}$ is close to 10^{-7}, we cannot assume that $[H^+] > [OH^-]$. Since HCl is a strong acid, it will be almost completely ionized. Thus, an initial assumption is that $C_{T,HCl} \cong [Cl^-] = 10^{-8}$. The charge balance becomes

$$[H^+] = \frac{K_W}{[H^+]} + 10^{-8}$$

Rearranging gives a quadratic equation:

$$[H^+]^2 - 10^{-8}[H^+] - 10^{-14} = 0$$

Solution involves the use of the quadratic formula:

$$[H^+] = \frac{-(-10^{-8}) \pm \sqrt{(-10^{-8})^2 - 4(1)(-10^{-14})}}{2} = 1.05 \times 10^{-7}$$

Thus, pH = 6.98. We now need to check our assumptions. We can use α values to do this.

$$[HCl] = \alpha_0(10^{-8}) \qquad \text{and} \qquad [Cl^-] = \alpha_1(10^{-8})$$

At pH = 6.98,

$$\alpha_0 = \frac{1}{1 + \dfrac{10^3}{10^{-6.98}}} = 1.05 \times 10^{-10}$$

$$\alpha_1 = \frac{1}{1 + \dfrac{10^{-6.98}}{10^3}} = 1.00$$

Thus, $[HCl] = 1.05 \times 10^{-18}$ M and $[Cl^-] = 10^{-8}$, and our assumption was a good one.

EXAMPLE 4.7

Using simplifying assumptions, calculate the equilibrium pH of a solution made by adding acetic acid to water to give a concentration of 10^{-2} M at 25°C. Ignore activity corrections.

In this case, C_T is 10^{-2}, which is much greater than 10^{-7}. Since it is the acid form, then it is reasonable to assume, ignoring activity corrections, that $[H^+]$ is greater than $[OH^-]$ and that [HAc] is greater than $[Ac^-]$.

Equilibrium relationships of interest are (noting that molar concentration equals activity for this problem):

$$[H^+][OH^-] = K_W = 10^{-14}$$

$$\frac{[H^+][Ac^-]}{[HAc]} = K_A = 1.8 \times 10^{-5}$$

The mass balance on acetic acid is

$$[HAc] + [Ac^-] = 0.01$$

The charge balance and proton condition, with H_2O and HAc being the reference species, are identical:

$$[H^+] = [OH^-] + [Ac^-]$$

From the assumption noted above ($[H^+] \gg [OH^-]$), the charge balance reduces to

$$[H^+] \cong [Ac^-]$$

If we let $x = [H^+] = [Ac^-]$ and note from the mass balance on acetic acid that [HAc] = $0.01 - x$, we can substitute into the equilibrium relationship for acetic acid:

$$\frac{x^2}{0.01 - x} = 1.8 \times 10^{-5}$$

Multiplying through and rearranging terms yields

$$x^2 + 1.8 \times 10^{-5}x - 1.8 \times 10^{-7} = 0$$

This equation can be solved using the quadratic formula:

$$x = \frac{-(1.8 \times 10^{-5}) \pm \sqrt{(1.8 \times 10^{-5})^2 - 4(1)(-1.8 \times 10^{-7})}}{2}$$

$$= 4.15 \times 10^{-4}$$

Thus, $[H^+] = 4.15 \times 10^{-4}$, or pH = 3.38. This is the same answer obtained in Example 4.2 when simplifying assumptions were not used.

It is clear that since pH = 3.38, our assumption that $[H^+]$ is much greater than $[OH^-]$ is a good one. The second assumption that $[HAc]$ is greater than $[Ac^-]$ is also a good one since $[Ac^-] = 4.14 \times 10^{-4}$. From the mass balance on acetic acid:

$$[HAc] = 0.01 - 4.14 \times 10^{-4} = 9.59 \times 10^{-3}$$

Rework Example 4.7 assuming that the solution now contains 0.01 M NaCl and use activity corrections. The student should note that NaCl completely ionizes upon addition to water.

EXAMPLE 4.8

For this problem, the equilibrium relationships are

$$\{H^+\}\{OH^-\} = K_W = 10^{-14}$$

$$\frac{\{H^+\}\{Ac^-\}}{\{HAc\}} = K_A = 1.8 \times 10^{-5}$$

The mass balance relationships are

$$C_{T,HAc} = 0.01 = [HAc] + [Ac^-]$$
$$C_{T,Na} = 0.01 = [Na^+]$$
$$C_{T,Cl} = 0.01 = [Cl^-]$$

The charge balance is

$$[H^+] + [Na^+] = [OH^-] + [Ac^-] + [Cl^-]$$

The proton condition, with H_2O and HAc being the reference species, is

$$[H^+] = [OH^-] + [Ac^-]$$

Note that in mass balance, charge balance, and proton condition relationships, molar concentrations and not activities are required.

The ionic strength must now be calculated so that activity corrections can be made. Since HAc is a weak acid, its ionization can be ignored ($[Ac^-] \cong 0$) and we can assume that the ionic strength is due entirely to the added NaCl.

$$\mu = \tfrac{1}{2}\sum_i C_i Z_i^2 = 0.5[0.01(1)^2 + 0.01(-1)^2] = 0.01 \text{ M}$$

Using the Güntelberg approximation [Eq. (4.2)]:

$$\log \gamma = -0.5Z^2 \frac{\sqrt{0.01}}{1 + \sqrt{0.01}} = -0.0455Z^2$$

Thus, for the ions whose $Z = 1$, we have $\log \gamma = -0.0455$ and $\gamma = 0.90$.

The same assumption used in Example 4.7, $\{H^+\}$ is much greater than $\{OH^-\}$ since the acid form is being added, can be made here. Now, since $[Na^+] = [Cl^-] = 0.01$, the charge balance reduces to

$$[H^+] \cong [Ac^-]$$

as was the case in Example 4.7. We can again let $x = [H^+] = [Ac^-]$. Recognizing that HAc is an un-ionized species, and that for solutions with ionic strengths less than 0.1 M, $\{HAc\} = [HAc]$, we see from the mass balance on acetic acid that $[HAc] = 0.01 - x$. Substituting into the equilibrium expression gives

$$\frac{(0.9x)(0.9x)}{0.01 - x} = 1.8 \times 10^{-5}$$

Multiplying through and rearranging terms yields

$$x^2 + 2.22 \times 10^{-5}x - 2.22 \times 10^{-7} = 0$$

This equation can be solved using the quadratic formula:

$$x = \frac{-(2.22 \times 10^{-5}) \pm \sqrt{(2.22 \times 10^{-5})^2 - 4(1)(-2.22 \times 10^{-7})}}{2}$$

$$= 4.60 \times 10^{-4} \, M = [H^+] = [Ac^-]$$

Thus, $\{H^+\} = 0.9(4.60 \times 10^{-4})$, or pH $= 3.38$. This is the same answer obtained in Example 4.7 when activity corrections were not made. Even though the pH did not change significantly, values for [HAc] and [Ac$^-$] did change by approximately 0.5 and 10 percent, respectively.

EXAMPLE 4.9

Calculate the equilibrium pH of a 25°C solution made by adding enough sodium acetate (NaAc) to water to give a concentration of 0.01 M. Include activity corrections. The student should note that NaAc completely dissolves upon addition to water, and the sodium exists there only as the ion, as with all sodium salts.

Since it is the base form being added to water, the equilibrium relationships are

$$\{H^+\}\{OH^-\} = K_W = 10^{-14}$$

$$\frac{\{OH^-\}\{HAc\}}{\{Ac^-\}} = K_B = 5.56 \times 10^{-10}$$

The mass balance relationships are

$$C_{T,HAc} = 0.01 = [HAc] + [Ac^-]$$
$$C_{T,Na} = 0.01 = [Na^+]$$

The charge balance is

$$[H^+] + [Na^+] = [OH^-] + [Ac^-]$$

The proton condition, with H_2O and Ac^- being the reference species, is

$$[H^+] + [HAc] = [OH^-]$$

The ionic strength must now be calculated so that activity corrections can be made. Since NaAc, a weak base, is added to water, we could assume that $[HAc] \cong 0$ relative to $[Ac^-]$ and that the ionic strength is due entirely to the added $[Na^+]$ and $[Ac^-]$. Thus,

$$\mu = \tfrac{1}{2}\sum_i C_i Z_i^2 = 0.5(0.01(1)^2 + 0.01(-1)^2) = 0.01 \text{ M}$$

Using the Güntelberg approximation, we have $\log \gamma = -0.0455$ and $\gamma = 0.90$ as was the case in Example 4.8.

Since it is the base form that is added, we can assume that $\{OH^-\}$ is much greater than $\{H^+\}$, and from the proton condition:

$$[HAc] \cong [OH^-]$$

We can now set $x = [HAc] = [OH^-]$. From the mass balance, $[Ac^-] = 0.01 - x$. We can now substitute in the equilibrium expression, correcting for activity, and obtain

$$\frac{(x)(0.9x)}{0.9(0.01 - x)} = 5.56 \times 10^{-10}$$

Multiplying through and rearranging terms yields

$$x^2 + 5.56 \times 10^{-10}x - 5.56 \times 10^{-12} = 0$$

This equation can be solved using the quadratic formula:

$$x = \frac{-(5.56 \times 10^{-10}) \pm \sqrt{(5.56 \times 10^{-10})^2 - 4(1)(-5.56 \times 10^{-12})}}{2}$$
$$= 2.36 \times 10^{-6} \text{ M} = [OH^-] = [HAc]$$

Solving for pH:

$$\{OH^-\} = 0.90(2.36 \times 10^{-6}) = 2.12 \times 10^{-6} \text{ M}$$
$$\{H^+\} = 4.72 \times 10^{-9} \text{ M}$$
$$pH = 8.33$$

We must now check our two major assumptions. The results obtained show that $\{OH^-\}$ is indeed much larger than $\{H^+\}$. Our assumption that the ionic strength was due to Na^+ and Ac^- was a good one since $[HAc]$ is 2.36×10^{-6} M, which is much smaller than 0.01 M.

Examples 4.3 to 4.9 used monoprotic acids and bases, and the assumptions made were relatively simple. Example 4.10 with a triprotic acid is a bit more complex but is approached in a similar manner.

EXAMPLE 4.10

Rework Example 4.5 using simplifying assumptions.

The equilibrium relationships of interest are

$$[H^+][OH^-] = 10^{-14} \qquad \text{at } 25°C \qquad (1)$$

$$\frac{[H^+][H_2PO_4^-]}{[H_3PO_4]} = K_{A1} = 7.5 \times 10^{-3} \qquad \text{at } 25°C \qquad (2)$$

$$\frac{[H^+][HPO_4^{2-}]}{[H_2PO_4^-]} = K_{A2} = 6.2 \times 10^{-8} \qquad \text{at } 25°C \qquad (3)$$

$$\frac{[H^+][PO_4^{3-}]}{[HPO_4^{2-}]} = K_{A3} = 4.8 \times 10^{-13} \qquad \text{at } 25°C \qquad (4)$$

The mass balance relationships are

$$C_{T,PO_4} = 10^{-4} = [H_3PO_4] + [H_2PO_4^-] + [HPO_4^{2-}] + [PO_4^{3-}] \qquad (5)$$
$$C_{T,Na} = 2 \times 10^{-4} = [Na^+] \qquad (6)$$

The charge balance is

$$[H^+] + [Na^+] = [OH^-] + [H_2PO_4^-] + 2[HPO_4^{2-}] + 3[PO_4^{3-}] \qquad (7)$$

The proton condition, with H_2O and HPO_4^{2-} being the reference species, is

$$[H^+] + [H_2PO_4^-] + 2[H_3PO_4] = [OH^-] + [PO_4^{3-}] \qquad (8)$$

Since $C_{T,PO_4} = 10^{-4}$ M, a relatively low ionic strength solution, we can ignore activity corrections. From Eqs. (1) to (8), we see that we have six unknown species (H_3PO_4, $H_2PO_4^-$, HPO_4^{2-}, PO_4^{3-}, H^+, and OH^-) and seven equations to use to solve the problem. The first task is to make some assumptions to simplify the problem.

Using the same approach as in Examples 4.6 to 4.8, first compare expected $[H^+]$ with $[OH^-]$ to see if one can be ignored. When HPO_4^{2-} is added to water, it can act as an acid or a base:

$$HPO_4^{2-} + H_2O \rightleftharpoons H_2PO_4^- + OH^- \qquad (pK_{B2} = 6.79)$$
$$HPO_4^{2-} \rightleftharpoons PO_4^{3-} + H^+ \qquad (pK_{A3} = 12.32)$$

Since pK_{B2} is less than pK_{A3}, HPO_4^{2-} is a stronger base than it is an acid and we can assume that $[OH^-]$ is greater than $[H^+]$. [Also note that pK_{B2} is less than pK_{A2} (7.21) indicating that HPO_4^{2-} is a stronger base than $H_2PO_4^-$ is an acid.] Next, try to eliminate some of the phosphate forms from the mass balance.

Since HPO_4^{2-} is the form added, it is unlikely that H_3PO_4 will be present in significant quantities since it is not in direct equilibrium with HPO_4^{2-}. Finally, since HPO_4^{2-} is a stronger base than it is an acid, it is likely that the concentration of $H_2PO_4^-$ will be greater than the concentration of PO_4^{3-}. Using these assumptions the proton condition [Eq. (8)] becomes

$$[H_2PO_4^-] \cong [OH^-]$$

and the mass balance becomes

$$10^{-4} \cong [H_2PO_4^-] + [HPO_4^{2-}]$$

We can now set $x = [H_2PO_4^-] = [OH^-]$ and substitute in the base equilibrium expression (with $pK_{B2} = 6.79$), ignoring activity corrections, and obtain

$$\frac{(x)(x)}{(0.0001 - x)} = 1.62 \times 10^{-7}$$

Multiplying through and rearranging terms yields

$$x^2 + 1.62 \times 10^{-7}x - 1.62 \times 10^{-11} = 0$$

This equation can be solved using the quadratic formula:

$$x = \frac{-(1.62 \times 10^{-7}) \pm \sqrt{(1.62 \times 10^{-7})^2 - 4(1)(-1.62 \times 10^{-11})}}{2}$$
$$= 3.94 \times 10^{-6}\,M = [OH^-] = [H_2PO_4^-]$$

Solving for pH:

$$[H^+] = 2.54 \times 10^{-9}\,M$$
$$pH = 8.60$$

This compares well with the "exact" solution of Example 4.5 (pH = 8.59).

We must now check our assumptions. The results obtained show that $[OH^-]$ is indeed much larger than $[H^+]$. Using equilibrium expressions and knowing that $[H^+] = 2.54 \times 10^{-9}\,M$ and $[H_2PO_4^-] = 3.94 \times 10^{-6}\,M$, we can calculate the following:

$$[HPO_4^{2-}] = 9.62 \times 10^{-5}\,M$$
$$[H_3PO_4] = 1.33 \times 10^{-12}\,M$$
$$[PO_4^{3-}] = 1.82 \times 10^{-8}\,M$$

From these values, we see that $[H_3PO_4]$ is very small, as we assumed, and $[H_2PO_4^-]$ is greater than $[PO_4^{3-}]$, also as we assumed. If we want to calculate the value of the error introduced by our assumptions, we can use the mass balance, charge balance, and proton condition with the following equation:

$$\text{Percent error} = \left(1 - \frac{\text{right side}}{\text{left side}}\right)100$$

By substituting concentrations into the mass balance, charge balance, or proton condition, sums can be obtained for the right and left sides of these equations. For example, a check of the mass balance equation gives

$$10^{-4} = 1.33 \times 10^{-12} + 3.94 \times 10^{-6} + 9.62 \times 10^{-5} + 1.82 \times 10^{-8}$$

Using the percent error equation, the calculated error is −0.16 percent. The errors for the charge balance and proton condition are −0.17 percent and −0.40 percent, respectively. If the errors so calculated are deemed to be unacceptable, the pH can be adjusted slightly, new values for all species present can be calculated, and the error checked again. For most calculations, errors of less than 5 to 10 percent are considered acceptable.

Logarithmic Concentration Diagrams

Another technique for solving acid-base problems uses logarithmic concentration diagrams. In this graphical procedure, pH is used as the master variable and is plotted as the abscissa. The logarithm of the concentration of each constituent of importance in a given case is plotted on the ordinate. A log concentration–pH diagram represents a mass balance; at every pH on the diagram, the mass balance is solved. The general idea is to develop equations for the concentration of a given species as a function of pH and known quantities (e.g., K_W, K_A, C_T). The procedure for drawing a log concentration–pH diagram will be demonstrated using a solution containing 0.02 M acetic acid. The diagram is shown in Fig. 4.1. It should also be noted that spreadsheet software such as Excel can be used to draw log concentration–pH (log C–pH) diagrams.

Logarithmic Concentration Diagram for a Weak Monoprotic Acid   The relationships given by Eqs. (4.4) to (4.7) for acetic acid apply for all monoprotic acids. The line for $[H^+]$ is obtained from the relationship

$$\log[H^+] = -pH \tag{4.26}$$

That for $[OH^-]$ is obtained from the relationship

$$\log[OH^-] = pH - pK_W \tag{4.27}$$

The concentration of acetic acid and acetate can be obtained by combining Eqs. (4.5) and (4.6) with the definitions for α_0 and α_1.

$$[HAc] = \alpha_0 C_T = \frac{C_T[H^+]}{K_A + [H^+]} \tag{4.28}$$

and

$$[Ac^-] = \alpha_1 C_T = \frac{C_T K_A}{K_A + [H^+]} \tag{4.29}$$

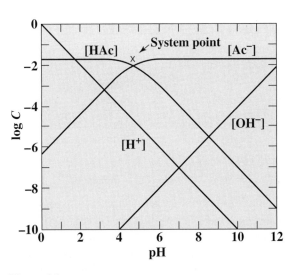

Figure 4.1
Logarithmic concentration diagram for 0.02 M acetic acid.

The logarithms of [HAc] and [Ac⁻] appear as straight lines connected by a short curve. The intersection of the two straight lines is called the system point and is located where pH $=$ pK_A: To the left of this point, [H⁺] is greater than K_A, so the logarithms of the concentrations from Eqs. (4.28) and (4.29) reduce to

$$\log[\text{HAc}] \cong \log C_T \tag{4.30}$$

and
$$\log[\text{Ac}^-] \cong \log C_T - \text{p}K_A + \text{pH} \tag{4.31}$$

The first equation represents a horizontal line and the second a diagonal line with a slope of $+1$, both passing through the system point.

To the right of the system point, [H⁺] is less than K_A, and the logarithms reduce to

$$\log[\text{HAc}] \cong \log C_T + \text{p}K_A - \text{pH} \tag{4.32}$$

and
$$\log[\text{Ac}^-] \cong \log C_T \tag{4.33}$$

The first equation represents a line with slope -1 and the second a horizontal line passing through the system point.

Just below the system point, $[\text{HAc}] = [\text{Ac}^-] = \frac{1}{2}C_T$ [from Eq (4.6)]. The log of $\frac{1}{2}C_T$ equals $\log C_T + \log 0.5$ or $\log C_T - 0.3$. Thus, the curves connecting the diagonal and horizontal lines intersect at 0.3 of a logarithmic unit below the system point.

Construction of Log Concentration Diagrams The construction of a log concentration–pH diagram is straightforward. Draw a horizontal line representing log C_T. Locate the system point at pH $=$ pK_A, and draw 45° lines sloping to the left and right through the system point. Locate a point 0.3 logarithmic units below the system point and connect the horizontal and 45° lines with short curves passing through this point. The lines for [H⁺] and [OH⁻] are drawn as 45° lines that intersect where pH equals 7 and log C equals -7. Thus, no involved numerical calculations are required. Changing the concentration of the solution simply shifts the [HAc] and [Ac⁻] curves up or down. Through use of transparent overlays, solutions to acid and base equilibria can be obtained rapidly.

Logarithmic Concentration Diagram for a Monoprotic Weak Base Figure 4.2 shows graphically the effect of pH on [NH₃] and [NH₄⁺] for a 0.01 M solution. In essence, the construction is similar to that for a weak acid except the system point is found from the relationship

$$\text{pH} = \text{p}K_W - \text{p}K_B \tag{4.34}$$

Since for ammonia, pK_B is 4.74, the pH of the system point is $14 - 4.74$ or 9.26.

Logarithmic Concentration Diagram for a Weak Acid and a Weak Base A logarithmic concentration diagram for a mixture containing 0.1 M ammonium hydroxide and 0.1 M acetic acid is shown in Fig. 4.3. These constituents are commonly found in natural waters and wastewaters. They may exist in relatively high

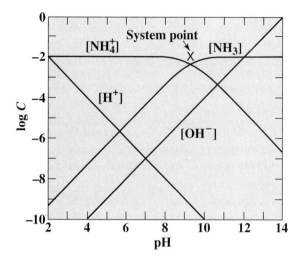

Figure 4.2
Logarithmic concentration diagram for 0.01 M ammonia
solution.

concentration in anaerobic digesters. The logarithmic concentration diagram is
made by superimposing the curves for each material on a single diagram.

Logarithmic Concentration Diagram for Polyprotic Acids and Bases   Logarithmic diagrams become even more useful for solution of equilibrium problems involving polyprotic acids and bases. The carbonic acid system, which is most impor-

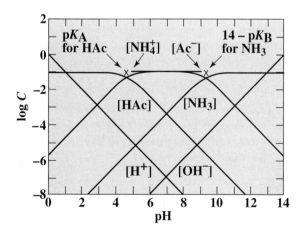

Figure 4.3
Logarithmic concentration diagram for a mixture of 0.1 M
acetic acid and 0.1 M ammonia.

tant for regulating the pH of most natural waters, will be used for illustration. The logarithmic diagram for a solution containing 0.01 M H_2CO_3 is illustrated in Fig. 4.4. The appropriate equations that apply are as follows:[5]

$$\frac{[H^+][HCO_3^-]}{[H_2CO_3^*]} = K_{A1} = 4.3 \times 10^{-7} \qquad pK_{A1} = 6.37$$

$$\frac{[H^+][CO_3^{2-}]}{[HCO_3^-]} = K_{A2} = 4.7 \times 10^{-11} \qquad pK_{A2} = 10.33$$

$$[H_2CO_3^*] + [HCO_3^-] + [CO_3^{2-}] = C_T = 10^{-2}$$

and

$$[H^+][OH^-] = K_W = 10^{-14} \qquad pK_W = 14$$

Solution of these equations for the individual carbonic acid species gives

$$[H_2CO_3^*] = \alpha_0 C_T = \frac{C_T}{1 + K_{A1}/[H^+] + K_{A1}K_{A2}/[H^+]^2} \qquad (4.35)$$

$$[HCO_3^-] = \alpha_1 C_T = \frac{C_T}{1 + [H^+]/K_{A1} + K_{A2}/[H^+]} \qquad (4.36)$$

$$[CO_3^{2-}] = \alpha_2 C_T = \frac{C_T}{1 + [H^+]/K_{A2} + [H^-]^2/K_{A1}K_{A2}} \qquad (4.37)$$

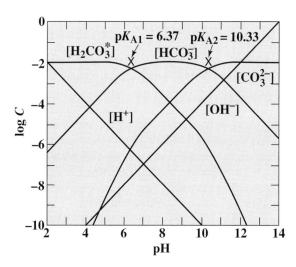

Figure 4.4

Logarithmic concentration diagram for 0.01 M carbonic acid.

---

[5]By convention, $[H_2CO_3^*]$ is taken to equal the sum of the actual carbonic acid concentration $[H_2CO_3]$ plus the dissolved carbon dioxide concentration $[CO_2(aq)]$.

It will be noted that the logarithmic diagram is constructed just as for monoprotic acids and bases, except that the slope of the line for $[CO_3^{2-}]$ changes from $+1$ to $+2$ when the pH drops below pK_{A1}. Also, the slope of the line for $[H_2CO_3^*]$ changes from -1 to -2 when the pH becomes greater than pK_{A2}. The reason for these slope changes is as follows. First, in reference to Eq. (4.37), when the pH drops below pK_{A1} ($[H^+] \gg K_{A1}$), the last term in the denominator becomes dominant, so

$$[CO_3^{2-}] \cong \frac{C_T}{[H^+]^2/K_{A1}K_{A2}}$$

and

$$\log [CO_3^{2-}] \cong \log C_T + \log K_{A1} + \log K_{A2} - 2 \log [H^+]$$

or

$$\log [CO_3^{2-}] \cong \log C_T - pK_{A1} - pK_{A2} + 2pH$$

Thus, the slope of the line for $\log [CO_3^{2-}]$ versus pH is 2. Similarly, when the pH exceeds pK_{A2}, the last term in the denominator of Eq. (4.35) dominates and

$$[H_2CO_3^*] \cong \frac{C_T}{K_{A1}K_{A2}/[H^+]^2}$$

The logarithmic form of this equation is

$$\log [H_2CO_3^*] = \log C_T + pK_{A1} + pK_{A2} - 2pH$$

The slope of the curve for $\log [H_2CO_3^*]$ versus pH is thus -2.

The construction of a logarithmic diagram for a diprotic acid or base is straightforward. System points are located at intersections between a horizontal line representing $\log C_T$ and pH values equal to pK_{A1} and pK_{A2} (or $pK_W - pK_{B1}$ and $pK_W - pK_{B2}$ for a base). Diagonal lines with slopes of $+1$ and -1 are drawn from each system point to a point located just below an adjacent system point. Beyond the adjacent system point, the slope of the line changes from -1 to -2, or $+1$ to $+2$, as the case may be.

For polyprotic acids or bases with more than two exchangeable protons, the procedure is similar except that the slope of a diagonal line drawn to the left of a system point changes from $+1$ to $+2$ when it reaches below the first adjacent system point and from $+2$ to $+3$ when and if it passes below a second adjacent system point. Proof of this is left to the reader. The logarithmic concentration diagram for phosphoric acid, a triprotic acid is illustrated in Fig. 4.5.

Example Problems Using Logarithmic Concentration Diagrams   The utility of logarithmic concentration diagrams in solving acid-base problems will be demonstrated using several examples. The student should remember that a log concentration–pH diagram is a picture of the mass balance and includes the equilibria involved (system points). In Examples 4.11 to 4.15, activity corrections will not be made. If activity corrections are deemed necessary, they will impact the location of the system point. Log concentration–pH diagrams are used with the charge balance and proton condition to solve acid-base problems.

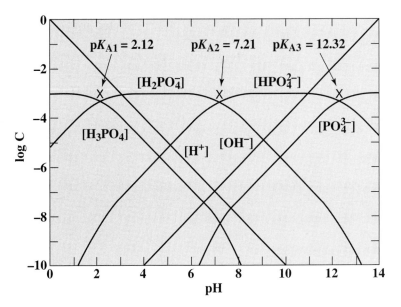

Figure 4.5
Logarithmic concentration diagram for 10^{-3} M phosphoric acid.

Estimate the equilibrium pH of a 25°C solution made by adding acetic acid to water to give a concentration of 0.02 M. The logarithmic concentration diagram shown in Fig. 4.1 can be used to solve this problem. The charge balance and proton condition (H_2O and HAc are the reference species) are identical:	**EXAMPLE 4.11**

$$[H^+] = [OH^-] + [Ac^-]$$

The method of solution is to locate a point on the log concentration–pH diagram where this equation is solved. Perhaps the easiest way to do this is to look at intersections of lines in the diagram involving the species in the charge balance. Inspection of the intersection of the $[H^+]$ and $[OH^-]$ line (pH = 7) indicates that the charge balance will not be solved at this pH since $[Ac^-]$ is much greater than 1×10^{-7}. The solution is then where the $[H^+]$ and $[Ac^-]$ lines intersect, which is at a pH of approximately 3.2. Please note that the mass balance between [HAc] and $[Ac^-]$ is solved at any pH on the diagram.

Estimate the equilibrium pH of a 25°C solution made by adding potassium acetate to water to give a concentration of 0.02 M. Ignore ionic strength corrections. The logarithmic concentration diagram shown in Fig. 4.1 can once again be used to solve this problem. As a matter of fact, Fig. 4.1 can be used to solve any problem where the combina-	**EXAMPLE 4.12**

tion of acetic acid and potassium acetate added totals 0.02 M. The charge balance for this problem is

$$[H^+] + [K^+] = [OH^-] + [Ac^-]$$

where $[K^+]$ is 0.02 M. The proton condition, with H_2O and Ac^- as the reference species, is

$$[H^+] + [HAc] = [OH^-]$$

In this case, the method of solution is to locate a point on the log concentration–pH diagram where the proton condition is solved and then check this solution using the charge balance. Inspection of the $[H^+]$ and $[OH^-]$ lines indicates that the proton condition will not be satisfied where these lines intersect at pH 7 since [HAc] there is much greater than 1×10^{-7}. The solution is then where the $[OH^-]$ and [HAc] lines intersect, which is at a pH of approximately 8.5. Checking this solution using the charge balance indicates that it is a good solution.

EXAMPLE 4.13

Estimate the equilibrium pH of a 25°C solution made by adding ammonium chloride to water to give a concentration of 0.01 M. Ignore ionic strength corrections. The logarithmic concentration diagram shown in Fig. 4.2 can be used to solve this problem. The charge balance for this problem is

$$[H^+] + [NH_4^+] = [OH^-] + [Cl^-]$$

where $[Cl^-]$ is 0.01 M. the proton condition, with H_2O and NH_4^+ as the reference species, is

$$[H^+] = [OH^-] + [NH_3]$$

In this case, the method of solution is to locate a point on the log concentration–pH diagram where the proton condition is solved and then check this solution using the charge balance. Inspection of the $[H^+]$ and $[OH^-]$ lines intersection at pH 7 indicates that the proton condition will not be satisfied here since $[NH_3]$ is much greater than 1×10^{-7}. The solution is then where the $[H^+]$ and $[NH_3]$ lines intersect, which is at a pH of approximately 5.8. Checking this solution using the charge balance indicates that it is a good solution.

EXAMPLE 4.14

Estimate the equilibrium pH of a 25°C solution made by adding ammonium chloride and acetic acid to water to give a concentration of 0.10 M for each. Ignore ionic strength corrections. The logarithmic concentration diagram shown in Fig. 4.3 can be used to solve this problem. The charge balance for this problem is

$$[H^+] + [NH_4^+] = [OH^-] + [Ac^-] + [Cl^-]$$

where $[Cl^-]$ is 0.10 M. The proton condition, with H_2O, HAc, and NH_4^+ as the reference species, is

$$[H^+] = [OH^-] + [Ac^-] + [NH_3]$$

In this case, the method of solution is to locate a point on the log concentration–pH diagram where the proton condition is solved and then check this solution using the charge balance. Inspection of the $[H^+]$ and $[OH^-]$ lines intersection at pH 7 indicates that the proton condition will not be solved here since $[Ac^-]$ and $[NH_3]$ are much greater than 1×10^{-7}. Similarly, the intersection of $[H^+]$ and $[NH_3]$ will not work since $[Ac^-]$ is too large. The solution is then where the $[H^+]$ and $[Ac^-]$ lines intersect, which is at pH of approximately 2.8. Checking this solution using the charge balance indicates that it is a good solution. The student should note that for this solution, the ionic strength is approximately 0.10 M, giving a γ of 0.76, which may be large enough to change the equilibrium pH. To check if this is true, the system point as defined by pK_A would need to be modified by using activity corrections and the pH scale changed to be $-\log [H^+]$. Alternatively, the problem could be worked using the algebraic methods described earlier with a pH of 2.8 as the initial estimate.

EXAMPLE 4.15

Estimate the equilibrium pH of a 25°C solution made by adding sodium bicarbonate ($NaHCO_3$) to water to give a concentration of 0.01 M. Ignore ionic strength corrections. The logarithmic concentration diagram shown in Fig. 4.4 can be used to solve this problem. The charge balance for this problem is

$$[H^+] + [Na^+] = [OH^-] + [HCO_3^-] + 2[CO_3^{2-}]$$

where $[Na^+]$ is 0.01 M. The proton condition, with H_2O and HCO_3^- as the reference species, is

$$[H^+] + [H_2CO_3^*] = [OH^-] + [CO_3^{2-}]$$

In this case, the method of solution is to locate a point on the log concentration–pH diagram where the proton condition is solved and then check this solution using the charge balance. Inspection of the $[H^+]$ and $[OH^-]$ lines (pH = 7) indicates that the proton condition will not be solved at this intersection since $[H_2CO_3^*]$ and $[CO_3^{2-}]$ are much greater than 1×10^{-7} and not equal to each other. Three other intersections can be checked. The solution is where the $[H_2CO_3^*]$ and $[CO_3^{2-}]$ lines intersect, which is at a pH of approximately 8.3. Checking this solution using the charge balance indicates that it is a good solution.

Open Versus Closed Systems

Up to now, we have been dealing with *closed* systems, that is, situations where we assume there is no interchange with the gas phase. For some acids and bases, such considerations are irrelevant; for others they are very important. For example, phosphoric acid (H_3PO_4) does not partition with the gas phase. On the other hand, carbon dioxide (CO_2), ammonia (NH_3), and hydrogen sulfide (H_2S) are gases that can partition between the gas and aqueous phases. In such cases, Henry's law is used to describe the gas-water equilibrium. In the following, CO_2 is used to show how such problems are solved. It is chosen because of the ubiquitous nature of the carbonate species and their importance in determining the pH of natural waters and waters being treated in engineered reactors.

The following are the equilibria (ignoring activity correction) and the mass balance of interest:

$$H_2CO_3^*(aq) \rightleftharpoons CO_2(g) + H_2O \qquad K_H = \frac{P_{CO_2}}{[H_2CO_3^*]} = 31.6 \frac{atm}{M} \qquad at\ 25°C$$

$$H_2CO_3^* \rightleftharpoons HCO_3^- + H^+ \qquad K_{A1} = 4.3 \times 10^{-7}$$

$$HCO_3^- \rightleftharpoons CO_3^{2-} + H^+ \qquad K_{A2} = 4.7 \times 10^{-11}$$

$$C_T = [H_2CO_3^*] + [HCO_3^-] + [CO_3^{2-}]$$

In a *closed* system, C_T remains constant with pH as shown in Fig. 4.4. In an *open* system, there is interchange between the gas and aqueous phases and at equilibrium $[H_2CO_3^*]$ is determined by the concentration of carbon dioxide in the gas phase ($[CO_2(g)]$), and is independent of pH:

$$[H_2CO_3^*] = P_{CO_2}/K_H$$

It is clear that while $[H_2CO_3^*]$ remains constant with changes in pH, C_T changes. As pH increases, both $[HCO_3^-]$ and $[CO_3^{2-}]$ increase. Thus, C_T increases with increasing

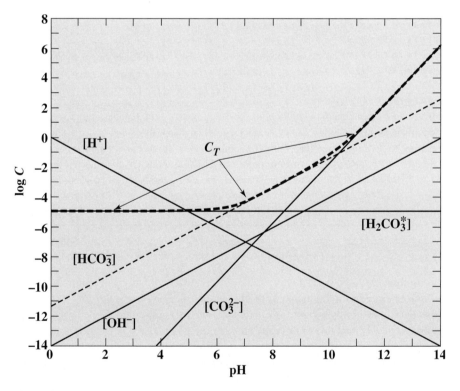

Figure 4.6
Logarithmic concentration diagram for carbonate species in an open system [$K_H = 31.6$ atm/M; $P_{CO_2} = 10^{-3.5}$ atm (316 ppm CO_2); $T = 25°C$].

pH. This is demonstrated in the log concentration–pH diagram for an open system shown in Fig. 4.6. The same solution techniques described for closed systems are used to solve equilibrium problems for open systems.

Calculate the pH of clean rain in equilibrium with an atmospheric CO_2 partial pressure of $10^{-3.5}$ atm (recent estimates indicate the average CO_2 content is 350 ppm or $10^{-3.46}$ atm). Ignore activity corrections.

EXAMPLE 4.16

The charge balance for this problem is, assuming clean rain has no other cations or anions,

$$[H^+] = [OH^-] + [HCO_3^-] + [CO_3^{2-}]$$

Since this is an open system, we can calculate $[H_2CO_3^*]$ from the following:

$$[H_2CO_3^*] = P_{CO_2}/K_H = 10^{-3.5}/31.6 = 1.00 \times 10^{-5} \text{ M}$$

We can now develop expressions for substituting into the charge balance:

$$[HCO_3^-] = \frac{K_{A1}[H_2CO_3^*]}{[H^+]} = \frac{(4.3 \times 10^{-7})(1.00 \times 10^{-5})}{[H^+]} = \frac{4.30 \times 10^{-12}}{[H^+]}$$

$$[CO_3^{2-}] = \frac{K_{A2}[HCO_3^-]}{[H^+]} = \frac{K_{A1}K_{A2}[H_2CO_3^*]}{[H^+]^2} = \frac{2.02 \times 10^{-22}}{[H^+]^2}$$

We can substitute these into the charge balance to develop an equation with a single unknown, $[H^+]$.

$$[H^+] = \frac{K_W}{[H^+]} + \frac{4.30 \times 10^{-12}}{[H^+]} + \frac{2.02 \times 10^{-22}}{[H^+]^2}$$

This equation can be solved using a spreadsheet as given here.

pH	[H⁺]	[OH⁻]	[HCO₃⁻]	[CO₃²⁻]	Left side	Right side	% error
5.000	1.00E-05	1.00E-09	4.30E-07	20.2E-12	1.00E-05	4.31E-07	95.69
6.000	1.00E-06	1.00E-08	4.30E-06	2.02E-10	1.00E-06	4.31E-06	−331.02
5.600	2.51E-06	3.98E-09	1.71E-06	3.20E-11	2.51E-06	1.72E-06	31.69
5.700	2.00E-06	5.01E-09	2.16E-06	5.07E-11	2.00E-06	2.16E-06	−8.26
5.670	2.14E-06	4.68E-09	2.01E-06	4.42E-11	2.14E-06	2.02E-06	5.71
5.680	2.09E-06	4.79E-09	2.06E-06	4.63E-11	2.09E-06	2.06E-06	1.26
5.690	2.04E-06	4.90E-09	2.11E-06	4.85E-11	2.04E-06	2.11E-06	−3.39
5.681	2.08E-06	4.80E-09	2.06E-06	4.65E-11	2.08E-06	2.07E-06	0.81
5.682	2.08E-06	4.81E-09	2.07E-06	4.67E-11	2.08E-06	2.07E-06	0.35
5.683	2.07E-06	4.82E-09	2.07E-06	4.69E-11	2.07E-06	2.08E-06	−0.11
5.6825	2.08E-06	4.81E-09	2.07E-06	4.68E-11	2.08E-06	2.07E-06	0.12
5.6828	**2.08E-06**	**4.82E-09**	**2.07E-06**	**4.69E-11**	**2.08E-06**	**2.08E-06**	**−0.02**

Thus, the pH of clean rain is approximately 5.68. In Fig. 4.6, this pH is represented by the intersection of the $[H^+]$ and $[HCO_3^-]$ lines. If $P_{CO_2} = 10^{-3.46}$ atm is used, the equilibrium pH decreases to 5.66.

EXAMPLE 4.17

A groundwater sample is taken from a geologic formation containing various carbonate minerals [e.g., $CaCO_3(s)$, $MgCO_3(s)$, $CaCO_3 \cdot MgCO_3(s)$] without exposing the water sample to the atmosphere. Analysis of the sample indicates that C_{T,CO_3} is 10^{-3} M, the pH is 7.6, and there are no other weak acids or bases in the sample. Ignore activity corrections.

(a) What is the acid neutralization capacity (ANC) of this water?

(b) This groundwater sample (no solids present) is then brought into equilibrium with the atmosphere ($P_{CO_2} = 10^{-3.5}$ atm). What is the resulting pH?

(c) ANC (also called *alkalinity*—more detail is contained in Chap. 18) can be determined from the following equation. The units of ANC are eq/L.

$$ANC = [OH^-] - [H^+] + \sum_{i=1}^{n} \text{eq/L conjugate bases of weak acids}$$

Here, the ANC is determined by OH^-, H^+, HCO_3^-, and CO_3^{2-} as follows:

$$ANC = [OH^-] - [H^+] + [HCO_3^-] + 2[CO_3^{2-}]$$
$$ANC = [OH^-] - [H^+] + C_{T,CO_3}(\alpha_1 + 2\alpha_2)$$
$$[H^+] = 10^{-7.6} = 2.51 \times 10^{-8} \text{ M}$$
$$[OH^-] = 10^{-6.4} = 3.98 \times 10^{-7} \text{ M}$$
$$C_{T,CO_3} = 10^{-3} \text{ M}$$

For pH = 7.6, α_1 [Eq.(4.36)] = 0.943 and α_2 [Eq.(4.37)] = 1.76×10^{-3}.

$$ANC = 3.98 \times 10^{-7} - 2.51 \times 10^{-8} + 10^{-3}(0.943 + 2(1.76 \times 10^{-3}))$$
$$ANC = 9.47 \times 10^{-4} \text{ eq/L} = 47 \text{ mg/L as } CaCO_3$$

(Equivalent weight of $CaCO_3$ is 50,000 mg per equivalent, see Chapter 18.)

(b) As always, we use the charge balance to develop the master equation:

$$C_B + [H^+] = [OH^-] + [HCO_3^-] + 2[CO_3^{2-}] + C_A$$

C_B is the sum of the strong-base cations (e.g., Na^+, Ca^{2+}), and C_A is the sum of the strong acid anions (eg., Cl^-, NO_3^-).

$$C_B - C_A = ANC = 9.47 \times 10^{-4} \quad \text{from solution to part } (a) \text{ before exposure to } CO_2$$

Please note that $C_B - C_A$ does not change with exposure to CO_2. So, the new charge balance becomes

$$9.47 \times 10^{-4} + [H^+] = [OH^-] + [HCO_3^-] + 2[CO_3^{2-}]$$

When the water is exposed to the atmosphere and at equilibrium:

$$[H_2CO_3^*] = P_{CO_2}/K_H = 10^{-3.5}/31.6 = 10^{-5} \text{ M}$$

It is also true that $[H_2CO_3^*] = \alpha_0 C_{T,CO_3}$

So, $C_{T,CO_3} = \dfrac{10^{-5}}{\alpha_0}$

Making the appropriate substitutions in the charge balance gives the following:

$$9.47 \times 10^{-4} = \frac{10^{-14}}{[H^+]} - [H^+] + \frac{10^{-5}}{\alpha_0}(\alpha_1 + 2\alpha_2)$$

This equation is solved using a spreadsheet and gives pH = 8.33. Apparently, CO_2 leaves the system and pH increases to maintain a constant ANC ($C_B - C_A$).

pH	[H$^+$]	α_0	α_1	α_2	Left side	Right side	% error
8	1.00E-08	2.26E-02	9.73E-01	4.57E-03	9.47E-04	4.35E-04	54.06
9	1.00E-09	2.22E-03	9.53E-01	4.48E-02	9.47E-04	4.71E-03	−397.80
8.3	5.01E-09	1.14E-02	9.79E-01	9.18E-03	9.47E-04	8.76E-04	7.49
8.4	3.98E-09	9.07E-03	9.79E-01	1.16E-02	9.47E-04	1.11E-03	−17.01
8.33	4.68E-09	1.07E-02	9.80E-01	9.84E-03	9.47E-04	9.40E-04	0.75
8.34	4.57E-09	1.04E-02	9.80E-01	1.01E-02	9.47E-04	9.62E-04	−1.61
8.332	4.66E-09	1.06E-02	9.80E-01	9.89E-03	9.47E-04	9.44E-04	0.28
8.333	4.65E-09	1.06E-02	9.80E-01	9.91E-03	9.47E-04	9.47E-04	0.04
8.334	4.63E-09	1.06E-02	9.80E-01	9.93E-03	9.47E-04	9.49E-04	−0.19

We can obtain a similar result from Fig. 4.6. First, we recognize for the charge balance that [H$^+$], [OH$^-$], and [CO$_3^{2-}$] would be small relative to [HCO$_3^-$], so that [HCO$_3^-$] about equals ANC. Thus, we find the location where log[HCO$_3^-$] = logANC = log(9.47 × 10^{-4}) = −3.02. This occurs approximately at pH 8.3.

Acid or Base Additions to Solution

When an acid or base is added to a solution, interactions between the different chemical species present will occur in such a way as to establish chemical equilibrium. The nature of the changes that will occur depends upon the nature and concentration of the chemical species present. These changes can be predicted by following the procedures outlined in Sec. 4.4. However, in addition to being cumbersome to carry out, such calculations often do not give an intuitive feel for the interactions that occur. In order to help understand the processes, the following discussion on acid-base titrations is presented. A *titration* is the procedure by which a measured amount of chemical or reagent is added to a solution in order to bring about a desired and measured change. Several examples will be given for illustration. In addition, simplifying assumptions to help solve the acid-base equilibria involved will be presented.

Titration of Strong Acids and Bases   When a strong base is titrated with a strong acid, the initial pH of the base is very high, usually in the range of 12 to 13. As acid is added, the pH changes very little at first and then slowly declines to a pH of about 10. From then on, the pH falls very rapidly until a pH of about 4 is reached; then changes become much more gradual. A plot of such data, commonly called a titration curve, is shown in Fig. 4.7. It may be noted that the curve is essentially vertical between pH values of 10 and 4; therefore, the stoichiometric end point or equivalence point lies between these values and, for all practical purposes, anywhere between them. The *equivalence point* is reached when the equivalents of base (or acid) added equal the equivalents of acid (or base) initially pre-

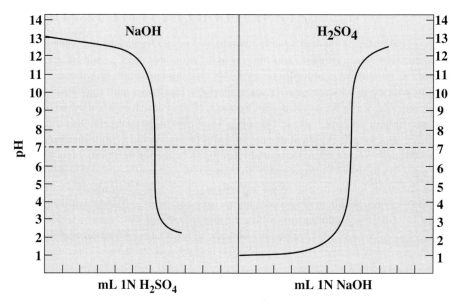

Figure 4.7
Titration curves for strong bases and acids.

sent. When a strong acid is titrated with a strong base, a curve quite similar in character is produced but is, of course, a mirror image of the former, as shown also in Fig. 4.7.

This discussion indicates the qualitative nature of strong acid and base titrations. It is desirable to illustrate also the quantitative changes that take place during the titration of a strong acid or base. Such calculations can be readily made if the strengths of the solutions are known. These calculations will not give exact values of pH when concentrations of ions rather than activities are used, but they are sufficiently close for most practical purposes.

Titration problems can be solved using the charge balance. Consider the case when the strong acid HCl is titrated with the strong base NaOH. The neutralization reaction in this case would be

$$NaOH + HCl \rightarrow H_2O + Na^+ + Cl^- \tag{4.38}$$

The charge balance is

$$[Na^+] + [H^+] = [OH^-] + [Cl^-]$$

If we let C_0 represent the molar concentration of HCl, V_0 be the volume of HCl to be titrated, C represent the molar concentration of NaOH, and V be the volume of NaOH added, the charge balance becomes

$$\frac{VC}{V_0 + V} + [H^+] = [OH^-] + \frac{V_0C_0}{V_0 + V} \tag{4.39}$$

Equation (4.39) is sometimes rearranged as follows:

$$\frac{V_0 C_0}{V_0 + V}\left(\frac{CV}{C_0 V_0} - 1\right) = [OH^-] - [H^+] \qquad (4.40)$$

$$\frac{V_0 C_0}{V_0 + V}(f - 1) = \frac{K_W}{[H^+]} - [H^+] \qquad (4.41)$$

where f equals $CV/C_0 V_0$ and is called the *equivalence fraction*. At the equivalence point, $f = 1$.

In many cases C is chosen to be much greater than C_0 so that $V_0 + V$ is about equal to V_0 in Eqs. (4.39) to (4.41). Example 4.18 illustrates the use of these equations.

EXAMPLE 4.18

Suppose 500 mL of 0.01 M HCl is titrated with 0.5 M NaOH. Calculate the pH values at different points in the titration to show the relative changes in pH with respect to the volume of titrating solution added. The neutralization reaction for this titration is given by Eq. (4.38). Equation (4.40) can be modified to calculate the pH as a function of the volume of NaOH added:

$$\frac{500(0.01)}{500 + V}\left(\frac{0.5V}{500(0.01)} - 1\right) = [OH^-] - [H^+]$$

At the beginning of the titration, the $[H^+]$ concentration of the 0.01 M acid solution is 0.01. The pH is equal to $-\log[H^+]$ or 2. If V is considered negligible relative to 500 mL, and $[OH^-]$ is assumed to be negligible relative to $[H^+]$, it is seen that when 5.0 mL of base is added, 50 percent of the HCl is neutralized ($f = 0.5$), and thus $[H^+] = 5 \times 10^{-3}$. The pH then would equal 2.3. By the time 9.0 mL of base is added, only 10 percent of the acid remains, and thus $[H^+] = 10^{-3}$ and the pH has increased only one unit to 3.

To reach the equivalence point ($f = 1$), 10 mL of base must be added. Here, the solution contains only NaCl and water. Since NaCl is a neutral salt, the pH value of the solution is 7.

Table 4.3 | Calculated values of pH in 500 mL of 0.01 M HCl to which different amounts of 0.5 M NaOH have been added

mL 0.5 M NaOH		HCl neutralized, %	Moles per liter		
Added	Excess		$[H^+]$	$[OH^-]$	pH
0.00	0.0	0	1.0×10^{-2}	1.0×10^{-12}	2.0
5.00	0.0	50	5.0×10^{-3}	2.0×10^{-12}	2.3
9.00	0.0	90	1.0×10^{-3}	1.0×10^{-11}	3.0
9.90	0.0	99	1.0×10^{-4}	1.0×10^{-10}	4.0
9.99	0.0	99.9	1.0×10^{-5}	1.0×10^{-9}	5.0
9.999	0.0	99.99	1.0×10^{-6}	1.0×10^{-8}	6.0
10.00	0.0	100	1.0×10^{-7}	1.0×10^{-7}	7.0
10.10	0.1	100	1.0×10^{-10}	1.0×10^{-4}	10.0
11.00	1.0	100	1.0×10^{-11}	1.0×10^{-3}	11.0
20.00	10.0	100	1.0×10^{-12}	1.0×10^{-2}	12.0

If the equivalence point is passed, $[OH^-]$ is larger than $[H^+]$, and Eq. (4.41) becomes

$$\frac{500(0.01)}{500 + V}\left(\frac{0.5V}{500(0.01)} - 1\right) \cong \frac{K_W}{[H^+]}$$

To illustrate, if 11 mL of 0.5 M NaOH are added, $[H^+]$ is 10^{-11} and the pH equals 11.

These and other values for this titration are listed in Table 4.3. By the time the pH has increased to 4, we see that 99 percent of the acid has been neutralized, and so for practical purposes the titration can be considered complete. After this point is reached, it requires only an additional 0.2 mL of NaOH to increase the pH to 10.

Titration of Weak Acids and Bases   When weak acids are titrated with strong bases, the character of the titration curve depends upon whether the acid is monobasic or polybasic, i.e., whether it yields one or more hydrogen ions. The initial pH of solutions of poorly ionized acids depends largely upon the degree of ionization, and the titration curves vary markedly, as shown in Fig. 4.8. Since the pH of natural waters if influenced mainly by weak acids and their salts, it is good to have a close familiarity with their chemistry. For this reason, it is desirable to develop the mathematical relationships which describe the influence of addition of strong acids or bases on the pH of water containing weak acids and bases.

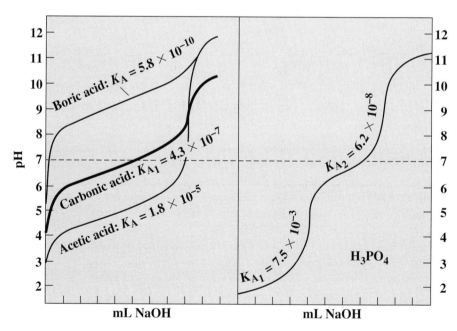

Figure 4.8
Titration curves for weak acids.

These same relationships will also indicate the factors affecting the pH of the equivalence point of a titration.

Consider first the titration of a weak monobasic acid with a strong base. The acid ionizes as follows:

$$HA \rightleftharpoons H^+ + A^- \tag{4.42}$$

The equilibrium relationships (ignoring ionic strength effects) that must be considered in evaluating the pH during the titration are those for the acid and for water:

$$\frac{[H^+][A^-]}{[HA]} = K_A \tag{4.43}$$

$$[H^+][OH^-] = K_W \tag{4.44}$$

When the weak acid is titrated with a strong base, the following neutralization takes place:

$$HA + B^+ + OH^- \rightarrow B^+ + A^- + H_2O \tag{4.45}$$

Here B^+ represents the cation associated with the strong base. For strong bases such as NaOH and KOH, the cations Na^+ and K^+ remain ionized in solution. The charge balance for this situation is

$$[H^+] + [B^+] = [A^-] + [OH^-] \tag{4.46}$$

The $[H^+]$ at any point during the titration of a monoprotic weak acid with a strong base can be determined by a simultaneous solution of Eqs. (4.43), (4.44), and (4.46).

The approach used for strong acid–strong base titrations can be applied here also. If C_0 and V_0 represent the molar concentration and initial volume for the weak acid and C and V represent the molar concentration and volume for the strong base being added, the following relationships can be written.

The mass balance on the weak acid is

$$[HA] + [A^-] = \frac{C_0 V_0}{V_0 + V} \tag{4.47}$$

The mass balance on the strong base cation is

$$[B^+] = \frac{CV}{V_0 + V} \tag{4.48}$$

Additionally, we can define the following by combining the mass balance and equilibrium expression for the weak acid:

$$\alpha_0 = \frac{[HA](V_0 + V)}{C_0 V_0} = \left(1 + \frac{K_A}{[H^+]}\right)^{-1} \tag{4.49}$$

$$\alpha_1 = \frac{[A^-](V_0 + V)}{C_0 V_0} = \left(1 + \frac{[H^+]}{K_A}\right)^{-1} \tag{4.50}$$

The charge balance now becomes

$$\frac{CV}{V_0 + V} + [H^+] = [OH^-] + \alpha_1\left(\frac{C_0 V_0}{V_0 + V}\right) \tag{4.51}$$

Rearranging gives

$$\frac{CV}{C_0 V_0} = f = \alpha_1 + \left(\frac{K_W}{[H^+]} - [H^+]\right)\left(\frac{V_0 + V}{C_0 V_0}\right) \tag{4.52}$$

This equation can now be used to generate the titration curve. Simplifying assumptions similar to those used for Examples 4.7 through 4.15 can be used here also.

EXAMPLE 4.19

Consider the titration of 500 mL of 0.01 M acetic acid with 0.5 M NaOH. Calculate the pH at different points in the titration to show the relative changes in pH with respect to the volume of titrating solution added. Ignore activity corrections.

At the beginning of the titration the problem is the same as Examples 4.2 and 4.7. The assumption made in Example 4.7 was that $[H^+]$ was much greater than $[OH^-]$ and a quadratic equation was developed and solved, giving an equilibrium pH of 3.38.

After 1 mL of 0.5 M NaOH is added, $f = 0.1$ and, ignoring V relative to V_0, Eq. (4.52) becomes

$$0.1 = \alpha_1 + \left(\frac{K_W}{[H^+]} - [H^+]\right)\left(\frac{1}{0.01}\right)$$

Recognizing that $[H^+]$ is likely to be greater than $[OH^-]$, the following equation can be developed:

$$0.001 = 0.01\alpha_1 - [H^+]$$

Substituting for α_1 and rearranging this equation gives

$$[H^+]^2 + 1.02 \times 10^{-3}[H^+] - 1.62 \times 10^{-7} = 0$$

Solving this quadratic equation gives a pH of 3.87.

When 10 mL of 0.5 M NaOH have been added, $f = 1.0$, and the equivalence point has been reached. Here the solution can now be considered to be a 0.01 M solution of sodium acetate. As was the case in Example 4.9, we can now assume that $[OH^-]$ is greater than $[H^+]$, and Eq. (4.52) becomes

$$0.0098 = 0.0098\alpha_1 + \frac{K_W}{[H^+]}$$

Solving this equation for $[H^+]$ gives a pH of 8.37.

pH values associated with different volumes of 0.5 M NaOH can be similarly calculated. A summary of such calculations is given here.

V, mL	f	pH
0.0	0.0	3.38
1.0	0.1	3.87
2.5	0.25	4.37
5.0	0.5	4.75
7.5	0.75	5.22
9.0	0.9	5.70
10	1.0	8.37
10.1	1.01	10.0
11	1.1	11.0
12	1.2	11.3
20	2.0	12.0

If certain additional approximations are made for Example 4.19, the calculations can be further simplified. This will be illustrated in the estimation of the pH at the beginning, the midpoint, and the equivalence point in the titration of a solution containing C_0 moles per liter of a weak acid, HA, with a strong base, $B^+ + OH^-$.

Beginning of Titration   In the initial solution containing the weak acid alone, $[B^+] = 0$. Also, the pH will be low so that $[OH^-] \ll [H^+]$. Therefore, from the charge balance [Eq. (4.46)], we see that $[H^+]$ and $[A^-]$ are approximately equal:

$$[H^+] \cong [A^-] \tag{4.53}$$

Substituting $[H^+]$ for $[A^-]$ in Eq. (4.43) and solving for $[H^+]$ yields

$$[H^+] \cong \sqrt{K_A[HA]} = \sqrt{K_A(C_0 - [A^-])} \tag{4.54}$$

For many cases, at the beginning of the titration, $[A^-] \ll C_0$, and the following approximation is sufficient:

$$[H^+] \cong \sqrt{K_A C_0} \tag{4.55}$$

From this the pH becomes

$$pH = -\log [H^+] \cong -\tfrac{1}{2}\log K_A - \tfrac{1}{2}\log C_0$$

or $\qquad\qquad pH \cong \tfrac{1}{2}(pK_A - \log C_0) \tag{4.56}$

EXAMPLE 4.20

Calculate the pH of a 0.01 M (600 mg/L) acetic acid solution.
    By approximate Eq. (4.56),

$$pH \cong \tfrac{1}{2}(pK_A - \log C_0) = \tfrac{1}{2}(4.74 + 2) = 3.37$$

Check: $[Ac^-] = [H^+] = 0.0004 \ll 0.01$; therefore, the approximation is good. (In the case of weak acids, the $[H^+]$ must be significantly less than the molar concentration of the acid for the approximation to be good. A comparable situation exists for bases.)

EXAMPLE 4.21

Calculate the pH of a 0.01 M phosphoric acid solution.

By approximate Eq. (4.56),

$$pH \cong \tfrac{1}{2}(pK_A - \log C_0) = \tfrac{1}{2}(2.12 + 2) = 2.06$$

Check: $[H_2PO_4^-] = [H^+] \cong 0.01 = C_0$; therefore, the approximation is poor. Use the better approximation given by Eq. (4.54) as follows:

$$[H^+] = [H_2PO_4^-] = \sqrt{K_A(0.01 - [H_2PO_4^-])}$$
$$[H_2PO_4^-]^2 = (7.5 \times 10^{-5}) - (7.5 \times 10^{-3})[H_2PO_4^-]$$
$$[H_2PO_4^-] = \frac{-7.5 \times 10^{-3} \pm \sqrt{(7.5 \times 10^{-3})^2 + 4(7.5 \times 10^{-5})}}{2}$$
$$= 5.7 \times 10^{-3} \text{ mol/L}$$
$$[H^+] = [H_2PO_4^-] = 5.7 \times 10^{-3} \text{ mol/L}$$
$$pH = 2.25$$

The error from use of approximate Eq. (4.56) was 0.2 of a pH unit. However, the error from using approximate Eq. (4.54) is less than 0.01 pH unit.

Midpoint of Titration   As the titration proceeds, $[A^-]$ increases and $[HA]$ decreases, as indicated by Eq. (4.45). When the neutralization is 50 percent complete, $[B^+] \cong \tfrac{1}{2}C_0$. Also, $[B^+]$ and $[A^-]$ become quite significant in concentration, and thus $[H^+] \ll [B^+]$ and $[OH^-] \ll [A^-]$. Therefore, from the charge balance [Eq. (4.46)],

$$[B^+] \cong [A^-] \cong \tfrac{1}{2}C_0 \qquad (4.57)$$

Also, since

$$[HA] + [A^-] = C_0$$

we have

$$[HA] \cong \tfrac{1}{2}C_0$$

Placing the above values of $[HA]$ and $[A^-]$ in Eq. (4.43) gives

$$\frac{[H^+](\tfrac{1}{2}C_0)}{\tfrac{1}{2}C_0} \cong K_A$$

Thus,

$$[H^+] \cong K_A \qquad (4.58)$$

and

$$pH \cong pK_A \qquad (4.59)$$

Equation (4.59) indicates the value of expressing the ionization constant in the pK_A form. The pK_A indicates the pH that will result when a weak acid is half neutralized. The weaker the acid, the higher the pK_A value, and thus the higher the pH at the half neutralization point.

Equivalence Point of Titration   At the equivalence point of the titration, the equivalents of base added equal the equivalents of acid in the original solution ($f =$

1), and thus $[B^+] = C_0$. Also, the pH is usually sufficiently high that $[H^+] \ll [OH^-]$. Therefore, from the charge balance [Eq. (4.46)],

$$C_0 \cong [A^-] + [OH^-] \qquad (4.60)$$

and

$$C_0 - [A^-] \cong [OH^-]$$

but

$$C_0 - [A^-] = [HA]$$

and thus

$$[HA] \cong [OH^-] = \frac{K_W}{[H^+]} \qquad (4.61)$$

Substituting Eq. (4.61) in Eq. (4.43) results in

$$\frac{1}{[H^+]} = \left[\frac{[A^-]}{K_W K_A} \right]^{1/2} \qquad (4.62)$$

Since at the end of the titration, $[A^-] \cong C_0$, the following approximate solution for the equivalence point pH results:

$$pH \cong \tfrac{1}{2}(\log C_0 + pK_A + pK_W) \qquad (4.63)$$

This equation indicates that the pH at the equivalence point of the titration depends on the concentration of the acid as well as on its equilibrium constant. Frequently for practical work, a pH for the equivalence point of a titration can be chosen that will give sufficiently accurate results over a wide range of concentrations. In other cases, it may be necessary to have an idea of the concentration so that the best pH can be chosen. It is evident from Eq. (4.63) that the weaker the acid and the higher its concentration, the higher the equivalence point pH.

In the titration of a weak base with a strong acid, the pH at various points can be determined in a fashion similar to that for a weak acid. It should be obvious, however, that such a titration is the mirror image of the acid titration and that the equations for the acid titration are applicable if pOH is substituted for pH and if pK_B is substituted for pK_A. A summary of the approximate equations for pH for both weak acid and weak base titrations is given here.

Weak acid pH during titration with strong base:

Beginning of titration	$pH \cong \tfrac{1}{2}(pK_A - \log C_0)$	(4.56)
Midpoint of titration	$pH \cong pK_A$	(4.59)
Equivalence point of titration	$pH \cong \tfrac{1}{2}(\log C_0 + pK_A + pK_W)$	(4.63)

Weak base pH during titration with strong acid:

Beginning of titration	$pH \cong pK_W - \tfrac{1}{2}pK_B + \tfrac{1}{2}\log C_0$	(4.64)
Midpoint of titration	$pH \cong pK_W - pK_B$	(4.65)
Equivalence point of titration	$pH \cong \tfrac{1}{2}(pK_W - pK_B - \log C_0)$	(4.66)

The inflection at the equivalence point in the titration of weak acids or bases with ionization constants less than about 10^{-7} or pK values greater than 7 is not sharp and so is difficult to detect accurately.

This is illustrated in Fig.4.8, showing the titration curves for three different weak acids with significantly different ionization constants. Solutions of acetic acid normally have an initial pH of about 3. During titration with a strong base the pH increases slowly to about 7, and then rapidly until pH 10 is reached. The equivalence point is usually between pH 8 and 9, on the rapidly rising portion of the curve. In the case of carbonic acid, the ionization constant of 10^{-7} is much smaller than the 10^{-5} for acetic acid, and so the initial pH is much higher. Also, for the same reason, the midpoint and equivalence point for carbonic acid are much higher. The pH for the carbonic acid titration does not break sharply until a pH of 8 is reached. All weak acids with ionization constants greater than 10^{-7} and in the concentrations normally encountered in practice show inflections in the curve or equivalence points at a pH of about 8.5. Because carbonic acid has a second ionization constant, its titration curve breaks away to the right at pH 8.5 from the near vertical line followed by acetic acid. The titration curve for such polyprotic weak acids will be described in more detail for phosphoric acid.

Boric acid is a good example of a very weak acid with an ionization constant much less than 10^{-7}. This acid is not completely neutralized until a pH of about 11 is reached. The titration curve for boric acid indicates that the inflection point at this pH is barely detectable, and an accurate measurement of the boric acid concentration by such a titration is not feasible.

The titration curve for phosphoric acid illustrates very well the behavior during titration of a polyprotic weak acid, yielding more than one measurable hydrogen ion (proton). Reference to Fig. 4.8 will show that the first ionization of phosphoric acid is similar to that of a strong acid and that the resulting hydrogen ions are neutralized by the time sufficient base has been added to reach a pH of about 4. The hydrogen ion resulting from the second step of ionization is neutralized by the time the pH has been raised to about 8.5. The third hydrogen ion of phosphoric acid ($K_{A3} = 4.8 \times 10^{-13}$), like the second of carbonic acid, is not measurable by ordinary titrations. Calculations of pH during the titration of polyprotic acids are more complex than for monoprotic acids, as the equilibrium relationships for each ionization must be considered simultaneously for an exact solution. Without going into the details, it is sufficient to indicate that the pH of the equivalence point for each ionization is approximately equal to the average of the pK_A value for that ionization and the pK_A value for the following ionization. For example, with phosphoric acid, $pK_{A1} = 2.12$ and $pK_{A2} = 7.21$, and the equivalence point pH between the first and second ionization equals about (2.12 + 7.21)/2 or 4.67.

Typical titration curves for weak bases are shown in Fig. 4.9. Weak bases with ionization constants greater than 10^{-7} and in the concentrations normally encountered in practice have equivalence points at a pH of 4 or higher. Salts of strong bases and weak acids, such as sodium carbonate and sodium acetate, are alkaline in character and behave like bases during titration (see Fig.4.9). When considered as such, their ionization can be considered to take place as shown here for acetate:

$$CH_3COO^- + H_2O \rightleftharpoons CH_3COOH + OH^-$$

and the equilibrium relationship becomes

$$\frac{[OH^-][CH_3COOH]}{[CH_3COO^-]} = K_B$$

It follows that K_B is equal to K_W/K_A, where K_A is the ionization constant for acetic acid. It also follows that pK_B for the salt formed from a weak acid and a strong base equals $pK_W - pK_A$. With this notation, Eqs. (4.64) to (4.66) for titration of a weak base may be used to evaluate the pH during titration of salts of weak acids. Titration curves for salts of weak acids are simply acid titration curves in reverse, and the curve is a mirror image of that for the acid. This is illustrated by comparison of the titration curves for acetic acid and carbonic acid (Fig. 4.8) with that for their salts, sodium acetate and sodium bicarbonate, respectively (Fig. 4.9).

The titration curve for sodium carbonate is characteristic of a base with two ionizations. The initial pH of its solution is rather high, and during titration with a strong acid, neutralization occurs in two steps, corresponding to each ionization. Addition of strong acid results in a gradual drop in pH, with a poorly defined inflection in the curve at a pH of about 8.5. This corresponds to the equivalence point for the conversion of carbonate ion to bicarbonate ion, as follows:

$$CO_3^{2-} + H^+ \rightarrow HCO_3^-$$

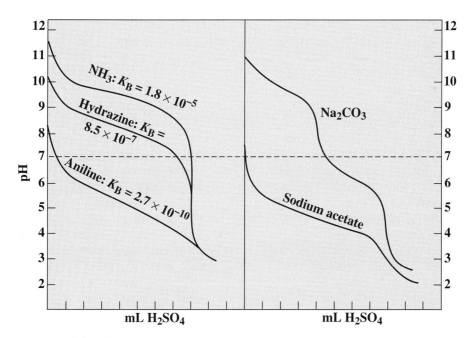

Figure 4.9
Titration curves for weak bases and for salts of weak acids.

Further addition of acid results in a gradual lowering of the pH until a value of about 5 is reached. The curve then passes through an inflection at a pH of about 4.5. This corresponds to the equivalence point for the conversion of bicarbonate ion to carbonic acid:

$$HCO_3^- + H^+ \rightarrow H_2CO_3$$

EXAMPLE 4.22

A solution containing 34 mg/L (0.001 M) H_2S is titrated with a strong NaOH solution. Calculate the pH at the beginning, the midpoint, and the equivalence point for the first ionization of the acid. Assume no H_2S interchange with the atmosphere.

The initial pH, by approximate Eq. (4.56), is

$$pH \cong \tfrac{1}{2}(pK_A - \log C_0) = \tfrac{1}{2}(7.04 + 3.0) = 5.02$$

Check: $[H^+] = [HS^-] = 10^{-5} \ll 10^{-3}$; therefore, the approximation is good.
The midpoint pH, by approximate Eq. (4.59), is

$$pH \cong pK_{A1} = 7.04$$

The equivalence point pH, since this is a diprotic acid, can be found as follows:

$$pH \cong \tfrac{1}{2}(pK_{A1} + pK_{A2}) = \tfrac{1}{2}(7.04 + 12.89) = 9.97$$

Compare with the value given by Eq. (4.63),

$$pH \cong \tfrac{1}{2}(\log C_0 + pK_A + pK_W) = \tfrac{1}{2}(-3 + 7.04 + 14) = 9.02$$

Since pH 9.02 would be reached before pH 9.97, pH 9.0 would be satisfactory for use as the equivalence point.

EXAMPLE 4.23

Calculate the pH at the beginning, midpoint, and equivalence point for the titration of a solution containing 10^{-4} mol/L of sodium propionate with H_2SO_4.

Since sodium propionate is the salt of a weak acid, CH_3CH_2COOH, and a strong base, NaOH, it acts like a weak base,

$$pK_B = pK_W - pK_A = 14 - 4.89 = 9.11$$

The initial pH, by Eq.(4.64), is

$$pH \cong pK_W - \tfrac{1}{2}pK_B + \tfrac{1}{2}\log C_0 = 14 - \tfrac{1}{2}(9.11) + \tfrac{1}{2}(-4) = 7.45$$

The midpoint pH, by Eq. (4.65), is

$$pH \cong pK_W - pK_B = 14 - 9.11 = 4.89$$

The equivalence point pH, using Eq. (4.66), is

$$pH \cong \tfrac{1}{2}(pK_W - pK_B - \log C_0) = \tfrac{1}{2}(14 - 9.11 + 4.0) = 4.45$$

A check reveals that all these approximate equations are satisfactory for use.

Use of Log Concentration Diagrams for Titrations   The graphical approach can be used to determine the equivalence point for acid-base titrations, and the pH at intermediate points during a titration. We will use Fig. 4.1 as an example. At all points, Eqs. (4.26) to (4.29) must be satisfied. In addition, a charge balance must be maintained:

$$[B^+] + [H^+] = [OH^-] + [Ac^-] \qquad (4.67)$$

Since a strong base is completely ionized, $[B^+]$ will equal the molar concentration of the base added if a monoprotic base such as NaOH is used for the titration. Initially, before NaOH is added, the condition will be

$$[H^+] = [OH^-] + [Ac^-] \qquad (4.68)$$

Since for acid conditions, $[OH^-]$ will be very small compared with $[H^+]$, Eq. (4.68) can be approximated as

$$[H^+] \cong [Ac^-] \qquad (4.69)$$

Thus, the initial pH of the solution is approximated in Fig. 4.1 by the intersection between the curves for $[H^+]$ and $[Ac^-]$. This occurs at pH 3.2. As base is added, $[B^+]$ soon becomes much greater than $[H^+]$ or $[OH^-]$, so Eq. (4.67) can be approximated as

$$[B^+] \cong [Ac^-] \qquad (4.70)$$

and the pH corresponding to a given base addition can be found by locating the point on the graph where Eq.(4.70) is satisfied. At the midpoint in the titration, $[B^+] = \frac{1}{2}C_0 = [Ac^-] = [HAc]$. Thus, the midpoint is located at pK, the system point pH. Finally, the equivalence point is reached when $[B^+] = C_0 = [HAc] + [Ac^-]$, which when substituted into Eq. (4.67) gives

$$[HAc] + [H^+] = [OH^-] \qquad (4.71)$$

Under the more basic conditions, $[OH^-]$ is much greater than $[H^+]$, so the equivalence point pH is approximated by the intersection between the $[HAc]$ and $[OH^-]$ lines. From Fig. 4.1, the equivalence point occurs at a pH of approximately 8.5.

Using this approach, the initial point, midpoint, and equivalence point for a titration are represented approximately by the intersection of curves and can readily be found. The effect of changes in the concentration of weak acid on these different points can be observed by moving the horizontal $[HAc]$ and $[Ac^-]$ curves up or down. Moving the curves up represents an increase in weak acid concentration and results in a decrease in the initial pH and an increase in the equivalence point pH. The pH of the midpoint of the titration, of course, does not change.

Using ammonium hydroxide (ammonia) as an example (Fig 4.2), the initial pH and equivalence points for titration of a weak base with a strong acid are seen to be analogous to those for a weak acid and strong base. The initial pH of a 0.01 M ammonium hydroxide solution is given by the intersection of the $[NH_4^+]$ and $[OH^-]$ lines and would equal about 10.6. The equivalence point of a titration with strong acids is given by the intersection of $[NH_3]$ and $[H^+]$ and would equal about 5.7.

Figures 4.4 and 4.5 illustrate the influence of pH on the species present in solutions of two different polyprotic weak acids. For example, the pH resulting from carbonic acid alone is approximated by the intersection between the lines for $[HCO_3^-]$ and $[H^+]$. If this solution is titrated with NaOH, then two equivalence points are reached. The first after the addition of an equivalent of 0.01 M of Na^+ is represented by the intersection of the $[H_2CO_3^*]$ and $[CO_3^{2-}]$ lines at pH 8.3. The second is reached after an additional 0.01 M Na^+ is added and is represented approximately by the intersection of the $[OH^-]$ and the $[HCO_3^-]$ lines at pH 11.1. The pH values for all intermediate points in the titration are represented by points that satisfy the charge-balance relationship

$$[Na^+] + [H^+] = [HCO_3^-] + 2[CO_3^{2-}] + [OH^-]$$

Solutions for intermediate points can be found by trial and error, using the logarithmic concentration diagram as a direct aid.

More complicated titration problems can also be solved using log concentration diagrams. Consider Fig. 4.3 as representing a solution containing 0.1 M ammonium hydroxide and 0.1 M acetic acid. By itself, the diagram illustrates the concentration of the various species present if the pH is varied by the addition of a strong base (NaOH) or a strong acid (HCl). The particular condition that will result from base or acid addition can be obtained, as usual, by imposing the charge-balance condition:

$$[H^+] + [NH_4^+] + [Na^+] = [OH^-] + [CH_3COO^-] + [Cl^-]$$

where $[Na^+]$ and $[Cl^-]$ represent the molar concentrations of NaOH and HCl added, respectively.

For the initial condition when acetic acid and ammonium hydroxide are added to distilled water, $[Na^+]$ and $[Cl^-]$ are zero, and $[H^+]$ and $[OH^-]$ are very small relative to $[NH_4^+]$ and $[CH_3COO^-]$. The resulting pH is approximated by the point where the $[NH_3]$ and $[HAc]$ curves intersect and equals about 7.0.

More complex systems involving several weak acids and bases can be solved in a similar manner. The respective curves are superimposed on one another and then the charge-balance condition is imposed in order to obtain a solution to a particular problem. A trial-and-error solution through direct use of the logarithmic concentration diagram is relatively rapid.

4.6 | BUFFERS

Buffers may be defined as *substances in solution that offer resistance to changes in pH* as acids or bases are added to or formed within the solution. There are many occasions in practice when it is desirable to use buffered solutions to closely maintain a desired pH level. Buffer solutions usually contain mixtures of weak acids and their salts *(conjugate bases)* or weak bases and their salts *(conjugate acids)*. The fundamental basis for an understanding of buffer action was developed in Sec. 4.5. Figs. 4.8 and 4.9 illustrate that at the midpoint in the titration of a weak acid or base, the slope of the titration curve is at a minimum. Thus, at this point the smallest change in pH occurs for a given volume of titrant added, and hence at this point the *buffering capac-*

ity is the greatest. At the midpoint the solution contains equal quantities of the ionized and the un-ionized acid or base. As indicated by Eqs. (4.59) and (4.65), the pH at the midpoint in the titration is given by pK_A for weak acids and by $pK_W - pK_B$ for weak bases. Thus, the pK values listed in Tables 4.1 and 4.2 can be used to indicate the pH at which the various acids and bases have the most effective buffering capacity.

Although weak acids and bases and their salts are most effective as buffers at a pH near the pK value, they may also be used effectively within ± 1 pH unit of the pK value by changing the relative concentration of acid to salt or base to salt. To illustrate, consider the equilibrium relationship for a weak acid (ignoring activity corrections),

$$K_A = \frac{[H^-][A^-]}{[HA]}$$

By rearranging, we obtain

$$\frac{1}{[H^+]} = \frac{1}{K_A} \frac{[A^-]}{[HA]}$$

Thus,

$$pH = pK_A + \log \frac{[salt]}{[acid]} \tag{4.72}$$

Equation (4.72) indicates that the pH of a buffer solution made of a weak acid and its salt depends upon the *ratio* of salt concentration to acid concentration. Thus, solutions of weak acids and their salts will have the same pH regardless of concentration as long as the *ratio* of concentrations remains the same. However, from the standpoint of buffering capacity, the larger the actual concentrations of salt and acid, the less they will be numerically changed by the addition of a given amount of acid or base and the more effective is the buffering action.

The student should also note that Eq. (4.72) can be derived by combining the equilibrium expression for the weak acid and its conjugate base (ignoring activity corrections), the mass balance, and the charge balance, and assuming that the concentrations of the weak acid and conjugate base added are likely to be much higher than the concentrations of $[H^+]$ and $[OH^-]$. Thus, Eq. (4.72) is a good one for concentrated buffers.

The pK value for the second ionization of phosphoric acid is very near 7 (7.21), and hence salts of phosphoric acid are quite useful substances to buffer solutions near a neutral pH. They are commonly used in analytical tests. Of special interest is the use of phosphate salts to maintain a neutral pH in the biochemical oxygen demand (BOD) test (Chapter 23). Here the monobasic potassium salt (KH_2PO_4) is used as the acid, and both the dibasic potassium salt (K_2HPO_4) and sodium salt (Na_2HPO_4) are used as the salt of the acid. These salts ionize in solution to establish the following equilibrium:

$$H_2PO_4^- \rightleftharpoons H^+ + HPO_4^{2-}$$

Although the buffering capacity of natural waters is largely due to salts of carbonic acid, one should remember that alkalinity determinations measure the buffering capacity of all salts of weak acids (see Chap. 18). This is particularly pertinent in the analysis of industrial wastes. The best way of evaluating the buffer capacity of an industrial waste is to perform an electrometric titration, using a standard acid or base. If observations of pH versus titrant additions are made, curves can be plotted that show its capacity and the pH range over which the buffer is especially effective.

Buffer capacity is an important characteristic of wastes that are submitted to biological treatment. The oxidation of neutral compounds—sugar, for example—results in the production of organic acids as intermediates. If the buffering capacity is not sufficient, the pH may fall to levels that inhibit the action of the bacteria. In an instance in which formaldehyde was involved, the pH was reduced to 4.5 within a matter of minutes, and the process was considered a failure until adequate buffer was applied. The initiation of anaerobic digestion to produce methane from sludge is hampered by limitations of buffer capacity. This process frequently results in the formation of considerable quantities of organic acids, and liming of digesters is often practiced to maintain favorable pH conditions. In effect, the lime combines with carbon dioxide and water to form calcium bicarbonate, a salt of carbonic acid. Hence, an understanding of the buffering action of carbonic acid and its salts is of value in digester control.

In the oxidation of ammonia, nitrous and nitric acids are formed. These must be neutralized to maintain a favorable environment for nitrifying bacteria. Acid rain inputs to natural waters need to be neutralized to prevent adverse environmental effects (e.g., fish kills). When chlorine is used for drinking water disinfection, it can react with ammonia to produce acids which need to be neutralized. The bicarbonates in natural waters, along with other salts of weak acids, serve this purpose. If they are not present in sufficient amounts, alkaline materials such as lime or sodium hydroxide must be added to maintain favorable pH.

EXAMPLE 4.24

Calculate the pH of a buffer solution containing 0.01 M acetic acid and 0.01 M sodium acetate. Then calcuate the pH after enough HCl is added to give a concentration of 0.001 M.

From Eq. (4.72),

$$\text{pH} = pK_A + \log\frac{[0.01]}{[0.01]} = 4.74 + 0 = 4.74$$

When the HCl is added, and we assume that all the HCl dissociates, Eq. (4.72) then becomes

$$\text{pH} = 4.74 + \log\frac{0.01 - 0.001}{0.01 + 0.001} = 4.74 - 0.09 = 4.65$$

A check of this solution using the charge balance shows that it is a good solution.

4.7 | BUFFER INDEX

The buffering capacity of a solution can be indicated quantitatively by the *buffering index* (sometimes called buffer intensity) β, which is defined as the slope of a titration curve of pH versus moles of strong base added (C_B) or moles of strong acid added (C_A):

$$\beta = \frac{dC_B}{d\text{pH}} = -\frac{dC_A}{d\text{pH}} \tag{4.73}$$

The buffer index indicates the number of moles of acid or base required to produce a given change in pH. While β can be readily determined from a titration curve, it can also be calculated if the composition of a solution is known. For example, for a solution containing a total concentration C_T of a monoprotic acid, Eqs. (4.28) and (4.29) apply. If C_B moles of base are added to the solution, then a charge balance as given by Eq. (4.67) must also hold, or

$$C_B + [H^+] = [OH^-] + [A^-] \tag{4.74}$$

Combining with Eqs. (4.28) and (4.29) gives

$$C_B = \frac{K_W}{[H^+]} - [H^+] + \frac{C_T K_A}{K_A + [H^+]} \tag{4.75}$$

Also,

$$\frac{dC_B}{dpH} = \frac{dC_B}{d[H^+]} \frac{d[H^+]}{dpH} \tag{4.76}$$

Since

$$pH = -\log[H^+] = -\ln[H^+]/2.303$$

then

$$d[H^+]/dpH = -2.303[H^+]$$

and

$$\beta = dC_B/dpH = -2.303[H^+](dC_B/d[H^+]) \tag{4.77}$$

Differentiating C_B in Eq. (4.75) with respect to $[H^+]$ and substituting into Eq. (4.77) yields

$$\beta = 2.303 \left[\frac{K_W}{[H^+]} + [H^+] + \frac{C_T K_A [H^+]}{(K_A + [H^+])^2} \right] \tag{4.78}$$

Thus, if we know the pH for a given buffer solution and the molar concentration of weak acid plus its conjugate base, the buffer index can be determined by direct substitution into Eq. (4.78). When dealing with a weak base, Eq.(4.78) can also be used by substituting K_W/K_B for K_A. Solutions for β for polyprotic acids or bases are also available.[6,7]

| A buffer solution has been prepared by adding 0.2 mol/L of acetic acid and 0.1 mol/L of acetate. The pH of the solution has been adjusted to 5.0 by addition of NaOH. How much NaOH (mol/L) is required to increase the pH to 5.1? | **EXAMPLE 4.25** |

$$C_T = 0.2 + 0.1 = 0.3 \text{ mol/L}$$
$$[H^+] = 10^{-5}$$

From Eq. (4.78),

$$\beta = 2.303 \left[\frac{10^{-14}}{10^{-5}} + 10^{-5} + \frac{0.3(1.8 \times 10^{-5})(10^{-5})}{(1.8 \times 10^{-5} + 10^{-5})^2} \right]$$
$$= 0.16 \text{ mol/L of base per pH unit}$$

NaOH required $= (5.1 - 5.0)(0.16) = 0.016 \text{ mol/L}$

---

[6]J. N. Butler, "Ionic Equilibrium," Addison-Wesley, Reading, MA, 1964.

[7]W. Stumm and J. J. Morgan, "Aquatic Chemistry," 3rd ed., Wiley-Interscience, New York, 1996.

4.8 | COMPLEX FORMATION

Background information and general nomenclature for complexes and complex formation are given in Sec. 2.13. The effect of complex formation on solubility of salts will be discussed in Sec. 4.9. Most metals form complexes with a variety of ligands, resulting in a number of negatively, neutral, or positively charged species. Because of differences such as charge, size and shape, and mobility, the various complexes of a metal behave differently chemically. For a given metal, some complexes are more toxic to organisms although in most cases it is the free metal that is most toxic. Some complexes reduce the availability of nutrient metals to living organisms, and some are removed more readily by chemical flocculation, activated carbon adsorption, and ion exchange. A knowledge of complex formation can aid in the understanding of metal behavior and fate in natural water systems, and can aid in designing treatment systems for metal removal.

Mononuclear Complexes

A mononuclear complex consists of a single, central metal ion to which is bound a number of neutral or anionic ligands. The ligands that attach to the central ion at only one site on the ligand are termed *mondentate* ligands. The number of ligands attached to the central ion is called the *coordination number.* Metal ions in solution are never in the uncomplexed state; they always are bound with solvent molecules. Thus, Cu^{2+} exists in solution as the hydrated $Cu(H_2O)_6^{2+}$, and Al^{3+} as $Al(H_2O)_6^{3+}$. Formation of other complexes in aqueous solution can be thought to result from the replacement of bound water by other ligands. Thus, formation of the hydroxide complex of copper can be thought to occur as follows:

$$Cu(H_2O)_6^{2+} + H_2O \rightleftharpoons Cu(H_2O)_5(OH)^+ + H_3O^+$$

For ease in writing, water can be dropped from the complex to give the abbreviated equation:

$$Cu^{2+} + H_2O \rightleftharpoons CuOH^+ + H^+ \tag{4.79}$$

However, Eq. (4.79) is more commonly written as

$$Cu^{2+} + OH^- \rightleftharpoons CuOH^+ \tag{4.80}$$

Generally in this book, abbreviated equations will be used.

Table 4.4 lists stepwise formation (stability) constants for a number of mononuclear metal ions and ligands. Table 4.5 lists values for hydroxide complexes. Reported values can vary widely among different literature sources, as indicated in Table 4.5 for hydroxide complexes. Such differences are especially common for organic ligands. However, when such constants are available, calculations of complex ion concentration under a given set of conditions is relatively straightforward. Consider the complexes of copper with ammonia,

$$Cu^{2+} + NH_3 \rightleftharpoons CuNH_3^{2+} \tag{4.81}$$

$$CuNH_3^{2+} + NH_3 \rightleftharpoons Cu(NH_3)_2^{2+} \tag{4.82}$$

$$Cu(NH_3)_2^{2+} + NH_3 \rightleftharpoons Cu(NH_3)_3^{2+} \tag{4.83}$$

$$Cu(NH_3)_3^{2+} + NH_3 \rightleftharpoons Cu(NH_3)_4^{2+} \tag{4.84}$$

Table 4.4 | Stepwise formation constants for various ligand and metal complexes

Ligand	Metal ion	Logarithm of constants				Source
		K_1	K_2	K_3	K_4	
Cl^-	Ag^+	3.45	2.22	0.33	0.04	a
	Cd^{2+}	2.0	0.6	−0.2	−0.7	b
	Fe^{2+}	0.36	0.04			a
	Fe^{3+}	1.48	0.65	−1.0		a
	Hg^{2+}	6.72	6.51	1.00	0.97	a
	Pb^{2+}	1.60	0.18	−0.1	−0.3	a
	Zn^{2+}	0.43	0.02	0.05	−0.3	c
F^-	Ag^+	0.4				b
	Al^{3+}	7.0	5.6	4.1	2.4	b
	Cd^{2+}	1.0	0.4			b
	Cr^{3+}	5.2	4.0	2.8		b
	Fe^{3+}	6.0	4.6	3.1		b
	Mg^{2+}	1.82				a
	Pb^{2+}	2.0	1.4			b
CN^-	Cd^{2+}	6.0	5.1	4.6	2.2	b
	Hg^{2+}	17.0	15.8	3.5	2.7	b
	Zn^{2+}	5.7	5.4	5.0	3.5	b
HS^-	Ag^+	14.05	4.4			c
	Cd^{2+}	10.17	6.36	2.18	2.19	c
	Pb^{2+}	15.27	1.3			c
	Zn^{2+}	14.94	1.16			c
HCO_3^-	Ca^{2+}	11.59				b
	Mg^{2+}	11.49				b
CO_3^{2-}	Ca^{2+}	3.2				b
	Cu^{2+}	6.7	3.5			b
	Mg^{2+}	3.4				b
SO_4^{2-}	Ag^+	1.3				a
	Al^{3+}	3.73				a
	Ca^{2+}	2.3				a
	Cd^{2+}	2.3	0.9	−0.5		b
	Cr^{3+}	3.0				b
	Cu^{2+}	2.4				b
	Fe^{2+}	2.3				a
	Fe^{3+}	4.04	1.30			a
	Pb^{2+}	3.7				a
	Zn^{2+}	2.1	1.0			b
NH_3	Ag^+	3.32	3.92			a
	Cd^{2+}	2.51	1.96	1.30	0.79	a
	Cu^{2+}	3.99	3.34	2.73	1.97	a
	Hg^{2+}	8.8	8.7	1	0.78	a
	Ni^{2+}	3.0	2.18	1.64	1.16	a
	Zn^{2+}	2.18	2.25	2.31	1.96	a

(continued)

Table 4.4 | (continued)

Ligand	Metal ion	Logarithm of constants				Source
		K_1	K_2	K_3	K_4	
Acetate	Ag^+	0.7	−0.1			b
	Cu^{2+}	2.2	1.4			b
	Cd^{2+}	1.9	1.3			b
	Cr^{3+}	5.4	3.0	2.8		b
	Fe^{2+}	1.4				b
	Fe^{3+}	4.0	3.6	2.0		b
	Hg^{2+}	6.1	4.0	4.0	3.5	b
	Pb^{2+}	2.7	1.4			b
	Zn^{2+}	1.6	0.2			b

Sources:
a. L. G. Sillén and A. E. Martell, Stability Constants, *Spec. Publs.* 17, 1964, and 25, 1971, Chemical Society, London.
b. F. M. M. Morel and J. J. Hering, "Principles and Applications of Aquatic Chemistry," Wiley-Interscience, New York, 1993.
c. M. M. Benjamin, "Water Chemistry," McGraw-Hill, New York, 2002.

Table 4.5 | Stepwise formation constants for various mononuclear hydroxide complexes

Metal ion	Logarithm of constants				Source
	K_1	K_2	K_3	K_4	
Ag^+	2.3	1.9			a
Ba^{2+}	0.64				a
Ca^{2+}	1.51				a
	1.15				b
Cd^{2+}	6.08	2.62	−0.32	0.04	a
	3.9	3.7			b
Cr^{3+}	10.0	8.3	5.70	4.6	b
Cu^{2+}	6.0	7.18	1.24	0.14	a
Fe^{2+}	4.5	2.9	3.6		b
Fe^{3+}	11.5	9.3			a
Hg^{2+}	10.3	11.4			a
	10.6	11.2	−0.90		b
Mg^{2+}	2.60				a
Mn^{2+}	3.4				a
	3.4	2.4	1.4	0.5	b
Ni^{2+}	4.0				a
	4.1	4.9	3.0		b
Pb^{2+}	7.82	3.06	3.06		a
	6.3	4.6	3.0		b
Zn^{2+}	4.15	6.00	4.11	1.26	a
	5.0	6.1	2.5	1.2	b

Sources:
a. L. G. Sillén and A. E. Martell, Stability Constants, *Spec. Publs.* 17, 1964, and 25, 1971, Chemical Society, London.
b. F. M. M. Morel and J. J. Hering, "Principles and Applications of Aquatic Chemistry," Wiley-Interscience, New York, 1993.

The stepwise equilibrium relationships for Eqs. (4.81) to (4.84) are

$$\frac{[CuNH_3^{2+}]}{[Cu^{2+}][NH_3]} = K_1 = 9.8 \times 10^3 \tag{4.85}$$

$$\frac{[Cu(NH_3)_2^{2+}]}{[CuNH_3^{2+}][NH_3]} = K_2 = 2.2 \times 10^3 \tag{4.86}$$

$$\frac{[Cu(NH_3)_3^{2+}]}{[Cu(NH_3)_2^{2+}][NH_3]} = K_3 = 5.4 \times 10^2 \tag{4.87}$$

$$\frac{[Cu(NH_3)_4^{2+}]}{[Cu(NH_3)_3^{2+}][NH_3]} = K_4 = 9.3 \times 10 \tag{4.88}$$

The significance of the various ammonia complexes of copper in aqueous systems can perhaps best be illustrated with a logarithmic concentration diagram. Here, the NH_3 ligand is taken as the master variable and $pNH_3(-\log [NH_3])$ is plotted as the abscissa as illustrated in Fig. 4.10. The calculations are considerably simplified if the concentration of the uncomplexed Cu^{2+} is arbitrarily held constant at some value, say 10^{-7} M. Once the concentration is fixed, Eqs. (4.85) to (4.88)

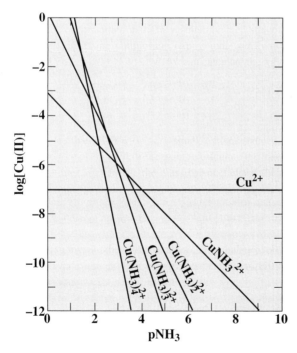

Figure 4.10
Logarithmic concentration diagram for illustrating effect of ammonia concentration on relative concentrations of various ammonia complexes of copper.

become straight lines on a logarithmic concentration diagram. The logarithm of Eq. (4.85) is

$$\log [CuNH_3^{2+}] = \log [Cu^{2+}] + \log K_1 + \log [NH_3] \qquad (4.89)$$

or since $\log K_1 = 3.99$, $\log [NH_3] = -pNH_3$, and $\log [Cu^{2+}] = -7$,

$$\log [CuNH_3^{2+}] = -3.01 - pNH_3 \qquad (4.90)$$

This line has a slope of -1 in Fig. 4.10 and intersects the $pNH_3 = 0$ ordinate at -3.01. The negative slopes for each of the succeeding complexes increase, as indicated by the respective logarithmic equations:

$$\log [Cu(NH_3)_2^{2+}] = 0.33 - 2pNH_3 \qquad (4.91)$$

$$\log [Cu(NH_3)_3^{2+}] = 3.06 - 3pNH_3 \qquad (4.92)$$

$$\log [Cu(NH_3)_4^{2+}] = 5.03 - 4pNH_3 \qquad (4.93)$$

From Fig. 4.10 it can readily be seen which complex dominates at a given ammonia concentration. If $[Cu^{2+}]$ were chosen to be some other value such as 10^{-9}, the relative concentrations of the various species would not change. The curves would all simply shift together two logarithmic units downward.

Ammonia is a base that ionizes in water to form NH_4^+. The distribution between NH_3 and NH_4^+ as a function of pH is illustrated in Fig. 4.2. At pH 7, only about 0.5 percent of the ammonia is in the NH_3 form, and this percentage increases to about 5 percent at a pH of 8. Under these conditions in natural waters, pNH_3 would generally range from about 5 to 7, and from Fig. 4.10, Cu^{2+} would be the predominant species. In municipal wastewaters, pNH_3 would vary from about 3.5 to 5.5. Over this range, any one of the species, Cu^{2+}, $CuNH_3^{2+}$, or $Cu(NH_3)_2^{2+}$ may dominate. During anaerobic digestion of sewage sludge, ammonia is released in relatively high concentration such that pNH_3 may range from 2.5 to 4. Under these conditions ammonia complexes of copper would generally predominate over Cu^{2+}. When pNH_3 equals 2.5, only about 0.1 percent of the soluble copper in the digester would be in the Cu^{2+} form. Consideration of the effect of complexes on the solubility of heavy metals in anaerobic systems and on potential toxicity to microorganisms would appear appropriate.

Another method of graphically presenting complex formation calculations is through a *distribution diagram,* which shows the relative concentration of each species as a function of ligand concentration. A distribution diagram for ammonia complexes of copper is shown in Fig. 4.11. This diagram can be constructed by first assuming an arbitrary value for $[Cu^{2+}]$ such as 1, and then through use of equations similar to Eqs. (4.90) to (4.93), the concentration of each species at a given pNH_3 can be determined. The total molar concentration C_T of Cu(II) is then

$$C_T = [Cu^{2+}] + [CuNH_3^{2+}] + [Cu(NH_3)_2^{2+}]$$
$$+ [Cu(NH_3)_3^{2+}] + [Cu(NH_3)_4^{2+}] \qquad (4.94)$$

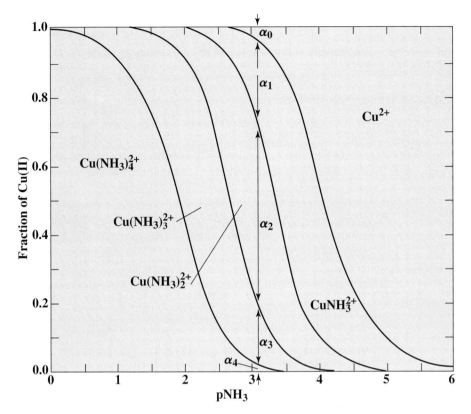

Figure 4.11
Distribution diagram for ammonia complexes of copper.

The fraction of copper present as each species at a given pNH_3 is then determined as follows:

$$\alpha_0 = \frac{[Cu^{2+}]}{C_T} \tag{4.95}$$

$$\alpha_1 = \frac{[CuNH_3^{2+}]}{C_T} \tag{4.96}$$

$$\alpha_2 = \frac{[Cu(NH_3)_2^{2+}]}{C_T} \tag{4.97}$$

$$\alpha_3 = \frac{[Cu(NH_3)_3^{2+}]}{C_T} \tag{4.98}$$

$$\alpha_4 = \frac{[Cu(NH_3)_4^{2+}]}{C_T} \tag{4.99}$$

From such calculations at different values of pNH$_3$ a distribution diagram like Fig. 4.11 can be constructed. Perhaps the relative importance of the different species can be discerned more rapidly from distribution diagrams than from logarithmic concentration diagrams.

Mixed Ligand Complexes

Natural water systems and wastewaters frequently contain several ligands such as Cl$^-$, NH$_3$, S^{2-}, and OH$^-$, and each will compete for complex formation with the metal ions present. The solution to this kind of a problem is not complex mathematically, but requires that all stepwise equilibria for each ligand and the metal be considered. As before, a concentration for the uncomplexed ion is first assumed, and then the concentration of each complex is determined for given values of each ligand. The results can perhaps best be displayed graphically with a *predominance area diagram* as illustrated in Fig. 4.12 for complexes between mercury and the ligands Cl$^-$ and OH$^-$. The abscissa is used to represent OH$^-$ where pH = 14 + log [OH$^-$]. The other ligand, chloride, is represented as the ordinate in the form of pCl. From calculations similar to those discussed previously, the complex which predominates under a given set of ligand concentrations can be determined. The lines

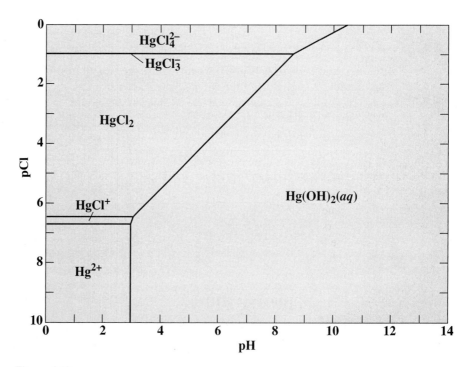

Figure 4.12
Predominance area diagram illustrating effect of pH and chloride concentration on chloride and hydroxide complexes of mercury.

drawn in Fig. 4.12 represent the locations where a change in predominance from one complex to another occurs.

Figure 4.12 illustrates the distribution of soluble mercury species in water when the total molar concentration of mercury is less than 10^{-4}. At higher mercury concentrations, precipitation of HgO might occur at high pH values. For fresh waters, pCl would normally lie between 2 and 4. With pH greater than 7, the predominant species is the neutral $Hg(OH)_2$, while below 7 it is $HgCl_2$. In seawater, pCl lies between 0 and 1, and $HgCl_4^{2-}$ would be the predominant species. At low pH, mercury may occur primarily as a positively charged, a neutral, or a negatively charged complex, depending upon chloride concentration.

When more than two ligands are involved, computations for solution are similar to those described above for the ammonia complexes, but involve more equations. Graphical presentation, however, becomes more difficult. Three-dimensional graphs can be used for three-ligand systems. However, it is rare for a given water or wastewater that more than two ligands will compete for predominance. The problem is to determine which two ligands this may be. Computer methods are also available for solution of these complex equilibrium problems.

Polynuclear complexes of metals may also form in solution. An important example is the dimerization of Fe(III),

$$2Fe(H_2O)_5OH^{2+} \rightleftharpoons \left[(H_2O)_4Fe \overset{\displaystyle \overset{H}{\underset{}{O}}}{\underset{\displaystyle \underset{H}{O}}{}} Fe(H_2O)_4 \right]^{4+} + 2H_2O \qquad (4.100)$$

which can be simplified as

$$2FeOH^{2+} \rightleftharpoons Fe_2(OH)_2^{4+} \qquad (4.101)$$

Polynuclear species of iron and aluminum may predominate in natural water and are believed to be important in coagulation of water and wastewater. Further discussion of mixed ligand and polynuclear species of importance in natural waters can be found elsewhere.[8,9]

The solubility of metal ions is increased by the presence of "chelating agents." Many of these chelators are *multidentate* ligands, meaning that the central metal ion may attach to more than one site on the ligand. These substances have the ability to seize or "sequester" metal ions and hold them in a clawlike grip (the word is from the Greek *chele,* meaning claw). Like a claw, a chelating molecule forms a ring in which the metal ion is held by a pair of pincers, so it is not free to form an insoluble salt. The pincers of a chelating molecule consist of "ligand" atoms (usually nitrogen, oxygen, or sulfur), each of which donates two electrons to form a "coordinate" bond with the ion. There are many natural chelates, such as hemoglo-

---

[8]J. N. Butler, "Ionic Equilibrium."

[9]Stumm and Morgan, "Aquatic Chemistry."

bin (containing iron), vitamin B-12 (containing cobalt), and chlorophyll (containing magnesium). Many well-known substances such as aspirin, citric acid, adrenaline, and cortisone can act as chelating agents; EDTA (ethylenediaminetetraacetic acid) is a chelating agent that has a remarkable affinity for calcium and is used for the determination of water hardness (Chap. 19). Natural organic material present in waters, such as humic substances (Sec. 5.32), may chelate metals and govern their fate and transport in the environment. Bacteria and algae produce and excrete organic compounds that will chelate metals; in some cases this chelation increases the bioavailability of nutrient metals, in others it will decrease the toxicity of metals by making them less bioavailable. Additional discussion about chelates can be found in the references listed at the end of this chapter.

Solving Equilibrium Problems Involving Complexes

Equilibrium problems involving complexes are solved in much the same fashion as other equilibrium problems. The general steps involved include

1. Write all reactions of interest.
2. List all species present at equilibrium.
3. Write equilibrium expressions using appropriate equilibrium (stability) constants.
4. Write mass-balance equations for the metal(s) ($C_{T,\text{metal}}$) and ligand(s) $C_{T,\text{ligand}}$) of interest.
5. If necessary, make assumptions based on
   (a) The relative magnitude of $C_{T,\text{metal}}$ and $C_{T,\text{ligand}}$,
   (b) The magnitude of the stability constants.
6. If necessary and appropriate, write charge-balance equation.

Example 4.26 illustrates these concepts. Problems can also be solved using graphical techniques such as those given in Figs. 4.10 to 4.12.

EXAMPLE 4.26	A water has a total soluble lead concentration of 1 mg/L and a total soluble chloride concentration of 100 mg/L. What are the concentrations of all the lead-and chloride-containing species? Ignore activity corrections and any lead complexes with hydroxide (generally not a good idea unless the pH is low enough).

$$C_{T,\text{Pb}} = \frac{1 \text{ mg/L}}{207{,}200 \text{ mg/mol}} = 4.83 \times 10^{-6} \text{ M}$$

$$C_{T,\text{Pb}} = [\text{Pb}^{2+}] + [\text{PbCl}^+] + [\text{PbCl}_2] + [\text{PbCl}_3^-] + [\text{PbCl}_4^{2-}] = 4.83 \times 10^{-6} \text{ M}$$

$$C_{T,\text{Cl}} = \frac{100 \text{ mg/L}}{35{,}453 \text{ mg/mol}} = 2.82 \times 10^{-3} \text{ M}$$

$$C_{T,\text{Cl}} = [\text{Cl}^-] + [\text{PbCl}^+] + 2[\text{PbCl}_2] + 3[\text{PbCl}_3^-] + 4[\text{PbCl}_4^{2-}] = 2.82 \times 10^{-3} \text{ M}$$

The equilibria involved, ignoring activity corrections, are given below. For this problem, it is more convenient to use overall formation constants (β). The student should recall from Sec. 2.13 that $\beta_1 = K_1, \beta_2 = K_1 K_2, \beta_3 = K_1 K_2 K_3$, and $\beta_4 = K_1 K_2 K_3 K_4$.

$$Pb^{2+} + Cl^- \rightleftharpoons PbCl^+ \qquad \beta_1 = \frac{[PbCl^+]}{[Pb^{2+}][Cl^-]} \qquad \beta_1 = 10^{1.60}$$

$$Pb^{2+} + 2Cl^- \rightleftharpoons PbCl_2(aq) \qquad \beta_2 = \frac{[PbCl_2(aq)]}{[Pb^{2+}][Cl^-]^2} \qquad \beta_2 = 10^{1.78}$$

$$Pb^{2+} + 3Cl^- \rightleftharpoons PbCl_3^- \qquad \beta_3 = \frac{[PbCl_3^-]}{[Pb^{2+}][Cl^-]^3} \qquad \beta_3 = 10^{1.68}$$

$$Pb^{2+} + 4Cl^- \rightleftharpoons PbCl_4^{2-} \qquad \beta_4 = \frac{[PbCl_4^{2-}]}{[Pb^{2+}][Cl^-]^4} \qquad \beta_4 = 10^{1.38}$$

Solve $C_{T,Pb}$ for $[Pb^{2+}]$:

$$4.83 \times 10^{-6} = [Pb^{2+}](1 + \beta_1[Cl^-] + \beta_2[Cl^-]^2 + \beta_3[Cl^-]^3 + \beta_4[Cl^-]^4)$$

Solve $C_{T,Cl}$ in terms of $[Pb^{2+}]$ and $[Cl^-]$:

$$2.82 \times 10^{-3} = [Cl^-] + \beta_1[Pb^{2+}][Cl^-] + 2\beta_2[Pb^{2+}][Cl^-]^2 + 3\beta_3[Pb^{2+}][Cl^-]^3 + 4\beta_4[Pb^{2+}][Cl^-]^4)$$

To solve this problem "exactly" would involve some type of computer solution (e.g., use of trial and error by spreadsheet or numerical method). Or, we could try to make some simplifying assumptions based on what we know about the species involved.

We can compare $C_{T,Pb}$ with $C_{T,Cl}$ to see if we can simplify either of the mass-balance equations. We note that $C_{T,Cl} \gg C_{T,Pb}$, such that $[Cl^-]$ is not likely to be depleted much by all of the Pb-Cl complexes. So, we can assume $[Cl^-] = 2.82 \times 10^{-3}$ and solve the $C_{T,Pb}$ equation for $[Pb^{2+}]$:

$$4.83 \times 10^{-6} = [Pb^{2+}](1 + 10^{1.6}(2.82 \times 10^{-3}) + 2(10^{1.78})(2.82 \times 10^{-3})^2 + 3(10^{1.68})(2.82 \times 10^{-3})^3 + 4(10^{1.38})(2.82 \times 10^{-3})^4)$$

From which,

$$[Pb^{2+}] = 4.34 \times 10^{-6} \, M = 0.90 \, mg/L \, Pb$$
$$[PbCl^+] = 4.87 \times 10^{-7} \, M = 0.10 \, mg/L \, Pb$$
$$[PbCl_2] = 2.08 \times 10^{-9} \, M = 4.31 \times 10^{-4} \, mg/L \, Pb$$
$$[PbCl_3^-] = 4.66 \times 10^{-12} \, M = 9.65 \times 10^{-7} \, mg/L \, Pb$$
$$[PbCl_4^{2-}] = 6.58 \times 10^{-15} \, M = 1.36 \times 10^{-9} \, mg/L \, Pb$$

Please note that our assumption that $C_{T,Cl} \cong [Cl^-]$ is a good one since the concentration of chloride tied up in lead complexes is seen to be only slightly more than $4.87 \times 10^{-6} \, M$ (about a 0.2% error!).

4.9 | SOLUBILITY OF SALTS

Basic concepts of salt solubility and the concept of solubility product were discussed in Sec. 2.13. Consideration of solubility relationships can aid in understanding the natural forces that dissolve rocks and other minerals, bringing minerals into aqueous solution. Variations in mineral characteristics of water can also be understood to some degree by considering the factors affecting solubility of minerals. Knowledge of solubility relationships can also help to devise methods for treating water supplies and wastewaters to rid them of hardness-causing constituents, heavy-metal contaminants, orthophosphate (PO_4^{3-}), and some organic materials. Also, since a variety of pollutants (e.g., heavy metals and some organic compounds) sorb to the surface of solids, an understanding of solubility concepts is needed to explain and help describe these phenomena. For purposes of the following discussion, we will define the solubility S of a metal as the sum of all dissolved species of the metal:

$$S = C_{T,\text{metal}} = [\text{free metal } (aq)] + \sum_{i=1}^{n} \text{dissolved metal complexes} \quad (4.102)$$

Logarithmic Concentration Diagrams for Simple Solubility Determinations

Logarithmic concentration diagrams are useful for illustrating solubility as well as acid-base relationships. Figure 4.13 shows the relationship between carbonate concentration and the saturation concentration for various cations based upon the solubility product constants from Table 2.5, and the solubility product relationship (ignoring activity corrections)

$$[M^{2+}] [CO_3^{2-}] = K_{sp} \quad (4.103)$$

which yields straight lines of slope $+1$ when written and plotted from the form

$$\log [M^{2+}] = pCO_3 + \log K_{sp} \quad (4.104)$$

In Fig. 4.13 $[CO_3^{2-}]$ is represented as a function of pCO_3 by a line having a slope of -1. Of the cations shown, Pb^{2+} is the least soluble and Mg^{2+} is the most soluble. Carbonate, even at relatively low concentration, can be quite effective in removal of most divalent cations from solution. Calcium hardness removal by carbonate precipitation is a commonly used water treatment process. Heavy metals are also removed quite effectively in this way by some advanced wastewater treatment processes.

The intersection of the CO_3^{2-} line with a cation solubility product line indicates the condition for the solubility of that particular salt in pure water,

$$[M^{2+}] = [CO_3^{2-}] = \text{solubility} \quad (4.105)$$

If the cation and anion do not have the same charge, then the solubility is displaced slightly from the point of intersection since the cation molar concentration would not equal the anion molar concentration. The solubility of the salts represented in

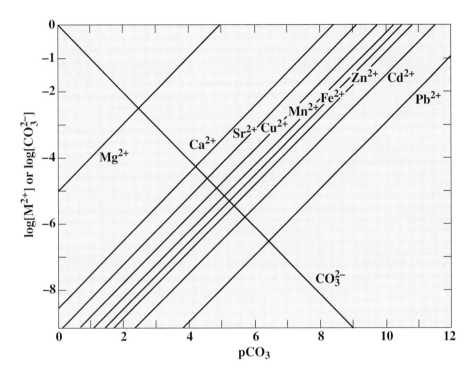

Figure 4.13
Logarithmic concentration diagram showing the solubility of various metallic carbonates at 25°C.

Fig. 4.13 is in the same order as that of the solubility products, which is not generally the case as discussed in Sec. 2.13. This occurs because the cations shown are all divalent. If monovalent or trivalent cations were shown, the slopes of their solubility product lines would have been $+\frac{1}{2}$ and $+\frac{3}{2}$, respectively. Use of logarithmic diagrams as in Fig. 4.13 may be no simpler than solubility product calculations for simple solubilities of the type shown. However, when complexes are present, such diagrams can give a complete and compact picture of all the equilibria involved. This is illustrated later in this section.

Determine the solubility of Ca^{2+} in a water sample containing 10^{-3} molar carbonate at 25°C.

$$pCO_3 = -\log 10^{-3} = 3$$

From Fig. 4.13, when $pCO_3 = 3$, $\log [Ca^{2+}] = -5.3$; therefore,

$$S = [Ca^{2+}] = 10^{-5.3} = 5 \times 10^{-6} \text{ M}$$

EXAMPLE 4.27

Complex Solubility Relationships

Solubility relationships are generally much more complex than indicated in Sec. 2.13 or in the preceding dicussion. Generally, in natural waters or wastewaters, several other factors must be considered in order to make realistic solubility calculations. First, as discussed in Sec. 4.3, the ionic strength of the solution affects ion activity and must be considered if more exact calculations are desired.

Perhaps more important, other equilibria besides the solubility product affect the concentration of the ions present. Reactions of the cation or anion with water to form hydroxide complexes or protonated anion species are common. In addition, the cations or anions may form complexes with other materials in solution, thus reducing their effective concentration. Finally, other ions may form salts with less solubility than the one under consideration. If these are ignored, then solubility predictions may be considerably in error.

Effect of pH on Solubility of Hydroxide Salts   First, consider the effect of solution pH on the solubility of cations. Assuming that no other materials exist in solution to react with the cations, the solubility product between any cation and OH^- would be

$$[M^{z+}][OH^-]^z = K_{sp} \tag{4.106}$$

and

$$\log [M^{z+}] = \log K_{sp} - z \log [OH^-] \tag{4.107}$$

However, $\log[OH^-]$ is a function of pH,

$$pH = pK_W - pOH = pK_W + \log [OH^-] \tag{4.108}$$

so that

$$\log [M^{z+}] = \log K_{sp} - z(pH - pK_W) \tag{4.109}$$

This relationship between pH and $\log [M^{z+}]$ is illustrated in Fig. 4.14 for several cations. The slope of the lines is equal to $-z$ as indicated by Eq. (4.109). Use is made of the effect of pH on cation solubility in water and wastewater treatment. Magnesium hardness can be reduced by raising the pH to 11 or greater and precipitating $Mg(OH)_2$. Copper, zinc, and chromium [Cr(III)] are often removed from metal-containing wastewaters by precipitation at a pH of 7 or above.

Effect of Weak Acids and Bases   The pH of water not only affects solubility of metal hydroxides as illustrated in Fig. 4.14, but also affects other equilibria in water, which in turn can affect solubility. For example, the solubility of salts of weak acids will be influenced by solution pH. The solubility of calcium carbonate will be used for illustration. It dissolves in water to give its ions,

$$CaCO_3(s) \rightleftharpoons Ca^{2+} + CO_3^{2-} \tag{4.110}$$

the equilibrium being controlled by the solubility product,

$$[Ca^{2+}][CO_3^{2-}] = K_{sp} = 5 \times 10^{-9} \tag{4.111}$$

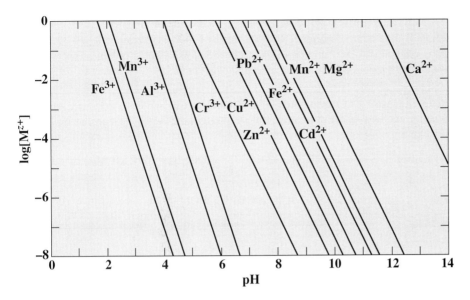

Figure 4.14
Logarithmic concentration diagram showing the solubility of various metallic hydroxides.

Carbonate is an anion of the weak diprotic acid, $H_2CO_3^*$, which ionizes in water as follows:

$$H_2CO_3^* \rightleftharpoons H^+ + HCO_3^- \tag{4.112}$$

$$HCO_3^- \rightleftharpoons H^+ + CO_3^{2-} \tag{4.113}$$

Equilibrium expressions for Eqs. (4.112) and (4.113) are

$$\frac{[H^+][HCO_3^-]}{[H_2CO_3^*]} = K_{A1} = 4.3 \times 10^{-7} \tag{4.114}$$

$$\frac{[H^+][CO_3^{2-}]}{[HCO_3^-]} = K_{A2} = 4.7 \times 10^{-11} \tag{4.115}$$

Consider a solution formed by the addition of carbon dioxide to distilled water. Next, the pH is adjusted by addition of NaOH, while maintaining the total molar concentration of carbon and species containing it at its initial value (that is, the system is considered closed to the atmosphere), where

$$CO_2 + H_2O \rightarrow H_2CO_3 \tag{4.116}$$

and

$$[H_2CO_3^*] = [CO_2] + [H_2CO_3] \tag{4.117}$$

Thus, the total concentration of soluble inorganic carbon in solution, C_T can be written as

$$C_T = [H_2CO_3^*] + [HCO_3^-] + [CO_3^{2-}] \tag{4.118}$$

Now, if at a given pH, $CaCl_2$ is added to the solution, at what $[Ca^{2+}]$ will the solution just become saturated with respect to $CaCO_3(s)$? How will this saturation value change as a function of pH and C_T? For this set of problems, Eqs. (4.111), (4.114), (4.115), and (4.118) apply. In addition,

$$[H^+][OH^-] = K_W = 10^{-14} \qquad (4.119)$$

must be satisfied. In order to solve this problem in a general sense, a charge balance must be maintained,

$$2[Ca^{2+}] + [Na^+] + [H^+] = [HCO_3^-] + 2[CO_3^{2-}] + [OH^-] + [Cl^-] \qquad (4.120)$$

In general, this equation can be solved for $[H^+]$ (e.g., with a spreadsheet). If other cations or anions are present in the solution, they must also be accounted for in the charge balance. However, for the particular problems at hand, the pH is given and so Eq. (4.120) is not required. From Eq. (4.37) $[CO_3^{2-}]$ can be found as a function of C_T and $[H^+]$ (and hence pH). $[Ca^{2+}]$ can then be determined by substituting the value for $[CO_3^{2-}]$ into Eq. (4.111).

The relationship between pH, C_T, and the saturation value for $[Ca^{2+}]$ as determined by the procedure just discussed is illustrated graphically in Fig. 4.15.

EXAMPLE 4.28

If $[Ca^{2+}] = 10^{-4}$ and $C_T = 10^{-2}$, at what pH will the solution become saturated with respect to $CaCO_3(s)$?

From Fig. 4.15, for $\log [Ca^{2+}] = -4$ and $C_T = 10^{-2}$, the pH of saturation equals 8.0. If the pH is raised above this value, the water will be supersaturated such that $CaCO_3(s)$ may begin to precipitate.

EXAMPLE 4.29

What is the maximum solubility of calcium in water with a pH of 10 and $C_T = 10^{-2}$?

From Fig. 4.15 $\log [Ca^{2+}] = -5.8$, so $[Ca^{2+}] = 10^{-5.8} = 1.6 \times 10^{-6}$ M.

EXAMPLE 4.30

A water has an initial $[Ca^{2+}]$ of 4×10^{-3} and C_T of 10^{-2}. It is desired to reduce $[Ca^{2+}]$ to 10^{-4} by precipitation of $CaCO_3(s)$. Based upon equilibrium relationships, what would be the final concentration of C_T, and how high should the solution pH be maintained in order to achieve the desired calcium removal?

The quantity of Ca^{2+} precipitated is $4 \times 10^{-3} - 10^{-4}$ or 3.9×10^{-3} mol per liter of solution. Since for each mole of calcium precipitated, 1 mol of CO_3^{2-} is removed from solution,

$$C_{T,(final)} = 10^{-2} - 3.9 \times 10^{-3} = 6.1 \times 10^{-3} \text{ M}$$

From Fig. 4.15 when $[Ca^{2+}] = 10^{-4}$ and $C_T = 6.1 \times 10^{-3}$, the pH is about 9.0. Thus, the pH would need to be raised to at least 9.0 to obtain the desired removal. Practically, a higher pH would be used in order to ensure that sufficient removal was obtained.

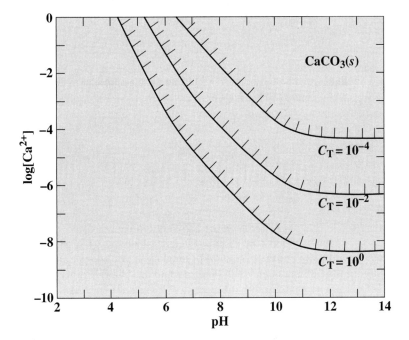

Figure 4.15
Logarithmic concentration diagram showing the relationship between pH, C_T (mol/L of inorganic carbon), and the equilibrium concentration of Ca^{2+} with respect to $CaCO_3(s)$.

The situation changes when the system is open to the atmosphere. Consider a situation where $CaCO_3(s)$ is in equilibrium with water and the atmosphere, that is, the system is open. When in equilibrium with the atmosphere, the total concentration of dissolved inorganic carbon [Eq. (4.118)] is not constant with changing pH as is the case with a closed system (if no precipitation occurs). Rather, it is $[H_2CO_3^*]$ that is constant according to the following equilibrium:

$$[H_2CO_3^*] = P_{CO_2}/K_H \tag{4.121}$$

where K_H is Henry's law constant (31.6 atm/M at 25°C) and P_{CO_2} is the partial pressure of CO_2 in the atmosphere expressed in units of atmospheres. Values for $[HCO_3^-]$, $[CO_3^{2-}]$, and C_T can then be determined from Eqs. (4.114), (4.115), and (4.118), respectively, if P_{CO_2} and pH are known.

If pH is unknown, the charge balance is needed to solve for solution pH:

$$2[Ca^{2+}] + [H^+] = [OH^-] + [HCO_3^-] + 2[CO_3^{2-}] \tag{4.122}$$

Equation (4.122) can be solved for $[H^+]$ by trial and error or an appropriate numerical method using Eqs. (4.111), (4.114), (4.115), (4.118) and (4.119). Once again, if other cations or anions are present, they must be included in the charge balance.

EXAMPLE 4.31

Consider a water in equilibrium with $CaCO_3(s)$ at 25°C. Ignoring activity corrections and any complexes with Ca^{2+} or CO_3^{2-}, determine the equilibrium pH for (a) a closed system and (b) a system open to an atmosphere with $P_{CO_2} = 10^{-3.5}$ atm.

As noted, these kinds of problems are solved using the charge balance, the equation of which is the same for both closed and open systems:

$$2[Ca^{2+}] + [H^+] = [OH^-] + [HCO_3^-] + 2[CO_3^{2-}]$$

or

$$2[Ca^{2+}] + [H^+] = \frac{K_W}{[H^+]} + C_{T,CO_3}(\alpha_1 + 2\alpha_2)$$

and

$$S = C_{T,Ca} = [Ca^{2+}]$$

(a) In a closed system, when 1 mol of $CaCO_3(s)$ dissolves, 1 mol of carbonate (C_{T,CO_3}) results. So, in this case

$$S = [Ca^{2+}] = C_{T,CO_3}$$

From the solubility product relationship, we can develop the following:

$$K_{sp} = 5 \times 10^{-9} = [Ca^{2+}][CO_3^{2-}] = S(\alpha_2 S)$$

$$S = [Ca^{2+}] = \sqrt{\frac{5 \times 10^{-9}}{\alpha_2}}$$

Making appropriate substitutions in the charge balance gives

$$2\sqrt{\frac{5 \times 10^{-9}}{\alpha_2}} + [H^+] = \frac{K_W}{[H^+]} + \sqrt{\frac{5 \times 10^{-9}}{\alpha_2}}(\alpha_1 + 2\alpha_2)$$

This is an equation with $[H^+]$ as the only unknown. It can be solved using a spreadsheet as demonstrated in previous examples.

The equilibrium pH for the closed system is 9.96.

(b) In an open system, $[H_2CO_3^*]$ is determined by P_{CO_2} as given in Eq. (4.121) and is independent of pH.

$$[H_2CO_3^*] = 10^{-3.5}/31.6 = 10^{-5} \text{ M}$$

Thus, in an open system, C_{T,CO_3} is determined by P_{CO_2} and α_0:

$$C_{T,CO_3} = \frac{10^{-5}}{\alpha_0}$$

The expression for $[Ca^{2+}]$ becomes

$$[Ca^{2+}] = \frac{5 \times 10^{-9}}{\alpha_2 C_{T,CO_3}} = \frac{\alpha_0(5 \times 10^{-9})}{\alpha_2(10^{-5})}$$

Making substitutions in the charge balance gives

$$\frac{2\alpha_0(5 \times 10^{-9})}{\alpha_2(10^{-5})} + [H^+] = \frac{K_W}{[H^+]} + \frac{10^{-5}}{\alpha_0}(\alpha_1 + 2\alpha_2) \tag{4.140}$$

This is again an equation with $[H^+]$ as the only unknown. It can be solved using a spreadsheet as demonstrated in previous examples.

The equilibrium pH for the open system is 8.35. It would be expected that the pH for the open system would be less than for the closed system since CO_2 is an acid gas.

It should be noted that the effect of adding acid (or base) to these systems could be accounted for by adjusting the charge balance. For example, if HCl were added $[Cl^-]$ would be included in the right side of the equation.

Effect of Complexes Complex formation also affects the solubility of salts. For example, zinc is commonly removed from acid-plating wastewaters by adding base to increase the pH to form the insoluble $Zn(OH)_2(s)$. However, if excess base is added, zinc will form soluble complexes with OH^- and will return to or remain in solution. The appropriate equations governing this behavior of zinc are as follows:

Solubility product:

$$Zn(OH)_2(s) \rightleftharpoons Zn^{2+} + 2OH^- \tag{4.123}$$

Complex formation:

$$Zn^{2+} + OH^- \rightleftharpoons ZnOH^+ \tag{4.124}$$

$$ZnOH^+ + OH^- \rightleftharpoons Zn(OH)_2(aq) \tag{4.125}$$

$$Zn(OH)_2(aq) + OH^- \rightleftharpoons Zn(OH)_3^- \tag{4.126}$$

$$Zn(OH)_3^- + OH^- \rightleftharpoons Zn(OH)_4^{2-} \tag{4.127}$$

Water ionization:

$$H_2O \rightleftharpoons H^+ + OH^- \tag{4.128}$$

Also, using values from Table 2.5 and from Table 4.5 reported by Sillen and Martell, equilibrium relationships for Eqs. (4.123) to (4.128) are:

$$[Zn^{2+}][OH^-]^2 = K_{sp} = 8 \times 10^{-18} \tag{4.129}$$

$$[ZnOH^+]/[Zn^{2+}][OH^-] = K_1 = 1.4 \times 10^4 \tag{4.130}$$

$$[Zn(OH)_2(aq)]/[ZnOH^+][OH^-] = K_2 = 1 \times 10^6 \tag{4.131}$$

$$[Zn(OH)_3^-]/[Zn(OH)_2(aq)][OH^-] = K_3 = 1.3 \times 10^4 \tag{4.132}$$

$$[Zn(OH)_4^{2-}]/[Zn(OH)_3^-][OH^-] = K_4 = 1.8 \times 10^1 \tag{4.133}$$

$$[H^+][OH^-] = K_W = 1 \times 10^{-14} \tag{4.134}$$

In addition, a mass balance for zinc in solution must be maintained:

$$C_{T,Zn} = [Zn^{2+}] + [ZnOH^+] + [Zn(OH)_2(aq)] + [Zn(OH)_3^-] + [Zn(OH)_4^{2-}] \quad (4.135)$$

It should be noted that $C_{T,Zn}$, which represents the total soluble concentration of zinc, is also termed the solubility (S) of zinc. Figure 4.16 is a logarithmic concentration diagram showing the relationships between pH and the various zinc species for a solution saturated with respect to $Zn(OH)_2(s)$. If for a given pH, the concentration of zinc in solution were found sufficiently high to be in the $Zn(OH)_2(s)$ area of the diagram, then the solution would be supersaturated, and $Zn(OH)_2(s)$ precipitation would be likely to occur.

Figure 4.16 was constructed as follows. For a given pH, $[OH^-]$ was calculated from Eq. (4.134) and used in Eq. (4.129) to determine $[Zn^{2+}]$. The latter value was in turn used in Eq. (4.130) to calculate $[ZnOH^+]$, and so on until the concentration of each complex was known. These values were then inserted in Eq. (4.135) to determine the total soluble concentration of zinc present. The total values so calculated were plotted as a function of pH and are represented by the curve bordered by cross-hatches in Fig. 4.16.

The reader should note that log C–pH equations can be developed for each Zn species shown in Fig. 4.16. For example, taking the log of both sides of Eq. (4.129) gives

$$\log [Zn^{2+}] = \log K_{sp} - 2 \log [OH^-] \quad (4.136)$$

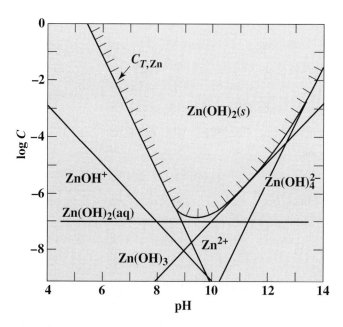

Figure 4.16
Solubility of $Zn(OH)_2(s)$ as a function of solution pH.

From Eq. (4.134):

$$\log [H^+] + \log [OH^-] = -14 \qquad (4.137)$$

Combining Eqs. (4.136) and (4.137), and recognizing that $pH = -\log [H^+]$,

$$\log [Zn^{2+}] = 10.9 - 2pH \qquad (4.138)$$

Taking the log of both sides of Eq. (4.130) gives

$$\log [ZnOH^+] = \log K_1 + \log [Zn^{2+}] + \log [OH^-] \qquad (4.139)$$

Now, substituting Eqs. (4.137) and (4.138) into Eq. (4.139) gives

$$\log [ZnOH^+] = 1.05 - pH \qquad (4.140)$$

Similar equations can be developed for the remaining Zn-hydroxide complexes.

It is possible for zinc to form polynuclear complexes containing two or more zinc atoms, and these would add to the total concentration of zinc in solution. However, for this example the polynuclear species were not considered, and in fact, would not have affected the equilibrium concentration significantly.

Figure 4.16 illustrates that at low pH, Zn^{2+} is the predominant soluble species present, followed by $ZnOH^+$. At very high pH, $Zn(OH)_4^{2-}$ is the predominant species followed by $Zn(OH)_3^-$. Quite obviously, if complex formation were ignored and zinc solubilities were calculated solely by use of Eq. (4.129), erroneous conclusions could be drawn.

At what pH does zinc have its minimum solubility in water? | **EXAMPLE 4.32**

From Fig. 4.16, minimum solubility occurs at pH 9.4. At this pH the total equilibrium zinc solution concentration equals $10^{-6.8}$ or 1.6×10^{-7} M. The predominant soluble species present is $Zn(OH)_2(aq)$ at 1.1×10^{-7} M, followed by $Zn(OH)_3^-$ at 3.6×10^{-8} M and Zn^{2+} at 1.3×10^{-8} M.

A plating waste has a zinc concentration of 10^{-3} mol/L. If lime is added to the solution, above what pH will zinc just begin to precipitate? What minimum pH should be used to decrease the zinc concentration below 10^{-5} M? | **EXAMPLE 4.33**

From Fig. 4.16, the pH for a maximum equilibrium solution concentration of zinc of 10^{-3} ($\log C = -3$) is 6.9. At a pH above this value, zinc should begin to precipitate from solution.

In order to reduce the zinc concentration to 10^{-5} ($\log C = -5$), the pH from Fig. 4.16 must equal 7.9. In practice, a higher pH (but not above 9.4) should be maintained to ensure adequate zinc removal.

Solubility of metals can be further complicated by the presence of ligands other than hydroxide. The effect of ammonia complexes on the solubility of $Zn(OH)_2(s)$ will serve as an example. Inspection of Table 4.4 shows that there are four Zn-ammonia complexes:

$$\frac{[Zn(NH_3)^{2+}]}{[Zn^{2+}][NH_3]} = K_1 = 151.4 \tag{4.141}$$

$$\frac{[Zn(NH_3)_2^{2+}]}{[Zn(NH_3)^{2+}][NH_3]} = K_2 = 177.8 \tag{4.142}$$

$$\frac{[Zn(NH_3)_3^{2+}]}{[Zn(NH_3)_2^{2+}][NH_3]} = K_3 = 204.2 \tag{4.143}$$

$$\frac{[Zn(NH_3)_4^{2+}]}{[Zn(NH_3)_3^{2+}][NH_3]} = K_4 = 91.2 \tag{4.144}$$

The mass balance for zinc is now

$$C_{T,Zn} = [Zn^{2+}] + [ZnOH^+] + [Zn(OH)_2] + [Zn(OH)_3^-] + [Zn(OH)_4^{2-}]$$
$$+ [ZN(NH_3)^{2+}] + [Zn(NH_3)_2^{2+}] + [Zn(NH_3)_3^{2+}] + [Zn(NH_3)_4^{2+}] \tag{4.145}$$

The mass balance for ammonia is

$$C_{T,N} = [NH_4^+] + [NH_3] + [Zn(NH_3)^{2+}] + 2[Zn(NH_3)_2^{2+}]$$
$$+ 3[Zn(NH_3)_3^{2+}] + 4[Zn(NH_3)_4^{2+}] \tag{4.146}$$

If it is assumed that the concentration of the Zn-ammonia complexes is small relative to $[NH_4^+] + [NH_3]$, then the mass balance for ammonia can be approximated as

$$C_{T,N} = [NH_4^+] + [NH_3] \tag{4.147}$$

These equations can be used along with Eqs. (4.129) to (4.134) to construct Fig. 4.17 for a total ammonia concentration ($C_{T,N}$) of 0.1 M. The concentration of $[NH_3]$ as a function of pH can be calculated using Eq. (4.147) and the equilibrium relationship given in Table 4.1. The procedure is then much the same as that used to calculate the concentrations of the Zn-hydroxide complexes. The concentration of $[Zn^{2+}]$ determined from Eq. (4.129) is used with $[NH_3]$ to calculate $[Zn(NH_3)^{2+}]$ from Eq. (4.141), which in turn is used to determine $[Zn(NH_3)_2^{2+}]$ from Eq. (4.142), and so on. These values are then substituted in Eq. (4.145) to determine the total soluble concentration of zinc. These values are plotted as a function of pH and are represented by the curve bordered by cross-hatches in Fig. 4.17.

The reader should again note that log C–pH equations for Zn-ammonia complexes can be developed in the same manner as those for the Zn-hydroxide complexes. For example, from Eq. (4.141),

$$\log[Zn(NH_3)^{2+}] = \log K_1 + \log[Zn^{2+}] + \log[NH_3] \tag{4.148}$$

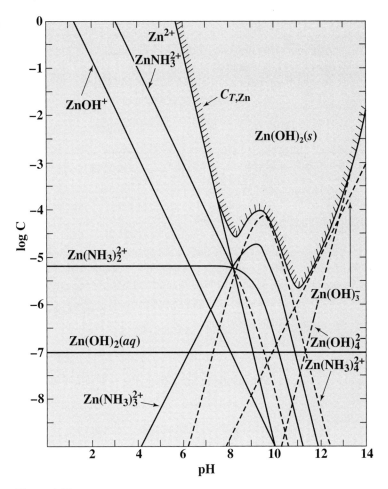

Figure 4.17
Solubility of $Zn(OH)_2(s)$ as a function of pH for a solution containing 0.1 M total ammonia.

Substituting Eq. (4.138) into Eq. (4.148) gives

$$\log[Zn(NH_3)^{2+}] = 13.08 - 2pH + \log[NH_3] \tag{4.149}$$

Values for $\log [NH_3]$ as a function of pH can be determined using Eq. (4.147) and the equilibrium expression from Table 4.1. Equations for the remaining Zn-ammonia complexes can be developed in a similar fashion:

$$\log [Zn(NH_3)_2^{2+}] = 15.33 - 2pH + 2\log[NH_3] \tag{4.150}$$

$$\log [Zn(NH_3)_3^{2+}] = 17.64 - 2pH + 3\log[NH_3] \tag{4.151}$$

$$\log [Zn(NH_3)_4^{2+}] = 19.60 - 2pH + 4\log[NH_3] \tag{4.152}$$

Inspection of Figs. 4.16 and 4.17 indicates that the presence of ammonia at 0.1 M has a significant impact on the solubility of zinc, especially for pH values between 7 and 11. It must be emphasized that a new logarithmic concentration diagram must be drawn for a different total ammonia concentration ($C_{T,N}$). Obviously, if $C_{T,N}$ is low enough, there will be no impact on zinc solubility.

EXAMPLE 4.34

Using Fig. 4.17, rework Examples 4.32 and 4.33 for a situation where $C_{T,N}$ is 0.1 M.

The minimum zinc solubility occurs at a pH of approximately 11. At this pH, the total soluble zinc ($C_{T,Zn} = S$) is approximately $10^{-5.7}$ or 2×10^{-6} M. These are significantly different than the values in the absence of NH_3 (pH = 9.4 and solubility of 1.6×10^{-7} M).

For a plating waste with a zinc concentration of 10^{-3} M, zinc will just begin to precipitate at a pH of about 6.9, the same as when ammonia is absent. In order to reduce the total soluble zinc to 10^{-5} M, a pH of 10.3 is required, which is significantly different than the required pH of 7.9 in the absence of ammonia complexes.

A final metal solubility example is given for a nonhydroxide salt.

EXAMPLE 4.35

What is the solubility of silver (in mg/L of Ag) in a solution at 25°C and in equilibrium with AgCl(s), 10^{-3} M total ammonia, 10^{-2} M total chloride, and a pH of 9.0? Ignore activity corrections.

Ag^+ forms complexes with OH^-, Cl^-, and NH_3 (see Tables 4.4 and 4.5). Thus,

$$C_{T,Ag} = S = [Ag^+] + [AgOH(aq)] + [Ag(OH)_2^-] + [AgCl(aq)] + [AgCl_2^-] + [AgCl_3^{2-}] + [AgCl_4^{3-}] + [AgNH_3^+] + [Ag(NH_3)_2^+]$$

$$K_{sp} = 3 \times 10^{-10} = [Ag^+][Cl^-] \qquad [Ag^+] = \frac{3 \times 10^{-10}}{[Cl^-]}$$

We need to know [Cl^-]. The mass balance for chloride is

$$C_{T,Cl} = [Cl^-] + [AgCl(aq)] + 2[AgCl_2^-] + 3[AgCl_3^{2-}] + 4[AgCl_4^{3-}]$$

We could assume here that [Cl^-] $>>$ sum of all Ag-Cl complexes. That is, [Cl^-] = 0.01 M.

So,
$$[Ag^+] = \frac{3 \times 10^{-10}}{[Cl^-]} = \frac{3 \times 10^{-10}}{0.01} = 3 \times 10^{-8} \text{ M}$$

Using data from Table 4.4 for Ag-Cl complexes,

$$[AgCl(aq)] = 3 \times 10^{-8}(0.01)(10^{3.45}) = 8.46 \times 10^{-7} \text{ M}$$
$$[AgCl_2^-] = 8.46 \times 10^{-7}(0.01)(10^{2.22}) = 1.40 \times 10^{-6} \text{ M}$$
$$[AgCl_3^{2-}] = 1.40 \times 10^{-6}(0.01)(10^{0.33}) = 3.00 \times 10^{-8} \text{ M}$$
$$[AgCl_4^{3-}] = 3.00 \times 10^{-8}(0.01)(10^{0.04}) = 3.29 \times 10^{-10} \text{ M}$$

It appears our assumption that $[Cl^-] >>$ sum of all Ag-Cl complexes is a good one!
Using data from Table 4.5 for complexes with OH^- for pH = 9.0,

$$[AgOH(aq)] = 3 \times 10^{-8}(10^{-(14-9)})(10^{2.3}) = 5.99 \times 10^{-11} \text{ M}$$
$$[Ag(OH)_2^-] = 5.99 \times 10^{-11}(10^{-(14-9)})(10^{1.9}) = 4.75 \times 10^{-14} \text{ M}$$

In order to determine the concentrations of the $Ag\text{-}NH_3$ complexes, we need to determine $[NH_3]$ for pH = 9.0. The mass balance for total ammonia is

$$C_{T,NH_3} = 0.01 = [NH_4^+] + [NH_3] + [AgNH_3^+] + 2[Ag(NH_3)_2^+]$$

We could assume here that the $Ag\text{-}NH_3$ complexes are negligible compared to the sum of NH_4^+ and NH_3.

$$C_{T,NH_3} = 0.01 = [NH_4^+] + [NH_3]$$

From Eq. (4.22) and definitions following Eq. (4.20)

$$[NH_3] = \alpha_1 C_{T,NH_3} = \left(\frac{1}{[H^+]/K_{A1} + 1}\right)(0.01) = \left(\frac{10^{-9}}{10^{-9.26}} + 1\right)^{-1}(0.01) = 3.55 \times 10^{-3} \text{ M}$$

Now, using equilibrium constant data from Table 4.4,

$$[AgNH_3^+] = 3 \times 10^{-8}(3.55 \times 10^{-3})(10^{3.32}) = 2.23 \times 10^{-7} \text{ M}$$
$$[Ag(NH_3)_2^+] = 2.23 \times 10^{-7}(3.55 \times 10^{-3})(10^{3.92}) = 6.57 \times 10^{-6} \text{ M}$$

From these calculations we see that our assumption that $Ag\text{-}NH_3$ complexes contribute little to C_{T,NH_3} is a good one.
Substituting into the solubility equation gives

$$S = 3 \times 10^{-8} + 5.99 \times 10^{-11} + 4.75 \times 10^{-14} + 8.46 \times 10^{-7} + 1.40 \times 10^{-6}$$
$$+ 3.00 \times 10^{-8} + 3.29 \times 10^{-10} + 2.23 \times 10^{-7} + 6.57 \times 10^{-6}$$
$$= 9.10 \times 10^{-6} \text{ M} = 0.98 \text{ mg/L of Ag}$$

Adsorption at the Solid-Water Interface

The chemistry of solid-water interactions is more complex than the simple equilibrium between ions in solution and a solid as described here. Adsorption at the solid-water interface is a many-faceted process that plays a major role in the fate and transport of metals, some ions, and organic compounds in natural waters and engineered reactors. Adsorption was discussed in Sec. 3.13 with an emphasis on its importance in engineered reactors. Adsorption of organic compounds is discussed in Sec. 5.34. Related aspects of colloid chemistry are presented in Chap. 7. What follows is a brief introduction to some basic concepts of the chemistry of solid-water interfaces. For a detailed treatise, the student is referred to texts by Stumm[10] and Benjamin.[11]

Reactions of metals, acids, and bases with metal oxides (for example, Al_2O_3, Fe_2O_3, FeOOH, SiO_2) will be used to introduce the concepts involved. Equilibrium expressions can be developed much in the same fashion as the equilibrium expressions presented earlier in this chapter. Consider the neutral surface of an iron oxide being represented by $\equiv FeOH$ (most metal oxides, when present in water, have surface hydroxide groups). This surface group can accept or donate a proton as described by the following acid-base equilibria:

$$\equiv FeOH_2^+ \rightleftharpoons \equiv FeOH + H^+ \tag{4.153}$$

$$\equiv FeOH \rightleftharpoons \equiv FeO^- + H^+ \tag{4.154}$$

Equilibrium constants for these expressions are (ignoring activity corrections)

$$K_{A1} = \frac{(\equiv FeOH)[H^+]}{(\equiv FeOH_2^+)} \tag{4.155}$$

$$K_{A2} = \frac{(\equiv FeO^-)[H^+]}{(\equiv FeOH)} \tag{4.156}$$

where ( ) represents the concentration of surface species in moles per kilogram of iron oxide. It should be noted that these concentrations can be converted to M (mol/L) by knowing the concentration of surface sites (moles per area of solid), the specific surface area of the solid (area/kg of solid), and the total concentration of solid in solution (kg/L). These equilibrium expressions are very similar to those presented in Sec. 4.5 for ionization of acids (for example, carbonic acid in Table 4.1).

Metals can adsorb to these surfaces according to the following equilibrium:

$$\equiv FeOH + M^{2+} \rightleftharpoons \equiv FeOM^+ + H^+ \tag{4.157}$$

Such metal-oxide complexes are commonly called *surface complexes*. The corresponding equilibrium constant for this surface-metal complex is

$$K_{SM} = \frac{(\equiv FeOM^+)[H^+]}{(\equiv FeOH)[M^{2+}]} \tag{4.158}$$

This surface complexation approach can be used to develop the Langmuir isotherm (Sec. 3.12) as well as more complicated models of adsorption behavior. These equilibrium expressions, along with mass balance equations and the charge balance, can be used to solve problems involving adsorption of metals (and other ions) to metal oxides and other similar mineral surfaces. A simple example follows.

---

[10]W. Stumm. "Chemistry of the Solid-Water Interface." Wiley-Interscience, New York, 1992.

[11]M. M. Benjamin, "Water Chemistry," McGraw-Hill, New York, 2002.

EXAMPLE 4.36

Consider the adsorption of copper ion (Cu^{2+}) by the iron oxide geothite (FeOOH). Estimate the percentage of copper sorbed to the surface of FeOOH for the following conditions. Ignore Cu-OH complexes for this example.

$$K_{A1} = \frac{[\equiv FeOH][H^+]}{[\equiv FeOH_2^+]} \qquad pK_{A1} = 6.0$$

$$K_{A2} = \frac{[\equiv FeO^-][H^+]}{[\equiv FeOH]} \qquad pK_{A2} = 8.8$$

$$K_{SCu} = \frac{[\equiv FeOCu^+][H^+]}{[\equiv FeOH][Cu^{2+}]} \qquad pK_{SCu} = -8.0$$

Note that molar concentrations (M) of the surface species in these expressions are used.

$$\text{Total iron oxide surface concentration} = 10^{-6}\,M$$
$$\text{Total copper concentration} = 10^{-6}\,M$$
$$pH = 7.0$$

The mass balance equations for the concentration of surface and copper species are

$$\text{Total iron oxide surface species} = 10^{-6}\,M = [\equiv FeOH_2^+] + [\equiv FeOH]$$
$$+ [\equiv FeO^-] + [\equiv FeOCu^+]$$
$$\text{Total copper species} = 10^{-6}\,M = [Cu^{2+}] + [\equiv FeOCu^+]$$

Using the equilibrium expressions and making appropriate substitutions gives

$$10^{-6} = [\equiv FeOH]\left(\frac{[H^+]}{K_{A1}} + 1 + \frac{K_{A2}}{[H^+]} + \frac{K_{SCu}[Cu^{2+}]}{[H^+]}\right)$$

$$10^{-6} = [Cu^{2+}]\left(1 + K_{SCu}\frac{[\equiv FeOH]}{[H^+]}\right)$$

Solving these two equations and making the appropriate substitutions gives the following:

$$[\equiv FeOH] = 2.99 \times 10^{-11}\,M$$
$$[\equiv FeOH_2^+] = 2.99 \times 10^{-13}\,M$$
$$[\equiv FeO^-] = 4.74 \times 10^{-13}\,M$$
$$[\equiv FeOCu^+] = 1.00 \times 10^{-6}\,M$$
$$[Cu^{2+}] = 3.34 \times 10^{-11}\,M$$

In this example, with a pH of 7.0, essentially 100 percent of the copper is adsorbed to the iron oxide surface. This indicates the potential significance of surface complexes in the speciation of metals in natural waters and engineered reactors. pH has a significant impact on sorption. For example, if the pH for this example were 2.0, approximately 38 percent of the copper would be sorbed to the iron oxide with the remaining 62 percent being dissolved Cu^{2+}.

The chemistry of the solid-water interface is further complicated by the potential interactions of dissolved complexes with hydroxide (metal hydrolysis) and other ligands. In addition ligands such as organic acids and humic acids may inhibit or enhance the formation of surface complexes. Electrostatic effects may be significant. The surface may catalyze oxidation-reduction reactions. Computer models are usually needed to solve such complicated problems. A thorough discussion of these effects is beyond the scope of this text.

4.10 | OXIDATION-REDUCTION REACTIONS

Oxidation-reduction reactions are among the most important with which environmental engineers and scientists deal. Many reactions of interest in wastewater treatment such as organic oxidation and methane fermentation, nitrification, and denitrification are of this type and are mediated by bacteria. Oxidation-reduction reactions are important in the solubilization and precipitation of iron and manganese. Oxidants such as chlorine and ozone are added to water and wastewater to bring about desired inorganic and organic transformations as well as to disinfect. The fate of materials introduced into the environment frequently depends upon the redox environment to which they become subjected. Also, many environmental analyses depend upon oxidation-reduction reactions.

As with acid-base, solubility, or complex formation, oxidation-reduction reactions tend toward a state of equilibrium. An understanding of oxidation-reduction equilibria can help to indicate whether a particular reaction is possible under given environmental conditions. It will not tell whether the reaction will in fact occur, or how fast it will occur, but nevertheless such an understanding is helpful in evaluating how conditions might best be changed to encourage desirable transformations or to prevent undesirable ones. Oxidation-reduction reactions can be quite complex. Graphical approaches can be helpful to reduce this complexity and to illustrate the significant factors involved for a particular case.

Equilibrium Relationships

In Sec. 2.7 methods for balancing oxidation-reduction (typically abbreviated redox) reactions were presented. A brief introduction to equilibrium in redox reactions was given in Sec. 3.9. These concepts are expanded in this section.

Redox reactions involve electron transfer as seen in the following generic half-reaction:

$$\text{Ox} + z e^- \rightleftharpoons \text{Red} \tag{4.159}$$

In this general format, "Ox" represents the oxidized species and "Red" represents the reduced species. At equilibrium, the following holds:

$$K = \frac{\{\text{Red}\}}{\{\text{Ox}\}\{e^-\}^z} \tag{4.160}$$

Solving for $\{e^-\}$ gives

$$\{e^-\} = \left(\frac{\{\text{Red}\}}{K\{\text{Ox}\}} \right)^{1/z} \tag{4.161}$$

The electron activity $\{e^-\}$ is expressed in two ways: (1) as a measured potential equal to $-2.3RT\log\{e^-\}/zF$ with the units being volts (or millivolts) and using the symbol E, or (2) using a logarithmic scale where $pE = -\log\{e^-\}$.[12] Although pE seems analogous to pH ($-\log\{H^+\}$) in that the logarithm of an activity is used, the electron is not an individual species like H^+ which can be measured. It is true, however, that like acid-base reactions where H^+ are exchanged, redox reactions do involve the exchange (transfer) of e^- from one species to another. E is usually called the redox potential. These terms are also directly related to the free energy for the system. A half reaction for the reduction of NO_3^- to NH_4^+ will be used for illustration,

$$\tfrac{1}{8}NO_3^- + \tfrac{5}{4}H^+ + e^- \;\rightleftharpoons\; \tfrac{1}{8}NH_4^+ + \tfrac{3}{8}H_2O \qquad (4.162)$$

$$\Delta G_f^0 \quad \tfrac{1}{8}(-111.34) \quad 0 \qquad 0 \qquad \tfrac{1}{8}(-79.37) \quad \tfrac{3}{8}(-237.18)$$

The standard free energies of formation for each of the reactants and products are obtained from Table 3.1 or from the Appendix. By convention, ΔG_f^0 for H^+ and e^- are zero. The standard free energy for the reaction as written is thus

$$\Delta G^0 = \tfrac{1}{8}(-79.37) + \tfrac{3}{8}(-237.18) - \tfrac{1}{8}(-111.34) = -84.95 \text{ kJ}$$

The standard electrode potential for the reaction can be calculated from ΔG^0 using Eq. (3.44),

$$E^0 = -\frac{\Delta G^0}{zF} = -\frac{-84.95}{1(96.485)} = 0.880 \text{ volts}$$

where 96.485 kJ/volt-eq is the Faraday, the conversion factor necessary to make the units consistent (see Sec. 3.9). As indicated by Eqs. (3.14) and (3.44), the equilibrium condition for the oxidation-reduction reaction can be determined from either ΔG^0 or E^0:

$$\ln K = 2.3 \log K = -\frac{\Delta G^0}{RT} = \frac{zFE^0}{RT}$$

Thus, for Eq. (4.162) the following relationship between concentrations must hold at equilibrium (using molar concentration except for $\{e^-\}$):

$$\log \frac{[NH_4^+]^{1/8}}{[NO_3^-]^{1/8}[H^+]^{5/4}\{e^-\}} = -\frac{\Delta G^0}{2.3\,RT} = \frac{zFE^0}{2.3RT} \qquad (4.163)$$

This equation can be converted into the following forms for $T = 25°C$:

$$E = E^0 - 0.059 \log \frac{[NH_4^+]^{1/8}}{[NO_3^-]^{1/8}[H^+]^{5/4}} \qquad (4.164)$$

---

[12]A note here about symbols. The symbols E and E_H are typically used for redox potential (units are volts or millivolts). The subscript "H" is used to indicate that the reference for measuring this potential is the half reaction for the $H_2(g)/H^+$ couple, which is assigned a value of 0 volts. The symbols pE, pe, and $pε$ have been used for "$-\log\{e^-\}$." In this text we will use E and pE.

or,
$$pE = pE^0 - \log \frac{[NH_4^+]^{1/8}}{[NO_3^-]^{1/8}[H^+]^{5/4}} \tag{4.165}$$

where
$$zpE^0 = -\frac{\Delta G^0}{2.3RT} = \frac{zFE^0}{2.3RT} \tag{4.166}$$

For a temperature of 25°C, $pE^0 = 16.9E^0/$volt and $pE = 16.9 E/$volt. Table 2.4 lists values of pE^0 and ΔG^0 for various half reactions of interest in water quality.

The redox potential of a system can be indicated by pE. When pE is large, the electron activity is low and the system tends to be an oxidizing one: i.e., half reactions tend to be driven to the left. When pE is small, the system is reducing and reactions tend to be driven to the right.

For a given system to be at equilibrium, E or pE for all possible redox half reactions for constituents present must be the same. In other words, for the "whole system," $E = pE = 0$ at equilibrium. This concept is demonstrated in Example 4.37.

EXAMPLE 4.37

What is the concentration of ferrous iron ($[Fe^{2+}]$) in a water that is in equilibrium with $Fe(OH)_3(s)$ and the atmosphere ($P_{O_2} = 0.21$ atm), and has a pH of 5.0?

There are a number of ways to work this problem. We will demonstrate two. The following will be useful.

$$
\begin{array}{llll}
Fe(OH)_3(s) + 3H^+ + e^- = Fe^{2+} + 3H_2O & E^0 = 0.947 \text{ volts} & pE^0 = 16.00 \\
\frac{1}{4}O_2(g) + H^+ + e^- = \frac{1}{2}H_2O & E^0 = 1.229 \text{ volts} & pE^0 = 20.77
\end{array}
$$

Method 1: Here we will make use of the redox potential E. At equilibrium, $E = \Delta G = 0$. A balanced redox reaction can be created by reversing the first reaction and adding it to the second reaction as was shown in Sec. 3.9.

$$
\begin{array}{ll}
Fe^{2+} + 3H_2O = Fe(OH)_3(s) + 3H^+ + e^- & E^0 = -0.947 \text{ volts} \\
\frac{1}{4}O_2(g) + H^+ + e^- = \frac{1}{2}H_2O & E^0 = 1.229 \text{ volts} \\
\hline
Fe^{2+} + 0.25O_2(g) + 2.5H_2O = Fe(OH)_3(s) + 2H^+ & E^0 = 0.282 \text{ volts}
\end{array}
$$

From Eq. (3.45) we can calculate the equilibrium constant for this reaction:

$$\log K = 16.9zE^0 = 16.9(1)(0.282) = 4.77$$

From this, we have, ignoring activity corrections;

$$K = 10^{4.77} = \frac{[Fe(OH)_3(s)][H^+]^2}{[Fe^{2+}](P_{O_2})^{0.25}[H_2O]^{2.5}} = \frac{(1)(10^{-5})^2}{[Fe^{2+}](0.21)^{0.25}(1)^{2.5}}$$

Solving gives $[Fe^{2+}] = 2.51 \times 10^{-15}$ M. From this it is clear that $Fe(OH)_3(s)$ is the dominant species when O_2 is present.

Method 2: Here we will make use of pE. If the system is at equilibrium, then by definition overall $pE = 0$. This also means that pE for the two half reactions involved must be equal (both must be written in the reduction direction). We can develop equations analogous to Eq. (4.165) for the two half reactions given.

$$pE_{Fe} = pE_{Fe}^0 - \log \frac{[Fe^{2+}][H_2O]^3}{[Fe(OH)_3(s)][H^+]^3} = 16.00 - \log \frac{[Fe^{2+}](1)^3}{(1)(10^{-5})^3}$$

$$pE_{O_2} = pE_{O_2}^0 - \log \frac{[H_2O]^{1/2}}{(P_{O_2})^{1/4}[H^+]} = 20.77 - \log \frac{(1)^{1/2}}{(0.21)^{1/4}(10^{-5})} = 20.77 - 5.17 = 15.60$$

Now, at equilibrium, $pE_{Fe} = pE_{O_2}$:

$$15.60 = 16.00 - \log \frac{[Fe^{2+}](1)^3}{(1)(10^{-5})^3} = 16.00 - \log [Fe^{2+}] + 3 \log (10^{-5})$$

$$\log [Fe^{2+}] = 16.00 - 15.60 - 15 = -14.60$$

$$[Fe^{2+}] = 2.51 \times 10^{-15} \text{ M} \qquad \text{(same as method 1)}$$

Logarithmic Concentration Diagrams

A logarithmic concentration diagram can be used to illustrate the relative equilibrium concentrations for NH_4^+ and NO_3^- as a function of pE or E, as shown in Fig. 4.18. This figure was constructed for 25°C and using Eq. (4.165). The value for pE° was obtained from Table 2.4. For pH = 7, $[H^+] = 10^{-7}$; thus,

$$pE = 14.88 - \tfrac{1}{8} \log [NH_4^+] + \tfrac{1}{8} \log [NO_3^-] + \tfrac{5}{4} \log (10^{-7})$$

which reduces to

$$pE = 6.13 - \tfrac{1}{8} \log [NH_4^+] + \tfrac{1}{8} \log [NO_3^-] \qquad (4.167)$$

It was also assumed that $C_T = [NH_4^+] + [NO_3^-] = 10^{-3}$ M.

Figure 4.18 is constructed similar to an acid-base logarithmic concentration diagram. A horizontal line representing $\log C_T$ is drawn. A system point is located at $pE = 6.13$. Two lines are drawn through the system point, one to the left with a slope of $\tfrac{1}{8}$ representing NO_3^-, and one to the right with a slope of $-\tfrac{1}{8}$ representing NH_4^+. Since when $pE = 6.13$, $[NH_4^+] = [NO_3^-] = C_T/2$, the two diagonal lines must intersect at $\log \tfrac{1}{2}$ or -0.3 logarithmic units below the system point. Curved lines through a point 0.3 units below the system point and connecting the straight lines completes the figure.

Two additional lines are shown in Fig. 4.18 and represent respective values for pE at which water is reduced and oxidized:

Reduction:

$$H^+ + e^- \rightleftharpoons \tfrac{1}{2}H_2(g)$$

$$pE = 0.0 - \log \frac{[H_2]^{1/2}}{[H^+]} \qquad (4.168)$$

At pH = 7,

$$pE = -7 - \tfrac{1}{2} \log [H_2] \qquad (4.169)$$

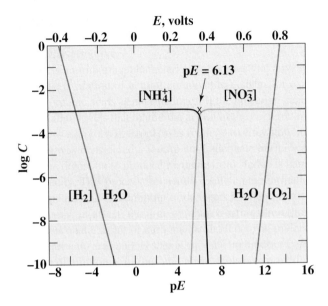

Figure 4.18
Logarithmic concentration diagram showing the relationship between $[NH_4^+]$ and $[NO_3^-]$ as a function of pE for pH = 7 and $[NH_4^+] + [NO_3^-] = 10^{-3}$ mol/L.

Oxidation:

$$\tfrac{1}{4}O_2(g) + H^+ + e^- \rightleftharpoons \tfrac{1}{2}H_2O$$

$$pE = 20.77 - \log \frac{1}{[O_2]^{1/4}[H^+]} \tag{4.170}$$

At pH = 7,

$$pE = 13.77 + \tfrac{1}{4}\log[O_2] \tag{4.171}$$

If pE for a given aqueous system were to the left of the $[H_2]$ line, the system would be highly reducing such that water would tend to decompose with the evolution of H_2. With pE to the right of the $[O_2]$ line, the system would be highly oxidizing with the result that water would tend to decompose with O_2 evolution. Natural aquatic systems are characterized by pE values between these two extremes. The student should recall that since H_2 and O_2 are gases, the unit of concentration for these species is partial pressure in atmospheres.

pE–pH Diagrams

Figure 4.18 represents the conditions for a neutral pH. As indicated by Eq. (4.165), the pE in the $NH_4^+ - NO_3^-$ system is also a function of $[H^+]$ and hence pH. This is true for any oxidation-reduction reaction in which H^+ or OH^- is involved. A predominance area or pE–pH diagram is frequently used to illustrate the relationship between pH, pE, and the most stable species in a given oxidation-reduction system. A pE–pH diagram for the nitrogen system is illustrated in Fig. 4.19. In order to construct Fig. 4.19, relationships are needed for the following couples: NH_4^+/NO_3^-, NH_4^+/NO_2^-, NO_2^-/NO_3^-, NH_3/NO_3^-, NH_3/NO_2^- and NH_4^+/NH_3. Equation (4.165) represents the equation for the NH_4^+/NO_3^- couple. Half reactions from Table 2.4 can be used to construct the remaining relationships. These equations are now developed.

The general form of the equation to be used is

$$pE = pE^0 - \frac{1}{z}\log \frac{(reduced)}{(oxidized)} \tag{4.172}$$

where (*reduced*) represents the species on the reduced side of the half reaction and (*oxidized*) represents the species on the oxidized side. If the half reaction is written for $1e^-$, then $z = 1$. In order to generate a pE–pH equation, $[H^+]$ must be separated from the log term in Eq. (4.172). Equation (4.165) then becomes

$$pE = 14.88 - \frac{5}{4}pH - \frac{1}{8}\log \frac{[NH_4^+]}{[NO_3^-]} \tag{4.173}$$

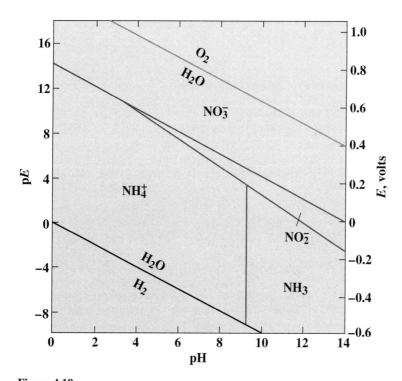

Figure 4.19
pE–pH diagram illustrating predominant nitrogen forms at equilibrium in an aqueous system.

Similarly, half reaction 19 from Table 2.4 becomes

$$pE = 15.08 - \tfrac{4}{3}pH - \tfrac{1}{6}\log \frac{[NH_4^+]}{[NO_2^-]} \tag{4.174}$$

Table 2.4 does not contain a half reaction for the NO_2^-/NO_3^- couple. Thus, a new half reaction must be developed. This can be done by combining half reactions 19 and 20 or 21 and 22 from Table 2.4. We will combine half reactions 19 and 20 to eliminate NH_4^+. By multiplying inverted half reaction 19 by 6

$$NH_4^+ + 2H_2O = NO_2^- + 8H^+ + 6e^-$$

and adding it to half reaction 20 multiplied by 8

$$NO_3^- + 10H^+ + 8e^- = NH_4^+ + 3H_2O$$

gives

$$NO_3^- + 2H^+ + 2e^- = NO_2^- + H_2O$$

or

$$\tfrac{1}{2}NO_3^- + H^+ + e^- = \tfrac{1}{2}NO_2^- + \tfrac{1}{2}H_2O \tag{4.175}$$

Since E^0 is given on a per electron-equivalent basis, calculation of E^0 for Eq. (4.175) is

$$E^0 = \frac{6(-0.892) + 8(0.880)}{2} = 0.844 \text{ volt}$$

From this, $pE^0 = 16.9E^0 = 14.26$. The pE–pH equation for the NH_2^-/NO_3^- couple is then

$$pE = 14.26 - pH - \tfrac{1}{2}\log \frac{[NO_2^-]}{[NO_3^-]} \tag{4.176}$$

In order to generate equations for the NH_3/NO_3^- and NH_3/NO_2^- couples, the following is needed:

$$K_A = \frac{[NH_3][H^+]}{[NH_4^+]} = 10^{-9.26} \tag{4.177}$$

The logarithmic form of Eq. (4.177) is

$$\log [NH_4^+] = \log [NH_3] - pH + 9.26 \tag{4.178}$$

Substitution of Eq. (4.178) into Eqs. (4.173) and (4.174) gives

$$pE = 13.72 - \tfrac{9}{8}pH - \tfrac{1}{8}\log \frac{[NH_3]}{[NO_3^-]} \tag{4.179}$$

$$pE = 13.54 - \tfrac{7}{6}pH - \tfrac{1}{6}\log \frac{[NH_3]}{[NO_2^-]} \tag{4.180}$$

The lines in Fig. 4.19 represent points of transition from predominance by one nitrogen form to predominance by another. The lines represent the location where the concentrations of the two most predominant nitrogen forms are equal. Thus, the equation for the line between NH_4^+ and NO_3^- is given by the result from setting $[NH_4^+]$ equal to $[NO_3^-]$ in Eq. (4.173),

$$pE = 14.88 - \tfrac{5}{4}pH \qquad\qquad (4.181)$$

The more reduced form, NH_4^+, is predominant in the area below the line while the oxidized form is predominant in the area above the line. The other lines in Fig. 4.19 are obtained from Eqs. (4.174) to (4.180) in a similar manner. The lines representing the boundaries where water is reduced or oxidized are obtained from Eqs. (4.168) and (4.170) by setting $[H_2]$ and $[O_2]$ equal to 1 atm. The stable range for water occurs between these two lines.

From Fig. 4.19 at high pE, the system tends to be oxidizing and the most oxidized form, NO_3^-, is dominant. If ammonia were introduced into such a system, it would not be stable and would tend to be oxidized to nitrate. The rate of this conversion at temperatures of natural waters, however, would be extremely slow if it were not for the nitrifying bacteria that catalyze the oxidation and obtain some of the energy released for growth. If ammonia were introduced into a system with low pE, it would be stable and could not be oxidized chemically or biologically. Thus, definition of a given system in terms of pE and pH can aid in understanding which reactions may occur when materials are added to the system.

Figure 4.20 is a pE–pH diagram illustrating zones of importance in natural water systems. It was constructed using half reactions from Table 2.4 for carbon dioxide reduction to methane, for sulfate reduction to hydrogen sulfide, and for nitrogen transformations including that between N_2 and nitrate. These oxidation-reduction reactions are commonly brought about through bacterial action at normal water temperatures. The aerobic zone represents conditions in aerated waters where the partial pressure of oxygen $[O_2]$ exceeds about 10^{-3} atm. This is the case for most natural streams, rivers, lake surfaces, and the ocean. Also represented are aerobic treatment processes such as activated sludge.

The anaerobic zone in the lower portion of Fig. 4.20 represents conditions where sulfate is reduced to hydrogen sulfide and organic materials are reduced to methane gas. Such conditions are common in organic laden sediments, the lower zones in polluted lakes, sanitary landfills, and anaerobic digesters.

A transition zone lies between the above two extremes. Reduction of nitrate to nitrogen gas through bacterial denitrification can occur in the lower portion of this zone, a condition common at natural water-sediment interfaces, the interior of bacterial flocs in otherwise aerobic treatment processes, saturated zones in soils, and in some biological treatment processes, specifically designed for this purpose. In the upper portion of the transition zone, nitrates are stable and oxidation of ammonia to nitrate by bacterial nitrification is possible.

Figure 4.20 illustrates that pE values associated with the different redox zones are dependent upon pH. At neutral pH, aerobic conditions are associated with a pE of about 13, and anaerobic by a pE below -3. For each unit decrease in pH, the respective values for pE increase about one unit.

An idea of the transformations likely to occur in natural water systems when other materials such as iron, manganese, mercury, and chlorine are introduced, can be obtained by comparing their pE–pH diagrams with the zones illustrated in Fig.

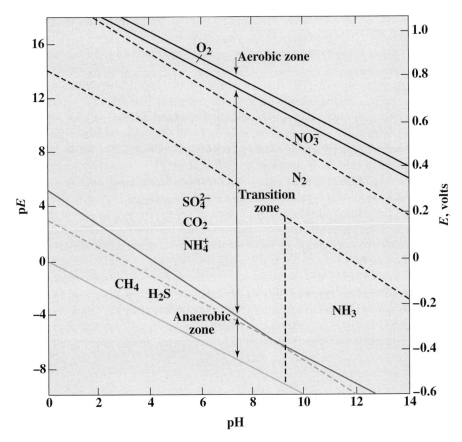

Figure 4.20
pE–pH diagram illustrating major oxidation-reduction zones in aqueous systems.

4.20. Pictorial representation significantly reduces the apparent complexity of such systems and can give environmental engineers and scientists a better intuitive feel for the systems with which they work.

4.11 | COMPUTER METHODS FOR SOLVING EQUILIBRIUM PROBLEMS

The chemical equilibrium problems presented in this chapter are relatively simple ones involving only a few species and equilibrium expressions. They have been solved using equilibrium constants, mass balances, charge balances, and sometimes proton conditions, and spreadsheets or by making simplifying assumptions and generating quadratic equations or by drawing log concentration diagrams. However, for more complex problems where many species are involved (say more than 15 to 20), where several solids may be present, and where equilibrium between the atmosphere and the water must be considered, computer solutions are required. Most available computer programs are based on the principles developed in this chapter. Basically, all use mass balances and equilibrium expressions that must be solved simultaneously. The numerical methods used for solving the equations are different for different computer programs. Additionally, there may be some constraints that are specific to the system being analyzed (for example, oxidized versus reduced environment, solids likely to be present, open or closed to the atmosphere). Mass balances must be written in a format that the computer program will recognize; these formats are different for the different computer programs. Matrices called tableaus are used by most for input data. Description of these formats is beyond the scope of this text. Most of the available programs come with databases that include equilibrium constants and other thermodynamic data for acid-base equilibria, complex-ion equilibria, and solubility equilibria.

The most commonly used software are based on MINEQL.[13] Examples include MINTEQA2,[14] Visual MINTEQ,[15] and MINEQL+.[16] Some of these software packages can be downloaded (or purchased) from the internet. More detailed descriptions of the format used for these programs is given by Benjamin[17] and Morel and Hering.[18]

PROBLEMS

For the following, assume that the temperature is 25°C. Ignore activity corrections unless otherwise noted.

4.1   Calculate the activity coefficient and activity of each ion in a solution containing 300 mg/L $NaNO_3$ and 150 mg/L $CaSO_4$.

Answer: Na^+ and NO_3^-, 0.91 and 3.21×10^{-3}; Ca^{2+} and SO_4^{2-}, 0.69 and 0.76×10^{-3}

4.2   Calculate the activity coefficient and activity of each ion in a solution containing 75 mg/L Na^+, 25 mg/L Ca^{2+}, 10 mg/L Mg^{2+}, 125 mg/L Cl^-, 50 mg/L HCO_3^-, and 48 mg/L SO_4^{2-}.

4.3   A solution is prepared by diluting 10^{-3} mol of propionic acid to 1 liter with distilled water. Calculate the equilibrium concentration for each chemical species in the water.

Answer: $[H^+] = 1.08 \times 10^{-4}$, $[OH^-] = 9.29 \times 10^{-11}$, $[HPr] = 8.92 \times 10^{-4}$, $[Pr^-] = 1.08 \times 10^{-4}$

4.4   A solution is prepared by diluting 10^{-2} mol of ammonia to 1 liter with distilled water. Calculate the equilibrium concentration for each chemical species in the water.

[13]J. C. Westall, J. L. Zachary, and F. M. M. Morel, "MINEQL, A Computer Program for the Calculation of Chemical Equilibrium Composition of Aqueous Systems," Technical Note 18, Parsons Laboratory, MIT, Cambridge, MA, 1976.

[14]U.S. Environmental Protection Agency.

[15]http://www.lwr.kth.se/english/OurSoftware/vminteq/index.htm

[16]MINEQL+ is a registered trademark of Environmental Research Software.

[17]M. M. Benjamin, "Water Chemistry," McGraw-Hill, New York, 2002.

[18]F. M. M. Morel and J. G. Hering, "Principles and Applications of Aquatic Chemistry," Wiley-Interscience, New York, 1993.

4.5 Calculate the equilibrium pH of a solution containing (*a*) 10^{-3} M H_2SO_4; (*b*) 10^{-8} M H_2SO_4.

  Answer: (a) 2.70; *(b)* 6.96

4.6 Calculate the equilibrium pH of NaOH solutions containing (*a*) 10^{-4} M NaOH; (*b*) 10^{-8} M NaOH.

4.7 Calculate the pH of solutions containing 100 mg/L of each of the following weak acids or weak bases: (*a*) acetric acid; (*b*) hypochlorous acid; (*c*) ammonia; (*d*) hydrocyanic acid.

  Answer: (a) 3.78; *(b)* 5.13; *(c)* 10.50; *(d)* 5.88

4.8 Calculate the pH of a solution containing 50 mg/L of each of the following weak acids or salts of weak acids: (*a*) carbonic acid; (*b*) sodium acetate; (*c*) sodium hypochlorite; (*d*) phosphoric acid.

4.9 Calculate the ratio of ammonia nitrogen in the NH_3 form to that in the NH_4^+ form in a solution with a pH of 7.4.

  Answer: 0.014

4.10 Calculate the ratio of hypochlorous acid to hypochlorite ion in solutions with the following pH values:

  (*a*) 6.0

  (*b*) 7.0

  (*c*) 8.0

4.11 Un-ionized hydrogen cyanide (HCN) is toxic to fish. Assume that the toxic level of HCN for a given species of fish is 10^{-6} M. For a total cyanide concentration of 10^{-5} M, determine at what pH HCN reaches toxic levels for

  (*a*) Ionic strength approximately 0

  (*b*) Ionic strength $= 0.1$ M

  Answer: 10.27; 10.15

4.12 What is the pH of a solution containing 10^{-2} M propionic acid?

4.13 What is the pH of a solution containing 10^{-2} M sodium propionate?

  Answer: 8.44

4.14 A solution is made by adding $Ca(OCl)_2$ to water to yield a concentration of 0.01 M. Using the algebraic approximation method,

  (*a*) What is the pH of the solution if activity corrections are ignored?

  (*b*) What is the pH of the solution if activity corrections are included?

  Answer: 9.92; 9.83

4.15 A solution is made by adding NH_4Cl to water to yield a concentration of 0.02 M. Using the algebraic approximation method,

  (*a*) What is the pH of the solution if activity corrections are ignored?

  (*b*) What is the pH of the solution if activity corrections are included?

4.16 Aspirin is produced from salicylic acid (HOC_6H_4COOH). This diprotic acid has $pK_{A1} = 2.97$ and $pK_{A2} = 13.70$ at 25°C. A solution is made by adding enough salicylic acid to water to give a concentration of 10^{-3} M.

  (*a*) Determine the equilibrium pH using a spreadsheet.

  (*b*) Determine the equilibrium pH using the algebraic approximation method.

  (*c*) What do you think this pH does to your stomach?

  Answer: 3.20; 3.20

4.17 Oxalic acid (HOOCCOOH) is a diprotic acid with $pK_{A1} = 1.25$ and $pK_{A2} = 4.28$. Using a spreadsheet, determine the equilibrium pH of a solution made by adding enough oxalic acid to water to give a concentration of 10^{-2} M.

4.18 A solution is made by adding enough sodium phthalate ($Na_2C_8O_4H_4$) to water to give a concentration of 10^{-3} M. Assume the sodium phthalate completely dissolves. At 25°C, $K_{A1} = 2.95$ and $K_{A2} = 5.41$.

(*a*) Determine the equilibrium pH using a spreadsheet.

(*b*) Determine the equilibrium pH using the algebraic approximation method.

4.19 Sufficient Na_2CO_3 is added to water to yield a concentration of 0.01 M.

(*a*) What is the pH of the solution at 25°C if activity corrections are ignored?

(*b*) What is the pH of the solution at 25°C if activity corrections are included?

(*c*) What is the pH of the solution at 15°C if activity corrections are included? (At 15°C, $pK_W = 14.34$, $pK_{A1} = 6.41$, and $pK_{A2} = 10.42$. These values were calculated using relationships developed in Chap. 3.)

4.20 Arsenic acid (H_3AsO_4) is a triprotic acid with $pK_{A1} = 2.22$, $pK_{A2} = 6.98$, and $pK_{A3} = 11.53$. KH_2AsO_4 is added to water to give a total concentration of 0.001 M.

(*a*) Write the mass balance for this solution.

(*b*) Write the charge balance for this solution.

(*c*) Write the proton condition for this solution.

(*d*) Using the algebraic approximation method, calculate the equilibrium pH.

(*e*) Determine the equilibrium pH using a spreadsheet.

4.21 Calculate the concentration of each chemical species in a solution made by dissolving enough NaH_2PO_4 in water to yield a concentration of 100 mg/L.
   Answer: pH = 5.17

4.22 A solution is made by adding acetic acid to yield a concentration of 0.01 M and sodium acetate to yield a concentration of 0.01 M. Using the algebraic approximation method,

(*a*) What is the pH of the solution if activity corrections are ignored?

(*b*) What is the pH of the solution if activity corrections are included?
   Answer: 4.74; 4.69

4.23 A solution is made by adding HCN to yield a concentration of 0.01 M and KCN to yield a concentration of 0.01 M. Using the algebraic approximation method,

(*a*) What is the pH of the solution if activity corrections are ignored?

(*b*) What is the pH of the solution if activity corrections are included?

4.24 (*a*) Draw a logarithmic concentration diagram for a 10^{-3} M solution of hydrogen sulfide. Assume a closed system.

(*b*) From the diagram, determine the pH for solutions that contain the following (1) 10^{-3} M H_2S, (2) 10^{-3} M Na_2S, (3) 0.5×10^{-3} M HS^- and 0.5×10^{-3} M S^{2-}, and (4) 0.5×10^{-3} M H_2S and 0.5×10^{-3} M HS^-.
   Answer: (1) 5.0; (2) 10.9; (3) 12.9; (4) 7.0

4.25 (*a*) Draw a logarithmic concentration diagram for a 10^{-2} M solution of hydrogen sulfide. Assume a closed system.

(*b*) From the diagram, determine the pH for solutions that contain the following (1) 10^{-2} M H_2S, (2) 10^{-2} M Na_2S, (3) 0.5×10^{-2} M HS^- and 0.5×10^{-2} M S^{2-}, and (4) 0.5×10^{-2} M H_2S and 0.5×10^{-2} M HS^-.

4.26   Using a logarithmic concentration diagram, determine the pH of a solution containing 10^{-2} M H_2CO_3 and 2×10^{-2} M Na_2CO_3.

4.27   Using a logarithmic concentration diagram, determine the pH of a solution containing 10^{-2} M acetic acid and 2×10^{-2} M sodium acetate.
Answer: 5.0

4.28   Using a logarithmic concentration diagram, determine the pH of a solution containing 10^{-3} M acetic acid and 10^{-3} M hydrochloric acid.

4.29   Using a logarithmic concentration diagram, determine the pH of a solution containing 10^{-3} M Na_2CO_3 and 10^{-3} M NaOH.
Answer: 11.1

4.30   Using a logarithmic concentration diagram, find the pH of a solution containing 10^{-2} M acetic acid and 10^{-2} M ammonia.

4.31   Using a logarithmic concentration diagram, find the pH of a solution containing 10^{-1} M acetic acid and 2×10^{-1} M sodium bicarbonate.
Answer: 6.4

4.32   Using a logarithmic concentration diagram, find the pH of a solution formed from mixing equal volumes of 10^{-3} M acetic acid and 10^{-2} M ammonium bicarbonate.

4.33   Using a logarithmic concentration diagram, find the pH of a solution formed from mixing equal volumes of 10^{-2} M propionic acid, 10^{-2} M acetic acid, and 10^{-1} M sodium bicarbonate.
Answer: 7.0

4.34   A solution is made by adding ammonium acetate to water to give a concentration of 0.01 M. Using a logarithmic concentration diagram, determine the equilibrium pH.
Answer: 7.0

4.35   A solution is made by adding ammonium carbonate to water to give a concentration of 0.01 M. Using a logarithmic concentration diagram, determine the equilibrium pH.

4.36   The pH of clean rain was calculated to be 5.68 in Example 4.16. Acid rain, defined here as rain with pH less than 5.68, can be formed when sulfur dioxide (SO_2) is discharged into the atmosphere. Assume that the atmosphere contains 20 ppb by volume of SO_2 ($10^{-7.7}$ atm). Develop an equation that will allow determination of the pH of rainwater in equilibrium with CO_2 in air @ $10^{-3.5}$ atm (see Example 4.16) and SO_2. Use a spreadsheet to calculate this pH. Comment on the relative impacts of SO_2 and CO_2 on the pH of rain. The following should be used.

$$H_2SO_3(aq) = SO_2(g) + H_2O \qquad K_{H,SO_2} = 0.80 \text{ atm}-\text{L/mol}$$
$$H_2SO_3 = HSO_3^- + H^+ \qquad pK_{A1} = 1.77$$
$$HSO_3^- = SO_3^{2-} + H^+ \qquad pK_{A2} = 7.21$$

4.37   Calculate the pH of the equivalence point in the titration of solutions containing the following concentrations of acetic acid with sodium hydroxide: (*a*) 100 mg/L; (*b*) 1000 mg/L; (*c*) 10,000 mg/L.
Answer: (*a*) 8.0; (*b*) 8.5; (*c*) 9.0

4.38   Calculate the pH of the equivalence point in the titration of each of the following concentrations of sodium bicarbonate with sulfuric acid: (*a*) 10 mg/L; (*b*) 100 mg/L; (*c*) 1000 mg/L. Assume a closed system.

4.39   Calculate the equivalence point pH for both ionizations in the titration of 150 mg/L of sodium carbonate with sulfuric acid. Assume a closed system.
Answer: 8.4, 4.6

4.40 Calculate the equivalence point pH for both ionizations in the titration of 150 mg/L of sodium sulfide (Na_2S) with sulfuric acid.

4.41 A 1-liter solution contains 100 mg of HCl. Calculate the following:

(*a*) Initial pH of the solution

(*b*) pH after addition of 1 mL of 1 N NaOH

(*c*) pH after addition of 2 mL of 1 N NaOH

(*d*) pH after addition of 3 mL of 1 N NaOH

(*e*) mL of 1 N NaOH required to reach the equivalence point
Answer: (*a*) 2.6; (*b*) 2.8; (*c*) 3.1; (*d*) 10.4; (*e*) 2.74 mL

4.42 A 500-mL solution contains 100 mg of NaOH. Calculate the following:

(*a*) Initial pH of the solution

(*b*) pH after addition of 2 mL of 1 N H_2SO_4

(*c*) pH after addition of 4 mL of 1 N H_2SO_4

(*d*) mL of 1 N H_2SO_4 required to reach the equivalence point

4.43 (*a*) Using the approximating equations developed in the text, sketch the titration curve for titrating a solution of 0.001 M sodium propionate with 0.1 M HCl.

(*b*) At what pH is the buffer intensity (buffer index) maximum? What is the value of the buffer intensity here?

(*c*) At what pH is the buffer intensity minimum? What is the value of the buffer intensity here?
Answer: (b) 4.89, 6.05×10^{-4} M; (*c*) 3.95, 4.71×10^{-4} M

4.44 A solution contains 10^{-3} M total inorganic carbon and has a pH of 7.2. No weak acids or bases are present other than the carbonate species, but other cations and anions are present. Neglecting ionic strength effects, plot the titration curve for titration of 1 liter of this water with 0.1 M HCl. Assume no interchange with the atmosphere occurs.

4.45 A solution contains 10^{-3} M dissolved inorganic carbon and has a pH of 8.23. Assume a closed system.

(*a*) How much strong acid (HCl in moles per liter of solution) would be required to titrate this solution to a pH of 6.3?

(*b*) How much strong acid (HCl in moles per liter of solution) would be required to titrate this solution to its equivalence point?
Answer: 5×10^{-4} M; 10^{-3} M

4.46 A water is in equilibrium with the atmosphere (partial pressure of $CO_2 = 5.0 \times 10^{-4}$ atm) at 25°C, and has a pH of 8.1. The total phosphate concentration is 0.001 M. No other weak acids or bases are present. How many milliliters of 0.5 M H_2SO_4 are needed to decrease the pH of 1 liter of this water to 6.0?

4.47 How many milliliters of 1 N NaOH must be added to a 500-mL solution containing 500 mg of acetic acid to increase the pH to 5?

4.48 (*a*) Draw a logarithmic concentration diagram for a 10^{-4} M solution of hypochlorous acid (25°C).

(*b*) From the diagram obtained in part (*a*), determine (1) the equilibrium pH for 10^{-4} M hypochlorous acid, (2) the pH after 0.5×10^{-4} mol of NaOH has been added to a liter of solution, and (3) the pH after 10^{-4} mol of NaOH has been added per liter of solution.
Answer: (1) 5.75; (2) 7.5; (3) 8.7

4.49 (a) Draw a logarithmic concentration diagram for a 10^{-4} M solution of propionic acid.

(b) From the diagram obtained in part (a), determine (1) the equilibrium pH for 10^{-4} M propionic acid, (2) the pH after 0.5×10^{-4} mol of NaOH has been added to a liter of solution, and (3) the pH after 10^{-4} mol/L of NaOH has been added to the solution.

4.50 Calculate the pH of a buffer solution prepared by mixing 2.4 g of acetic acid (CH_3COOH) and 0.73 g of sodium acetate (CH_3COONa) in 1 liter of water (equivalent to buffer solution for amperometric chlorine determination).

Answer: 4.0

4.51 Calculate the pH of a buffer solution prepared by mixing 8.5 g KH_2PO_4 with 43.5 g K_2HPO_4 in 1 liter of water (equivalent to buffer solution for BOD dilution water).

4.52 Calculate the pH (a) before and (b) after adding 20 mg/L HCl to a buffer solution containing 100 mg/L KH_2PO_4 and 200 mg/L K_2HPO_4.

Answer: (a) 7.40; (b) 6.88

4.53 Calculate the pH of a buffer solution prepared with 500 mg/L acetic acid (CH_3COOH) and 250 mg/L sodium acetate (CH_3COONa) under the following conditions:

(a) Initially

(b) After 20 mg/L of HCl is added to the solution

(c) After 20 mg/L of NaOH is added to the solution

4.54 Calculate the pH of a 200-mL buffer solution containing 20 mg/L carbonic acid and 50 mg/L bicarbonate ion, under the following conditions. Assume a closed system.

(a) Initially

(b) After 3 mL of 0.02 N H_2SO_4 is added

(c) After 3 mL of 0.02 N NaOH is added

Answer: (a) 6.8; (b) 6.3; (c) 8.1

4.55 Find the pH and buffer index for the following:

(a) 0.1 M acetic acid plus 0.1 M sodium acetate

(b) 0.19 M acetic acid plus 0.01 M sodium acetate

(c) 0.02 M acetic acid plus 0.18 M sodium acetate

4.56 Find the pH and buffer index for the following:

(a) 0.005 M HOCl plus 0.005 M NaOCl

(b) 0.001 M HOCl plus 0.009 M NaOCl

(c) 0.008 M HOCl plus 0.002 M NaOCl

Answer: (a) 7.5, 5.8×10^{-3}; (b) 8.45, 2.2×10^{-3}; (c) 6.9, 1.26×10^{-3}

4.57 Draw a curve of buffer index as a function of pH for 0.1 M acetic acid.

4.58 Draw a curve of buffer index as a function of pH for 0.1 M ammonium chloride.

4.59 You wish to develop a buffer that will maintain the pH of a laboratory reactor in the range 7.0 ± 0.3. The biological reaction of interest is nitrification, which is described by the following:

$$NH_4^+ + 2O_2 \rightarrow NO_3^- + 2H^+ + H_2O$$

(a) Assume that the nitrogen is initially present as NH_4Cl and that no other weak acids or bases exist in the solution. Select an appropriate acid/conjugate base pair to serve as the buffer and calculate the concentration of acid and conjugate base required for treatment of 50 mg/L of N in the form of NH_4^+.

(b) Calculate the buffer intensity of your buffer system.

4.60 A buffer is made by combining monosodium oxalate (NaC_2O_4H) and disodium oxalate ($Na_2C_2O_4$) to give concentrations of 0.01 M for NaC_2O_4H and 0.02 M for $Na_2C_2O_4$. pK_{A1} is 1.25 and pK_{A2} is 4.28.

(a) What is the initial pH of this buffer?

(b) What is its buffer intensity?

(c) What is the pH of this buffer after the addition of 0.001 M NaOH?

  Answer: (a) 4.58; (b) 1.54×10^{-2} M; (c) 4.65

4.61 You wish to construct a buffer that will resist pH changes during biological denitrification. During this process 1 mol of OH^- is produced for each mole of NO_3^- removed. You wish to remove 10^{-3} M of NO_3^- and maintain the pH at 7.5 ± 0.5. Select an appropriate buffer system and calculate the minimum quantities of chemicals required to maintain the desired pH range.

4.62 (a) What is the pH of a buffer made by adding 2.91×10^{-2} M $NaHCO_3$ and 4.30×10^{-3} M Na_2CO_3 to water in a *closed* system?

(b) What is the pH of the buffer after addition of 10^{-3} M H_2SO_4? The system is still *closed.*

(c) If the buffer in part (a) is *opened* to an atmosphere with $P_{CO_2} = 10^{-3.46}$ atm, what is the resulting pH? Does C_{T,CO_3} increase or decrease? By how much?

4.63 From a distribution diagram for $Cu-NH_3$ complexes, what is the predominant Cu(II) species when the ammonia concentration is (a) 0.1 mg/L, (b) 1 mg/L, (c) 10 mg/L, and (d) 100 mg/L?

4.64 Using a logarithmic concentration diagram, determine the concentration of various ammonia complexes of copper if $[Cu^{2+}] = 10^{-5}$ mol/L and $[NH_3] = 10^{-4}$ mol/L.

  Answer: $Cu(NH_3)^{2+}$, 10^{-5}; $Cu(NH_3)_2^{2+}$, 2.5×10^{-6}; $Cu(NH_3)_3^{2+}$, 10^{-7}; $Cu(NH_3)_4^{2+}$, 10^{-12}.

4.65 Using a logarithmic concentration diagram, determine the concentration of various ammonia complexes of copper if $[Cu^{2+}] = 10^{-4}$ mol/L and $[NH_3] = 10^{-3}$ mol/L.

4.66 Draw a logarithmic concentration diagram illustrating the effect of chloride concentration on the relative concentrations of various chloride complexes of mercury, assuming $[Hg^{2+}] = 10^{-7}$ mol/L. Which complex predominates when the chloride concentration equals (a) 0.1 mg/L, (b) 1 mg/L, (c) 10 mg/L, (d) 100 mg/L?

  Answer: $HgCl_2$ under all conditions

4.67 Draw a logarithmic concentration diagram illustrating the effect of fluoride concentration on the relative concentrations of various fluoride complexes of aluminum, assuming $[Al^{3+}] = 10^{-4}$ mol/L. Which complex predominates when the fluoride concentration equals (a) 0.1 mg/L, (b) 1 mg/L, (c) 10 mg/L?

4.68 Draw a distribution diagram for the chloride complexes of mercury.

4.69 Draw a distribution diagram for the fluoride complexes of aluminum.

4.70 From a predominance area diagram, determine which species of Hg(II) is likely to predominate (among hydroxide and chloride complexes only) when the chloride concentration is 35 mg/L and pH is (a) 6, (b) 7, (c) 8, and (d) 9.

  Answer: (a) $HgCl_2$; (b) $Hg(OH)_2$; (c) $Hg(OH)_2$; (d) $Hg(OH)_2$

4.71 Draw a predominance area diagram illustrating the effect of pH and fluoride concentration on fluoride and hydroxide complexes of Fe(III). Assume that the Fe(III) concentration is sufficiently low so that precipitation of iron does not occur.

4.72 Mercury (Hg) is a toxic heavy metal. Typically the "free metal ion," Hg^{2+}, is considered the most toxic inorganic form. For a natural water with a pH = 7.8 and containing 0.5 mg/L of total soluble Hg(II) and 50 mg/L of total soluble chloride (Cl^-) species, calculate the concentration of *all* soluble Hg(II) species. Which Hg(II) species is most prevalent? Assume there are no solids present, that the major ligands present are Cl^- and OH^-, and ignore activity corrections. If you make additional assumptions to solve this problem, clearly state what they are and be sure to check to see if they are reasonable.

4.73 Mercury (Hg) and Cadmium (Cd) are toxic heavy metals; it is usually the Hg^{2+} and the Cd^{2+} forms that are considered the most toxic inorganic forms. For a natural water containing 10^{-6} M of total soluble Hg(II), 10^{-6} M of total soluble Cd(II) and 10^{-3} M of total soluble chloride (Cl^- species), calculate the concentrations of all Hg(II) and Cd(II) species. Which Hg(II) species is the most prevalent? Which Cd(II) species is most prevalent? For this problem, ignore complexes of Hg(II) and Cd(II) with OH^-, assume no solids are present, and ignore activity corrections. Use the equilibrium constants given in Table 4.4 for the chloride complexes of Hg(II) and Cd(II).

4.74 Rework Prob. 4.73 with the same assumptions except this time include the OH^- complexes of Hg(II) and Cd(II). Assume the pH of the water is 6.5. Consult Table 4.5 for appropriate values of K.

4.75 Calculate the concentrations of all species (i.e., HCN, CN^-, Ni^{2+}, $Ni(CN)_4^{2-}$) in a solution with a pH of 9.0 having a total cyanide concentration of 10^{-3} M and a total nickel concentration of 2.0×10^{-4} M. State and verify all assumptions. Ignore other Ni complexes. The following equilibrium will be useful:

$$Ni^{2+} + 4CN^- = Ni(CN)_4^{2-} \qquad \beta_4 = 10^{+30}$$

Answer: $[Ni^{2+}] = 1.13 \times 10^{-17}$ M

4.76 From a logarithmic concentration diagram, estimate the minimum pH to which a water or wastewater need be raised to effect the precipitation as a metallic hydroxide of all but 10^{-4} mol/L of each of the following (*a*) Cr^{3+}, (*b*) Cu^{2+}, (*c*) Zn^{2+}, (*d*) Mg^{2+}, and (*e*) Ca^{2+}.

4.77 From a logarithmic concentration diagram, estimate the minimum pH to which a water or wastewater need be raised to effect the precipitation as a metallic hydroxide of all but 10^{-5} mol/L of each of the following (*a*) Fe^{3+}, (*b*) Fe^{2+}, (*c*) Mn^{2+}, (*d*) Al^{3+}, (*e*) Mg^{2+}, and (*f*) Cu^{2+}.
Answer: (*a*) 3.2; (*b*) 9.2; (*c*) 10.0; (*d*) 5.0; (*e*) 10.8; (*f*) 7.2

4.78 From a logarithmic concentration diagram, estimate the minimum concentration in moles per liter of CO_3^{2-} required to precipitate as a metallic carbonate all but 10^{-4} mol/L of (*a*) Mg^{2+}, (*b*) Ca^{2+}, (*c*) Sr^{2+}, (*d*) Zn^{2+}, and (*e*) Pb^{2+}. Ignore complexes.

4.79 From a logarithmic concentration diagram, estimate the minimum concentration in moles per liter of CO_3^{2-} required to precipitate as a metallic carbonate all but 10^{-5} mol/L of (*a*) Ca^{2+}, (*b*) Cu^{2+}, (*c*) Fe^{2+}, (*d*) Cd^{2+}, and (*e*) Pb^{2+}. Ignore complexes.
Answer: (*a*) 4×10^{-4}; (*b*) 2×10^{-5}; (*c*) 5×10^{-6}; (*d*) 5×10^{-7}; (*e*) 1.6×10^{-8}

4.80 From a logarithmic concentration diagram, determine the minimum pH at which each of the following concentrations of Ca^{2+} would be at equilibrium with $CaCO_3$ precipitate if the total concentration of inorganic carbon in solution equals 10^{-2} mol/L: (*a*) 10^{-3} mol/L, (*b*) 10^{-4} mol/L, (*c*) 10^{-5} mol/L. Ignore complexes.

4.81 Do Prob. 4.80, but assume that the total concentration of inorganic carbon in solution equals 10^{-4} mol/L. Ignore complexes.
    Answer: (a) 9.1; 10.4; *(c)* Not saturated with $CaCO_3(s)$

4.82 Construct a logarithmic concentration diagram showing the relationship between pH and the equilibrium concentration of Ca^{2+} with respect to $CaCO_3(s)$, assuming that the total concentration of inorganic carbon in solution, C, equals 10^{-1} mol/L. Ignore complexes.

4.83 A water has an initial Ca^{2+} concentration of 2×10^{-3} mol/L, and the total concentration of inorganic carbon in solution equals 2×10^{-2} mol/L. It is desired to reduce $[Ca^{2+}]$ to 2×10^{-4} mol/L by precipitation of $CaCO_3(s)$. What minimum pH would be required to effect this removal, and what would be the final molar concentration of inorganic dissolved carbon? Ignore complexes.

4.84 Draw a diagram that shows the solubility of $Cd(OH)_2(s)$ as a function of solution pH, and that also shows the concentration of other cadmium hydroxide complexes in a saturated solution. At what pH does $Cd(OH)_2(s)$ have minimum solubility? K_{sp} for $Cd(OH)_2(s) = 2 \times 10^{-14}$.

4.85 It has been reported that Americans ingest a total of approximately 8 tons/day of lead. The drinking water standard for lead is 0.05 mg/L total soluble lead. One method for removing heavy metals such as lead is pH adjustment to precipitate the metal hydroxide. Using a logarithmic concentration diagram, determine the minimum solubility (S) of lead (in mg/L as Pb) in water. Be sure to include lead complexes with hydroxide. At what pH does the minimum solubility occur? K_{sp} for $Pb(OH)_2(s) = 2.5 \times 10^{-16}$.
    Answer: 11.7 mg/L, pH $= 11.0$

4.86 A common way of removing Cr^{3+} from solution is precipitation of $Cr(OH)_3(s)$. An environmental engineer tells you that adjusting the pH to 8.5 will decrease the total soluble Cr(III) to below 1 mg/L. Is this true? In other words, is the solubility of Cr(III) in the presence of $Cr(OH)_3(s)$ at a pH of 8.5 less than 1 mg/L? The pK_{sp} of $Cr(OH)_3(s)$ is 30.22.

4.87 What is the solubility of iron (in mg/L Fe) in a solution in equilibrium with $Fe(OH)_3(s)$, 10^{-2} M total sulfate, and a pH of 7.0?

4.88 An industrial wastewater contains cadmium, ammonia, and chloride, and is in equilibrium with $Cd(OH)_2(s)$.
    (a) Write the general mass balance equation for soluble Cd.
    (b) Write the general mass balance equation for ammonia-N.
    (c) Write the general mass balance equation for chloride.
    (d) Write the equation for the solubility (S) of Cd.

4.89 A solution contains 0.01 M of each of the following metals: Ag^+, Ba^{2+}, Ca^{2+}, Pb^{2+}, and Sr^{2+}. You are to titrate this solution with Na_2SO_4. Ignore complexes.
    (a) Which metal will precipitate first?
    (b) Which metal will precipitate last?
    Be sure to give adequate justification for your answers. The pK_{sp} for $Ag_2SO_4(s)$ is 4.80, for $BaSO_4(s)$ it is 9.96, for $CaSO_4(s)$ it is 4.70, for $PbSO_4(s)$ it is 7.80, and for $SrSO_4(s)$ it is 6.55.

4.90   A water is in equilibrium with $Ca_3(PO_4)_2(s)$. The solution contains other cations and anions, but no other weak acids or bases or sources of Ca and PO_4, and has a pH of 8.6. Ignore complexes.

(a) What is the solubility (S) of Ca in this water?

(b) Will S increase, decrease, or remain constant if HCl is added to this water? Why?

Answer: $S = 1.46 \times 10^{-4}$ M; increase

4.91   An industrial wastewater has the following characteristics: temperature = 25°C, pH = 2.8, total soluble ammonia = 2000 mg/L as N, total soluble nickel = 490 mg/L as Ni, total inorganic carbon = 0. Using a logarithmic concentration diagram, to what pH must this water be adjusted to reduce the total soluble nickel (S) to 1 mg/L? After the pH has been raised, is the solution well buffered? Why?

4.92   Consider a water in equilibrium with $Ca_3(PO_4)_2(s)$. Assume there are no other solids and no complexes. Develop an equation that will allow you to solve for the pH of this water; that is, your equation will have only one unknown, $[H^+]$. Solve this equation for equilibrium pH using a spreadsheet.

Answer: pH = 9.59

4.93   A water is in equilibrium with CO_2 in the atmosphere (partial pressure is 3.16×10^{-4} atm) and with $CaCO_3(s)$. The pH is 8.1. Ignoring Ca complexes, what is the solubility (S) of Ca?

4.94   Consider a water in equilibrium with $CaCO_3(s)$ and 10^{-4} M NH_4Cl. Assume the system is closed and ignore complexes.

(a) Using a spreadsheet, determine the equilibrium pH of this system.

(b) How many milliliters of 0.1 M HCl will be required to decrease the pH to 6.0?

Answer: pH = 9.72; 0.32 mL

4.95   Pure water is brought into equilibrium with $Mg(OH)_2(s)$ and $CaCO_3(s)$ in a closed system at 25°C.

(a) Using a spreadsheet, determine the equilibrium pH of this water.

(b) The solids are removed by filtration and the system is opened to an atmosphere and allowed to come to equilibrium with $P_{CO_2} = 10^{-3.5}$ atm. How many milliliters of 0.1 N HCl are required to reduce the pH of this water to 7.0?

4.96   Iron oxide will adsorb radium ions (Ra^{2+}) to form a 1:1 surface complex. Calculate as a function of pH the *percent* of radium that will be adsorbed by 10 mg/L of iron oxide. Plot your results for pH values of 3 to 11. Assume no competition with other metal ions (other than protons). You might want to compare the total soluble radium concentration with available surface sites on the iron oxide to see if appropriate, simplifying assumptions can be made. The following data are to be used.

$$C_{T,Ra} = 10^{-10} \text{ M}$$

For the iron oxide: $pK_{A1} = 7.0$ $pK_{A2} = 9.0$ $pK_{SRa} = 2.0$
Surface site density = 10^{-4} mol/mg of iron oxide

This problem, with minor modifications, was contributed by Prof. Richard Valentine of the University of Iowa.

4.97   Draw a logarithmic concentration diagram showing the relationship between $[CO_2]$ and $[CH_4]$ as a function of pE for pH = 7 and $[CO_2] + [CH_4] = 1$ atm.

4.98 Draw a logarithmic concentration diagram showing the relationship between $[SO_4^{2-}]$ and $[H_2S]$ as a function of pE for pH $= 7$ and $[SO_4^{2-}] + [H_2S] = 10^{-3}$ mol/L.

4.99 Draw a pE–pH diagram illustrating predominant iron forms (Fe^{3+}, Fe^{2+}, Fe) in an aqueous system.

4.100 Draw a pE–pH diagram illustrating predominant manganese forms (Mn^{2+}, MnO_2, MnO_4^-) in an aqueous system.

4.101 From a pE–pH diagram, estimate the pE range for aerobic conditions in an aqueous system at pH equal to (*a*) 4, (*b*) 7, and (*c*) 10.
Answer: (*a*) 16.0 to 16.8; (*b*) 13.0 to 13.8; (*c*) 10.0 to 10.8

4.102 From a pE–pH diagram, estimate the pE range that is typical for sulfide and methane production in an aqueous system at pH equal to (*a*) 4, (*b*) 7, and (*c*) 10.

4.103 Develop the pE–pH equation for the SO_4^{2-}/H_2S line in Fig. 4.20.

4.104 Develop the pE–pH equation for the CO_2/CH_4 line in Fig. 4.20.

4.105 Consider the oxidation of Mn^{2+} to $MnO_2(s)$ by molecular oxygen [$O_2(g)$].
(*a*) Using appropriate half reactions in Table 2.4, write the pE–pH equations for reduction of $MnO_2(s)$ to Mn^{2+} and the reduction of $O_2(g)$ to H_2O.
(*b*) For a pH of 7.0 and a total soluble Mn concentration of 10^{-2} M, what partial pressure of oxygen (in atm) is required for Mn^{2+} to be the dominant form of Mn in the system?
(*c*) For a pH of 7.0, which form of Mn [Mn^{2+} or $MnO_2(s)$] is thermodynamically favored under normal atmospheric conditions (oxygen partial pressure of 0.21 atm)?
Answer: (*b*) 1.10×10^{-24} atm; (*c*) $MnO_2(s)$

4.106 Using data in Table 2.4, construct a balanced half reaction for the reduction of HOCl to Cl^-. What is E^0 for this half reaction?

4.107 Using data in Table 2.4;
(*a*) Construct a half reaction for the reduction of thiosulfate ($S_2O_3^{2-}$) to elemental sulfur ($S(s)$).
(*b*) Generate the pE–pH equation for this half reaction for a pH of 7.0.

4.108 Can ozone [$O_3(g)$] be used to oxidize NH_4^+ to NO_3^-? Show all calculations necessary to justify your answer. The following is useful:

$$O_3(g) + 2H^+ + 2e^- = O_2(g) + H_2O \qquad E^0 = 2.07 \text{ volts}$$

Answer: Yes

4.109 Some engineers are recommending addition of the strong oxidant potassium permanganate ($KMnO_4$) for remediating aquifers contaminated with chlorinated solvents such as TCE. In many of these sites, we also find trivalent chromium (Cr(III)), which is the less toxic, less mobile form of Cr. Addition of $KMnO_4$ may oxidize Cr(III) to its more toxic, more mobile hexavalent form (Cr(VI)). Using the data given, answer the following question: *Under standard conditions, can MnO_4^- oxidize Cr(III)?* Give justification for your answer.

$$HCrO_4^- + 4H^+ + 3e^- = Cr(OH)_3(s) + H_2O \qquad E^0 = 1.20 \text{ volts}$$
$$MnO_4^- + 8H^+ + 5e^- = Mn^{2+} + 4H_2O \qquad E^0 = 1.491 \text{ volts}$$

4.110 Consider the possibility of oxidation of Fe^{2+} to Fe^{3+} with aqueous chlorine.

(a) Write a balanced equation for this overall reaction in water.

(b) Is this reaction thermodynamically possible? Justify your answer.

4.111 Consider the following expression:

$$Au^{3+} + 3e^- = Au(s) \qquad E^0 = 1.42 \text{ volts}$$

(a) Pure gold [$Au(s)$] is placed in water with a pH of 7.0 and in equilibrium with the atmosphere (partial pressure of $O_2 = 0.21$ atm). What will be the concentration of Au^{3+} at equilibrium?

(b) Given your answer to part (a), what volume of water would be required to dissolve 1 g of gold? For comparison, the entire volume of the earth is approximately 10^{24} liters.

(This problem courtesy of Dr. James Gossett of Cornell University.)

Answer: (a) 6.04×10^{-32} M; (b) 8.4×10^{28} liters

4.112 A "globule" of elemental mercury (Hg^0, which is a liquid at room temperature) is placed in water that is in equilibrium with an atmosphere containing 10^{-2} atm O_2 and has a pH of 7.0. What will be the concentration of Hg^{2+} under these conditions? Assume the Hg^0 is in equilibrium with the water. Use data available in Table 2.4.

4.113 The major species of lead (Pb) that are dominant in natural systems include $PbO_2(s)$, Pb^{2+}, $Pb(OH)_3^-$, and $Pb^0(s)$.

(a) What are E^0, pE^0, and ΔG^0 for the reductive dissolution of $PbO_2(s)$?

$$PbO_2(s) + 4H^+ + 2e^- = Pb^{2+} + 2H_2O \qquad \log K = 49.2$$

(b) Given the reaction for the dissolution of $PbO_2(s)$, derive the pE–pH equation for the reduction of $PbO_2(s)$ to $PbO(s)$.

$$PbO(s) + 2H^+ = Pb^{2+} + H_2O \qquad \log K = 12.7$$

(c) What is the concentration of Pb^{2+} in a lake that has a pH of 6.0 and is in equilibrium with $PbO_2(s)$ and atmospheric oxygen?

(This problem, with minor modifications, was contributed by Prof. Michelle Scherer of the University of Iowa.)

Answer: (c) 10^{-4} M

4.114 Chlorinated solvents such as carbon tetrachloride (CCl_4) can be degraded in a process called reductive dechlorination where Cl atoms are removed and replaced with H atoms (see Chap. 6). For example,

$$CCl_4 + H^+ + 2e^- = CHCl_3 + Cl^- \qquad E^0 = 0.67 \text{ volt}$$

In in-situ bioremediation an electron donor such as acetate is typically added and is oxidized to provide the electrons for the reduction of CCl_4. In groundwaters devoid of oxygen, nitrate typically serves as the elctron acceptor. Using data from Table 2.4,

(a) Is it possible that CCl_4 could serve as an electron acceptor for acetate?

(b) Which electron acceptor, CCl_4 or NO_3^-, would be thermodynamically preferred if both were at the same concentration?

4.115 The following describes the equilibrium between Fe^{2+} and $Fe(OH)_3(s)$:

$$Fe(OH)_3(s) + 3H^+ + e^- = Fe^{2+} + 3H_2O \qquad E^0 = 1.06 \text{ volts}$$

(a) Develop a pE–pH equation for this half reaction assuming a total soluble iron concentration of 10^{-7} M.

(b) Many Midwestern groundwaters are devoid of oxygen but contain significant concentration of NO_3^-. Using the equation developed in part (a) and Fig. 4.20, what is the predominant form of iron present at neutral pH in groundwaters devoid of oxygen but containing NO_3^-?

REFERENCES

Benjamin, M. M.: "Water Chemistry," McGraw-Hill, New York, 2002.

Butler, J. N.: "Ionic Equilibrium, A. Mathematical Approach," Addison-Wesley, Reading, MA, 1964.

Chang, R.: "Chemistry," 7th ed., McGraw-Hill, New York, 2002.

Dzombak, D. A., and F. M. M. Morel: "Surface Complexation Modeling," Wiley-Interscience, New York, 1990.

Martell, A. E., and R. M. Smith: "Critical Stability Constants," vol. 1. Amino Acids, 1974 and vol. 3. Other Organic Ligands, 1977, Plenum, New York.

Morel, F. M. M.: "Principles of Aquatic Chemistry," Wiley, New York, 1983.

Morel, F. M. M., and J. G. Hering: "Principles and Applications of Aquatic Chemistry," Wiley-Interscience, New York, 1993.

Pankow, J. F.: "Aquatic Chemistry Concepts," Lewis, Chelsea, MI, 1991.

Sillén, L. G.: Graphic Presentation of Equilibrium Data, chap. 8, "Treatise on Analytical Chemistry," vol. 1, I. M. Kolthoff, P. J. Elving, and E. B. Sandell (eds.), "The Interscience Encyclopedia," New York, 1959.

Sillén, L. G., and A. E. Martell: Stability Constants of Metal-Ion Complexes, *Spec. Publ.* 17, 1964, and *Spec. Publ.* 25, 1971, Chemical Society, London.

Smith, R. M., and A. E. Martell: "Critical Stability Constants," vol. 2. Amines, 1975 and vol. 4. Inorganic Ligands, 1976, Plenum, New York.

Snoeyink, V. L., and D. Jenkins: "Water Chemistry," Wiley, New York, 1980.

Stumm, W.: "Chemistry of the Solid-Water Interface," Wiley-Interscience, New York, 1992.

Stumm, W., and J. J. Morgan: "Aquatic Chemistry," 3rd ed., Wiley-Interscience, New York, 1996.

5

Basic Concepts from Organic Chemistry

5.1 | INTRODUCTION

The fundamental information that environmental engineers and scientists need concerning organic chemistry differs considerably from that which the organic chemist requires. This difference is due to the fact that chemists are concerned principally with the synthesis of compounds, whereas environmental engineers and scientists are concerned, in the main, with how the organic compounds in liquid, solid, and gaseous wastes can be destroyed and how they react in the environment. Another major difference lies in the fact that the organic chemist is usually concerned with the product of the reaction: the by-products of a reaction are of little interest to him or her. Since few organic reactions give better than 85 percent yields, the amount of by-products and unreacted raw materials that represent processing wastes is of considerable magnitude. In addition, many raw materials contain impurities that do not enter the desired reaction and, of course, add to the organic load in waste streams. A classical example is formaldehyde, which normally contains about 5 percent of methanol unless special precautions are taken in its manufacture. Unfortunately, organic chemists have presented very little information on the nature of the by-products of reactions to aid environmental engineers and scientists in solving industrial and hazardous waste problems. Fortunately, this is changing because of the large liabilities that companies now face from discharge of environmental pollutants. Awards are now being given for "green chemistry," that is, for changing the ways chemicals are produced in order to reduce the environmental harm they or their production cause.

The environmental engineer and scientist, like the biochemist, must have a fundamental knowledge of organic chemistry. It is not important for either to know a multiplicity of ways of preparing a given organic compound and the yields to be expected from each. Rather, the important consideration is how the compounds react in the atmosphere, in the soil, in water, and in treatment reactors, especially when serving as a source of energy for living organisms. It is from this viewpoint that organic chemistry will be treated in this chapter, and considerations will be from the viewpoint of classes rather than individual compounds.

History

Organic chemistry deals with the compounds of carbon. The science of organic chemistry is considered to have originated in 1685 with the publication by Lémery[1] of a chemistry book that classified substances according to their origin as mineral, vegetable, or animal. Compounds derived from plants and animals became known as *organic* and those derived from nonliving sources were inorganic.

Until 1828 it was believed that organic compounds could not be formed except by living plants and animals. This was known as the *vital-force theory,* and belief in it severely limited the development of organic chemistry. Wöhler,[2] in 1828, by accident, found that application of heat to ammonium cyanate, an inorganic compound, caused it to change to urea, a compound considered organic in nature. This discovery dealt a death blow to the vital-force theory, and by 1850 modern organic chemistry became well established. Today about 13 million organic compounds are known.[3] Many of these are products of synthetic chemistry, and similar compounds are not known in nature. Approximately 70,000 organic chemicals are in commercial use.

Elements

All organic compounds contain carbon in combination with one or more elements. The hydrocarbons contain only carbon and hydrogen. A great many compounds contain carbon, hydrogen, and oxygen, and they are considered to be the major elements. Minor elements in naturally occurring compounds are nitrogen, phosphorus, and sulfur, and sometimes halogens and metals. Compounds produced by synthesis may contain, in addition, a wide variety of other elements.

Properties

Organic compounds, in general, differ greatly from inorganic compounds in seven respects:

1. Organic compounds are usually combustible.
2. Organic compounds, in general, have lower melting and boiling points.
3. Organic compounds are usually less soluble in water.
4. Several organic compounds may exist for a given formula. This is known as isomerism.
5. Reactions of organic compounds are usually molecular rather than ionic. As a result, they are often quite slow.
6. The molecular weights of organic compounds may be very high, often well over 1000.
7. Most organic compounds can serve as a source of food for bacteria.

---

[1]Nicholas Lémery (1645–1715), French physician and chemist.

[2]Freidrich Wöhler (1800–1882), German chemist.

[3]R. Chang, "Chemistry," 7th ed., McGraw-Hill, New York, 2002.

Sources

Organic compounds are derived from three sources:

1. *Nature:* fibers, vegetable oils, animal oils and fats, alkaloids, cellulose, starch, sugars, and so on.
2. *Synthesis:* A wide variety of compounds and materials prepared by manufacturing processes.
3. *Fermentation:* Alcohols, acetone, glycerol, antibiotics, acids, and the like are derived by the action of microorganisms upon organic matter.

The wastes produced in the processing of natural organic materials and from the synthetic organic and fermentation industries constitute a major part of the industrial and hazardous waste problems that environmental engineers and scientists are called upon to solve.

The Carbon Atom

A question commonly asked is: How is it possible to have so many compounds of carbon? There are two reasons. In the first place, carbon normally has four covalent bonds (four electrons to share). This factor alone allows many possibilities, but the most important reason is concerned with the ability of carbon atoms to link together by covalent bonding in a wide variety of ways They may be in a continuous open chain,

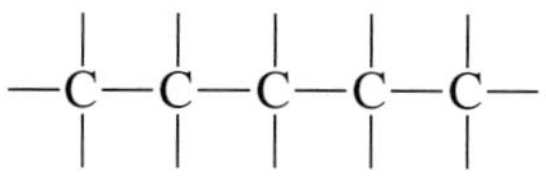

$$-\overset{|}{\underset{|}{C}}-\overset{|}{\underset{|}{C}}-\overset{|}{\underset{|}{C}}-\overset{|}{\underset{|}{C}}-\overset{|}{\underset{|}{C}}-$$

or a chain with branches,

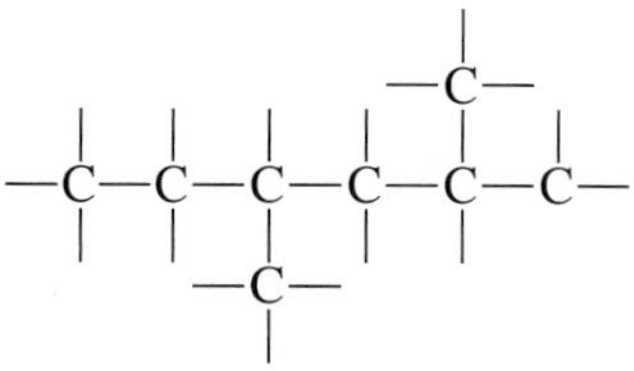

$$-\overset{|}{\underset{|}{C}}-\overset{|}{\underset{|}{C}}-\overset{|}{\underset{|}{C}}-\overset{|}{\underset{|}{C}}-\overset{\overset{\displaystyle -\overset{|}{C}-}{|}}{\underset{\underset{\displaystyle -C-}{|}}{C}}-\overset{|}{\underset{|}{C}}-$$

or in a ring,

$$-\overset{|}{C}\overset{\displaystyle \overset{|}{C}}{\underset{\displaystyle \underset{|}{C}}{}}\overset{|}{C}-$$

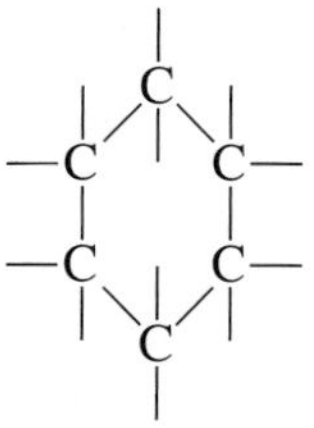

or in chains or rings containing other elements,

These examples will serve to show the tremendous number of possibilities that exist.

Isomerism

In inorganic chemistry, a molecular formula is specific for one compound. In organic chemistry, most molecular formulas do not represent any particular compound. For example, the molecular formula $C_3H_6O_3$ represents at least four separate compounds and therefore is of little value in imparting information other than that the compound contains carbon, hydrogen, and oxygen. Four compounds having the formula $C_3H_6O_3$ are

Compounds having the same molecular formula are known as *isomers*. In the case just cited, the first two isomers are hydroxy acids, the third is an ester of a hydroxy acid, and the fourth is a methoxy acid. To the organic chemist, each of the formulas represents a chemical compound with definite physical and chemical properties. The term *structural* formulas is applied to molecular representations as drawn for the four compounds. They are as useful to a chemist as blueprints are to an engineer.

In many cases structural formulas may be simplified as *condensed* formulas so as to use only one line. Thus, the formula

may be written as

$$CH_3-CH_2-CH_2OH \quad \text{or} \quad CH_3CH_2CH_2OH$$

thereby saving a great deal of space.

There are three major types of organic compounds, the *aliphatic, aromatic,* and *heterocyclic.* The *aliphatic* compounds are those in which the characteristic groups are linked to a straight or branched carbon chain. The *aromatic* compounds have these groups linked to a particular type of six-member carbon ring that contains three alternating double bonds. Such rings have peculiar stability and chemical character and are present in a wide variety of important compounds. The *heterocyclic* compounds have a ring structure with or without the alternating double-bond structure of aromatic compounds, and in which at least one member is an element other than carbon.

ALIPHATIC COMPOUNDS

5.2 | HYDROCARBONS

The hydrocarbons are compounds of carbon and hydrogen. There are two types, saturated and unsaturated. Saturated hydrocarbons are those in which adjacent carbon atoms are joined by a single covalent bond and all other bonds are satisfied by hydrogen.

$$
\begin{array}{ccccc}
& H & H & H & \\
& | & | & | & \\
H\!-\!\!&C\!-\!\!&C\!-\!\!&C\!-\!\!&H \\
& | & | & | & \\
& H & H & H &
\end{array}
$$

A saturated compound

Unsaturated hydrocarbons have at least two carbon atoms that are joined by more than one covalent bond and all remaining bonds are satisfied by hydrogen.

$$
\begin{array}{ccccc}
& H & H & H & \\
& | & | & | & \\
H\!-\!\!&C\!-\!\!&C\!=\!\!&C\!-\!\!&H \qquad HC\!\equiv\!CH \\
& | & & & \\
& H & & &
\end{array}
$$

Unsaturated compounds

Saturated Hydrocarbons

The saturated hydrocarbons form a whole series of compounds starting with one carbon atom and increasing one carbon atom, stepwise. These compounds are called *alkanes,* or the *methane series.* The principal source is petroleum. Gasoline is a mixture containing several of them; diesel fuel is another such mixture.

The hydrocarbons serve as feedstocks for the preparation of a wide variety of organic chemicals. This knowledge serves as the basis of the great petrochemical industry within the petroleum industry. Saturated hydrocarbons are quite inert toward most chemical reagents. For this reason they were termed "paraffins" by early chemists (from the Latin *parum affinis,* meaning "little affinity").

Methane (CH_4) is the simplest hydrocarbon. It is a gas of considerable importance to environmental engineers and scientists since it is a major end product of the anaerobic treatment process as applied to sewage sludge and other organic waste materials. It is a component of marsh gas and of natural gas and, in a mixture with air containing from 5 to 15 percent methane, it is highly explosive. This property allows its use as fuel for gas engines. Methane is commonly called "firedamp" by miners and makes their work particularly hazardous. Methane is also considered to be an important greenhouse gas; its concentration in the stratosphere affects the earth's heat balance, and thus temperature. On a per molecule basis, it is 21 times more effective at trapping heat in the atmosphere than carbon dioxide, the primary greenhouse gas.

Ethane (CH_3—CH_3) is the second member of the series.

Propane (CH_3—CH_2—CH_3) is the third member of the series.

Butane (C_4H_{10}) is the fourth member of the series and is of interest because it occurs in two isomeric forms:

n-Butane   Isobutane

Pentane (C_5H_{12}) is the fifth member of the series and exists in three isomeric forms:

| *n*-Pentane | Isopentane | Neopentane |
| bp, 36.2°C | bp, 28°C | bp, 9.5°C |

The third isomer of pentane might also be called tetramethylmethane or dimethylpropane, as the reader will shortly recognize.

As the number of carbon atoms increases in the molecule, the number of possible isomers increases accordingly. There are five possible isomers of *hexane* (C_6H_{14}) and 75 possible isomers of *decane* ($C_{10}H_{22}$).

Physical Properties   Table 5.1 lists the names and physical constants of the *normal* saturated hydrocarbons of 1 to 10 carbon atoms per molecule. The term "normal" applies to the isomer that has all its carbon atoms linked in a *straight chain*. The others are referred to as *branched-chain* compounds. The branched form of butane and the simplest branched form of pentane are commonly given the prefix *iso-*.

The saturated hydrocarbons are colorless, practically odorless, and quite insoluble in water, particularly those with five or more carbon atoms. They dissolve read-

Table 5.1 | Physical constants of some normal alkanes

Name	Formula	Mp, °C	Bp, °C	Sp. gr., 20°/4°*	Calcd. no. of isomers
Methane	CH_4	−182.4	−161.5	$0.423^{-162°}$	1
Ethane	C_2H_6	−182.8	−88.6	$0.545^{-89°}$	1
Propane	C_3H_8	−187.6	−42.1	$0.493^{25°}$	1
Butane	C_4H_{10}	−138.2	−0.5	$0.573^{25°}$	2
Pentane	C_5H_{12}	−129.7	36.0	0.626	3
Hexane	C_6H_{14}	−95.3	68.7	0.655	5
Heptane	C_7H_{16}	−90.6	98.5	0.684	9
Octane	C_8H_{18}	−56.8	125.6	$0.699^{25°}$	18
Nonane	C_9H_{20}	−53.5	150.8	0.718	35
Decane	$C_{10}H_{22}$	−29.7	174.1	0.730	75

*Density of compound at 20°C (or at the temperature noted by superscript) over the maximum density of water, which is at 4°C.

Source: D. R. Linde (ed.): "Handbook of Chemistry and Physics," 82nd ed., CRC Press LLC, Boca Raton, 2001.

ily in many organic solvents. At room temperature all members through C_5 are gases, those from C_6 to C_{17} are liquids, and those above C_{17} are solids. Data in Table 5.1 show that as molecular size increases, the melting and boiling points of alkanes increase. Solubility in water in general decreases with increasing size. Such relationships are important in understanding the behavior of organic compounds in the environment and in engineered reactors (Sec. 5.34).

Homologous Series   It will be noted from Table 5.1 that each successive member of the series differs from the previous member by CH_2. When the formulas of a series of compounds differ by a common increment, such as CH_2, the series is referred to as being a *homologous series*. Such compounds can be expressed by a general formula. That for the alkane series is C_nH_{2n+2}.

Radicals   The inert character of the alkanes has been mentioned; however, they may be made to react under the proper conditions, and a wide variety of compounds results. It becomes necessary, therefore, to establish some form of nomenclature to identify the products formed. When one hydrogen is replaced from a molecule of an alkane, the *-ane* ending is dropped and a *-yl* is added. The names of some radicals are shown in Table 5.2. The system serves quite well for the normal compounds but is of little value in naming derivatives of the isomers.

Nomenclature   The alkanes are characterized by names ending in *-ane*. The straight-chain compounds are termed *normal* compounds. The branched-chain

Table 5.2 | Names of alkane-series radicals (alkyl groups)

Parent compound	Radical	Formula
Methane	Methyl	$CH_3—$
Ethane	Ethyl	$C_2H_5—$
Propane	*n*-Propyl	$C_3H_7—$
Propane	Isopropyl	$(CH_3)_2CH—$
n-Butane	*n*-Butyl	$C_4H_9—$

compounds and the derivatives of both straight- and branched-chain compounds are difficult to name with any degree of specificity. The IUPAC system, as proposed by the International Union of Pure and Applied Chemistry, is commonly used. In this system the compounds are named in terms of the longest continuous chain of carbon atoms in the molecule. A few examples will illustrate the method.

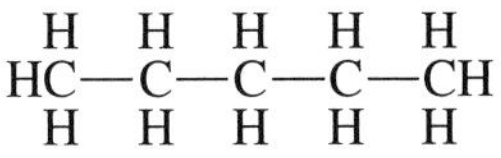

$$
\begin{array}{ccccc}
H & H & H & H & H \\
HC- & C- & C- & C- & CH \\
H & H & H & H & H
\end{array}
$$
n-Pentane

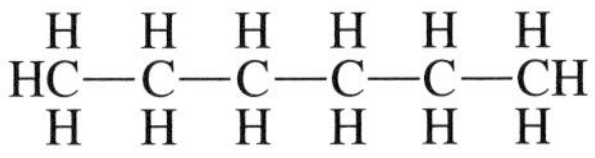

$$
\begin{array}{cccccc}
H & H & H & H & H & H \\
HC- & C- & C- & C- & C- & CH \\
H & H & H & H & H & H
\end{array}
$$
n-Hexane

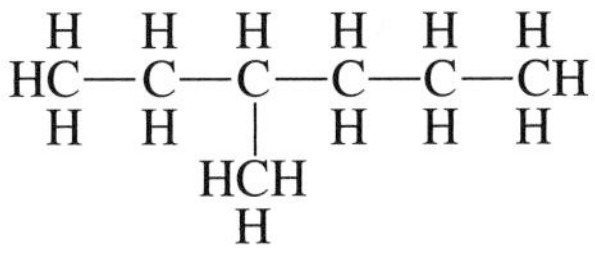

$$
\begin{array}{cccccc}
H & H & H & H & H & H \\
HC- & C- & C- & C- & C- & CH \\
H & H & | & H & H & H \\
 & & HCH & & & \\
 & & H & & &
\end{array}
$$
3-Methylhexane
(an isomer of heptane)

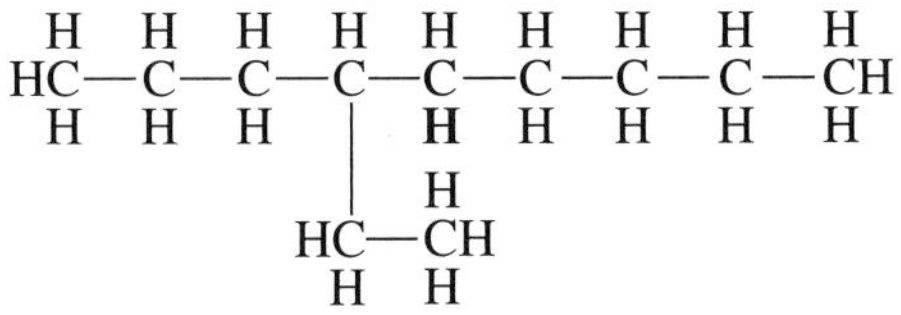

$$
\begin{array}{ccccccccc}
H & H & H & H & H & H & H & H & H \\
HC- & C- & C- & C- & C- & C- & C- & C- & CH \\
H & H & H & | & H & H & H & H & H \\
 & & & H & & & & & \\
 & & HC- & CH & & & & & \\
 & & H & H & & & & &
\end{array}
$$
4-Ethylnonane
(an isomer of undecane)

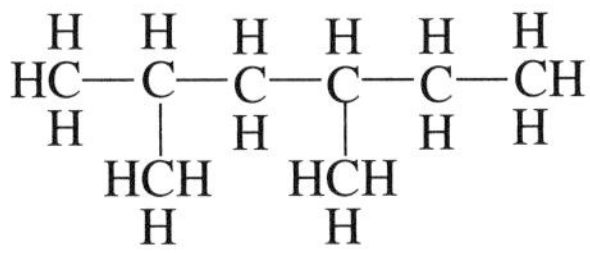

$$
\begin{array}{cccccc}
H & H & H & H & H & H \\
HC- & C- & C- & C- & C- & CH \\
H & | & H & | & H & H \\
 & HCH & & HCH & & \\
 & H & & H & &
\end{array}
$$
2,4-Dimethylhexane
(an isomer of octane)

It will be noted that a chain is numbered from the end nearest the attached radical. The rule is to make the numbers as small as possible. The IUPAC system is applied to other compounds as well as to hydrocarbons.

Chemical Reactions Strong bases, acids, or aqueous solutions of oxidizing agents do not react with saturated hydrocarbons at room temperature. At elevated temperatures, strong oxidizing agents, such as concentrated sulfuric acid, oxidize the compounds to carbon dioxide and water. Other reactions of importance are as follows:

1. *Oxidation with oxygen or air:*

$$CH_4 + 2O_2 \xrightarrow{\Delta} CO_2 + 2H_2O \tag{5.1}$$

2. *Substitution of hydrogen by halogens:*

$$CH_4 + Cl_2 \rightarrow HCl + CH_3Cl \tag{5.2}$$

This reaction does not ordinarily occur in aqueous solutions and therefore is of little significance in environmental engineering and science.

3. *Pyrolysis or cracking:* High-molecular-weight hydrocarbons may be broken into smaller molecules by heat treatment. The process is used in the petroleum industry to increase the yield of light boiling fractions, suitable for sale as gasoline or for chemical synthesis. Heat treatment results in disruption of the large molecules as follows:

$$\left.\begin{array}{c}\text{High-mol.-wt} \\ \text{alkanes}\end{array}\right\} \xrightarrow[\text{pressure}]{\text{heat}} \begin{array}{l} \text{alkanes of lower mol. wt} \\ + \text{ alkenes} \\ + \text{ hydrogen } + \text{ naphthenes} \\ + \text{ carbon}\end{array} \tag{5.3}$$

4. *Biological oxidation:* Hydrocarbons are oxidized by certain bacteria under aerobic conditions. The oxidation proceeds through several steps. The first step is very slow biologically and involves conversion to alcohols with attack occurring on terminal carbon atoms, i.e., omega oxidation.

$$\underset{\text{Hydrocarbon}}{2CH_3CH_2CH_3} + O_2 \xrightarrow{\text{bact.}} \underset{\text{Alcohol}}{2CH_3CH_2CH_2OH} \tag{5.4}$$

Through additional oxidative steps, which will be discussed in Chap. 6, microorganisms convert the hydrocarbon to carbon dioxide and water and derive energy in the process.

$$CH_3CH_2CH_3 + 5O_2 \xrightarrow{\text{bact.}} 3CO_2 + 4H_2O \tag{5.5}$$

Such reactions, particularly the intermediate steps, are of great interest to environmental engineering and science.

Unsaturated Hydrocarbons

The unsaturated hydrocarbons are usually separated into four classes.

Alkenes   Each member of the alkane group except methane can lose hydrogen to form an unsaturated compound or alkene. The alkenes all contain one double bond between two adjacent carbon atoms,

$$\begin{array}{cc} \overset{\text{H H}}{\underset{|\ \ |}{\text{H—C=C—H}}} & \overset{\text{H H H H}}{\underset{\overset{|}{\text{H}}\ \ \ \ \overset{|}{\text{H}}}{\text{HC—C=C—CH}}} \\ \text{Ethylene} & \text{Butylene} \\ \text{or} & \text{or} \\ \text{ethene} & \text{2-butene} \end{array}$$

and their names all end in *-ylene* (older nomenclature) or *-ene*. The alkenes are also called *olefins*. Alkenes, particularly ethene, propene, and butene are formed in great quantities during the cracking or pyrolysis of petroleum.

The names, formulas, and physical constants of a number of important alkenes are given in Table 5.3. In naming specific alkenes, the IUPAC system

Table 5.3 | Physical constants of selected alkenes

IUPAC name	Formula	Mp, °C	Bp, °C	Sp. gr., 20°C/4°C*	Calcd. no. of isomers
Ethene	$CH_2{=}CH_2$	−169	−103.7	$0.568^{-104°}$	1
Propene	$CH_2{=}CHCH_3$	−185.2	−47.6	$0.505^{25°}$	1
1-Butene	$CH_2{=}CHCH_2CH_3$	−185.3	−6.2	$0.588^{25°}$	3
1-Pentene	$CH_2{=}CH(CH_2)_2CH_3$	−165.2	29.9	0.640	5
1-Hexene	$CH_2{=}CH(CH_2)_3CH_3$	−139.7	63.4	0.673	13
1-Heptene	$CH_2{=}CH(CH_2)_4CH_3$	−119.7	93.6	0.697	27
1-Octene	$CH_2{=}CH(CH_2)_5CH_3$	−101.7	121.2	0.715	66
1-Nonene	$CH_2{=}CH(CH_2)_6CH_3$	−81.3	149.9	$0.725^{25°}$	153
1-Decene	$CH_2{=}CH(CH_2)_7CH_3$	−66.3	170.5	0.741	377

*Density of compound at 20°C (or at the temperature noted by superscript) over the maximum density of water, which is at 4°C.

Source: D. R. Linde (ed.): "Handbook of Chemistry and Physics," 82nd ed., CRC Press LLC, Boca Raton, 2001.

must be employed on all compounds with over three carbon atoms. The nomenclature becomes quite complicated with branched-chain isomers. Fortunately, there is little reason to differentiate between normal and branched-chain compounds in this series.

Diolefins   When aliphatic compounds contain two double bonds in the molecule, they are called alkadienes, some times dienes for short. The compound 1,3-butadiene is an important example:

$$CH_2{=}CHCH{=}CH_2$$

It has been used to make polymers.

Alkadienes   Some organic compounds contain more than two double bonds per molecule. The red coloring matter of tomatoes, lycopene, and the yellow coloring matter of carrots are examples.

Lycopene ($C_{40}H_{56}$)

These compounds are of interest because of their occurrence in industrial wastes produced in the preparation of vegetables for canning. The oxidant (e.g., chlorine) demand of such wastes is extremely high.

Alkynes   The alkynes have a triple bond between adjacent carbon atoms.

$$H{-}C{\equiv}C{-}H$$

These compounds are found to some extent in industrial wastes from certain industries, particularly those from the manufacture of some types of synthetic rubber.

Chemical Reactions   Unsaturated linkages occur in many types of organic compounds and exhibit many properties in common, regardless of the type of compound in which they exist. Unsaturated compounds undergo several reactions with relative ease.

1. *Oxidation:* The compounds are easily oxidized in aqueous solution by oxidizing agents such as potassium permanganate. A glycol is the normal product.

2. *Reduction:* Under special conditions of temperature, pressure, and catalysis, hydrogen may be caused to add at double or triple bonds. This reaction is of considerable importance commercially in the conversion of vegetable oils to solid fats. Many vegetable shortenings are made by this process.

3. *Addition:* Halogen acids, hypochlorous acid, and halogens will add across unsaturated linkages.

$$CH_3-C=C-H + HOCl \rightarrow CH_3-\overset{OH}{\underset{H}{C}}-\overset{Cl}{\underset{H}{C}}-H \qquad (5.6)$$

The reaction with hypochlorous acid is most important. Industrial wastes containing appreciable amounts of unsaturated compounds exhibit high chlorine-demand values because of such reactions.

4. *Polymerization:* Molecules of certain compounds having unsaturated linkages are prone to combine with each other to form polymers of higher molecular weight.

$$nCH_2=CH_2 \xrightarrow[\text{and pressure}]{\text{high temperature}} (C_2H_4)n \qquad n = 70 \text{ to } 700 \qquad (5.7)$$
$$\text{Polyethylene}$$

Similar reactions serve as the basis for many industrial products, e.g., synthetic resins, synthetic fibers, synthetic rubber, and synthetic detergents. Industrial wastes from such industries can be expected to contain a wide variety of polymers and usually exhibit a high chlorine demand. Historically such wastes were treated by chlorination. This practice is currently discouraged because the chlorinated organic compounds so formed have adverse impacts to human health and the environment.

5. *Bacterial oxidation:* It is generally considered that organic compounds possessing unsaturated linkages are more prone to bacterial oxidation than corresponding saturated compounds because of the ease of oxidation at the double bonds.

5.3 | ALCOHOLS

Alcohols are considered the primary oxidation product of hydrocarbons.

$$\underset{\text{Methane}}{CH_4} + \tfrac{1}{2}O_2 \rightarrow \underset{\text{Methyl alcohol or methanol}}{CH_3OH} \qquad (5.8)$$

$$\underset{\text{Propane}}{CH_3-CH_2-CH_3} + \tfrac{1}{2}O_2 \rightarrow \underset{\text{n-Propyl alcohol or propanol}}{CH_3-CH_2-CH_2OH} \qquad (5.9)$$

They cannot be prepared in this manner, however, because the reaction cannot be stopped with alcohols as the end product. Nevertheless, the reaction illustrates an initial step in biological degradation of hydrocarbons under aerobic conditions.

Alcohols may be considered as *hydroxy alkyl* compounds. For convenience, the alkyl group in alcohols and other organic compounds is often represented by R—, and the general formula for alcohols is R—OH. The OH group does not easily ionize; consequently, alcohols are neutral in reaction. The chemistry of alcohols is related entirely to the OH group.

Classification

Alcohols are classified into three groups: primary, secondary, and tertiary, depending upon where the OH group is attached to the molecule. If the OH group is on a terminal (primary) carbon atom, it is a *primary* alcohol.

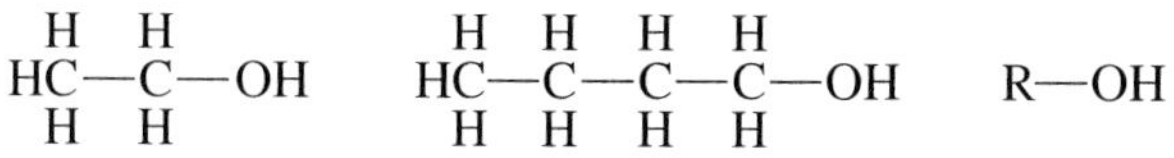

Primary alcohols

If the OH group is attached to a carbon atom that is joined to two other carbon atoms, it is a *secondary* alcohol, and the carbon atom to which it is attached is a *secondary carbon atom.*

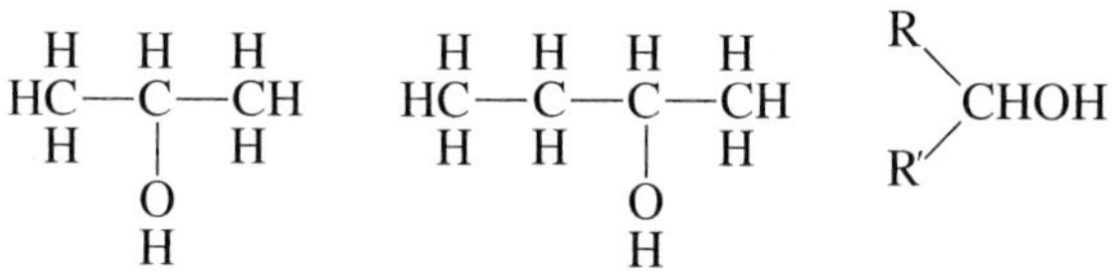

Secondary alcohols

If the OH group is attached to a carbon atom that is joined to three other carbon atoms, it is a *tertiary* alcohol, and the carbon atom to which it is attached is a *tertiary carbon atom.*

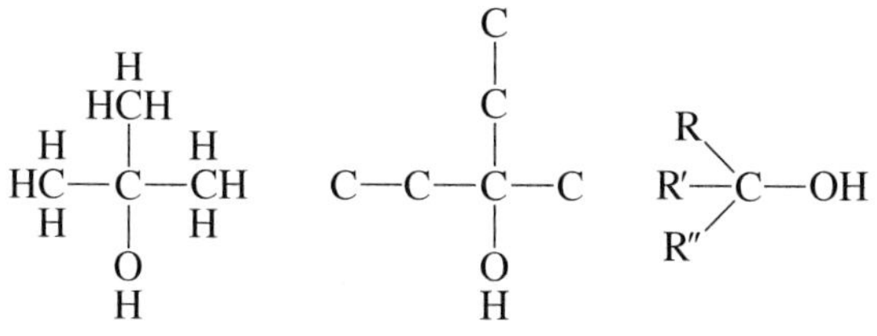

Tertiary alcohols

The chemistry of the primary, secondary, and tertiary alcohols differs considerably (e.g., bacterial oxidation of tertiary alcohols is much more difficult than of primary alcohols). Therefore, it is important to know how to differentiate among them.

Common Alcohols

The alcohols of greatest commercial importance are methyl, ethyl, isopropyl, and *n*-butyl.

Methanol (CH₃OH)   Methanol (also called methyl alcohol) is used to a considerable extent for synthesis of organic compounds. It has been used as an antifreeze for automobiles. It is prepared mainly by synthesis from natural gas and steam as follows:

$$CH_4 + H_2O \xrightarrow[\text{catalyst}]{\Delta \text{ pressure}} CH_3OH + H_2 \tag{5.10}$$

but may be manufactured from carbon monoxide and hydrogen.

Ethanol (CH₃CH₂OH)   Ethanol (ethyl alcohol) is used for the synthesis of organic compounds, the production of beverages, and the manufacture of medicines. Recently it has found use as a fuel oxygenate for reformulated gasoline (Sec. 3.8). It is prepared largely by fermentation processes. Alcohol intended for beverages is often manufactured by the fermentation of starch derived from a variety of materials, such as corn, wheat, rye, rice, and potatoes. The reactions involved are as follows:

$$\text{Starch} + \text{water} \xrightarrow[\text{of malt}]{\text{enzyme}} \text{maltose} \tag{5.11}$$

$$\underset{\text{Maltose}}{C_{12}H_{22}O_{11}} + H_2O \xrightarrow[\text{of yeast}]{\text{enzyme}} 2 \text{ glucose} \tag{5.12}$$

Fermentation of the glucose yields carbon dioxide and alcohol.

$$\underset{\text{Glucose}}{C_6H_{12}O_6} \xrightarrow{\text{fermentation}} 2CO_2 + 2C_2H_5OH \tag{5.13}$$

Wine is produced from the fermentation of sugars in grapes. Industrial alcohol is produced largely from the fermentation of solutions containing sugars that are difficult to reclaim, such as molasses and, sometimes, even spent sulfite liquor.

$$C_{12}H_{22}O_{11} + H_2O \xrightarrow{\text{invertase}} \underset{\text{Glucose}}{C_6H_{12}O_6} + \underset{\text{Fructose}}{C_6H_{12}O_6} \tag{5.14}$$

$$\underset{\substack{\text{Glucose} \\ \text{Fructose}}}{C_6H_{12}O_6} \xrightarrow{\text{yeast}} 2CO_2 + 2C_2H_5OH$$

The residues remaining after distillation of the desired product, ethanol, constitute some of the most potent industrial wastes in terms of strength with which the environmental engineer has to deal.

Isopropanol (CH₃CHOHCH₃)   Isopropanol (isopropyl alcohol) is widely used in organic synthesis, and considerable amounts are sold as "dry gas" to prevent separation of water in the fuel tanks of automobiles. It is prepared by hydration of propylene derived from the cracking of petroleum.

***n*-Butanol (CH₃CH₂CH₂CH₂OH)**   Normal butanol (*n*-butyl alcohol) is used to prepare butyl acetate, an excellent solvent. It is often referred to as "synthetic banana oil" because of its odor which resembles natural banana oil, amyl acetate. Normal butanol is prepared from cornstarch by a fermentation process utilizing a particular microorganism, *Clostridium acetobutylicum.* Considerable amounts of acetone and some ethanol and hydrogen are produced during the fermentation. The liquid wastes remaining after distillation of the desired products are classed as industrial wastes, and their treatment and ultimate disposal usually fall to the lot of the environmental engineer. They are similar in character to the residues from the production of ethanol but offer less promise of by-product recovery. While methanol, ethanol, and isopropanol are completely miscible with water, *n*-butanol has a somewhat limited solubility in water.

Physical Properties of Alcohols

The short-chain alcohols are completely soluble in water due in part to the increased polarity caused by the hydroxyl group. As with alkanes and alkenes, as alcohol molecular size increases, melting and boiling points increase while water solubility decreases. The miscibility of C_1–C_3 alcohols makes them useful as cosolvents for remediation of contaminated subsurface environments (Sec. 3.8). Alcohols with more than 12 carbon atoms are colorless waxy solids and very poorly soluble in water. The physical constants of several alcohols are given in Table 5.4.

Nomenclature

The alcohols of commercial significance are usually called by their common names. The IUPAC system must be employed, however, to differentiate among isomers and to name the higher members, such as hexadecanol. In this terminology, the names of all alcohols end in *-ol*. The formulas, common names, and IUPAC names of several alcohols are given in Tables 5.4 and 5.5. In this system, the longest carbon chain containing the hydroxyl group determines the name. The locations of alkyl groups and the hydroxyl group are described by number; e.g., isobutanol is described as 2-methyl-1-propanol.

Table 5.4 | Physical constants of normal primary alcohols

Name of radical	IUPAC name of alcohol	Formula	Mp, °C	Bp, °C	Sp. gr., 20°/4°*	Calcd. no. of isomers
Methyl	Methanol	CH_3OH	−97.6	64.6	0.791	1
Ethyl	Ethanol	C_2H_5OH	−114.1	78.2	0.789	1
Propyl	1-Propanol	C_3H_7OH	−126.1	97.2	$0.800^{25°}$	2
Butyl	1-Butanol	C_4H_9OH	−89.8	117.7	0.810	4
Pentyl	1-Pentanol	$C_5H_{11}OH$	−78.9	137.9	0.814	8
Hexyl	1-Hexanol	$C_6H_{13}OH$	−44.6	157.6	0.814	17
Heptyl	1-Heptanol	$C_7H_{15}OH$	−34	176.4	0.822	39
Octyl	1-Octanol	$C_8H_{17}OH$	−15.5	195.1	$0.826^{25°}$	89
Nonyl	1-Nonanol	$C_9H_{19}OH$	−5	213.3	0.827	211[†]
Decyl	1-Decanol	$C_{10}H_{21}OH$	6.9	231.1	0.830	507

*Density of compound at 20°C (or at the temperature noted by superscript) over the maximum density of water, which is at 4°C.
[†]These numbers are for all the isomers of a given carbon content.
Source: D. R. Linde (ed.): "Handbook of Chemistry and Physics," 82nd ed., CRC Press LLC, Boca Raton, 2001.

Table 5.5 | Nomenclature of alcohols

Formula	Common name	IUPAC name
CH_3OH	Methyl alcohol	Methanol
C_2H_5OH	Ethyl alcohol	Ethanol
$CH_3CH_2CH_2OH$	*n*-Propyl alcohol	1-Propanol
CH_3 $\diagdown$CHOH $\diagup$ CH_3	Isopropyl alcohol	2-Propanol
$CH_3CH_2CH_2CH_2OH$	*n*-Butyl alcohol	1-Butanol
CH_3 $\diagdown CHCH_2OH$ $\diagup$ CH_3	Isobutyl alcohol	2-Methyl-1-propanol
CH_3 $\diagdown$CHOH $\diagup$ C_2H_5	*sec*-Butyl alcohol	2-Butanol
$(CH_3)_3COH$	*tert*-Butyl alcohol	2-Methyl-2-propanol

Polyhydroxy Alcohols

Those alcohols having two hydroxyl groups per molecule are known as *glycols.* The principal glycol of commercial significance is *ethylene glycol* (1,2-ethanediol), which is prepared from ethene. Ethene adds hypochlorous acid to form ethylene chlorohydrin.

$$CH_2{=}CH_2 + HOCl \rightarrow \underset{\text{Ethylene chlorohydrin}}{\begin{array}{cc} H & H \\ HC & \!\!-\!\!CH \\ | & | \\ Cl & OH \end{array}} \qquad (5.15)$$

and treatment of the chlorohydrin with sodium bicarbonate produces ethylene glycol.

$$\begin{array}{cc} H_2C & \!\!-\!\!CH_2 \\ | & | \\ Cl & OH \end{array} + NaHCO_3 \rightarrow NaCl + CO_2 + \underset{\text{Ethylene glycol}}{\begin{array}{cc} H_2C & \!\!-\!\!CH_2 \\ | & | \\ OH & OH \end{array}} \qquad (5.16)$$

It is used extensively as a nonevaporative, radiator antifreeze compound.

Glycerol or glycerin is a trihydroxy alcohol (1,2,3-trihydroxypropane).

$$\underset{\text{Glycerol}}{\begin{array}{ccc} H_2 & H & H_2 \\ C & \!\!-C-\!\! & C \\ | & | & | \\ OH & OH & OH \end{array}}$$

It was formerly produced in large quantities in the soap industry through saponification of fats and oils. Presently, considerable amounts are produced by chemical synthesis. Glycerol is used in a wide variety of commercial products: foods, cosmetics, medicines, tobaccos, and so on. It is used for the manufacture of nitroglycerin, an important component of dynamite.

Chemical Reactions of Alcohols

Alcohols undergo two types of reactions of special interest.

Ester Formation   Alcohols react with acids, both inorganic and organic, to form esters. Inorganic hydroxy acids yield "inorganic" esters:

$$ROH + H_2SO_4 \rightleftharpoons H_2O + ROSO_3H \qquad (5.17)$$

Organic acids yield organic esters:

$$ROH + R_1CO_2H \rightleftharpoons H_2O + R_1CO_2R \qquad (5.18)$$

Organic esters are discussed in Sec. 5.6.

Oxidation   Most alcohols are readily oxidized by strong oxidizing agents and by many microorganisms. The product of the oxidation depends upon the class of alcohol involved.

Primary alcohols are oxidized to aldehydes. The general equation is

$$\underset{\substack{\text{Primary}\\\text{alcohol}}}{RCH_2OH} + \tfrac{1}{2}O_2 \rightarrow H_2O + \underset{\text{An aldehyde}}{RC\overset{H}{=}O} \qquad (5.19)$$

Care must be used in selecting the oxidizing agent, or the aldehyde may be oxidized still further to an acid.

Secondary alcohols are oxidized to ketones.

$$\underset{\text{Isopropyl alcohol}}{H_3C-\underset{\underset{H}{|}}{\overset{\overset{OH}{|}}{C}}-CH_3} + \tfrac{1}{2}O_2 \rightarrow H_2O + \underset{\substack{\text{Acetone}\\\text{(a ketone)}}}{H_3C-\overset{\overset{O}{\|}}{C}-CH_3} \qquad (5.20)$$

The ketones are not easily oxidized and can usually be recovered completely.

Tertiary alcohols are not oxidized by ordinary agents in aqueous solution. When attacked by very strong oxidizing agents, they are converted to carbon dioxide and water.

Microorganisms oxidize primary and secondary alcohols readily under aerobic conditions. The end products are carbon dioxide and water, but aldehydes and ketones are believed to exist as intermediates. Present evidence indicates that tertiary alcohols are very resistant to microbial degradation and are oxidized initially through terminal methyl groups, as with the hydrocarbons.

5.4 | ALDEHYDES AND KETONES

Aldehydes are the oxidation products of primary alcohols (ROH). *Ketones* are the oxidation products of secondary alcohols.

Aldehydes

The oxidation of primary alcohols to aldehydes is as follows:

$$RCH_2OH + \tfrac{1}{2}O_2 \rightarrow R-\overset{H}{\underset{}{C}}{=}O + H_2O \qquad (5.21)$$

All aldehydes have the characteristic *carbonyl* group, $-\overset{H}{C}{=}O$. The general structural formula for an aldehyde is R—CHO, where R represents any alkyl group: CH_3-, C_2H_5-, and so on.

Aldehydes can also be formed from unsaturated hydrocarbons by ozonation. The hydrocarbons are first converted to an ozonide by ozone.

$$RCH{=}CHR' + O_3 \rightarrow RCH\underset{O-O}{\overset{O}{\diagup\diagdown}}CHR'$$

An ozonide (5.22)

The ozonides react readily with water to form aldehydes.

$$RCH\underset{O-O}{\overset{O}{\diagup\diagdown}}CHR' + H_2O \rightarrow RCHO + R'CHO + H_2O_2 \qquad (5.23)$$

These reactions are particularly significant in air pollution, where aldehydes are formed when unsaturated hydrocarbons discharged in automobile exhausts combine with ozone catalytically produced by reactions of oxygen with sunlight in the presence of oxides of nitrogen. The aldehydes so formed cause eye irritation, one of the most serious problems associated with air pollution. The oxides of nitrogen required for catalyzing ozone formation are produced in great quantities during high-temperature combustion of fossil fuels in steam power plants and internal combustion engines. Automobile and truck engine exhaust gases are a particularly significant source because of wide distribution at ground level where contact with human, animal, and plant life is most probable.

Although a wide variety of aldehydes can be formed from primary alcohols, only a few are of commercial importance. Aldehydes can also be produced by reduction of carboxylic acids.

Formaldehyde Formaldehyde is formed by the oxidation of methanol.

$$CH_3OH + \tfrac{1}{2}O_2 \rightarrow H\overset{H}{\underset{}{C}}{=}O + H_2O$$

Formaldehyde (5.24)

It is used extensively in organic synthesis. It is very toxic to microorganisms, and, because of this property, it is used in embalming fluids and fluids used for the preservation of biological specimens. Industrial wastes containing formaldehyde were considered at one time to be too toxic for treatment by biological methods.

Through dilution of such wastes to reduce the concentration of formaldehyde below 1500 mg/L, it was found that microorganisms could use the formaldehyde as food and oxidize it to carbon dioxide and water. This experience has led to the concept of *toxicity thresholds* in industrial waste treatment practice. It means that below certain concentrations all materials are nontoxic. The completely mixed activated sludge system is designed to take advantage of this concept.

Acetaldehyde   Acetaldehyde is formed by the oxidation of ethanol.

$$CH_3CH_2OH \ + \ \tfrac{1}{2}O_2 \ \rightarrow \ CH_3\!-\!\overset{\text{H}}{C}\!=\!O \ + \ H_2O \qquad (5.25)$$
Acetaldehyde

It is used extensively in organic synthesis. A major industrial use involves its condensation with formaldehyde to produce pentaerythritol $[C(CH_2OH)_4]$, an important intermediate for the production of a wide variety of products, including aldehyde resin paints. Development of a biological treatment process for the formaldehyde-bearing industrial wastes from the manufacture of pentaerythritol led to the concept of toxicity threshold mentioned earlier.

A wide variety of aldehydes are of commercial interest. The names and formulas of several of them are given in Table 5.6. The IUPAC names of all aldehydes end in *-al*.

Ketones

Ketones are prepared by the oxidation of secondary alcohols.

$$R\!-\!\overset{\text{O}}{\underset{\text{H}}{\overset{\text{H}}{C}}}\!-\!R' \ + \ \tfrac{1}{2}O_2 \ \rightarrow \ R\!-\!\overset{\text{O}}{\overset{\|}{C}}\!-\!R' \ + \ H_2O \qquad (5.26)$$
Ketone

Ketones have two alkyl groups attached to the carbonyl group, $-\overset{\text{O}}{\overset{\|}{C}}\!-$, while aldehydes have one R group and a hydrogen atom. The R groups in ketones may be the same or different.

Table 5.6 | Common aldehydes

Common name	IUPAC name	Formula
Formaldehyde	Methanal	HCHO
Acetaldehyde	Ethanal	CH_3CHO
Propionaldehyde	*n*-Propanal	C_2H_5CHO
Butyraldehyde	*n*-Butanal	C_3H_7CHO
Valeraldehyde	*n*-Pentanal	C_4H_9CHO
Caproaldehyde	*n*-Hexanal	$C_5H_{11}CHO$
Heptaldehyde	*n*-Heptanal	$C_6H_{13}CHO$
Acrolein		$CH_2\!=\!CHCHO$
Citral		$C_9H_{15}CHO$
Citronellal		$C_9H_{17}CHO$

Acetone   Acetone (dimethyl ketone) is the simplest ketone and is produced by the oxidation of isopropyl alcohol (2-propanol).

$$CH_3-\underset{\underset{H}{|}}{\overset{\overset{H}{|}}{\underset{}{C}}}-CH_3 + \tfrac{1}{2}O_2 \rightarrow \underset{\text{Acetone}}{CH_3-\overset{\overset{O}{\|}}{C}-CH_3} + H_2O \qquad (5.27)$$

Methyl ethyl ketone   Methyl ethyl ketone is prepared by the oxidation of 2-butanol.

$$CH_3-\underset{\underset{H}{|}}{\overset{\overset{H}{\underset{|}{O}}}{C}}-CH_2CH_3 + \tfrac{1}{2}O_2 \rightarrow \underset{\text{Methyl ethyl ketone}}{CH_3-\overset{\overset{O}{\|}}{C}-C_2H_5} + H_2O \qquad (5.28)$$

Ketones are used as solvents in industry and for the synthesis of a wide variety of products. The names of a few ketones are given in Table 5.7.

Chemical Properties of Aldehydes and Ketones

Aldehydes and ketones differ in ease of oxidation.

1. Aldehydes are easily oxidized to the corresponding acids.

$$R-\overset{\overset{H}{|}}{C}=O + \tfrac{1}{2}O_2 \rightarrow R-\overset{\overset{O}{\|}}{C}-OH \qquad (5.29)$$

2. Ketones are more difficult to oxidize. This is because there is no hydrogen attached to the carbonyl group. As a result, further oxidation must initiate in one of the alkyl groups, the molecule is cleaved, and two or more acids are produced.

$$\underset{\text{Acetone}}{CH_3\overset{\overset{O}{\|}}{C}CH_3} + 2O_2 \rightarrow CO_2 + H_2O + \underset{\text{Acetic acid}}{CH_3COOH} \qquad (5.30)$$

Table 5.7 | Common ketones

Common name	IUPAC name
Acetone	Propanone
Methyl ethyl ketone	Butanone
Diethyl ketone	3-Pentanone
Methyl propyl ketone	2-Pentanone
Methyl isopropyl ketone	3-Methyl-2-butanone
n-Butyl methyl ketone	2-Hexanone
Ethyl propyl ketone	3-Hexanone
Dipropyl ketone	4-Heptanone
Dibutyl ketone	5-Nonanone

In the case of acetone, carbon dioxide and acetic acid are formed. Theoretically, formic acid should be formed, but it is so easily oxidized that it is generally converted under the prevailing conditions to carbon dioxide and water. Higher ketones such as diethyl ketone (3-pentanone) are oxidized as follows:

$$\underset{\underset{H}{\overset{H}{|}}}{R}-\overset{H}{\underset{H}{\overset{|}{C}}}-\overset{O}{\overset{\parallel}{C}}-R' + \tfrac{3}{2}O_2 \rightarrow R-\overset{O}{\overset{\parallel}{C}}-OH + R'-\overset{O}{\overset{\parallel}{C}}-OH \qquad (5.31)$$

Oxidation of both aldehydes and ketones to organic acids is accomplished readily by many microorganisms. However, since the organic acids serve as a good food supply, the ultimate end products under aerobic conditions are carbon dioxide and water.

5.5 | ACIDS

Acids represent the highest oxidation state that an organic compound can attain. Further oxidation results in the formation of carbon dioxide and water, which are classed as inorganic compounds, and the organic compound is considered completely destroyed.

$$CH_4 \rightarrow CH_3OH \rightarrow H_2C{=}O \rightarrow HCOOH \rightarrow H_2O + CO_2 \qquad (5.32)$$

Hydro-   Alcohol   Aldehyde   Acid   Products of
carbon   complete
                                        oxidation

Organic acids typically contain the $-\overset{O}{\overset{\parallel}{C}}-OH$ group. This is called the *carboxyl* group and is commonly written $-COOH$. Acids with one carboxyl group are known as *monocarboxylic* acids, and those with more than one are *polycarboxylic* acids. In addition, derivatives of phenol (Sec. 5.14) can act as acids. The acids may be saturated or unsaturated. Some contain hydroxy groups within the molecule.

Saturated Monocarboxylic Acids

A wide variety of saturated monocarboxylic acids occur in nature as constituents of fats, oils, and waxes. Unsaturated acids are also found in these materials, and, as a result, both types are commonly known as *fatty* acids. The majority of the fatty acids derived from natural products have an even number of carbon atoms and usually have a straight-chain or normal structure.

Physical Properties The first nine members, C_1 to C_9, are liquids. All the others are greasy solids. Formic, acetic, and propionic acid have sharp penetrating odors;

the remaining liquid acids have disgusting odors, particularly butyric and valeric. Butyric acid gives rancid butter its characteristic odor. Industrial wastes from the dairy industry must be treated with considerable care to prevent formation of butyric acid and consequent odor problems. The "volatile acids" (C_1 to C_5) are quite soluble in water.

The names, formulas, and physical constants of the important saturated acids are given in Table 5.8. All the acids are considered weak acids from the viewpoint of ionization. Formic acid is the strongest of all.

Nomenclature   The common names are usually used for most of the acids, except for those with 7, 8, 9, and 10 carbon atoms. The IUPAC names are given in Table 5.8. In naming derivatives of acids, the IUPAC system is frequently abandoned for a system using Greek letters to identify the carbon atoms. In this system the carboxyl group is the reference point, and carbon atoms are numbered from it as follows:

$$\overset{5}{C}H_3\overset{4}{C}H_2\overset{3}{C}H_2\overset{2}{C}H_2\overset{1}{C}OOH$$
$$\quad\quad\delta\quad\;\gamma\quad\;\beta\quad\;\alpha$$

The carbon atom next to the carboxyl group is *alpha,* the next *beta,* then *gamma, delta,* and so on. The terminal carbon atom is also referred to as being in the *omega* position. The α-amino acids are particularly important compounds and are discussed in Sec. 5.22.

Unsaturated Monocarboxylic Acids

The principal unsaturated monocarboxylic acids are as follows:

Acrylic Acid (CH_2=CHCOOH)   Acrylic acid is used extensively because of its ability to polymerize, a characteristic of many compounds with unsaturated link-

Table 5.8 | Physical constants of some normal monocarboxylic acids

Common name	IUPAC name	Formula	Mp, °C	Bp, °C	Sp. gr., 20°/4°*	pKa at 25°
Formic	Methanoic	HCOOH	8.3	101	1.220	3.75
Acetic	Ethanoic	CH_3COOH	16.6	117.9	$1.045^{25°}$	4.756
Propionic	Propanoic	C_2H_5COOH	−20.7	141.1	0.993	4.87
Butyric	Butanoic	C_3H_7COOH	−5.7	163.7	0.958	4.83
Valeric	Pentanoic	C_4H_9COOH	−34	186.1	0.939	$4.83^{20°}$
Caproic	Hexanoic	$C_5H_{11}COOH$	−3	205.2	0.927	4.85
Enanthic	Heptanoic	$C_6H_{13}COOH$	−7.5	222.2	0.918	4.89
Caprylic	Octanoic	$C_7H_{15}COOH$	16.3	239	0.911	4.89
Pelargonic	Nonanoic	$C_8H_{17}COOH$	12.3	254.5	0.905	4.96
Capric	Decanoic	$C_9H_{19}COOH$	31.9	268.7	$0.886^{40°}$	
Palmitic	Hexadecanoic	$C_{15}H_{31}COOH$	63.1	351.5	$0.853^{62°}$	
Stearic	Octadecanoic	$C_{17}H_{35}COOH$	68.8	232	0.941	

*Density of compound at 20°C (or at the temperature noted by superscript) over the maximum density of water, which is at 4°C.
Source: D. R. Linde (ed.): "Handbook of Chemistry and Physics," 82nd ed., CRC Press LLC, Boca Raton, 2001.

ages. Derivatives of the acid are used to form colorless plastics such as Lucite and Plexiglas.

Oleic Acid [$CH_3(CH_2)_7CH=CH(CH_2)_7COOH$]

Linoleic Acid [$CH_3(CH_2)_4CH=CHCH_2CH=CH(CH_2)_7COOH$]

Linolenic Acid [$CH_3(CH_2CH=CH)_3CH_2(CH_2)_6COOH$]

Oleic, linoleic, and linolenic acids are normal constituents of the glycerides of most fats and oils. A glyceride is the ester (Sec. 5.6) formed by combining glycerol (Sec. 5.3) with these acids. Oleic acid is considered to be an essential acid in the diet of humans and animals. Linoleic and linolenic acids as glycerides are important constituents of linseed and other drying oils. Their value for this purpose is dependent upon the multiple double bonds that they possess.

Chemical Properties of Acids   The chemical properties of acids are determined largely by the carboxyl group. All form metallic salts that have a wide range of commercial use. In addition, the unsaturated acids have chemical properties characterized by the double bond, as described under unsaturated hydrocarbons in Sec. 5.2. The unsaturated acids may be reduced with hydrogen to give corresponding saturated acids.

Organic acids serve as food for many microorganisms and are oxidized to carbon dioxide and water. Ease and rate of oxidation are believed to be enhanced by the presence of unsaturated linkages. The rate of biological attack on high-molecular-weight fatty acids is often limited by their solubility in water. This is a particular problem in sludge digesters where fatty materials tend to float and segregate themselves in a scum layer.

Polycarboxylic Acids

The most important of the polycarboxylic acids are those that have two carboxyl groups, one on each end of a normal chain of carbon atoms. The most important acids are listed in Table 5.9.

Adipic acid is of some interest because it is used in the manufacture of nylon fiber and may be expected to occur in the industrial wastes of that industry.

Table 5.9 | Dicarboxylic acids

Name	Formula
Oxalic	$(COOH)_2$
Malonic	$CH_2(COOH)_2$
Succinic	$(CH_2)_2(COOH)_2$
Glutaric	$(CH_2)_3(COOH)_2$
Adipic	$(CH_2)_4(COOH)_2$
Pimelic	$(CH_2)_5(COOH)_2$
Suberic	$(CH_2)_6(COOH)_2$

Hydroxy Acids

Hydroxy acids have OH groups attached to the molecule other than in the carboxyl group. Thus, they act chemically as acids and as alcohols. A number of the hydroxy acids have special names. Some examples are

$HOCH_2COOH$	Hydroxyacetic acid, glycolic acid
$CH_3CHOHCOOH$	α-Hydroxypropionic acid, lactic acid
$HOCH_2CH_2COOH$	β-Hydroxypropionic acid, hydracrylic acid
$HOCH_2CH_2CH_2COOH$	γ-Hydroxybutyric acid

Lactic acid, α-hydroxypropionic acid, is of special interest since it is formed during bacterial fermentation of milk and therefore is a normal constituent of industrial wastes from the dairy industry. Whey from cheese making contains considerable amounts of lactic acid. It is the principal acid in *sauerkraut* juice and prevents spoilage of the sauerkraut.

Lactic acid is also of interest because it possesses the property of *optical activity*. One form of lactic acid is *levorotatory,* which rotates polarized light in the counterclockwise direction (levo means "to the left"). The other form is *dextrorotatory,* which rotates polarized light in a clockwise fashion (dextro means "to the right"). In the past, *l* was used to name the levorotatory form while *d* was used to name the dextrorotatory form. Current convention is to use a minus sign ($-$) for the *l* form and a plus sign ($+$) for the *d* form.

Many organic compounds are optically active. In order to be optically active, a compound must generally possess at least one *asymmetric* carbon atom. An asymmetric carbon atom is one having four dissimilar groups attached to it, as follows:

$$d-\underset{\underset{c}{|}}{\overset{\overset{a}{|}}{C}}-b$$

Such compounds are not superposable on their mirror images. Many other objects lack symmetry and cannot be superposed on their mirror image. For example, the left and right hands are mirror images of each other, but cannot be made to coincide at every point with each other. Objects, including molecules, that are nonsuperposable on their mirror image are said to be *chiral*. With molecules, the asymmetric carbon atom is termed the *chiral center.* Chiral compounds may have more than one chiral center. The different chiral forms of a molecule are called *enantiomers.* Chiral compounds are of environmental significance in that their enantiomers can differ in biological activity or biodegradability. The most sensational case is that of the drug thalidomide. The ($+$) enantiomer was employed as a sedative and antinausea drug around 1960. As is the case with most commercially available chiral molecules, thalidomide was sold as a *racemic* mixture [approximately equal amounts of the ($+$) and ($-$) enantiomers; such mixtures carry the symbol ($\pm$)]. Unfortunately, the ($-$) enantiomer caused serious birth defects in children born to pregnant women taking the drug during pregnancy.

Environmental engineers and scientists have recently begun to use the properties of chiral molecules to better understand the fate of organic pollutants.[4]

Lactic acid has one chiral center, and two optically active enantiomers are possible.

$$
\begin{array}{cc}
\text{COOH} & \text{COOH} \\
| & | \\
\text{H}-\text{C}-\text{OH} & \text{HO}-\text{C}-\text{H} \\
| & | \\
\text{CH}_3 & \text{CH}_3 \\
\text{D-(–)-Lactic Acid}^5 & \text{L-(+)-Lactic Acid}^5
\end{array}
$$

It is the L form of lactic acid that is commonly produced in living organisms.

Hydroxy Polycarboxylic Acids

There are several hydroxy polycarboxylic acids.

$$
\begin{bmatrix}
\text{H}_2\text{CCOOH} \\
| \\
\text{HOCCOOH} \\
| \\
\text{H}_2\text{CCOOH}
\end{bmatrix} \cdot \text{H}_2\text{O}
\qquad
\begin{array}{c}
\text{COOH} \\
| \\
\text{HCOH} \\
| \\
\text{HOCH} \\
| \\
\text{COOH}
\end{array}
$$

Citric acid   L-(+)-Tartaric acid

Tartaric acid occurs in many fruits, especially grapes, and is present in canning and winery wastes. Citric acid is the major acid of all citrus fruits: oranges, lemons, limes, and grapefruit. It is, of course, a major component of the liquid wastes of the citrus industry.

5.6 | ESTERS

Esters are compounds formed by the reaction of acids and alcohols, similar to the reactions of acids and bases to form salts in inorganic chemistry. The reaction between low-molecular-weight organic acids and alcohols is never complete. Hydrolysis occurs and a reversible reaction results. The reaction may be represented by the general equation

$$\text{RCO}-\boxed{\text{OH} + \text{H}}-\text{OR}_1 \rightleftharpoons \text{H}_2\text{O} + \text{RCOOR}_1 \qquad (5.33)$$

---

[4]J. J. Ridal et al., Enantiomers of α-Hexachlorocyclohexane as Tracers of Air-Water Gas Exchange in Lake Ontario, *Env. Sci. Tech.*, **31**: 1940–1945, 1997.

[5]The symbols D and L are used to describe the configuration of the groups surrounding the asymmetric carbon and are not to be confused with the + or − designations. Additional configuration information is conveyed by the use of R and S symbols. Details concerning this terminology are beyond the scope of this text. The reader should refer to the organic chemistry texts listed at the end of this chapter for the finer points of chiral nomenclature.

The general formula of an ester is

$$
\underset{\text{R}-\overset{\displaystyle O}{\overset{\|}{\text{C}}}-\text{O}-\text{R}'}{}
$$

A wide variety of esters are used in chemical manufacturing. Most esters have highly pleasing odors. Butyl acetate smells like banana oil (amyl acetate) and is used for solvent purposes. Many esters are used in flavoring extracts and perfumes.

Esters have been used to some extent as immiscible solvents in the separation and purification of antibiotics. Considerable quantities often reach the sewer system and become an industrial waste problem. Enzymes liberated by many microorganisms hydrolyze esters to yield the corresponding acid and alcohol.

$$
\text{R}-\overset{\displaystyle O}{\overset{\|}{\text{C}}}-\text{OR}' + \text{HOH} \xrightarrow[\text{enzyme}]{} \text{RCOOH} + \text{R}'-\text{OH} \tag{5.34}
$$

The acid and alcohol serve as bacterial food and are oxidized to carbon dioxide and water, as discussed in Secs. 5.3 and 5.5.

5.7 | ETHERS

Ethers are formed by treatment of alcohols with strong dehydrating agents. In the reaction, one molecule of water is removed from two molecules of alcohol.

$$
\text{RO\underline{H} + \underline{HO}R'} \xrightarrow[\substack{\text{dehydrating} \\ \text{agent}}]{\Delta} \underset{\text{Ether}}{\text{R}-\text{O}-\text{R}'} + \text{H}_2\text{O} \tag{5.35}
$$

The two fragments of the alcohol join to form an ether. The alkyl groups are joined through an oxygen atom; thus, a carbon-to-oxygen-to-carbon bond,

$$
-\overset{|}{\underset{|}{\text{C}}}-\text{O}-\overset{|}{\underset{|}{\text{C}}}- ,
$$

is established.

Ethers are used widely as solvents. The low-molecular-weight ethers are highly flammable. When left exposed to air, they are prone to form peroxides that are extremely explosive, particularly when recovery by distillation is practiced and the distillation is allowed to go to dryness. Diethyl ether has been used widely as an anesthetic.

Ethers are generally resistant to biological oxidation. Fortunately, most are relatively insoluble in water and can be separated from industrial wastes by flotation or decantation procedures.

An important ether is methyl *tert*-butyl ether (MTBE).

$$
\text{H}_3\text{C}-\text{O}-\overset{\displaystyle \text{CH}_3}{\overset{|}{\underset{|}{\underset{\displaystyle \text{CH}_3}{\text{C}}}}}-\text{CH}_3
$$

MTBE was first used in the late 1970s as an octane enhancer when lead additives to gasoline were being phased out. More recently it has been added to gasoline as a fuel oxygenate to help reduce carbon monoxide emissions from gasoline combustion. MTBE is very soluble in water, volatile, not strongly sorbed to particulate matter, and relatively nonbiodegradable. Thus, upon release to the environment, it is very mobile. Through leaking storage tanks, pipelines, and other uncontrolled releases to the environment, MTBE has been found in a large number of surface waters and groundwaters. Although the health effects associated with MTBE exposure are not yet completely known, action levels as low as 20 μg/L have been set due to taste and odor considerations.[6] Because of these problems, there is a proposed ban on the use of MTBE as a fuel additive. Ethanol is an alternative for MTBE as a fuel additive. However, it has its own set of potential problems (Sec. 3.8).

5.8 | ALKYL HALIDES AND OTHER HALOGENATED ALIPHATIC COMPOUNDS

The *alkyl halides* are aliphatic compounds in which a halogen is attached to an alkyl carbon. These are to be distinguished from other halogenated aliphatics where the halogen is attached to a double-bonded carbon. Both types of compounds are discussed here. Halogenated aliphatic compounds are used extensively in organic synthesis, and a few of them have important industrial uses. Many are toxic to humans and are of significant environmental concern. They are among the most frequently found hazardous chemicals at abandoned waste sites, refuse disposal areas, and industrial and municipal wastewaters. The U.S. federal government has set drinking water maximum contaminant levels (MCLs) for some of these compounds and is proposing MCLs for others. These are listed in Table 34.1. Most of the halogenated aliphatic compounds are volatile and thus may be significant air pollutants.

Simple Alkyl Halides

The alkyl halides are of great value in organic synthesis because they react with potassium cyanide to form compounds with an additional carbon atom.

$$R \cdot \overline{|\underline{I + K}|} \cdot CN \rightarrow KI + R \cdot CN \qquad (5.36)$$
A nitrile
(alkyl cyanide)

The nitrile formed can be hydrolyzed to an acid and then reduced to an alcohol, if desired. The alcohol can be converted to an alkyl halide and the process repeated. In this manner the organic chemist can increase the length of carbon-chain compounds one atom at a time.

---

[6]R. Johnson, et al., MTBE—To What Extent Will Past Releases Contaminate Community Water Supply Wells, *Env. Sci. Tech.,* **34**(9): 210A–217A (2000).

Methyl chloride (CH$_3$Cl), also called chloromethane, and ethyl chloride (C$_2$H$_5$Cl), also called chloroethane, were used extensively as refrigerants in the past. Chloroethane is used in the manufacture of tetraethyl lead, great quantities of which were previously used in the production of high-octane, antiknock gasolines.

$$4C_2H_5Cl + 4NaPb \text{ (alloy)} \xrightarrow{40-60°} 4NaCl + 3Pb + \underset{\substack{\text{Tetraethyl} \\ \text{lead}}}{(C_2H_5)_4Pb} \qquad (5.37)$$

During combustion the lead is converted to lead oxide and is emitted in the exhaust gases, creating a lead pollution problem. The lead also fouls catalytic devices being used to reduce the quantity of unburned fuel discharged to the atmosphere. For these reasons use of leaded gasoline has been eliminated in the United States and is being eliminated in Europe.

Vinyl chloride (CH$_2$=CHCl), also called monochloroethene, is used in large quantities to produce a wide variety of polyvinylchloride (PVC) products. It is also formed biologically through reduction of polyhalogenated ethenes. Through this route it is frequently found in contaminated groundwaters and in gases emanating from sanitary landfills. This compound is very volatile and is a known carcinogen. A drinking water MCL of 2 μg/L has been established (Table 34.1). Because of its volatility, vinyl chloride is a potentially significant air pollutant.

Polyhalogen compounds

A wide variety of *polyhalogen* compounds are used for industrial purposes. Because of a variety of public and environmental health problems, use of several of these compounds is being discouraged. The challenge is to find environmentally friendly substitutes.

1,2-Dibromoethane   This compound has also been called ethylene dibromide (EDB), although it is not actually an alkene. 1,2-Dibromoethane is formed from ethene by addition of bromine.

$$\underset{}{\overset{\text{H H}}{HC=CH}} + Br_2 \rightarrow \underset{\text{Br Br}}{\overset{\text{H H}}{HC-CH}} \qquad (5.38)$$

1,2-Dibromoethane

1,2-Dibromoethane has many industrial uses and has been used as a pesticide in agricultural areas and as an ingredient in ethyl gasoline. In the early 1980s, it was discovered in fumigated fruits and vegetables, and several grain products, causing some commercial products to be removed from stores. It is a commonly detected groundwater contaminant. Because of its potential health effects, a drinking water standard of 0.05 μg/L has been established (Table 34.1).

Dichloromethane (CH$_2$Cl$_2$)   Dichloromethane, sometimes called methylene chloride, is a volatile liquid that has been used as an industrial solvent, in paint strippers, to decaffeinate coffee, and in the making of polyurethane products. It can be produced from the chlorination of methane, as can all the chlorinated methanes. Dichloromethane is quite soluble in water, more dense than water, and is fairly volatile. A drinking water MCL of 2 μg/L has been established.

Chloroform (CHCl$_3$)   Chloroform (trichloromethane) was one of our first anesthetics (1847) and was widely used until about 1920. It is used in industry as a solvent for oils, waxes, and so on. It is nonflammable. Chloroform has been found present in microgram per liter to milligram per liter concentrations in drinking water supplies as a result of chlorination for disinfection. It is formed through reaction of chlorine with organics of biological origin commonly present in natural waters. Because it is a potential carcinogen, an MCL in drinking water has been established for the sum total of chloroform and other trihalomethanes (THMs) of 80 μg/L.

Carbon Tetrachloride (CCl$_4$)   Carbon tetrachloride (tetrachloromethane) has been widely used as a fire extinguisher in small units (Pyrene) and as a solvent. It is a toxic compound, and its use should be restricted to well-ventilated areas. Its use as a fire extinguisher is fraught with some danger. In contact with hot iron and oxygen, it is converted to phosgene (COCl$_2$), a highly toxic gas. For this reason, trained fire fighters use other types of fire-fighting equipment. An MCL for carbon tetrachloride has been established at 5 μg/L.

Halogenated Ethanes   Several halogenated ethanes, particularly some chlorinated ethanes, are of environmental significance. For the most part, these compounds are used as industrial solvents. They are volatile, heavier than water, and have varying solubilities in water. Some are important because they are products formed during the transformation (chemical and biological) of other halogenated compounds. These aspects are further discussed in Sec. 5.34 and Chap. 6.

1,1,1-Trichloroethane (CCl$_3$CH$_3$) is a widely used industrial solvent and is a commonly detected groundwater contaminant. 1,2-Dichloroethane (CH$_2$ClCH$_2$Cl) is also an industrial solvent. Drinking water MCLs are in place for both of these compounds (200 μg/L and 5 μg/L, respectively), and for 1,1,2-trichloroethane (CHCl$_2$CH$_2$Cl), another industrial solvent (5 μg/L).

Halogenated Ethenes   As with the halogenated ethanes, it is the chlorinated ethenes that are of most environmental significance. They are volatile solvents, heavier than water, and have varying solubilities in water. Transformation reactions can convert parent chlorinated ethenes into other ethenes. A classical example is the reductive dechlorination of trichloroethene to dichloroethene, and then to vinyl chloride.

Trichloroethene (TCE: CCl$_2$=CHCl), also called trichloroethylene, is a nonflammable, volatile liquid that has been used as an industrial solvent and degreaser, a dry-cleaning solvent, and in many household products. In the past it was used to decaffeinate coffee. It is one of the most commonly found groundwater pollutants. 1,1-Dichloroethene (CCl$_2$=CH$_2$), also called 1,1-dichloroethylene, is also a solvent and has been found in the environment as a transformation product of both TCE and 1,1,1-trichloroethane. There are drinking water MCLs for both TCE (5 μg/L) and 1,1-dichloroethene (7 μg/L).

The chlorinated ethenes with the formula CHCl=CHCl are examples of a special type of isomerism called *geometric isomers*. In *cis*-1,2-dichloroethene the chlorines are on the same side of the molecule while in *trans*-1,2-dichloroethene the chlorines are on the opposite side of the molecule:

cis-1,2-Dichloroethene   trans-1,2-Dichloroethene

It is important to note that the physical properties can be different for geometric isomers and thus their behavior in the environment and in engineered reactors can be different. For example, the following are given for the cis and trans isomers of 1,2-dichloroethene (values are for 20°C except for boiling point):

Property	cis isomer	trans isomer
Boiling point, °C	60.3	47.5
Density, g/cm^3	1.2837	1.2565
Vapor pressure, atm	0.24	0.37
Henry's constant, atm/M	7.4	6.7
Water solubility, mg/L	3500	6300

Both *cis*- and *trans*-1,2-dichloroethene have been found in the environment as biological transformation products of TCE. Drinking water MCLs have been established for both (70 and 100 μg/L, respectively).

Tetrachloroethene (CCl$_2$=CCl$_2$), also called perchloroethene (PCE), is a widely used industrial solvent and is a commonly found groundwater pollutant. It is a regulated drinking water contaminant with an MCL of 5 μg/L.

Dibromochloropropane (DBCP)   Although there are nine possible isomers of dibromochloropropane (C$_3$H$_5$Br$_2$Cl), it is 1,2-dibromo-3-chloropropane (CH$_2$BrCH BrCH$_2$Cl) that is most environmentally significant. This isomer has been widely used as a pesticide and is frequently detected in groundwater. It is thought to cause cancer and adverse reproductive outcomes, and for these reasons a drinking water MCL has been established (0.2 μg/L).

Chlorofluorocarbons (CFCs)   Chlorofluorocarbon compounds are one- and two-carbon compounds containing chlorine and fluorine. They are sometimes called by their trade name, Freons. For many years these compounds have been widely used as refrigerants, solvents, and aerosol propellants. They were used because of their stability and their nonflammable and nontoxic properties. The most commonly used CFCs are CCl$_3$F (CFC-11), CCl$_2$F$_2$ (CFC-12), CCl$_2$FCClF$_2$ (CFC-113), CClF$_2$CClF$_2$ (CFC-114), and CClF$_2$CF$_3$ (CFC-115). Recently, CFCs have been shown to be involved in the destruction of the stratospheric ozone layer, the "ozone hole" effect, via the following reactions with ultraviolet radiation.

$$CF_2Cl_2 + h\nu \rightarrow Cl\cdot + CF_2Cl\cdot$$
(Cl atom)

$$Cl\cdot + O_3 \rightarrow ClO\cdot + O_2$$
(ozone) (free radical chlorine oxide)

$$ClO\cdot + O \rightarrow Cl\cdot + O_2$$
(O atom)

$$O + O_3 \rightarrow O_2 + O_2$$

Chlorofluorocarbons are also known to be greenhouse gases. Because of their role in ozone destruction and the greenhouse effect, manufacture and use of chlorofluorocarbons is being eliminated worldwide.

Haloacetic Acids (HAA) Another important group of halogenated aliphatic compounds is the haloacetic acids. They are produced during chlorination of drinking waters that contain natural organic matter. An MCL of 60 μg/L has been established for the sum of five haloacetic acids (HAA5), mono-, di-, and trichloroacetic acids and mono- and dibromoacetic acids.

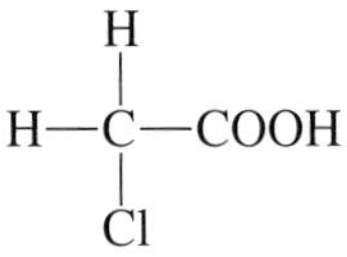

Monochloroacetic acid 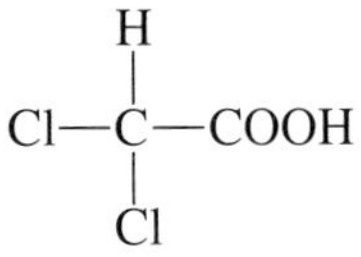   Dichloroacetic acid 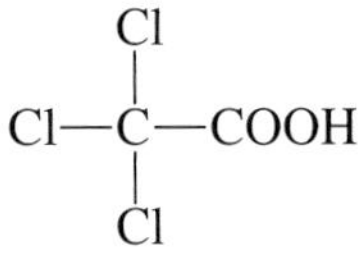   Trichloroacetic acid

Perfluorinated Compounds Another class of halogenated compounds of emerging concern are the *perfluorinated* compounds. Here, all C—H bonds are replaced by C—F bonds. These bonds are very stable and tend to persist in the environment. Perfluorinated compounds have been used in fabric protectors and in a variety of products ranging from paper plates to semiconductor coatings to airplane hydraulic fluid. A particular group of these compounds that are of emerging concern are those that either use perfluoro-octane-sulfonate (PFOS) in their manufacture or break down to this compound.

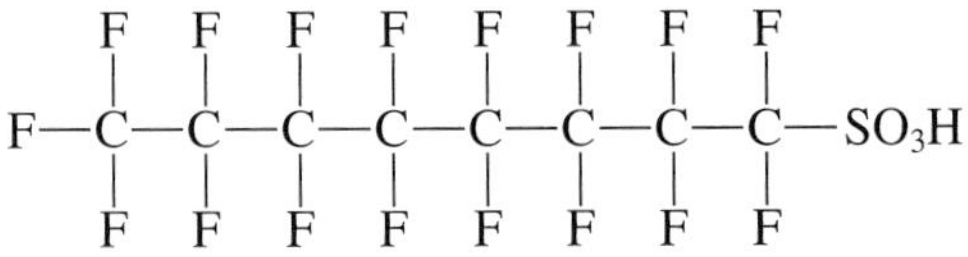

Perfluoro-octane-sulfonate (PFOS)

These compounds are known to bioaccumulate and are being found in blood samples from humans and animals far from their known sources.[7] Laboratory toxicity

---

[7]J. P. Giesy and K. Kannan, Global Distribution of Perfluorooctane Sulfonate in Wildlife, *Env. Sci. Tech.*, **35:** 1339–1342 (2001).

studies are showing that these compounds are toxic, causing a variety of effects in the livers of test animals, and perhaps carcinogenic.[8,9]

5.9 | SIMPLE COMPOUNDS CONTAINING NITROGEN

The simple aliphatic compounds containing nitrogen are of three types: amines, amides, and nitriles (cyanides). Other nitrogen-containing compounds of environmental significance include nitrosamines and isocyanates.

Amines

The amines are alkyl derivatives of ammonia. They are of three types: primary, secondary, and tertiary.

$$
\text{R—NH}_2 \qquad \begin{array}{c} \text{R} \\ \diagdown \\ \diagup \\ \text{R}' \end{array}\!\!\text{NH} \qquad \begin{array}{c} \text{R} \\ \diagdown \\ \diagup \\ \text{R}' \end{array}\!\!\text{N—R}''
$$

Primary amine	Secondary[10] amine	Tertiary[10] amine

In *primary amines,* one hydrogen atom of ammonia is replaced by an alkyl group such as CH_3—, C_2H_5—, and so on. In *secondary amines,* two hydrogen atoms of ammonia are replaced by alkyl groups, and in *tertiary amines,* all three hydrogens are replaced. The amines, like ammonia, are all basic in reaction. The basicity increases from primary to tertiary.

The amines are found in certain industrial wastes, particularly those from the fish and beet-sugar industries. It is well known that deamination reactions (removal of ammonia) are easily accomplished with primary amines, and somewhat less easily with secondary amines.

Tertiary amines combine with alkyl halides to form *quaternary ammonium salts* as follows:

$$
\begin{array}{c} \text{R} \\ | \\ \text{R—N} \\ | \\ \text{R} \end{array} + \text{RCl} \xrightarrow{\Delta} \left[\begin{array}{c} \text{R} \\ | \\ \text{R—N—R} \\ | \\ \text{R} \end{array}\right]^{+} + \text{Cl}^{-} \tag{5.39}
$$

The compounds formed are actually chloride salts and ionize to form a quaternary ammonium ion and a chloride ion. The quaternary ammonium salts have bactericidal properties that can be enhanced by the proper choice of the R groups. They

---

[8]R. Renner, Growing Concern Over Perfluorinated Chemicals, *Env. Sci. Tech.,* **35:** 154A–160A (2001).

[9]R. Renner, Perfluorinated Compounds Linked to Carcinogenicity in Vitro," *Env. Sci. Tech.,* **35:** 180A (2001).

[10]R, R′, R″ represent alkyl groups. They may all be different or all alike.

are therefore of interest to public health professionals, who find them useful as disinfecting agents in food- and beverage-dispensing establishments. They are also used as disinfectants in the laundering of babies' diapers to control infections of bacteria responsible for the rapid hydrolysis of urea. Solutions of the quaternary ammonium salts are sold for disinfecting purposes under a variety of trade names.

Amides

The amides may be considered as being derived from organic acids and ammonia under special conditions. The ordinary reaction between ammonia and an organic acid, of course, produces an ammonium salt.

$$RCOOH + NH_3 \rightarrow RCOO^- + NH_4^+ \qquad (5.40)$$

Under special conditions, an amide results.

$$\underset{}{R-\overset{O}{\overset{\|}{C}}-\overline{OH + H}-NH_2} \rightarrow R-\overset{O}{\overset{\|}{C}}-NH_2 + H_2O \qquad (5.41)$$
$$\text{Amide}$$

Amides are of considerable significance to organic chemists in synthetic work. When they are caused to react with a halogen (Hofmann reaction), an atom of carbon is lost from the amide, and an amine with one less carbon atom is formed.

$$CH_3CONH_2 + Br_2 + 4NaOH \rightarrow 2NaBr + Na_2CO_3 + 2H_2O + CH_3NH_2 \qquad (5.42)$$

This constitutes a method of reducing a length of a carbon chain by one atom.

Amides as such are of little importance to the environmental engineer, but the *amide group* $\left(\overset{O}{\overset{\|}{-C}}-NH_2\right)$ is related to an important group of compounds containing the *peptide linkage* $\left(\overset{O}{\overset{\|}{-C}}-\overset{H}{\underset{}{N}}-\right)$, as discussed in Sec. 5.22.

Urea,

$$\underset{NH_2}{\overset{NH_2}{\diagup}} C=O$$

is an amide of considerable importance because of its many commercial uses and because it is a normal constituent of urine. It is a constituent of many agricultural fertilizers and is used in the manufacture of synthetic resins.

Although urea was originally considered an organic compound and its accidental production from ammonium cyanate by Wöhler is considered to have initiated the study of organic chemistry, it is in effect an inorganic compound since it cannot be used by saprophytic bacteria as a source of energy. In aqueous solutions containing soil bacteria, urea is hydrolyzed to carbon dioxide and ammonia. These combine to form ammonium carbonate in the presence of water.

$$C{=}O \begin{array}{c} \diagup NH_2 \\ \\ \diagdown NH_2 \end{array} + H_2O \xrightarrow[\text{enzymes}]{\text{bact.}} CO_2 + 2NH_3 \tag{5.43}$$

and

$$CO_2 + 2NH_3 + H_2O \rightleftharpoons (NH_4)_2CO_3 \tag{5.44}$$

The penetrating odor at latrines, privies, and some urinals is due to bacterial infections and their action on urea with subsequent release of free ammonia to the atmosphere. The use of disinfectants will control the decomposition of urea and, thereby, control odors.

Nitriles

Nitriles, or organic cyanides, are important compounds of industry. They have the general formula R—CN, and the R group may be saturated or unsaturated. The names and formulas of a few nitriles of industrial importance are given in Table 5.10.

The nitriles are used extensively in the manufacture of synthetic fibers and can be expected to be present in industrial wastes of that industry. Some are quite toxic to microorganisms.

Nitrosamines

Nitrosamines, or *N*-nitroso compounds, have the general formula R_2—N—N$=$O and are environmentally significant because they have been shown to cause cancer. They can be formed in the environment through the action of microorganisms and have been detected in beer and whiskey as well as a variety of foods. In the past these compounds had been used as industrial solvents. Of particular recent concern is *N*-nitrosodimethylamine (NDMA).

$$O{=}N{-}N \begin{array}{c} \diagup CH_3 \\ \\ \diagdown CH_3 \end{array}$$

N-Nitrosodimethylamine (NDMA)

NDMA has been used in rocket fuel production and for a variety of industrial purposes. It has been found in a variety of foods and can also be formed naturally

Table 5.10 | Important nitriles

Name as nitrile	Name as cyanide	Formula
Acetonitrile	Methyl cyanide	CH_3—CN
Propionitrile	Ethyl cyanide	CH_3—CH_2—CN
Succinonitrile	—	CN—CH_2—CH_2—CN
Acrylonitrile	Vinyl cyanide	$CH_2{=}CH$—CN

in the environment.[11] It is classified as a "probable" human carcinogen. NDMA has been detected in surface waters and treated wastewaters and drinking waters. Recent evidence suggests that NDMA can be formed as a disinfection by-product during chloramination with dimethylamine (DMA) as the precursor.[12]

Isocyanates

Isocyanates have the general formula R—N=C=O and are widely used industrial chemicals. Methyl isocyanate (CH_3—N=C=O) was the material involved in the Bhopal, India, chemical plant explosion in which over 1000 people were killed and many thousands were affected to various degrees. The volatility of these compounds makes them potent potential air pollutants.

5.10 | CYCLIC ALIPHATIC COMPOUNDS

A number of *cyclic aliphatic* hydrocarbons are known. Many of these occur in petroleum and are known as naphthenes.

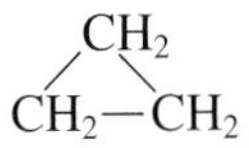

Cyclopropane 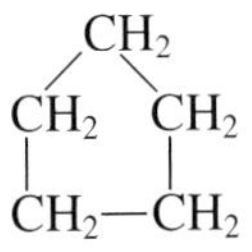   Cyclopentane 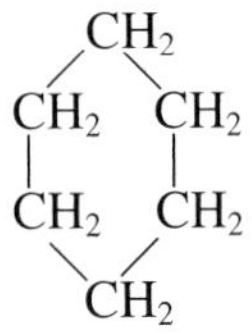   Cyclohexane

They are characterized by having two atoms of hydrogen attached to each carbon in the ring; i.e., they are saturated.

A wide variety of cyclic alcohols and ketones are known. Examples are cyclohexanol and cyclohexanone.

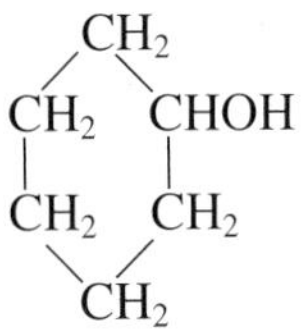

Cyclohexanol 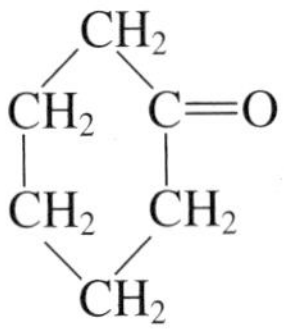   Cyclohexanone

5.11 | MERCAPTANS OR THIOALCOHOLS

Mercaptans or *thioalcohols* are aliphatic compounds that contain sulfur. They have a structure similar to alcohols, except that oxygen is replaced by sulfur.

ROH   RSH

Alcohol   Mercaptan

---

[11]M. Alexander, "Biodegradation and Bioremediation," 2nd ed., Academic Press, San Diego, 1999.

[12]I. Najm and R. R. Trussell, NDMA Formation in Water and Wastewater, *J. American Water Works Assoc.,* **93**(2): 92–99 (2001).

Mercaptans are noted for their disagreeable odor and are found in certain industrial wastes, particularly those from the pulping of wood by the Kraft or sulfate process. They are considered to be quite toxic to fish. The odor of skunks is largely due to butyl mercaptan.

AROMATIC COMPOUNDS

5.12 | INTRODUCTION

The *aromatic* organic compounds are all ring compounds or have cyclic groups of aromatic nature in their structure. The carbon atoms in these ring compounds have only one covalent bond, in contrast to those in aliphatic compounds with two.

Aliphatic ring   Aromatic ring

The simplest aromatic ring is made up of six carbon atoms and is known as the benzene ring. *Benzene* (C_6H_6) is known as the *parent compound* of the aromatic series. Hydrocarbons based on the benzene ring as a structural unit are also called *arenes*. The benzene ring is usually represented by the Kekule formula.

Kekule   Simplified
benzene formula   formulas

This formula shows double bonds between alternate carbon atoms in the ring. The double bonds, however, are not like those in the aliphatic series. For example, halogens will not add to such bonds. For purposes of simplicity, most chemists represent the benzene ring as one of the simplified formulas shown.

Nomenclature

It is important to note that carbon atoms are not shown in the simplified benzene formula. Also, each carbon atom in a ring is like all others, and therefore, when substitution occurs on one carbon atom, the same compound is formed as though substitution had occurred on any of the other five carbon atoms. Thus, for mono-chlorobenzene (C_6H_5Cl) there is only one compound, no matter how the structural formula is written.

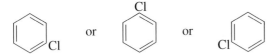

When substitution occurs on two or more carbon atoms of a benzene ring, it becomes necessary to establish some system of nomenclature. Two systems are used.

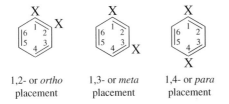

| 1,2- or *ortho* | 1,3- or *meta* | 1,4- or *para* |
| placement | placement | placement |

Di-substituted compounds, such as dichlorobenzene, are commonly referred to as *ortho, meta,* or *para,* depending on the point of substitution. If substitution is on adjacent carbon atoms, the term ortho is used; if on carbon atoms once removed, the term meta is used; and if on carbon atoms opposite each other, the term para is used. Tri- and other poly-substituted compounds must be named by another system. In this system, the carbon atoms of the benzene ring are numbered in a clockwise manner. Examples are

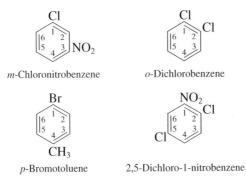

When the benzene ring is attached to aliphatic compounds, the products are also called *phenyl* derivatives, the phenyl group being C_6H_5—. Thus ethyl benzene is also phenyl ethane.

5.13 | HYDROCARBONS

Two series of homologous aromatic hydrocarbons are known: the benzene and the polyring series.

Benzene Series

The benzene series of homologous compounds is made up of alkyl substitution products of benzene. They are found along with benzene in coal tar and in many crude petroleums. Table 5.11 lists the benzene-series hydrocarbons of commercial importance. *Toluene,* or methylbenzene, is the simplest alkyl derivative of benzene. *Xylene* is a dimethyl derivative of benzene.

Toluene   Ethylbenzene   *o*-Xylene   *m*-Xylene   *p*-Xylene

It exists in three isomeric forms: ortho-xylene, meta-xylene, and para-xylene. All are isomeric with ethylbenzene in that they have the same general formula, C_8H_{10}. Together with benzene, these compounds are commonly referred to as the BTEX group.

The benzene-series hydrocarbons are used extensively as solvents and in chemical synthesis, and are common constituents of petroleum products (e.g., gasoline). Although they are relatively insoluble in water, wastewaters and leachates containing 10 to as high as 1000 mg/L for the BTEXs have been observed. These compounds are frequently detected in groundwaters, with a major source being leaking underground gasoline storage tanks.

The benzene-series compounds have been implicated in several human health effects, most notably cancer. Benzene is known to cause leukemia. The current drinking water MCLs are 5 μg/L for benzene, 700 μg/L for ethylbenzene, 1 mg/L for toluene, and 10 mg/L for the sum total of the xylenes.

Styrene (phenyl ethene), a benzene derivative, is an environmentally significant compound. It is used as a monomer in the production of a wide variety of poly-

Table 5.11 | Benzene-series hydrocarbons

Name	Formula	Mp, °C	Bp, °C	Sp. gr., 20°/4°
Benzene	C_6H_6	5.5	80.0	0.877
Toluene	$C_6H_5CH_3$	−94.9	110.6	0.867
o-Xylene	$C_6H_4(CH_3)_2$	−25.2	144.5	$0.880^{10°}$
m-Xylene		−47.8	139.1	0.864
p-Xylene		13.2	138.3	0.861
Ethylbenzene	$C_6H_5C_2H_5$	−94.9	136.1	0.867

Source: D. R. Linde (ed.): "Handbook of Chemistry and Physics," 82nd ed., CRC Press LLC, Boca Raton, 2001.

styrene products (e.g., plastics, synthetic rubber). A drinking water MCL of 100 μg/L has been set for styrene.

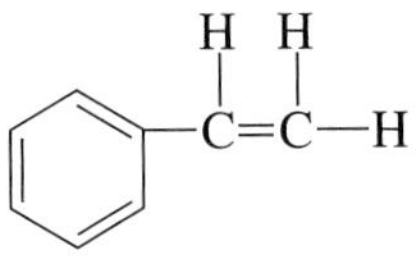

Styrene

Polyring Hydrocarbons

A wide variety of polycyclic aromatic hydrocarbons (PAHs) are known. A few examples will illustrate the possibilities.

Naphthalene ($C_{10}H_8$)   *Naphthalene* is a white crystalline compound derived from coal tar and was formerly used to produce mothballs.

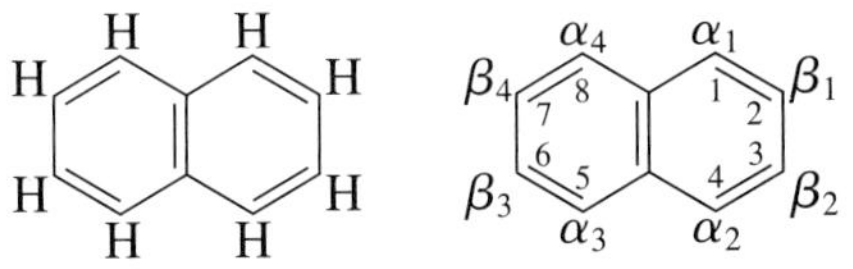

Naphthalene

It has been displaced largely from this market by paradichlorobenzene. A new system of nomenclature is applied to this type of compound. Carbon atoms adjacent to those shared in common by the two rings are known as α-carbon atoms and the others are known as β-carbon atoms. The carbon atoms shared by the two rings do not have hydrogen attached to them and so are given no designation. The specific name, naphthalene, should not be confused with naphthene (Sec. 5.10).

Anthracene ($C_{14}H_{10}$) and Phenanthrene ($C_{14}H_{10}$)   *Anthracene* and *phenanthrene* are isomers.

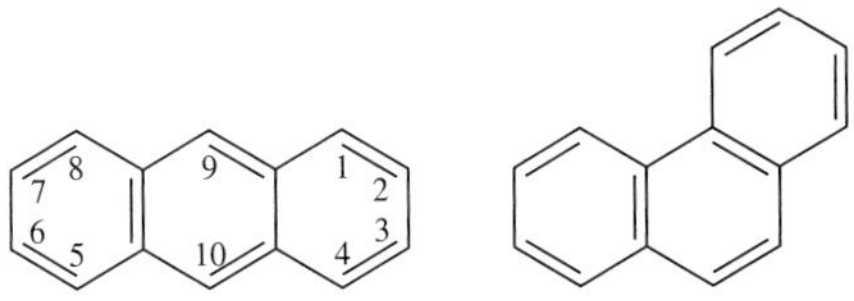

Anthracene   Phenanthrene

Their formulas illustrate the possible ways in which polycyclic aromatic hydrocarbons may occur. Many other more complex compounds, such as *benzo(a)pyrene* and *picene,* are known. Benzo(a)pyrene is a potent carcinogen, and has a drinking water MCL of 0.2 μg/L.

Benzo(a)pyrene   Picene

It should be remembered that hydrogen atoms occur on all carbon atoms of these compounds that are not common to two rings.

Naphthalene and anthracene are widely used in the manufacture of dye-stuffs. The phenanthrene nucleus is found in important alkaloids, such as morphine, vitamin D, sex hormones, and other compounds of great biological significance.

PAHs are associated with combustion products, and are believed to be the cause of the first recognized chemically related cancer that was found in chimney sweeps in the late eighteenth century. They are common residual contaminants at sites where coal was used to manufacture consumer gas (also called coal gas or town gas) in the early twentieth century, and where creosote, which is manufactured from coal, has been used to preserve wood. The larger compounds (five or more aromatic rings) tend not only to be carcinogenic, but are also the more difficult ones for bacteria to degrade.

Chlorinated Aromatic Hydrocarbons

There are many chlorinated aromatic hydrocarbons (also called aryl halides) of importance. They are common industrial chemicals that through their widespread use and relative environmental persistence have become significant environmental problems, much in the same manner as the chlorinated aliphatic compounds. In general, as chlorine atoms are added to the benzene ring, water solubility decreases hydrophobicity increases, and the molecule's vapor pressure decreases. Two of the most important classes are the chlorinated benzenes and the polychlorinated biphenyls.

Chlorinated Benzenes   Chlorinated benzenes are benzenes with one or more of the hydrogens replaced with chlorine. They are widely used industrial chemicals that have solvent and pesticide properties. Like the chlorinated aliphatics, they have been found at abandoned waste sites and in many wastewaters and leachates. They are fairly volatile, slightly to moderately soluble, and hydrophobic.

Chlorobenzene (monochlorobenzene) has a drinking water MCL of 100 μg/L. There are three dichlorobenzene isomers: 1,2-dichlorobenzene (ortho isomer), 1,3-dichlorobenzene (meta isomer), and 1,4-dichlorobenzene (para isomer). The current MCLs are 75 μg/L for *p*-dichlorobenzene and 600 μg/L for *o*-dichlorobenzene; the meta isomer is not regulated at this time. There are also three trichlorobenzene isomers: 1,2,3-trichlorobenzene, 1,2,4-trichlorobenzene, and 1,3,5-trichlorobenzene. A drinking water MCL of 70 μg/L has been established for 1,2,4-trichlorobenzene.

Hexachlorobenzene (C_6Cl_6) has been used in the synthesis of several chlorinated benzenes and is a commonly found groundwater contaminant. Its MCL is 1 μg/L.

Polychlorinated Biphenyls (PCBs)   Polychlorinated biphenyls are chlorinated benzenes with the general structure:

where X can be either a chlorine or a hydrogen. There are 209 possible PCBs (called congeners). Groups of congeners with similar average numbers of chlorine atoms are typically described with the name Aroclor and a corresponding number, for example, Aroclor 1260. PCBs are very stable compounds of low vapor pressure and high dielectric constants. They were widely used as coolants in transformers and capacitors, plasticizers, solvents, and hydraulic fluids. PCBs are very hydrophobic and tend to bioconcentrate. Significant concentrations of PCBs have been found in higher levels of the food chain (e.g., fish and birds). Because of human health and environmental effects associated with PCBs, their manufacture in the United States was banned in 1977. The drinking water MCL for total PCBs is 0.5 μg/L. PCBs have recently been implicated as *endocrine disruptors*. Endocrine-disrupting chemicals (EDC), also called hormonally active agents, are those chemicals that have been reported to be associated with adverse reproductive and developmental effects in humans and wildlife. Additional discussion of EDCs is given in Sec. 5.32. The disposal of PCB wastes is strictly controlled and continues to be problematical. Some PCB congeners have been shown to be degraded by microorganisms; the more highly chlorinated congener ones by anaerobic microorganisms, and the less chlorinated ones by aerobic microorganisms. Transformation rates, however, tend to be very slow such that the compounds are generally considered to be highly resistant to biodegradation.

5.14 | PHENOLS

The phenols are among the most important of the aromatic compounds.

Monohydroxy Phenols

There are several monohydroxy phenols of interest.

Phenol (C_6H_5OH)   The monohydroxy derivative of benzene is known as *phenol*.

Phenol

Its formula and name indicate that it might correspond in the aromatic series to alcohols in the aliphatic series. This is not the case, however. Phenol is also known as carbolic acid. It ionizes to yield H^+ to a limited extent ($K_A = 1.2 \times 10^{-10}$), and in concentrated solution is quite toxic to bacteria. It has been used widely as a germicide, and disinfectants have been rated in terms of "phenol coefficients," i.e., relative disinfecting power with respect to phenol. The system is considered archaic at the present time.

Phenol is recovered from coal tar, and considerable amounts are manufactured synthetically. It is used extensively in the synthesis of organic products, particularly phenolic-type resins. It occurs as a natural component in industrial wastes from the coal-gas, coal-coking, and petroleum industries as well as in a wide variety of industrial wastes from processes involving the use of phenol as a raw material.

Biological treatment of wastes containing more than 25 mg/L of phenol was considered impossible some years ago. However, research and much practice have demonstrated that phenol will serve as food for aerobic bacteria without serious toxic effects at levels as high as 500 mg/L. Studies with it and with formaldehyde have established the concept of toxicity thresholds. At levels below the threshold, bacteria use the material as food, but above the threshold they find it too toxic for use as food and reproduction of the organisms. Phenol has also been shown to be highly degradable even by anaerobic bacteria, and at levels of up to 2000 mg/L.[13]

Cresols and Other Alkylphenols   The next higher homologs of phenol are cresols.

o-Cresol	*m*-Cresol	*p*-Cresol

They are found in coal tar and have a higher germicidal action than phenol. They are less toxic to humans. *Lysol* is a mixture of cresols which is sold as a household and sickroom disinfectant. Cresols are the major constituents of "creosote" which is used extensively for the preservation of wood.

Industrial wastes containing cresols are somewhat difficult to treat by biological methods. Research with aerobic[14] and anaerobic[15] microorganisms has shown the pure cresols to be relatively nontoxic at concentrations of 250 mg/L. The toxicity of crude cresols is believed, therefore, to be due to other compounds.

[13]D. J. W. Blum, R. Hergenroeder, G. F. Parkin, and R. E. Speece, Anaerobic Treatment of Coal Conversion Wastewater Constituents: Biodegradability and Toxicity, *J. Water Pollution Control Federation,* **58:** 122–131 (1986).

[14]R. E. McKinney, H. D. Tomlinson, and R. L. Wilcox, Metabolism of Aromatic Compounds in Activated Sludge, *Sewage and Ind. Wastes,* **28:** 547–557 (1956).

[15]Blum et al., Anaerobic Treatment of Coal Conversion Wastewater Constituents: Biodegradability and Toxicity, *J. Water Pollution Control Federation,* **58:** 122–131 (1986).

Alkylphenols have recently been implicated as endocrine disrupters. Alkylphenols are combined with ethylene oxide (Secs. 5.17 and 5.25) to produce alkylphenol ethoxylates, which are used as surfactants in products such as detergents, paints, and pesticides. Action by naturally occurring bacteria degrade these molecules releasing the alkylphenols. These compounds have been found in effluents from wastewater treatment plants, wastewater plant sludges, and natural waters.[16,17] Nonylphenol is believed to be one of the more potent potential endocrine disruptors.[18]

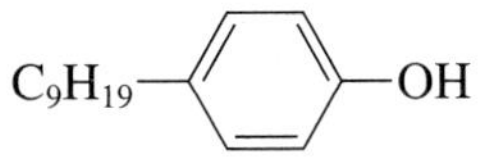

Nonylphenol

The nonyl group may be highly branched, making the compound resistant to biodegradation.

Another phenol that has been implicated as an important endocrine disrupter is bisphenol A.[19] Bisphenol A is an important intermediate in the production of a wide variety of industrial products including polymers such as polycarbonate plastics, dyes, epoxy coatings, and flame retardants.

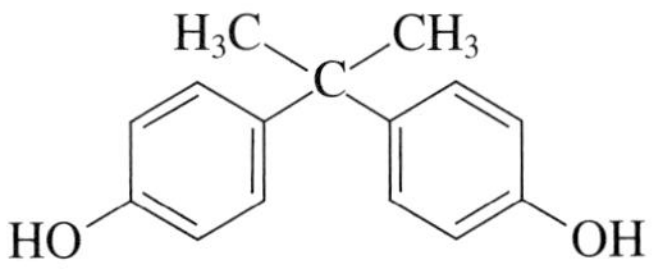

Bisphenol A

Chlorinated Phenols Chlorinated phenols have been used as wood preservatives. The most commonly used phenol for this purpose is pentachlorophenol.

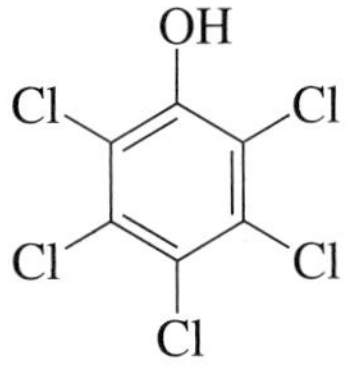

Pentachlorophenol

The drinking water MCL for pentachlorophenol is 1 μg/L.

---

[16]B. E. Erickson, Akylphenols in Sewage Sludge Applied to Land, *Env. Sci. Tech.,* **36**(1): 10A–11A, 2002.

[17]C. Maczka et al., Evaluating Impacts of Hormonally Active Agents in the Environment, *Env. Sci. Tech.,* **34**(5): 136A–141A, 2000.

[18]Maczka et al., Evaluating Impacts of Hormonally Active Agents in the Environment.

[19]Ibid.

Polyhydroxy Phenols

Three isomeric dihydroxy phenols are known. All have been shown[20] to be readily oxidized by properly acclimated microorganisms.

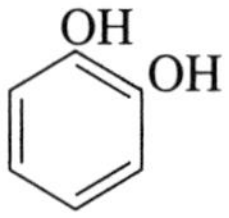

Benzene-1,2-diol,
pyrocatechol,
catechol

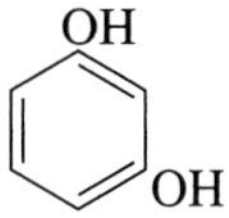

Benzene-1,3-diol,
resorcinol

Benzene-1,4-diol,
hydroquinone

Pyrogallol   *Pyrogallol,* 1,2,3-trihydroxybenzene, is known as *pyrogallic acid.*

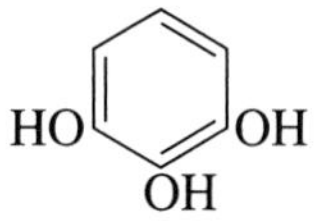

Pyrogallol

It is easily oxidized and serves as a photographic developer. Pyrogallol is a minor constituent of spent tan liquors. When they are discharged to streams containing iron, inky black ferric pyrogallate is formed.

5.15 | ALCOHOLS, ALDEHYDES, KETONES, AND ACIDS

The aromatic alcohols, aldehydes, ketones, and acids are all formed from alkyl derivatives of benzene or one of its homologs. The active group is always in the alkyl group, and therefore the chemistry of the aromatic alcohols, aldehydes, ketones, and acids is very similar to that of the corresponding aliphatic compounds. Common names are usually employed with these compounds, but they may be more properly named as phenyl derivatives of aliphatic compounds.

Alcohols

The aromatic alcohols compose a homologous series. They are phenyl methyl, phenyl ethyl, phenyl *n*-propyl, phenyl isopropyl, and so on.

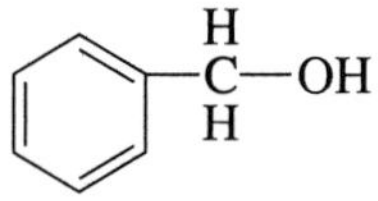

Benzyl alcohol,
phenyl methyl alcohol

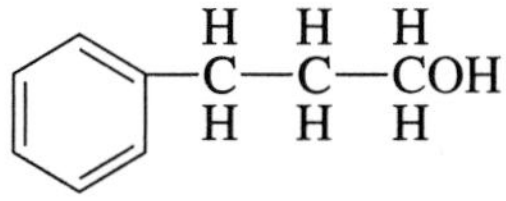

Phenyl *n*-propyl alcohol

---

[20]Blum et al., Anaerobic Treatment of Coal Conversion Wastewater Constituents: Biodegradability and Toxicity, *J. Water Pollution Control Federation,* **58:** 122–131 (1986).

The aromatic alcohols are subject to chemical and biological oxidation. Oxidation of primary alcohols produces aldehydes and oxidation of secondary alcohols produces ketones.

Aldehydes

The aromatic aldehydes are important compounds in chemical synthesis.

Benzaldehyde   Phenyl propyl aldehyde

They are easily oxidized to the corresponding acids. Many of the more complex aldehydes have fragrant odors: coumarin, anisaldehyde, vanillin, and so on.

Ketones

The aromatic ketones are of two types: those that have one phenyl group attached to the carbonyl group and those that have two.

Acetophenone   Benzophenone

Chemical and biological oxidation results in disruption of the molecule, with the formation of lower-molecular-weight acids and, probably, carbon dioxide.

Acids

A wide variety of aromatic, monocarboxylic acids is known. Oxidation of benzaldehyde produces benzoic acid. Sodium benzoate is used as a food preservative. Salicylic acid is used to prepare aspirin.

Benzoic acid   Salicylic acid

Oxidation of naphthalene produces an important dicarboxylic acid, phthalic acid.

Phthalic acid   Phthalic anhydride

It and its anhydride are important in the manufacture of a variety of organic compounds. Phthalates are used as plasticizers in the production of a wide variety of plastics. A drinking water MCL of 6 μg/L has been set for di (2-ethylhexyl) phthalate. Phthalates have also been implicated as endocrine disruptors.

Phenolphthalein, used as a pH indicator in the laboratory, is another example.

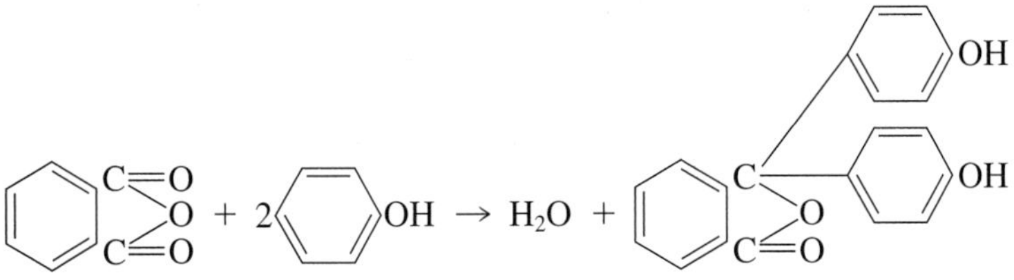

Phenolphthalein

Most aromatic acids are subject to biological oxidation. The normal end products are carbon dioxide and water. However, some phthalates are resistant to biodegradation.

5.16 | SIMPLE COMPOUNDS CONTAINING NITROGEN

The aromatic compounds containing nitrogen are derivatives either of ammonia or of nitric acid. The former are called *amines,* and the latter are called *nitro* compounds. In addition, many complex nitrogen compounds exist that are outside the scope of this book.

Amines

The aromatic amines are of two types: those in which the phenyl or other aromatic group is attached directly to nitrogen and those in which the nitrogen occurs in an attached alkyl group. There are three phenyl derivatives of ammonia: primary, secondary, and tertiary amines.

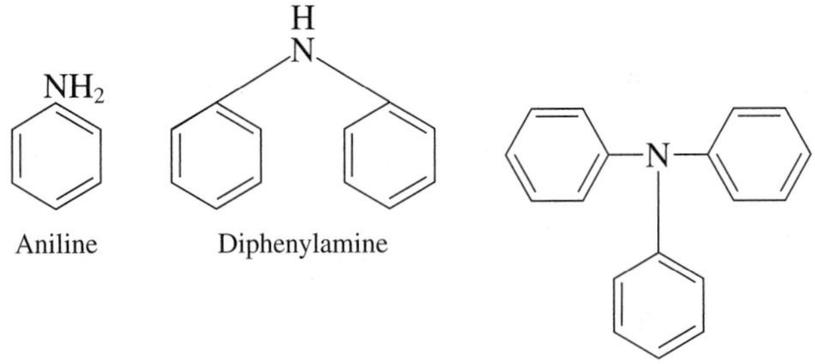

Aniline   Diphenylamine

Triphenylamine

The primary form is called aniline and the secondary form is diphenylamine. They are both basic in character, react with strong acids to form salts, and are important compounds in organic synthesis. Aniline dyes are derived from aniline. Sulfanilic acid, used in the colorimetric determination of nitrites, is made from aniline.

$$\underset{\substack{\text{Aniline hydrogen}\\\text{sulfate}}}{\text{C}_6\text{H}_5\overset{\text{H}_3}{\text{N}}\text{—OSO}_3\text{H}} \xrightarrow{180°} \text{H}_2\text{O} + \underset{\substack{\text{Sulfanilic acid}}}{\text{C}_6\text{H}_4(\text{NH}_2)(\text{SO}_3\text{H})} \tag{5.46}$$

1-Naphthylamine HCl is an important compound in water analysis.

NH₃Cl

A derivative, N-(1-naphthyl)-ethylenediamine dihydrochloride is used in conjunction with sulfanilic acid in the determination of nitrite nitrogen.

Benzylamine ($\text{C}_6\text{H}_5\text{CH}_2\text{NH}_2$) is an example of an aromatic amine that has the —NH₂ group attached to the aliphatic part of the molecule. These compounds are not important commercially.

Nitro Compounds

Nitric acid reacts with benzene and other aromatic compounds to form *nitro* compounds. The reaction is as follows:

$$\text{C}_6\text{H}_5(\text{H} + \text{HO})\text{NO}_2 \xrightarrow{\text{H}_2\text{SO}_4} \underset{\text{Nitrobenzene}}{\text{C}_6\text{H}_5\text{—NO}_2} + \text{H}_2\text{O} \tag{5.47}$$

A dehydrating agent, usually sulfuric acid, must be present to remove the water that is formed. One additional nitro group may be added under proper conditions. The principal product is *m*-dinitrobenzene,

m-Dinitrobenzene   Trinitrotoluene (TNT)

as the presence of the first nitro group directs the second into the meta position. Trinitrobenzene is very difficult to prepare.

Nitration of toluene results in the formation of trinitrotoluene, or TNT. The first nitro group is directed into the ortho position by the methyl group, and additional nitro groups attach in the meta position with respect to the first nitro group. TNT is widely used as an ingredient of military explosives. It is relatively resistant to aerobic biological degradation. However, it is fairly easily reduced to the corresponding amine under reducing conditions.

HETEROCYCLIC COMPOUNDS

5.17 | HETEROCYCLIC COMPOUNDS

Heterocyclic compounds have one other element in the ring in addition to carbon. A wide variety of compounds exists; some are aliphatic in character, and some are aromatic. Many are of great biological importance, and several are of significance in environmental engineering and science.

Furaldehyde, or *furfural,* is an example of an aliphatic heterocyclic compound having a five-membered ring containing oxygen. It is produced from pentose sugars by dehydration. Commercially, it is made from oat hulls and corn cobs, waste products of the cereal industry.

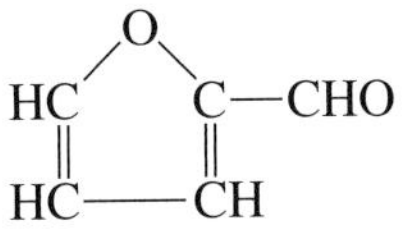

Furaldehyde

Both were formerly disposed of by burning. Manufacture of furfural yields some liquid wastes of concern to environmental engineers. Furfural has been shown to be biodegradable.

Epoxides are three-membered rings where oxygen is bonded with two carbons. Several epoxides are environmentally significant. For example, ethylene oxide,

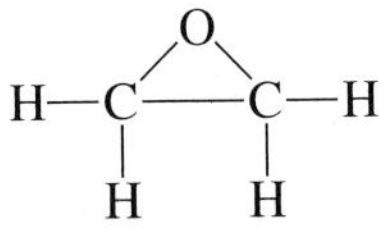

Ethylene oxide

is a toxic epoxide that has been used in chemical synthesis [e.g. alkylphenol ethoxylates (Sec. 5.14) and synthetic detergents (Sec. 5.26)] and as a sterilizing agent and pesticide. It is a regulated air pollutant. In general, epoxides are very reactive. They have been shown to be important intermediates in the biotransformation of several organics, for example, trichloroethene.

Pyrrole and *pyrrolidine* are examples of heterocyclic compounds having five-membered rings containing nitrogen.

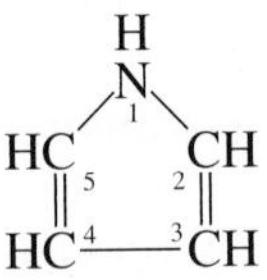

Pyrrole 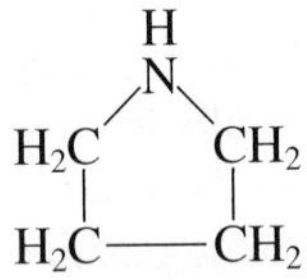   Pyrrolidine

The pyrrole or pyrrolidine ring occurs in the structure of many important natural compounds, e.g., nicotine, cocaine, chlorophyll, and hemoglobin.

Pyridine is an example of a six-membered aromatic heterocyclic compound with nitrogen contained in the ring.

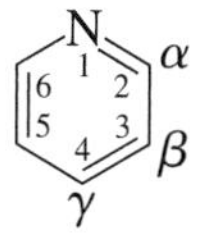

Pyridine

It is an especially vile-smelling liquid. It is used as a denaturant in ethyl alcohol, to make it unpalatable, and in chemical synthesis. It is weakly basic in character and forms salts with strong acids. *Nicotinic acid* is a derivative of pyridine, having a carboxyl group in the β position. It is a key component of a coenzyme, NAD, which is present in all cells and involved in oxidation of organic matter. Nicotinic acid is a vitamin required by humans and many other mammals, a deficiency in the diet of humans causing pellagra.

Purine and *pyrimidine* are two other ring compounds containing nitrogen and of immense biological importance.

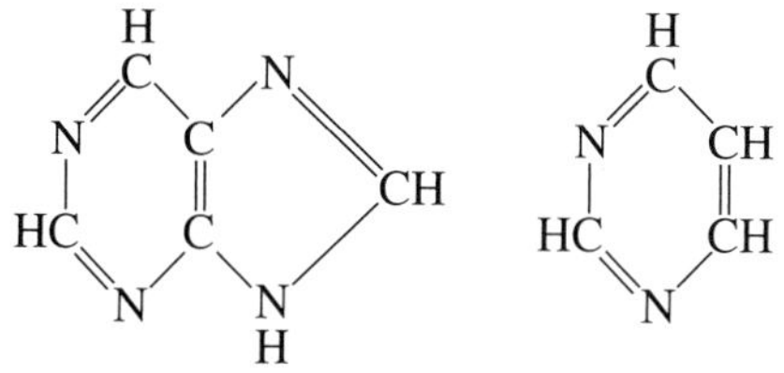

Purine   Pyrimidine

Important derivatives of purine are *adenine* and *guanine*, and of pyrimidine are *cytosine, uracil,* and *thymine.* These compounds or *bases* form the major components of nucleic acids, which carry the genetic information for all life (see Sec. 6.14). In addition, they are components of key biological molecules such as ATP, the primary carrier of chemical energy in all cells, and coenzyme A, which is necessary for fatty acid degradation.

Indole and *skatole* are examples of heterocyclic compounds that possess a benzene nucleus condensed with a pyrrole nucleus.

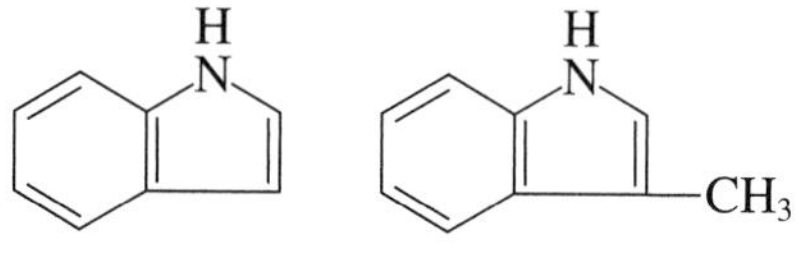

Indole   Skatole

Both possess unpleasant odors and are produced during the putrefaction of protein matter. Under controlled conditions, such as exist in well-operated sludge digesters, very little indole or skatole is formed.

Other heterocyclic nitrogen compounds of interest include the explosives RDX (hexahydro-1,3,5-trinitro-1,3,5-triazine) and HMX (octahydro-1,3,5,7-tetranitro-

1,3,5,7-tetrazocine). These compounds have been found as soil or groundwater contaminants at a number of U.S. Department of Defense facilities. They are slightly soluble in water but are not particularly hydrophobic; they have low vapor pressures and are not very biodegradable under most conditions found in the natural environment. They have been detected in a number of groundwater supplies and present environmental engineers and scientists with challenging remediation problems.

RDX
(hexahydro-1,3,5-trinitro-1,3,5-triazine)

HMX
(octahydro-1,3,5,7-tetranitro-1,3,5,7-tetrazocine)

5.18 | DYES

The subject of *dyes* is of such magnitude and complexity that a discussion of various types will not be presented here. The environmental engineer and scientist concerned with the treatment of textile wastes, and possibly a few others, will be confronted with the need to learn more about these materials. Recourse for information should be made to standard organic chemistry texts or treatises on dyes. The sulfur dyes are noted for their toxic properties.

THE COMMON FOODS AND RELATED COMPOUNDS

5.19 | GENERAL

The term *food* is applied to a wide variety of organic materials that can serve as a source of energy for living organisms. In the case of bacteria, these compounds range from hydrocarbons through various oxidation products, including organic acids. In the case of higher animals and humans, the principal or common foods are restricted to *carbohydrates, fats,* and *proteins.* Other organic compounds such as ethanol, certain aldehydes, and many acids serve as food or energy sources also. The latter are sometimes referred to as exotic foods, as they are not considered part of an essential diet but are added to increase palatability or for other reasons.

5.20 | CARBOHYDRATES

The term *carbohydrate* is applied to a large group of compounds of carbon, hydrogen, and oxygen in which the hydrogen and oxygen are in the same ratio as in water, i.e., two atoms of hydrogen for each atom of oxygen. The processing of carbohydrate materials occurs in the lumber, paper, and textile industries, as well as in the food industry. Wastes from these industries are major problems and tax the ingenuity of environmental engineers and scientists to find satisfactory solutions.

Carbohydrates may be grouped into three general classifications, depending upon the complexity of their structure: (1) simple sugars, or *monosaccharides;* (2) complex sugars, or *disaccharides;* (3) *polysaccharides.* In general, the *-ose* ending is used to name carbohydrates.

Simple Sugars, or Monosaccharides

The simple sugars, or *monosaccharides,* all contain a carbonyl group in the form of an aldehyde or a keto group. Those with aldehyde groups are known as *aldoses* and those with keto groups are known as *ketoses.* They are also glycols, as they possess several OH groups. Two series of simple sugars are of importance commercially: the *pentoses* are five-carbon-atom sugars and the *hexoses* are six-carbon-atom sugars.

Pentoses   Pentoses have the general formula $C_5H_{10}O_5$. Two pentoses are of commercial importance, and both are *aldopentoses. Xylose* is formed by the hydrolysis of pentosans, which are commonly found in waste organic materials such as oat hulls, corn cobs, and cottonseed hulls. Considerable amounts of xylose are formed in the pulping of wood through the hydrolysis of hemicellulose. *Arabinose* is produced by the hydrolysis of gum arabic or wheat bran.

$$
\begin{array}{cc}
\begin{array}{c}
\text{H} \\
\text{C}=\text{O} \\
| \\
\text{H}-\text{C}-\text{OH} \\
| \\
\text{HO}-\text{C}-\text{H} \\
| \\
\text{H}-\text{C}-\text{OH} \\
| \\
\text{CH}_2\text{OH}
\end{array}
&
\begin{array}{c}
\text{H} \\
\text{C}=\text{O} \\
| \\
\text{HO}-\text{C}-\text{H} \\
| \\
\text{H}-\text{C}-\text{OH} \\
| \\
\text{H}-\text{C}-\text{OH} \\
| \\
\text{CH}_2\text{OH}
\end{array}
\\
\text{D(–)-Xylose[21]} & \text{D(–)-Arabinose[21]}
\end{array}
$$

Both xylose and arabinose are used in bacteriological work in media used to differentiate among various bacteria. Certain bacteria can ferment one but not the other, and vice versa. Mixed cultures of bacteria, such as those derived from the soil or waste, convert both sugars to carbon dioxide and water. The pentose sugars are not fermented by yeast under anaerobic conditions; therefore, they cannot be used to produce ethanol. They do serve as an energy source for yeast under aerobic conditions, however, and advantage is taken of this fact in one method of treating spent sulfite liquors from the pulping of wood.

Hexoses   There are four important hexose sugars with the general formula $C_6H_{12}O_6$. *Glucose, galactose,* and *mannose* are all *aldoses,* and *fructose* is a *ketose.*

---

[21]All sugars are optically active. The nomenclature is somewhat complicated, however, and details should be obtained from a standard text on organic chemistry.

Glucose Glucose is the most common of the aldohexose sugars. It is found naturally in fruit juices and in honey. It is manufactured in great quantity by the hydrolysis of corn starch. It is the principal component of corn syrup. Both corn syrup and glucose are used extensively in candy manufacture. Glucose is much less sweet than ordinary sugar and replaces it for many purposes.

$$
\begin{array}{c}
\text{H} \\
\text{C}=\text{O} \\
| \\
\text{H}-\text{C}-\text{OH} \\
| \\
\text{HO}-\text{C}-\text{H} \\
| \\
\text{H}-\text{C}-\text{OH} \\
| \\
\text{H}-\text{C}-\text{OH} \\
| \\
\text{CH}_2\text{OH}
\end{array}
$$

D-(+)-Glucose

Glucose is the only hexose sugar that can be prepared in relatively pure form by the hydrolysis of disaccharides or polysaccharides. All the other hexose sugars occur in combination with glucose.

Fructose Fructose is the only significant ketohexose and occurs naturally in honey. When cane or beet sugar is hydrolyzed, one molecule of fructose and one molecule of glucose are formed from each molecule of sucrose.

Galactose and Mannose Galactose and mannose do not occur in free form in nature. Galactose is produced by the hydrolysis of lactose, more commonly called milk sugar. Glucose is formed simultaneously. Mannose is produced by the hydrolysis of ivory nut, and glucose is formed at the same time.

$$
\begin{array}{ccc}
\begin{array}{c}
\text{CH}_2\text{OH} \\
| \\
\text{C}=\text{O} \\
| \\
\text{HO}-\text{C}-\text{H} \\
| \\
\text{H}-\text{C}-\text{OH} \\
| \\
\text{H}-\text{C}-\text{OH} \\
| \\
\text{CH}_2\text{OH}
\end{array}
&
\begin{array}{c}
\text{H} \\
\text{C}=\text{O} \\
| \\
\text{H}-\text{C}-\text{OH} \\
| \\
\text{HO}-\text{C}-\text{H} \\
| \\
\text{HO}-\text{C}-\text{H} \\
| \\
\text{H}-\text{C}-\text{OH} \\
| \\
\text{CH}_2\text{OH}
\end{array}
&
\begin{array}{c}
\text{H} \\
\text{C}=\text{O} \\
| \\
\text{HO}-\text{C}-\text{H} \\
| \\
\text{HO}-\text{C}-\text{H} \\
| \\
\text{H}-\text{C}-\text{OH} \\
| \\
\text{H}-\text{C}-\text{OH} \\
| \\
\text{CH}_2\text{OH}
\end{array} \\
\text{D-(–)-Fructose} & \text{D-(+)-Galactose} & \text{D-(+)-Mannose}
\end{array}
$$

Glucose and galactose are of particular interest. Glucose is always one of the products and may be the sole product when di- or polysaccharides are hydrolyzed. It is therefore found in a wide variety of industrial wastes. Galactose is formed from the hydrolysis of lactose, or milk sugar, and is found in wastes from the dairy industry. Both sugars are readily oxidized by aerobic bacteria to form acids, and the oxidation may stop at that point because of the unfavorable pH conditions produced by the acids unless precautions are taken to control the pH by means of buffers or alkaline materials. Lactic acid is an important intermediate in the oxidation of galactose. Both sugars are fermented rapidly under anaerobic conditions, with acid formation.

Complex Sugars, or Disaccharides

There are three important sugars with the general formula $C_{12}H_{22}O_{11}$: *sucrose, maltose,* and *lactose.* All disaccharides may be considered as consisting of two hexose sugars hooked together in one molecule. Hydrolysis results in cleavage of the molecule and formation of the hexoses.

Sucrose Sucrose is the common sugar of commerce. It is derived largely from sugarcane and sugar beets. The sap of such trees as the sugar maple contains considerable sucrose. Hydrolysis of the sucrose molecule results in the formation of one molecule of glucose and one molecule of fructose.

Sucrose ($C_{12}H_{22}O_{11}$)

Maltose Maltose is made by the hydrolysis of starch, induced by *diastase,* an enzyme present in barley malt. The starch may be derived from a wide variety of sources, and its hydrolysis by diastase results in commercial maltose, which is used in infant foods and in malted milk.

Maltose ($C_{12}H_{22}O_{11}$)

Maltose is readily hydrolyzed to yield two molecules of glucose.

Alcohol production by fermentation processes uses starch from a wide variety of sources. The starch is converted to maltose by the enzyme from barley malt. Enzymes from the yeast hydrolyze maltose to glucose and convert the glucose to alcohol and carbon dioxide (Sec. 5.3).

Lactose   Lactose, or milk sugar, occurs in the milk of all mammals. Upon hydrolysis, the molecule is split to yield a molecule of glucose and a molecule of galactose.

Lactose ($C_{12}H_{22}O_{11}$)

Lactose is used in infant foods and in candy making. Dried skimmed-milk solids contain about 60 percent lactose.

Polysaccharides

The polysaccharides are all condensation products of hexoses or other monosaccharides. Glucose and xylose are the most common units involved. Three polysaccharides are of interest: *starch, cellulose,* and *hemicellulose.* None of them have the characteristic sweet taste of sugar because of their insolubility and complex molecular structure.

Starch   Starch has the general formula $(C_6H_{10}O_5)_x$. It occurs in a wide variety of products grown for food purposes (corn, wheat, potatoes, rice, etc.). It is the cheapest foodstuff and serves mainly in human nutrition as a source of energy. Starch is used in fermentation industries to produce a wide variety of products. The structure of the starch molecule is not known definitely. Its hydrolysis yields glucose as the only monosaccharide, and its general formula may be indicated as follows:

Starch (amylose, x = about 100 to 1000)

Starch consists of two major fractions. One fraction, consisting of glucose units connected in a straight chain, is termed *amylose,* and the other fraction, consisting of glucose units attached to form branched chains, is termed *amylopectin.* The amylose molecule contains 100 to 1000 glucose units. Amylose is soluble in water and absorbs up to 20 percent of its own weight in iodine to form the blue complex used as an indicator in iodimetric analysis (Sec. 11.4). The amylopectin molecule, not shown, is much larger and contains about 500 to 5000 glucose units. It is not as soluble in water as amylose.

The glucose units in starch are connected by what is termed an *alpha* linkage. This linkage is readily hydrolyzed by enzymes common to all mammals as well as to microorganisms, and hence they are able to use starch as food.

The industrial wastes produced from the manufacture of starch, from the processing of carbohydrate foods, and from the industrial uses of starch can be readily treated by biological processes.

Cellulose   Cellulose forms the structural fiber of many plants. Cotton is essentially pure cellulose. High-grade cellulose can be produced from wood through the sulfite and sulfate pulping processes. Like starch, cellulose consists of glucose subunits. However, these units are connected by what is termed *beta* linkage, and mammals, including humans, do not have enzymes capable of promoting the hydrolysis of this linkage. Therefore, cellulose passes through the digestive tract unchanged. Certain animals, especially ruminants (cud-chewing animals) such as the cow, have bacteria in the digestive tract that can hydrolyze the beta link. This is a convenient arrangement, for the animal can digest the bacterial fermentation products and thus derive nourishment indirectly from cellulose. The beta linkage for cellulose is indicated as follows:

Portion of cellulose molecule

Industrial wastes from the paper industry usually contain considerable amounts of cellulose in suspension. This is particularly true of the wastes from the manufacture of low-grade papers involving the reuse of waste paper. Most of the wastes from other industries processing cellulose contain very little cellulose. The principal contaminants are inorganic compounds, derivatives of cellulose, and other organic compounds. Since certain bacteria can hydrolyze cellulose, biological treatment of cellulose containing wastes is possible. However, treatment by aerobic processes is slow. Since most of the cellulose will settle to produce a sludge, preliminary treatment by sedimentation is practiced, and the sludges produced are disposed of by anaerobic digestion or by physical methods such as filtration, centrifugation, and incineration.

Hemicelluloses   The hemicelluloses are compounds that have characteristics somewhat like cellulose. They are composed of a mixture of hexose and pentose units, however, and upon hydrolysis yield glucose and a pentose, usually xylose.

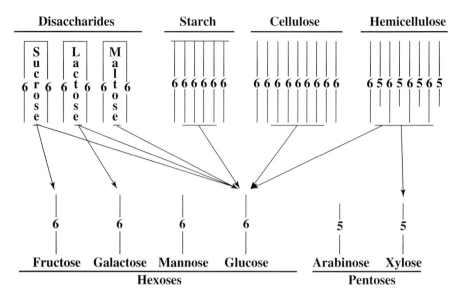

Figure 5.1
Summary of hydrolytic behavior of carbohydrates.

Most natural woods contain cellulose, hemicellulose, and lignin, along with resins, pitch, and so on. In the pulping process, the lignin, hemicellulose, resins, and so on, are dissolved, leaving cellulose as the product. As a result, spent pulping liquors contain considerable amounts of glucose and xylose as well as other organic substances, principally derivatives of lignin. The lignin derivatives, which have highly complex aromatic structures, are very resistant to biological degradation. Glucose, xylose, and other organic substances are converted to carbon dioxide and water by yeast or bacteria under aerobic conditions. Yeast may be used to ferment the glucose to alcohol under anaerobic conditions, but the xylose, a pentose, is not fermentable to alcohol.

Summary of Hydrolytic Behavior of Carbohydrates

The hydrolytic behavior of carbohydrates is presented in a simplified graphic form in Fig. 5.1. All di- and polysaccharides yield glucose. Sucrose yields fructose, lactose yields galactose, and hemicellulose yields xylose in addition to glucose.

5.21 | FATS, OILS, AND WAXES

Fats, oils, and *waxes* are all esters. Fats and oils are esters of the trihydroxy alcohol, glycerol, while waxes are esters of long-chain monohydroxy alcohols. All serve as food for humans, as well as bacteria, since they can be hydrolyzed to the corresponding fatty acids and alcohols.

Fats and Oils

Fats and oils are both glycerides of fatty acids. The fatty acids are generally of 16- or 18-carbon atoms, although butyric, caproic, and caprylic acids are present to a significant extent as components of the esters of butterfat. The acids may also be unsaturated. Oleic and linoleic are important acids in cottonseed oil. Linseed oil contains large amounts of linoleic and linolenic acids. The glycerides of fatty acids that are liquid at ordinary temperatures are called *oils* and those that are solids are called *fats*. Chemically they are quite similar. The oils have a predominance of short-chain fatty acids or fatty acids with a considerable degree of unsaturation, such as linoleic or linolenic.

Glyceryl (or glycerol) tributyrate, tributyrin

Glyceryl (or glycerol) tristearate, tristearin

The fatty acids in a given molecule of a glyceride may be all the same, as just shown, or they may all be different.

Glyceryl (or glycerol) oleo-butyropalmitate (found in butter fat)

β-Palmito-α,α′-distearin

The principal acids composing the glycerides of fats and oils are shown in Table 5.12.

The relative amounts of the major fatty acids contained in various fats and oils are shown in Table 5.13.

Fats and oils undergo three types of chemical reactions of interest: hydrolysis, addition, and oxidation.

Table 5.12 | Acids of fats and oils

Name	Formula	Mp, °C	Source
Butyric	C_3H_7COOH	−5.7	Butter
Caproic	$C_5H_{11}COOH$	−3	Butter, coconut oil
Caprylic	$C_7H_{15}COOH$	16.3	Palm oil, butter
Capric	$C_9H_{19}COOH$	31.9	Coconut oil
Lauric	$C_{11}H_{23}COOH$	43.2	Coconut oil, spermaceti
Myristic	$C_{13}H_{27}COOH$	53.9	Nutmeg, coconut oil
Palmitic	$C_{15}H_{31}COOH$	63.1	Palm oil, animal fats
Stearic	$C_{17}H_{35}COOH$	69.6	Animal and vegetable fats, oils
Arachidic	$C_{20}H_{40}O_2$	76.5	Peanut oil
Behenic	$C_{22}H_{44}O_2$	81.5	Ben oil
Oleic	$C_{18}H_{34}O_2$	13.4	Animal and vegetable fats, oils
Erucic	$C_{22}H_{42}O_2$	34.7	Rape oil, mustard oil
Linoleic	$C_{18}H_{32}O_2$	−12	Cottonseed oil
Linolenic	$C_{18}H_{30}O_2$	−11	Linseed oil

Source: D. R. Linde (ed.): "Handbook of Chemistry and Physics," 82nd ed., CRC Press LLC, Boca Raton, 2001.

Table 5.13 | Acid content of fats and oils (percent)

Name	Oleic	Linoleic	Linolenic	Stearic	Myristic	Palmitic	Arachidic
Butter[a]	27.4			11.4	22.6	22.6	
Mutton tallow	36.0	4.3		30.5	4.6	24.6	
Castor oil[b]	9	3		3			
Olive oil	84.4	4.6		2.3	Trace	6.9	0.1
Palm oil	38.4	10.7		4.2	1.1	41.1	
Coconut oil[c]	5.0	1.0		3.0	18.5	7.5	
Peanut oil[d]	60.6	21.6		4.9		6.3	3.3
Corn oil[e]	43.4	39.1		3.3		7.3	0.4
Cottonseed oil	33.2	39.4		1.9	0.3	19.1	0.6
Linseed oil	5	48.5	34.1				
Soybean oil[f]	32.0	49.3	2.2	4.2		6.5	0.7
Tung oil[g]	14.9			1.3		4.1	

[a]Contains caproic, 1.4 percent; caprylic, 1.8 percent; capric, 1.8 percent; butyric, 3.2 percent; lauric, 6.9 percent.
[b]Contains about 85 percent of ricinoleic acid, 12-hydroxy-9-octadecenoic acid (mp, 17°), $CH_3(CH_2)_5CHOHCH_2CH{=}CH(CH_2)_7COOH$.
[c]Contains caprylic, 9.5 percent; capric, 4.5 percent; lauric, 51 percent.
[d]Contains 2.6 percent lignoceric acid.
[e]Contains 0.2 percent lignoceric acid.
[f]Contains 0.1 percent lignoceric acid.
[g]Contains 79.7 percent eleostearic acid.

Source: From E. Wertheim and H. Jeskey, "Introductory Organic Chemistry," 3rd ed., McGraw-Hill, New York, 1956. Table reproduced by permission of the authors.

Hydrolysis   Since fats and oils are esters, they undergo hydrolysis with more or less ease. The hydrolysis may be induced by chemical means, usually by treatment with NaOH, or by bacterial enzymes that split the molecule into glycerol plus fatty acids. Hydrolysis with the aid of NaOH is called *saponification*. Hydrolysis by bacterial action may produce *rancid* fats or oils and renders them unpalatable. Rancid butter and margarine are notorious for their bad odor.

Addition   The fats and oils containing unsaturated acids add chlorine at the double bonds, as other unsaturated compounds do. This reaction is often slow because of the relative insolubility of the compounds. It may represent a significant part of the chlorine demand of some wastes, and chlorinated organics will be produced during chlorination, some of which may be of health concern.

Oils that contain significant amounts of oleic and linoleic acids may be converted to fats by the process of *hydrogenation.* In this process hydrogen is caused to add at the double bonds, and saturated acids result. Thus, low-priced oils such as soybean and cottonseed can be converted into margarine, which is acceptable as human food. Many cooking fats or shortenings are made in the same manner. The hydrogenation can be controlled to produce any degree of hardness desired in the product.

Oxidation   The oils with appreciable amounts of linoleic and linolenic acids or other highly unsaturated acids, such as linseed and tung oil, are known as *drying oils*. In contact with the air, oxygen adds at the double bonds and forms a resinlike material. The drying oils are the major component in all oil-based paints.

Waxes

Waxes, with the exception of paraffin wax, are esters of long-chain acids and alcohols of high molecular weight. *Beeswax* is an ester of palmitic acid and myricyl alcohol ($C_{15}H_{31}COOC_{31}H_{63}$). It also contains cerotic acid ($C_{25}H_{51}COOH$). Spermaceti is obtained from the heads of sperm whales and is principally an ester of palmitic acid and cetyl alcohol ($C_{15}H_{31}COOC_{16}H_{33}$). Cetyl esters of lauric and myristic acids are also present to a limited extent.

5.22 | PROTEINS AND AMINO ACIDS

Proteins are complex compounds of carbon, hydrogen, oxygen, and nitrogen. Phosphorus and sulfur are present in a few. They are among the most complex of the organic compounds produced in nature and are widely distributed in plants and animals. They form an essential part of protoplasm and enzymes, and are a necessary part of the diet of all higher animals, in which they serve to build and repair muscle tissue. Like polysaccharides, which may be considered to be made up of glucose units, proteins are formed by the union of α-amino acids. Since more than 20 different amino acids are normally found present in proteins, the variety of proteins is considerable.

Amino Acids

The *α-amino acids* are the building blocks from which proteins are constructed. Most plants and bacteria have the ability to synthesize the amino acids from which they build proteins. Animals are unable to synthesize certain of the amino acids and must depend upon plants to supply them in the form of proteins. Such amino acids are considered to be indispensable.

The amino acids that occur in proteins all have an amino group attached to the alpha carbon atom and are therefore called *α-amino acids*.

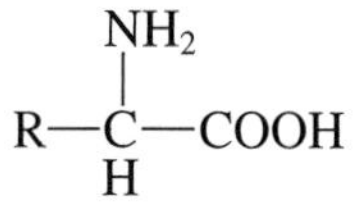

$$\begin{array}{c} NH_2 \\ | \\ R-C-COOH \\ | \\ H \end{array}$$

α-Amino acid

Chemistry of Amino Acids The free amino acids behave like acids and also like bases because of the amino group that they contain. Thus, they are *amphoteric* in character and form salts with acids or bases.

Salt formation with an acid:

$$H_2NCH_2COOH + HCl \rightarrow Cl^- + {}^+H_3NCH_2COOH \tag{5.48}$$

Salt formation with a base:

$$H_2NCH_2COOH + NaOH \rightarrow H_2O + H_2NCH_2COO^- + Na^+ \tag{5.49}$$

The amino acids having one amino and one carboxyl group are essentially neutral in aqueous solution. This is considered to be due to a case of self-neutralization in which the hydrogen ion of the carboxyl group migrates to the amino group and a positive-negative (dipolar) ion known as a *zwitterion* results.

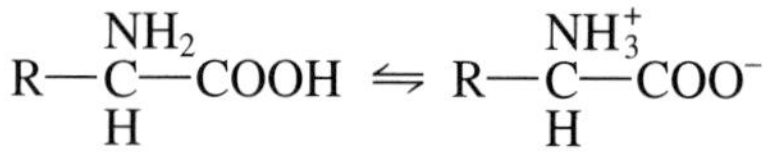

$$\begin{array}{c} NH_2 \\ | \\ R-C-COOH \\ | \\ H \end{array} \rightleftharpoons \begin{array}{c} NH_3^+ \\ | \\ R-C-COO^- \\ | \\ H \end{array} \tag{5.50}$$

Zwitterion

In Sec. 5.9 it was shown that organic acids can react with ammonia to form amides. The amino and carboxyl groups of separate amino acid molecules can react in the same manner.

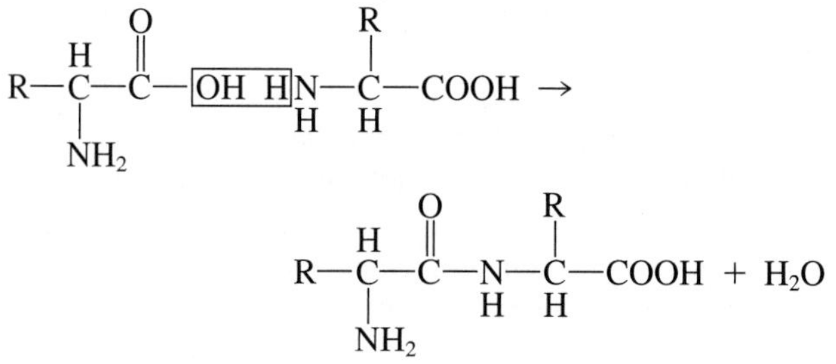

$$\begin{array}{c} \quad\quad O \\ \quad H \quad \| \quad\quad\quad R \\ R-C-C-\boxed{OH\ H}N-C-COOH \rightarrow \\ \quad | \quad\quad\quad H \ \ H \\ \quad NH_2 \end{array}$$

$$\begin{array}{c} \quad\quad O \quad\quad R \\ \quad H \quad \| \quad\quad | \\ R-C-C-N-C-COOH + H_2O \\ \quad | \quad\quad H \ \ H \\ \quad NH_2 \end{array} \tag{5.51}$$

A dipeptide

It is this ability to form linkages between the amino and carboxyl groups that allows the large complex molecules of proteins to be formed. In the example given in Eq. (5.51), the resulting molecule contains one free amino and one free carboxyl group. Each can combine with another molecule of an amino acid. In turn, the resulting molecule will contain free amino and carboxyl groups, and the process can be repeated, presumably, *ad infinitum*. Biochemical processes, however, direct the synthesis to produce the type and size of protein molecules desired.

The molecule formed by the union of two molecules of amino acids is known as a *dipeptide;* if there are three units, the name is *tripeptide;* if more than three units, the compound is called a *polypeptide.* The particular linkage formed when amino acids join is called the *peptide link* and is formed by loss of water between an amino and a carboxyl group.

$$
\underset{\text{Peptide link}}{\overset{O}{\underset{}{\overset{\|}{-C}}}-[OH\ H]-\overset{H}{N}-\ \rightarrow\ -\overset{O}{\overset{\|}{C}}-\overset{H}{N}-\ +\ H_2O} \tag{5.52}
$$

Classes of Amino Acids   About 20 different α-amino acids can generally be isolated by the hydrolysis of protein matter. The simplest have one amino group and one carboxyl group per molecule. Some have sulfur in the molecule. Some have two amino groups and one carboxyl group and consequently are basic in reaction. Some have one amino group and two carboxyl groups and are acidic in reaction. Others have aromatic or heterocyclic groups.

All the amino acids, except glycine, are optically active. In the following list, those marked with an asterisk are considered indispensable in human nutrition. The generally used abbreviation for each amino acid is noted after the name.

Monoamino monocarboxy acids

H_2NCH_2COOH	Glycine (Gly)
CH_3CHNH_2COOH	Alanine (Ala)

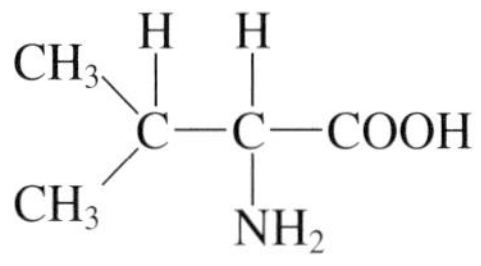

*Valine (Val)

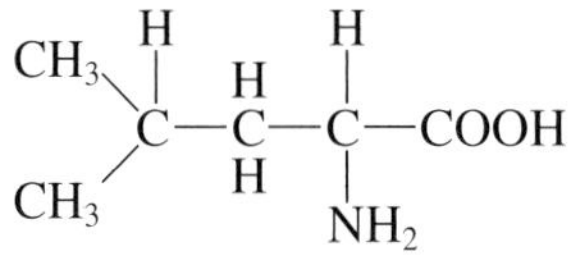

*Leucine (Leu)

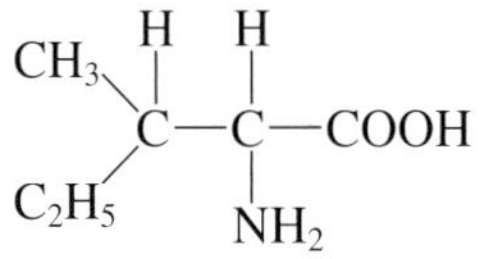

*Isoleucine (Ile)

Monocarboxy diamino acids

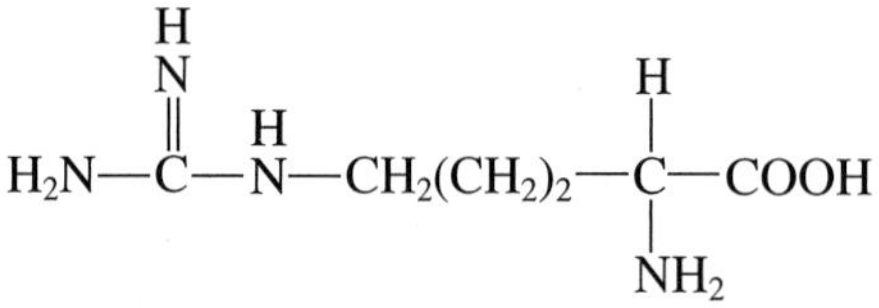

$$H_2N—\overset{\overset{\displaystyle H}{\overset{\displaystyle N}{\|}}}{C}—\overset{\displaystyle H}{N}—CH_2(CH_2)_2—\overset{\overset{\displaystyle H}{|}}{\underset{\underset{\displaystyle NH_2}{|}}{C}}—COOH \quad \text{Arginine (Arg)}$$

$$H_2N—\overset{\overset{\displaystyle O}{\|}}{C}—CH_2—\overset{\underset{\underset{\displaystyle NH_2}{|}}{}}{CH}—COOH \quad \text{Asparagine (Asn)}$$

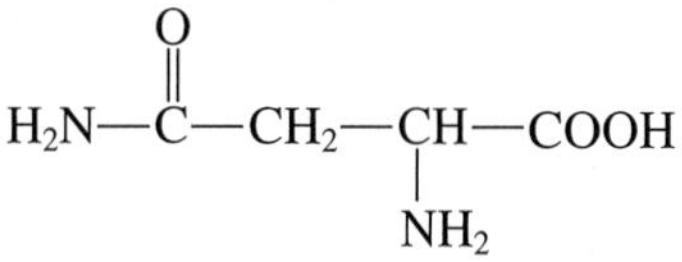

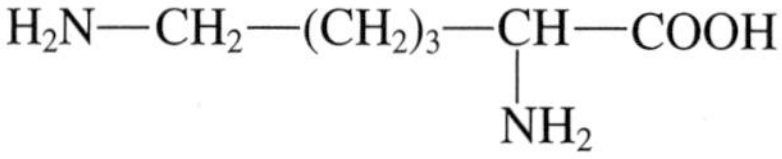

$$H_2N—CH_2—(CH_2)_3—\overset{\underset{\underset{\displaystyle NH_2}{|}}{}}{CH}—COOH \quad \text{*Lysine (Lys)}$$

$$H_2N—\overset{\overset{\displaystyle O}{\|}}{C}—(CH_2)_2—\overset{\underset{\underset{\displaystyle NH_2}{|}}{}}{CH}—COOH \quad \text{Glutamine (Gln)}$$

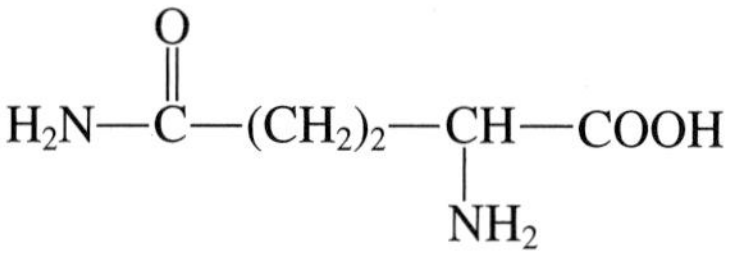

Aromatic homocyclic acids

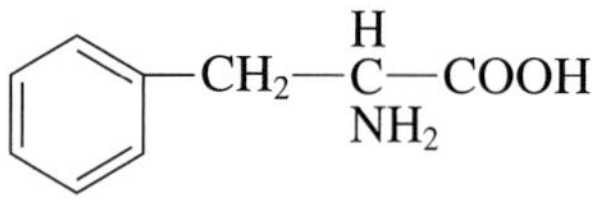

$$\text{⬡}—CH_2—\overset{\overset{\displaystyle H}{|}}{\underset{\underset{\displaystyle NH_2}{}}{C}}—COOH \quad \text{*Phenylalanine (Phe)}$$

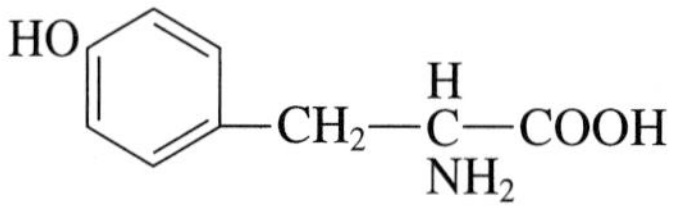

$$HO\text{⬡}—CH_2—\overset{\overset{\displaystyle H}{|}}{\underset{\underset{\displaystyle NH_2}{}}{C}}—COOH \quad \text{Tyrosine (Tyr)}$$

$$HO\overset{\overset{\displaystyle I}{}}{\underset{\underset{\displaystyle I}{}}{\text{⬡}}}—O—\overset{\overset{\displaystyle I}{}}{\underset{\underset{\displaystyle I}{}}{\text{⬡}}}\overset{\overset{\displaystyle H\ \ H}{|\ \ \ |}}{\underset{\underset{\displaystyle H\ \ NH_2}{}}{C—C}}—COOH \quad \text{Thyroxine (Thy)}$$

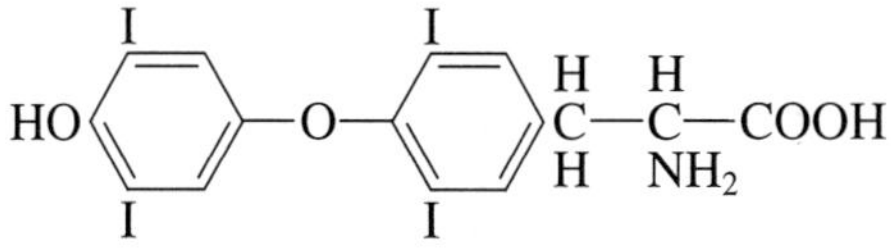

Monoamino monocarboxy monohydroxy acids

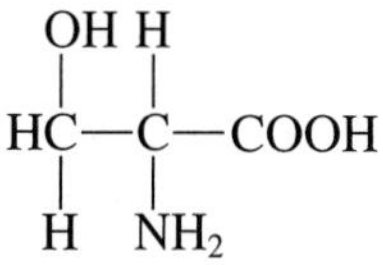

$$\overset{\overset{\displaystyle OH\ H}{|\ \ \ |}}{\underset{\underset{\displaystyle H\ \ \ NH_2}{}}{HC—C}}—COOH \quad \text{Serine (Ser)}$$

$$CH_3—\overset{\overset{\displaystyle H\ \ H}{|\ \ \ |}}{\underset{\underset{\underset{\displaystyle H}{O}}{|}\ \ \underset{\displaystyle NH_2}{}}{C—C}}—COOH \quad \text{*Threonine (Thr)}$$

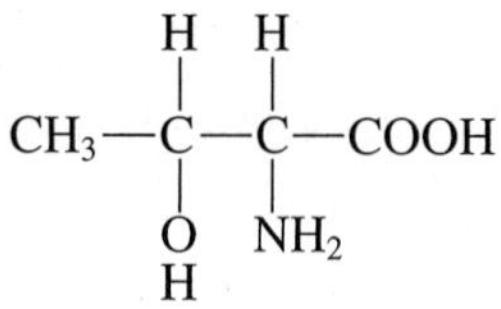

Sulfur-containing acids

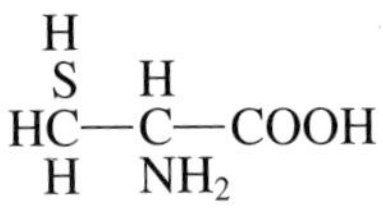

$$
\begin{array}{c}
\text{H} \\
\overset{|}{\text{S}} \quad \text{H} \\
\text{HC}\!-\!\overset{|}{\text{C}}\!-\!\text{COOH} \\
\overset{|}{\text{H}} \quad \overset{|}{\text{NH}_2}
\end{array}
\qquad \text{Cysteine (Cys)}
$$

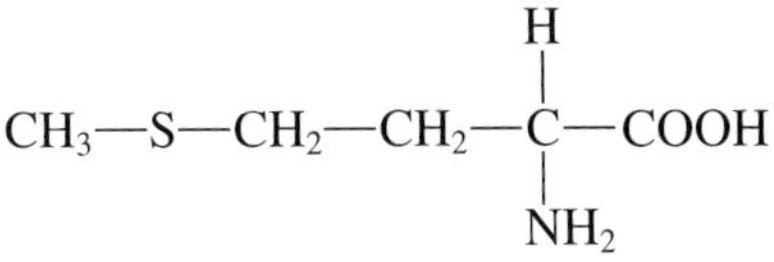

$$
\text{CH}_3\!-\!\text{S}\!-\!\text{CH}_2\!-\!\text{CH}_2\!-\!\overset{\overset{\text{H}}{|}}{\underset{\underset{\text{NH}_2}{|}}{\text{C}}}\!-\!\text{COOH}
\qquad \text{*Methionine (Met)}
$$

Dicarboxy monoamino acids

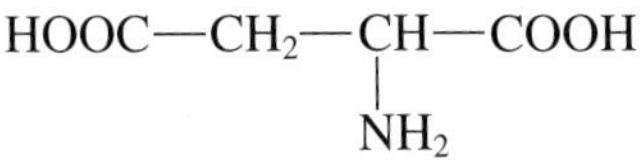

$$
\text{HOOC}\!-\!\text{CH}_2\!-\!\underset{\underset{\text{NH}_2}{|}}{\text{CH}}\!-\!\text{COOH}
\qquad \text{Aspartic acid (Asp)}
$$

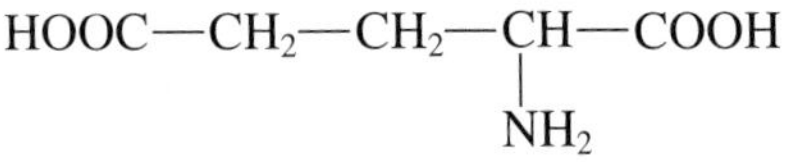

$$
\text{HOOC}\!-\!\text{CH}_2\!-\!\text{CH}_2\!-\!\underset{\underset{\text{NH}_2}{|}}{\text{CH}}\!-\!\text{COOH}
\qquad \text{Glutamic acid (Glu)}
$$

Heterocyclic acids

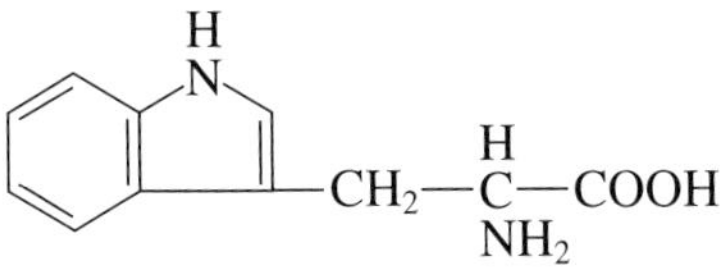

*Tryptophan (Trp)

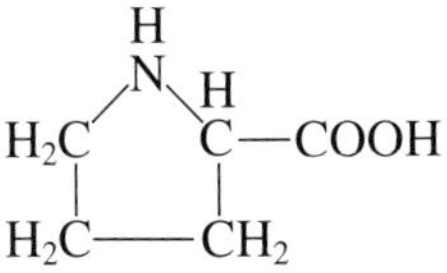

Proline (Pro)

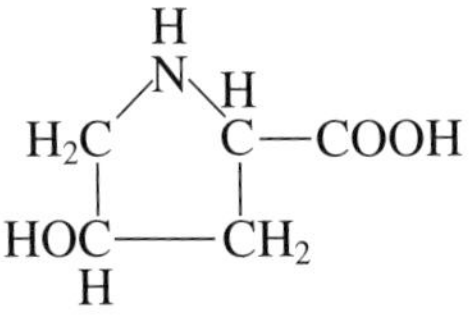

Hydroxyproline (Hypro)

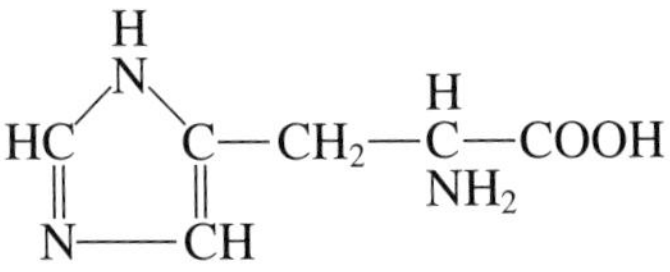

Histidine (His)

Proteins from different sources yield varying amounts of the different amino acids upon hydrolysis. The protein from a given source, however, normally yields the same amino acids and in the same ratio. All proteins yield more than one amino acid.

Proteins

Proteins constitute a very important part of the diet of humans, particularly in the form of meats, cheeses, eggs, and certain vegetables. The processing of these materials, except for eggs, results in the production of industrial wastes that can generally be treated by biological processes.

Properties of Proteins   Protein molecules are very large and have complex chemical structures. Insulin, which contains 51 amino acid units per molecule, was the first protein for which the precise order of the atoms in the molecule was discovered. For this significant achievement, Frederick Sanger at the University of Cambridge received the Nobel prize in 1958. Among the large number of proteins for which the order of atoms is now known are ribonuclease with 124 amino acids, tobacco mosaic virus protein with 158 amino acids, and hemoglobin with 574 amino acid units. Hemoglobin has a molecular weight of 64,500 and contains 10,000 atoms of hydrogen, carbon, nitrogen, oxygen, and sulfur, plus 4 atoms of iron. The iron atoms are more important than all the rest as they give blood its ability to combine with oxygen. Some protein molecules are thought to be 10 to 50 times larger than that of hemoglobin.

All proteins contain carbon, hydrogen, oxygen, and nitrogen. Regardless of their source, be it animal or vegetable, the ultimate analysis of all proteins falls within a very narrow range, as shown here:

	Percentage
Carbon	51–55
Hydrogen	6.5–7.3
Oxygen	20–24
Nitrogen	15–18
Sulfur	0.0–2.5
Phosphorus	0.0–1.0

The nitrogen content varies from 15 to 18 percent and averages about 16 percent. Since carbohydrates and fats do not contain nitrogen, advantage is taken of this fact in food analysis to calculate protein content. The value for nitrogen as determined by the Kjeldahl digestion procedure (Sec. 25.3), when multiplied by the factor 100/16 or 6.25, gives an estimate of the protein content. This procedure is sometimes used to estimate the protein content of domestic and industrial wastes and of sludges from their organic nitrogen (Sec. 25.3) content. There are several other nitrogen-containing organics in such wastes, however, and so such estimates should be considered only as crude approximations.

Biological Treatment of Protein Wastes   In general, satisfactory treatment of wastes containing significant amounts of proteins requires the use of biological processes. In these processes, the first step in degradation of the protein is considered to be hydrolysis, induced by hydrolytic enzymes. The hydrolysis is considered to progress in steps in reverse manner to those in which proteins are synthesized.

Hydrolysis Products of Proteins

$$\text{Protein} \rightarrow \text{polypeptides}$$
$$\downarrow \qquad\qquad (5.53)$$
$$\alpha\text{-amino acids} \leftarrow \text{dipeptides}$$

The α-amino acids are then deaminated by enzymic action, and free fatty and other acids result. The free acids serve as food for the microorganisms, and they are converted to carbon dioxide and water.

DETERGENTS

5.23 | DETERGENTS

The term *detergent* is applied to a wide variety of cleansing materials used to remove soil from clothes, dishes, and a host of other things. The basic ingredients of detergents are organic materials that have the property of being "surface active" in aqueous solution and are called surface-active agents or *surfactants*. All surfactants have rather large polar functional groups. One end of the molecule is particularly soluble in water and the other is readily soluble in oils. The solubility in water is due to carboxyl, sulfate, hydroxyl, or sulfonate groups. The surfactants with carboxyl, sulfate, and sulfonate groups are all used as sodium or potassium salts.

$$
\begin{array}{ll}
\textit{Oil-soluble part} & \textit{Water-soluble part} \\[6pt]
 & -COO^-Na^+ \\
\text{Organic group} \left. \vphantom{\begin{array}{c} \\ \\ \\ \\ \end{array}} \right\} & -SO_4^-Na^+ \\
 & -SO_3^-Na^+ \\
 & -OH
\end{array}
$$

The nature of the organic part of the molecule varies greatly with the various surfactant types.

5.24 | SOAPS

Ordinary *soaps* are derived from fats and oils by *saponification* with sodium hydroxide. Saponification is a special case of hydrolysis in which an alkaline agent is present to neutralize the fatty acids as they are formed. In this way the reaction is caused to go to completion.

$$
\begin{array}{ll}
\text{H}_2\text{COOCC}_{17}\text{H}_{35} & \text{H}_2\text{COH} \\
\quad | & \quad | \\
\text{H}-\text{COOCC}_{17}\text{H}_{35} + 3\text{NaOH} \rightarrow \text{H}-\text{COH} + 3\text{C}_{17}\text{H}_{35}\text{COONa} \qquad (5.54) \\
\quad | & \quad | \\
\text{H}_2\text{COOCC}_{17}\text{H}_{35} & \text{H}_2\text{COH} \\
\qquad \text{Stearin} & \qquad \text{Glycerol}
\end{array}
$$

The fats and oils are split into glycerol and sodium soaps. The nature of the soap depends upon the type of fat or oil used. Beef fat and cottonseed oil are used to produce low-grade, heavy-duty soaps. Coconut and other oils are used in the production of toilet soaps.

All sodium and potassium soaps are soluble in water. If the water is hard, the calcium, magnesium, and any other ions causing hardness precipitate the soap in the form of metallic soaps.

$$2C_{17}H_{35}COONa(aq) + Ca^{2+} \rightarrow (C_{17}H_{35}COO)_2Ca(s) + 2Na^+ \tag{5.55}$$

Soap must be added to precipitate all the ions causing hardness before it can act as a surfactant, usually indicated by the onset of frothing upon agitation.

5.25 | SYNTHETIC DETERGENTS

Since 1945 a wide variety of synthetic detergents have been accepted as substitutes for soap. Their major advantage is that they do not form insoluble precipitates with the ions causing hardness. Most commercially available products contain from 20 to 30 percent surfactant (active ingredient) and 70 to 80 percent *builders.* The builders are usually sodium sulfate, sodium tripolyphosphate, sodium pyrophosphate, sodium silicate, and other materials that enhance the detergent properties of the active ingredient. The use of phosphate has been curtailed because of its role in eutrophication. The synthetic surfactants are of three major types: *anionic, nonionic,* and *cationic.*

Anionic Detergents

The anionic detergents are all sodium salts and ionize to yield Na^+ plus a negatively charged, surface-activated ion. The common ones are all sulfates and sulfonates.

Sulfates   Long-chain alcohols when treated with sulfuric acid produce sulfates (inorganic esters) with surface-active properties. Dodecyl or lauryl alcohol is commonly used.

$$\underset{\substack{\text{Lauryl} \\ \text{alcohol}}}{C_{12}H_{25}OH} + H_2SO_4 \rightarrow C_{12}H_{25}-O-SO_3H + H_2O \tag{5.56}$$

The sulfated alcohol is neutralized with sodium hydroxide to produce the surfactant.

$$C_{12}H_{25}-O-SO_3H + NaOH \rightarrow \underset{\text{Sodium lauryl sulfate}}{C_{12}H_{25}-O-SO_3Na} + H_2O \tag{5.57}$$

The sulfated alcohols were the first surfactants to be produced commercially. The sulfated alcohols are used in combination with other synthetic detergents to produce blends with desired properties.

Sulfonates The principal sulfonates of importance are derived from esters, amides, and alkylbenzenes.

The esters and amides are of organic acids with 16 or 18 carbon atoms. In the past the alkylbenzene sulfonates (ABS) were derived largely from polymers of propylene, and the alkyl group, which averaged 12 carbon atoms, was highly branched. These materials are now made largely from normal (straight-chain) paraffins, and thus the alkane chain is not branched and the benzene ring is attached primarily to secondary carbon atoms. These latter materials have been labeled LAS (linear alkyl sulfonate).

Nonionic Detergents

The *nonionic* detergents do not ionize and have to depend upon groups in the molecule to render them soluble. All depend upon polymers of ethylene oxide (C_2H_4O) (polyethoxylates) to give them this property.

Nonylphenol (Sec. 5.14) and octylphenol with a variable number of ethylene oxides are important members of the aryl class.

Cationic Detergents

The *cationic* detergents are salts of quaternary ammonium hydroxide. In quaternary ammonium hydroxide, the hydrogens of the ammonium ion have all been replaced with alkyl groups. The surface-active properties are contained in the cation.

$$
\left[
\begin{array}{c}
R' \\
| \\
R—N—R'' \\
| \\
R'''
\end{array}
\right]^{+} Cl^{-}
$$

A cationic detergent

The cationic detergents are noted for their disinfecting (bactericidal) properties. They are used as sanitizing agents for dishwashing where hot water is unavailable or undesirable. They are also useful in the washing of babies' diapers, where sterility is important. If diapers are not sterilized by some means, bacterial infestations may occur that release enzymes that will hydrolyze urea to produce free ammonia [Eq. (5.43)]. The high pH resulting is harmful to the tender skin of babies, and the odor of free ammonia is unpleasant to all.

Biological Degradation of Detergents

Detergents vary greatly in their biochemical behavior, depending on their chemical structure.[22] Common soaps and the sulfated alcohols are readily used as bacterial food. The synthetic detergents with ester or amide linkages are readily hydrolyzed. The fatty acids produced serve as sources of bacterial food. The other hydrolysis product may or may not serve as bacterial food, depending upon its chemical structure. The synthetic detergents prepared from polymers of ethylene oxide appear susceptible to biological attack. However, recent evidence has demonstrated they are only partially transformed, leaving an alkyl aromatic compound that can be chlorinated or brominated during chlorine disinfection. As noted previously, the alkylphenols are suspected endocrine disruptors. The alkyl-benzene sulfonates derived from propene were highly resistant to biodegradation, and their persistence resulted in excessive foaming in rivers and groundwaters in the 1950s. This presented some of the first evidence of the potential harmful environmental consequences of synthetic organic chemicals. For this reason, the detergent manufacturing industry changed to the production of LAS surfactants. LAS is readily degradable under aerobic conditions, and its use has helped relieve the most serious problems of detergent foaming. However, unlike common soap, it is resistant

[22]C. N. Sawyer and D. W. Ryckman, Anionic Detergents and Water Supply Problems, *J. Amer. Water Works Assoc.,* **49:** 480 (1957).

to degradation under anaerobic conditions. Currently, there is a United States EPA secondary standard of 0.5 mg/L for foaming agents. Enforcement of secondary standards by states is optional.

PESTICIDES

5.26 | PESTICIDES

Pesticides are materials used to prevent, destroy, repel, or otherwise control objectionable insects, rodents, plants, weeds, or other undesirable forms of life. Common pesticides can be categorized chemically into three general groups, inorganic, natural organic, and synthetic organic. They may also be classified by their biological usefulness, viz., insecticides, herbicides, algicides, fungicides, and rodenticides.

The synthetic organic pesticides gained prominence during World War II, and since then their numbers have grown into the thousands, while the total annual production has increased to about 1.5 billion pounds of active ingredients. They are used mainly for agricultural purposes. The synthetic organic pesticides are best classified according to their chemical properties, since this more readily determines their persistence and behavior when introduced into the environment. The major types of synthetic pesticides are the chlorinated hydrocarbons, the organic phosphorus pesticides, and the carbamate pesticides. An additional category of current interest is the s-triazine pesticides. There are many other types of synthetic pesticides.

5.27 | CHLORINATED PESTICIDES

Chlorinated pesticides are of many types and have been widely used for a variety of purposes. DDT proved to be an extremely versatile insecticide during World War II when it was used mainly for louse and mosquito control. It is still one of the major pesticides used internationally, although use in the United States has been banned for quite some time for environmental reasons. DDT is a chlorinated aromatic compound with the following structure:

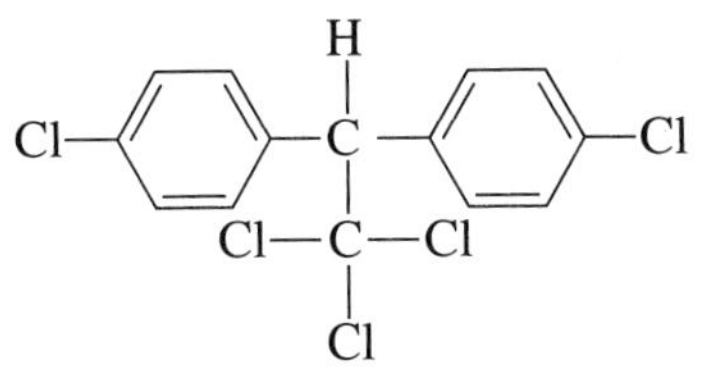

DDT

Technical grade DDT contains three isomers, the above isomer representing about 70 percent of the total. It has the long technical name 2,2-bis(*p*-chlorophenyl)-1,1,1-trichloroethane.

When benzene and chlorine react in direct sunlight, the addition product benzene hexachloride or BHC is formed. Several stereoisomers are produced, but the gamma isomer called *lindane* is by far the most effective as an insecticide.

γ-Benzene hexachloride (lindane)

Two other formerly widely used chlorinated insecticides are endrin and dieldrin. They are isomers, and their structural formulas appear the same:

Dieldrin or endrin

However, their spatial configurations are significantly different, as are their insecticidal properties.

Chlorinated pesticides are also used as herbicides. Two of the most common are 2,4-D and 2,4,5-T.

2,4-Dichlorophenoxyacetic acid
(2,4-D)

2,4,5-Trichlorophenoxyacetic acid
(2,4,5-T)

These two herbicides are effective in destroying certain broad-leaf plants while not killing grasses. They have also been used for aquatic-plant control in lakes, ponds, and reservoirs. Dioxin, an extremely toxic organic to humans, is a side product contaminant in 2,4,5-T, which has led to restrictions in its use. Dioxins are also formed during the combustion of chlorinated organic compounds. The most toxic dioxin is 2,3,7,8-tetrachlorodibenzo-*p*-dioxin (TCDD):

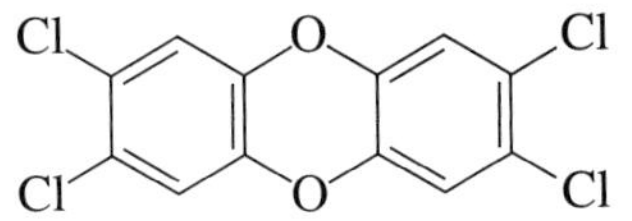

2,3,7,8-Tetrachlorodibenzo-*p*-dioxin (TCDD)

Because of its extreme toxicity, a very low drinking water MCL of 3×10^{-5} μg/L has been established for this dioxin.

The family of compounds called the chloroacetamides are also commonly used as herbicides for control of broad-leaf weeds. Two of the most common examples are alachlor [2-chloro-2′,6′-diethyl-N-(methoxymethyl)-acetanilide) and meto-lachlor [2-chloro-6′-ethyl-N-(2-methoxy-1-methyl-ethyl)acet-o-toluidine]. These herbicides are used primarily for weed control in the production of corn and soy-beans.

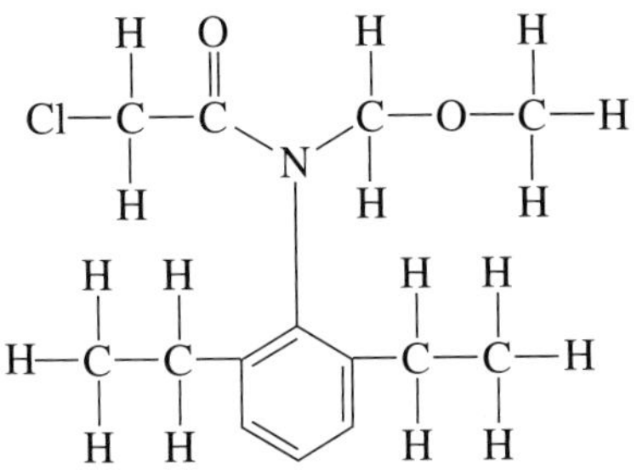

Alachlor

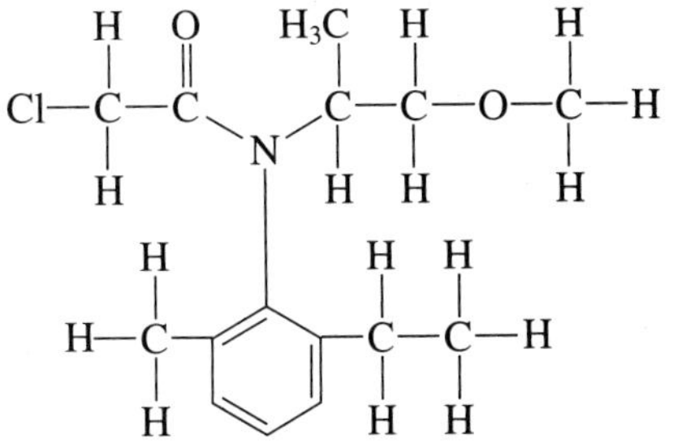

Metolachlor

Other chlorinated or halogenated pesticides of significance are aldrin, chlor-dane, toxaphene, heptachlor, methoxychlor, DDD, EDB, DBCP, and 1,2-dichloropropane; all have been used as insecticides, fungicides, or nematocides. All halogenated pesticides are considered to be of significant concern because of their persistence and high potential for creating harm to humans and the environ-ment. For this reason aldrin and dieldrin are now banned from use in the United States, and many of the other chlorinated pesticides have been greatly restricted in usage.

5.28 | ORGANIC PHOSPHORUS PESTICIDES

The organic phosphorus pesticides became important as insecticides after World War II. These compounds were developed in the course of chemical warfare re-search in Germany and in general are quite toxic to humans as well as to pests. *Parathion* is an important pesticide, which was introduced into the United States from Germany in 1946. It is an aromatic compound and contains sulfur and nitrogen as well as phosphorus in its structure, so do many of the organic phosphorus pesticides.

O,O-Diethyl-*O*-*p*-nitrophenylthiophosphate
(parathion)

Parathion has been particularly effective against certain pests such as the fruit fly. However, it is also quite toxic to humans and extreme caution must be exercised in its use.

Malathion is highly toxic to a variety of insects, but unlike many organic phosphorus pesticides, it has low toxicity to mammals. This has made it particularly successful and it is widely used.

S-(1,2)-Dicarbethoxyethyl)-*O,O*-dimethyldithiophosphate
(malathion)

Other organic phosphorus pesticides of significance are methyl parathion, glyphosate, demeton, guthion, systox, metasystox, chlorthion, disyston, and dicapthon.

5.29 | CARBAMATE PESTICIDES

Carbamate pesticides are amides having the general formula RHNCOOR′. One that has received wide usage is isopropyl *N*-phenylcarbamate (IPC).

Isopropyl *N*-phenylcarbamate
(IPC)

IPC is a herbicide that is effective for the control of grasses, without affecting broad-leaf crops. Other carbamates of importance are aldicarb, carbaryl (Sevin®), carbofuran, ferbam, and captan. Carbamates in general appear to have low toxicity to mammals.

5.30 | s-TRIAZINES

The class of compounds called the s-triazines are primarily used as herbicides in agricultural regions and are of three general types: (1) chloro s-triazines, (2) methylthio s-triazines, and (3) methoxy s-triazines:

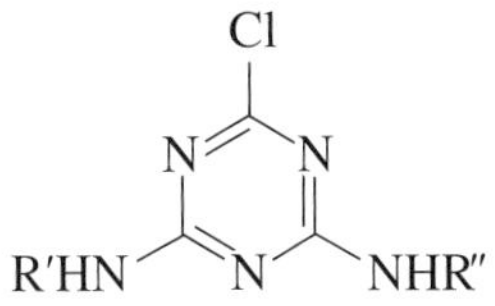

Cl SCH_3 OCH_3

R'HN — N — NHR''   R'HN — N — NHR''   R'HN — N — NHR''

Chloro-s-triazine 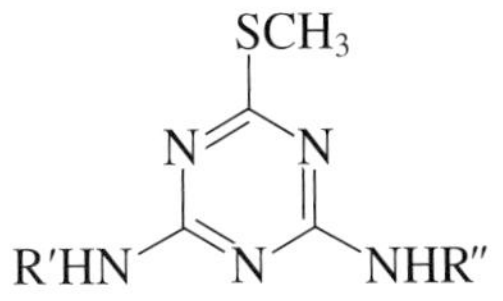   Methylthio-s-triazine 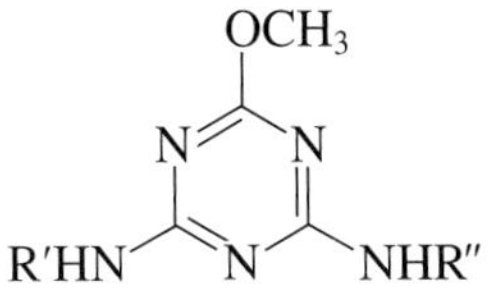   Methoxy-s-triazine

Two of the most commonly used s-triazines are atrazine (2-chloro-4-ethyl-amino-6-isopropylamino-s-triazine) and cyanazine [2-chloro-4-ethylamino-6-(1-cyano-1-methylethylamino)-s-triazine].

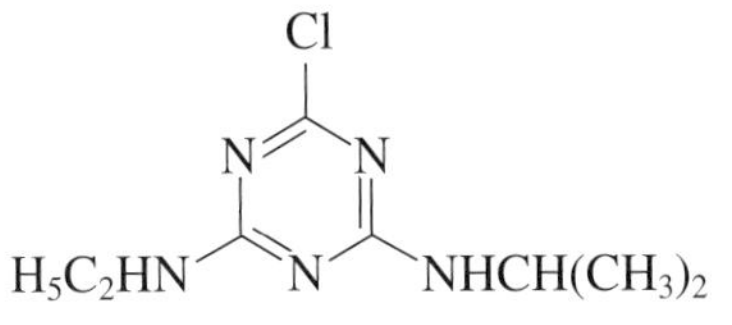

Cl   Cl

H_5C_2HN — N — $NHCH(CH_3)_2$ H_5C_2HN — N — $NHC(CH_3)$—CN with CH_3

Atrazine 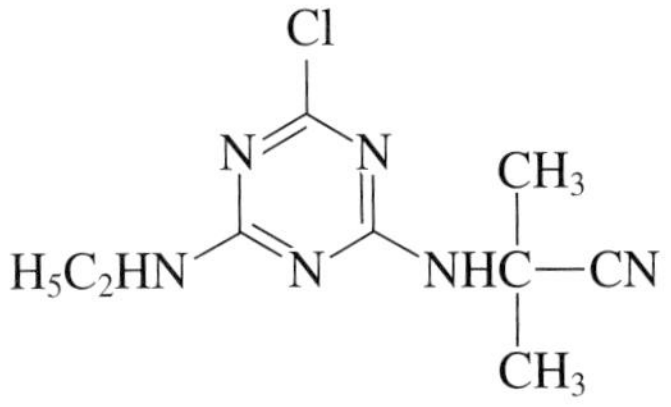   Cyanazine

Atrazine is currently one of the most widely used herbicides, is difficult to degrade biologically, and is a commonly detected groundwater and surface water contaminant in agricultural areas.

5.31 | BIOLOGICAL PROPERTIES OF PESTICIDES

Pesticides may gain access to groundwater and surface water supplies through direct application or through percolation and runoff from treated areas. Several recent surveys have shown that pesticides are present in many groundwater and surface water supplies. For example, one report indicated that 46 different pesticides were found in groundwaters in 26 states.[23] Concentrations are typically higher in surface waters than in groundwaters. Some pesticides are toxic to fish and other aquatic life at only a small fraction of a milligram per liter. They also tend to concentrate in aquatic plants and animals to values several thousand times that occurring in the water in which they live. Also, some pesticides are quite resistant to biological degradation and persist in soils and water for long periods of time. A major concern with all pesticides is the potential for incomplete transformation of the parent compound into metabolites that may be more or less toxic.

---

[23]D. J. Munch et al., Methods Development and Implementation for the National Pesticide Survey. *Env. Sci. Tech.,* **24:** 1445–1451 (1990).

In general, the chlorinated pesticides are the most resistant to biological degradation and may persist for months or years following application. Many are also highly toxic to aquatic life or to birds that feed on aquatic life. This persistence and great potential for harm has led to restrictions on chlorinated pesticide usage. Several chlorinated pesticides (e.g., DDT, dieldrin, methoxychlor, toxaphene) are suspected endocrine disruptors. As a group, the organic phosphorus pesticides are not too toxic to fish life and have not caused much concern in this respect, except when large accidental spills have occurred. Also, they tend to hydrolyze rather quickly at pH values above neutral, thus losing their toxic properties. Under proper conditions such as dryness, however, some have been observed to persist for many months. Their main potential for harm is to farm workers who become exposed both from contact with previously sprayed foliage as well as from direct contact during organic phosphorus pesticide application. Some of these pesticides are readily absorbed through the skin and affect the nervous system by inhibiting the enzyme cholinesterase, which is important in the transmission of nerve impulses. The carbamates are noted for their low toxicity and high susceptibility to degradation. The s-triazines are in general quite resistant to environmental degradation and are known endocrine disruptors.

Because of the several potential health problems that have been associated with pesticides, drinking water MCLs have been established for several pesticides that are still in somewhat common usage. These are listed in Table 34.1.

5.32 | PHARMACEUTICALLY ACTIVE AND ENDOCRINE-DISRUPTING CHEMICALS

Two groups of chemicals of emerging concern are *pharmaceutically active chemicals* (PhACs) and *endocrine-disrupting chemicals* (EDCs). These compounds are being found in surface waters and groundwaters and in wastewater treatment plant effluents in the ng/L (parts per trillion) to μg/L concentration range. Understanding the fate and effects of these chemicals in the environment and removing them from drinking water supplies and wastewaters present difficult challenges for environmental engineers and scientists.

Pharmaceutically Active Chemicals (PhACs)

Pharmaceuticals are generally defined as chemicals used for the treatment or prevention of illness. As such, they can range from compounds used for cancer treatment (chemotherapy) and birth control to antibiotics used to combat infection to compounds used to relieve pain (e.g., aspirin and ibuprofen). They are found in personal care products such as fragrances, disinfectants and antiseptics, sunscreen agents, and preservatives. Pharmaceuticals are also used in veterinary health care (e.g., antibiotics and growth hormones). Very little is known with respect to the effect of PhAC on human and wildlife health. However, there is potential for adverse effects and the reasoning goes like this. Most of these compounds are lipophilic ("fat-loving"; synonymous with hydrophobic) and their activity is slow to decay (that is, they remain pharmaceutically active for an extended time). Thus, bioconcentration is possible. While the concentration of individual PhACs in water supplies is low (generally less than 0.5 μg/L with many in the ng/L range), the presence of numerous drugs with similar modes of action could lead to measurable effects. Finally, exposure can be chronic because PhACs are continually introduced into the environment via human wastewater treatment, livestock production, and similar activities. One special concern with antibiotics is the development of antibiotic-resistant strains of pathogenic bacteria due to overuse of these compounds. In general this is because a large portion of the administered antibiotic leaves the body (humans and animals) via urine and feces as a mixture of parent compound and metabolites.

PhACs come from a wide variety of organic chemical classes. Examples of PhACs that have been found in surface waters and wastewater treatment effluents include pain killers such as ibuprofen (a chiral compound), acetaminophen, acetylsalicylic acid (aspirin), and codeine; antibiotics in the sulfonamide, tetracycline, fluoroquinolone, and macrolide classes; chlofibric acid [environmental metabolic of the lipid regulator clofibrate (helps lower cholesterol)], and caffeine. Some chemical structures are given here.

Ibuprofen   Acetaminophen   Acetylsalicylic acid
(aspirin)

Codeine   Sulfanilamide

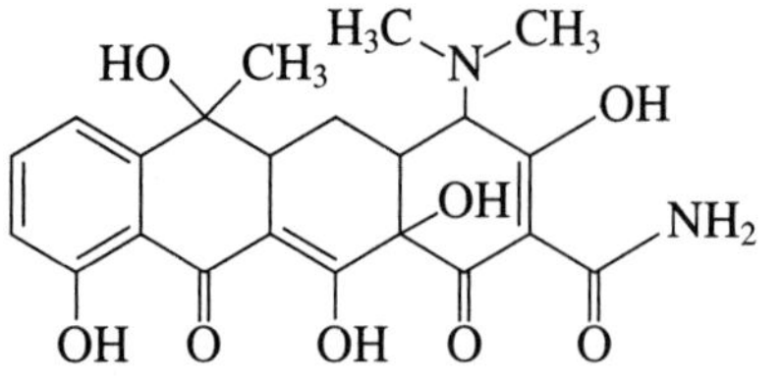

Tetracycline

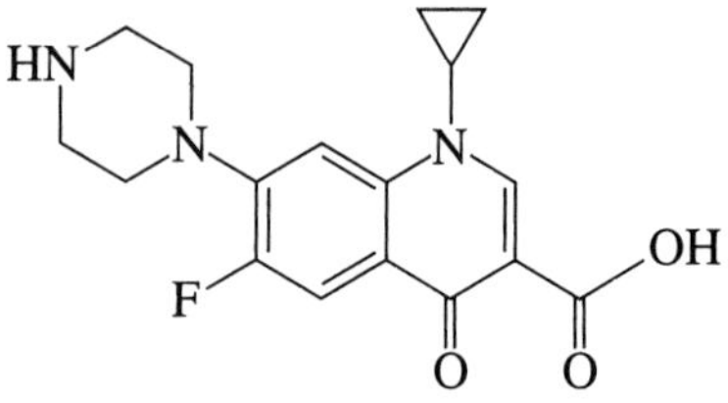

Fluoroquinolone

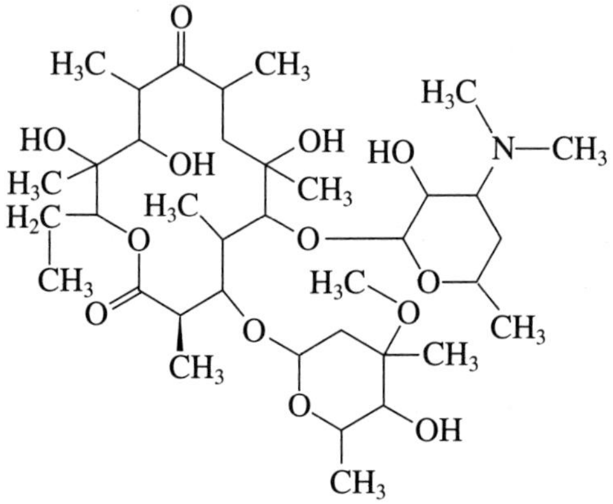

Erythromycin (a macrolide antibiotic)

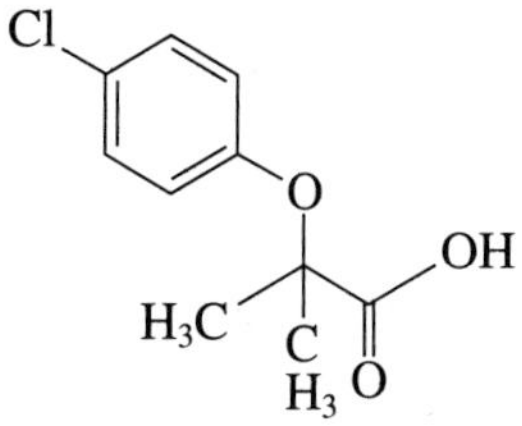

Clofibric acid

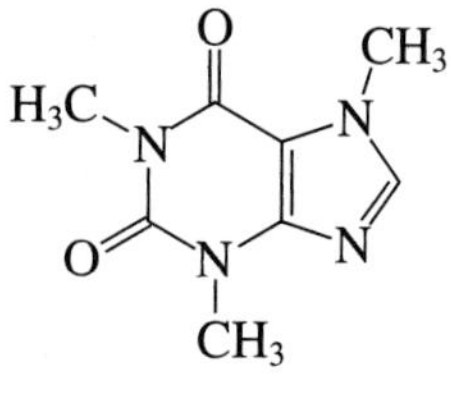

Caffeine

Endocrine-Disrupting Chemicals (EDCs)

The human endocrine system, which includes a variety of glands (e.g., thyroid, pituitary, pineal, ovaries, and testes) and the hormones they produce (e.g., adrenaline, estrogens, and testosterone), regulates growth, development, reproduction, and behavior. Chemical structures for testosterone and estradiol, an estrogen, are given here.

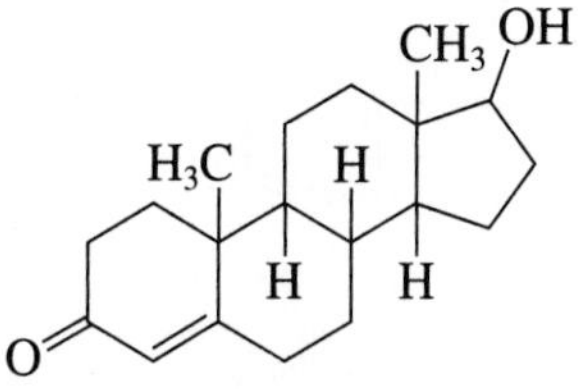

Testosterone

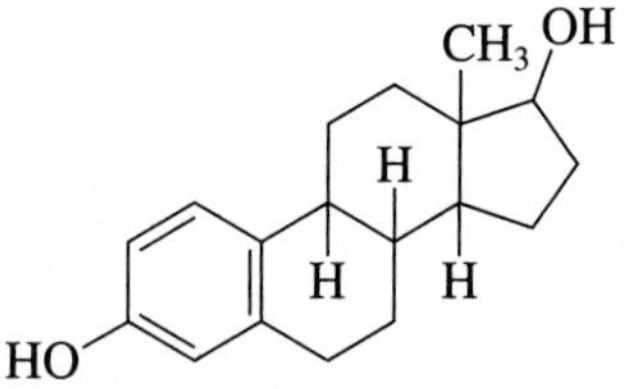

Estradiol (an estrogen)

There is concern that some synthetic organic chemicals released into the environment can interfere with or disrupt this system. In general, it is thought that such compounds mimic the body's natural hormones, binding with receptor molecules in the body, and thus interfering with the regulation of the endocrine system. An excellent review of how EDCs can interfere with the endocrine system is given by Trussell.[24] There is still some controversy as to whether observed effects of EDCs are due to their hormonal properties or some other toxicological mechanism.[25] For this and other reasons, a National Research Council committee recommends use of the term *hormonally active agent* (HAA) instead of EDC.[26] It should be noted that some naturally produced organic compounds (e.g., phytoestrogens produced by plants such as soybeans) and inorganic chemicals (e.g., lead, mercury) can also act as EDCs. Some PhACs are also EDCs (for example, growth hormones used in livestock production).

EDCs have been reported to have a variety of effects on humans and wildlife; most are developmental abnormalities, although some neurological and immunological effects have been reported. A widely reported example is the effect of diethylstilbestrol (DES), a synthetic estrogen used in the 1950s to prevent miscarriages. Children of mothers taking DES during pregnancy experienced a wide variety of problems including cancers of the vagina and cervix, malformed reproductive organs, undescended testicles, and abnormal sperm. These effects were associated with high exposure to DES. Recent reports indicate effects in wildlife associated with exposure to lower levels of EDCs. For example, developmental abnormalities in alligators in Florida and male fish living near outfalls from wastewater treatment plants, and declines in fish and bird populations in the Great Lakes have been associated with EDC exposure. It must be stressed that much more needs to be learned about the effects of EDCs on adverse health outcomes in humans and wildlife.

A wide variety of chemicals have been implicated as EDCs. These include several pesticides, organochlorine compounds including PCBs, phenolic compounds, phthalates, PAHs, synthetic steroids, and triazines. Specific examples include malathion, dieldrin, methoxychlor, DDT and its metabolites, dioxin (TCDD), pentachlorophenol, bisphenol A, nonylphenol, butyl benzyl phthalate, and atrazine. Phytoestrogens include a variety of flavones and isoflavones. Chemical structures for PCBs, malathion, dieldrin, DDT, dioxin, pentachlorophenol, bisphenol A, nonylphenol, and atrazine were given in Secs. 5.13 to 5.30. Structures

---

[24]R. R. Trussell, Endocrine Disrupters and the Water Industry, *J. American Water Works Assoc.,* **93**(2): 58–65, 2001.

[25]National Research Council, "Hormonally Active Agents in the Environment," National Academy Press, Washington, D.C., 1999.

[26]Use of the acronym HAA for hormonally active agent may cause confusion. As noted in Sec. 5.8, HAA is also used by environmental engineers and scientists to describe the haloacetic acids produced as disinfection by-products and regulated as drinking water contaminants.

for methoxychlor, butyl benzyl phthalate, and genistein, an isoflavone in soybeans are given here.

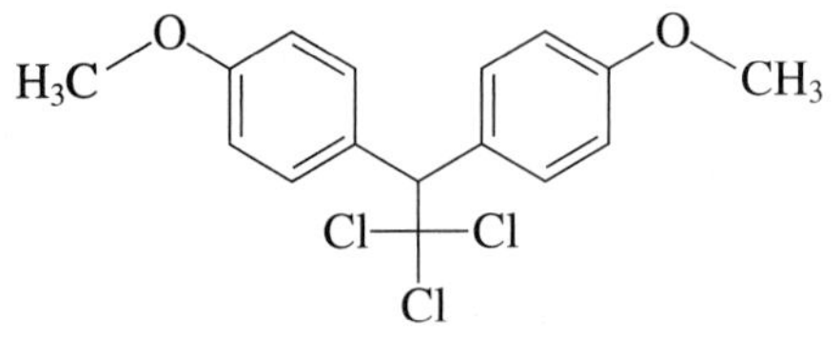

Methoxychlor

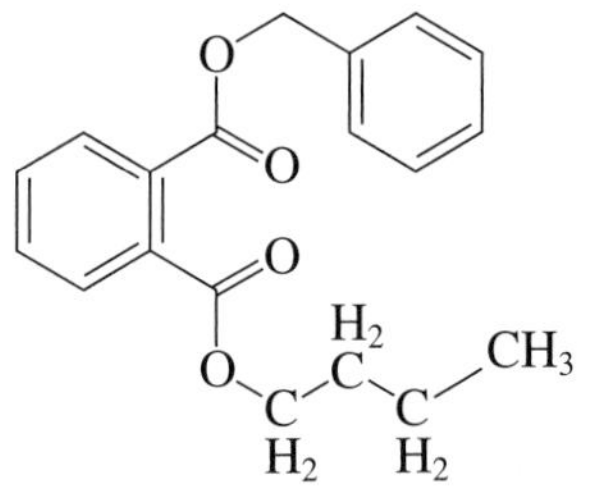

Butyl benzyl phthalate

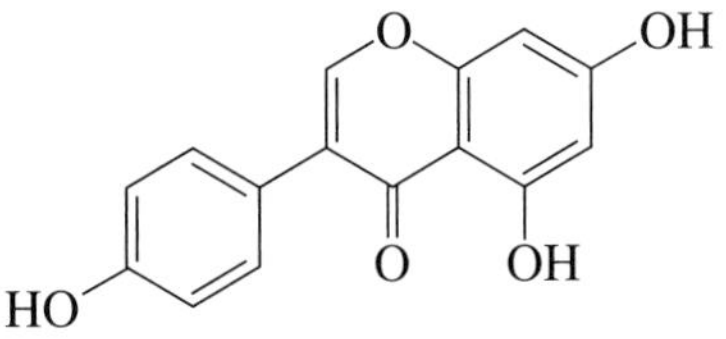

Genistein

BEHAVIOR OF ORGANICS IN THE ENVIRONMENT AND IN ENGINEERED SYSTEMS

5.33 | INTRODUCTION

It is important for the environmental engineer and scientist to have knowledge of the properties, both physicochemical and structural, of the different types and classes of organic compounds to aid in understanding and predicting the fate, effects, and potential of engineered processes for removal or control of these compounds. In the previous sections classes of organic compounds were described with respect to the functional group or groups characteristic of each class of compound. These characteristic functional groups also manifest themselves in other important properties that aid in understanding the behavior of organics in the environment and in engineered reactors. Physicochemical properties include, but are not limited to, solubility, hydrophobicity, polarity, volatility, density, and energy content. Solubility, hydrophobicity, and polarity are somewhat related and are useful in understanding the tendency of organics to partition between phases (i.e., solid-liquid or liquid-gas partitioning). Volatility, which can be quantified using Henry's constant or vapor pressure (Chap. 2), is useful in understanding partitioning between the gas and liquid phases. Density is useful in understanding physical separation potential and transport behavior. Energy content is useful in predicting bacterial yields (Chap. 6). Structural characteristics such as molecular connectivity and molecular surface area can be related through structure- and property-activity relationships (SARs and PARs) that can be used to predict compound behavior and fate. Detailed discussion of these and other properties is beyond the scope of this text; additional material can be found in the references listed at the end of the chapter.

5.34 | FATE OF ORGANICS

Important processes involved in the movement and fate of organics in the environment and in engineered systems are listed in Table 5.14. Processes especially important in understanding the fate and removal of organics found at contaminated sites and in industrial wastes and leachates are volatilization, sorption, and transformation reactions.

Volatilization

The concepts relevant to volatilization were introduced in Sec. 2.9 and are reviewed here. The equation for Henry's law constants (K_H) for chemical equilibrium between gaseous and aqueous phases is repeated here:

$$K_H = \frac{\text{concentration in gas}}{\text{concentration in water}} = \frac{P_{gas}}{C_{equil}} \qquad (2.15)$$

K_H can be thought of as a partition coefficient between water and the atmosphere. It can take several forms. The most common units are partial pressure (atm)/molar concentration (mol/L) or atm/M. The so-called unitless Henry's constant H results when the concentrations in both water and gas are expressed in mol/vol. Conversion can be made using the following:

$$H = \frac{K_H}{RT} \qquad (5.58)$$

where R is 0.08206 atm/M-K and T is the temperature (K). Henry's constant has a strong temperature dependence; organic compounds are less volatile at lower temperatures. Values of Henry's constant for some environmentally significant organic compounds are given in Table 5.15. The higher the value of K_H, the more likely the compound is to partition to the atmosphere.

Table 5.14 | Processes involved in the fate of organics

Process	Air	Water	Soil
Meteorological transport	X		
Condensation	X		
Settling	X	X	
Volatilization		X	X
Runoff (erosion)			X
Filtration			X
Sorption	X	X	X
Desorption	X	X	X
Precipitation	X	X	X
Chemical transformation			
Photolysis	X	X	
Hydrolysis	X	X	X
Oxidation	X	X	X
Reduction	X	X	X
Bio-uptake (bioconcentration)		X	X
Biodegradation (biotransformation)		X	X

Table 5.15 | Values of K_{ow}, water solubilities and Henry's law constants for selected organic compounds

Compound	log K_{ow}	Water solubility, mg/L	K_H, atm/M
Data from Yaws for 25°C			
Halogenated aliphatic compounds			
Methanes			
Chloromethane	0.91	5,900	8.2
Dichloromethane	1.25	19,400	2.5
Chloroform	1.97	7,500	4.1
Bromoform	2.4	3,100	0.59
Carbon tetrachloride	2.83	790	29
Dichlorodifluoromethane	2.16	18,800	390
Ethanes			
Chloroethane	1.43	9,000	6.9
1,1-Dichloroethane	1.79	5,000	5.8
1,2-Dichloroethane	1.48	8,700	1.18
1,1,1-Trichloroethane	2.49	1,000	22
1,1,2-Trichloroethane	1.89	4,400	0.92
Hexachloroethane	3.91	8	25
Ethenes			
Vinyl chloride	1.62	2,700	22
1,1-Dichloroethene	2.13	3,400	23
1,2-*cis*-Dichloroethene	1.86	3,500	7.4
1,2-*trans*-Dichloroethene	2.09	6,300	6.7
Trichloroethene	2.42	1,100	11.6
Tetrachloroethene	3.4	150	26.9
Aromatic compounds			
Hydrocarbons			
Benzene	2.13	1,760	5.6
Toluene	2.73	540	6.4
Ethylbenzene	3.15	165	8.1
Styrene	2.95	322	2.6
o-Xylene	3.12	221	4.2
m-Xylene	3.2	174	6.8
p-Xylene	3.15	200	6.2
1,2,3-Trimethylbenzene	3.66	36	7.4
1,2,4-Trimethylbenzene	4.02	35	
Napthalene	3.3	32	0.46 (20°C)
Phenanthrene	4.46	1.18	
Anthracene	4.45	0.053	
Fluorene	4.18	1.89	
Other aromatic compounds			
Chlorobenzene	2.84	390	4.5
1,2-Dichlorobenzene	3.43	92	2.8
1,3-Dichlorobenzene	3.53	123	3.4
1,4-Dichlorobenzene	3.44	80	
1,2,4-Trichlorobenzene			3.0
Hexachlorobenzene	5.73	0.0047	

Table 5.15 | (continued)

Compound	log K_{ow}	Water solubility, mg/L	K_H, atm/M
Data from Yaws for 25°C			
Other aromatic compounds			
Nitrobenzene	1.85	1,940	0.021
3-Nitrotoluene	2.45	500	0.075
Phenol	1.46	80,000	0.00076
Diethyl phthalate	2.47	1,000	0.00014
2-Chlorophenol	2.15	25,000	0.037
3-Chlorophenol	2.5	25,000	0.00204
Dibenzofuran	4.12		
Other aliphatic compounds			
Methyl *t*-butyl ether	0.94	51,000	0.54
Methyl ethyl ketone	0.29	250,000	0.030
Data from Schnoor et al. for 20°C			
2-Nitrophenol	1.75	2,100	
Benzo(a)pyrene	6.06	0.0038	0.00049
Acrolein	0.01	210,000	0.0038
Alachlor	2.92	240	
Atrazine	2.69	33	
Pentachlorophenol	5.04	14	
DDT	6.91	0.0055	0.038
Lindane	3.72	7.52	0.0048
Dieldrin	3.54	0.2	0.0002
2,4-D	1.78	900	0.00000172

Sources: C. L. Yaws, "Chemical Properties Handbook." McGraw-Hill, New York, 1999. Schnoor et al., "Processes, Coefficients, and Models for Simulating Toxic Organics and Heavy Metals in Surface Waters," U.S. Environmental Protection Agency, EPA/600/3-87/015, June 1987.

In general, it is considered that if K_H is less than 0.2 atm/M or H is less than about 0.01, the compound will not be efficiently removed from water by air stripping in an engineered reactor (e.g., stripping tower, aeration pond). The larger the value of K_H (and H), the easier the compound can be removed by air stripping. It should be noted that values of K_H reported in the literature should be used with some caution, since methods and conditions used in determining K_H vary widely.

When designing an engineered system to remove organics by volatilization, the following relationship from Sec. 3.10 is often used:

$$\frac{dC}{dt} = K_L a (C_{equil} - C) \tag{3.52}$$

where
C = concentration in water (mass/vol)

C_{equil} = concentration in water that would be in equilibrium with the concentration in air (mass/vol)

$K_L a$ = overall mass-transfer rate coefficient (time^{-1})

Often the air concentration is near zero, so C_{equil} becomes zero. Different engineered systems will give different values for K_La, and this rate coefficient will be different for different organics as well. It is also a strong function of temperature, increasing with an increase in temperature.

EXAMPLE 5.1

Gossett[27] developed relationships describing the effect of temperature on Henry's constant for a variety of chlorinated aliphatic compounds. The relationship between K_H (atm-m^3/mol) and T for trichloroethene (CHCl$_2$=CCl$_2$) is given here.

$$\ln K_H = 11.37 - \frac{4780}{T}$$

Calculate the value for K_H (atm/M) for temperatures of 10°C (283 K) and 25°C (298 K), which might represent the extremes in groundwater temperatures in the midwest.

For 10°C: $\ln K_H = 11.37 - \dfrac{4780}{283} = -5.52$ $K_H = 0.0040$ atm-m^3/mol or 4 atm/M

For 25°C: $\ln K_H = 11.37 - \dfrac{4780}{298} = -4.67$ $K_H = 0.0094$ atm-m^3/mol or 9.4 atm/M

For TCE, K_H more than doubled due to the 15°C temperature increase. This means it will be much easier to remove TCE from contaminated groundwater by air stripping in the summer months. It may be noted that the 25°C K_H value of 9.4 atm/M is somewhat less than the 11.6 atm/M value in Table 5.15, reflecting the different sources of the data.

EXAMPLE 5.2

Removing organics from water by volatilization can be thought of as forcing the organic to partition into the air. An important parameter is the air-to-water ratio, the volume of air exposed to a given volume of contaminated water. An easy way to view the importance of this ratio and of K_H is by using a simple equilibrium model in which a given volume of air is mixed with a given volume of water. A mass balance yields the following expression:

$$\frac{C}{C_0} = \frac{1}{1 + \left(\dfrac{V_a}{V_w}\right)\left(\dfrac{K_H}{RT}\right)}$$

where C = equilibrium concentration in water

C_0 = initial concentration in water

V_a/V_w = volume of air divided by volume of water (the air-to-water ratio)

T = temperature (absolute)

R = universal gas constant

---

[27]J. J. Gossett, Measurements of Henry's Law Constants for C1 and C2 Chlorinated Hydrocarbons, *Env. Sci. Tech.*, **21**: 202–208, 1987.

Consider the two chlorinated solvents carbon tetrachloride and trichlorothene. What air-to-water ratio is required to remove 95 percent of each from water? For an air-to-water ratio of 100 (typical values for air stripping towers are 10 to 100), what is the percent removal of each compound? Assume a temperature of 25°C.

For 25°C, RT is 24.5 atm/M. From Table 5.15, K_H is 29 atm/M for carbon tetrachloride and 11.6 atm/M for trichloroethene.

For 95 percent removal, C/C_0 is 0.05. Solving the given expression for V_a/V_w yields

$$\frac{V_a}{V_w} = \left(\frac{C_0}{C} - 1\right)\left(\frac{RT}{K_H}\right)$$

Substituting the proper K_H yields a V_a/V_w of 16 for carbon tetrachloride and 40 for trichloroethene.

A V_a/V_w of 100 yields a C/C_0 of 0.0084 (99.2 percent removal) for carbon tetrachloride and 0.0207 (97.9 percent removal) for trichloroethene.

Solubility

Water solubility of an organic compound is generally defined as the concentration (mass/vol or mol/vol) resulting when the water is in equilibrium with the pure compound [gas (1 atm), liquid, or solid]. There are many factors affecting compound solubility. They include, but are not limited to, size of the molecule; the nature, number, and location of functional groups in the molecule; temperature; pH; dissolved salt concentration; and the presence of other phases (i.e., organic liquids, solids, gases). In general, as molecular size increases, solubility decreases. Polar functional groups (e.g., —OH) tend to increase solubility. Addition of Cl atoms or NO_2 groups in general decreases solubility. As temperature increases, the solubility of solids generally increases as does volatility (vapor pressure and K_H). A detailed discussion of the thermodynamics of solubility is beyond the scope of this text. The interested reader is referred to the Schwarzenbach et al. text listed at the end of this chapter. Values of solubility for some environmentally significant compounds are given in Table 5.15. Caution should be exercised when using literature values for solubility, since methods and conditions used to determine them vary widely.

A word about the term *hydrophobicity* is warranted. The octanol-water partition coefficient (K_{ow}—described in Sorption/Partitioning) is often used to describe hydrophobicity. However, since the term hydrophobic refers to "water hating," aqueous solubility is a truer measure of the hydrophobicity of a compound (how much it "hates" water). In general, compounds with high K_{ow} have low aqueous solubility. For example, DDT has a K_{ow} of $10^{6.91}$ and an aqueous solubility of 0.0055 mg/L (at 20°C). In some cases, however, compounds with low solubility do not have particularly high values of K_{ow}. For example, the explosive HMX (Sec. 5.17) has a solubility of 5 mg/L at 20°C and its K_{ow} is $10^{0.20}$.

Sorption/Partitioning

Adsorption was discussed in Sec. 3.12 with a primary emphasis on its use in engineered reactors. Adsorption of metals on surfaces was discussed in Sec. 4.9. Sorption is also an important process affecting the fate and transport of organic compounds in surface waters and groundwaters. The general term sorption is often used for the natural process rather than adsorption or absorption because the exact manner in which partitioning to solids occurs is often not known. Partitioning of an organic compound between solids and water (e.g., aquifer solids in groundwater, particulates and sediments in surface water systems) can be understood and predicted to some degree using physicochemical properties of organic compounds such as the relative partitioning between the liquid solvent n-octanol and water (octanol-water partition coefficient—K_{ow}) and water solubility. K_{ow} has long been used as a surrogate for living tissue and natural organic matter. For example, in the pharmaceutical industry, the K_{ow} of a drug is often correlated to its activity (higher K_{ow}, higher activity). Bioconcentration of organic compounds in fatty tissues of living organisms has been correlated to K_{ow} (Sec. 5.35). Values of K_{ow} for some environmentally significant compounds are given in Table 5.15. As with K_H and solubility, values of K_{ow} should be used with some caution, since methods and conditions used in determining them vary widely.

Sorption in most surface water and groundwater systems can be described by the Freundlich equation (Sec. 3.12). In some situations, concentration of the organic is low and the number of surface sites is quite high, so the value of n approaches 1 and a linear isotherm results (Fig. 5.2). In this case, K [Eq. (3.84)] is commonly called the partition coefficient K_p. The units of K_p are typically mg organic sorbed/kg of solid divided by mg organic/liter of water, or liter of water/kg of solid.

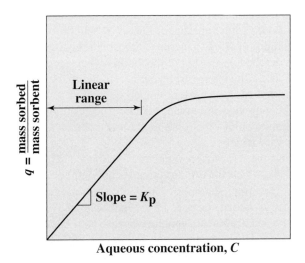

Figure 5.2
Linear isotherm for sorption of organics.

It has been found that sorption of hydrophobic organic compounds is often related to the organic content of the solids, indicating that the process is an organic-organic partitioning, or absorption, rather than true adsorption.[28] Thus, partitioning would be expected to be a function of the nature of the natural organic matter present. An organic-carbon-normalized partition coefficient K_{oc} is sometimes calculated by dividing K_p by the mass fraction of organic carbon in the solids. It should be noted that in some cases, equilibrium between solute and sorbent may take days, months, or even years, and may be more complicated than the linear isotherm given here.[29,30] Desorption (partitioning back into the water) can be a very slow process. For hydrophobic compounds in contact with environmental solids for extended periods of time (sometimes called "aging"), sorption may be for all intents and purposes, irreversible. This behavior significantly affects our ability to remediate such soils.

Values for K_{oc} have been estimated from K_{ow} using the following relationship developed for a series of polycyclic aromatics and chlorinated hydrocarbons.[31]

$$K_{oc} = 0.63(K_{ow}) \tag{5.59}$$

where K_{oc} is the organic-carbon-normalized partition coefficient in liters per kilogram of organic carbon. Here K_{ow} serves as a surrogate for natural organic matter. Values for K_{oc} have also been related to water solubility.[32]

$$\log K_{oc} = -0.54 \log S + 0.44 \tag{5.60}$$

where S is the water solubility of the compound expressed as mole fraction. For the compounds tested, use of K_{ow} provided a better estimate of K_{oc} than S.

One of the convenient uses of these concepts is in the understanding of retardation of groundwater contaminants (movement of compounds relative to movement of water) due to sorption. The following equation has been used.[33]

$$t_r = 1 + \rho K_p/e \tag{5.61}$$

where t_r = retardation factor or ratio of water movement rate to chemical movement rate

   p = bulk density of soil, kg/L

   K_p = partition coefficient, L/kg

   e = void fraction or porosity of soil

[28]S. W. Karickhoff, D. S. Brown, and T. A. Scott, Sorption of Hydrophobic Pollutants on Natural Sediments, *Water Research,* **13:** 241–248 (1979).

[29]W. J. Weber, Jr., P. M. McGinley, and L. E. Katz, A Distributed Reactivity Model for Sorption by Soils and Sediments. 1. Conceptual Basis and Equilibrium Assessments, *Env. Sci. Tech.,* **26:** 1955–1962 (1992).

[30]R. P. Schwarzenbach, P. M. Gschwend, and D. M. Imboden, "Environmental Organic Chemistry," Wiley-Interscience, New York, 1993.

[31]Karickhoff et al., Sorption of Hydrophobic Pollutants on Natural Sediments.

[32]Ibid.

[33]P. V. Roberts, M. Reinhard, and P. L. McCarty, Organic Contaminant Behavior During Groundwater Recharge, *J. Water Pollution Control Fed.,* **52:** 161–172 (1980).

EXAMPLE 5.3

Compare the retardation of chloroform, chlorobenzene, and DDT for the following groundwater conditions:

$\rho = 2$ kg/L
f_{oc} = fraction organic carbon = 0.01 (soil contains 1 percent organic carbon)
$e = 0.20$

The following equations are used to solve this problem:

$$K_p = K_{oc}f_{oc}$$
$$K_{oc} = 0.63(K_{ow})$$
$$t_r = 1 + \rho K_p/e$$

Using the data from Table 5.15:

Compound	log K_{ow}	K_{ow}	K_{oc}	K_p	t_r
Chloroform	1.97	93.3	58.8	0.588	6.88
Chlorobenzene	2.84	692	436	4.36	44.6
DDT	6.91	8.13×10^6	5.12×10^6	5.12×10^4	5.12×10^5

These calculations indicate that chloroform will move slower than water by a factor of 7, chlorobenzene by a factor of 45, and DDT by a factor of about 500,000. Of these three compounds, chloroform is thus expected to move much more rapidly with groundwater flow whereas DDT is not likely to move far from its source.

Transformation Reactions

Transformation reactions important in the fate and removal of organics are listed in Table 5.14 and include photolysis, hydrolysis, oxidation, reduction, and biotransformation. Hydrolysis, oxidation, and reduction can be chemical (abiotic) or mediated by microorganisms (biotic). An additional transformation reaction important with halogenated organics is elimination (dehydrohalogenation). In many cases, microbially mediated transformation reactions are much more rapid than abiotic reactions. Photochemical transformations are sometimes quite rapid. Biotransformation reactions are discussed in more detail in Chap. 6.

Photochemical Transformations   Photochemical transformations are important fate processes for organics in the near-surface aquatic environment as well as in the upper atmosphere. Enhanced photochemical processes are also being used for the treatment of some hazardous wastes. Inorganic pollutants can also be transformed by photochemical reactions; the best known such case is the production of photochemical smog. There are four major photochemical reactions of environmental significance: direct photolysis, indirect photolysis, oxidation, and free-radical oxidation. In direct photolysis, the organic absorbs light energy (photons) and is converted into an excited state which then releases this energy in conjunction with conversion into a product (different) compound. In indirect photolysis, a nontarget compound (for example, dissolved organic material such as humic substances) absorbs the photons and becomes excited. This energized molecule then transmits its energy to the pollutant (target organic) causing it to be transformed.

In oxidation and free-radical oxidation, light energy is typically absorbed by an intermediate compound such as dissolved organic matter, nitrate, or Fe(III), with the resultant production of oxidants such as H_2O_2 and O_3, free radicals such as hydroxyl (·OH) and peroxy (·OOR), and/or other reactive species such as singlet oxygen $^1O_2$. These oxidants are then available to oxidize a wide variety of organics. Chlorinated compounds are particularly susceptible to oxidation by these species. Mixtures of ultraviolet light and ozone or hydrogen peroxide are used in engineered systems for treatment of trace levels of organics in gases and water, and rely on production of these activated chemical species. As described in Sec. 5.8, free-radical reactions are involved in the depletion of the ozone layer.

Hydrolysis and Other Nucleophilic Substitutions   Hydrolysis reactions are nucleophilic substitution reactions where water acts as a nucleophile and attacks an organic bond. *Nucleophiles* are electron-rich ions (H_2O is an exception) that are often called "nucleus-liking, positive-charge-liking" species. They typically like to attack saturated bonds. Other common nucleophiles include OH^-, NO_3^-, SO_4^{2-}, HS^-, HCO_3^-, and HPO_4^{2-}. There is no change in oxidation state of the organic compound during these transformations. In aquatic systems, the most common reactions are those in which water is added, and these reactions may be abiotic or biotic. Abiotic reactions are a strong function of pH. Organic compounds susceptible to hydrolysis and nucleophilic substitution include alkyl halides (halogenated aliphatics), amides, amines, carbamates, epoxides, esters, nitriles, and phosphoric acid esters. A few examples, mostly with H_2O serving as the nucleophile, are given here.

1. Alkyl halides (X = halogen)

$$R—CH_2X + H_2O \rightarrow R—CH_2OH + H^+ + X^-$$
$$R—CH_2X + HS^- \rightarrow R—CH_2SH + X^-$$

Iron sulfide minerals [e.g., mackinawite, typically abbreviated FeS(s)] can also facilitate nucleophilic substitution reactions for halogenated organic compounds. They can also bring about reductive dehalogenation and other electron-transferring processes. Halogenated aromatics (aryl halides) may also undergo hydrolysis. For example,

Chlorobenzene   Phenol

2. Esters

Methyl acetate   Acetic acid   Methanol

3.   Amides

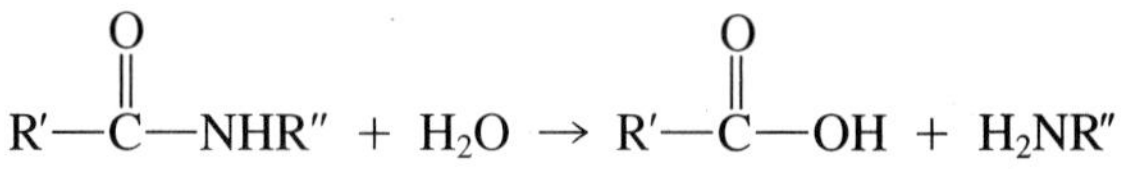

$$R'-\overset{\overset{\displaystyle O}{\|}}{C}-NHR'' + H_2O \rightarrow R'-\overset{\overset{\displaystyle O}{\|}}{C}-OH + H_2NR''$$

Amide   Acid   Amine

4.   Carbamates are converted into CO_2, an amine, and other compounds (for example, alcohols).

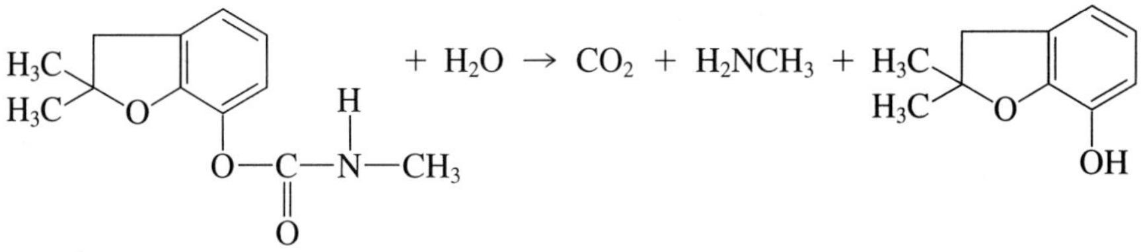

$$+ H_2O \rightarrow CO_2 + H_2NCH_3 +$$

Furadan, a pesticide   Amine

5.   Phosphoric esters

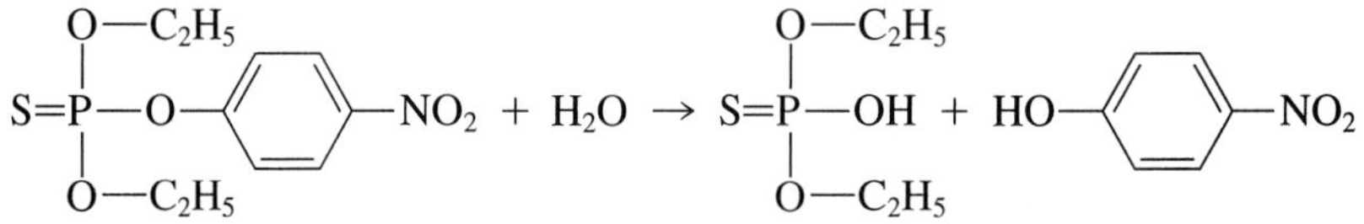

Parathion

Nucleophilic substitution reactions are typically expressed as second-order or pseudo-first-order reactions:

$$\text{Rate} = -k_h[C][N] \qquad (5.62)$$

where k_h = hydrolysis rate coefficient
            $[C]$ = concentration of organic
            $[N]$ = concentration of nucleophile

The minus sign represents the disappearance of the organic compound. When water is the nucleophile, or when the concentration of the nucleophile is constant or unknown, a pseudo-first-order expression may be used:

$$\text{Rate} = -k_h[C] \qquad (5.63)$$

Hydrolysis reactions can be acid or base catalyzed, in which case the rate expressions can be a bit more complicated:

$$\text{Rate} = -(k_a[H^+] + k_n + k_b[OH^-])[C] \qquad (5.64)$$

where k_a = acid-catalyzed rate coefficient
            k_n = neutral rate coefficient
            k_b = base-catalyzed rate coefficient

The overall hydrolysis rate coefficient can then be expressed as

$$k_h = k_a[H^+] + k_n + k_b[OH^-] \qquad (5.65)$$

Care must be taken to ensure that the units used in these rate expressions are consistent.

The concept of half-life is often used to describe the persistence of an organic in the environment. Half-life ($t_{1/2}$) was defined in Sec. 3.10 as $0.693/k$, where k is a first-order rate coefficient. With hydrolysis, k is represented by k_h or $k_h[N]$. Hydrolysis reactions may be quite slow (half-lives of years) or quite rapid (half-lives of seconds) depending on the organic and the environmental conditions.

EXAMPLE 5.4

The following are hydrolysis rate coefficients reported by Jeffers and co-workers.[34]

Compound	k_h, min^{-1}
Chloroform	7.13×10^{-10}
Carbon tetrachloride	3.26×10^{-8}
1,1,1-Trichloroethane	1.24×10^{-6}
Pentachloroethane	1.31×10^{-4}
Trichloroethene	1.07×10^{-12}
Tetrachloroethene	1.37×10^{-15}

From these data, calculate the half-lives, in years, of these compounds in the environment assuming that no other transformation reactions occur.

$$\text{Half-life} = t_{1/2} = \frac{\ln 2}{k_h} = \frac{0.693}{(k_h)(60)(24)(365)}$$

Compound	$t_{1/2}$, years
Chloroform	1849
Carbon tetrachloride	40
1,1,1-Trichloroethane	1.1
Pentachloroethane	0.010 (3.7 days)
Trichloroethene	1.23×10^6
Tetrachloroethene	9.62×10^8

These data indicate that chlorinated aliphatics are in general very slowly hydrolyzed. Other data indicate that the environmental half-lives of these compounds are sometimes much shorter, indicating that other transformation processes, perhaps biotic, are more important in controlling their fate.

---

[34]P. Jeffers, L. Ward, L. Woytowitch, and L. Wolfe, Homogeneous Hydrolysis Rate Constants for Selected Chlorinated Methanes, Ethanes, Ethenes, and Propanes. *Env. Sci. Techn.,* **23:** 965–969 (1989).

Elimination (Dehydrohalogenation)   Elimination reactions may occur abiotically or be microbially mediated. Although these reactions have been shown to occur with many organics, the ones of most importance to environmental engineers and scientists are those with halogenated organics. The most common of these reactions are characterized by the release of HX (dehydrohalogenation) and the formation of a double bond.

$$Cl\text{—}\underset{\underset{Cl}{|}}{\overset{\overset{Cl}{|}}{C}}\text{—}\underset{\underset{H}{|}}{\overset{\overset{H}{|}}{C}}\text{—}H \rightarrow Cl\text{—}\overset{\overset{Cl}{|}}{C}=\overset{\overset{H}{|}}{C}\text{—}H + HCl$$

1,1,1-Trichloroethane   1,1-Dichloroethene

DDT   DDE

Oxidation   Oxidation occurs when an organic loses (donates) one or more electrons to an oxidizing agent (electron acceptor). Abiotic oxidation reactions are typically very slow or nonexistent under environmental conditions. The most common natural oxidants are molecular oxygen, Fe(III), and Mn(III/IV). These are to be contrasted with the potentially rapid oxidations that occur with photochemically produced oxidants such as H_2O_2, O_3, the free radicals such as hydroxyl ($\cdot OH$) and peroxy ($\cdot OOR$), and other reactive species such as singlet oxygen $^1O_2$. Strong oxidizing agents such as H_2O_2, O_3, MnO_4^- (permanganate), and various chlorine species (e.g., Cl_2, ClO_2) can oxidize organics during water and wastewater treatment. Biological oxidations will be covered in more detail in Chap. 6.

Rates of abiotic oxidation reactions are often described using a second-order rate expression:

$$\text{Rate} = -k_o[C][Ox] \tag{5.66}$$

where k_o = oxidation rate coefficient
            $[C]$ = concentration of organic
            $[Ox]$ = concentration of oxidant

Reduction   Reduction occurs when an organic gains (accepts) one or more electrons from a reducing agent (electron donor). Highly oxidized organics (e.g., highly chlorinated organics and nitroaromatics) are susceptible to reduction. Abiotic reduction reactions are quite common, especially under reducing conditions (absence of oxygen). In many cases it is difficult to distinguish between truly abiotic reductions and those mediated by microorganisms. For example, HS^- can take part in nucleophilic substitution reactions (see Hydrolysis and Other Nucleophilic Substitutions) and can be involved in reduction reactions where it is oxidized to elemental sulfur

(S(s)) with the reduction of the organic. In anaerobic environments HS^- is present primarily due to microbial reduction of sulfate; without the microorganisms, the reduction would not take place. Reducing agents found in the environment include HS^-; Fe(II) compounds such as iron sulfides, iron carbonates; iron oxides; iron and other porphyrins; and other reduced organic compounds (natural organic matter).

Rates of abiotic reduction reactions are often described using a second-order rate expression:

$$\text{Rate} = -k_r[C][Red] \qquad (5.67)$$

where

$$k_r = \text{reduction rate coefficient}$$
$$[C] = \text{concentration of organic}$$
$$[Red] = \text{concentration of reductant.}$$

Many types of organics are susceptible to abiotic or biotic reduction. A few examples are given here.

1. Reductive dehalogenation.

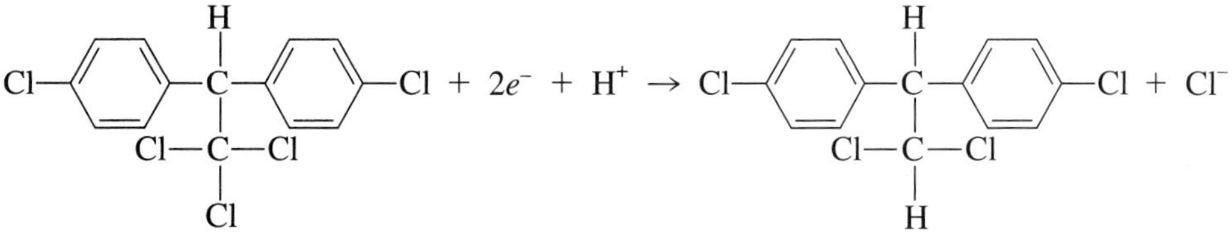

DDT → DDD

The biological reductive dehalogenation of tetrachloroethene (PCE) and trichloroethene (TCE) illustrate the potential for transformation reactions to result in more problematic products (vinyl chloride in this case). These reductions proceed with an overall transfer of $2e^-$ per molecule and replacement of one Cl atom with a H atom:

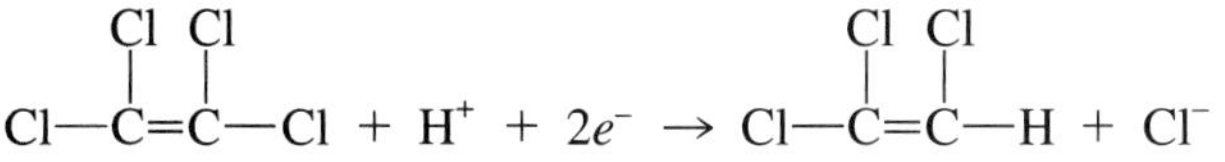

Tetrachloroethene → Trichloroethene

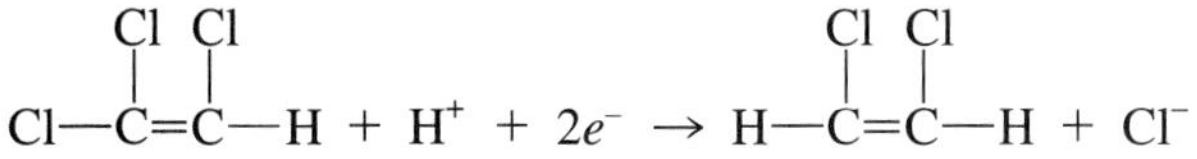

Trichloroethene → cis-1,2-Dichloroethene

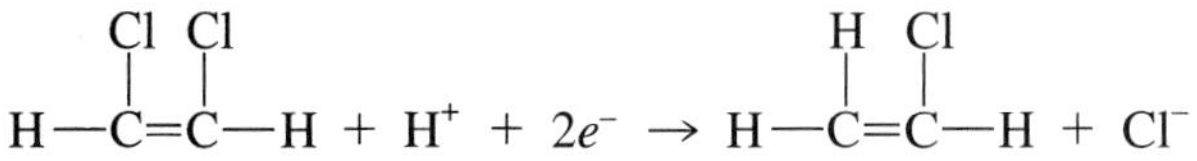

cis-1,2-Dichloroethene → Chloroethene (vinyl chloride)

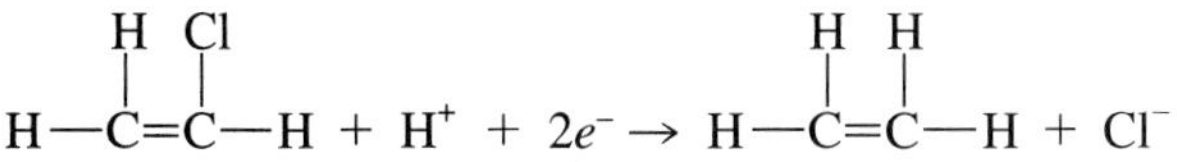

Vinyl chloride → Ethene

2. Reductive elimination

$$Cl-\underset{\underset{Cl}{|}}{\overset{\overset{Cl}{|}}{C}}=\underset{\underset{Cl}{|}}{\overset{\overset{Cl}{|}}{C}}-Cl + 2e^- \rightarrow Cl-\overset{\overset{Cl}{|}}{C}=\overset{\overset{Cl}{|}}{C}-Cl + 2Cl^-$$

Hexachloroethane   Tetrachloroethene

Treatment of contaminated groundwaters using *permeable reactive barriers* (PRB) containing iron metal (Fe(0)) is based on reduction. As Fe(0) is oxidized (corrodes), e^- are released which can be used to reduce highly oxidized pollutants:

$$Fe^0 \rightarrow Fe^{2+} + 2e^-$$

These Fe(0)-PRB have been used to treat groundwaters contaminated with chlorinated aliphatics such as perchloroethene and trichloroethene and highly oxidized metals such as Cr(VI). In the case of Cr(VI), the less mobile and less toxic trivalent form is produced. Example stoichiometries are given here for trichloroethene ($CCl_2 = CHCl$) reduction to ethene ($CH_2 = CH_2$) and chromate ($HCrO_4^-$) reduction to trivalent Cr ($Cr(OH)_3(s)$):

$$3Fe^0 \rightarrow 3Fe^{2+} + 6e^-$$
$$CCl_2 = CHCl + 3H^+ + 6e^- \rightarrow CH_2 = CH_2 + 3Cl^-$$
$$\overline{}$$
$$3Fe^0 + CCl_2 = CHCl + 3H^+ \rightarrow 3Fe^{2+} + CH_2 = CH_2 + 3Cl^-$$

$$1.5Fe^0 \rightarrow 1.5Fe^{2+} + 3e^-$$
$$HCrO_4^- + 4H^+ + 3e^- \rightarrow Cr(OH)_3(s) + H_2O$$
$$\overline{}$$
$$1.5Fe^0 + HCrO_4^- + 4H^+ \rightarrow 1.5Fe^{2+} + Cr(OH)_3(s) + H_2O$$

As discussed in Chap. 6, the electrons necessary for accomplishing reduction can also be provided by microbial oxidation of a wide variety of electron donors.

3. Reduction of nitro groups to the corresponding amino groups.

$$NO_2 \qquad \qquad NH_2$$
$$+ 6e^- + 6H^+ \rightarrow \qquad + 2H_2O$$

Nitrobenzene   Aniline

4. Reduction of disulfides to mercaptans.

$$R'-S-S-R'' + 2e^- + 2H^+ \longrightarrow R'-SH + R''-SH$$

5.35 | STRUCTURE- AND PROPERTY-ACTIVITY RELATIONSHIPS

There is considerable interest in developing methods for predicting the properties (e.g., solubility) and activities (e.g., fate, toxicity) of chemicals, especially organics, in the environment and in engineered systems. Physical, chemical, and structural (molecular) characteristics of organic compounds are used in this regard, and correlations have been developed for a wide variety of structures, properties, and activities. For the purposes of this discussion, *structure* refers to molecular characteristics. *Activity* refers to a biological effect such as toxicity, biotransformation, or bioconcentration (to be distinguished from the definition of activity given in Sec. 2.12). It may also refer to the potential for transformation of organic compounds via abiotic reactions. *Property* is a more all-encompassing term that includes such environmentally significant characteristics as solubility, volatility, and liquid-liquid or solid-liquid partitioning. It is the purpose of this section to introduce the student to the nature and usefulness of these relationships for understanding and predicting the behavior of organic compounds. The reader should be aware that the definition of the terms used herein may differ slightly with those used by others in this evolving field.

Relationships that have been developed include *structure-activity, property-activity, structure-property,* and *property-property* relationships. All are commonly referred to in the literature as *quantitative structure-activity relationships (QSAR).* The general goal of any QSAR is to quantitatively relate, using a variety of statistical correlation techniques, a property or activity of a given set of chemicals to characteristics (structure or property), sometimes called descriptors, of these chemicals. QSARs were initially used in drug design and formulation of pesticides, and later were extended for use in environmental toxicology. QSARs are of current interest because they can help understand and predict the impact of organic chemicals for which there is very little environmental data. It has been estimated that there are about 70,000 synthetic chemicals currently being used, with 500–1000 new ones being added each year. Data about environmental fate and effects exist for only about 20% of these chemicals.[35] Although QSARs are empirical relationships, they do provide important information that can help interpret causes and mechanisms involved. Some caution must be exercised when attempting to use QSARs for compounds with properties (e.g., chemical structure) that are significantly different from the compounds used to develop the QSAR.

Relationships that are developed using basic structural (molecular) characteristics such as molecular connectivity, molar surface area, intrinsic molar volume, atomic charge, and others have some advantages in that they can be determined without need for experimental data. However, the environmental engineer or scientist must have access to the structural characteristics and calculation methods necessary to determine values for these parameters. Some of this infor-

---

[35]S. Postel, in "State of the World 1987," L. R. Brown, et al., eds., Worldwatch Institute Report, W. W. Norton, New York, 1987.

mation is currently available in the literature, and more is becoming increasingly available in software packages that can be purchased.[36] Detailed discussion of these parameters and the methods used to determine them is beyond the scope of this text.

Prediction of Properties of Organic Chemicals

Relationships have been developed that allow for prediction of environmentally significant properties such as solubility, K_H, K_{ow}, and K_{oc}. An example of a structure-property relationship for solubility is[37]

$$\log S = 1.543 + 1.638^{\circ}\chi - 1.374^{\circ}\chi^{v} + 1.003\overline{\Phi} \tag{5.68}$$

where S = aqueous solubility, mol/L

$^{\circ}\chi$ = zero-order, simple molecular connectivity index

$^{\circ}\chi^{v}$ = valence molecular connectivity index

Φ' = modified polarizability parameter.

Values for $^{\circ}\chi$, $^{\circ}\chi^{v}$, and Φ' can be calculated from molecular structure without need for experimental data. Table 5.16 contains a sample of the type of data used to calculate solubility using this relationship. The relationship in Eq. (5.68) was developed from a data set of 145 different organic chemicals, including alcohols, halogenated aliphatics, and aromatics.

Many *property-property* relationships have been developed. Equation (5.59), relating K_{oc} to K_{ow}, and Eq. (5.60), relating K_{oc} to S, are examples. An additional example relates K_{ow} to S (mol/L) for a set of 36 different organic compounds.[38]

$$\log K_{ow} = -0.862 \log S + 0.710 \tag{5.69}$$

Table 5.16 | $^{\circ}\chi$, $^{\circ}\chi^{v}$, and Φ' values for selected organic compounds

Compound	$^{\circ}\chi$	$^{\circ}\chi^{v}$	Φ'
Phenol	4.38	3.83	−4.11
2-Chlorophenol	5.30	4.89	−4.71
Pentachlorophenol	9.00	9.46	−7.12
Naphthalene	5.61	5.61	−6.72
Benzo(a)pyrene	10.92	10.92	−12.00

Source: Data taken from N. N. Nirmalakhandan and R. E. Speece, Prediction of Aqueous Solubility of Organic Chemicals Based on Molecular Structure. 2. Application to PNAs, PCBs, PCDDs, etc., *Env. Sci. Tech.,* **23:** 708–713 (1989).

---

[36]M. Reinhard and A. Drefahl, "Handbook for Estimating Physiochemical Properties of Organic Compounds," John Wiley & Sons, Inc., New York, 1999.

[37]N. N. Nirmalakhandan and R. E. Speece, Prediction of Aqueous Solubility of Organic Chemicals Based on Molecular Structure. 2. Application to PNAs, PCBs, PCDDs, etc., *Env. Sci. Tech.,* **23:** 708–713 (1989).

[38]C. T. Chiou, Partition Coefficient and Water Solubility in Environmental Chemistry, in "Assessment of Chemicals," vol. 1, Academic Press, pp. 117–153, 1981.

Prediction of Activities of Organic Chemicals

Activity in the context of QSARs usually refers to a biological effect such as toxicity, biotransformation, or bioconcentration. Many such relationships have been developed;[39] a few are described here.

It has long been recognized that bioaccumulation or bioconcentration (roughly defined as the ratio of the concentration of the organic compound in a target organism to the concentration of chemical in water) can be related to K_{ow}. This is expected since bioaccumulation is thought to be a partitioning process. An example of such a property-activity relationship is[40]

$$\log \text{BF} = \log K_{ow} - 1.32 \tag{5.70}$$

where BF is the bioconcentration factor. Equation (5.70) was developed from a variety of aquatic organisms (fish, algae, and mussels) and a data set of 51 organic chemicals.

Attempts have been made to develop structure-activity relationships, such as biodegradability versus chemical structure. An example is[41]

$$\text{BOD} = 1015(|\Delta\delta|_{x-y}) + 1.193 \tag{5.71}$$

where BOD is the percentage of theoretical BOD achieved in 5 days and $|\Delta\delta|_{x-y}$ is the modulus of atomic charge differences across common bonds (e.g., C—O). This relationship was developed from 79 organic compounds, including amines, phenols, aldehydes, carboxylic acids, halogenated hydrocarbons, and amino acids.

One of the most promising uses of QSARs is in the understanding and prediction of toxicity of organic chemicals to organisms of environmental significance. Recent studies have related the toxicity of 50 to 130 organic chemicals to a variety of bacteria. Examples of *structure-activity* and *property-activity* relationships, respectively, for aerobic, heterotrophic bacteria are given here.[42]

$$\log \text{IC}_{50} = 5.24 - 4.15(V_i/100) + 3.71\beta_m - 0.41\alpha_m \tag{5.72}$$

$$\log \text{IC}_{50} = 5.12 - 0.76 \log K_{ow} \tag{5.73}$$

where
IC_{50} = concentration (mg/L) at which the biological reaction rate is reduced by 50 percent

V_i = intrinsic molar volume

β_m and α_m = measures of ability to participate in hydrogen bonding as a hydrogen donor or acceptor, respectively

V_i, β_m, and α_m are commonly termed linear solvation energy relationship (LSER) parameters. Table 5.17 contains a sample of values of these parameters needed to use Eq. (5.72).

[39]Reinhard and Drefahl, "Handbook for Estimating Physiochemical Properties of Organic Compounds."

[40]D. Mackay, Correlation of Bioconcentration Factors, *Env. Sci. Tech.,* **16:** 274–278 (1982).

[41]J. C. Dearden, and R. M. Nicholson, The Prediction of Biodegradability by the Use of Quantitative Structure-Activity Relationships: Correlation of Biological Oxygen Demand with Atomic Charge Difference, *Pesticide Science,* **17:** 305–310 (1986).

[42]D. J. W. Blum and R. E. Speece, Determining Chemical Toxicity to Aquatic Species. *Env. Sci. Tech.,* **24:** 284–293 (1990).

Table 5.17 | Values of $V_i/100$, β_m, and α_m for selected organic compounds

Compound	$V_i/100$	β_m	α_m
Dichloromethane	0.336	0.05	0.13
Chloroform	0.427	0	0.20
1,1,1-Trichloroethane	0.519	0	0
Trichloroethene	0.492	0.03	0.12
Benzene	0.491	0.14	0
Toluene	0.592	0.14	0
Chlorobenzene	0.581	0.09	0
Phenol	0.536	0.33	0.60

Source: Data taken from D. J. W. Blum, Chemical Toxicity to Environmental
Bacteria: Quantitative Structure Activity Relationships and Interspecies Correlations
and Comparisons, Ph.D. Dissertation, Drexel University, Philadelphia (1989).

One of the best-known property-activity relationships is commonly termed a
linear free-energy relationship (LFER) and correlates reaction rates with equilibrium constants. This is an attempt to relate the kinetics of a chemical reaction to its
thermodynamics. An example is given here for the base-catalyzed hydrolysis of a
series of 16 N-phenylcarbamates.[43]

$$\log k_b = -1.15 pK_A + 13.6 \tag{5.74}$$

where k_b is the first-order hydrolysis rate ($M^{-1}s^{-1}$) and pK_A is the pK_A of the alcohol, or leaving group. Recall that hydrolysis of carbamates yields an amine, an alcohol, and CO_2. Leaving groups included nitrophenolates, chlorophenolates, and halogenated aliphatic compounds, among others. An example is given here for
4-nitrophenyl-N-phenylcarbamate ($k_b = 2.7 \times 10^5\ M^{-1}s^{-1}$):

Leaving group:
p-nitrophenylate

4-Nitrophenyl-*N*-phenylcarbamate

$+\ H^+$ $pK_A = 7.1$

p-nitrophenol   *p*-nitrophenylate

[43]R. P. Schwarzenbach and P. M. Gschwend, Chemical Transformations of Organic Pollutants in the
Aquatic Environment, in "Aquatic Chemical Kinetics," W. Stumm, ed., Wiley-Interscience, New York,
1990.

Another example of a LFER relates the surface-area normalized rate coefficient (k_{SA}, L/m^2-h) for reduction of 48 chlorinated alkanes and alkenes by Fe(0) to calculated estimates of the lowest unoccupied molecular orbital energies (E_{LUMO}, eV) for these compounds.[44]

$$\log k_{SA} = -5.603 - 1.49 E_{LUMO} \qquad (5.75)$$

For example, E_{LUMO} for tetrachloroethene is estimated to be -1.689 eV. Using Eq. (5.75) gives a rate coefficient of 8.2×10^{-4} L/m^2-h. If the surface-area concentration in a Fe(0) permeable reactive barrier were 2 m^2/mL (2000 m^2/L), then the pseudo-first-order rate coefficient would be 1.64 h^{-1}.

PROBLEMS

5.1 Why is it possible to have so many compounds of carbon?

5.2 Define homologous series, homolog, hydrocarbon, and alkyl radical.

5.3 Define isomerism and illustrate with structural formulas of compounds with the molecular formula C_6H_{14}.

5.4 In general, how do organic compounds differ from inorganic compounds?

5.5 How many grams of oxygen are required to furnish just enough oxygen for the complete oxidation of 20 g of butane?

5.6 What is the difference between an aliphatic and an aromatic compound?

5.7 Define with suitable illustrations the terms primary alcohol, secondary alcohol, and tertiary alcohol, and indicate their relative ease of biodegradation.

5.8 Show the structures and names of the intermediate compounds formed in the biological oxidation of *n*-butane to butyric acid.

5.9 Why are oxidation reactions of organic compounds important to environmental engineers and scientists?

5.10 What might be formed from CH_3—CH_2—CH=CH—CH_3 in the atmosphere in the presence of sunlight, oxides of nitrogen, and water droplets?

5.11 Which of the following compounds have optical isomers?

$$CH_2OH-\underset{\underset{OH}{|}}{CH}-CHO \qquad CH_3-\underset{\underset{H}{|}}{C}=\underset{\underset{CH_3}{|}}{C}-H \qquad CH_3-CH_2-\underset{\underset{OH}{|}}{CH}-CH_3$$

$$CHO-\underset{\underset{OH}{|}}{CH}-\underset{\underset{OH}{|}}{CH}-CH_2OH \qquad CH_3-\underset{\underset{CH_3}{|}}{CH}-COOH \qquad CH_3-\underset{\underset{H}{|}}{C}=\underset{\underset{CH_3}{|}}{CH}-CH_3$$

5.12 Show a general formula for fats and oils. What is the difference between a fat and an oil?

5.13 What is the difference between a fat and a wax?

5.14 How many asymmetric carbon atoms are there in glucose?

---

[44]M. M. Scherer, B. A. Balko, D. A. Gallagher, and P. G. Tratnyek, Correlation Analysis of Rate Constants for Dechlorination by Zero-Valent Iron, *Env. Sci. Tech.*, **32:** 3026–3033 (1998).

5.15 Write a possible structural formula for a ketopentose and for an aldopentose.

5.16 What is the major factor contributing to the difference in biodegradability between starch and cellulose?

5.17 Show a general formula for the building blocks of proteins and illustrate how these building blocks are connected to form proteins.

5.18 If a sample of food waste contains 2.5 percent of organic nitrogen, what approximate percentage of the sample is protein?

5.19 What functional group is characteristic of each of the following: alkenes, alcohols, aldehydes, ketones, acids, amines, amides, ethers, esters, and aromatic compounds?

5.20 What monosaccharides are formed upon hydrolysis of the following: cellulose, starch, and hemicellulose?

5.21 Name the three general classes of synthetic detergents, and give an example of each.

5.22 Name the four general classes of synthetic pesticides, and describe how the pesticides in each class differ in degree of biodegradability.

5.23 Explain why pharmaceutically active chemicals (PhACs) are of concern to environmental engineers and scientists. Give two examples of PhACs.

5.24 Explain why endocrine-disrupting chemicals (EDCs) are of concern to environmental engineers and scientists. Give two examples of EDCs.

5.25 Give the structural formula for each of the following:

    (*a*) 3-Nonene

    (*b*) *n*-Octyl phenol

    (*c*) Dimethyl ether

    (*d*) 1,1,2-Trichloroethane

    (*e*) Any PCB

    (*f*) *n*-Pentanoic acid

    (*g*) Isopropanol

    (*h*) 2,4,6-Trinitrotoluene (TNT)

5.26 Give the structural formula for the following:

    (*a*) 1,2-Dichloroethene

    (*b*) Diethyl ketone

    (*c*) Pentachlorophenol

    (*d*) Butyric acid

    (*e*) *m*-Chlorophenol

    (*f*) A poly-cyclic aromatic hydrocarbon (PAH)

    (*g*) 3-Pentanol

    (*h*) 1-Octene

5.27 Given the following structural formulas, name (IUPAC or common) the compound.

    (*a*) $CH_3CHOH\ CH_2CH_3$

    (*b*)

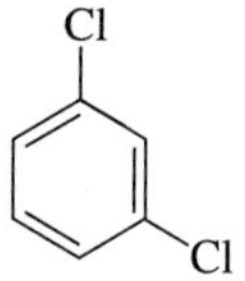

(c) $CHCl_2CH_2Br$

(d) $CH_3OCH_2CH_3$

(e) CH_3CHO

(f) CCl_2CHCl

(g) $CH_3CH_2CH_2COCH_2CH_3$

(h)

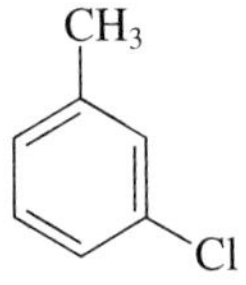

(i)

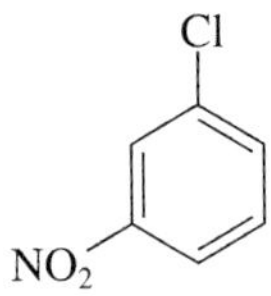

5.28 Name the following:

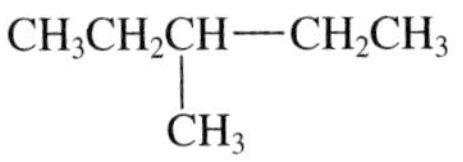

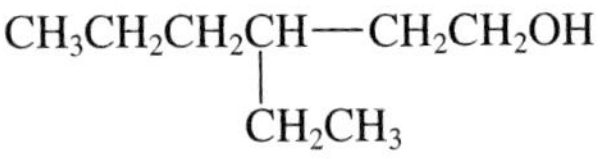

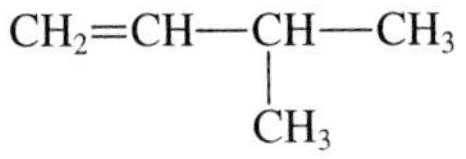

$CH_3CH_2CH{-}CH_2CH_3$ $CH_3CH_2CH_2CH{-}CH_2CH_2OH$

       $|$ $|$

      CH_3 CH_2CH_3

$CH_2{=}CH{-}CH{-}CH_3$ $CH_3CH_2CH_2COOH$

             $|$

            CH_3

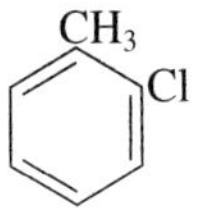

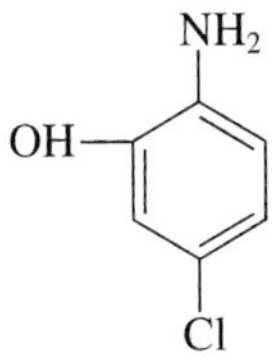

5.29 A contaminated groundwater contains 50 $\mu g/L$ chlorobenzene, 25 $\mu g/L$ dichloromethane, and 85 $\mu g/L$ ethylbenzene. Using the simplified equilibrium model developed in Sec. 5.34, estimate the air-to-water ratio required to reduce the concentration of each of these compounds below 5 $\mu g/L$. Assume a temperature of 25°C. If you were to design a stripping tower to remove these contaminants, would you expect your design air-to-water ratio to be larger, smaller, or the same? Why?

    Answer: 48.9; 39.1; 48.3

5.30 Regarding air stripping for the removal of volatile organic compounds, what impact on removal efficiency would an increase in the following parameters have?

(a) H

(b) V_a/V_w

(c) K_La

(d) Temperature

5.31 Rank the following compounds from most volatile (most likely to partition into the atmosphere) to least volatile. Include appropriate data and reference the data to substantiate your ranking.

(a) Benzene

(b) Phenol

(c) Chloroform

(d) Bromoform

(e) Trichloroethene

(f) 1,1,1-Trichloroethane

(g) Tetrachloroethene (also called perchloroethene)

5.32 An industry has dumped a total of 100 pounds of trichloroethene (TCE) into a well-capped landfill (that is, assume the landfill is closed to the atmosphere). This chemical landfill is 100 feet long by 100 feet wide and 10 feet deep. The void volume is 30 percent and 70 percent of the voids are filled with water. Assuming that aqueous solubility is not exceeded and that there is no transformation or sorptive removal, estimate the concentration of TCE in the aqueous phase. Assume a temperature of 25°C. [Think about how complex this problem would be if biodegradation (vinyl chloride can be produced under anaerobic conditions!) and sorption occurred, and if the landfill were not capped and/or a substantial amount of methane were produced and vented to the atmosphere.]

  Answer: 63 mg/L

5.33 A wastewater is saturated with trichloroethene giving a concentration of 1100 mg/L. One million liters of this waste is put into a *closed* holding tank (total tank volume = 2 million liters). What percent removal of TCE *from the liquid* would occur from this action? Assume a temperature of 25°C.

  Answer: 32 percent

5.34 Using values from Table 5.15, estimate K_{oc} using octanol-water partition coefficients and water solubilities for the following compounds:

(a) Chloroform

(b) 1,1,1-Trichloroethane

(c) Benzene

(d) Chlorobenzene

(e) Phenol

(f) Pentachlorophenol

(g) Benzo(a)pyrene

(h) Atrazine

5.35 Using data from Table 5.15, determine which of the following will move faster with groundwater. By what degree? Show all calculations necessary to justify your answer. Assume the aquifer has a void fraction of 0.20, a bulk density of 2 kg/L, and contains 1 percent organic carbon.

(a) Benzene

(b) Toluene

(c) Trichloroethene

(d) 1,1,1-Trichloroethane

    (*e*) Acrolein

    (*f*) Naphthalene

        Answer: acrolein is fastest; naphthalene slowest

5.36 Assuming a temperature of 25°C, estimate the degree to which tetrachloroethene (PCE) will be retarded in a groundwater aquifer with

    (*a*) A void fraction of 0.30, a bulk density of 2000 kg/m^3, and an organic carbon content of 0.25 percent.

    (*b*) A void fraction of 0.30, a bulk density of 2000 kg/m^3, and an organic carbon content of 1.0 percent.

5.37 Which parameter, H, water solubility, or K_{ow}, would you expect to be a better predictor for whether a compound will sorb strongly to particulate material in soils and sediments? Why?

5.38 The organic compounds listed here are contained in a combined domestic-industrial wastewater. Rank the compounds in increasing order of susceptibility to removal by settling of wastewater suspended solids during primary sedimentation. Provide suitable justification for your rankings.

    (*a*) Alachlor

    (*b*) Phenol

    (*c*) Pentachlorophenol

    (*d*) Acrolein

    (*e*) Carbon tetrachloride

    (*f*) Nitrobenzene

    (*g*) Ethanol

        Answer: pentachlorophenol most susceptible

5.39 Groundwater contamination problems have been caused because of underground storage tanks leaking gasoline. In most situations the compounds causing the most concern are MTBE and the BTEX compounds. Using data contained in Table 5.15, answer the following.

    (*a*) Which compound would you expect to move fastest in the groundwater? Why?

    (*b*) Which compound would be easiest to remove by air stripping? Why?

    (*c*) Which compound would be easiest to remove by adsorption? Why?

5.40 The following are types of transformation reactions. Give an adequate definition of each by writing an example reaction with a specific organic for each. Do *not* use the example reactions listed in the text, generate your own.

    (*a*) Hydrolysis

    (*b*) Elimination

    (*c*) Oxidation

    (*d*) Reduction

    (*e*) Nucleophilic substitution

5.41 Using example reactions, explain the difference between dehydrohalogenation, reductive dehalogenation, and reductive elimination.

5.42 The pesticides shown here are subject to transformation in the environment. Give the class of pesticide and suggest a transformation reaction that it may possibly undergo in the environment. Balance the reaction as best you can and tell under what conditions it might occur (i.e., aerobic, anaerobic).

(*a*)

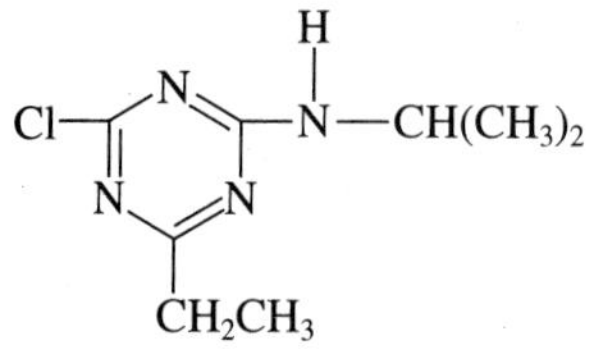

(*b*)

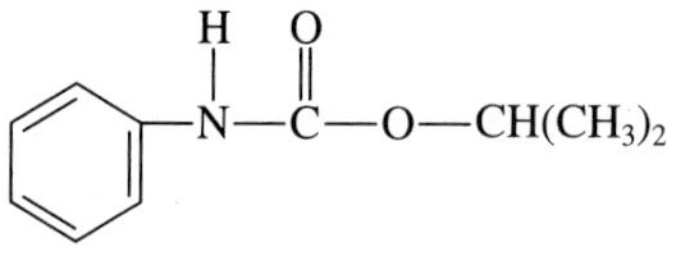

(*c*)

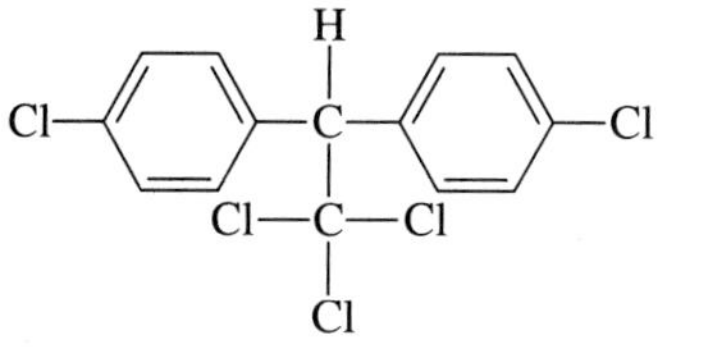

(*d*)

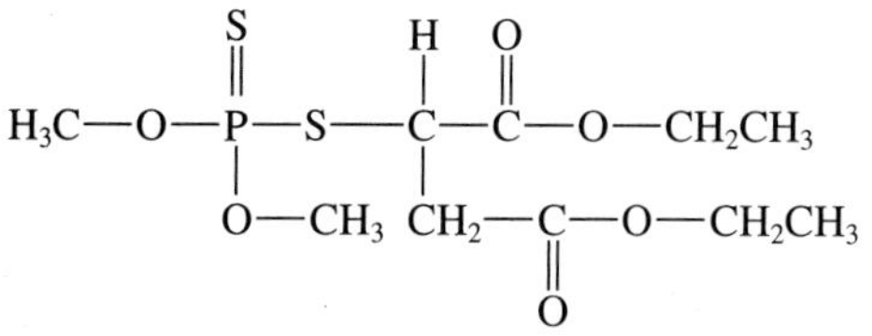

5.43 The following are pseudo-first-order rate coefficients for the pesticide parathion:

$k_p = 0.10 \text{ day}^{-1}$ photolysis
$k_h = 0.028 \text{ day}^{-1}$ hydrolysis
$k_o = 0.005 \text{ day}^{-1}$ oxidation

Calculate the environmental half-life for each of the transformation reaction types and the overall half-life of parathion. (*Note:* Rate coefficients for this problem were provided by Prof. J. L. Schnoor of the University of Iowa.)
    Answer: 6.9 d; 24.8 d; 139 d; 5.2 d

5.44 A groundwater plume contains trichloroethene (TCE). TCE is reduced by iron metal (Fe(0)) following first-order kinetics with respect to TCE concentration. A permeable reactive barrier (PRB) containing Fe(0) is proposed to intercept and treat the contaminated groundwater. The following information is to be used to solve this problem.

TCE concentration entering the PRB $= 100 \ \mu g/L$
$k_{TCE} = 0.002 \text{ h}^{-1}$

For the PRB,

$f_{oc} = 0.01$ $\rho = 2.8 \text{ kg/L}$ porosity $= 0.45$

Linear groundwater velocity $= 0.3$ ft/d

Since the PRB contains organic carbon, TCE will be retarded as it passes through the PRB as well as reduced by the Fe(0). Will a 3-foot-wide PRB be adequate to reduce the concentration of TCE below its drinking water MCL (5 μg/L)? Assume a temperature of 25°C. *Hint: Calculate the "retarded" velocity of TCE and note that TCE removal by reduction is described by first-order kinetics.* (Note: This problem is courtesy of Prof. Michelle Scherer of the University of Iowa.)

5.45 Using Eq. (5.68), and assuming a temperature of 25°C, estimate the solubility of the compounds listed in Table 5.16. Compare these values with those given in Table 5.15.

5.46 Using Eq. (5.69) and K_{ow} data in Table 5.15, estimate the solubility of the compounds listed in Table 5.16. Compare these values with your answers for Prob. 5.45.

5.47 Using K_{ow} data from Table 5.15, rank the following compounds on the basis of lowest to highest degree of bioconcentration by estimating the bioconcentration factor for each.

(*a*) Chloroform

(*b*) Trichloroethene

(*c*) Benzene

(*d*) Toluene

(*e*) Phenol

(*f*) Benzo(a)pyrene

(*g*) Acrolein

(*h*) DDT

5.48 Using K_{ow} data from Table 5.15, rank the following compounds on the basis of least to most toxic to aerobic heterotrophic bacteria by estimating the IC_{50} using Eqs. (5.72) and (5.73).

(*a*) Dichloromethane

(*b*) Chloroform

(*c*) 1,1,1-Trichloroethane

(*d*) Trichloroethene

(*e*) Benzene

(*f*) Toluene

(*g*) Chlorobenzene

(*h*) Phenol
    Answer: least, phenol or dichloromethane

REFERENCES

Brezonik, P. L.: Principles of Linear Free-Energy and Structure-Activity Relationships and Their Applications to the Fate of Chemicals in Aquatic Systems, in "Aquatic Chemical Kinetics," W. Stumm, ed., Wiley-Interscience, New York, 1990.

Carey, F. A.: "Organic Chemistry," 4th ed., McGraw-Hill, New York, 2000.

Loudon, G. C.: "Organic Chemistry," 3rd ed., Benjamin/Cummings, Menlo Park, CA, 1995.

Manahan, S. E.: "Environmental Chemistry," 5th ed., Lewis, Chelsea, MI, 1991.

Morrison, R. T., and R. N. Boyd: "Organic Chemistry," 4th ed., Allyn and Bacon, Boston, 1983.

Reinhard, M., and A. Drefahl, "Handbook for Estimating Physiochemical Properties of Organic Compounds," John Wiley & Sons, Inc., New York, 1999.

Schwarzenbach, R. P., P. M. Gschwend, and D. M. Imboden: "Environmental Organic Chemistry," Wiley-Interscience, New York, 1993.

Schwarzenbach, R. P. and P. M. Gschwend: Chemical Transformations of Organic Pollutants in the Aquatic Environment, in "Aquatic Chemical Kinetics," W. Stumm, ed., Wiley-Interscience, New York, 1990.

6

Basic Concepts from Biochemistry

6.1 | INTRODUCTION

Many environmental engineers and scientists can be classified as biochemical engineers because they spend considerable time and effort designing and operating treatment facilities that utilize living organisms to bring about the destruction or transformation of waste organic and inorganic materials. Therefore, an important facet of their education is biochemistry. Such education has become even more important in understanding the biotransformation of xenobiotic chemicals (e.g., halogenated organics) in the environment and in engineered reactors. This chapter is meant to provide the reader with an introduction to important concepts in biochemistry. It is not meant to serve as an introduction to environmental microbiology, although some discussion is included. For additional detail on basic microbiology and biochemistry, the reader is referred to the Rittmann and McCarty reference and basic biochemistry and microbiology texts listed at the end of this chapter.

Biochemistry deals with chemical changes that are brought about by living organisms. The reactions may be *extracellular* or *intracellular*. Hydrolytic reactions (breaking chemical bonds by addition of water) are generally extracellular in character and necessarily so, because such reactions are often required to reduce the complexity of organic compounds to a point at which they can dialyze through the cell wall. The energy requirement for hydrolytic reactions is considered nil. Oxidative reactions occur intracellularly and produce energy in accordance with the free energy of the particular reaction involved.

Biochemical reactions occur at temperatures with a normal range from about 0 to 60°C. Organisms that thrive at 0 to 10, 10 to 40, and above 40° are classed as *psychrophilic, mesophilic,* and *thermophilic,* respectively. *Hyperthermophilic* organisms can thrive at temperatures between 65 and 110°C. The majority of the chemical reactions that these organisms bring about occur at far lower temperatures than would be needed in their absence. For this reason, catalysts that lower markedly the activation energy of the reactions are required. The catalysts are supplied by the living organisms as part of their life processes and are known as *enzymes.* They serve to initiate the reactions and also to control their speed in a manner that serves the best interests of the particular organism. This mechanism sometimes leads to a competition between groups of organisms when operating in mixed culture.

6.2 | ENZYMES

Enzymes have been defined as temperature-sensitive, organic catalysts produced by living cells and capable of action outside or inside the cell. Certain of the enzymes are secreted by the cell and are known as *extracellular* enzymes. Others are associated with the protoplasm of the cell and perform their function within the cell, and so they are called *intracellular* enzymes. Enzymes are also located in the cell wall and cell membrane. Enzymes may be *constitutive,* meaning they are produced continuously, or they may be *inducible,* being produced on demand in response to some externally applied stimulus. For example, inducible enzymes are likely to be involved when an organism is exposed to an organic it has not seen before.

Enzymes are proteinaceous in character. Some are simple proteins, whereas others are of a complex conjugated type. They are highly specific for the reactions that they catalyze. Enzymes are grouped into six major classes, depending on the nature of the reaction that they control (Table 6.1). For example, those that catalyze hydrolytic reactions are known as *hydrolases,* and those that catalyze oxidation-reduction reactions are oxido-reductases. In general, hydrolases are extracellular and the other enzymes are intracellular. The *-ase* ending is used to designate enzymes.

Some of these enzymes are of particular current interest because of additional reactions they catalyze. For example, methane monooxygenase, toluene monooxygenase, and toluene dioxygenase (Table 6.1) are involved in biochemical pathways responsible for the initial steps in the oxidation of methane, phenol, and toluene or benzene, respectively, by certain aerobic bacteria. These enzymes have also been shown to degrade trichloroethene and other chlorinated aliphatic compounds, a process called *cometabolism,* as the reactions are fortuitous and the microorganisms obtain little direct benefit from the transformations. Another important example is the discovery of *dehalorespiring* bacteria that can derive energy for growth through reduction of chlorinated compounds such as PCE and TCE. These organisms contain reductive dehalogenases that catalyze the removal of chlorine atoms from these compounds, replacing them with hydrogen. Understanding these phenomena is of value as they offer possibilities for treating waters and soils contaminated with such compounds.

A great number of enzyme and bacterial preparations have been marketed as agents capable of solving problems related to the operation of septic tanks, sludge digestion units, and other treatment facilities, and for degrading biologically resistant, hazardous compounds. Research has indicated that domestic wastewater and sludges generally contain bacteria capable of producing the necessary enzyme systems in adequate amounts and that no beneficial effects can be demonstrated by the addition of enzymes from outside sources. In the case of malfunctioning units, reports from field studies (usually conducted without controls) have often been favorable. Laboratory studies often fail to show any beneficial effects. Generally, little benefit can be obtained by adding most of the commercially available cultures that promise special powers for degrading xenobiotic chemicals. There are exceptions,

Table 6.1 | Classification of enzymes

Enzyme	Substrate	Products
Hydrolases		
1. Carbohydrases:		
a. Glycosidases (sugar splitters):		
Sucrase	Sucrose	Glucose + fructose
Maltase	Maltose	Glucose
Lactase	Lactose	Glucose + galactose
b. Amylases (starch splitters):		
Diatase ⎱	Starch	Maltose
Ptyalin ⎰		
c. Cellulase	Cellulose	Cellobiose
2. Esterases:		
a. Lipases:		
Lipase	Glycerides	Glycerol + fatty acids
b. Phosphatases	Phosphoric esters	H_3PO_4 + alcohols
3. Proteases:		
a. Proteinases:		
Pepsin ⎱	Proteins	Polypeptides
Trypsin ⎰		
b. Peptidases	Polypeptides	Amino acids
4. Amidases:		
a. Urease	Urea	$NH_3 + CO_2$
5. Deaminases:	Amino acids	NH_3 + organic acids
Oxido-reductases (oxidation-reduction reactions)		
1. Dehydrogenases		
2. Hydroxylase		
3. Reductive dehalogenase	PCE	TCE
4. Oxidases		
5. Oxygenases		
6. Methane monooxygenase	Methane	Methanol
7. Toluene monooxygenase	Phenol	Catechol
8. Toluene dioxygenase	Benzene	Dihydroxy benzene
	Toluene	Dihydroxy toluene
9. Ammonia monooxygenase	Ammonia	Hydroxylamine
Transferases (transfer of functional groups)		
1. Transaminases		
2. ATP		
Lyases (addition to double bonds)		
Isomerases (isomerization reactions)		
Ligases (formation of bonds by ATP)		

but practicing engineers and scientists need to be skeptical in the absence of sound, scientific, performance results. One successful application was the addition of a de-halorespiring enrichment to accelerate conversion of TCE and *cis*-PCE to ethene.[1]

Enzyme activity is dependent upon cofactor presence, temperature, pH, and macronutrient and micronutrient availability.

---

[1]D. E. Ellis, E. J. Lutz, J. M. Odon, R. J. Buchanan, Jr., C. L. Bartlett, M. D. Lee, M. R. Harkness, and K. A. Deweerd, Bioaugmentation for Accelerated In Situ Anaerobic Bioremediation. *Env. Sci. Tech.,* **34:** 2254–2260 (2000).

6.3 | COFACTORS

Some enzymes depend for their activity only upon their structure as proteins. Others require in addition nonprotein structures, or *cofactors.* The cofactor may be a metal ion, or it may be a complex organic molecule called a *coenzyme.* Sometimes both are required. Metal ions that sometimes serve as cofactors are zinc, magnesium, manganese, iron, copper, nickel, cobalt, potassium, and sodium. The organic structures of coenzymes are generally heat-stable whereas most enzyme proteins are not. Cofactors may be bound tightly or loosely to the enzyme protein.

A limited number of coenzymes are known. Some of the most important are as follows:

NAD. Nicotinamide adenine dinucleotide, an enzyme that transfers hydrogen atoms or electrons and participates in the oxidation of organic materials.

NADP. Nicotinamide adenine dinucleotide phosphate, which is similar in structure and function to NAD, but has three phosphorus atoms in the structure instead of two.

Coenzyme A. A derivative of pantothenic acid, it functions in fatty-acid metabolism (oxidation) and synthesis.

Flavoproteins. Flavin mononucleotide and flavin adenine dinucleotide (FAD) are known collectively as flavoproteins. They are important in the transport of hydrogen or electrons from metabolites to oxygen.

Coenzyme M. A coenzyme unique to methanogenic bacteria that is involved in the production of methane, serving as a methyl carrier.

F_{420}. Another coenzyme contained in methanogenic bacteria that is involved in the production of methane. It serves as an electron carrier, carrying two electrons per mole.

F_{430}. A novel nickel-containing coenzyme present in methanogenic bacteria and involved in the production of methane. This coenzyme can also facilitate dechlorination of chlorinated organic compounds.

6.4 | TEMPERATURE RELATIONSHIPS

Biochemical reactions, in general, follow the van't Hoff rule of a doubling of reaction rate for a 10°C increase in temperature, over a restricted temperature range. Classic studies with activated sludge have shown the reaction rate to be more than doubled for a 10°C rise in temperature, as shown in Fig. 6.1.[2]

In biochemistry, temperature relationships are often referred to as Q_{10} values, which are the ratio of the reaction rate at a particular temperature to the rate at 10°C lower. Table 6.2 shows Q_{10} values from studies conducted many years ago for aerobic processes involving activated sludge[3] and conventional anaerobic digestion.[4] These general relationships still hold.

---

[2] C. N. Sawyer and G. A. Rohlich, *Sewage Works J.,* **11:** 946 (1939).

[3] Ibid.

[4] G. M. Fair and E. W. Moore, *Sewage Works J.,* **6:** 3 (1934).

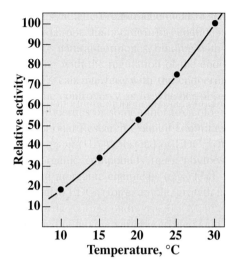

Figure 6.1
Effect of temperature on activity of activated
sludge as measured by oxygen requirements
per unit time.

The data in table 6.2 illustrate that the influence of temperature varies considerably, depending upon the range of temperature and also the nature of the reaction involved. A great deal of study has been given to the effect of temperature on the rate of enzyme-induced reactions. Since all biological reactions are dependent upon enzymes, the data are of considerable interest to environmental engineering and science. Table 6.3 shows the effect of temperature on a number of enzymes, and Fig. 6.2 shows the activity of amylase at several temperatures when hydrolyzing starch.

The data in Fig. 6.2 illustrate the point that increasing temperature has a favorable effect upon biochemical reactions, within limits. As the temperature is increased, eventually a point is reached at which the enzyme becomes less active. This change is considered to be due to *denaturation* of the enzyme, a changing of the enzyme's structure. In systems containing living organisms, the adverse effects

Table 6.2 | Effect of temperature on aerobic
and anaerobic biochemical processes

Biological system	Temperature, °C	Q_{10}
Activated sludge	10–20	2.85
	15–25	2.22
	20–30	1.89
Anaerobic sludge	10–20	1.67
	15–25	1.73
	20–30	1.67
	25–35	1.48
	30–40	1.0

Table 6.3 | Effect of temperature on the activity of
certain enzymes

Enzyme	Temperature, °C	Q_{10}
Amylase	10–20	1.34
	15–25	1.59
	20–30	1.44
	25–35	1.27
	30–40	1.17
Pepsin	0–10	2.60
	10–20	2.00
	20–30	1.80
	30–40	1.60
Steapsin	0–10	1.50
	10–20	1.34
	20–30	1.26

of high temperature may be explained by considering that the enzymes are denaturized or that the ability of the organisms to produce enzymes has been destroyed. The net effect is the same in either case.

The importance of temperature as a factor in determining the rate of biological reactions has been recognized for some time in connection with the design and operation of anaerobic biological processes, trickling filters, lagoons, and ponds. The

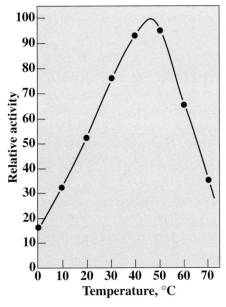

Figure 6.2
Effect of temperature on the action of malt
amylase when hydrolyzing starch to glucose.

significance of temperature in activated sludge treatment is a somewhat more complicated matter as it involves three major factors, all of which are temperature-dependent: (1) compound degradation rate, (2) oxygen transfer rate, and (3) oxygen solubility.

6.5 | pH

Hydrogen-ion concentration is one of the most important factors that influences the speed of biochemical reactions, and since such reactions are induced and controlled by enzymes, it is necessary to have a knowledge of how pH affects enzyme activity. The range of pH through which a particular enzyme can act effectively is usually quite narrow. Some enzymes act best at low pH levels; others require high pH; but the majority are most effective in neutral solutions. The optimum pH and the effective range of pH of a few enzymes are given in Fig. 6.3.

Most biological processes employed in environmental engineering practice involve the use of soil organisms operating in mixed culture. The enzyme systems of these organisms are adapted to operating in essentially neutral solutions; therefore, it is important that pH be controlled over a rather narrow range of about 6 to 9. Control of pH is best accomplished by means of a buffer system, as discussed in Sec. 4.6.

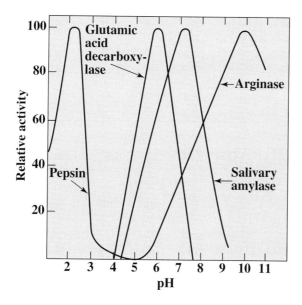

Figure 6.3
Effect of pH on enzyme activity, illustrating the rather
narrow range of optimum action.

6.6 | MAJOR AND TRACE ELEMENTS

Bacteria are among the simplest forms of living matter, and like all living matter, they must reproduce in order to survive. Many bacteria can thrive upon very simple substrates, such as sugar, provided that certain essential elements are present in the form of ions or inorganic salts. Bacterial cell tissue has an empirical formula that approximates $C_5H_7O_2N$, corresponding to about 12 percent nitrogen.

It is known that bacteria are capable of synthesizing protein from a wide variety of carbonaceous materials. If bacterial growth is to occur, sufficient nitrogen must be present to supply the 12 percent fraction of nitrogen in their cells. Because of its relatively high fraction in cells, nitrogen is considered a major nutrient element ("macronutrient") in bacterial nutrition. Phosphorus, generally required at about one-fifth the nitrogen requirement (i.e., 0.2 kg of phosphorus required for every kilogram of nitrogen), and sulfur are other elements essential to the formation of some conjugated proteins and are considered major nutrient elements.

Certain other elements are needed in trace amounts ("micronutrients") for cell metabolism. Many are known to be important in enzyme function or in other physiological capacity. Calcium, cobalt, copper, iron, magnesium, manganese, potassium, selenium, and zinc are probably essential for most bacteria. Other elements are necessary for certain bacteria. Molybdenum, for example, is required by nitrogen-fixing bacteria. Nickel is required by methanogenic bacteria. Many of these metals serve as cofactors essential for enzyme activity.

Environmental engineers and scientists may for the most part assume that domestic wastewater will provide all the major and trace elements needed for its treatment by bacteria. This may or may not be the case with industrial wastes. Many industrial wastes are deficient in the major nutrient elements nitrogen and phosphorus. Some may be deficient in sulfur and in trace elements, depending upon the nature of the source water. Similarly, biological treatment of contaminated groundwater and leachates from hazardous waste sites may be limited by major and/or trace nutrients. Careful waste characterization and laboratory studies are needed to identify potential nutrient limitations.

6.7 | BIODEGRADATION

There are two general types of biodegradation. *Mineralization* occurs when organic compounds are converted under aerobic conditions by living organisms to mineral (nonorganic) end products:

$$\underset{\text{(organic)}}{C_aH_bO_cN_dP_eS_fCl_g} + hO_2 \longrightarrow iCO_2 + jH_2O + kNH_4^+ + lPO_4^{3-} + mSO_4^{2-} + nCl^-$$

Energy is produced during mineralization. *Biotransformation* occurs when parent organic compounds are not completely mineralized, a portion is converted into other organics. An example is the reductive dechlorination of trichloroethene to produce dichloroethene:

$$CCl_2{=}CHCl + H^+ + 2e^- \longrightarrow CHCl{=}CHCl + Cl^-$$

Fermentations all represent biotransformations. An example is the fermentation of sugars such as glucose to ethanol. Energy is often obtained from fermentations, but may not be with all biotransformations.

Some general rules can be made about biotransformations. While there are many exceptions, such general rules are useful for understanding the environmental fate of organic compounds or their treatment.

1. Simple carbohydrates and amino acids are very biodegradable; fats and oils may be more difficult to degrade because of solubility limitations.

2. Hydrocarbons are more difficult to oxidize than alcohols, aldehydes, or acids. Aldehydes are generally more toxic than alcohols or acids. Unsaturated (containing double bonds) aliphatic compounds are more readily degraded than saturated aliphatics.

3. Ketones are more difficult to degrade than aldehydes.

4. Ethers are difficult to biodegrade.

5. Tertiary and quaternary carbons and nitrogens are much more difficult to degrade than primary or secondary carbons or nitrogens. This explains the difference in degradability of ABS and LAS synthetic detergents (Sec. 5.25). Quaternary nitrogen compounds are very toxic to bacteria.

6. Hydrolysis of esters, amides, and carbamates, among others, is generally fast and easily carried out by microorganisms.

7. Adding a chlorine atom or a nitro group ($-NO_2$) to a benzene ring increases its resistance to biodegradation and its toxicity.

8. Meta substitution (1,3 position) to a benzene ring generally makes it more difficult to degrade than to the ortho (1,2) or the para (1,4) positions.

9. Polycyclic aromatic hydrocarbons (PAHs) with more than 3 rings are very resistant to biodegradation.

Biotransformation of halogenated organics is somewhat more complicated. It had been felt that as the number of chlorine atoms in a molecule increases, it becomes more toxic and more difficult to degrade. However, such behavior is organism and environment specific. For example, the more highly chlorinated compounds, such as PCBs, carbon tetrachloride, and tetrachloroethene, are more readily transformed under anaerobic conditions, while those compounds with fewer chlorines, such as monochlorinated biphenyls and chloroethene (vinyl chloride), are more readily biotransformed aerobically.

The following is a list of transformation reactions that can be mediated by bacteria. Some have been discussed previously in Sec. 5.34 and some are discussed in more detail in the sections that follow.

1. *Oxidation.* Release of electrons during transformation

2. *Reduction.* Addition of electrons during transformation

3. *Hydrolysis.* Addition of water

4. *Substitution.* Exchange of one group for another (e.g., OH^- or HS^- for Cl^-)

5. *Elimination.* Removal of atoms form adjacent carbons, leaving a double bond between them

6. *Dealkylation.* Removal of an alkyl group

7. *Deamination.* Removal of an NH_2 group

8. *Condensation.* Production of a larger molecule from smaller molecules

9. *Isomerization.* Conversion of one isomer into another (e.g., conversion of *cis*-1,2-dichloroethene to *trans*-1,2-dichloroethene)

10. *Ring cleavage.* Opening of an organic ring structure, generally for the purpose of further biotransformation

6.8 | BIOCHEMISTRY OF CARBOHYDRATES

The primary use of carbohydrates in higher animals is to serve as a source of energy. With microscopic organisms, however, the differentiation of foods for particular purposes is not a rigid matter. Bacteria, for example, utilize carbohydrates for the synthesis of fats and proteins as well as for energy. In addition, of course, carbohydrates are also used in building cell tissue and may be stored as polysaccharide inside or outside the cell wall.

The mechanisms by which bacteria and other microorganisms transform carbohydrates are believed to be essentially the same as those occurring in plants and animals. The first stage in carbohydrate metabolism involves hydrolysis (see Fig. 5.1). This degradation must progress to at least the disaccharide stage before transfer through the cell wall can occur. Once within the cell wall, the simple sugars are used for energy or synthesis.

The pathway by which bacteria metabolize simple sugars for energy depends upon whether the conditions are aerobic or anaerobic. In either case, the initial conversion is to pyruvic acid as follows for a simple hexose:

$$\underset{\text{Hexose}}{C_6H_{12}O_6} \xrightarrow[\text{enzymes}]{\text{bacteria}} \underset{\text{Pyruvic acid}}{2CH_3\overset{\overset{\text{O}}{\|}}{C}COOH} + 4H \tag{6.1}$$

Under an aerobic conditions, the pyruvic acid and hydrogen are oxidized to carbon dioxide and water for energy:

$$2CH_3\overset{\overset{\text{O}}{\|}}{C}COOH + 4H + 6O_2 \xrightarrow[\text{enzymes}]{\text{bacteria}} 6CO_2 + 6H_2O \tag{6.2}$$

Under anaerobic conditions, however, this oxidation is not possible, and so the bacteria discharge various fermentation end products back into solution. The nature of the end products depends upon the bacterial species involved, as well as the environmental conditions imposed. A few possibilities are indicated below:

$$
2CH_3CCOOH + 4H
\begin{cases}
\xrightarrow{\text{bacteria enzymes}} \underset{\substack{|\\ OH}}{2CH_3CHCOOH} \\[2mm]
\xrightarrow[\text{enzymes}]{\text{bacteria}} CH_3CH_2COOH + CH_3COOH + HCOOH \\[2mm]
\xrightarrow{\text{bacteria enzymes}} 2CH_3CH_2OH + 2CO_2
\end{cases}
$$

(6.3)

Other acids, alcohols, and ketones may also be formed under anaerobic conditions.

The large quantities of organic acids formed from anaerobic carbohydrate metabolism may exceed the buffering capacity of a waste, resulting in a low pH and inhibition or cessation of biological activity. This is a distinct possibility even with aerobic systems, which may suddenly be overloaded with a carbohydrate waste, resulting in temporary buildup of acid intermediates and a resulting drop in pH.

6.9 | BIOCHEMISTRY OF PROTEINS

Proteins are essential in the diets of higher animals and are used to build and repair muscle tissue. Amounts in excess of these requirements may be consumed for energy or converted to carbohydrates and fats. Bacteria are much less demanding in their protein requirements. Most of them are capable of synthesizing protein from inorganic nitrogen and non-protein-containing organics, such as carbohydrates, fats, and alcohols. This ability is most fortunate, for many industrial wastes, including some from the food industry, have very low protein content. It would add greatly to the cost of biological treatment if proteinaceous matter had to be added.

The first step in biological utilization of proteins involves their hydrolysis, which progresses in steps as shown in Sec. 5.22. It is fairly well established that hydrolysis must yield α-amino acids before passage through the cell wall is possible. Within the cell, *deamination* of the amino acids occurs. The nature of the deamination reactions varies under aerobic and anaerobic conditions.

Deamination under Aerobic Conditions

Bacteria deaminize amino acids under aerobic conditions to produce saturated acids with one less carbon atom,

$$
\underset{\substack{|\\ H}}{\overset{\substack{NH_2\\ |}}{R-C-COOH}} + O_2 \xrightarrow[\text{enzymes}]{\text{bacteria}} R-COOH + CO_2 + NH_3 \qquad (6.4)
$$

or hydroxy acids with the same number of carbon atoms,

$$R\overset{\overset{\displaystyle NH_2}{|}}{\underset{\underset{\displaystyle H}{|}}{C}}-COOH + H_2O \xrightarrow[\text{enzymes}]{\text{bacteria}} R\overset{\overset{\displaystyle OH}{|}}{\underset{\underset{\displaystyle H}{|}}{C}}-COOH + NH_3 \qquad (6.5)$$

Deamination under Anaerobic Conditions

Bacterial deamination under anaerobic conditions may proceed with or without reduction to form the corresponding saturated or unsaturated acids.

$$R\overset{\overset{\displaystyle NH_2}{|}}{\underset{\underset{\displaystyle H}{|}}{C}}-COOH + H_2 \xrightarrow[\text{enzymes}]{\text{bacteria}} R-CH_2COOH + NH_3 \qquad (6.6)$$

$$R-CH_2-\overset{\overset{\displaystyle NH_2}{|}}{\underset{\underset{\displaystyle H}{|}}{C}}-COOH \xrightarrow[\text{enzymes}]{\text{bacteria}} R-CH{=}CH-COOH + NH_3 \qquad (6.7)$$

The acids formed under aerobic or anaerobic conditions submit to further oxidation, as discussed in Secs. 6.10 and 6.11. Anaerobic degradation of proteins may produce large quantities of alkalinity in the form of ammonium bicarbonate (NH_4HCO_3). pH may increase significantly if protein concentrations are high.

6.10 | BIOCHEMISTRY OF FATS AND OILS

The degradation or assimilation of fatty materials is sometimes restricted because of their relative insolubility. This can be a serious problem in anaerobic sludge digestion units. Because of their low specific gravity, the fatty materials tend to float and complicate scum conditions. In the scum layer, these fatty materials may be rather remote from the bacteria that are capable of utilizing them. Good mixing in a digester can help reduce this problem.

The biological degradation of fatty materials in its initial phases is known to progress along similar lines under aerobic and anaerobic conditions. The first step is hydrolysis, with the production of glycerol and fatty acids, as discussed in Sec. 5.21. These fatty acids are further processed as will now be described.

Beta Oxidation of Fatty Acids

The free fatty acids derived from the hydrolysis of fatty materials, the deaminization of amino acids, carbohydrate fermentation, and omega oxidation (Sec. 5.2) undergo further breakdown by oxidation. Oxidation is believed to occur at the beta carbon atom in accordance with *Knoop's theory,* sometimes called the *beta oxidation theory.* According to this theory, oxidation proceeds in a series of steps. Coenzyme A is known to be active in these transformations. Oxidation is accomplished by enzymatic hydrogen (and electron) removal and water addition, which are facilitated by the electron carriers, FAD and NAD:

$$R-CH_2-\underset{\underset{\text{β-carbon}}{\overset{|}{\underset{H}{}}}}{\overset{\overset{H}{|}}{C}}-\overset{\overset{O}{\parallel}}{C}-OH + HSCoA \xrightarrow{\text{enzyme}} R-CH_2-\overset{\overset{H}{|}}{\underset{\underset{H}{|}}{C}}-\overset{\overset{O}{\parallel}}{C}-SCoA + H_2O$$

$$\text{β-carbon} \qquad \text{Coenzyme A}$$

$$(6.8)$$

$$R-CH_2-\overset{\overset{H}{|}}{\underset{\underset{H}{|}}{C}}-\overset{\overset{O}{\parallel}}{C}-SCoA + FAD^+ \xrightarrow{\text{enzyme}} R-\overset{\overset{H}{|}}{C}=\overset{\overset{H}{|}}{C}-\overset{\overset{O}{\parallel}}{C}-SCoA + FADH + H^+$$

$$(6.9)$$

$$R-\overset{\overset{H}{|}}{C}=\overset{\overset{H}{|}}{C}-\overset{\overset{O}{\parallel}}{C}-SCoA + H_2O \xrightarrow{\text{enzyme}} R-\overset{\overset{OH}{|}}{\underset{\underset{H}{|}}{C}}-\overset{\overset{H}{|}}{\underset{\underset{H}{|}}{C}}-\overset{\overset{O}{\parallel}}{C}-SCoA \quad (6.10)$$

$$R-\overset{\overset{OH}{|}}{\underset{\underset{H}{|}}{C}}-\overset{\overset{H}{|}}{\underset{\underset{H}{|}}{C}}-\overset{\overset{O}{\parallel}}{C}-SCoA + NAD^+ \xrightarrow{\text{enzyme}} R-\overset{\overset{O}{\parallel}}{C}-\overset{\overset{H}{|}}{\underset{\underset{H}{|}}{C}}-\overset{\overset{O}{\parallel}}{C}-SCoA + NADH + H^+$$

$$(6.11)$$

$$R-\overset{\overset{O}{\parallel}}{C}-\overset{\overset{H}{|}}{\underset{\underset{H}{|}}{C}}-\overset{\overset{O}{\parallel}}{C}-SCoA + 2H_2O \xrightarrow{\text{enzyme}} R-\overset{\overset{O}{\parallel}}{C}-OH + CH_3COOH + HSCoA$$

$$(6.12)$$

In the final step, rupture of the molecule occurs, with formation of one molecule of acetic acid, and the original molecule of acid appears as a new acid derivative with two less carbon atoms. Thus, by successive oxidations at the beta carbon atom, long-chain fatty acids are "whittled" into fragments consisting of acetic acid. During this oxidation four hydrogen atoms, with corresponding electrons, are removed for each acetic acid unit produced. These electrons are contained in NADH and FADH. For example, hydrolysis of animal fat may yield palmitic acid ($C_{15}H_{31}COOH$). β-oxidation of this acid would proceed by "chopping off" two carbons at a time in the form of acetic acid. Since the acid has a total of 16 carbons, a total of 8 mol of acetic acid will be formed for each mole of palmitic acid. By this stoichiometry, the net result would be the following:

$$C_{15}H_{31}COOH + HSCoA + 7FAD^+ + 7NAD^+ + 14H_2O \rightarrow$$
$$8CH_3COOH + HSCoA + 7FADH + 7NADH + 14H^+ \qquad (6.13)$$

Under aerobic conditions, the electrons now carried by FADH and NADH are used to reduce molecular oxygen to water and produce energy. Under anaerobic conditions, however, it is not possible for the bacteria to rid themselves of the electrons in this fashion, and another scheme must be used. Methods by which this may be accomplished are indicated in Sec. 6.11.

6.11 | GENERAL BIOCHEMICAL PATHWAYS

Sections 6.8 to 6.10 have indicated pathways used by microorganisms for the oxidation of carbohydrates, proteins, and fats in order to obtain energy for their life processes. Oxidation involves the transfer of electrons from a reduced substance termed the *electron donor* to an oxidizing material termed the *electron acceptor.* Generally, we think of the electron donor as being the "food" for the organism. Organic matter is generally used as food by bacteria and fungi as well as by animals. However, with some bacteria, reduced inorganic materials such as ammonia, sulfide, molecular hydrogen, and ferrous iron may also serve as electron donors and thus as energy sources. Bacteria are typically classified on the basis of the carbon sources used for cell synthesis and how they generate energy. Organisms that obtain energy from sunlight are termed *photosynthetic* or *phototrophic.* Organisms that obtain energy from oxidation of organic compounds are termed *chemoorganotrophs,* while those that obtain energy from the oxidation of inorganic compounds are termed *chemolithotrophs.* Bacteria that use organic compounds for cell synthesis are termed *heterotrophs,* and those that use inorganic carbon are termed *autotrophs.* These different classifications are typically used interchangeably. For example, chemolithotrophic bacteria are usually autotrophs and chemoorganotrophs are usually heterotrophs.

Energy is transferred from the electron donor to the organism for synthesis and maintenance by a complex series of enzymatic reactions, often starting with the formation of the electron carriers FADH and NADH as illustrated in Eqs. (6.9) and (6.11). A key compound in subsequent energy transfer is the nucleotide, adenosine diphosphate (ADP) which has the following structure:

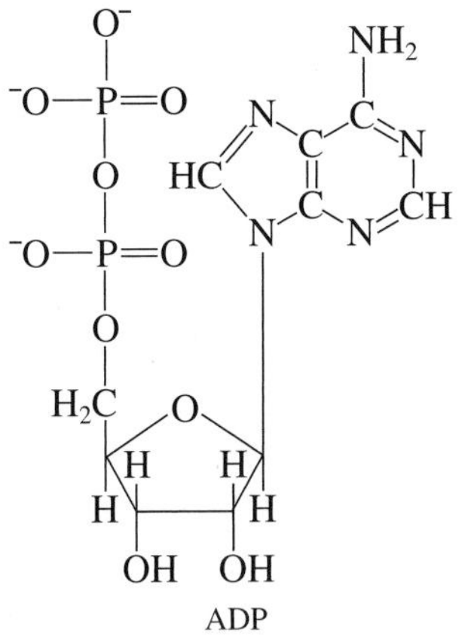

ADP

ADP uses the energy released from oxidation to form a bond with phosphate to form another nucleotide, adenosine triphosphate (ATP):

$$ADP + H_3PO_4 = ATP + H_2O \qquad \Delta G^{0\prime} = 32 \text{ kJ/mol} \qquad (6.14)$$

Here, $\Delta G^{0\prime}$ represents the free energy of formation at pH $= 7.0$. The ATP so formed can travel through the cell to give up its energy for cell synthesis, maintenance, or movement by the reverse of Eq. (6.14). A general schematic of these processes is given in Fig. 6.4.

The process by which the energy contained in NADH (and other electron carriers such as FADH and NADPH) can be converted to ATP is complex and involves the development of a proton (H^+) gradient commonly called the *proton motive force*. This process is called *oxidative phosphorylation* and is closely tied to *respiration* (the reduction of the terminal electron acceptor). When NADH is oxidized to release its electrons, a proton is released and discharged outside the cell membrane:

$$NADH = NAD^+ + H^+ + 2e^- \qquad \Delta G^0 = -62 \text{ kJ/mol} \qquad (6.15)$$

Some have used the analogy of charging a car battery to describe this process. The energy stored in this gradient is used to make ATP, and the amount of energy made depends on the terminal electron acceptor available. For example, under aerobic conditions, a complex series of reactions can be summarized as:

$$NADH + 0.5O_2 + H^+ + 3ADP + 3H_3PO_4 \rightarrow NAD^+ + 4H_2O + 3ATP \qquad (6.16)$$

Under iron (Fe(III)) reduction and denitrifying (NO_3^- is the terminal electron acceptor) conditions slightly less than 3 mol of ATP are produced for each mole of

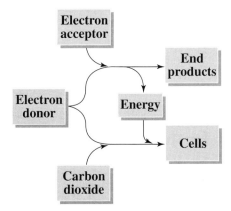

Figure 6.4
Schematic diagram of biological oxidation of an electron donor for energy and transfer of the energy for cell synthesis. Either an organic electron donor or carbon dioxide may provide the cellular need for carbon.

NADH oxidized to NAD^+. With strictly anaerobic conditions, less than 0.5 mol of ATP may be produced per mole of NADH. Thus more energy is produced from each mole of NADH under aerobic, iron-reducing, and denitrifying conditions than under anaerobic conditions.

Such thermodynamic considerations can be easily seen using an "electron tower" such as that shown in Fig. 6.5. For example, the electrons carried in NADH "fall" further when used to reduce O_2 to H_2O ($\Delta G^{0\prime} = -78.72 - 31 = -109.72$ kJ/eq)[5] than when used to reduce sulfate to sulfide ($\Delta G^{0\prime} = 20.85 - 31 = -10.15$ kJ/eq). Thus, approximately 3 mol of ATP are formed from each mole of NADH when O_2 serves as the terminal electron acceptor while approximately 0.3 mol of ATP may be formed per mole of NADH when SO_4^{2-} is the terminal electron acceptor. Additional detail is contained in the references listed at the end of this chapter.

A greatly simplified and general scheme for oxidation and fermentation of organic materials and transfer of energy to ATP is illustrated in Fig. 6.6. Carbohydrates, proteins, and fats enter the overall biochemical scheme at different locations. Not all bacteria have the capability to oxidize all organic compounds, but most are able to at least carry out some portion of the overall conversions illustrated. With organic compounds other than carbohydrates, proteins, and fats, bacteria still need to convert them to a limited number of precursor molecules that can enter pathways common to all bacteria (for example, conversion to pyruvate for entry into the tricarboxylic acid cycle, also known as Krebs cycle). Organic oxidation can be thought of generally as removal of electrons (or hydrogen atoms) from organic molecules with the aid of the coenzyme NAD^+, which in turn is reduced and converted to NADH. In oxidative phosphorylation the electrons are transferred from NADH through a series of coenzymes (termed the *electron transport system*) and ultimately to the terminal electron acceptor (e.g., O_2 in an aerobic system). During these transfers, a portion of the energy available is transferred to ADP to make ATP as previously indicated. Another portion is lost as heat because of inefficiencies in energy transfer. Some ATP may be formed directly from organic transformations such as during the conversion of glucose to pyruvate.

In aerobic systems, oxygen is the terminal electron acceptor and is reduced while organic or inorganic electron donors are being oxidized. In the absence of oxygen, other materials such as nitrate, Fe(III), Mn(IV), sulfate, and carbon dioxide may become electron acceptors. The use of sulfate and carbon dioxide requires strictly anaerobic conditions. Nitrate can be used, however, by facultative organisms living under intermediate conditions referred to as anoxic, which are characterized by end products of carbon dioxide, water, and nitrogen gas. All are inoffensive as opposed to products of methane and H_2S formed under strictly anaerobic conditions. Additionally, Fe(II) and Mn(II) formed from reduction of Fe(III) and Mn(IV) represent potential water quality problems (e.g., staining).

---

[5]eq= mole of e^-. For example, for Eq. (6.15), $\Delta G^{0\prime} = -62$ kJ/mol of NADH. Two moles of e^- are generated. Thus, $\Delta G^{0\prime} = -62\dfrac{\text{kJ}}{\text{mol NADH}}\left(\dfrac{1 \text{ mol NADH}}{2 \text{ mol } e^-}\right) = 31$ kJ/eq.

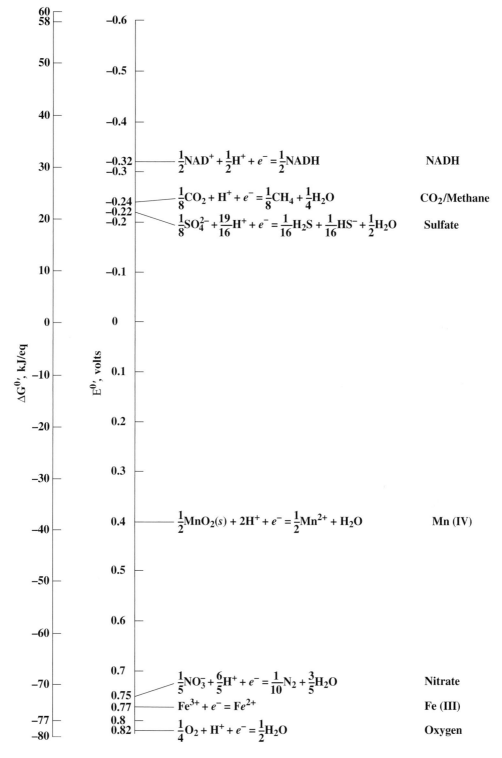

Figure 6.5

Electron tower comparing the energy content of common electron acceptors and the NADH/NAD$^+$ couple. *Note:* 1 volt $= -96.485$ kJ/eq. (*Adapted from Rittmann and McCarty Environmental Biotechnology,* McGraw-Hill, New York, 2001.)

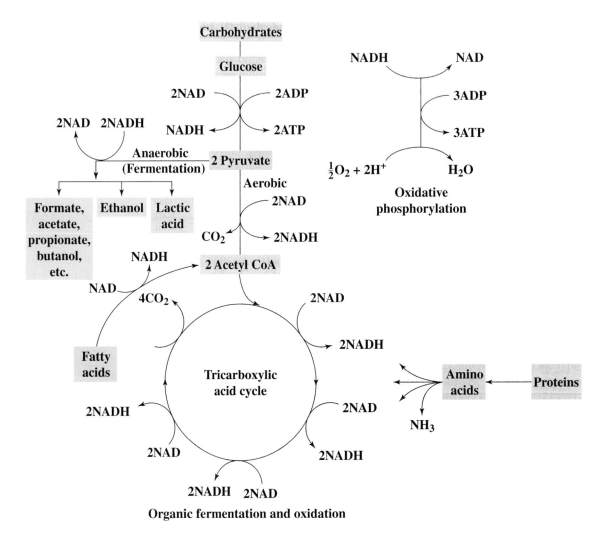

Figure 6.6
Generalized pathways for aerobic or anaerobic fermentation and oxidation of organics and transfer of energy released through NAD to form ATP.

Energy availability can also be assessed using free-energy considerations. Balanced reactions for oxidation of electron donors and reduction of terminal electron acceptors can be obtained by combining half reactions from Table 6.4. Values for $\Delta G^{0'}$ are calculated as described in Chap. 3 except that ΔG^0 for H^+ is taken to equal -39.87/electron-equivalent, the value it would have at pH 7 when $[H^+] = 10^{-7}$ M. The reverse of Reaction 14 in Table 6.4 can be combined with the electron acceptor half reactions to yield the following:

Reaction	$\Delta G^{0\prime}$ kJ/eq	
Aerobic:		
$\frac{1}{8}CH_3COO^- + \frac{1}{4}O_2 \qquad = \frac{1}{8}CO_2 + \frac{1}{8}HCO_3^- + \frac{1}{8}H_2O$	-106.13	(6.17)
Fe(III) reduction:		
$\frac{1}{8}CH_3COO^- + Fe^{3+} + \frac{3}{8}H_2O \quad = \frac{1}{8}CO_2 + \frac{1}{8}HCO_3^- + Fe^{2+} + H^+$	-101.68	(6.18)
Denitrification:		
$\frac{1}{8}CH_3COO^- + \frac{1}{5}NO_3^- + \frac{1}{5}H^+ \quad = \frac{1}{8}CO_2 + \frac{1}{8}HCO_3^- + \frac{1}{10}N_2 + \frac{9}{40}H_2O$	-99.61	(6.19)
Mn(IV) reduction:		
$\frac{1}{8}CH_3COO^- + \frac{1}{2}MnO_2(s) + H^+ = \frac{1}{8}CO_2 + \frac{1}{8}HCO_3^- + \frac{1}{2}Mn^{2+} + \frac{5}{8}H_2O$	-66.3	(6.20)
Anaerobic:		
$\frac{1}{8}CH_3COO^- + \frac{1}{8}SO_4^{2-} + \frac{3}{16}H^+ = \frac{1}{8}CO_2 + \frac{1}{8}HCO_3^- + \frac{1}{16}H_2S + \frac{1}{16}HS^- + \frac{1}{8}H_2O$ $\quad -6.56$		(6.21)
$\frac{1}{8}CH_3COO^- + \frac{1}{8}H_2O \qquad = \frac{1}{8}CH_4 + \frac{1}{8}HCO_3^-$	-3.89	(6.22)

The standard free energies for each reaction indicate, as stated previously, that more energy is available for biological growth from aerobic, metal-reducing, and denitrifying reactions than from anaerobic reactions. The last two reactions of sulfate reduction and methanogenesis release the least energy of all. These reactions can be carried out only by highly specialized groups of anaerobic bacteria and cannot generally proceed in the presence of O_2, Fe(III), NO_3^-, or Mn(IV).

Methanogenesis is the major process by which wastes are stabilized in anaerobic digestion and is an important part of the stabilization occurring in septic tanks, lagoons, and sanitary landfills. The methane gas released can be used as a fuel. Increased interest in this process has been shown in recent years because it offers an alternative to the dwindling supplies of energy. About 70 percent of the methane resulting from the complete methane fermentation of complex wastes results from fermentation of acetic acid,[6] according to Eq. (6.22). Acetic acid is formed as an intermediate in the anaerobic fermentation of carbohydrates, proteins, fats, and other organic compounds.

In the anaerobic fermentation of carbohydrates, an external electron acceptor is not required; the organic matter itself serves as the electron acceptor. As depicted in Fig. 6.6, electrons removed from glucose by NAD are transferred to pyruvate to form the variety of possible end products, as discussed previously. A knowledge of the biochemical pathways for organic degradation can aid in understanding what transformations are possible under different environmental conditions, a topic discussed in more detail in Sec. 6.13.

---

[6]J. S. Jeris and P. L. McCarty, The Biochemistry of Methane Fermentation Using C^{14} Tracers, *J. Water Pollut. Contr. Fed.,* **37:** 178 (1965).

Table 6.4 | Useful half reactions for bacterial systems

Reaction number	Half reaction	$\Delta G^{0\prime}*$ kJ/eq
	Reactions for bacterial cell synthesis (R_c)	
	Ammonia as nitrogen source:	
1	$\frac{1}{5}CO_2(g) + \frac{1}{20}HCO_3^- + \frac{1}{20}NH_4^+ + H^+ + e^- = \frac{1}{20}C_5H_7O_2N + \frac{9}{20}H_2O$	
	Nitrate as nitrogen source:	
2	$\frac{5}{28}CO_2(g) + \frac{1}{28}NO_3^- + \frac{29}{28}H^+ + e^- \qquad = \frac{1}{28}C_5H_7O_2N + \frac{11}{28}H_2O$	
	Reactions for electron acceptors (R_a)	
	Oxygen:	
3	$\frac{1}{4}O_2(g) + H^+ + e^- \qquad = \frac{1}{2}H_2O$	−78.72
	Ferric iron:	
4	$Fe^{3+} + e^- \qquad = Fe^{2+}$	−74.27
	Nitrate:	
5	$\frac{1}{5}NO_3^- + \frac{6}{5}H^+ + e^- \qquad = \frac{1}{10}N_2(g) + \frac{3}{5}H_2O$	−72.20
	Manganese:	
6	$\frac{1}{2}MnO_2(s) + 2H^+ + e^- = \frac{1}{2}Mn^{2+} + H_2O$	−38.89
	Sulfate:	
7	$\frac{1}{8}SO_4^{2-} + \frac{19}{16}H^+ + e^- \quad = \frac{1}{16}H_2S(aq) + \frac{1}{16}HS^- + \frac{1}{2}H_2O$	20.85
	Carbon dioxide:	
8	$\frac{1}{8}CO_2(g) + H^+ + e^- \quad = \frac{1}{8}CH_4(g) + \frac{1}{4}H_2O$	23.52
	Reactions for electron donors (R_d)	
	Organic donors	
	Domestic wastewater:	
9	$\frac{9}{50}CO_2(g) + \frac{1}{50}HCO_3^- + \frac{1}{50}NH_4^+ + H^+ + e^- = \frac{1}{50}C_{10}H_{19}O_3N + \frac{9}{25}H_2O$	32.80
	Protein (amino acids, proteins, nitrogenous organics):	
10	$\frac{8}{33}CO_2(g) + \frac{2}{33}NH_4^+ + \frac{31}{33}H^+ + e^- \qquad = \frac{1}{66}C_{16}H_{24}O_5N_4 + \frac{27}{66}H_2O$	31.66
	Carbohydrates (cellulose, starch, sugars):	
11	$\frac{1}{4}CO_2(g) + H^+ + e^- \qquad = \frac{1}{4}CH_2O + \frac{1}{4}H_2O$	41.25
	Grease (fats and oils):	
12	$\frac{4}{23}CO_2(g) + H^+ + e^- \qquad = \frac{1}{46}C_8H_{16}O + \frac{15}{46}H_2O$	27.02
	Benzoate:	
13	$\frac{1}{5}CO_2(g) + \frac{1}{30}HCO_3^- + H^+ + e^- \qquad = \frac{1}{30}C_6H_5COO^- + \frac{13}{30}H_2O$	27.34
	Acetate:	
14	$\frac{1}{8}CO_2(g) + \frac{1}{8}HCO_3^- + H^+ + e^- \qquad = \frac{1}{8}CH_3COO^- + \frac{3}{8}H_2O$	27.41
	Propionate:	

Table 6.4 | (continued)

Reaction number	Half reaction		$\Delta G^{0\prime}$* kJ/eq
	Reactions for electron donors (R_d)		
15	$\frac{1}{7}CO_2(g) + \frac{1}{14}HCO_3^- + H^+ + e^-$	$= \frac{1}{14}CH_3CH_2COO^- + \frac{5}{14}H_2O$	27.63
	Ethanol:		
16	$\frac{1}{6}CO_2(g) + H^+ + e^-$	$= \frac{1}{12}CH_3CH_2OH + \frac{1}{4}H_2O$	31.16
	Lactate:		
17	$\frac{1}{6}CO_2(g) + \frac{1}{12}HCO_3^- + H^+ + e^-$	$= \frac{1}{12}CH_3CHOHCOO^- + \frac{1}{3}H_2O$	32.29
	Pyruvate:		
18	$\frac{1}{5}CO_2(g) + \frac{1}{10}HCO_3^- + H^+ + e^-$	$= \frac{1}{10}CH_3COCOO^- + \frac{2}{5}H_2O$	35.10
	Methanol:		
19	$\frac{1}{6}CO_2(g) + H^+ + e^-$	$= \frac{1}{6}CH_3OH + \frac{1}{6}H_2O$	36.84
	Inorganic donors		
20	$Fe^{3+} + e^-$	$= Fe^{2+}$	−74.27
21	$\frac{1}{2}NO_3^- + H^+ + e^-$	$= \frac{1}{2}NO_2^- + \frac{1}{2}H_2O$	−41.65
22	$\frac{1}{8}NO_3^- + \frac{5}{4}H^+ + e^-$	$= \frac{1}{8}NH_4^+ + \frac{3}{8}H_2O$	−35.11
23	$\frac{1}{6}NO_2^- + \frac{4}{3}H^+ + e^-$	$= \frac{1}{6}NH_4^+ + \frac{1}{3}H_2O$	−32.93
24	$\frac{1}{6}SO_4^{2-} + \frac{4}{3}H^+ + e^-$	$= \frac{1}{6}S(s) + \frac{2}{3}H_2O$	19.15
25	$\frac{1}{8}SO_4^{2-} + \frac{19}{16}H^+ + e^-$	$= \frac{1}{16}H_2S(aq) + \frac{1}{16}HS^- + \frac{1}{2}H_2O$	20.85
26	$\frac{1}{4}SO_4^{2-} + \frac{5}{4}H^+ + e^-$	$= \frac{1}{8}S_2O_3^{2-} + \frac{5}{8}H_2O$	23.58
27	$H^+ + e^-$	$= \frac{1}{2}H_2$	39.87
28	$\frac{1}{2}SO_4^{2-} + H^+ + e^-$	$= \frac{1}{2}SO_3^{2-} + \frac{1}{2}H_2O$	50.30

*Reactants and products at unit activity except $[H^+] = 10^{-7}$ M. Values for $\Delta G^{0\prime}$ calculated using ΔG_f^0 values given in App. A. *Note:* 1 kcal = 4.184 kJ.

It should also be noted that other compounds can serve as terminal electron acceptors in biological systems. Notable among these are oxidized sulfur compounds (e.g., $S(s)$, $S_2O_3^{2-}$), and even halogenated organic compounds (chlorinated solvents and aromatic compounds are environmentally significant examples). These reactions are discussed in more detail in Sec. 6.13. Oxidation-reduction reactions with iron are particularly important in groundwater environments. Nitrite (NO_2^-) can also serve as a terminal electron acceptor; however, it is rarely found in significant concentrations in the environment.

6.12 | ENERGETICS AND BACTERIAL GROWTH

Microorganisms oxidize inorganic and organic materials in order to obtain energy for growth and maintenance. Heterotrophic organisms use a portion of organic material metabolized for energy, which in turn is used to convert another portion of the organic matter into cells. Autotrophic organisms on the other hand oxidize inorganic materials for energy, and in turn use the energy released to reduce carbon dioxide to form cellular organics. Electrons are needed to reduce the carbon dioxide and are obtained by oxidizing another portion of the inorganic electron donor. Thus, whether heterotrophic or autotrophic growth is considered, a portion of the electron donor is used for energy and a portion is used for synthesis. These two uses for the electron donor are shown schematically in Fig. 6.4.

In the design of biological systems for wastewater treatment and remediation of contaminated aquifers environmental engineers and scientists need to know all the transformations that take place. They need to know what portion of the electron donor will be converted for energy so that they can determine the quantity of oxygen, nitrate, or sulfate that will be required, or the quantity of methane that will be produced. They need to know the portion that will be converted to microbial cells since they will need to be sure sufficient nutrients such as nitrogen and phosphorus are available. They will also need to provide sludge-handling facilities to treat and dispose of the biological materials that the process produces. In other words, they need to make a mass balance for the system.

A balanced chemical reaction for the overall biological conversion can aid in making such a mass balance. The reaction should include terms for synthesis and for energy. For the synthesis portion an empirical formula for bacterial cells, $C_5H_7O_2N$, is commonly used.[7] A balanced overall reaction can be written through use of the half reactions listed in Table 6.4, and combining them in accordance with the following:[8]

$$R = f_s R_c + f_e R_a - R_d \qquad (6.23)$$

R_c represents the half reaction for synthesis of bacterial cells and would be either Reaction 1 or 2 from Table 6.4, depending on whether ammonia or nitrate was available, respectively, for satisfying the cell nitrogen requirements. R_a represents the half reaction for the electron acceptor, and R_d for the electron donor. The values f_s and f_e represent the portion of the electron donor used for synthesis and energy, respectively, and the sum of the two must equal 1.0. The plus sign indicates that the half reactions as written in Table 6.4 are added together, and the minus sign means that the half reaction is first switched so that the right-hand side of the equation in the table becomes the left-hand side, and then the half reaction is added.

Typical maximum values for f_s [$f_{s(max)}$] are given in Table 6.5, and are appropriate for young rapidly growing bacterial cells. In general, reactions between electron

---

[7]P. L. McCarty, Stoichiometry of Biological Reactions, *Progr. in Water Technol.*, **7**: 157, Pergamon, London (1975).

[8]Ibid.

Table 6.5 | Typical values for $f_{s(max)}$ for bacterial
reactions

Electron donor	Electron acceptor	$f_{s(max)}$
Heterotrophic reactions		
Carbohydrate	O_2	0.72
Carbohydrate	NO_3^-	0.60
Carbohydrate	SO_4^{2-}	0.30
Carbohydrate	CO_2	0.28
Protein	O_2	0.64
Protein	CO_2	0.08
Fatty acid	O_2	0.59
Fatty acid	SO_4^{2-}	0.06
Fatty acid	CO_2	0.05
Methanol	NO_3^-	0.36
Methanol	CO_2	0.15
Autotrophic reactions		
S	O_2	0.21
$S_2O_3^{2-}$	O_2	0.21
$S_2O_3^{2-}$	NO_3^-	0.20
NH_4^+	O_2	0.10
H_2	O_2	0.24
H_2	CO_2	0.04
Fe^{2+}	O_2	0.07

donors and acceptors that yield more energy result in higher values for $f_{s(max)}$. For old or slowly growing cultures, values for f_s would be less than $f_{s(max)}$ because a larger portion of the energy is used for cell maintenance than for synthesis. For very old cells, values for f_s may be as little as 20 percent of $f_{s(max)}$. The value for f_s also depends upon characteristics of the bacterial species, efficiencies of energy transfer, and environmental conditions.

(*a*) Write a balanced equation for the methane fermentation of glucose, assuming that f_s equals 0.28. (*b*) Based on this equation, what quantity of methane is produced from fermentation of 1000 g of glucose, what quantity of bacteria is produced in this conversion, and what quantity of ammonia nitrogen is needed to satisfy the growth requirements of the bacteria?

EXAMPLE 6.1

(*a*) $f_e = 1 - f_s = 0.72$

Assuming that ammonia is available for cell synthesis, R_c = Reaction 1, R_a = Reaction 8, and R_d = Reaction 11 from Table 6.4:

$f_e R_c$: $0.056CO_2 + 0.014HCO_3^- + 0.014NH_4^+ + 0.28H^+ + 0.28e^- = 0.014C_5H_7O_2N + 0.126H_2O$

$f_e R_a$: $0.09CO_2 + 0.72H^+ + 0.72e^- = 0.09CH_4 + 0.18H_2O$

$-R_d$: $\underline{0.25CH_2O + 0.25H_2O = 0.25CO_2 + H^+ + e^-}$

R: $0.25CH_2O + 0.014HCO_3^- + 0.014NH_4^+ = 0.104CO_2 + 0.056H_2O + 0.09CH_4 + 0.014C_5H_7O_2N$

(b)   Reaction R indicates that the metabolism of 0.25 mol or 7.5 g of CH_2O requires 0.014 mol or 0.196 g of ammonia nitrogen and results in the production of 0.09 mol or 2.02 liters (STP) of methane and 0.014 mol or 1.58 g of cells. Therefore, the fermentation of 1000 g of glucose will result in the following:

$$\text{Ammonia-N used} = 0.196(1000)/7.5 = 26.1 \text{ g}$$
$$\text{Methane produced} = 2.02(1000)/7.5 = 269 \text{ liters (STP)}$$
$$\text{Cells produced} = 1.58(1000)/7.5 = 211 \text{ g}$$

EXAMPLE 6.2

Write a balanced equation for the autotrophic oxidation of NH_4^+ to NO_2^-, using the value of $f_{s(max)}$ from Table 6.5.

From Table 6.5, $f_s = 0.10$, and $f_e = 1.0 - 0.1 = 0.9$. Since ammonia is available for cell synthesis, use R_c = Reaction 1, R_a = Reaction 3, and R_d = Reaction 23 from Table 6.4:

$f_s R_c$: $0.02CO_2 + 0.005HCO_3^- + 0.005NH_4^+ + 0.1H^+ + 0.1e^- = 0.005C_5H_7O_2N + 0.045H_2O$

$f_e R_a$: $0.225O_2 + 0.9H^+ + 0.9e^- \qquad\qquad\qquad\qquad = 0.45H_2O$
$-R_d$: $0.167NH_4^+ + 0.333H_2O \qquad\qquad\qquad\qquad = 0.167NO_2^- + 1.333H^+ + e^-$

R: $0.172NH_4^+ + 0.225O_2 + 0.005HCO_3^- + 0.02CO_2 = $
$\qquad\qquad\qquad\qquad 0.167NO_2^- + 0.005C_5H_7O_2N + 0.333H^+ + 0.162H_2O$

A similar method of writing balanced equations may be used for bacterial fermentations. Here, organics serve both as electron donor and acceptor, the donor portion is oxidized generally to carbon dioxide, and the acceptor portion is converted to a more reduced organic material. An example is the fermentation of glucose to ethanol:

$$C_6H_{12}O_6 = 2CH_3CH_2OH + 2CO_2 \qquad\qquad (6.24)$$

EXAMPLE 6.3

Write a balanced equation for the bacterial fermentation of glucose to ethanol, assuming that ammonia is present as a nutrient and that f_s equals 0.15.

Here, $f_e = 0.85$, R_c = Reaction 1, R_d = Reaction 11, and R_a = Reaction 16 (ethanol is the end product and conceptually can be thought to be formed through CO_2 reduction, even though this is not the case).

$f_s R_c$: $0.03CO_2 + 0.0075HCO_3^- + 0.0075NH_4^+ + 0.15H^+ + 0.15e^- = 0.0075C_5H_7O_2N + 0.0675H_2O$

$f_e R_a$: $0.142CO_2 + 0.85H^+ + 0.85e^- \qquad\qquad\qquad = 0.0708CH_3CH_2OH + 0.2125H_2O$

$-R_d$: $0.25CH_2O + 0.25H_2O \qquad\qquad\qquad\qquad = 0.25CO_2 + H^+ + e^-$

R: $0.25CH_2O + 0.0075HCO_3^- + 0.0075NH_4^+ = $
$\qquad\qquad\qquad 0.0708CH_3CH_2OH + 0.0075C_5H_7O_2N + 0.078CO_2 + 0.03H_2O$

In addition to providing estimates of electron donor, electron acceptor, and nutrient (N) requirements, and bacteria and end-product (e.g., CH_4) production, this approach gives additional insight. For instance, in Exs. 6.1 and 6.3, bicarbonate (HCO_3^-) is consumed and CO_2 is produced. Thus, there is the potential for a pH decrease if the system is not properly buffered. In Example 6.2, HCO_3^- is consumed and H^+ is produced. Again, without proper buffering, pH may decrease to a level that is toxic to the bacteria. In this case, a total of 0.338 eq of alkalinity (0.005 HCO_3^- plus 0.333 H^+) would be required to buffer the oxidation of 0.172 mol of NH_4^+ and prevent a pH decrease. A total of $\frac{0.338 \times 50}{0.172 \times 14}$ or 7.02 mg of alkalinity as $CaCO_3$ would be required per mg of NH_4^+-N oxidized.

6.13 | NOVEL BIOTRANSFORMATIONS

In this section, several novel biotransformation reactions are discussed. These transformations are important in understanding the fate and transport of xenobiotic and anthropogenic chemicals in the environment and in devising methods for their removal from waters, wastewaters, and contaminated aquifers and sediments. These biotransformations are termed novel here for two general reasons. First, in some cases the organic compounds are not being used as electron donors or electron acceptors for growth (primary substrates). Second, in some cases compounds such as tetrachloroethene, trichloroethene, 2-chlorophenol, Cr(VI), perchlorate (ClO_4^-), and chlorate (ClO_3^-) serve as electron acceptors supporting growth. Such capabilities were largely unknown prior to around 1990.

Xenobiotic chemicals are those that are foreign to natural biota. *Anthropogenic* compounds are those that are synthetic. Examples included in both categories are pesticides, PCBs, and chlorinated solvents. These compounds are considered hazardous and toxic. Some toxic chemicals are naturally occurring organics such as BTEXs, phenol, and PAHs, and inorganics such as perchlorate, chlorate, and Cr(VI). All these compounds are of concern because of the wide variety of human health and environmental problems that have been linked with their presence in the environment. Among these are several types of cancer, central nervous system disorders, adverse reproductive outcomes (e.g., birth defects), disruption of the endocrine system, and a range of specific organ disorders in humans and other species. As discussed in Chaps. 5 and 34, air and drinking water standards have been established for many of these compounds. These compounds are most often associated with wastewater discharges, air pollution, and pollution of groundwaters, surface waters, and sediments through accidental spills, poor waste management practices, and leaking storage tanks. Many of the organic compounds were discussed in general in Chap. 5, and important properties of specific compounds were given in Table 5.15.

Transformation reactions important in the fate and removal of toxic compounds were discussed in Sec. 5.34. Some of these reactions are mediated by microorganisms (biotic) and some can occur chemically (abiotic). In general, biotic transformation reactions are faster than abiotic reactions, but this is not always the case. The most important biotransformation reactions are thought to be hydrolysis, oxidation, and reduction. No dehydrohalogenation (elimination) reactions have yet been reported to be microbially mediated.[9]

Organic and inorganic compounds can be transformed by microorganisms through various mechanisms, and this applies to xenobiotic compounds as well. The best known way is when the compound serves as a *primary substrate.* A primary substrate is an inorganic or organic electron donor that provides the main energy source for the microorganism. Another way a compound can be transformed is while serving as an *electron acceptor* for energy production. Example compounds that can be transformed in this way are trichloroethene and perchlorate. A third way that a compound can be transformed is while acting as a *secondary substrate.* A secondary substrate is one with a concentration that is too low to provide net energy production for growth of the microorganisms. Under such conditions the compound by itself cannot be expected to be transformed. However, a secondary substrate can be transformed if it coexists with a primary substrate at sufficient concentration to provide net energy for growth. An example here is dichloromethane, which can be used as a primary substrate if at high enough concentration. However, if its concentration is only 1 μg/L, then the concentration is too low to support growth. But if acetate is also present at 1-mg/L concentration, which is sufficient under aerobic conditions for acetate to act as a primary substrate, then the dichloromethane can be oxidized simultaneously with the acetate as sufficient energy would then be available. A fourth way in which a compound can be transformed is through *cometabolism.* Here, the microorganism is unable to use the compound as a primary substrate or electron acceptor to satisfy its energy needs, regardless of concentration. However, the microorganism may be able to transform the compound if it has enzymes for other purposes that fortuitously interact with and transform the compound. The organism then needs some other compound to serve as a primary substrate for growth and production of the transforming enzyme. An example of trichloroethene cometabolism is given under **Oxidation** in the following.

Hydrolysis (Nucleophilic Substitution)

Hydrolysis is the most common substitution reaction catalyzed by microorganisms. Here, water serves as a nucleophile. Hydrolysis is an especially important transformation reaction with halogenated organic compounds. Brominated compounds are more susceptible to hydrolysis than chlorinated compounds since bromine is a better leaving group than chlorine. All the hydrolysis reactions listed in Sec. 5.34 can be catalyzed by microorganisms.

Bisulfide (HS^-) is an important nucleophile and may take part in a strictly abiotic transformation with halogenated organic compounds. However, HS^- is typically produced through microbial reduction of SO_4^{2-} (Sec. 6.12).

---

[9]T. M. Vogel, C. S. Criddle, and P. L. McCarty, Transformations of Halogenated Aliphatic Compounds, *Env. Sci. Tech.,* **21:** 722–736 (1987).

Oxidation

Microbial oxidation is a process where electrons are released through enzyme-catalyzed reactions. A terminal electron acceptor is then required to complete the process (Sec. 6.12). Some xenobiotics and other toxic compounds may serve as primary substrates and are mineralized (e.g., benzene, toluene, phenol, chlorobenzene, dichloromethane, nitrotoluene). As such, half reactions can be constructed for these compounds as described in Sec. 6.12 and $\Delta G^{0'}$ can be calculated provided the free energy of formation (ΔG_f^0) of the organic compound is known. In general, the more reduced the organic compound, the greater the energy availability through oxidation.

Xenobiotic compounds can also be oxidized via *cometabolism*. One important example is the oxidation of TCE by oxygenases such as methane monooxygenase, toluene monooxygenase, toluene dioxygenase, and ammonia monooxygenase. Toluene dioxygenase (TDO), for example, is involved in toluene oxidation, transforming toluene to *cis*-toluene dihydrodiol.[10]

$$\text{(6.25)}$$

This enzyme has been shown to oxidize TCE by cometabolism to a four-membered ring structure.[11]

$$\text{(6.26)}$$

Growth on the primary substrate toluene results in production of TDO, allowing for the biotransformation of TCE. Similarly, addition of methane stimulates the growth of methanotrophic bacteria and can result in TCE cometabolic transformation (as well as that of some other chlorinated organics) with methane monooxygenase. Here, a TCE epoxide is formed:

$$\text{(6.27)}$$

[10]G. J. Zylstra, L. P. Wackett, and D. T. Gibson, Trichloroethylene Degradation by *Escherichia coli* Containing the Cloned *Pseudomonas putida* F1 Toluene Dioxygenase Genes, *Appl. and Env. Micro.,* **55:** 3162–3166 (1989).

[11]S. Li, and L. P. Wackett, Trichloroethylene Oxidation by Toluene Dioxygenase, *Biochem. Biophys. Res. Commun.,* **185**(1): 443–451 (1992).

Molecular oxygen and energy (NADH), as noted, are required for these reactions to occur. The products of TCE cometabolism can be further degraded to CO_2, formic acid, and glyoxylate, among others. Such reactions have demonstrated potential for treating waters contaminated with some halogenated compounds.[12]

Aromatic rings are typically cleaved by oxidation reactions, although hydrolysis may also be involved in initial steps. For example, the *cis*-toluene dihydrodiol formed by the action of toluene dioxygenase [Eq. (6.25)] can be further oxidized and the aromatic ring cleaved to give a dicarboxylic acid:

$$\text{[structure]} + 2H_2O \rightarrow \text{[structure]} + 6H^+ + 6e^- \tag{6.28}$$

In a similar fashion, xenobiotic aromatic compounds such as *p*-nitrobenzoic acid may be oxidized to a dicarboxylic acid:

$$\text{[structure]} + 2H_2O \rightarrow \text{[structure]} + NO_2^- + 2H^+ + e^- \tag{6.29}$$

$$\text{[structure]} + 2H_2O \rightarrow \text{[structure]} + 4H^+ + 4e^- \tag{6.30}$$

With 2,4-D, hydrolysis and reductive dealkylation may precede oxidation to a dicarboxylic acid:

$$\text{[structure]} + 2H_2O + 2e^- \rightarrow \text{[structure]} + CH_3COOH + 2Cl^- \tag{6.31}$$

$$\text{(structure)} + 2H_2O \rightarrow \text{(structure)} + 4H^+ + 4e^- \qquad (6.32)$$

Side chains such as the —CH_2COOH group in 2,4-D are typically removed before the aromatic ring is cleaved. These carboxylic acids are eventually mineralized to CO_2.

As described, halogenated xenobiotic compounds can often be oxidized by bacteria. In general, the more chlorine atoms a compound has, the less susceptible it is to oxidation and the more susceptible it is to reduction. For example, hexachloroethane (C_2Cl_6) is very resistant to oxidation, but is readily reduced to form tetrachloroethene. However, 1,2-dichloroethane (CH_2ClCH_2Cl) is more susceptible to oxidation, and chloroethane is even more readily oxidized.

Reduction

Through reduction reactions, xenobiotic compounds may serve as electron acceptors. Generally, oxygen, Fe(III), nitrate, sulfate, and carbon dioxide serve as terminal electron acceptors. However, chlorate (ClO_3^-), perchlorate (ClO_4^-), Cr(VI), and halogenated compounds can serve this role as well. General reduction reactions were discussed in Sec. 5.34.

The potential for chlorate, perchlorate, Cr(VI), and halogenated aliphatic compounds to serve as electron acceptors can be evaluated using half-reaction reduction potentials or free-energy considerations. Half-reaction potentials and $\Delta G^{0\prime}$ values for these chemicals are listed in Table 6.6. These data indicate that chlorate, perchlorate, and hexachloroethane (CCl_3CCl_3) are stronger electron acceptors than O_2, Fe(III), and NO_3^-. Also, several chlorinated aliphatic compounds and 1,2-dibromoethane (CH_2BrCH_2Br) may serve as electron acceptors under denitrifying conditions. Cr(VI) is a more thermodynamically favorable electron acceptor than Mn(IV). All compounds listed in Table 6.6 can serve as electron acceptors under sulfate-reducing (SO_4^{2-}) and methanogenic (CO_2) conditions. The general trend is that the more chlorines a compound has (that is, the more oxidized the carbon), the stronger the tendency to serve as an electron acceptor. These thermodynamic considerations can also be visualized by entering the half reactions in the "electron tower" given in Fig. 6.5.

Many of these reductions are cometabolic; that is, they are not coupled with energy production for the microorganism. However, some of these compounds can serve as electron acceptors and generate energy for growth. For example, tetrachloroethene, trichloroethene, *cis*-1,2-dichloroethene, and chloroethene (vinyl chloride) have been shown to act in this capacity. The anaerobic bacteria that use chlorinated compounds in this fashion are termed *dehalorespiring* bacteria—they produce energy while using these compounds as terminal electron acceptors. Chlorate and

Table 6.6 | Half-reaction free energies and reduction potentials for some *novel* and common electron acceptors

Reaction number	Half reaction		$\Delta G^{0\prime}*$ kJ/eq	$E^{0\prime},^{\dagger}$ volts
1	$\frac{1}{6}ClO_3^- + H^+ + e^-$	$= \frac{1}{6}Cl^- + \frac{1}{2}H_2O$	-100.04	1.04
2	$\frac{1}{8}ClO_4^- + H^+ + e^-$	$= \frac{1}{8}Cl^- + \frac{1}{2}H_2O$	-94.06	0.97
3	$\frac{1}{2}CCl_3CCl_3(aq) + e^-$	$= \frac{1}{2}CCl_2{=}CCl_2(aq) + Cl^-$	-92.70	0.96
4	$\frac{1}{4}O_2(g) + H^+ + e^-$	$= \frac{1}{2}H_2O$	-78.72	0.82
5	$Fe^{3+} + e^-$	$= Fe^{2+}$	-74.27	0.77
6	$\frac{1}{5}NO_3^- + \frac{6}{5}H^+ + e^-$	$= \frac{1}{10}N_2(g) + \frac{3}{5}H_2O$	-72.20	0.75
7	$\frac{1}{2}CCl_4(aq) + \frac{1}{2}H^+ + e^-$	$= \frac{1}{2}CHCl_3(aq) + \frac{1}{2}Cl^-$	-56.42	0.58
8	$\frac{1}{2}CCl_3CCl_3(aq) + \frac{1}{2}H^+ + e^-$	$= \frac{1}{2}CHCl_2CCl_3(aq) + \frac{1}{2}Cl^-$	-55.04	0.57
9	$\frac{1}{2}CH_2BrCH_2Br + \frac{1}{2}H^+ + e^-$	$= \frac{1}{2}CH_3CH_2Br + \frac{1}{2}Br^-$	-52.49	0.54
10	$\frac{1}{2}CCl_2{=}CCl_2(aq) + \frac{1}{2}H^+ + e^-$	$= \frac{1}{2}CHCl{=}CCl_2(aq) + \frac{1}{2}Cl^-$	-46.81	0.49
11	$\frac{1}{2}CH_3CCl_3(aq) + \frac{1}{2}H^+ + e^-$	$= \frac{1}{2}CH_3CHCl_2(aq) + \frac{1}{2}Cl^-$	-45.54	0.47
12	$\frac{1}{2}CHCl_3(aq) + \frac{1}{2}H^+ + e^-$	$= CH_2Cl_2(aq) + \frac{1}{2}Cl^-$	-45.52	0.47
13	$\frac{1}{2}CHCl{=}CCl_2(aq) + \frac{1}{2}H^+ + e^-$	$= \frac{1}{2}cis\text{-}CHCl{=}CHCl(aq) + \frac{1}{2}Cl^-$	-44.52	0.46
14	$\frac{1}{2}CHCl{=}CCl_2(aq) + \frac{1}{2}H^+ + e^-$	$= \frac{1}{2}trans\text{-}CHCl{=}CHCl(aq) + \frac{1}{2}Cl^-$	-42.39	0.44
15	$\frac{1}{2}NO_3^- + H^+ + e^-$	$= \frac{1}{2}NO_2^- + \frac{1}{2}H_2O$	-41.65	0.43
16	$\frac{1}{6}HCrO_4^- + \frac{1}{6}CrO_4^{2-} + \frac{5}{2}H^+ + e^-$	$= \frac{1}{3}Cr^{3+} + \frac{4}{3}H_2O$	-39.61	0.41
17	$\frac{1}{2}MnO_2(s) + 2H^+ + e^-$	$= \frac{1}{2}Mn^{2+} + H_2O$	-38.89	0.40
18	$\frac{1}{8}NO_3^- + \frac{5}{4}H^+ + e^-$	$= \frac{1}{8}NH_4^+ + \frac{3}{8}H_2O$	-35.11	0.36
19	$\frac{1}{2}CH_2{=}CHCl(aq) + \frac{1}{2}H^+ + e^-$	$= \frac{1}{2}CH_2{=}CH_2(aq) + \frac{1}{2}Cl^-$	-34.82	0.36
20	$\frac{1}{2}trans\text{-}CHCl{=}CHCl(aq) + \frac{1}{2}H^+ + e^-$	$= \frac{1}{2}CH_2{=}CHCl(aq) + \frac{1}{2}Cl^-$	-31.92	0.33
21	$\frac{1}{2}cis\text{-}CHCl{=}CHCl(aq) + \frac{1}{2}H^+ + e^-$	$= \frac{1}{2}CH_2{=}CHCl(aq) + \frac{1}{2}Cl^-$	-29.79	0.31
22	$\frac{1}{2}CH_3CH_2Br(aq) + \frac{1}{2}H^+ + e^-$	$= \frac{1}{2}CH_3CH_3(aq) + \frac{1}{2}Br^-$	-15.84	0.16
23	$\frac{1}{8}SO_4^{2-} + \frac{19}{16}H^+ + e^-$	$= \frac{1}{16}H_2S(aq) + \frac{1}{16}HS^- + \frac{1}{2}H_2O$	20.85	-0.22
24	$\frac{1}{8}CO_2(g) + H^+ + e^-$	$= \frac{1}{8}CH_4(g) + \frac{1}{4}H_2O$	23.52	-0.24

*$\Delta G^{0\prime}$ calculated using values taken from Table 3.1 or App. A and assuming $\{H^+\} = 10^{-7}$. (*Note:* 1 kcal = 4.184 kJ.)
†Values for $E^{0\prime}$ obtained by dividing $\Delta G^{0\prime}$ by -96.485 kJ/volt-eq.

perchlorate have also been shown to serve as terminal electron acceptors for energy and growth. Reductase enzymes catalyze the reduction of perchlorate to chlorate, and then to chlorite (ClO_2^-). A fairly unique enzyme, a dismutase, then facilitates a disproportionation of ClO_2^- to form Cl^- and O_2.

Fate of Xenobiotic Compounds in Biotic Systems

The fate of xenobiotic chemicals (and other environmentally significant compounds) in the environment and in engineered reactors may be dependent upon the dominant reactions occurring from a variety of possible abiotic and biotic reactions. An example of a compound featuring a variety of possible transformation pathways is 1,1,1-trichloroethane, particularly under anaerobic conditions (Table 6.7). Thus, understanding the environmental fate of even a single xenobiotic chemical may be quite difficult and complex.

6.14 | MOLECULAR BIOLOGY AND GENETIC ENGINEERING

Advances in knowledge in the fields of biochemistry, microbiology, and molecular biology (the study of the transfer of genetic information) are aiding environmental engineers and scientists in their search for solutions to environmental problems. Examples include the exploitation of specific enzymatic pathways to create genetically engineered organisms for bioremediation, development of sensitive probes for detecting the presence of specific bacteria or specific degradation capability, and development of immunoassay techniques for analysis of specific pollutants such as pesticides. The purpose of this section is to briefly introduce the student to the ter-

Table 6.7 | Possible pathways for the transformation of 1,1,1-trichloroethane under anaerobic conditions

Pathway 1		
CH_3CCl_3	$\longrightarrow CH_2{=}CCl_2 + H^+ + Cl^-$	Abiotic elimination
$CH_2{=}CCl_2 + H^+ + 2e^-$	$\longrightarrow CH_2{=}CHCl + Cl^-$	Biotic reduction
$CH_2{=}CHCl + H^+ + 2e^-$	$\longrightarrow CH_2{=}CH_2 + Cl^-$	Biotic reduction
$CH_2{=}CHCl + 4H_2O$	$\longrightarrow 2CO_2 + 11H^+ + 10e^- + Cl^-$	Biotic oxidation
Pathway 2		
$CH_3CCl_3 + H^+ + 2e^-$	$\longrightarrow CH_3CHCl_2 + Cl^-$	Biotic reduction
$CH_3CHCl_2 + H^+ + 2e^-$	$\longrightarrow CH_3CH_2Cl + Cl^-$	Biotic reduction
$CH_3CH_2Cl + H_2O$	$\longrightarrow CH_3CH_2OH + Cl^- + H^+$	Abiotic hydrolysis
$CH_3CH_2OH + 3H_2O$	$\longrightarrow 2CO_2 + 12H^+ + 12e^-$	Biotic oxidation
Pathway 3		
$CH_3CCl_3 + 2H_2O$	$\longrightarrow CH_3COOH + 3Cl^- + 3H^+$	Abiotic hydrolysis
CH_3COOH	$\longrightarrow CH_4 + CO_2$	Methanogenesis
$CH_3COOH + 2H_2O$	$\longrightarrow 2CO_2 + 8H^+ + 8e^-$	Biotic oxidation

Note: Electrons generated from oxidation to CO_2 can be used to generate CH_4, to reduce SO_4^{2-} or NO_3^-, or to reduce additional chlorinated compounds.

Adapted from T. M. Vogel, C. S. Criddle, and P L. McCarty, Transformations of Halogenated Aliphatic Compounds, *Env. Sci. Tech.,* **21:** 722–736 (1987).

minology and principles involved, and the potential uses of these advances. Additional detail can be found in recent biochemistry and microbiology textbooks such as those listed at the end of this chapter.

Terminology and Principles

The guiding principle for these recent advances involves the manipulation of genetic information. The genetic information of all living organisms is contained in *deoxyribonucleic acid (DNA)*. DNA is a polymer made of several million nucleotides, each nucleotide being made up of the six-carbon sugar deoxyribose, a purine or pyrimidine base, and a phosphate. Adenine (A) and guanine (G) are the purine bases, and cytosine (C) and thymine (T) are the pyrimidine bases (Fig. 6.7). The individual nucleotides are hooked together via a phosphate-ester linkage to form a strand of DNA (Fig. 6.8). Most DNA exists as the so-called double helix in which two strands of DNA are bonded together in a complementary fashion where a purine or pyrimidine base on one strand is hydrogen-bonded to a complementary purine or pyrimidine base on the other strand (Fig. 6.9). Most often guanine hydrogen-bonds with cytosine and adenine with thymine. A *gene* is a segment of DNA that contains information ("codes") for the production (or replication) of a specific protein. The transfer of genetic information via DNA usually involves *ribonucleic acid (RNA)*. RNA is similar in structure to a single strand of DNA except that the sugar is ribose, and uracil (Fig. 6.7) replaces thyamine as the second pyrimidine base involved. RNA may be used in several different ways to convert genetic information from DNA into desired products.

A simplified way of looking at the transfer of genetic information from one generation to the next is as follows. A portion of the genetic information contained in DNA is converted into a complementary strand of RNA. This RNA is then transferred to a site in the cell where replication occurs (the ribosome). The sequence of bases in the DNA, now contained in this "messenger" RNA, tells what amino acids and thus what protein or enzyme is to be produced (or replicated). Several types of RNA are involved in the synthesis of a new protein in the ribosome.

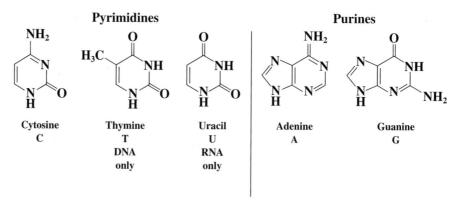

Figure 6.7
Nucleic acid bases of DNA and RNA.

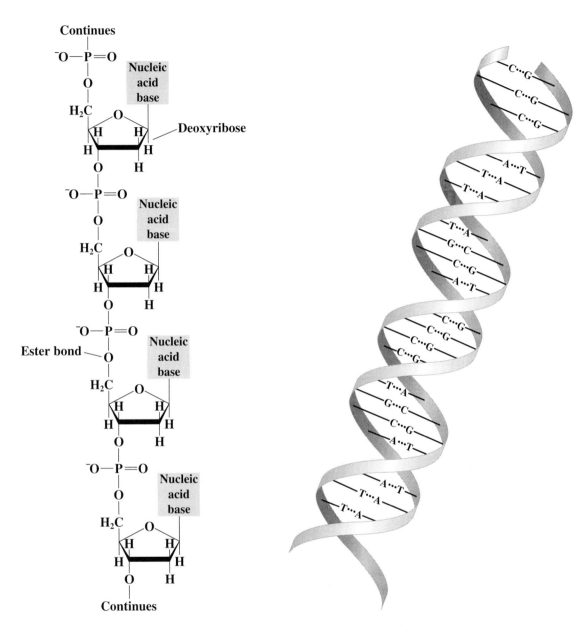

Figure 6.8
Partial structure of a DNA strand.

Figure 6.9
Representation of a portion of the DNA double-helix
structure.

Two forms of DNA are of interest. One is chromosomal DNA, which is the essential form that contains all the basic information required for normal cell metabolism, growth, and reproduction. The other is DNA contained in separate, relatively small, self-replicating circular fragments called *plasmids.* Plasmids contain genetic information that conveys additional capabilities to an organism such as resistance to toxic heavy metals such as mercury or to drugs such as penicillin. A plasmid may also contain genetic information for a key enzyme that allows an organism to initiate a reaction with an organic compound that would otherwise be non-biodegradable by the organism. Another interesting property of plasmids is that they can be exchanged between bacterial species, sometimes resulting in a particularly good combination of enzymes in one organism, allowing it to degrade a new organic chemical that is introduced into the environment. This is an important "adaptive" mechanism for bacteria. It is known that some environmentally significant reactions such as degradation of halogenated organic compounds are encoded on plasmids.[13]

Much of the information contained in DNA and plasmids is unique to a specific organism, a specific degradation pathway, or production of a specific product. Scientists now have the capability of separating and identifying the components (i.e., enzymes, genes, DNA, plasmids, etc.) responsible for these activities. With this capability comes the potential for manipulating, controlling, and understanding various biological reactions. Some specific examples are given next.

Example Applications

New genetic tools (sometimes referred to as "molecular biology tools") are being developed and introduced rapidly. They are having wider application in helping to understand and solve environmental contamination problems. The environmental engineer and scientist is well advised to keep abreast of new developments.

Genetic Engineering   *Genetic engineering,* sometimes called *recombinant-DNA technology,* involves the recombination of DNA from several sources. One successful application of genetic engineering is in the production of human insulin with genetically engineered bacteria. The genes involved in the production of insulin in the human pancreas were characterized, separated, recombined, and introduced into the bacterium *Escherichia coli.* This process is termed *gene cloning.* The *E. coli* then can produce large quantities of insulin precursors that are easily converted to human insulin by chemical means.

Another example involves the enzyme toluene dioxygenase described in Sec. 6.13. The genes responsible for production and activity of this enzyme, along with its ability to biodegrade chlorinated organic compounds such as TCE, were isolated from the bacterium *Pseudomonas putida* F1 and placed into an *E. coli* strain.[14] This

---

[13]B. E. Rittmann, B. F. Smets, and D. A. Stahl, The Role of Genes in Biological Processes: Part I, *Env. Sci. Tech.,* **24:** 23–29 (1990).

[14]G. J. Zylstra, L. P. Wackett, and D. T. Gibson, Trichloroethylene Degradation by *Escherichia coli* Containing the Cloned *Pseudomonas putida* F1 Toluene Dioxygenase Genes. *Appl. and Env. Micro.,* **55:** 3162–3166 (1989).

genetically engineered *E. coli* was then shown to degrade TCE as long as an inducing substrate (toluene or isopropyl-β-D-thiogalactopyranoside in this case) was present. There are indications that enzyme activity such as this can be induced by means other than chemical addition (e.g., temperature).

Genetic engineering has resulted in at least one patented bacterium for use in bioremediation. A. M. Chakrabarty developed a genetically engineered *Pseudomonas* strain by cloning four different plasmids into the organism, giving it the ability to biodegrade a wide variety of petroleum products.[15] Successes like this and the other ones described lead some engineers and scientists to believe that genetic engineering holds significant promise for remediating many hazardous and industrial wastes and that we can develop "designer genes" for the removal of specific pollutants.

The use of genetic engineering is not without controversy. There is concern with release of genetically altered organisms into the environment. A current example is the controversy surrounding use of genetically engineered corn. The bacterium *Bacillus thuringiensis* produces toxins that will kill some types of insects. These insecticidal genes have been cloned into several varieties of corn with the goal of protecting the corn without having to apply insecticides. One major concern is the development of insects that are resistant to these naturally produced toxins. There is also concern with gene transfer to other strains of corn and perhaps other plants. Finally there is concern about the potential human and environmental health effects of long-term exposure to these genetically engineered plants.

Molecular Tools   These tools are being used to assess both bacterial community structure (i.e., molecular ecology—what types of bacteria are present and in what numbers) and biodegradation capability (what types of compounds can be degraded by the organisms present). This field is advancing at a very rapid pace with methods becoming more standard and commercially available. The following is meant to be a brief introduction to some of the more common techniques being used today. All are based on the ability to extract, concentrate (amplify), and detect DNA and RNA that is specific to the organism or degradation capability of interest. Much more detail is contained in the references given at the end of this chapter.

Several methods in current use are based on the sequence of bases contained in the 16S segment of ribosomal RNA. Portions of the base sequence in 16S rRNA are common to most life, while other portions are unique to a given organism. *Gene probes,* today commonly called *oligonucleotide probes,* are single strands of DNA or RNA containing perhaps 15 to 25 bases that, when exposed to the complementary sequence in the target 16S rRNA (or DNA), hybridizes (combines) to form a compound that can be measured. Experimental conditions are controlled such that the probe DNA does not combine with cells that don't contain the complementary sequence of 16S rRNA because the sequences don't match. Detection is typically accomplished by attaching an appropriate label (e.g., radioisotope, fluorescent compound) that can be measured analytically. With some of these techniques, the RNA

---

[15]P. J. VanDenmark and B. L. Batzing, "The Microbes: An Introduction to Their Nature and Importance," Benjamin/Cummings, Menlo Park, CA, 1987.

and DNA must be extracted from the microbial population. Commercially available kits are often used. In one technique, *fluorescence in situ hybridization (FISH),* the probe is tagged with a fluorescent molecule and added to "intact" cells. Here, the probe is contained within the cells, which are fixed and viewed using a fluorescence microscope. In this fashion, the cells of interest can be viewed in contrast to other cells within the community being studied. For example, methanogenic bacteria made to fluoresce could be distinguished from other members of an anaerobic bacterial community converting a carbohydrate to methane.

Another technique involves extraction of DNA from a bacterial community followed by amplification using the *polymerase chain reaction (PCR).* This process is somewhat analogous to enriching for bacteria by adding a specific substrate (e.g., acetate). In general this is done because, at least with environmental samples, bacterial numbers will be low. With PCR, one objective might be to produce many more copies of genes that code for 16S rRNA. These can then be hybridized with oligonucleotide probes that are specific for organisms of interest. For example, a general probe might hybridize with 16S rRNA of all bacterial life in the community. More specific probes can target specific strains of bacteria (e.g., strains of methanogenic or sulfate-reducing bacteria). In this fashion, we can learn about the variety of bacteria in the community and perhaps how the community changes with time. A limitation of this technique is that it works only for those bacteria that have been isolated and for which the ribosomal DNA has been sequenced. It is generally believed that only a few percent of the different microbial strains on earth have been isolated.

A promising modification that allows quantification of bacterial density (or, alternatively, the number of copies of a specific gene of interest) is *real-time quantitative PCR (RTQ-PCR).* A specific sequence in the 16S ribosomal DNA of a target organism [or, alternatively, a gene responsible for specific enzyme activity (e.g., toluene dioxygenase)] is amplified using PCR and a special fluorescent oligonucleotide probe. During PCR, the intensity of fluorescence increases and can be measured and related to bacterial density (or degradative gene density).

Another promising method for quantification of bacterial numbers or, more specifically, the number of copies of a specific target gene is *competitive PCR.* A competitor DNA is designed so that it can be amplified by PCR using the same primers as used to amplify the target DNA of the cells. Primers are compounds added to react with specific sequences of DNA or RNA. Then a known amount of a competitor DNA is added to the DNA extracted from cells. A series of samples is prepared with different amounts of competitor added to each. Then, primers are added, and the target and competitor DNA are amplified together. The amplified target DNA and competitor DNA are then separated by electrophoresis, and the amount of each present is quantified and compared. What is sought is a particular sample in which the amplified amounts of target DNA and competitor DNA are the same. Here, the number of copies of target DNA is equal to the known amount of competitor DNA added to that sample. The procedure is relatively rapid and precise.

There are techniques being developed that allow study of communities containing bacteria that have not been isolated and sequenced. In one such technique, community DNA is extracted and selectively amplified using PCR. Primers are used for this selective amplification and are chosen to amplify for some general or specific function. For example, primers can be very general such as for the 16S rRNA of all bacteria or very specific such as for the 16S rRNA of the dehalorespiring bacteria *Dehalococcoides ethenogenes.* Primers can also be designed to select for genes encoding specific enzymes such as toluene dioxygenase. One way the amplified DNA can be analyzed is by using *denaturing gradient gel electrophoresis* (DGGE). Here, electrophoresis is used to separate amplified DNA based on differences in the base sequences. Distinct bands form, which can represent, for example, different species of bacteria present in a community. It is also possible to remove these bands from the gel for additional analysis. For example, after further amplification, the base sequence can be determined and compared with computer databases containing 16S rRNA sequences for known bacterial species. Such analyses are useful for identifying bacteria.

These methods have been successful in detecting a wide variety of microorganisms such as *Salmonella, Legionella,* sulfate-reducing bacteria, methanogens, and dehalorespirers.[16] They are finding increasing use in assessing community structure and biodegradation capabilities[17,18] and in detecting and quantifying specific bacterial strains.[19]

Immunochemical Techniques *Immunochemical* techniques are special types of molecular biology techniques that involve immunization. The two most common examples are *immunoassays* and *immunochemical probes.* Immunoassay is a term used to describe assays that detect and quantify specific chemicals such as pesticides and PCBs. Immunochemical probes are used to detect specific microorganisms, genes, or enzymes (or enzyme activity).

These methods take advantage of the fact that the immune systems of living organisms respond to the introduction of foreign organisms or chemicals. The general procedure involves immunizing an animal, usually a mouse or rabbit, with a target organism (e.g., methanogenic bacteria), gene, or chemical. These foreign substances, called *antigens,* cause the production of specific proteins called *antibodies* by the animal's immune system. There are many antibody producing cells in the animal and the suite of proteins produced, called *polyclonal antibodies,* meaning clones from many different cells, may respond to many different antigens. Screening procedures are then used to isolate the antibodies that are most useful. In many cases it is important to have *monoclonal antibodies,* that is, antibodies produced by

[16]B. G. Olson, Tracking and Using Genes in the Environment, *Env. Sci. Tech.,* **25:** 604–611, 1991.

[17]S. J. Flynn, F. E. Löffler, and J. M. Tiedje, Microbial Community Changes Associated with a Shift from Reductive Dechlorination of PCE to Reductive Dechlorination of *cis*-DCE and VC, *Env. Sci. Tech.,* **34:** 1056–1061, 2000.

[18]F. E. Löffler, Q. Sun, J. Li, and J. M. Tiedje, 16S rRNA Gene-Based Detection of Tetrachloroethene-Dechlorinating *Desulfuromonas,* and *Dehalococcoides* Species, *Appl. & Env. Microbiol.,* **66:** 1369–1374, 2000.

[19]K. R. Hristova, C. M. Lutenegger, and K. M. Scow, Detection and Quantification of Methyl *tert*-Butyl Ether-Degrading Strain PM1 by Real-Time TaqMan PCR, *Appl. & Env. Microbiol.,* **67:** 5154–5160, 2001.

one type of cell that are specific for a specific antigen. These antibodies are usually obtained from cells taken from the animal's spleen. Since it is very difficult to culture these cells, they are fused with tumor cells to create special organisms called *hybridomas,* which can easily be grown in mass culture to produce relatively large quantities of the specific monoclonal antibody.

Assay procedures involve reacting the antibodies with environmental samples containing the target organism, gene, enzyme, or chemical. In some cases we are just interested in whether the target species is present while in other cases we are interested in the concentration of the target species. A common method of detection is by enzyme linked immunosorbent assays (ELISAs). Here an enzyme is linked to the antibody and its activity is detected and calibrated by the release of a colored product that can be measured spectrophotometrically (Chap. 12). Other methods of detection include tagging antibodies with radioactivity or fluorescent compounds. Immunochemical probes have been developed for organisms such as methanogens and sulfate-reducing bacteria. Major potential advantages of immunochemical techniques are ease of use, low cost, and specificity. For example, immunochemically based home pregnancy tests are widely used.

In the future immunoassay methods may make it possible for homeowners to accurately and inexpensively determine the concentration of chemicals such as atrazine in their drinking water. There are several commercially available immunoassay kits in common use today. Examples include kits for atrazine, cyanazine, alachlor, metolachlor, and 2,4-D, among others.[20] Detection limits are reported to be as low as 0.04 μg/L for some compounds. These kits are semi-quantitative and are currently used primarily as a screening tool; that is, the test kit will determine if the concentration is above or below the detection limit. If an accurate determination of concentration is desired, standard, approved analytical methods (e.g., extraction followed by gas chromatography) must be used. Such methods are described in Part 2 of this text.

6.15 | BIOCHEMISTRY OF HUMANS

Since environmental engineers and scientists are concerned with the disposal of human wastes, it is important that they be familiar with the major changes that organic matter, taken as food, undergoes in its passage through the body.

Carbohydrates

Much of the carbohydrates consumed by humans is utilized by the body. The remainder, consisting of undigestible matter, is eliminated in the feces. Most of the rejected carbohydrate matter is cellulose and other higher polysaccharides for which the human body does not provide enzymes to accomplish its hydrolysis, or for which the detention time in the intestine is too short to complete hydrolysis. The

---

[20]M. Vanderlaan, B. E. Watkins, and L. Stanker, Environmental Monitoring by Immunoassay, *Env. Sci. Tech.,* **22:** 247–254 (1988).

short detention time is aggravated by improper chewing of food and by diarrhetic conditions.

The carbohydrate matter that is assimilated into the blood stream is used for energy, stored as glycogen (animal starch) in muscle tissue and the liver, or converted to fat and stored as fatty tissue. The carbohydrates that are oxidized to produce energy are converted to carbon dioxide and water. The carbon dioxide is carried away, by the blood, from the cells where it is formed. The blood is buffered to such an extent that it can carry considerable amounts of CO_2 and release it to the air in the lungs, in accordance with the principles of Henry's law.

The human body contains a remarkable mechanism for controlling the amount of sugar (glucose) in the blood stream. If excessive amounts accumulate, the excess is released into the urine. This is a part of the kidney function. Persons with diabetes suffer from improper metabolism of sugar. As a result, blood sugar exceeds the amount acceptable to the kidney (renal threshold), and the excess is separated in the kidneys and escapes in the urine. The urine of diabetics shows the presence of glucose consistently. If the carbohydrate intake of a diabetic exceeds the capacity of the kidneys to excrete sugar, blood-sugar levels build up to a point at which the person may pass into a coma.

Fats

Crude fatty materials contain certain substances that are not hydrolyzed in the human alimentary system. These materials and some of the undigested fats are passed in the feces. Fats are hydrolyzed to a considerable extent by lipase in the stomach. Further hydrolysis occurs in the intestine, where the reaction is facilitated through the emulsifying properties of the bile salts. The fatty acids that enter the blood stream are oxidized to produce energy or are stored in fatty tissue for future use. The end products of oxidation are principally carbon dioxide and water, but some ketones, principally acetone, are formed. The carbon dioxide is expelled by the lungs. The ketones are excreted in the urine. Ketones are found in unusual amounts in the urine of diabetics and people suffering from faulty fat metabolism.

Proteins

Hydrolysis of proteins is started in the stomach and continues in the intestine. Amino acids, when released by hydrolysis, are absorbed into the blood stream. Fractions that are not completely hydrolyzed are excreted in the feces. The amino acids are used mainly for the building and repair of muscle tissue, and in these capacities they become fixed in body tissues.

The end products of protein metabolism that require excretion as waste products result principally from two processes: the "wearing" of muscle tissue and oxidation of amino acids to obtain energy. Deamination of amino acids precedes their use as energy sources. The ammonia is released principally as urea, but small amounts of NH_4^+ are normally present. Excretion is by way of the urine. The major function of the kidneys is to separate waste nitrogen compounds from the blood. That protein metabolism involves a variety of complicated processes may be deduced from the considerable number of nitrogenous compounds present in urine. Creatine, creatinine, uric acid, hippuric acid, and traces of purine bases are normally present, in addition to urea and ammonium ion.

Vitamins

Vitamins are very potent organic substances that occur in minute quantities in natural foodstuffs. They must be supplied in the diet of animals if they are not synthesized naturally within the animal from essential dietary or metabolic precursors. Some function as precursors of enzymes; with others, the function is not well understood. In general, they exert a hormone-like or enzymatic action in the control of specific chemical reactions in the animal body, and the absence or lack of a sufficient supply of certain ones leads to the development of vitamin-deficiency diseases, e.g., beriberi, rickets, pellagra, and scurvy.

A wide variety of vitamins are known. They are generally classified into two groups, the *fat-soluble* and the *water-soluble,* as shown in Table 6.8.

The role of vitamins in biological processes employed by environmental engineers and scientists has not been explored. Several of the vitamins are recovered from industrial wastes, particularly those from the fermentation industry, and their economic value has been an important factor in helping to solve the waste-disposal problem in the distilling industry. Activated sludge has been found to be a rich source of vitamin B_{12}. Methanogenic bacteria have also been shown to produce vitamin B_{12}, and this vitamin has been shown to facilitate reductive dehalogenation of halogenated aliphatic hydrocarbons through cometabolism. Environmental engineers and scientists should thus be informed on the subject of vitamins and their significance.

Table 6.8 | Classification and function of some important vitamins

Vitamin		Good sources	Function
Fat-soluble:			
A		Butter, liver oils	Eye health
D		Liver oils, egg	Ca metabolism, i.e., antirachitic
E		Cottonseed oil, cereals	Prevents sterility
K		Green plants, egg yolk	Clotting of blood
Water-soluble:			
B_1	Thiamine	Pork, whole wheat, peanuts	Antiberiberi
B_2	Riboflavin	Eggs, liver, cereals, milk	General health
	Nicotinic acid	Meat, whole wheat, yeast	Antipellagra
B_6	Pyridoxine	Egg yolk, liver, yeast	Skin tone
	Biotin	Egg yolk, liver, yeast	Skin tone
	Pantothenic acid	Egg yolk, liver, milk	Skin tone, growth
	Folic acid group	Green leafy vegetables	Antianemia
	Inositol	Fruits, vegetables	Hair, growth
B_{12}	Cobalamin	Liver, activated sludge	Antianemia
C	Ascorbic acid	Citrus fruits, apples	Antiscurvy

PROBLEMS

6.1 What role do enzymes play in living organisms?

6.2 What terms are used to describe enzymes with respect to (*a*) where their action occurs, and (*b*) the nature of the reaction that they control?

6.3 How does the environmental engineer make use of temperature and pH relationships of biochemical reactions in the design and operation of biological waste treatment facilities?

6.4 Explain how the enzyme methane monooxygenase might be used by environmental engineers and scientists.

6.5 Disposal of the large quantities of bacteria produced during waste treatment is one of the most significant (and expensive!) problems in environmental engineering. Explain why this particular problem might be minimized by anaerobic rather than aerobic waste treatment?

6.6 (*a*) How many acetic acid molecules are produced during the complete beta oxidation of a stearic acid molecule?

   (*b*) How many hydrogen atoms are removed from stearic acid during the beta oxidation of part (*a*)?

6.7 (*a*) Show how *n*-octane is degraded step by step first by omega oxidation (discussed in Chap. 5) and then by beta oxidation. Balance all reactions.

   (*b*) How many acetic acid molecules are produced during the complete beta oxidation of *n*-octane?

   (*c*) How many hydrogen atoms are removed from *n*-octane during this beta oxidation?
   Answers: (*b*) 4; (*c*) 12

6.8 Use of methanol has been proposed to rid a wastewater of nitrate by biological denitrification to N_2.

   (*a*) Write a balanced overall equation for nitrate removal with methanol, using $f_{s(max)}$ and assuming nitrate serves as the nutrient source for bacteria.

   (*b*) If the nitrate (NO_3^-) concentration in the wastewater equals 100 mg/L, what concentration of methanol must be added for complete nitrate removal?

   (*c*) What percentage of the nitrate-nitrogen is used for cell synthesis?

6.9 Acetic acid is a common fatty acid in wastewaters.

   (*a*) Write a balanced overall equation for aerobic oxidation of acetic acid, assuming ammonia is available as a nitrogen source and that $f_{s(max)}$ applies.

   (*b*) How many grams of oxygen are required to oxidize 100 g of acetic acid?

   (*c*) How many grams of bacteria are produced per gram of acetic acid metabolized?
   Answers: (*a*) $0.125CH_3COO^- + 0.0295NH_4^+ + 0.1025O_2 = 0.0295C_5H_7O_2N + 0.007CO_2 + 0.0955H_2O + 0.0955HCO_3^-$; (*b*) 44.5; (*c*) 0.45

6.10 The biological oxidation of Fe^{2+} in mine drainage waters can result in red discoloration in streams from subsequent Fe^{3+} precipitation.

   (*a*) Name the type of organism responsible for this process.

   (*b*) Is this reaction spontaneous? Show calculations to justify your answer.

   (*c*) Write a balanced overall equation for biological oxidation of Fe^{2+}, assuming $f_{s(max)}$ and the presence of ammonia.

   (*d*) How many grams of oxygen are required per gram of Fe^{2+} oxidized?

6.11   Denitrification of wastewaters can be obtained through autotrophic biological reactions.

(a)   Write a balanced equation for denitrification of NO_3^- to N_2 using $S_2O_3^{2-}$ (which is oxidized to SO_4^{2-}), using $f_{s(max)}$ and assuming the bacteria use ammonia as a nitrogen source for cell synthesis.

b)   How many grams of $Na_2S_2O_3$ would need to be added to the wastewater per gram of NO_3^- reduced?
Answers: (a) $0.125S_2O_3^{2-} + 0.04CO_2 + 0.01NH_4^+ + 0.01HCO_3^- + 0.16NO_3^- + 0.055H_2O = 0.25SO_4^{2-} + 0.01C_5H_7O_2N + 0.08N_2 + 0.09H^+$; (b) 1.99

6.12   Answer the following regarding the biodegradation of phenol.

(a)   Construct a half reaction for the oxidation of phenol to $CO_2(g)$.

(b)   Calculate $\Delta G^{0\prime}$ for this reaction, given that the free energy of formation for phenol(*aq*) is -47.5 kJ/mol.

(c)   For anaerobic conditions and pH = 7, is the conversion of phenol to $CH_4(g)$ thermodynamically favorable?

(d)   For anaerobic conditions and $f_s = 0.10$, construct a balanced reaction for the conversion of phenol to methane.

(e)   For a waste containing 1000 mg/L of phenol, what volume of methane is produced with 98 percent phenol utilization?

6.13   Biological nitrification is a process where ammonium (NH_4^+) is converted to (NO_3^-) by aerobic bacteria.

(a)   Write a balanced overall reaction for nitrification using $f_{s(max)}$.

(b)   How much dissolved oxygen is required to remove 25 mg/L of $NH_4^+ - N$?

(c)   What concentration of bacteria is produced?

(d)   What concentration of alkalinity (as $CaCO_3$) is consumed?
Answers: (a) $0.13NH_4^+ + 0.225O_2 + 0.005HCO_3^- + 0.02CO_2 = 0.125NO_3^- + 0.005C_5H_7O_2N + 0.25H^+ + 0.12H_2O$; (b) 98.9; (c) 7.8; (d) 175

6.14   Approximately 2500 dry tons of dewatered wastewater treatment plant sludge (chemical formula is $C_{10}H_{19}O_3N$) have been disposed in a sanitary landfill. Assuming anaerobic conditions, how many cubic feet of methane can be formed from the biological degradation of this sludge? You will have to assume an f_s to work this problem; justify your assumption.

6.15   At midnight on January 1, 1992, an employee of the M. T. Head Waste Disposal Corporation empties 2 million gallons of an industrial wastewater containing 1000 mg/L of benzoate (benzoic acid) and 500 mg/L propionate (propionic acid) from a tank truck into Innocent Lake. The lake has 500 million liters of water containing 7.5 mg/L dissolved oxygen and no nitrate. Assuming no atmospheric reaeration takes place, that sufficient ammonia-N is available for bacterial growth, and that the lake is completely mixed, will Innocent Lake go anaerobic due to the biological activity stimulated by discharge of this wastewater?
Answer: yes

6.16   Recently anaerobic "dehalorespiring" bacteria have been found that will use tetrachloroethene (PCE) as an electron acceptor in support of growth. These organisms have been shown to use molecular hydrogen (H_2) as an electron donor for growth.

(a)   Develop a balanced half reaction for the reduction of PCE(*aq*) to ethene(*aq*) and calculate $\Delta G^{0\prime}$ for your half reaction. You will need to use data contained in App. A.

(b) Assuming $f_s = 0.05$ (comment on why this is a reasonable choice for f_s), determine how much $H_2(g)$ is required to convert 50 g of PCE to ethene.

(c) Calculate the mass (g) of bacteria produced.

(d) Does the pH increase, decrease, or remain constant as a result of this transformation?

6.17 Perchlorate (ClO_4^-) is an oxyanion that has been found in some groundwaters especially near defense department sites. It has been used as a solid rocket fuel ("oxygen source"). Recently, several bacteria have been isolated that can use perchlorate as an electron acceptor for growth, reducing ClO_4^- to Cl^-. Typically, perchlorate degradation by these organisms requires addition of a carbon and energy source. One such compound that can serve both purposes is acetate.

(a) Is the combination of perchlorate as an electron acceptor and acetate as an electron donor thermodynamically favorable? Justify your answer.

(b) If so, and if you wanted to develop a balanced biochemical reaction (that is, R), what value would you use for f_s and why? Consult Table 6.5 in determining your answer.

(c) Using your value for f_s calculate how many mg/L of acetate will be required to biodegrade 50 mg/L of ClO_4^-

6.18 Given the following six pairs of compounds, for each pair, which of the two compounds is likely to be the easiest to degrade. Why? Be specific!

(a)

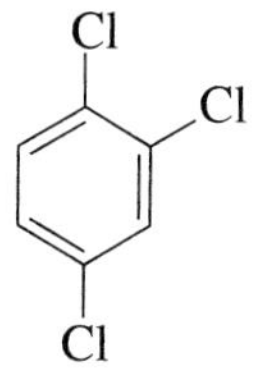

or

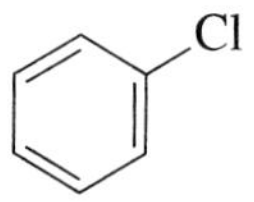

(b)

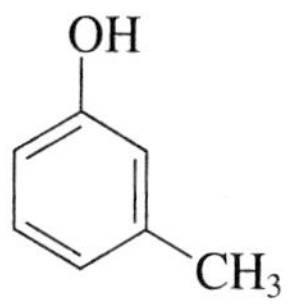

or

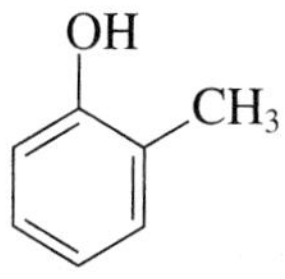

(c)

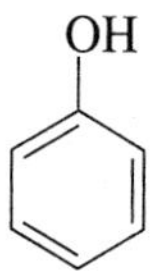

or

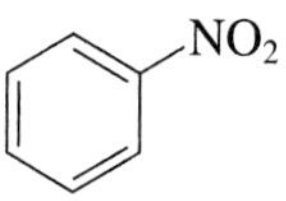

(d) $RCOOCH_2R'$ or RCH_2OCH_2R'

(e) RCH_2CHO or RCH_2COCH_3

(f) $RCH_2C(CH_3)_2CH_2COOH$ or $RCH_2CH(CH_3)CH_2COOH$

6.19 An organic chemist has developed a new pesticide with the hypothetical structure shown. Based on your expertise of degradation pathways, suggest a likely means (pathway) whereby this compound could be completely mineralized. Include appropriate environmental conditions (e.g., aerobic, anaerobic). Balance your reactions to the extent possible.

$$\text{Cl}-\underset{}{\bigcirc}-\text{NH}-\overset{\overset{\text{O}}{\|}}{\text{C}}-(\text{CH}_2)_4-\text{CH}_3$$

6.20 Biotransformation of 1,1,1-trichloroethane (CH_3CCl_3) in anaerobic biological systems can be accomplished by oxidation or reduction.

(a) Construct a half reaction for the reduction of CH_3CCl_3 to inorganic chloride.

(b) In a system fed acetate, construct a half reaction that could produce the electrons required for part (a).

(c) Add half reactions from parts (a) and (b) to yield a balanced reaction for reductive dechlorination of CH_3CCl_3 in an anaerobic biological system fed acetate.

(d) Construct a half reaction for the oxidation of CH_3CCl_3 to carbon dioxide and inorganic chloride.

(e) For an anaerobic biological treatment system, propose a half reaction that will use the electrons generated in part (d).

(f) Add half reactions from parts (d) and (e) to yield a balanced reaction for the oxidation of CH_3CCl_3 in an anaerobic biological system.

6.21 Rank the following organic compounds in order from easiest to oxidize to most resistant to oxidation. Provide justification for your answer.

(a) Carbon tetrachloride

(b) Dichloromethane

(c) Hexachloroethane

(d) Trichloroethene

(e) Vinyl chloride

6.22 A contaminated groundwater contains the following six xenobiotic chemicals: chlorobenzene, dichloromethane, hexachloroethane, pentachlorophenol, tetrachloroethene, and vinyl chloride. Propose a bioremediation scheme, based on terminal electron acceptor or acceptors to be used, that will remove all compounds to the maximum extent feasible.

6.23 Explain the difference between DNA and RNA.

6.24 Why are plasmids of importance to environmental engineers and scientists?

6.25 What is a genetically engineered microorganism? How might an environmental engineer make use of such an organism?

6.26 Explain in general how oligonucleotide probes work. What is a limitation of using such probes?

6.27 Explain how FISH might be used to determine whether a specific microbial species is present in a bacterial community.

6.28 What is the polymerase chain reaction?

6.29 Explain how DGGE might be used by environmental engineers and scientists.

6.30 What are monoclonal antibodies?

6.31 What is an immunoassay? How do environmental engineers and scientists use this technique?

REFERENCES

Berg, J. M., J. L. Tymoczko, and L. Stryer: "Biochemistry," 5th ed., W.H. Freeman, New York, 2002.

Denniston, K. J., J. J. Topping, and R. L. Caret: "General, Organic, and Biochemistry," 3rd ed., McGraw-Hill, New York, 2000.

Madigan, M. T., J. M. Martinko, and J. Parker: "Brock Biology of Microorganisms," 9th ed., Prentice-Hall, Inc., Upper Saddle River, NJ, 2000.

Maier, R. M., I. L. Pepper, and C. P. Gerba: "Environmental Microbiology," Academic Press, San Diego, 2000.

Mathews, C. K., K. E. van Holde, and K. G. Ahern: "Biochemistry," 3rd ed., Benjamin Cummings, San Francisco, 2000.

Nelson, D. L., and M. M. Cox: "Lehninger Principles of Biochemistry," 3rd ed., Worth Publishers, New York, 2000.

Rittmann, B. E., and P. L. McCarty: "Environmental Biotechnology, Principles and Applications," McGraw-Hill, New York, 2001.

CHAPTER 7

Basic Concepts from Colloid Chemistry

7.1 | INTRODUCTION

Colloid chemistry is concerned with dispersions. These may exist in solids, liquids, or gases. Of most environmental importance are those that occur in liquids or in gases. Eight classes of colloidal dispersions are known, as shown in Table 7.1. Classes 4 to 8 of this table are commonly encountered in environmental engineering and science, and classes 5 to 8 are readily recognized by the public by their common names.

Size

Colloidal dispersions consist of discrete particles that are separated by the dispersion medium. The particles may be aggregates of atoms, molecules, or mixed materials that are considered larger than individual atoms or molecules but are small enough to possess properties greatly different from coarse dispersions. Colloidal particles normally range in size from about 1 to 1000 nanometers (nm), or 0.001 to 1 micrometer (μm), and for the most part are not visible even with the aid of the ordinary optical microscope. Their relation to other dispersions is as follows:

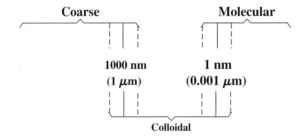

Colloidal dispersions may be considered as ultrafine dispersions and occupy a size range between fine and molecular. The boundaries between fine, colloidal and molecular dispersions are by no means hard-and-fast values. In fact, many consider colloids in water to be those particles with diameters less than 10 μm.

Table 7.1 | Classes of colloidal dispersions

Class	Dispersed phase	Dispersion medium	Common name
1	Solid	Solid	
2	Liquid	Solid	
3	Gas	Solid	
4	Solid	Liquid	Sol
5	Liquid	Liquid	Emulsions
6	Gas	Liquid	Foams
7	Solid	Gas	Smokes
8	Liquid	Gas	Fogs

Methods of Formation

Any material that is reasonably insoluble in the dispersion medium can be caused to form a colloidal dispersion. Colloidal-sized particles can be produced by grinding coarse materials. Devices designed for such purposes are called *colloid mills.* Colloidal particles are formed in considerable amounts in hard rock drilling and blasting operations. Colloidal-sized particles may be formed from ions that react to form insoluble compounds. Under the proper conditions, aggregates of molecules result that do not grow into crystals of a size large enough to settle or be filtered out. This often happens in gravimetric analysis, as discussed under crystal growth in Sec. 3.6. Natural, physical weathering of minerals such as iron oxides and silicates can also produce colloids. Microorganisms, living and dead, such as viruses, bacteria, and algae, when not clumped together, can form colloidal dispersions.

Certain organic substances and compounds that are considered soluble in water do not form true solutions; instead they form colloidal dispersions. Soap, starch, gelatin, agar-agar, gum arabic, and albumin are examples. Bentonite, a volcanic clay, is an example of an inorganic material that acts likewise. In these cases the dispersion medium, water, has the ability to disintegrate the material sufficiently to carry it into colloidal suspension, but it does not necessarily have the ability to complete the dispersion into molecular particles. In certain cases, such as with gelatin, gum arabic, albumin, and humic materials, the individual molecules may be so large as to fall into the colloidal range, even though dispersion might be complete.

General Properties

When discussing colloidal dispersions the term *stability* is often used. In this context, stability can be thought of as the resistance of the colloid to removal by settling or filtration. The stability of colloids is due to their size and electrical properties and is affected by the chemical nature of the colloid and the chemistry of the dispersion medium (e.g., the ionic strength, pH, and organic content of water). In order to remove colloids, they must somehow be destabilized. This is discussed further in Sec. 7.2.

Because colloidal particles are so small, their surface area in relation to mass is very great. Some concept of this relation can be obtained by consideration of how the surface area of a cube 1 cm in side length increases when it is reduced to colloidal-sized cubes. If colloidal-sized cubes of 10 nm are formed, the surface area is increased from 6 cm² to 600 m², or about $\frac{1}{7}$ acre. It is difficult to conceive of such a small mass of material having such a tremendous surface area. As a result of this large area, surface phenomena predominate, and control the behavior of colloidal suspensions, so much so that colloidal chemistry is often considered synonymous with surface chemistry. The mass of colloidal particles is so small that gravitational effects are unimportant.

Electrical Properties   All colloidal particles are electrically charged. The charge varies considerably in its magnitude with the nature of the colloidal material and may be positive or negative, as shown in Fig. 7.1. Many colloidal dispersions are dependent upon the electrical charge for their stability. Like charges repel, and as a result, similarly charged colloidal particles cannot come close enough together to agglomerate into larger particles.

The *electrokinetic* properties of colloids are of great importance. As destabilization and removal of colloids depends upon a knowledge of them. More detailed discussion is given in Sec. 7.2.

When colloidal particles are placed in an electric field, the particles migrate toward the pole of opposite charge. This phenomenon is known as *electrophoresis* and is used extensively to determine the nature of the charge on the colloidal particle and other properties.

Brownian Movement   Colloidal particles are bombarded by molecules of the dispersion medium, and because of their small mass, the colloids move about under the impetus of the bombardment in a helter-skelter manner. It was originally thought to be a characteristic of living matter, but a botanist, Robert Brown, showed in 1827 that nonliving material exhibited this same phenomenon. The term *Brownian movement* has been used to describe this action, whether the particles are living or inanimate.

Tyndall Effect   Because some colloidal particles have dimensions greater than the average wavelength of visible light, they interfere with the passage of light. Light that strikes them may be reflected. As a result, a beam of light passing through such a colloidal suspension is visible to an observer who is at or near right angles to the beam of light. This phenomenon is called the *Tyndall effect* in honor of the English physicist who studied it extensively. This test is often used to prove the presence of a colloid, as true solutions and coarse suspensions do not produce the phenomenon. Oftentimes rays of sunlight piercing between clouds are seen when the atmosphere

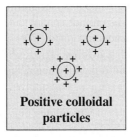
Positive colloidal particles

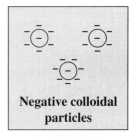
Negative colloidal particles

Figure 7.1
Positive and negative colloidal particles.

is charged with colloidal dust particles. Students often see the Tyndall effect illustrated in classrooms when chalk or other dust is present in colloidal form in the air. The Tyndall effect is used as one basis of determining turbidity in water.

Adsorption Colloids have tremendous surface area and, of course, great adsorptive powers. Adsorption is normally preferential in nature, some ions being chosen and others excluded. This selective action yields charged particles and is the fundamental basis of the stability of many colloidal dispersions. In addition to this primary role adsorption plays in determining the surface charge of colloids, the large surface area available also represents a great potential for adsorption of environmentally significant ions and molecules. Aspects of adsorption to surfaces were discussed in Secs. 3.12, 4.9 (adsorption at the solid-water interface), and 5.34 (sorptional-partitioning).

Effect on Freezing and Boiling Point Colloidal dispersions affect the freezing and boiling points of liquids as do dispersions of other particles; see Raoult's law (Sec. 2.10). Their effect, however, is not measurable with ordinary instruments, because the actual number of particles is so very few as compared with Avogadro's number. This is so because colloidal suspensions normally fall into the realm of very dilute solutions, and, in addition, each particle is made up of perhaps hundreds or even thousands of molecules.

Dialysis Colloids, because of their large particle size, do not pass through ordinary semipermeable membranes. Thus, a separation of crystalloids and colloids can be accomplished by dialysis, as discussed in Sec. 3.7.

Environmental Significance of Colloids

The traditional focus on colloids has been the removal of turbidity (Chap. 13) from drinking waters and the dewatering of sludges produced during water and wastewater treatment. A more recent focus is the role of colloids in the fate, transformation, and transport of pollutants in the aqueous environment and in engineered reactors. Metals and other ions, a wide variety of organic compounds, and radionuclides can be adsorbed by colloids. Various transformation reactions such as oxidation and reduction can be catalyzed by colloidal surfaces. The small size of colloids may allow them to be transported relatively long distances from their origin, or carried out with the effluent in engineered reactors. This is particularly true in the subsurface environment where contaminants thought to be immobilized on aquifer solids can be transported via colloids. Colloids have been shown to be of major importance in the fate, transformation, and transport of metals,[1] toxic organic compounds,[2,3] viruses, and radionuclides.[4] Recent advances in "nanotechnology" may

[1] W. Stumm, "Chemistry of the Solid-Water Interface," Wiley-Interscience, New York, 1992.

[2] Y. P. Chin and P. M. Gschwend, Partitioning of Polycyclic Aromatic Hydrocarbons to Marine Porewater Organic Colloids, *Env. Sci. Tech.,* **26:** 1621 (1992).

[3] R. P. Schwarzenbach, P. M. Gschwend, an D. M. Imboden, "Environmental Organic Chemistry," Wiley-Interscience, New York, 1993.

[4] J. F. McCarthy and J. M. Zachara, Subsurface Transport of Contaminants, *Env. Sci. Tech.,* **23:** 496 (1989).

allow the use of commercially produced colloidal nanoparticles (5 to 200 nm diameter) for catalytic destruction of environmental pollutants.[5]

Although most of this discussion emphasizes the importance of colloids in water, colloidal dispersions are similarly important in determining the fate, transformation, and transport of metals, organic compounds, radionuclides, biological particles, and other environmentally significant compounds in the atmosphere.

Nomenclature

The nomenclature applied to colloidal systems varies considerably with the type; consequently, there are few terms generally applicable.

7.2 | COLLOIDAL DISPERSIONS IN LIQUIDS

Colloidal dispersions of solids, liquids, and gases in liquids are commonly encountered in environmental engineering and science. The nomenclature and behavior of each type differ somewhat; consequently, discussions of each will be given.

Solids in Liquids

Colloidal dispersions of solids in liquids are generally of two types, those that bind strongly with the liquid and those that do not. Dispersions binding strongly with the liquid are generally more stable and difficult to separate from the liquid than those that do not. Colloids that bind strongly with water are termed *hydrophilic* (water loving), and those that do not are termed *hydrophobic* (water hating). Colloidal dispersions of solids in liquids are often referred to as *sols* or *suspensoids*.

Hydrophobic Colloids   Hydrophobic colloids are all electrically charged. The primary or surface charge may be developed in several ways and may be positive or negative. The sign and magnitude of the primary charge is a function of the character of the colloid, and the pH and general ionic characteristics of the water. A lower water pH tends to make colloids more positive or less negative.

In the realm of liquids, the greatest concern is with the removal of colloidal solids from water and wastewater. These colloids are not always well-defined hydrophobic sols, but they do lend themselves to separation when treatment designed to remove hydrophobic sols is used. The natural coloring matter of surface waters and the colloidally suspended matter of domestic wastewaters are examples of quasi hydrophobic-hydrophilic, negatively charged colloids encountered in environmental engineering and science practice.

Electrokinetic Properties   The stability of hydrophobic colloids depends upon the electrical charge they possess. This primary charge results from charged groups within the particle surface in combination with that gained by adsorption of a layer of ions from the surrounding medium, as illustrated in Fig. 7.2.

---

[5]D. W. Elliot and W. X. Zhang, "Field Assessment of Nanoscale Bimetallic Particles for Groundwater Treatment," *Env. Sci. Tech.,* **35:** 4922–4926 (2001).

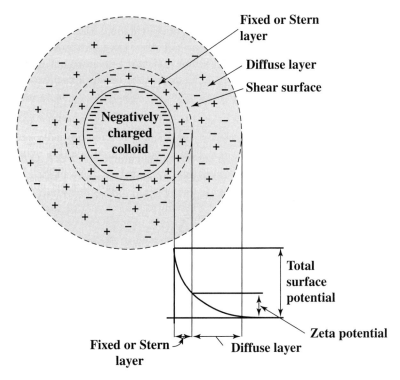

Figure 7.2
Electrical double layer of a negatively charged colloid.

A sol considered as a whole cannot have a net charge, so the charge that a given particle may possess must be counterbalanced by ions of opposite charge *(counter-ions)* near the surface and in the solution phase. This need for electroneutrality results in an *electric double layer* comprised of (1) a *fixed layer* (sometimes termed the Stern layer) of oppositely charged ions adsorbed to the colloid surface and (2) a *diffuse layer* of a mixture of charged ions as shown in Fig. 7.2. A shear surface separates the fixed and diffuse layers and defines the mobile portion of the colloid. The fixed layer, which includes bound water, moves with the colloid. Although the counter-ions of the fixed layer are attracted electrostatically and are thus concentrated in the interfacial region, they are rather loosely held and may diffuse away in response to thermal agitation, to be replaced by other ions. These competing forces (electrical attraction and diffusion) spread the charge over the electrical double layer such that the concentration of counter-ions is greatest at the surface and decreases gradually with distance from the surface. When the water contains a high concentration of ions (high ionic strength), the electrical double layer would obviously be compacted. It would thus occupy a smaller volume and would extend less far into solution.

Because of the primary charge on the particle, an electric potential exists between the surface of the particle and the bulk of the solution. The charge is a maximum at the particle surface and decreases with distance from the surface. When two similar colloidal particles with similar primary charge approach each other, their electrical double layers begin to interact. As they come closer, the similar primary charges they possess result in repulsive forces. The closer the particles approach, the stronger the repulsive forces.

These repulsive forces which keep particles from aggregating are counteracted to some degree by an intermolecular attractive force termed *van der Waals' force*. All molecules and colloidal particles possess this attractive force regardless of charge and composition. The magnitude of this force is a function of the composition and density of the colloid but is independent of the composition of the aqueous phase. The van der Waals' force decreases rapidly with increasing distance between the particles.

Figure 7.3 illustrates the effect of separating distance between two particles on the net force that exists between them. As two similar particles approach each other, the repulsive electrostatic forces increase to keep them apart. However, if they can be brought sufficiently close together to get past this energy barrier, the attractive van der Waals' force will predominate, and the particles will remain together. If it is desired to destabilize and coagulate colloidal particles, then they must be given sufficient kinetic energy to overcome the energy barrier that exists, or else the energy barrier must be lowered by some means. This theory that colloids are stable because of these attractive and repulsive forces is termed the DLVO theory after the work of Derjagin, Landau, Verwey, and Overbeek. Although it appears to do an adequate

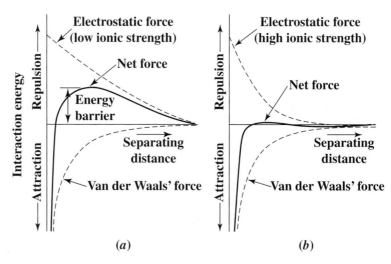

Figure 7.3
The effect of liquid ionic strength and separating distance between colloids on the forces of interaction between them.

job of describing colloidal stability, recent work has shown that it does not adequately explain the kinetics of chemical destabilization of colloids.[6,7]

In order to understand colloidal stability and what conditions will be necessary for destabilization, it is useful to know the surface charge, surface potential, and the point of zero charge. Surface charge (typical units are coulombs/m^2) can be determined experimentally using techniques such as potentiometric titrations. Surface potential (in volts or millivolts) cannot be directly measured. However, it can be estimated by at least two methods. One method is to calculate it using measured values of surface charge and electrical-double-layer theory as described in detail by Stumm.[8] A second method is to estimate it using measurements of colloid movement in an electric field (called electrophoretic mobility) to calculate the *zeta potential*, ζ. As indicated in Fig. 7.2, the zeta potential is the potential at the shear surface and is lower than the true surface potential of the colloid.

Zeta potential is defined by the equation

$$\zeta = \frac{4\pi\delta q}{D} \tag{7.1}$$

where q = charge at shear surface

δ = thickness of diffuse layer

D = dielectric constant of liquid

Electrophoretic mobility is measured as a velocity in an applied electric field; zeta potential can then be calculated from these measurements. For example, at 25°C

$$\zeta = \frac{12.9v}{V_F} \tag{7.2}$$

where ζ = zeta potential, mV

v = measured velocity, μm/s

V_F = applied electric field, volts/cm

Measurements of zeta potential can give an indication of the effectiveness of added electrolytes in lowering the energy barrier between colloids (that is, reducing surface potential and electrical-double-layer thickness), and thus can serve to guide the selection of optimum conditions for coagulation.

The *point of zero charge* is defined as that condition where the surface charge is zero.[9] It is often called the *isoelectric point.* pH is often the parameter governing surface charge and the pH at the point of zero charge is termed pH$_{pzc}$. The effect of

---

[6]C. R. O'Melia, Kinetics of Colloid Chemical Processes in Aquatic Systems, in, W. Stumm (ed.), "Aquatic Chemical Kinetics," Wiley-Interscience, New York, 1990.

[7]M. Elimelech and C. R. O'Melia, Kinetics of Deposition of Colloidal Particles in Porous Media, *Env. Sci. Tech.,* **24:** 1528 (1990).

[8]Stumm "Chemistry of the Solid-Water Interface."

[9]G. A. Parks, Surface-Energy and Adsorption at Mineral-Water Interfaces: An Introduction, *Rev. in Mineralogy,* **23,** 133–175, 1990.

pH on surface charge for three colloids of significance is shown in Fig. 7.4. As can be seen, clay particles, which can impart turbidity to water supplies, have a different charge behavior than do iron and aluminum colloids. Salts of iron and aluminum are often added as coagulants to water and wastewater to destabilize clay and other turbidity associated colloids by processes described in the following section on Destabilization and Coagulation. As indicated in the figure, the point of zero charge for the clay colloid is near pH 4, while that for the coagulants is much higher. That for iron (α-FeOOH) colloids is almost 8 and for aluminum oxide (δ-Al$_2$O$_3$) colloids about 9. In many cases colloids are least stable; that is, they tend to aggregate, at pH$_{pzc}$. The value of pH$_{pzc}$ for colloids such as metal oxides can have a significant impact on the adsorption of metals, ions, and other molecules.

Destabilization and Coagulation   Colloidal particles are too small to be removed by gravitational settling alone. However, if the colloids are destabilized by causing them to aggregate or coagulate into larger particles, they can be effectively re-moved. There are four basic mechanisms by which colloids can be coagulated: (1) double-layer compression, (2) charge neutralization, (3) entrapment in a precipi-tate, and (4) interparticle bridging.

If a high concentration of an electrolyte is added to a sol, the concentration of ions within the electrical double layer will increase, and hence the thickness of this layer will decrease. The addition of counter-ions with higher charge, such as Ca^{2+} instead of Na$^+$, will have a similar effect. This will result in a greater decrease in charge with distance from the particle interface, resulting in a decrease or perhaps an elimination of the potential barrier, as illustrated in Fig. 7.3. With a reduced or eliminated energy barrier, particles can come together and aggregate.

The charge on a colloid can sometimes be neutralized by addition of molecules of opposite charge that have the ability to adsorb onto the colloid. For example, pos-itively charged organic molecules such as dodecylammonium ion, C$_{12}$H$_{25}$NH$_3^+$, tend to be hydrophobic and readily adsorb to negatively charged colloidal particles. The

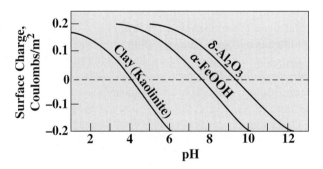

Figure 7.4
Effect of pH on effective surface charge of a typical clay colloid, and iron and aluminum colloids (after W. Stumm and J. J. Morgan, "Aquatic Chemistry, Chemical Equlibria and Rates in Natural Waters," 3rd ed., John Wiley & Sons, New York, 1996.)

opposite charges of the organic and the colloid cancel each other out, and coagulation results. An overdose of dodecylammonium ion, however, can result in charge reversal and the formation of a stable but positively charged particle.

When sufficient quantities of Al(III) and Fe(III) salts are added to a solution, they will form hydroxide precipitates [e.g., $Al(OH)_3(s)$ and $Fe(OH)_3(s)$]. Colloidal particles may provide condensation sites where the precipitates may form and hence the colloids become entrapped in the precipitate and settle with it. The settling precipitate can also entrap colloids that it passes, bringing them down and removing them from water.

Finally, long-chained charged synthetic and natural polymers *(polyelectrolytes)* can act to destabilize colloids by forming a bridge between one colloid and another. One charge site on the long polymer can attach or adsorb to a site on one colloid, while the remainder of the polymer molecule extends out into solution. If the extended portion becomes attached to another colloid, then the two colloids are effectively tied together. Commonly, a negatively charged polymer is most effective in bridging between negative colloids. This is believed to result from interaction between the polymer and specific sites on the colloid, some of which may be positive even though the overall charge of the colloid is negative.

There are essentially four different ways in which the above mechanisms may be applied in the destruction of colloids. Only two are applied to a significant extent in environmental engineering practice. The four methods are (1) boiling, (2) freezing, (3) addition of electrolytes, and (4) mutual precipitation by addition of a colloid of opposite charge.

Boiling of a hydrophobic colloidal suspension often results in coagulation of the colloidal particles. This action is not usually attributed to a reduction in the surface potential but rather to modification in the degree of hydration of the particles, or to increased kinetic velocities, permitting the particles to overcome the energy barrier that separates them (Fig. 7.3). Chemists often boil materials to accomplish coagulation of colloids, but boiling is generally too expensive for application in environmental engineering practice.

Freezing is another method by which colloids may be coagulated. During the freezing process, crystals of relatively pure water form. Thus, the colloidal and crystalloidal materials are forced into a more and more concentrated condition. Two additive effects cause coagulation to occur. As the colloidal suspension becomes more concentrated, opportunity for close contact increases. At the same time, the concentration of electrolytes increases, resulting in a decrease in the diffuse layer thickness. The net result is coagulation of the colloid.

Freezing has been proposed as a practical means of destroying the colloidal character of sludges in preparation for dewatering. However, except perhaps in cold climates, it is usually less costly to condition sludges of filtration by the use of chemicals.

The most common method of destabilizing hydrophobic colloids is by the addition of electrolytes. One way in which electrolytes act is by double-layer compression. A sufficient concentration of monovalent ions, such as NaCl, can bring about coagulation in this way. However, it has been noted that salts having divalent ions of charge opposite to that of the colloidal particle exert far greater coagulation powers. Salts having trivalent ions of opposite charge are even more effective. The significance of the relation between ionic charge and precipitating power was first pointed out by Schulze and verified by Hardy. Their findings are usually called the *Schulze-Hardy rule,* which states: *The coagulation of a colloid is effected by that ion of an added electrolyte that has a charge opposite in sign to that of the colloidal particle, and the effect of such ion increases markedly with the number of charges it carries.*

Table 7.2 lists a number of electrolytes and gives their relative coagulating powers for positive and negative colloids. From the data given in Table 7.2, it becomes obvious why aluminum and iron salts are so widely used as coagulants in environmental engineering practice. The greater value of the sulfate salts as compared with the chloride salts is not readily apparent but will be discussed shortly. Multivalent ions of opposite charge are considered able to penetrate the diffuse layer of colloidal particles and in this way neutralize, in part, the charge on the colloid.

The salts of Al(III) and Fe(III) used in coagulation of water to remove colloidal color and turbidity can function as coagulants by several mechanisms. When Al(III) and Fe(III) salts are added to water, they ionize to yield the trivalent, free metallic ions (Al^{3+}, Fe^{3+}), the amount and life of which are a function of water characteristics such as pH and alkalinity. Some of the trivalent metal ions undoubtedly reach the target and neutralize the charge on some of the colloidal particles. The majority of the trivalent metal ions, however, combine with hydroxide ions to give a variety of hydroxide complexes, including some polynuclear complexes [e.g., $Al_{13}(OH)_{34}^{5+}$], most of which carry a positive charge and can adsorb onto the colloid, causing charge neutralization and/or double-layer compression. These destabilized colloids can now come together to form larger particles that settle readily. If the amount of positively charged metal-hydroxide complexes is much in excess of the amount needed to react with the negatively charged color or turbidity particles, charge reversal can occur in much the same manner as described previously for dodecylammonium ion. The excess metallic hydroxides, some of which may be colloids, must

Table 7.2 | Relative coagulating power of several electrolytes

Electrolyte	Relative power* of coagulation	
	Positive colloids	Negative colloids
NaCl	1	1
Na_2SO_4	30	1
Na_3PO_4	1000	1
$BaCl_2$	1	30
$MgSO_4$	30	30
$AlCl_3$	1	1000
$Al_2(SO_4)_3$	30	>1000
$FeCl_3$	1	1000
$Fe_2(SO_4)_3$	30	>1000

*Values given are approximate and are for solutions of equivalent ionic strength.

be removed in some manner. This is where the negative ion of the Al(III) and Fe(III) salts may be of importance. If sulfate salts are used instead of chloride salts, the divalent sulfate ions act to compress the diffuse layer of the metal hydroxide colloid and thereby help to complete the coagulation of the colloidal system (i.e., the coming together to form particles large enough to settle in a timely fashion).

At higher doses of Al(III) and Fe(III) salts, hydroxide precipitates are rapidly formed. The mechanism of removal here is termed "sweep floc" as colloidal particles are enmeshed in the metal precipitate and thus "swept" from solution. These reactions are quite complicated. Figure 7.5 illustrates the effect of pH on the solubility of $Al(OH)_3(s)$. Because of the many complexes that aluminum (and iron) can form with hydroxide, aluminum (iron also) is more soluble at both higher and lower pH values. Coagulation is usually best carried out at the pH of lowest solubility. As noted in the previous paragraph, other polynuclear hydroxide complexes are formed, which further affect the solubility of aluminum (and iron). Addition of organic polyelectrolytes alone or in conjunction with metallic electrolytes can enhance the overall coagulation process by mechanisms already discussed. It should be noted that coagulation can also be optimized to remove natural organic materials that serve as precursors to the formation of trihalomethanes (THMs) during chlorine disinfection; such an application is termed *enhanced coagulation.*

Mutual precipitation occurs when colloids of opposite charge are mixed. If they are added in essentially equivalent amounts, in terms of electrostatic charge, coagulation occurs and is quite complete. This method is not used per se in environmental engineering practice because of the large volumes of water that would be needed to carry the second colloid and because of the relatively long time required for

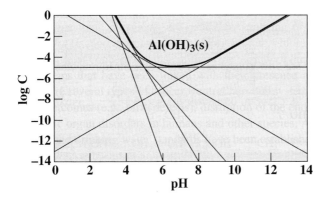

Figure 7.5
Solubility of $Al(OH)_3(s)$. Drawn as described in Chap. 4 with $\log K_{sp} = -32$ and five complexes with hydroxide: $AlOH^{2+}$, $Al(OH)_2^+$, $Al(OH)_3(aq)$, $Al(OH)_4^-$, and $Al_3(OH)_4^{5+}$. Stability constants for the hydroxide complexes taken from F. M. M. Morel and J. G. Hering, "Principles and Applications of Aquatic Chemistry," John Wiley & Sons, New York, 1993.

flocculation of the colloidal dispersions. Positively charged colloids formed when trivalent salts of iron and aluminum are added to water, however, may to some extent act in this manner to remove negatively charged colloids.

Hydrophilic Colloids   A wide variety of hydrophilic colloidal materials are known. Most of them are products of plant or animal life and therefore are of considerable concern to environmental engineers and scientists. Soap, soluble starch, soluble proteins, protein degradation products, blood serum, agar-agar, gum arabic, pectins, synthetic detergents, bacteria, and viruses are examples. These materials occur in natural waters, domestic wastewater, and in many industrial wastes. Soap, however, is usually precipitated by calcium and magnesium ions and does not often occur as a colloidal suspension except in laundry wastes.

The hydrophilic colloids are readily dispersed in water, and their stability usually depends more upon their love for the solvent rather than upon the slight charge (usually negative) that they possess. This property makes it difficult to remove them from aqueous suspension. The stability of hydrophilic colloids is strongly influenced by solution pH and the nature of the ionizable groups on the colloid surface (e.g., carboxyl, amino, phenolic, hydroxyl). Certain of them, such as the proteins and protein degradation products, form heavy metal salts that are insoluble; thus, their removal is effected by aluminum and ferric salts. Proteins, proteoses, peptones, polypeptides, and amino acids have a minimum solubility at their isoelectric point. The isoelectric point (pH_{pzc} for colloids and the pH where the molecule is neutral for dissolved species) varies from about pH 4.0 to 6.5 for the majority; consequently, it is usually uneconomical to attempt their removal by pH adjustment in wastes that are well buffered with bicarbonates.

Most hydrophilic colloids serve in a protective capacity for hydrophobic colloids. When acting in this capacity, they are called *protective colloids.* It is believed that they envelop the hydrophobic colloid in a manner to shield it from the action of electrolytes. Coagulation of such systems requires rather drastic treatment with massive doses of coagulant salts, often 10 to 20 times the amount used in conventional water treatment. Much less is known about the coagulation of hydrophilic colloids than is known about coagulation of hydrophobic colloids.

Liquid-in-Liquid Systems

Colloidal systems involving the dispersion of one liquid in another are known as *emulsions.* Obviously, the two liquids must be immiscible in each other. The emulsions of most interest are composed of oil and water. The oil may be dispersed in the water (the usual case), or water may be dispersed in oil. Most emulsions depend upon a third component, an *emulsifying agent,* for their stability.

Soap and synthetic detergents are excellent emulsifying agents, as would be expected from their common use for laundering, dishwashing, and other cleaning purposes. As discussed in Sec. 5.23, detergents are rather large polar molecules with one end generally a hydrophobic hydrocarbon group that is soluble in fats and oils, and the other an ionized group or highly oxygenated hydrophilic group that prefers water. When detergent molecules come in contact, say with an oil droplet, the hydrophobic end dissolves in the oil droplet, while the hydrophilic end remains in the water, imparting water stability to the droplet that is characteristic of emulsions. Many natural materials such as proteins, protein degradation products, egg yolk, lanolin, saponin, and gum arabic act as emulsifying agents. Egg yolk serves as the emulsifying agent in salad dressings, as in mayonnaise, for example. Kaolin, fuller's earth, colloidal clay, and lampblack have been found to act as good emulsifiers of mineral oil in water.

Water-in-Oil Emulsions Water-in-oil emulsions are quite common in the petroleum industry. They can be readily broken by heating, and such treatment is practical because of the value of the oil that can be recovered.

Oil-in-Water Emulsions Oil-in-water emulsions are usually milky white in appearance, and their destruction depends upon treatment to inactivate the emulsifying agent. This inactivation may be accomplished in a variety of ways, but economics usually dictates the use of chemical coagulating agents such as aluminum, ferric, and calcium salts.

Gas-in-Liquid Systems

Dispersions of gas bubbles in liquids are considered to be colloidal in character regardless of the bubble size; therefore, foams fall in this category. They are of concern in the treatment of certain industrial wastes such as those from the wood-pulping, textile, and meat-packing industries. Foams are sometimes a problem in anaerobic treatment, especially under unbalanced conditions.

Foams are normally stabilized by hydrophilic colloidal materials that are highly surface-active and tend to concentrate at air-water interfaces. Foams have been aptly described as a collection of interfaces separated by air bubbles. Destruction is usually accomplished in either of two ways. Water sprays are used to break the foam by dilution and mechanical action, or antifoaming materials may be added. The antifoaming material, to be effective, must lower the surface tension more than the hydrophilic colloid. It will then displace the colloid and cause the foam to collapse.

It is worth noting that dissolved-air flotation is a solid-liquid separation process that depends on a dispersion of a gas (air) in a liquid for its success. The process has been used for concentrating biological sludges.

7.3 | COLLOIDAL DISPERSIONS IN AIR

Dispersions of liquids and solids in air can cause significant air pollution problems and may have a role in global climate change. The term *aerosol* is sometimes used for describing both liquid and solid colloidal dispersions. Of particular interest are fogs, smog, smokes, and other particulate aerosols.

Fog and Smog

Fog consists of a colloidal dispersion of a liquid in air. The environmental engineer and scientist has no special interest in ordinary fog. However, photochemical and other reactions may occur in the polluted atmosphere near the ground to produce artificial fogs. In areas where large quantities of oxides of sulfur and nitrogen are present, *acid fogs* containing sulfuric and nitric acids may be formed. The pH of these fogs may be as low as 2.0. Fog produced from photochemical reactions is commonly referred to as *photochemical smog*. The term smog originated as a combination of the terms smoke and fog. Photochemical smog is formed by the reaction between hydrocarbons and oxides of nitrogen (NO and N_2O) in the presence of sunlight. Ozone, nitro derivatives of the hydrocarbons, and other organic compounds such as formaldehyde are formed in the reaction. The nitro compounds are suspected of condensing water from the atmosphere to produce the colloidal dispersion called smog. As with fog, the amounts involved are so tremendous that destruction is out of the question. Control appears to rest on limitation of the hydrocarbon compounds and oxides of nitrogen allowed to reach the atmosphere. Automobile exhaust gases are considered the major source. Photochemical smog is known to cause respiratory problems. Fogs are also known to concentrate other pollutants such as volatile organic compounds and pesticides.

Smoke and Other Particulate Aerosols

Fine particles of smog formed in the atmosphere through chemical reactions of sulfur dioxide, nitrogen oxides, and organic compounds are not the only particles of concern in the atmosphere. Atmospheric *Particulate Matter* (PM) in general is of increasing concern because of better knowledge about the adverse health and environmental effects it can cause. This is a special problem with the very small particulate matter that escapes the body's normal defense mechanisms during breathing and can reach deep into the lungs. Fine particulate matter (0.1 to 1 μm) also causes *haze,* affecting air visibility. There are several additional kinds of atmospheric particulate matter, such as dust, dirt, soot, and smoke. Sources include factories, incinerators, power plants, cars, trucks, buses, construction sites, unpaved roads, burning of wood and agricultural wastes and fossil fuels, tilling of fields, and crushing of stone. Diesel engines have become of special concern in recent years with their growth in usage in automobiles. Particulate matter from fossil fuel combustion that is large or dark enough to be readily seen is generally referred to as soot or smoke. It can be removed by electrostatic precipitators. Some particulate matter is so small that it can only be seen with an electron microscope. Small particulate matter does not settle readily and can remain in the air for long periods of time.

Coarser particles can accumulate in the respiratory system and aggravate asthma and other respiratory health problems. More recent scientific studies have indicated that finer particles, those smaller than 2.5 μm in size, are the ones that can penetrate deeply into the lungs, resulting in premature mortality and emergency hospital admissions. The problem is particularly severe with elderly individuals with cardiopulmonary disease and children with asthma. Such fine particles can result in decreased lung function and alterations in lung tissue and structure and in respiratory tract defense mechanisms.

Because of the recent findings of adverse health impacts of fine particulate matter, the United States Environmental Protection Agency enacted in 1997 revised air particulate matter standards. There is an annual $10\text{-}\mu\text{m}$ or smaller particulate size standard (PM_{10}) of $50\ \mu\text{g/m}^3$, with a 24-h standard of $150\ \mu\text{g/m}^3$, which is similar to the former standard. They added a new $PM_{2.5}$ standard for very small particles, those of $2.5\text{-}\mu\text{m}$ size or smaller. This is an annual concentration of $50\ \mu\text{g/m}^3$ and a 24-h standard of $65\ \mu\text{g/m}^3$. A separate program addresses regional haze.

There are several other particulate aerosols of environmental significance. Examples include inorganic lead aerosols produced primarily from combustion of lead-containing fuels and inorganic sulfates such as ammonium sulfate that comprise the dry deposition component of atmospheric acid deposition to natural waters (the other component being acid rain).

PROBLEMS

7.1 Define the following: sol, hydrophilic, emulsion, protective colloid, smog, smoke, Brownian movement, and Tyndall effect.

7.2 What gives stability to hydrophobic colloids in water?

7.3 What factors affect the primary charge on a colloid?

7.4 By what mechanism does addition of an electrolyte such as NaCl bring about destabilization of colloids in water?

7.5 By what different mechanisms can Fe(III) and Al(III) bring about colloid destabilization?

7.6 Explain the possible mechanism by which the addition of small quantities of some materials results in destabilization of sols, but addition of large quantities of the same material does not.

7.7 List the mechanisms that may be involved in destabilization of sols by (1) boiling, (2) freezing, and (3) addition of electrolytes.

7.8 What is the zeta potential? Explain how it can be used to assess the effectiveness of various coagulation strategies.

7.9 A set of experiments was conducted to assess the removal of colloid-caused turbidity by adding $Al_2(SO_4)_3$. During these experiments it was noted that significant removal of the metal radium occurred. Give two potential mechanisms for the observed radium removal.

REFERENCES

Adamson, A. W.: "Physical Chemistry of Surfaces," 6th ed., John Wiley & Sons, New York, 1997.

O'Melia, C. R.: Kinetics of Colloid Chemical Processes in Aquatic Systems, in "Aquatic Chemical Kinetics," W. Stumm (ed.), John Wiley & Sons, New York, 1990.

Stumm, W.: "Chemistry of the Soil-Water Interface," John Wiley & Sons, New York, 1992.

Stumm, W., and J. J. Morgan: "Aquatic Chemistry, Chemical Equilibria and Rates in Natural Waters," 3rd ed., John Wiley & Sons, New York, 1996.

Basic Concepts from Nuclear Chemistry

8.1 | INTRODUCTION

The science of *nuclear chemistry* deals with transformations in the nucleus of atoms. Atoms are the smallest particles of chemical elements. Some atoms are naturally stable, while others are not. When the instability of an atom leads to its transformation into a different, perhaps more stable form, the phenomenon of radioactivity results. Radioactivity is the ejection of particles or radiation from the nucleus. Environmental engineers and scientists have an interest in nuclear chemistry as the radioactivity emanating from changes in unstable elements can result in hazards to health. Radioactive elements may also be used as tracers for measuring the flow of water or for investigating the fate of materials in the environment.

Interest in radioactivity may be considered to date from 1895, when Roentgen discovered a new form of radiation from cathode-ray tubes. The radiations caused certain salts to become luminescent and also affected photographic plates. They are called *roentgen rays* or *X rays*. With a few modifications, the cathode-ray tube became the modern roentgen or X-ray tube that has been used so extensively in medical and industrial applications. Gamma rays released by radioactive materials and X rays are both electromagnetic waves; the gamma rays usually having somewhat shorter wavelengths.

Roentgen's discovery of a new radiation that affected photographic plates stimulated a great deal of testing of materials for similar characteristics. Becquerel and his father had been interested in phosphorescence for some time prior to 1891. They had noted that potassium uranyl sulfate [$K_2UO_2(SO_4)_2 \cdot 2H_2O$] exhibited pronounced phosphorescence when excited by ultraviolet light. It was natural the Becquerel would want to test his uranyl salts for emanation of X rays. He found them to do so, and subsequent observations on a wide variety of salts and materials containing uranium showed them to produce X rays in proportion to their content of uranium.

In 1898 Pierre and Marie Curie concluded that the X rays from uranium were an atomic phenomenon characteristic of the element, and they introduced the term *radioactivity*. The Curies pursued their studies of radioactive materials with much vigor. They found that compounds of thorium emitted radiation similar to that of uranium. They also noted that certain ores of uranium were more radioactive than uranium itself. This led to a search for other materials in the residues remaining after uranium extraction. Two new radioactive elements were isolated, polonium and radium. Radium is several thousand times more radioactive than uranium.

8.2 | ATOMIC STRUCTURE

Modern concepts of atomic structure are largely the result of knowledge gained from the behavior of radioactive materials. It is difficult, therefore, to discuss one without considering the other. Prior to the discovery of radioactivity, atoms were considered to be indivisible. With discovery that radioactive elements emitted positively and negatively charged particles, the foundation was laid for new concepts.

Nuclear Theory

By 1900 it was realized that atoms are not indivisible. However, it was not until 1911 that Rutherford proposed the nuclear concept of the atom. This theory held that atoms were composed of a small positively charged nucleus, containing most of the mass of the atom, with a cloud of negatively charged electrons surrounding it.

Electron Orbits

Bohr was the first to propose that the electrons about the nucleus of an atom are arranged in a methodical manner and revolve in orbits about the nucleus. Although his theory, issued in 1913, has undergone some refinements, it remains the basis of our modern-day knowledge. The present tendency is to think of the electrons as being arranged in shells about the nucleus. A major contribution to the Bohr theory was made by Sommerfeld, who has shown that the electrons within a given shell occur in several energy levels. Other contributions, particularly with respect to chemical properties, were made by Langmuir (octet theory), Mosely, G. N. Lewis, and W. Kossel.

The simplest atoms, hydrogen and helium, have one shell of electrons, and the most complex have seven. The shells or rings are designated as *K, L, M, N, O, P,* and *Q* in the order of their increasing remoteness from the nucleus. The maximum number of electrons that a given shell can have is shown by the formula $2n^2$ where n is the shell number, equaling 1 for the *K* shell and 7 for the *Q* shell. The arrangement of electrons for a number of elements is given in Table 8.1. Over the course of ensuing years, the positively charged nucleus was considered to consist of protons and electrons. This theory held that the protons were always in excess of the electrons in the nucleus, and this excess was equal to the planetary electrons; thus, the net charge on the atom was zero.

Table 8.1 | Arrangement of electrons for some common elements

Symbol	Atomic number	Number of electrons in shells						
		K	L	M	N	O	P	Q
H	1	1						
He*	2	2						
N	7	2	5					
Ne*	10	2	8					
Na	11	2	8	1				
Cl	17	2	8	7				
Ar*	18	2	8	8				
Ca	20	2	8	8	2			
Zn	30	2	8	18	2			
Br	35	2	8	18	7			
Kr*	36	2	8	18	8			
Ag	47	2	8	18	18	1		
Xe*	54	2	8	18	18	8		
Ba	56	2	8	18	18	8	2	
Hg	80	2	8	18	32	18	2	
Pb	82	2	8	18	32	18	4	
Rn*	86	2	8	18	32	18	8	
Ra	88	2	8	18	32	18	8	2

*Inert gases.

Neutron-Proton Concept of Nuclear Structure

In 1930 Bothe and Becker discovered a very penetrating secondary radiation when light elements, such as beryllium and lithium, were subjected to bombardment by alpha particles from polonium. The new radiation was first thought to be X rays of very short wavelength. In 1932 Chadwick showed this secondary radiation to be made up of neutral particles having a mass comparable to that of the proton. The new particles were given the name *neutrons,* and since their source was obviously the nucleus of the bombarded atoms, a new concept of nuclear structure evolved. From this came the present "practical" model of the nucleus. Since 1932 the nucleus has been found to consist of many other particles, such as mesons, muons, leptons, and quarks. However, these have little significance to the understanding of nuclear chemistry needed by the environmental engineer and scientist.

According to the simple, practical model we use, the nucleus of all atoms, except the simple hydrogen atom, consists of neutrons plus protons, both of which are called *nucleons.* The number of protons Z corresponds to the atomic number and is equal to the number of electrons about the nucleus of a neutral atom. The number of neutrons N is equal to the atomic weight A expressed as the nearest whole number less the number of protons: $N = A - Z$. The structure of the atom may be represented as shown in Fig. 8.1. The nucleus has a diameter on the order of 10^{-12} to 10^{-13} cm, and the atom a diameter of about 10^{-8} cm. The density of nuclear matter is tremendous. It is estimated that 1 cm^3 would weigh 10^{10} kg.

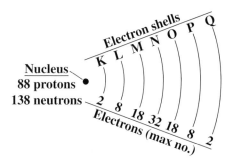

Figure 8.1
Structure of radium atom.

Nomenclature of Isotopes

An element is defined by its atomic number (number of protons, with an equivalent number of electrons). However, many elements have differing mass numbers; that is, they have a different number of neutrons. These different forms of the same element are called *isotopes*. In order to differentiate between isotopes, a new system of symbol writing had to be developed. The IUPAC nomenclature lists the atomic number as a subscript just before the symbol and the atomic weight, or mass number, as a superscript also before the symbol. In the IUPAC system, $^{204}_{82}$Pb, $^{206}_{82}$Pb, $^{207}_{82}$Pb, and $^{208}_{82}$Pb represent four isotopic forms of lead, each of which has 82 protons and 82 electrons. Since the atomic number of a given element is always the same, it is frequently eliminated when discussing isotopes. However, in radioactive changes involving transmutation of one element into another, such as in the conversion of ^{238}U to ^{206}Pb, the change is best shown by $^{238}_{92}$U $\rightarrow$ $^{206}_{82}$Pb. Anyone familiar with nuclear chemistry knows that such a change cannot occur in one step and that several intermediate steps arc involved (see Fig. 8.2).

8.3 | STABLE AND RADIOACTIVE NUCLIDES

The chemical identity of an atom is given by the number of protons in the nucleus Z. The stability of a nucleus can be empirically related to some degree with the ratio of the number of neutrons to the number of protons it contains: *N/Z*. For stable light nuclei *N/Z* is approximately 1, while for stable heavy nuclei the ratio approaches 2. When the nucleus is at variance with this pattern, instability results. The shakedown of an unstable nucleus to a more stable form results in the natural process of nuclear decay and the ejection of particles or radiation from the nucleus. Nuclear transformations or reactions can also be induced by bombarding the nucleus with particles or energy. Radioactive atoms may exist for times as short as 10^{-20} s or as long as 10^{18} yr. Those that decay too slowly to be measured are termed stable. All nuclides with atomic numbers greater than 83 or atomic weights greater than 200 are unstable. Table 8.2 lists isotopes for a few elements and illustrates some compositions that are stable and some that are radioactive.

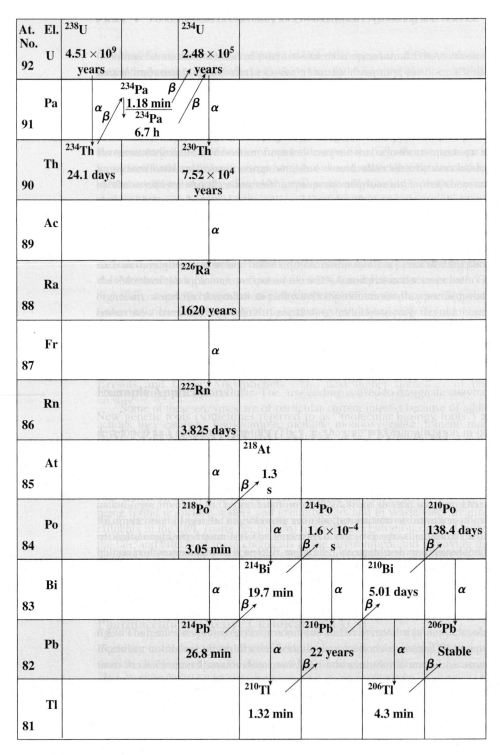

At. No.	El.	^{238}U 4.51×10^9 years		^{234}U 2.48×10^5 years			
91	Pa	α β	^{234}Pa β 1.18 min ^{234}Pa 6.7 h	β α			
90	Th	^{234}Th 24.1 days		^{230}Th 7.52×10^4 years			
89	Ac			α			
88	Ra			^{226}Ra 1620 years			
87	Fr			α			
86	Rn			^{222}Rn 3.825 days			
85	At			α	^{218}At β 1.3 s		
84	Po			^{218}Po 3.05 min	α	^{214}Po 1.6×10^{-4} β s	^{210}Po 138.4 days β
83	Bi				α	^{214}Bi 19.7 min β	α ^{210}Bi 5.01 days β α
82	Pb				^{214}Pb 26.8 min	α ^{210}Pb 22 years β	α ^{206}Pb Stable β
81	Tl				^{210}Tl 1.32 min		^{206}Tl 4.3 min

Figure 8.2
Steps in radioactive decay of U to stable Pb.

Table 8.2 | Isotopes of a few elements illustrating stable and radioactive forms

Element	Isotope				
Hydrogen		$^{1}_{1}H$	$^{2}_{1}H$	$^{3}_{1}H$	
		Stable	Stable	Radioactive	
Carbon		$^{11}_{6}C$	$^{12}_{6}C$	$^{13}_{6}C$	$^{14}_{6}C$
		Radioactive	Stable	Stable	Radioactive
Nitrogen	$^{12}_{7}N$	$^{13}_{7}N$	$^{14}_{7}N$	$^{15}_{7}N$	$^{16}_{7}N$
	Radioactive	Stable	Stable	Stable	Radioactive
Oxygen	$^{14}_{8}O$	$^{15}_{8}O$	$^{16}_{8}O$	$^{17}_{8}O$	$^{18}_{8}O$	$^{19}_{8}O$
	Radioactive	Radioactive	Stable	Stable	Stable	Radioactive
Phosphorus	$^{29}_{15}P$	$^{30}_{15}P$	$^{31}_{15}P$	$^{32}_{15}P$	$^{33}_{15}P$
	Radioactive	Radioactive	Stable	Radioactive	Radioactive
Chlorine	$^{34}_{17}Cl$	$^{35}_{17}Cl$	$^{36}_{17}Cl$	$^{37}_{17}Cl$	$^{38}_{17}Cl$
	Radioactive	Stable	Radioactive	Stable	Radioactive

Most of the presently existing atoms on earth are stable, but some are radioactive. Many of the radioactive isotopes present decay very slowly. Others are formed as short-lived intermediates following the decay of longer-lived isotopes. Bombardment of the earth with radiation from outer space also converts some stable atoms to radioactive ones. A good example is the production of $^{14}_{6}C$ from $^{14}_{7}N$ by cosmic rays. All living forms are kept uniformly radioactive with ^{14}C since atmospheric carbon that contains this produced material is the primary source of carbon for life.

The largest terrestrial sources of radiation are ^{40}K, ^{238}U, ^{235}U, and ^{232}Th. The total energy emission from ^{40}K in the earth's crust is estimated to be 4×10^{12} W. The heat generated by the decay of these elements has resulted in a much slower cooling of the earth than would have otherwise been the case. Decay of $^{40}_{19}K$ leads to the formation of the stable isotopes $^{40}_{20}Ca$ and $^{40}_{18}Ar$, each of which is formed by a different mechanism of decay.

The heavy-metal radioactive elements fall into three series: uranium, thorium, and actinium. The uranium series has ^{238}U as its parent substance and, after 14 successive transformations have occurred, the end product is ^{206}Pb. Thorium (^{232}Th) is the parent substance of the thorium series. After 10 transformations, it remains as ^{208}Pb. The parent element of the actinium series is ^{235}U and, after 11 transformations, it remains as ^{207}Pb. This series takes its name from the fact that ^{231}Pa (protactinium) and ^{227}Ac (actinium) are long-lived isotopes formed as steps in the transformation. The uranium series is sometimes called the radium series for the same reason. The steps in the radioactive decay of ^{238}U are shown in Fig. 8.2. ^{232}Th and ^{235}U decompose through similar steps.

Nature of Radiations

Early workers with radioactive materials were cognizant of the presence of only one form of radiation, and its properties were similar to those of X rays. Later investigations established the presence of three kinds of radiations designated as alpha, beta, and gamma. Separation and identification were accomplished by directing the radiation through a magnetic field, as shown in Fig. 8.3. Some of the radiation was bent slightly

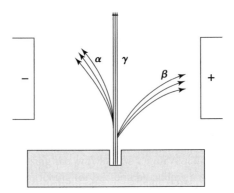

Figure 8.3
Effect of a magnetic field upon alpha, beta,
and gamma rays.

toward the negative pole. This phenomenon indicated that it had a positive charge and, probably, considerable mass. This is called alpha radiation. Other radiation was bent radically toward the positive pole, showing it to be negatively charged and, probably, of small mass. This was called beta radiation. A third radiation group was unaffected by the magnetic field. It does not have a charge and is called gamma radiation.

Alpha Radiation   Alpha radiation is not true electromagnetic radiation as are light and X rays. It consists of particles of matter. *Alpha particles* ($_2^4\alpha$) are actually doubly charged ions of helium with a mass of 4 (^{4}He). Although alpha particles are propelled from the nucleus of atoms at velocities ranging from 1.4 to 2×10^9 cm/s (about 10 percent of the speed of light), they do not travel much more than 10 cm in air at room temperature. They are stopped by an ordinary sheet of paper. The alpha particles emitted by a particular element are all released at the same velocity. The velocity varies, however, from element to element. The alpha particles have extremely high ionizing action within their range of travel.

Beta Radiation   Beta radiation consists of negatively charged particles moving at speeds ranging from 30 to 99 percent of the speed of light. *Beta particles (β)* are actually electrons, and the velocity of flight of individual electrons varies considerably for a given element, as well as for different elements. The penetrating power of beta particles varies with their speed. They normally travel several hundred feet in air. Shielding with aluminum sheeting a few millimeters thick will stop the particles. Because of the low mass, the ionizing power of beta radiation is much weaker than that of the alpha radiation.

Gamma Radiation   Gamma radiation *(γ)* is true electromagnetic radiation which travels with the speed of light. It is similar to X rays but has shorter wavelength and therefore greater penetrating power, which increases as the wavelength decreases. Proper shielding from gamma radiation requires several centimeters of lead or several feet of concrete. The unit of gamma radiation is the *photon.*

Energies of Radiations

It is important to know the energy of the various radiations produced by radioactive materials. Although alpha and beta particles have mass and gamma radiations do not, it is possible to establish a single system of expressing energies through use of Einstein's energy-mass equivalence formula

$$E = Mc^2 \tag{8.1}$$

With c as the velocity of light (2.998×10^{10} cm/s) and M the mass of the particle in grams, E has units of g $\cdot$ cm/s or ergs. However, a more convenient energy unit for nuclear particles is the *electron volt* (eV), which is the energy necessary to raise one electron through a potential difference of 1 volt. The conversion factor is 1 eV = 1.602×10^{-12} erg or 1.602×10^{-19} J. The energies of alpha and beta particles and of gamma photons range from several thousands up to several millions of electron volts. For this reason, energies are usually expressed as million electron volts (MeV).

Equation (8.1) indicates that potential energy is stored as mass in atoms and that in a given system a change in mass is accompanied by an equivalent change in energy:

$$\Delta E = \Delta Mc^2 \tag{8.2}$$

The spontaneous radioactive decay of unstable nuclides is accompanied by the release of energy as required by the laws of thermodynamics. The energy results from a loss in mass of the nuclide during its transformation. From this requirement it follows that spontaneous radioactive decay may take place by some mode if the masses of the products are lighter than the mass of the original nuclide. The excess mass is released as radiation during the disintegration. Thus, for the following disintegration of nuclide A into nuclide B plus radiation b,

$$A \rightarrow B + b \tag{8.3}$$

the change in mass

$$\Delta M = M_A - (M_B + M_b) \tag{8.4}$$

must be positive if the disintegration is to occur spontaneously. This does not mean the disintegration will occur, only that it can occur. If ΔM is negative, the radioactive decay cannot occur spontaneously; i.e., the nuclide is stable.

Atomic Changes Resulting from Release of Radiation

The change that atoms undergo when releasing alpha particles is considerably different from the change when beta particles are released. These changes are illustrated in Fig. 8.2 and were formulated into the so-called *displacement laws* by Fajans, Rutherford, and Soddy as follows.

Alpha-Particle Release   *When an element emits an alpha particle, the product has the properties of an element two places to the left of the parent in the periodic table.* In other words, emission of an alpha particle decreases the mass number by four units and the nuclear charge, or atomic number, by two units. An example is the decay of $^{238}_{92}U$ as shown in Fig. 8.2:

$$^{238}_{92}U \rightarrow {}^{234}_{90}Th + {}^{4}_{2}\alpha \tag{8.5}$$

Beta-Particle Release   *When an element emits a beta particle, the product has the properties of an element one place to the right of the parent in the periodic table.* Beta particles are released from an unstable nucleus when a neutron is transformed into a proton. In this change the mass remains the same and the atomic number increases one unit. For example, from Fig. 8.2

$$^{234}_{90}Th \rightarrow {}^{234}_{91}Pa + \beta \tag{8.6}$$

Gamma Radiations   Gamma radiation may accompany the release of alpha or beta particles and is a result of energy released by nuclear transformations or shifts of orbital electrons.

Units of Radioactivity

The unit of radioactivity is the *curie.* Formerly, it was considered to be the number of disintegrations occurring per second in 1 gram of pure radium. Since the constants for radium are subject to revision from time to time, the International Radium Standard Commission has recommended the use of a fixed value, 3.7×10^{10} disintegrations per second, as the *standard curie* (Ci).

The curie is used mainly to define quantities of radioactive materials. A curie of an alpha emitter is that quantity which releases 3.7×10^{10} alpha particles per second. A curie of a beta emitter is that quantity of material which releases 3.7×10^{10} beta particles per second, and a curie of a gamma emitter is that quantity of material which releases 3.7×10^{10} photons per second. The curie represents such a large number of disintegrations per second that the millicurie (mCi), microcurie (μCi), nanocurie (nCi), and picocurie (pCi), corresponding to 10^{-3}, 10^{-6}, 10^{-9}, and 10^{-12} curie, respectively, are more commonly used. The SI unit of radioactivity is the *becquerel* (Bq), which equals 1 disintegration per second. Thus, 1 Ci = 3.7×10^{10} Bq.

The *roentgen* (r) is a unit of gamma or X-ray radiation intensity. It is of value in the study of the biological effects of radiation that result from ionization induced within cells by the radiations. The roentgen is defined as the amount of gamma or X radiation that will produce in 1 cubic centimeter of dry air, at 0°C and 760 mm pressure, one electrostatic unit (esu) of electricity. This is equivalent to 1.61×10^{12} ion pairs per gram of air and corresponds to the absorption of 83.8 ergs of energy.

The roentgen is a unit of the total quantity of ionization produced by gamma or X rays, and dosage rates for these radiations are expressed in terms of roentgens per unit time.

With the advent of atomic energy involving exposure to neutrons, protons, and alpha and beta particles which also have effects on living tissue, it has become necessary to have other means of expressing ionization produced in cells. Three methods of expression have been used.

Table 8.3 | Half-lives of common radioactive isotopes

Atomic number	Nuclide	Half-life	Nature of radiation
1	^{3}H	12.3 yr	β
6	^{14}C	5730 yr	β
11	^{24}Na	15.0 h	β, γ
15	^{32}P	14.3 days	β
16	^{35}S	88 days	β
19	^{40}K	1.28×10^{9} yr	β
27	^{60}Co	5.3 yr	β, γ
35	^{78}Br	6.4 min	β, γ
38	^{90}Sr	28.1 yr	β
53	^{131}I	8.0 days	β, γ
55	^{137}Cs	30 yr	β
86	^{222}Rn	3.825 days	α
88	^{226}Ra	1620 yr	α, γ
92	^{238}U	4.51×10^{9} yr	α
94	^{239}Pu	24,360 yr	α, γ

The *roentgen-absorption-dose* (rad) is a unit of radiation corresponding to an energy absorption of 100 ergs per gram of any medium. It can be applied to any type and energy of radiation that leads to the production of ionization. Studies on the radiation of biological materials have shown that the roentgen is approximately equivalent to 100 ergs/g of tissue and can be equivalent to 90 to 150 ergs/g of tissue, depending on the energy of the X and γ radiation and type of tissue.

The rad represents such a tremendous radiation dosage, in terms of permissible amounts for human beings, that another unit has been developed specifically for humans. The term *roentgen-equivalent-man* (rem) is used. It corresponds to the amount of radiation that will produce an energy dissipation in the human body that is biologically equivalent to 1 roentgen of radiation of X rays, or approximately 100 ergs/g. The U.S. Occupational Safety and Health Administration (OSHA) standard for radiation workers is 5 rem/year.

Half-Lives

Radioactive decomposition is a true unimolecular reaction. The rate is constant over a wide variety of environmental conditions. Half-lives of the radioactive elements vary from fractions of a second to about 10^{20} yr. The half-lives of a number of elements are given in Table 8.3. The kinetics of unimolecular or first-order reactions are discussed in Sec. 3.10.

8.4 | ATOMIC TRANSMUTATIONS AND ARTIFICIAL RADIOACTIVITY

The experimental conversion of one element into another was accomplished by Rutherford in 1919. When alpha particles derived from radium C were passed through nitrogen gas, protons were detected. The collisions between alpha particles and nitrogen nuclei resulted in the formation of hydrogen and an isotope of oxygen as follows:

$$^{14}_{7}\text{N} + ^{4}_{2}\alpha \rightarrow ^{17}_{8}\text{O} + ^{1}_{1}\text{H} \tag{8.7}$$

By 1922 Rutherford and Chadwick had shown that all elements in the periodic table between boron and potassium, except carbon and oxygen, underwent similar transmutations when submitted to bombardment by alpha particles.

It was not until 1930 that radiation other than protons was detected when elements were subjected to alpha-particle bombardment. In that year, Bothe and Becker discovered a very penetrating, neutral, secondary radiation when beryllium or lithium was subjected to alpha particles from polonium. In 1932 Chadwick showed the particles to be neutrons, and the changes were described as follows:

$$^{7}_{3}\text{Li} + ^{4}_{2}\alpha \rightarrow ^{10}_{5}\text{B} + ^{1}_{0}n \tag{8.8}$$

$$^{9}_{4}\text{Be} + ^{4}_{2}\alpha \rightarrow ^{12}_{6}\text{C} + ^{1}_{0}n \tag{8.9}$$

where the neutron is represented by $^{1}_{0}n$.

The third important step in the transmutation of elements involved the discovery that a third particle was found in certain instances. In 1934 I. Curie and Joliot noted that when boron, magnesium, or aluminum was bombarded with alpha particles, the expected transmutation with neutron release occurred and that *positrons*[1] (positive electrons) were also produced. In addition, they found that positron emission continued after alpha bombardment was discontinued. The emission of positrons was shown to decrease in accordance with the decay law for radioactive materials. Through careful analysis of the materials produced, they were able to show that alpha bombardment of these elements had produce an atom with an unstable nucleus that underwent radioactive positron decay; thus, the production of artificial radioactive materials by alpha bombardment was established.

$$^{10}_{5}\text{B} + ^{4}_{2}\alpha \rightarrow ^{13}_{7}\text{N} + ^{1}_{0}n \tag{8.10}$$
$$\hookrightarrow ^{13}_{6}\text{C} + \beta^{+} \quad t_{1/2} = 10.0 \text{ min}$$

$$^{27}_{13}\text{Al} + ^{4}_{2}\alpha \rightarrow ^{30}_{15}\text{P} + ^{1}_{0}n \tag{8.11}$$
$$\hookrightarrow ^{30}_{14}\text{Si} + \beta^{+} \quad t_{1/2} = 2.5 \text{ min}$$

It soon became apparent that there is no real distinction between a nuclear reaction leading to stable products and one leading to unstable products. According to the Bohr concept, all bombardments result in an absorption of the bombarding particle by the nucleus to produce an unstable compound nucleus. The life of the compound nucleus is extremely short (10^{-12} to 10^{-14} s), and decomposition occurs to a set of products. The products may be stable or they may be unstable.

The discovery of the neutron and the fact that radioactive elements could be produced artificially set the stage for the developments of nuclear energy.

---

[1]The existence of positrons had been predicted and confirmed previously from cosmic-ray studies.

8.5 | NUCLEAR REACTIONS

Nuclear reactions may be induced by bombardment with a wide variety of particles. Most of the radioactive materials used in industry, research, and medicine today are produced by such bombardment, which may be done either in a nuclear reactor or in a particle accelerator.

Alpha-Induced Reactions

Because of the positive charge on the alpha particle, it has to overcome the repulsive forces of the positively charged nucleus of an atom before it can add to it. As the atomic number of elements increases, the repulsive force toward alpha particles increases. For this reason, alpha particles are unable to cause nuclear changes in elements of high atomic weight, and their use is restricted to action on the elements with light nuclei.

Alpha-induced reactions serve as the basis for the production of neutrons and therefore are extremely important.

Proton-Induced Reactions

Protons suffer from the same limitations as do alpha particles, even more so, because the ratio of mass to charge is only one-half that of the alpha particle. Therefore, they are repelled more easily by the positively charged nuclei.

Deuteron-Induced Reactions

Deuterons $[(^2_1H)^+]$ are probably the most effective of the positively charged particles, since they have only one charge, and the ratio of mass to charge is the same as for the alpha particle. Certain deuteron-induced reactions are excellent sources of neutrons.

Gamma-Induced Reactions

Gamma and X rays are inefficient in producing nuclear reactions.

Neutron-Induced Reactions

Neutrons, being neutral, are extremely efficient particles for bombarding the nuclei of all elements. By their use, all elements, with the exception of helium, have been transmuted into other elements.

Two neutron-induced reactions are commonly utilized. One is the n, gamma (n, γ) reaction such as used to produce ^{60}Co by bombarding ^{59}Co with neutrons,

$$^{59}_{27}\text{Co} + ^1_0n \rightarrow ^{60}_{27}\text{Co} + \gamma$$

or $\qquad ^{59}\text{Co}(n, \gamma)^{60}\text{Co} \qquad$ (8.12)

The other reaction, termed the n, p reaction, can be illustrated by the bombardment of ^{14}N with neutrons to produce ^{14}C and hydrogen,

$$^{14}_7\text{N} + ^1_0n \rightarrow ^{14}_6\text{C} + ^1_1\text{H}$$

or $\qquad ^{14}\text{N}(n, p)^{14}\text{C} \qquad$ (8.13)

Activation Analysis

An analytical technique that employs induced radioactivity for quantitative analysis of elements is termed *activation analysis*. Here an unknown sample is subjected to bombardment, generally with neutrons, although charged particles and photons may also be used. The irradiation is conducted for chosen lengths of time, and the elements of interest are identified and assayed by measurement of the characteristic radionuclides formed. Usually the irradiation must be followed by chemical separation of the desired radionuclides after the addition of a known quantity of appropriate carriers, which are nonradioactive isotopes of the radionuclide of interest. The carriers allow determination of separation efficiency. Such analyses are often sufficiently sensitive to measure trace elements in concentrations as low as parts per million or even parts per billion.

8.6 | NUCLEAR FISSION

Shortly after the discovery of the neutron, Fermi found neutron bombardment of some heavy metals to be followed by beta activity. Bombardment of uranium produced beta particles of four distinct activities. The activities could not be correlated with those of any of the elements with a mass in the range of uranium. The answer was found by Hahn and Stassmann, who conducted chemical analyses of the products. They found isotopes of barium, lanthanum, strontium, and yttrium, as well as an inert gas (Xe or Kr) and an alkali metal (Cs or Rb) present. From this information it was concluded that neutron capture by the uranium atom was followed by a rupture of the nucleus to form several elements of lower atomic weight. This is termed *nuclear fission*.

Nuclear fission is of interest because of the tremendous amounts of energy released as a result of the fission process. This release results from a conversion of some of the mass to energy. The energy released during the fission of one gram atom of ^{233}U, ^{235}U, or ^{239}Pu corresponds to 5.3×10^6 kWh. Since the gram atomic weights of these elements vary so little, the energy per kilogram of fissionable material is essentially as shown in Table 8.4.

Nuclear Explosions

Nuclear fission is initiated by neutron bombardment. During the fission of ^{235}U, an average of 2.5 neutrons are released for each atom undergoing fission; thus, the reaction can become self-perpetuating once fission is initiated. Fortunately, this had not been the case in early laboratory studies. Probability considerations, however, indicated that, if the mass of fissionable material were large enough, a self-sustaining chain reaction would occur. This has been proven many times, beginning with the first atomic bomb test at Alamagordo, New Mexico, in 1945.

Nuclear explosions are accompanied by release of tremendous amounts of radioactivity. The effect on the surrounding ground area depends upon the distance from the ground at which the explosion occurs. In any event, a great deal of the radioactive matter is projected into the upper atmosphere where it is carried around the world. It is constantly returning to the earth, particularly at times of rain and

Table 8.4 | Energy liberated per kilogram of fissionable material

2.0×10^{13} cal
8.2×10^{13} J
2.3×10^7 kWh
6.1×10^{13} ft-lb
7.8×10^{10} Btu

snowfall, as *fallout.* There are over 200 fission products from a nuclear explosion, although only 17 of these contribute most of the radioactivity present in the mixture. Of particular concern are those having a relatively long half-life, such as strontium 90 and cesium 137 (see Table 8.3). Also of concern are fission products that can be taken into the body and deposited at critical locations: for example, strontium 90, which is a bone seeker, and iodine 131, a thyroid seeker.

Nuclear Power

Control of fission reactions so that the great amount of energy released can be utilized for beneficial purposes has been the objective of a great deal of research, and nuclear-fueled power-generating plants are now in operation in many countries throughout the world. The reactors in which the fission occurs, of course, must contain the fissionable matter in excess of the "critical mass." To avoid an explosion, neutrons released in the fission process must be controlled in number to keep the chain reaction going at the desired speed. This is accomplished by means of neutron absorbers or moderators. Cadmium, boron, graphite, and deuterium oxide are excellent neutron absorbers.

In nuclear reactors, ^{235}U, ^{238}U, or ^{239}Pu are generally used as fuel to maintain the controlled chain reaction. During decay they function to produce neutrons that can also be used to transmute nonradioactive elements into fissionable or radioactive forms. The reactor at Oak Ridge, Tennessee, has been used largely to produce a wide variety of radioactive isotopes for use in medical, biological, and industrial research.

In conventional nuclear power plants, ^{235}U is bombarded with neutrons to produce ^{236}U, fission products, more neutrons, gamma radiation, and large quantities of energy:

$$^{235}U + {}_0^1n \rightarrow {}^{236}U + \textit{fission products} + (2 \text{ to } 3){}_0^1n + \gamma + \textit{energy} \quad (8.14)$$

Uranium ores are mostly ^{238}U, which is not fissionable, and contain very little ^{235}U. Thus, large quantities of uranium-containing ore must be mined and enriched for ^{235}U. Currently it is felt that the health risks associated with mining are much greater than those associated with generating energy from ^{235}U fission. The fission products, which may include ^{137}Cs, ^{131}I, and ^{90}Sr, among others, are radioactive, and attempts to dispose of spent fuel containing these materials (radioactive waste) is fraught with controversy.

Nuclear reactors require large quantities of water to dissipate the heat released by the nuclear fission that produces neutrons. Discharge of such cooling water to rivers has been of concern, not only because of possible thermal pollution effects, but also because of the induced radioactivity and possibly some radioactive fission products that it may contain. Cooling towers are constructed to remove heat from these waters prior to discharge. Concern has also been expressed over the safety of nuclear power plants from sabotage, earthquakes, and industrial accidents, and over the possible long-term fate of radioactive wastes removed during the reprocessing of fuel elements. The struggle between concern for uncertain risks and desire for potential benefits has been keenly evident in this area. In recent years, in the United

States, there has been a virtual halt in the construction of new nuclear power plants because of these concerns.

A *breeder reactor* takes advantage of the ability of the abundant isotope $^{238}_{92}\text{U}$ to absorb neutrons and eventually be transformed into $^{239}_{94}\text{Pu}$. This plutonium is then used as a fuel and bombarded with neutrons:

$$^{239}\text{Pu} + {}^1_0 n \rightarrow \textit{fission products} + (2 \text{ to } 3){}^1_0 n + \gamma + \textit{energy} \qquad (8.15)$$

The disadvantage of this process is that ^{239}Pu is very toxic and has a half-life of 24,360 years making disposal of the waste produced from breeder reactors very problematical. In addition, ^{239}Pu is a critical component of nuclear weapons.

8.7 | NUCLEAR FUSION

The fusion of two or more light atomic nuclei to form the nucleus of a heavier element is generally more productive of energy than the fission of heavy elements. The production of one atom of helium from the fusion of four atoms of hydrogen produces seven times as much energy per unit weight of material as the fission of ^{235}U or ^{239}Pu.

The hydrogen bomb depends upon *nuclear fusion* for its tremendous explosive power, resulting from the nuclear fusion of heavy hydrogen isotopes. Temperatures of over 40,000,000°C are needed to initiate the fusion process, and because of this the fusion process is known as a *thermonuclear reaction*. The high temperature required has been obtained by incorporating a fission device for ignition. Thus, development of the atomic bomb made the hydrogen bomb a possibility.

The usual fuel for the fusion reaction is deuterium (^2_1H) and tritium (^3_1H), which combine to produce ^4_2He with the release of a neutron.

$$^2_1\text{H} + {}^3_1\text{H} \xrightarrow{\Delta\Delta\Delta} {}^4_2\text{He} + {}^1_0 n \qquad (8.16)$$
$$\searrow \textit{energy}$$

Under these conditions, the neutron is converted to energy. In this reaction about 20 percent of the original mass of the hydrogen is converted to energy.

Nuclear fusion in itself does not release radioactive materials, and partly because of this a great deal of research is being conducted to produce a controlled fusion reaction for power production. At the International Conference on the Peaceful Uses of Atomic Energy, held in Geneva, Switzerland, in 1955, it was predicted that the fusion process would be harnessed to provide power for industrial uses within 20 years. This period has long since passed, and now there is question whether this development can take place in the near future. Controlled fusion power is the hope of many who otherwise see a dwindling of the world's fuel supplies and a decreasing standard of living for the human race.

8.8 | USE OF ISOTOPES AS TRACERS

Both stable and radioactive isotopes have been used as tracers in geology, anthropology, biology, chemistry, physics, medicine, and engineering. Compounds containing radioactive elements such as ^{14}C, ^{32}P, and ^{35}S, have been used extensively by researchers in the fields of biology, chemistry, and medicine to determine the course of chemical and biochemical reactions. Changes in the ratios of stable to radioactive isotopes of elements that occur over time are used to evaluate the age of minerals, water, and archeological artifacts. Stable isotope ratios of a given element can also change *(isotope fractionation)* as a result of physical, chemical, or biological processes affecting molecules containing the atoms[2,3]. Such changes are being used to evaluate large scale geological processes, to determine the sources of groundwater, to discover the origin of pollutants, and to assess the rates of chemical and biological transformations of contaminants. There are numerous potential applications for both stable and radioactive tracers for evaluating the effectiveness of treatment processes and for determining the movement and fate of chemicals in the environment.

Consideration has often been given to the use of radioactive tracers such as tritium for determining the flow-through time of sedimentation tanks or of reaches or stretches in rivers, or the direction and rate of flow of groundwater. However, such uses are generally confined to controlled laboratory experiments as the release of radioactive elements to the environment for such purposes is generally not allowed. The use of radioisotopes in chemical, biological, medical, and engineering research can create environmental problems, since some of the materials contained in wastewaters reach sewers and rivers. In sewers and waste treatment plants, certain of the isotopes, such as ^{131}I and ^{32}P, can accumulate in biological slimes and sludges. In rivers, radioactive material may be concentrated by microscopic forms that serve as food for fish and other forms of life consumed by humans. Many water supplies are derived from rivers; hence the disposal of radioactive wastes to rivers is a matter of concern to all the consumers. For this reason standards have been set for permissible levels of radioactive materials in water supplies and wastewater discharges. Those working with radioactive materials in the laboratory must follow strict guidelines for their use and disposal.

However, use as tracers for field studies may be made of radioactive elements that occur naturally in the environment or that have been inadvertently released to the environment by human activities. For example, ^{14}C is formed continuously in the atmosphere by cosmic rays from outer space through an *n, p* reaction with atmospheric ^{14}N. Since the half-life of ^{14}C is 5,720 years, its formation in the atmosphere is balanced by its decay, leading to a steady-state atmospheric ^{14}C concentration and relatively constant $^{14}C/^{12}C$ ratio. Plants growing on atmospheric carbon dioxide take on a similar $^{14}C/^{12}C$ ratio, but when they die and become buried or made into archeological artifacts such as wooden structures, the ratio decreases slowly with time since the ^{12}C isotope is stable. The $^{14}C/^{12}C$ ratio is essentially zero in very old carbon such as in fossil fuels. Thus the $^{14}C/^{12}C$ ratio can be used to distinguish between carbon dioxide produced from burning of recent biomass from that produced by fossil fuel combustion. This ratio is most widely used to determine

---

[2]E. M. Galimov, *The Biological Fractionation of Isotopes,* Academic Press, Orlando, Fl., (1985).

[3]D. Clark, P. Fritz, and P. Fritz, *Environmental Isotopes in Hydrogeology,* CRC Press, Boca Raton, FL, (1997).

the age of organic archeological artifacts that are up to several thousands of years of age as long as the ratio still has a measurable value above zero. Atmospheric testing of nuclear weapons during the 1950s and early 1960s significantly increased the ^{14}C concentration in the atmosphere, and thus upset the $^{14}C/^{12}C$ balance from what it had been for thousands of years.

Some of the radioactive elements formed from the period of atmospheric testing, such as ^{90}Sr, have been used in a positive way to evaluate the rate of sedimentation in rivers, lakes, bays and harbors. Such radioactive elements that became incorporated into material settling in natural bodies of water during the testing period have created a marker in sediments that is commonly used to evaluate the average rate of sedimentation that has occurred since them. The rate of sediment buildup can be an important piece of information in decisions of whether sedimentation is occurring fast enough to maintain a protective layer over hazardous contaminants such as mercury and PCBs that were deposited in the sediments in the past. The way radioactive elements from atmospheric testing are distributed vertically in sediments can also help in evaluating the significance of mixing processes resulting from burrowing organisms *(bioturbation)* that may also result in bringing hazardous materials to the sediment surface where they may cause harm to aquatic organisms.

The use of stable isotopes is not as limited for tracer studies as radioactive isotopes for they do not pose the same risks to human health and the environment. Stable isotopes of hydrogen (^2H), nitrogen (^{15}N), carbon (^{13}C), and oxygen (^{18}O) can generally be added in field studies to evaluate the movement and fate of contaminants under natural conditions. They are also widely used in the laboratory where the large quantity of dominate natural isotopes of these elements can otherwise make analyses of changes difficult to quantify. For example, the discovery of anaerobic benzene and toluene biodegradation[4] was seriously questioned by biochemists who thought atmospheric oxygen was required for their biodegradation. In order to overcome criticisms that atmospheric oxygen must be leaking into the culture vessels, the researchers added $H_2^{18}O$ to the culture vessel and determined that the oxygen in the degradation intermediates that were formed came from the water and not from O_2. This proved that atmospheric oxygen was not needed. As a result of such evidence it is now widely accepted that natural anaerobic biological removal of aromatic hydrocarbons in groundwater contaminated with gasoline does occur. Now, *natural attenuation* (a decrease in contaminant mass or concentration due to natural biological, chemical, and physical processes) is considered by regulators as a viable alternative for ridding groundwater of these contaminants.

Isotope fractionation of stable isotopes of an element that occurs during physical, chemical, and biological processes can also be used to help determine the movement and fate of chemicals in the environment. Small amounts of the stable isotopes ^{13}C(1.1%), ^{34}S(4.2%), ^{15}N(0.37%) and ^{18}O(0.20%) occur in nature com-

[4]T. M. Vogel and D. Grbic-Galic. Incorporation of Oxygen from Water into Toluene and Benzene during Anaerobic Fermentative Transformation. *Applied and Environmental Microbiology,* **5:** 200–202 (1986).

pared with the dominant isotopes ^{12}C, ^{32}S, ^{14}N, and ^{16}O. In biological transformations, the lighter isotopes are generally preferred over the heavier ones. Thus, for example in photosynthesis, plants growing on CO_2 become enriched in ^{12}C compared with the growth-medium CO_2, while the remaining growth-medium CO_2 becomes enriched in ^{13}C. Likewise, when methanogenic bacteria reduce CO_2 to CH_4, the methane formed contains much lighter carbon than was present in the original CO_2, and the sulfur that is reduced to sulfide during biological sulfate reduction is much lighter than in the original sulfate. During the physical process of evaporation of water from the oceans, the vapor contains less ^{18}O than the ocean water on a per unit mass basis. Thus, when the vapor condenses as rain over the land, the freshwater runoff is depleted in ^{18}O compared with that in sea water. This knowledge helps make it possible to evaluate the original sources of saline groundwaters.

Environmental engineers and scientists are attempting to evaluate when natural attenuation of organic and inorganic materials in groundwater is protective of human health and the environment. For this they need to determine how fast and to what extent such chemicals are being rendered harmless by natural chemical, physical, or biological processes. Isotope fractionation is an emerging tool that is helping in these efforts. For example, it is now known that anaerobic processes can result in the reductive dehalogenation of tetrachloroethene to trichloroethene, *cis*-1,2-dichloroethene, vinyl chloride, and then ethene (see Section 6.13). Because of the high cost of sampling groundwater at many locations, mass balances that would help to learn of the rate and extent of such transformations are very difficult to make. However, laboratory studies have demonstrated, as might be expected from the preceding discussion of isotope fractionation, that the carbon in the transformation products of tetrachloroethene are lighter in ^{13}C than the parent compound[5,6]. The isotope fractionation occurring during transformation of PCE to TCE and TCE to *c*-DCE is relatively small. However, the fractionation occurring from conversion of *c*-DCE to vinyl chloride and vinyl chloride to ethene is large. In these studies, while transformation was ongoing, the ethene had a lower $^{13}C/^{12}C$ ratio than the PCE, but as the transformation neared completion, the $^{13}C/^{12}C$ ratio increased in the ethene, and at completion, was the same as in the starting PCE, as would be expected from a mass balance of isotopes. All the carbon initially in the PCE at the end of transformation would be in the ethene. It is obvious that good judgment about the extent of transformation at intermediate points requires a good understanding of hydrogeology and kinetics, as well as about isotope fractionation.

Isotope fraction has also been used to evaluate the extent of reduction of Cr(VI) to Cr(III) in groundwater[7]. Cr(VI) in the form of chromate (CrO_4^{2-}) and dichromate

[5]D. Hunkeler, R. Aravena, and B. J. Butler, Monitoring Microbial Dechlorination of Tetrachloroethene (PCE) in Groundwater Using Compound-Specific Stable Carbon Isotope Ratios: Microcosm and Field Studies, *Environ. Sci. & Tech.,* **33:** 2733–2738 (1999).

[6]D. L. Song, M. E. Conrad, K. S. Sorenson, and L. Alvarez-Cohen, Stable Carbon isotope Fractionation during Enhanced In Situ Bioremediation of Trichloroethene, *Environ. Sci. Technol.,* **36:** 2262–2268 (2002).

[7]A. S. Ellis, T. M. Johnson, and T. D. Bullen, Chromium Isotopes and the Fate of Hexavalent Chromium in the Environment, *Science,* **295:** 2060–2062 (2002).

($Cr_2O_7^{2-}$) is toxic and very water soluble. However, when reduced to Cr(III), $Cr(OH)_3(s)$ forms at near neutral pH, removing chromium from solution and forming a stable non-toxic precipitate. Use is being made of chemical and biological reduction of chromium to protect down-gradient water users. The question again is how fast and to what extent is chromium reduction occurring at a given site. Chromium has four stable isotopes of masses 50 (4.35%), 52 (83.8%), 53 (9.50%), and 54 (2.37%). As Cr(VI) reduction takes place, the ratio of the most abundant isotopes ($^{53}Cr/^{52}Cr$) in the remaining Cr(VI) increases. The degree of increase can be used to indicate the extent of the reduction taking place. Isotope fractionation is thus an additional tool that can help solve complex movement and fate questions.

8.9 | EFFECT OF RADIATION ON HUMANS

Radiation effects on humans are classified as *somatic* or *genetic*. Somatic effects are those that cause damage to the individual and include anemia, fatigue, loss of hair, cataracts, skin rash, and cancer. Genetic effects include inheritable changes resulting from mutations in reproductive cells. It is widely held that even small dosages of radiation can have some adverse effects, genetic effects being of most concern. Humans are exposed to varying levels of natural radiation, especially from extraterrestrial sources. It is generally felt that radiation created through human activities should be kept well within the bounds of the natural background radiation. What should be the upper acceptable levels within these bounds is a subject of much debate.

Different types of radiation produce different effects in humans, and the effect is different for internal as compared with external exposure. External exposure to alpha particles represents a very small hazard since they have difficulty in penetrating the skin. However, alpha particles can be quite damaging if ingested since they can cause extensive ionization when they collide with matter making up the organs of the body. Beta particles are smaller and move faster than alpha particles and hence can penetrate to greater distances. They can penetrate from a few millimeters to a centimeter or so under the skin and so can be hazardous even with external exposure. Internally, beta particles are more hazardous than they are externally, but they are not as damaging as internal alpha particles. Gamma radiation is most dangerous since it has very high penetrating power and constitutes a hazard to the entire body. It can destroy tissue and inflict serious harm quite rapidly.

The interaction of radiation with biological material and with the water it contains results in the formation of a whole host of ionized species (such as H^+, H_2O^-, H_2O^+, e^-, e^+, H_3O^-, etc.), many of which are highly reactive. These go on to react with proteins and nucleic acids, to deactivate enzymes, to inhibit cell division, to disrupt cell membranes, and to otherwise damage cell performance. As knowledge of these effects has increased, so have efforts to prevent undue exposure to radiation. The risks versus benefits from use of radioactive materials should constantly be weighed in medical applications as well as in other uses that have already been discussed. With proper precautions, use of radioactive materials offers significant benefits to humanity.

Because of potential health effects associated with exposure to radiation, the U.S. Environmental Protection Agency (EPA) has set drinking water maximum contaminant limits (MCLs) for various types of radioactivity in drinking water. These include alpha particles (15 pCi/L), uranium (30 μ/L as of December 8, 2003), ^{226}Ra and ^{228}Ra (5 pCi/L combined), and beta particles and photon emitters (4 mrem per year). As noted in Sec. 8.3, rem is a unit that corresponds to the amount of radiation that will produce a certain energy dissipation in a human body. As such, it is not something that can be measured directly from a concentration in water alone. It depends upon the route of exposure, the contact time with the body, the rate of decay, and the energy released by decay. The former National Bureau of Standards (now the National Institute of Standards and Technology) published a report that related the rem with concentrations (pCi/L) of various radioactive materials in water (Publication 69, 1963). The EPA has used this publication as a basis for relating the concentrations of numerous radioactive elements in water that correspond to 4 mrem per year for humans drinking 2 liters of water per day.[8] This concentration in water is 20,000 pCi/L for tritium (^{3}H), 2000 pCi/L for ^{14}C, 500 pCi/L for ^{35}S, and 30 pCi/L for ^{32}P. If more than one beta particle or photon emitter is present in the water, then the total contribution cannot exceed 4 mrem/year. For example, if a water contained 12,000 pCi/L of tritium and 1000 pCi/L of ^{14}C, the total would be (12,000/20,000 + 1000/2000) $\times$ 4 or 4.4 mrem/year. This would exceed the MCL, even though it is not exceeded by either tritium or ^{14}C alone.

Radon

The EPA indicates that the average person is exposed to an effective ionizing radiation dose equivalent of approximately 360 mrem/year (whole body exposure) from all sources, of which about two-thirds is from natural background and radon and the remainder from other sources, especially medical diagnoses and treatment. The largest part of the natural radiation that we receive comes from radon, a naturally occurring radioactive gas. Radon is a fission product of ^{238}U, which is an abundant source of naturally occurring radioactivity. ^{238}U decays to ^{236}Ra, which decays by alpha emission to form ^{222}Rn. Radon has a short half-life of only 3.82 days, decaying by alpha emission to form a series of short-lived radioactive decay products (^{218}Po, ^{214}Pb, ^{214}Bi, and ^{214}Po; see Fig. 8.2). The damage to lungs comes from the decay products, and not from radon itself. Radon occurs in three forms, but the term radon is generally used specifically to refer to ^{222}Rn.

Radon is reported to be second only to smoking as a leading cause of lung cancer deaths in the United States. The National Research Council in a 1999 report[9] indicated that about 19,000 lung cancer deaths in the United States each year are radon related, and only 160 of these cancer deaths results from radon in drink-

[8]U.S. Environmental Protection Agency, "Implementation Guidance for Radionuclides," EPA 816-D-00-002, Office of Water Washington D. C. (January 2002).

[9]National Research Council, "Risk Assessment of Radon in Drinking Water," National Academy Press, Washington D.C., 1999.

ing water, most result from radon in indoor air. Most of the radon that enters buildings comes directly from soil that is in contact with or beneath the basement or foundation. Some comes from the building materials themselves. The average outdoor air concentration of radon over the United States is about 0.4 pCi/L, and the average indoor air concentration is about 1.24 pCi/L. Radon is also contained in well water and will enter the home whenever the water is used there. The National Research Council report[10] indicates that typical surface water concentrations of radon are less than 100 pCi/L, but the average concentration in groundwater sources used for drinking water is about 540 pCi/L, and some wells have concentrations 400 times the average. Radon enters the air when showering or washing with the water. Most risk from water comes from breathing the radon released to the air, rather than from drinking the water itself. The risk from radon is higher among smokers, the damage of the combination being much higher than from either alone. Indeed, most radon-related lung cancer could be eliminated through the elimination of smoking.

The EPA currently has no air or drinking water standard for radon, primarily because most exposure to radon occurs in the home environment where the EPA lacks jurisdiction. But the problem is a significant one. In order to address the issue, the EPA has proposed a multimedia framework that provides states with flexibility in how to limit exposure to radon by allowing them to focus their efforts on the area with greatest radon risks, that is on indoor air, while also reducing the risks from radon in drinking water. This proposal suggests as a first option that states address health risks from indoor air and combine this with having individual water systems reduce radon levels in drinking water to 4000 pCi/L or lower. The second option if a state chooses would be to focus only on drinking water, and in this case, reduce radon content there to 300 pCi/L.

PROBLEMS

8.1 What is the difference between alpha, beta, and gamma radiation?

8.2 What fraction of an initial amount of ^{14}C would be left after 1 year?

8.3 What is formed from the following disintegrations?

  (a) $^{230}_{90}Th$ emission of an alpha particle

  (b) $^{49}_{19}K$ emission of a beta particle

8.4 What is formed by the following induced reactions?

  (a) From ^{238}U by an n, gamma reaction

  (b) From ^1_1H by an n, gamma reaction

  (c) From ^6_3Li by an n, alpha reaction

  (d) From $^{10}_5B$ by an n, alpha reaction

8.5 The carbon in living plants and animals contains enough ^{14}C to yield about twelve ^{14}C disintegrations per minute per g C. Estimate the age of an entombed piece of wood that yields only four disintegrations per minute.

---

[10]Ibid.

8.6 How many millions of electron volts (MeV) are released by the complete conversion of 1 g of mass into energy?

8.7 Write balanced nuclear equations, in the same form as given in Secs. 8.3 and 8.4, for the noted decay of each of the following:

(a) ^{14}C, beta

(b) ^{22}Na, positron

(c) ^{226}Ra, alpha

(d) ^{32}P, beta

(e) ^{3}H, beta

8.8 How long would a waste containing ^{24}Na have to be stored for the concentration to be reduced to 0.1 percent of its initial value?

8.9 Assume you have 1 g of pure ^{239}Pu for disposal in a deep geologic formation. It has been estimated that 1×10^{-6} g of ^{239}Pu will cause cancer. How many years will pass before 1×10^{-6} g of of ^{239}Pu remains? (*Note:* It has been estimated that 1×10^{-3} g of ^{239}Pu is lethal to humans. Thus, for all the intervening years, the level of ^{239}Pu is very toxic!)

REFERENCES

Arnikar, H. J.: "Essentials of Nuclear Chemistry," John Wiley, New York, 1987.

Chopin, G. R., J. O. Liljenzin, and J. Rydberg: "Radiochemistry and Nuclear Chemistry," Butterworth-Heinemann, Boston, 2002.

Friedlander, G.: "Nuclear and Radiochemistry," John Wiley, New York, 1981.

Lieser, K. H.: "Nuclear and Radiochemistry, Fundamentals and Applications," Weinheim, New York, 1997.

Navrátil, O.: "Nuclear Chemistry," E. Horwood, New York, 1992.

Water and Wastewater Analysis

C H A P T E R

Introduction

9.1 | IMPORTANCE OF QUANTITATIVE MEASUREMENTS

Quantitative measurements of one sort or another serve as the keystone of engineering practice. Environmental engineering and science is perhaps most demanding in this respect, for it requires the use of not only the conventional measuring devices employed by engineers but, in addition, many of the techniques and methods of measurement used by chemists, physicists, and some of those used by biologists.

Every problem in environmental engineering and science must be approached initially in a manner that will define the problem. This approach necessitates the use of analytical methods and procedures, in the field and laboratory, that have been proved to yield reliable results in the hands of many people and on a wide variety of materials. Once the problem has been defined quantitatively, the engineer is usually in a position to design facilities that will provide a satisfactory solution.

After construction of the facilities has been completed and they have been placed in operation, usually constant supervision employing quantitative procedures is required to maintain economical and satisfactory performance. Records of performance are needed for reports that have to be made to supervisory personnel and regulatory agencies.

The increase in population density and new developments in industrial technology are constantly intensifying old problems and creating new ones. In addition, engineers are forever seeking more economical methods of solving old problems. Research is continuously under way to find answers to the new problems and better answers to old ones. Quantitative analysis will continue to serve as the basis for such studies.

9.2 | CHARACTER OF ENVIRONMENTAL ENGINEERING AND SCIENCE PROBLEMS

Most problems in environmental engineering and science involve relationships between living organisms and their environment. Because of this, the analytical procedures needed to obtain quantitative information are often a strange mixture of chemical and biochemical methods, and interpretation of the data is usually related to the effect on microorganisms or human beings. Also, many of the determinations used fall into the realm of microanalysis because of the small amounts of contaminants present in the samples. For these reasons, the usual course in quantitative analysis offered by schools of chemistry is of limited value to environmental engineers and scientists, other than to teach basic techniques. Specialized courses in environmental analytical chemistry have been developed to meet this particular need at nearly all universities that educate environmental engineers.

9.3 | STANDARD METHODS OF ANALYSIS

Concurrent with the evolution of environmental engineering practice, analytical methods have been developed to obtain the factual information required for the resolution and solution of problems. In many cases different methods were proposed for the same determination, and many of them were modified in some manner. As a result, analytical data obtained by analysts were often in disagreement. In cases involving litigation, judges often found it difficult to evaluate evidence based upon analytical methods. In an attempt to bring order out of chaos, the American Public Health Association (APHA) appointed a committee to study the various analytical methods available and published the recommendations of the committee as "Standard Methods of Water Analysis" in 1905. Since that time, the scope of "Standard Methods" has been enlarged to include wastewaters, and the American Water Works Association (AWWA) and the Water Environment Federation (WEF) have become collaborators in its preparation. The twentieth edition appeared in 1998.[1]

"Standard Methods" as published today is the product of the untiring effort of hundreds of individuals who serve on committees and subcommittees, testing and improving analytical procedures for the purpose of selecting those best suited for inclusion in "Standard Methods." Evidence that is obtained by qualified analysts based upon methods recommended in "Standard Methods" is normally accepted in the courts of the United States without qualification.

9.4 | SCOPE OF A COURSE IN ANALYSIS OF ENVIRONMENTAL SAMPLES

It would be impossible and unwise to attempt to teach a course in analysis dealing with all the determinations described in "Standard Methods." In the first place, space does not permit such treatment, and in the second place, many of the determi-

---

[1]L. S. Clesceri, A. E. Greenberg, and A. D. Eaton (eds.), "Standard Methods for the Examination of Water and Wastewater." 20th ed., American Public Health Association, New York, 1998.

nations are highly specific for certain industrial wastes. On the other hand, it is important that a good foundation in analytical procedures be established so that any contingencies that may arise during one's career can be met and handled with a reasonable degree of confidence.

The choice of determinations and the order in which they are included in a particular course depend greatly upon the interests of the instructor. The selection of topics for discussion in the following chapters covers items that the authors have found essential for the basic education of environmental engineers and scientists. The order of presentation is a matter of personal opinion but is based upon a natural sequence of dependence and increasing complexity.

A major objective of a course in the analysis of environmental samples should be to prepare the student for research, rather than as a technician. To this end, it is important that the student appreciate the nature and source of the materials under analysis, the limitations of the analytical methods, how to interpret the data, and how the information may be applied in environmental engineering practice.

9.5 | EXPRESSION OF RESULTS

Most materials subjected to analysis in the fields of water and wastewater fall into the realm of dilute solutions, and it is impractical to express results in terms of percent, as is the usual practice in analytical chemistry. Ordinarily, the amounts determined are a few milligrams per liter and oftentimes only a few micrograms. Water and wastewater samples are usually measured by volume, using a volumetric pipet; therefore, it is convenient and appropriate to express results in terms of milligrams per liter (mg/L). With air, soils, sludges, and semisolid samples, on the other hand, measurements are made by weight or volume, and thus the term *parts per million* (ppm) is here an appropriate method of expression. The term ppm is sometimes used for aqueous samples as well, but this frequently leads to misinterpretation and should not be used. The student should learn when it is appropriate to express measurements in mass per unit volume (mg/L) and in mass per unit mass (ppm).

Parts per Million

The term *parts per million* is a weight-to-weight ratio. Its use was more or less universal and unquestioned when analysis was principally concerned with water, because a liter of water weighs approximately 1000 g or 1,000,000 mg, and hence 1 mg/L was considered to be equal to 1 ppm. With the development and inclusion of methods for the analysis of polluted waters, such as domestic wastewater, the concept of the relation between parts per million and milligrams per liter did not change, because the specific gravity of domestic wastewater is essentially the same as that of water. As industrial wastes were included in the materials subjected to analysis, many of them were found to have specific gravities considerably different from that of water, and the close relationship between parts per million and milligrams per liter no longer applied. This discrepancy has led to abandonment of the term parts per million in water and wastewater analysis in favor of the use of milligrams per liter. However, as noted, when semisolid samples are analyzed and their weights rather than volumes are determined, ppm is then an appropriate method of expression. At times, very small concentrations of contaminants are measured in semisolid samples, and the terms parts per billion (ppb) or parts per trillion (ppt) may then be appropriate for expression of results.

| EXAMPLE 9.1 | What is the TCE concentration in g/m^3 for each of the following cases? |

> What is the TCE concentration in g/m^3 for each of the following cases?
>
> (*a*) A soil having a density of 2 g/cm^3 with TCE concentration of 4 ppm
> (*b*) An air sample at 20°C and 1 atm pressure with TCE concentration of 4 ppm
>
> (*a*) Since 4 ppm in soil equals 4 $g/10^6$ g soil,
>
> $$\frac{4\ g}{10^6\ g} \times \frac{2\ g}{cm^3} \times \frac{1000\ cm^3}{m^3} = 0.008\ g/m^3$$
>
> (*b*) Since in air, 4 ppm equals 4 mL (gaseous) TCE/10^6 mL air,
>
> $$\frac{4\ mL\ TCE}{10^6\ mL} \times \frac{1\ mol\ TCE}{22.4\ L\ (STP)} \times \frac{131.5\ g\ TCE}{mol\ TCE} \times \frac{10^3\ L}{m^3} \times \frac{273\ K}{(273 + 20)\ K} = 0.022\ g/m^3$$
>
> We see that the concentrations in g/m^3 are quite different for the soil and air samples, even though the concentrations when expressed in ppm are the same. This illustrates that we must understand well the general basis for mass expressions in different media.

Milligrams per Liter

Milligrams per liter is a weight-volume relationship and, when dealing with liquids, it offers a convenient basis for calculation. The expression

$$mg/L \times 8.34 = lb/million\ gal$$

is widely used and has general application. It replaces the original expression

$$ppm \times 8.34 = lb/million\ gal$$

which may be safely applied to problems involving water and wastewater and other liquids whose specific gravity is essentially 1.00, but may lead to serious errors with other liquids unless correction for specific gravity is applied. Specific gravity is the ratio of the mass of a substance to the mass of an equal volume of water at 4°C.

The use of milligrams per liter for aqueous samples eliminates any opportunity for misunderstanding and confusion. In the past many results have been reported in terms of parts per million with no reference to specific gravity. Obviously, the results should have been reported in milligrams per liter. Furthermore, milligrams per liter is directly applicable to the metric system.

$$mg/L = g/m^3$$

or
$$mg/L \times 10^{-3} = kg/m^3$$

An industry is found to be discharging 20 lb/day of benzene into a river with a flow of 10 billion gallons per day. What would be the resulting benzene concentration if the benzene is uniformly mixed with the river flow? Use SI units.

EXAMPLE 9.2

$$\text{Benzene concentration} = \frac{20 \text{ lb}}{10^9 \text{gallons}} \times \frac{1000 \text{ gallons}}{3.79 \text{ m}^3} \times \frac{\text{kg}}{2.2 \text{ lb}}$$

$$= 2.4(10^{-6}) \text{ kg/m}^3 \quad \text{or} \quad 2.4 \text{ } \mu\text{g/L}$$

SI Units

In order to develop a uniform method of reporting results that is applicable worldwide, the International Organization for Standardization (ISO) published in 1973 a document listing a recommended International System of Units (SI).[2] These recommendations were approved by Member Boards of ISO in 30 countries including the United States, and by the International Union of Pure and Applied Chemistry (IUPAC). Many engineering and scientific tests now use SI units exclusively, and many others present a portion of the problems and examples in SI units. In the United States increasing efforts are being made to familiarize the public with the metric system, which is the cornerstone of the SI system. Thus, it is prudent for students and practitioners to develop a working knowledge of this system. Fortunately, the metric system has generally been used in environmental chemistry and so adoption to the SI system requires little basic change.

The SI system is founded upon seven base units as listed in Table 9.1. There are a series of units that are derived from the base units through multiplication or division: for example, the SI unit for concentration would be kilograms per cubic meter (kg/m^3). For some derived SI units, special names and symbols exist. Those appropriate to the subject of this text are given in Table 9.2. In addition, the SI system recognizes that there are units outside the SI system that should be retained because of their practical importance or because of their use in specialized fields. A

Table 9.1 | Base units in the SI system

Quantity	Name of base SI unit	Symbol
Length	Meter	m
Mass	Kilogram	kg
Time	Second	s
Electric current	Ampere	A
Thermodynamic temperature	Kelvin	K
Amount of substance	Mole	mol
Luminous intensity	Candela	cd

---

[2]"International Standard 1000," International Organization for Standardization, 1973.

Table 9.2 | Derived units in the SI system

Quantity	Name of derived SI unit	Symbol	Expressed in terms of base or supplementary SI units or in terms of other derived SI units
Force	Newton	N	$1\ N = 1\ kg\text{-}m/s^2$
Pressure, stress	Pascal	Pa	$1\ Pa = 1\ N/m^2$
Energy, work, quantity of heat	Joule	J	$1\ J = 1\ N\text{-}m$
Power	Watt	W	$1\ W = 1\ J/s$
Electric charge, quantity of electricity	Coulomb	C	$1\ C = 1\ A\text{-}s$
Electric potential, potential difference, tension, electromotive force	Volt	V	$1\ V = 1\ J/C$
Electric capacitance	Farad	F	$1\ F = 1\ C/V$
Electric resistance	Ohm	Ω	$1\ \Omega = 1\ V/A$
Electric conductance	Siemens	S	$1\ S = 1\ \Omega^{-1}$

listing of applicable alternative units is given in Table 9.3. Thus, the expression of concentration in milligrams per liter is acceptable under the SI system.

There is a current interest in environmental chemistry in trace organics and metals occurring in less than 1 mg/L amounts. For expressing such low concentrations, it is convenient to have a system of expression that allows for a wide range of possible contaminant levels. This is accomplished in the SI system through the use of prefixes as listed in Table 9.4. The prefixes are used to form names and symbols for multiples of the SI units. For example, $\mu g/L$ is called micrograms per liter and is equivalent to 10^{-6} gram per liter. This is a concentration unit commonly used for trace materials in water. Measured pesticide concentrations in water are generally less than this and may be expressed in nanograms per liter (ng/L), which is equivalent to 10^{-9} gram per liter. Analytical instruments today are sufficiently sensitive so that picogram quantities of materials can often be detected. One pg/L of a pesticide is a concentration that would result from mixing about 4 mg of pesticide in the daily water supply for the city of Chicago. It is difficult to comprehend how such a small

Table 9.3 | Units outside the SI system that are acceptable

Quantity	Name	Symbol
Volume	Liter	L
Density	Grams per liter	g/L
Concentration	Moles per liter	M
Time	Day	d
Time	Hour	h
Time	Minute	min

Table 9.4 | Multiplication prefixes in SI system

Factor by which the unit is multiplied	Prefix	
	Name	Symbol
10^{12}	tera	T
10^9	giga	G
10^6	mega	M
10^3	kilo	k
10^2	hecto	h
10	deca	da
10^{-1}	deci	d
10^{-2}	centi	c
10^{-3}	milli	m
10^{-6}	micro	μ
10^{-9}	nano	n
10^{-12}	pico	p
10^{-15}	femto	f
10^{-18}	atto	a

concentration of anything could have public health or environmental significance. Ability to measure such small concentrations does indicate the danger of arbitrarily setting "zero" concentration limits for contaminants.

Other Methods of Expression

In certain determinations, such as color and turbidity, reference is made to arbitrary standards. In these cases results are expressed in units without designation, 0, 1, 5, and so on.

Sludges and some industrial wastes contain enough suspended or dissolved solids so that results can be best expressed in terms of percent. A rule of thumb often applied is this: When concentrations exceed 10,000 mg/L,[3] results are expressed as percent. Wherever practice has established a precedent in opposition to the rule, the rule is ignored.

Water chemists often prefer to express results in terms of milliequivalents (meq) per liter. This allows them to translate results directly in terms of other chemicals. Milliequivalents per liter are obtained by dividing milligrams per liter of the element or ion by its equivalent weight in grams. Equivalent weights are discussed in Chap. 11. Moles per liter are also frequently used because of need when working with equilibrium relationships. Thus, the environmental engineer and scientist becomes exposed to many methods of expression, and to be competent, he or she must develop a working knowledge of each. The need for a more uniform system of expression is evident!

---

[3] A figure of 10,000 mg/L is equivalent to 1 percent when the specific gravity is equal to 1.0.

9.6 | OTHER ITEMS

It is expected that environmental engineering and science students will become familiar with all the material in the Introduction to "Standard Methods." Close attention should be directed to the sections on quality assurance; data quality; collection and preservation of samples; laboratory apparatus, reagents, and techniques; and safety. The latter introductory material in particular should be required reading of all students prior to their working in the laboratory. Many of the chemicals used for analysis, as well as the samples being analyzed themselves, can be highly dangerous. The dangers and appropriate precautions to follow must be well understood by all laboratory workers. Disposal of samples after analysis and of waste hazardous chemical solutions must be carried out in an environmentally acceptable manner. Since costs for such disposal are increasing rapidly, good practice dictates that analytical procedures that minimize chemical usage be adapted. Also, chemicals should be purchased in sufficiently small quantities to satisfy only the immediate needs for them so that long-term storage and/or disposal of excess quantities can be avoided.

Chap. 10 addresses statistical methods of analysis. Those using analytical data often overlook the uncertainties inherent in such measurements, uncertainties that if not considered properly may lead to significant errors in judgment. Thus, all environmental engineers and scientists should understand as a minimum the basic statistical concepts discussed here. Chapters 11 and 12 contain discussions of some additional items of general importance and usefulness to analysis of environmental samples. It would be good for the student to become generally familiar with the contents of these chapters and to refer back to them for details of specific importance to individual analyses that are discussed in subsequent chapters.

PROBLEMS

9.1   What professional organizations collaborate in the preparation of "Standard Methods"?

9.2   What needs for environmental analysis are not generally met by the usual chemistry course in quantitative analysis?

9.3   The average carbon dioxide content in ambient air is 0.03 percent on a mole fraction basis. Express the concentration in ppm and mg/L, assuming standard conditions of 1 atm pressure and 0°C.

9.4   A sample of sludge has a total solids content of 3.42 percent. If the sludge specific gravity is 1.06, what is the total solids content in grams per liter?

9.5   If 1000 lbs of lime are added to 1.00 million gallons of water, what is the concentration of the lime in mg/L?

9.6   The PCB content of a soil is found to be 8 ppm. If the density of the soil is 2 kg/m^3, what is the PCB content in grams per cubic meter of soil?

9.7   How many grams of NaCl are there in a liter solution in which the NaCl concentration is 85 mM?

9.8   If the equivalent weight of potassium dichromate is one-sixth of the formula weight, how many grams would be present in a liter of solution containing one equivalent of potassium dichromate?

9.9   A water sample contains 20 ng/L. What is the concentration in kg/m^3?

9.10 What precautions should be taken in the purchase and disposal of hazardous chemicals?

9.11 If 50 liters of methane passes from an anaerobic digester into a room that is 20 m × 80 m × 5 m in size, what will be the methane concentration in ppm? Assume 1 atm of pressure at 20°C.

9.12 One liter of trichloroethene (TCE) leaks onto 5 m^3 of soil having a density of 2 g/cm^3. What is the average TCE concentration in the soil in ppm?

9.13 A drinking water supply contains 150 μg/L chloroform. How many pounds of chloroform are present in 150 million gallons?

9.14 The EPA maximum contaminant level (MCL) for trichloroethene (TCE) in drinking water is 5 μg/L. The specific gravity of pure TCE, which is a liquid, is 1.46. What volume of TCE in milliliters would need to be added to 200 million gallons of water (about the daily drinking water supply for 1 to 2 million people) to increase the TCE concentration by 5 μg/L?

Statistical Analysis of Analytical Data

10.1 | INTRODUCTION

All analytical procedures are subject to error. Thus, the measured concentration of a given constituent in water, air, or soil will generally be different from the true value. This raises several questions for the analytical chemist; for regulators relying on analytical results to determine the chemical or biological quality of water, air, or soil; for engineers evaluating the effectiveness of a given treatment process; and for those attempting to determine which of several potential processes provides the most efficient and reliable treatment for removal of chemicals of health or ecological concern. Because of the importance to decisions they must make, the uncertainties involved must be understood by those using analytical data. Statistics is the science that deals with uncertainties and how to measure and evaluate them. Analytical chemists, regulators, and practitioners should all be familiar with statistical analysis if they wish to make wise decisions about the analytical data they use. This chapter presents only a brief summary of statistical methods for analysis of analytical data. More thorough treatments can be found in textbooks on the subject. "Standard Methods"[1] also contains a more thorough treatment of some of the concepts presented here. In addition, the text by de Levie[2] is an excellent aid to the use of spreadsheets for statistical analysis.

Calculators and spreadsheets are essential for rapid statistical analysis of data. Their use has made it quite easy to conduct statistical calculations, thus encouraging their broader use and application. Indeed, many hand-held calculators and computer spreadsheets contain statistical functions that allow rapid calculation of

[1]L. S. Clesceri, A. E. Greenburg, and A. D. Eaton (eds.), "Standard Methods for the Examination of Water and Wastewater," 20th ed., American Public Health Association, New York, 1998.

[2]R. de Levie, "How to Use Excel® in Analytical Chemistry and in General Scientific Data Analysis," Cambridge University Press, Cambridge, MA, 2001.

such things as averages and standard deviations. However, it should be cautioned that the ready availability of such easy-to-use tools increases, rather than decreases, the need for one to thoroughly understand the fundamental basis behind the calculations. Otherwise the user is likely to carry out a statistical procedure by computer without knowing whether in fact it is appropriate for the data being analyzed. Blind application of statistical programs without adequate knowledge of the assumptions involved is likely to result in erroneous conclusions, rather than the more sophisticated analysis of data that may be desired. It is hoped that this chapter will form at least a beginning toward a good understanding of statistics and of the potential value of statistical analysis in environmental engineering and science.

10.2 | ROUNDING NUMERICAL DATA

We first address how to present analytical data so that proper information is conveyed to the reader about the analytical uncertainties involved. This is indicated by the number of *significant figures* that are reported for an analytical result. For example, a computer spreadsheet that calculates results for us based upon input from an analytical instrument might list 276.539 mg/L for sodium in a given sample. What number should we report to our client? Some possibilities are these: 276.539, 276.5, 276, or 280 mg/L. Which is correct? The answer depends upon how close the number is expected to represent the true value. Generally in environmental analyses, analytical measurements are seldom closer than ± 1 percent of the true value because of analytical uncertainties in sample collection, preservation, and preparation, as well as in the analytical measurements themselves. Frequently, they are only within ± 10 percent of the true value. This uncertainty as a minimum must be indicated in the number reported. Which of the listed possible numbers would reflect an uncertainty of 1 percent, and which would imply an uncertainty of 10 percent? The answers are 276 and 280 mg/L, respectively. To illustrate why this is the case, we see that 1 percent of 276 is 2.76. Thus, the true number is expected to lie somewhere between 273.24 and 278.76. If this is the case, then 276 is a better reflection of the uncertainty in the value than would be a number with more digits added to it. We gain no benefit by giving the client the spreadsheet value of 276.539, and to do so implies that we know the true value with more precision than is actually the case. If the uncertainty is 10 percent, then the true number might be expected to lie somewhere between 248 and 304 mg/L. Here, 280 mg/L is an adequate representation of the accuracy of the value. Calling it 276 mg/L instead implies again a precision that is not warranted by the uncertainty in the result.

What we mean by significant figures is the number of nonzero digits used to express the result. For example, the number 10,300, 273, 3.54, and 0.0976 all are reported to three significant figures and imply uncertainties of about 1 percent. Additionally, 24,000, 560, 0.42, and 0.00056 all represent values reported to two significant figures, representing uncertainties of about 10 percent. With computer-generated results being the rule rather than the exception today, those reporting analytical results must be extra careful in significant figure usage when they present their results in reports in order to avoid misleading others. As an added benefit, it is much easier to read results from tables when use of unnecessary significant figures is avoided. As an additional comment, it is okay and probably best not to round values to the appropriate significant figure level while computations of analytical results are being made with a computer, for if rounding is too severe, computational errors may result. The significant figure rules apply primarily when reporting results to others.

10.3 | DEFINITIONS

While reporting data using significant figures properly is helpful in conveying the uncertainties in information, there are much better statistical methods for doing this. Because of uncertainties in analyses and the possibility of mistakes in sample preparation and handling, the analyst generally carries two or more samples from the same treatment through an analytical procedure. Because of uncertainties in the analysis, the results are generally not all exactly the same. The analyst then generally uses some procedure for reporting the central tendency of the results and also for indicating how the analytical results vary from one another. A central value generally is more reliable than any of the individual results by themselves. The degree to which the individual results differ with one another provides a measure of the uncertainty in the results. The mean and median are two measures of central tendency.

Mean and Median

Mean, average, and *arithmetic mean* ($\bar{x}$) are terms used for the same measure of central tendency, which is obtained by dividing the sum of all the measurements by the number of measurements made:

$$\bar{x} = \frac{\sum_{i=1}^{n} x_i}{n} \tag{10.1}$$

where each x_i represent an individual measurement and n represents the total number of measurements made.

The median is the middle value in a series of values. If there is an odd number of values, then the median is the one directly in the middle, while if there is an even number of values, the median is the average of the two numbers in the middle.

EXAMPLE 10.1

Determine the mean and median concentration for the following measurements of effluent biochemical oxygen demand (BOD) for a wastewater treatment plant reported for the month of May: 15, 22, 10, 29, 5, 16, 19, 6, 28, 14, and 17 mg/L.

$$\text{Mean} = \bar{x} = (15 + 22 + 10 + 29 + 5 + 16 + 19 + 6 + 28 + 14 + 17)/11 = 16.45 = 16 \text{ mg/L}$$

Note that the answer is rounded to two significant digits only to match the significant figures of the individual measurements themselves. It is obvious from the spread in the numerical values of the measurements that no greater accuracy for the mean value should be implied.

For the median, arrange the numbers in numerical order (5, 6, 10, 14, 15, 16, 17, 19, 22, 28, 29). We see that the middle number (the sixth one) is 16. Thus, the median is 16 mg/L, and in this case happens to be the same as the mean. Thus, the two measures of central tendency agree with one another. This is not always the case.

Standard Deviation

The *standard deviation* is a measure of the spread of related values in a data set. It also can provide a measure of the uncertainty in a value or its precision, such as the concentration of a constituent in a water sample, when analyzed numerous times. The standard deviation s is defined as follows:

$$s = \sqrt{\frac{\sum(x_i - \bar{x})^2}{n - 1}} \tag{10.2}$$

which is numerically equivalent to

$$s = \sqrt{\frac{\sum x_i^2 - (\sum x_i)^2/n}{n - 1}} \tag{10.3}$$

Equation (10.2) indicates that the standard deviation is a measure of the relative differences between the individual values in a data set and their average. The greater the differences, the greater is the standard deviation. The value $n - 1$ in the denominator is referred to as the *degrees of freedom,* which in the case of a simple data set is the number of values in the set minus 1. The degrees of freedom will be different in other cases where data sets are combined as will be illustrated in Sec. 10.5. While Eq. (10.2) better shows the meaning of the standard deviation, Eq. (10.3) is more often useful for calculating the value from a table of data as the average value of the data does not have to be determined before the calculation is made.

The *variance* s^2 is useful at times in data analysis when standard deviations for various processes are to be combined. The variance is simply the square of the standard deviation:

$$s^2 = \frac{\sum(x_i - \bar{x})^2}{n - 1} \tag{10.4}$$

EXAMPLE 10.2

Determine the standard deviation and variance for the effluent BOD data given in Example 10.1.

The mean for the data in Example 10.1 was given as 16.45 mg/L, which was rounded to 16 mg/L as appropriate based upon significant figures. Let's use both Eqs. (10.2) and (10.3) to illustrate that they provide the same result:

x_i	$(x_i - \bar{x})^2$	x_i^2
15	2.12	225
22	30.75	484
10	41.66	100
29	157.39	841
5	131.21	25
16	0.21	256
19	6.48	361
6	109.30	36
28	133.30	784
14	6.02	196
17	0.30	289
$\Sigma = $ 181.00	618.73	3597
$\bar{x} = $ 16.45		

From Eq. (10.2),

$$ s = \sqrt{\frac{\sum (x_i - \bar{x})^2}{n - 1}} = \sqrt{\frac{618.73}{11 - 1}} = 7.87 \quad \text{or} \quad 8 \text{ mg/L using significant figures} $$

From Eq. (10.3),

$$ s = \sqrt{\frac{\sum x_i^2 - (\sum x_i)^2/n}{n - 1}} = \sqrt{\frac{3597 - (181^2/11)}{11 - 1}} = 7.87 \quad \text{or} \quad 8 \text{ mg/L as from Eq. (10.2)} $$

The results using Eq. (10.2) or Eq. (10.3) are the same. The variance is then simply s^2:

$$ s^2 = (7.87)^2 = 62 \ (\text{mg/L})^2 $$

Accuracy and Precision

Accuracy and precision are two important words that are often used incorrectly. They have specific meanings that should be clearly understood by those using analytical data. *Accuracy* refers to the closeness of a measurement or set of measurements to the true (or accepted) value. *Precision* refers to the closeness in agreement between two or more measured values. For example, assume three measurements for chloride concentration in water by a given analytical procedure result in the following values: 145, 147, and 143 mg/L. The numbers are quite close together with an average of 145 mg/L. Therefore, the analytical procedure might be said to give fairly precise results. However, if the accepted concentration based upon careful preparation of the sample that was analyzed is 155 mg/L, then the analytical procedure would be said to lack accuracy. Good precision in itself is not a sufficient test of the soundness of an analytical procedure. It must be accurate, that is, capable of giving results that are close to true or accepted values. In order to determine the accuracy of a given analytical procedure, standards against which the results of the analytical procedure can be evaluated are generally prepared using careful procedures and chemicals of known purity. Such standards then provide samples with the "accepted value" for the chemical constituent of interest.

Lack of accuracy in a measurement indicates that errors exist in the measurement. The absolute error E_a is given by the difference between the measured value and the accepted or true value.

$$E_a = x_i - x_t \tag{10.5}$$

where x_t is the accepted or true value. At times the relative error E_r is of more interest.

$$E_r = 100 \frac{x_i - x_t}{x_t} \tag{10.6}$$

For the data in the preceding paragraph the absolute error is $145 - 155$ or -10 mg/L, and the relative error is $100(145 - 155)/155$ or -6.5 percent.

Errors are of three types, *indeterminate* (or random) errors, *determinate* (or systematic) errors, and *gross* errors. Indeterminate errors cause data to be scattered in a more or less systematic fashion around the mean. These errors are a consequence of the many uncontrollable variables that are part of every analytical measurement. Determinate errors have a definite source that presumably can be identified. They may be instrument errors, methods errors, or personal errors. Instrument errors are inherent errors related to the measuring device used, whether this be an electronic instrument, a pipet, or a graduated cylinder. None are made perfect. Method errors are related to physical and chemical aspects of a particular method of analysis, such as impurities in the chemicals used to make standards, or incomplete chemical reactions or extractions that may be part of the analytical procedure. Personal errors relate to the carelessness or inherent limitations of the experimentalist, for example, the inability to inject a sample into a gas chromatograph in precisely the same manner every time. Gross errors generally are those that result from carelessness on the part of the experimenter, such as reading a scale wrongly, making calculation errors, failing to record results correctly, or failing to mix a solution properly. Analysts must consistently check their analytical procedures through proper use of standards to determine both the accuracy and precision of their measurements. If these do not measure up to that needed for the task at hand, then an exploration as to the particular source of the error should be made. By systematic evaluation of errors resulting from the analytical procedure itself, and those resulting from a combination of sample preparation and analysis, the analyst should be able to determine where the cause or causes of the errors lie and seek to correct them.

10.4 | DISTRIBUTION OF EXPERIMENTAL DATA

The distribution of data from most quantitative analytical experiments tends to approach that of a bell-shaped or *Gaussian curve*. This is termed a *normal distribution*. Such a curve is illustrated in Fig. 10.1. This is drawn to represent an infinite set of data distributed around a mean value, which is designated as μ, and for which the standard deviation is taken to be σ. These symbols are used to distinguish them from the symbols $\bar{x}$ and s used in Eqs. (10.1) and (10.2), which are generally used when the data set is relatively small or finite. Figure 10.1 indicates how the relative frequency of values in the infinite data set are distributed around the mean value μ, which is represented as 0 ($x = \mu$) on the z axis. The z and y axes are normalized, resulting in an area under the curve exactly equal to 1. The curve shown can be represented by the formula,

$$y = \frac{1}{\sqrt{2\pi}} e^{-z^2/2\sigma^2} \tag{10.7}$$

The maximum value of y when $z = \mu$ is 0.39894. Of great usefulness is the fact that 68.27 percent of the data lie within $\pm 1\sigma$ of the mean value, 95.45 percent within $\pm 2\sigma$ of the mean value, and 99.73 percent within $\pm 3\sigma$ of the mean value.

If data follow a normal distribution, then we would expect that the frequency distribution around the mean would tend to follow the bell-shaped distribution indicated in Fig. 10.1. In order to evaluate this, if sufficient data are available, one can make a *histogram* and compare the results with the normal distribution curve. For this purpose, a modification of Eq. (10.7) is useful,

$$Y = \frac{ni}{\sigma\sqrt{2\pi}} e^{-z^2/2\sigma^2} \tag{10.8}$$

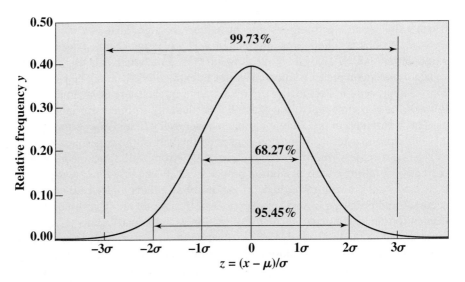

Figure 10.1
Gaussian frequency distribution of data.

where n represents the total number of data collected and i represents the interval within which the frequency of values is evaluated.

EXAMPLE 10.3

A series of trichloroethene concentrations ($\mu g/L$) obtained from analysis of groundwater samples taken from a given monitoring well over a several-day period are listed. Determine the averages (mean and median) and standard deviation for the data, draw a histogram showing a frequency distribution of the data, and compare this with a theoretical normal distribution.

33.2	35.0	25.5	23.6	41.8	27.3	31.8	31.4	26.4	33.4
38.3	37.9	26.4	23.2	39.1	36.4	30.0	29.9	27.9	24.4
33.2	35.2	30.6	22.5	24.7	25.5	28.7	33.6	24.6	29.3
36.3	30.7	28.9	28.0	30.0	9.0	42.2	29.9	25.4	24.0
40.6	28.8	33.7	23.7	35.1	9.4	30.9	33.3	28.8	23.5
33.1	36.6	28.1	21.9	33.9	29.5	22.1	25.8	26.0	25.2
37.5	27.0	35.2	12.9	35.7	37.3	36.8	28.3	27.0	23.3
46.2	29.1	24.0	12.6	33.3	28.9	40.6	37.1	25.2	26.9
33.0	33.2	25.6	14.7	36.0	27.5	31.4	35.9	27.7	27.5
37.8	29.0	32.8	13.3	30.5	7.7	29.0	26.1	24.6	26.1
41.6	30.2	23.8	17.9	32.1	30.6	5.9	25.7	24.8	
49.5	28.5	26.8	19.5	33.7	39.3	30.9	31.9	26.9	
36.0	34.8	23.3	36.9	30.5	29.3	33.2	28.3	23.1	
45.7	26.7	33.3	38.8	30.9	37.0	38.1	32.6	24.9	

First, the data are arranged in a series of increasing values from the lowest to the highest:

5.9	23.2	24.9	26.7	28.3	29.9	31.8	33.4	36.3	39.1
7.7	23.3	25.2	26.8	28.5	30.0	31.9	33.6	36.4	39.3
9.0	23.3	25.2	26.9	28.7	30.0	32.1	33.7	36.6	40.6
9.4	23.5	25.4	26.9	28.8	30.2	32.6	33.7	36.8	40.6
12.6	23.6	25.5	27.0	28.8	30.5	32.8	33.9	36.9	41.6
12.9	23.7	25.5	27.0	28.9	30.5	33.0	34.8	37.0	41.8
13.3	23.8	25.6	27.3	28.9	30.6	33.1	35.0	37.1	42.2
14.7	24.0	25.7	27.5	29.0	30.6	33.2	35.1	37.3	45.7
17.9	24.0	25.8	27.5	29.0	30.7	33.2	35.2	37.5	46.2
19.5	24.4	26.0	27.7	29.1	30.9	33.2	35.2	37.8	49.5
21.9	24.6	26.1	27.9	29.3	30.9	33.2	35.7	37.9	
22.1	24.6	26.1	28.0	29.3	30.9	33.3	35.9	38.1	
22.5	24.7	26.4	28.1	29.5	31.4	33.3	36.0	38.3	
23.1	24.8	26.4	28.3	29.9	31.4	33.3	36.0	38.8	

Next, an appropriate interval is selected that divides the data into 10 to 12, more or less, data sets. The given data set ranges from about 6 to 50, representing a difference of 44. Dividing this by 12 gives 3.7. We then might choose intervals of 3 or 4. Let us select

3. We then prepare a table in which we place the number of values within each three-unit interval:

Interval	Number of data
0.0–2.9	0
3.0–5.9	1
6.0–8.9	1
9.0–11.9	2
12.0–14.9	4
15.0–17.9	1
18.0–20.9	1
21.0–23.9	11
24.0–26.9	25
27.0–29.9	25
30.0–32.9	18
33.0–35.9	21
36.0–38.9	16
39.0–41.9	6
42.0–44.9	1
45.0–47.9	2
48.0–50.9	1

The number of data that lie within each interval is the frequency. This frequency for each interval versus the concentration interval is plotted in Fig. 10.2.

Figure 10.2

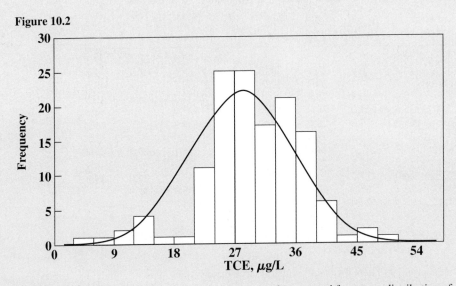

Comparison between a histogram and a Gaussian curve for a normal frequency distribution of TCE data (Example 10.2).

We now calculate the mean value $\bar{x}$ and standard deviation s of the data using Eqs. (10.1) and (10.3) and find that they equal 29.6 and 7.4, respectively. We can compare this mean value with the median, which is the middle value in the table of 136 ordered values. Since there is an even number of values, the median is the average of the 68th and 69th values [(29.3 + 29.5)/2], or 29.4. This is very close to the mean, as it should be if the data are normally spread.

We can now compare the frequency distribution for the data with the theoretical frequency we would have if $\mu = \bar{x}$ and $\sigma = s$. Then, using Eq. (10.8) with $n = 136$ and $i = 3$, we find the normal frequency distribution curve illustrated in Fig. 10.2. The fit is quite good, suggesting the data follow a normal distribution reasonably well. We can see that in order to make such a comparison, many data points are needed. The greater the number of data points we have, the more accurate will be the comparison.

Another comparison we can make is for the percentage of data points lying within given ranges as defined by the standard deviation and compare these with the theoretical values:

| | | Actual data in range | | Theoretical data |
Range around mean		Number	%	in range, %
$\pm s$	22.2–37.0	106	77.9	68.3
$\pm 2s$	14.8–44.4	125	91.9	95.5
$\pm 3s$	7.4–51.8	135	99.3	99.7

While the percentage of data in each range differs somewhat from the theoretical amount based upon a normal distribution, the difference is not large, again confirming that the data follow a normal distribution reasonably well. There are other more sophisticated methods for determining the goodness of fit to a normal distribution, but their description is beyond the scope of this chapter. Such methods are described in most of the statistical reference books listed at the end of this chapter.

10.5 | ERRORS

Improving Precision of Analytical Data

In spite of our best efforts to improve a given method of analysis, there will always be errors associated with sample collection, preparation, and analysis, so a given single analysis of a sample will likely deviate to some degree from the true value desired. Thus, if a given sample is analyzed repeatedly, a different value is likely to be obtained each time. It is obvious that by taking the mean of several measurements, we are likely to be closer to the true value than by using a single measurement alone. A question then is, how much improvement can we obtain by taking the average of multiple measurements, and how many replicate measurements is sufficient to obtain a desired improvement in precision? There are several measures that can be used to guide us.

Standard Error of the Mean

The standard deviation s provides us with a measure of the distribution of data around a mean value. If s for a given measurement is larger than desired, we can reduce this by taking more measurements of the same sample. The mean of multiple analyses will in general be closer to the true mean than if a single analysis is made. The *standard error of the mean* s_m is a function of the number of analyses used to determine the mean:

$$s_m = \frac{s}{\sqrt{n}} \tag{10.9}$$

EXAMPLE 10.4

Determine the standard error of the mean s_m for the rows of data given in Example 10.3 and compare the results with the standard deviation s of the mean values. Do this analysis separately for the rows having 10 values and the rows having 9 values.

The estimates of s_m for rows with n of 9 and 10, respectively, are

$$s_{m(n=9)} = \frac{s}{\sqrt{n}} = \frac{7.4}{\sqrt{9}} = 2.5 \qquad s_{m(n=10)} = \frac{s}{\sqrt{n}} = \frac{7.4}{\sqrt{10}} = 2.3$$

The mean values for each row from the upper table in Example 10.3 using Eq. (10.9) are summarized here.

Row	n	Row mean ($n = 10$)	Row mean ($n = 9$)
1	10	30.9	
2	10	31.4	
3	10	28.8	
4	10	28.4	
5	10	28.8	
6	10	28.2	
7	10	30.1	
8	10	30.4	
9	10	29.2	
10	10	25.7	
11	9		25.8
12	9		31.9
13	9		30.6
14	9		34.2
Mean of row means		29.2	30.6
Standard deviation of row means		1.6	3.5
Standard error of the mean [Eq. (10.9)]		2.3	2.5

Tabulated at the bottom of each column listing of row mean values is the mean value and standard deviation of the row means. The standard errors of the row means as calculated with Eq. (10.9) are also listed.

We see the improvement that has resulted from using several rather than a single value to determine the concentration. The standard deviations of row means are much lower than the standard deviation of 7.4 for the individual values of the original table. We also see that the standard deviations of the row means are reasonably close to the values calculated using Eq. (10.9). However, they do not equal the calculated values exactly. The standard deviation for the $n = 10$ row data of 1.6 is somewhat lower than the calculated standard error of the mean, while that of 3.5 for the $n = 9$ rows is somewhat higher. This illustrates that the calculated values also are not exact values; they too have uncertainties in them, which also can be represented by a standard deviation. This example serves to illustrate the normal variations that can be expected with random sets of data, approaches to analyzing the data, and some of the cautions that must be used in their application.

Standard Deviation from Pooled Data

Obtaining good estimates of s often requires much data, which can be time consuming and costly to obtain. However, one can obtain an improved estimate of s (s_{pooled}) by pooling replicate measurements from a series of different data sets. Here the deviation from the mean for each subset is first squared. The squares of all subsets are summed and then divided by an appropriate number of degrees of freedom. Finally, s_{pooled} is obtained by taking the square root of the quotient. One degree of freedom is lost for each subset. The number of degrees of freedom n used in this analysis is thus equal to the total number of measurements minus the number of subsets used.

Four different water samples were analyzed repeatedly for chloride. The results are listed. Determine s_{pooled} for the chloride analysis and compare with the standard deviation for each independent data set.

EXAMPLE 10.5

Sample number	Number of analyses	Chloride, mg/L	Mean, mg/L	Sum of squares of deviations from mean
1	4	2,220; 2,260; 2,230; 2,090	2,200	17,000
2	4	3,220; 3,200; 2,930; 3,100	3,110	52,675
3	7	2,460; 2,820 2,430; 2,540; 2,550; 2,560; 2,580	2,560	95,343
4	3	1,850; 1,810; 1,730	1,800	7,467
Total number of analyses	18	Total sum of squares		172,485

Sample number	Calculation of standard deviation	Standard deviation s, mg/L
1	$s = [17,000/(3 - 1)]^{0.5}$	75
2	$s = [53,675/(3 - 1)]^{0.5}$	133
3	$s = [95,343/(7 - 1)]^{0.5}$	126
4	$s = [7,467/(3 - 1)]^{0.5}$	61
Pooled	$s_{pooled} = [172,485/(18 - 4)]^{0.5}$	100

The pooled standard deviation of 100 mg/L is seen to lie between the individual sample standard deviations, which range from 61 to 133 mg/L. While the value of 100 mg/L is not far from the average of the four sample standard deviations (99 mg/L), the procedure used for computing the pooled value appropriately gives more weight to samples with more analyses than a simple average would.

Confidence Limits and Intervals

Often we are interested primarily in the true mean for a set of data. However, we never know the exact value of the mean because this would require an infinite number of measurements. However, using statistical theory we can estimate the range about an experimental mean where the true mean is expected to lie within desired levels of probability. The limits obtained in this manner are called the *confidence limits* (CL), and the interval they define is called the *confidence interval* (CI). Here, use is made of a statistical parameter t as follows:

$$\text{CL for } \mu = \bar{x} \pm \frac{ts}{\sqrt{n}} \qquad (10.10)$$

Table 10.1 | Values of Student's t

Degrees of freedom	Probability level				
	50%	**90%**	**95%**	**98%**	**99%**
1	1.000	6.314	12.71	31.821	63.66
2	0.816	2.920	4.303	6.965	9.925
3	0.765	2.353	3.182	4.541	5.841
4	0.741	2.132	2.776	3.747	4.604
5	0.727	2.015	2.571	3.365	4.032
6	0.718	1.943	2.447	3.143	3.707
7	0.711	1.895	2.365	2.998	3.499
8	0.706	1.860	2.306	2.896	3.355
9	0.703	1.833	2.262	2.821	3.250
10	0.700	1.812	2.228	2.764	3.169
12	0.695	1.782	2.179	2.681	3.055
14	0.692	1.761	2.145	2.624	2.977
16	0.690	1.746	2.120	2.583	2.921
18	0.688	1.734	2.101	2.552	2.878
20	0.687	1.725	2.086	2.528	2.845
25	0.684	1.708	2.060	2.485	2.787
30	0.683	1.697	2.042	2.457	2.750
40	0.681	1.684	2.021	2.423	2.704
50	0.679	1.676	2.009	2.403	2.678
60	0.679	1.671	2.000	2.390	2.660
80	0.678	1.664	1.990	2.374	2.639
∞	0.674	1.645	1.960	2.326	2.576

Here, μ is the true mean and $\bar{x}$ is the mean calculated from n data points. The coefficient t is commonly referred to as "Student's t" because the author of the original work on this concept, William S. Gosset, wrote under the pseudonym "Student." There are different values for t depending upon the level of probability desired and the degrees of freedom available. Table 10.1 contains a listing of t values for various levels of probability and degrees of freedom. We see that as the number of data points used to calculate $\bar{x}$ increases, the confidence interval decreases not only because the denominator in Eq. (10.10) increases, but because the numerator value t decreases. The more data used for determining a mean, the greater the certainty that the mean obtained is close to the true mean. One can select any particular level of probability one desires. As a general rule, the 95% probability is reported. This means that only 1 out of 20 times is the true mean likely to reside outside of the calculated confidence limits.

Determine the confidence limits at the 95% probability level for the mean of the BOD data given in Example 10.1.

EXAMPLE 10.6

The number of data points provided in Example 10.1 was 11, so the degrees of freedom are 10. The mean was calculated to be 16 mg/L in Example 10.1, and the standard deviation was 8 mg/L as indicated in Example 10.2. From Table 10.1, for 95% probability and 10 degrees of freedom, t equals 2.228. Thus,

$$\text{CI} = \frac{ts}{\sqrt{n}} = \frac{2.228(8)}{\sqrt{11-1}} = 5.6 \quad \text{or} \quad 6 \text{ mg/L when rounding to significant figures}$$

and, $\quad \text{CL} = 16 \pm 6 \text{ mg/L} \quad or \quad 10 \text{ to } 22 \text{ mg/L}$

This example illustrates that even with 11 data points, we are still not very sure of what the true mean is because the standard deviation for the data is so large. Even with 80 data points, the confidence interval of the average obtained would be ± 2 mg/L at the 95% probability level, which perhaps is surprisingly large considering the large amount of data involved. This fact should provide a caution to those wishing to jump to a strong conclusion with relatively little data in hand on which to support it.

Standard Deviation of Computed Results

At times, one wishes to add, subtract, multiply, or divide analytical results, each of which has a known standard deviation. The manner in which the standard deviation of the result is determined depends upon the mathematical operation used.

Standard Deviation of Sums and Differences   For mathematical operations involving sums and differences, the variance of the result is equal to the sum of the variances of the individual measurements. Thus, if y were a function of a, b, and c as in Eq. (10.11),

$$y(\pm s_y) = a(\pm s_a) + b(\pm s_b) - c(\pm s_c) \tag{10.11}$$

the standard deviation for the value y becomes

$$s_y = \sqrt{s_a^2 + s_b^2 + s_c^2} \tag{10.12}$$

EXAMPLE 10.7

An engineer wishes to estimate the mass of copper remaining in a sealed laboratory reactor. Based upon analytical measurements, he added $22.3(\pm 1.5)$ g initially and $13.5(\pm 1.0)$ g later. He subsequently removed $14.7(\pm 1.2)$ g at one time and $5.3(\pm 0.8)$ g later from the reactor.

Calculated amount remaining $= 22.3 + 13.5 - 14.7 - 5.3 = 15.8$ g

However, because of uncertainties in the measurements made, the actual amount remaining may be different. A calculation of standard deviation using Eq. (10.12) provides a measure of this uncertainty:

$$s_{15.8} = \sqrt{1.5^2 + 1.0^2 + 1.2^2 + 0.8^2} = 2.3 \text{ g}$$

Thus, the engineer could correctly report that $15.8(\pm 2.3)$ g remained in the reactor.

The Standard Deviation of Products and Quotients   With products and quotients, a somewhat different procedure is required to determine the standard deviation of a result. Here, the relative standard deviation must be used. Consider the following equation:

$$y(\pm s_y) = \frac{a(\pm s_a) \times b(\pm s_b)}{c(\pm s_c)} \tag{10.13}$$

The relative standard deviation for y or s_y/y is found as follows:

$$\frac{s_y}{y} = \sqrt{\left(\frac{s_a}{a}\right)^2 + \left(\frac{s_b}{b}\right)^2 + \left(\frac{s_c}{c}\right)^2} \tag{10.14}$$

EXAMPLE 10.8

An engineer wishes to know the efficiency of chemical oxygen demand (COD) mass removal in a reactor under steady-state conditions of operation. For this calculation she has measured the influent and effluent flow rates over time to be $1.5(\pm 0.05)$ m^3/s and $1.4(\pm 0.06)$m^3/s, the potential decrease being the result of water evaporation from the reactor. The influent and effluent COD values were found to be $195(\pm 22)$ mg/L and $9(\pm 3)$ mg/L, respectively.

The flow rate of COD mass into the reactor is given by the product of the influent flow rate and COD concentration, and similarly the flow rate of COD mass out of the reactor is given as the product of effluent flow rate and COD concentration. The percentage removal of COD is given by

$$\text{COD removal} = 100\left(1 - \frac{Q_{\text{out}}C_{\text{out}}}{Q_{\text{in}}C_{\text{in}}}\right)$$

We calculate the ratio from the information provided:

$$\frac{Q_{\text{out}}C_{\text{out}}}{Q_{\text{in}}C_{\text{in}}} = \frac{1.4 \times 9}{1.5 \times 195} = 0.043$$

The standard deviation of the value 0.043 is then found using Eq. (10.14),

$$\frac{s_{0.043}}{0.043} = \sqrt{\left(\frac{0.05}{1.5}\right)^2 + \left(\frac{3}{9}\right)^2 + \left(\frac{0.06}{1.4}\right)^2 + \left(\frac{22}{195}\right)^2} = 0.36$$

from which, $s_{0.043} = 0.043 \times 0.36 = 0.015$.

Thus, based on this mass balance, the overall removal efficiency is given correctly as

$$\text{COD removal} = 100[1 - 0.043(\pm 0.015)] = 95.7(\pm 1.5)\%$$

We see that this final calculation involves subtracting from the number 1 and multiplying by 100. However, these numbers are exact with no standard deviation. If we methodically went through the same procedures using 0 as the standard deviation for these numbers, we would find the same result or 95.7($\pm$1.5) percent.

The Standard Deviation of Exponentials and Logarithms Suppose we have a number carried to some exponential power, where the exponential is an exact number,

$$y(\pm s_y) = [a(\pm s_a)]^x \tag{10.15}$$

For this case we again use the relative standard deviation:

$$\frac{s_y}{y} = x\left(\frac{s_a}{a}\right) \tag{10.16}$$

For logarithms, we have,

$$y = \ln a \tag{10.17}$$

for which the derivative is

$$dy = \frac{da}{a} \tag{10.18}$$

We see then that a small change in the absolute value of y corresponds to a small change in the relative value of a; thus, the standard deviation in y corresponds to the relative standard deviation of a,

$$s_y = \frac{s_a}{a} \tag{10.19}$$

The rate of BOD exertion in the BOD test is generally considered to follow a first-order reaction as follows:

$$\ln \frac{L_t}{L} = -k't$$

where L_t/L represents the fraction of the ultimate BOD (L) remaining after time t. If the rate constant k were found to equal 0.28($\pm$0.05)day^{-1}, what would be the uncertainty in

EXAMPLE 10.9

the calculated fraction of BOD remaining in the standard 5-day BOD test expressed as a standard deviation?

First, we can measure time rather exactly, so we can consider the standard deviation on time to be 0. Then, in order to solve for the fraction of BOD remaining, L_t/L, we express Eq. (10.20) in exponential form:

$$\frac{L_t}{L} = e^{-k't} = e^{-0.28(5)} = 0.247$$

Relating Eqs. (10.17) and (10.20),

$$y = -kt \qquad \text{and} \qquad a = L_t/L$$

and thus from Eq. (10.19),

$$s_{-kt} = \frac{s_{L/L}}{L_t/L}$$

from which,

$$s_{L/L} = s_{-kt}\frac{L_t}{L}$$

To compute the standard deviation of the product $-kt$ we use Eq. (10.14),

$$\frac{s_{-kt}}{-kt} = \sqrt{\left(\frac{s_k}{-k}\right)^2 + \left(\frac{s_t}{t}\right)^2} = \frac{s_k}{-k}$$

Then, $s_{-kt} = ts_k = 5(0.05) = 0.25$

From this, we obtain $s_{L/L} = 0.25(0.247) = 0.062$

$$\frac{L_t}{L} = 0.247(\pm 0.062)$$

10.6 | HYPOTHESIS TESTING

The variable t was used previously to determine the confidence limits on means and standard deviations. It is also useful in determining whether or not two sets of data appear to be from the same pool of data for from different ones. For example, a comparison may be made between two different analytical methods to determine the concentration of a given constituent. If a set of measurements of a given constituent is taken by each of the methods, and the mean of each set is computed, it may be that the means are somewhat different from each other, and they may also be somewhat different from that of the known standard concentration. The question then is whether these differences are simply the result of the indeterminate errors that are a part of all measurements or due to a true difference between the procedures used. Another similar question arises if one wishes to determine whether some treatment has a significant impact on a process being evaluated. Measurements of results are obtained with and without the treatment, and then the results are compared. If a difference is noted, then the question again arises as to whether or not the difference is simply a result of the normal indeterminate errors associated with the analysis, or whether indeed the treatment had some significant impact on the result.

In order to obtain a statistical answer to these questions, we can resort to hypothesis testing. Here we use what is called the *null hypothesis.* This assumes that the numerical quantities that we are comparing are indeed the same. The probability that the differences being observed are the result of indeterminate errors is then computed from statistical theory. Generally, if the difference observed is one that would occur only 1 time out of 20 (that is only 5 percent of the time), then the null hypothesis is considered to be questionable and the difference is judged to be significant. We may apply a stricter standard if we wish, say a result that would occur only 1 out of 100 or 1 percent of the time. If this is the case, then we may be more sure that the difference observed is the result of a true difference between the two results. A less strict standard, say 10 percent, could also be chosen. However, the most common value selected for hypothesis testing is 5 percent. Let us examine some applications of hypothesis testing.

Experimental Mean versus True Value

A common way to evaluate an analytical procedure is to analyze the concentration of a given constituent in a solution where the concentration is truly known. We may make several measurements, compute the mean of the values obtained, and compare this mean with the known value. If the procedure has some variability in it, as most procedures do, then we would expect that the mean would not exactly equal the true value, that is, $\bar{x} - \mu$ is not 0. But is the difference significant? In order to answer this question, we determine what value of difference would occur due to indeterminate errors only 5 percent of the time. For this, we rewrite Eq. (10.10) in the form,

$$\bar{x} - \mu = \pm \frac{ts}{\sqrt{n}} \qquad (10.21)$$

The value s is the standard deviation of the n analyses that have been made with a mean value of $\bar{x}$. We select a value of t from Table 10.1 that has the desired level of probability, say 95% in this case, for $n - 1$ degrees of freedom. If the difference is greater than the absolute value of the right side of Eq. (10.21), then we say the null hypothesis is rejected, and the difference is significant at the 95% level of confidence. In other words, applying these criteria for our conclusion, we are likely to be wrong only 5 percent of the time.

An analytical procedure has been developed for measurement of chloride and has been calibrated against a standard solution containing 100 mg/L chloride. It is desired to determine how well the procedure calibrated at this level will work when measuring a standard solution containing only 25 mg/L chloride. Five analyses are made, giving measurements of 20, 22, 28, 24, and 25 mg/L. The average is 23.8 mg/L. Given the obvious variability in individual measurements, is this result satisfactory to confirm that the procedure is accurate at this lower level?

EXAMPLE 10.10

From Table 10.1, t at the 95% level for $n - 1 = 4$ is 2.78. The standard deviation is computed,

$$s = \sqrt{\frac{\begin{array}{c}(20 - 23.8)^2 + (22 - 23.8)^2 + (28 - 23.8)^2 \\ + (24 - 23.8)^2 + (25 - 23.8)^2\end{array}}{5 - 1}} = 3.0 \text{ mg/L}$$

Then, the value of the right side of Eq. (10.21) is determined,

$$\pm \frac{ts}{\sqrt{n}} = \pm \frac{2.78 \times 3.0}{\sqrt{5}} = \pm 3.7 \text{ mg/L}$$

The difference between $\bar{x}$ and μ is only -1.2 mg/L and thus is within the given range of 3.7 mg/L. This means we cannot reject the null hypothesis, and thus we accept the analytical procedure as being sufficiently accurate near the 25-mg/L concentration range even though it was standardized only at the 100-mg/L level.

Comparison Between Two Experimental Means

At times we wish to compare the results between two different analytical tests, or perhaps between two different treatment processes. Often we find the results are different. The question then arises as to whether there is a true difference between the two results, or whether the difference might be attributed to statistical variation in results. This problem has close similarities with the previous analysis of comparing an experimental result with the true value. The appropriate form of the equation here is

$$\bar{x}_1 - \bar{x}_2 = \pm \frac{ts}{\sqrt{\dfrac{n_1 n_2}{n_1 + n_2}}} \tag{10.22}$$

This procedure presupposes that the standard deviations for the two different samples are essentially the same. The appropriate t value is obtained from the t table, using $(n_1 + n_2 - 2)$ degrees of freedom. If the results of computation using the right side of the equation are greater than the left side of the equation, then we judge that the differences between the two values are not significant. If larger, then the difference is judged significant. Again here we might use the 95% probability level in our judgment.

EXAMPLE 10.11

We wish to compare two different analytical procedures using two different detectors for measuring the concentration of trichloroethene (TCE) in water. The question is whether the results are essentially the same or different. We are given the following analytical results from 10 analyses of a single sample by each procedure.

Detector	Measured TCE concentrations, mg/L										$\bar{x}$	s
FID	772	858	866	809	794	752	738	774	766	768	790	43
PID	812	778	764	710	733	691	689	724	692	711	730	42

The mean values of 790 and 730 mg/L are different by 60 mg/L, but is the difference significant? The standard deviation of 43 and 42 for the different detectors, respectively, are nearly the same. For this case, the degrees of freedom are 18 (10 + 10 − 2), and from Table 10.1, t at the 95% confidence level equals 2.101. We calculate the right side of Eq. (10.22) and compare with the difference between the means

$$\frac{ts}{\sqrt{\dfrac{n_1 n_2}{n_1 + n_2}}} = \frac{2.101(42.5)}{\sqrt{\dfrac{10(10)}{10 + 10}}} = 40 \text{ mg/L}$$

Since 60 mg/L is higher than 40 mg/L, we reject the null hypothesis and say that the differences are statistically significant at the 95% confidence level. We conclude with 95% confidence that the two detectors are giving different analytical results.

Comparison Between Two Standard Deviations

In Example 10.11, the standard deviations are about the same. But how can we compare them to determine whether or not the small difference is significant? Here, we can safely rely upon our judgment since the difference is so small. But what if the difference were greater? We might be interested in such a comparison, for example, if we were comparing two different analytical procedures and wanted to make a comparison between their precision. Here, we can rely upon another statistical procedure called the F test that is similar in concept to the comparison of two means. For this we use the following equation:

$$F = \frac{s_1^2}{s_2^2} \qquad s_1 > s_2 \tag{10.23}$$

The larger of the standard deviations is taken as s_1 so that the ratio in Eq. (10.23) is greater than 1. We then compare the calculated F value with the relevant F value from Table 10.2, which is for the 95% confidence level. Statistical handbooks have values for other confidence levels, if desired, but the 95% value is most commonly used.

Table 10.2 | F values at the 95% probability level

$n - 1$ for s_2	$n - 1$ for s_1											
	2	3	4	5	6	8	10	16	20	40	100	∞
2	19.00	19.16	19.25	19.30	19.33	19.37	19.39	19.43	19.44	19.47	19.49	19.50
3	9.55	9.28	9.12	9.01	8.94	8.84	8.78	8.69	8.66	8.60	8.56	8.53
4	6.94	6.59	6.39	6.26	6.16	6.04	5.96	5.84	5.80	5.71	5.66	5.63
5	5.79	5.41	5.19	5.05	4.95	4.82	4.74	4.60	4.56	4.46	4.40	4.36
6	5.14	4.76	4.53	4.39	4.28	4.15	4.06	3.92	3.87	3.77	3.71	3.67
8	4.46	4.07	3.84	3.69	3.58	3.44	3.34	3.20	3.15	3.05	2.98	2.93
10	4.10	3.71	3.48	3.33	3.22	3.07	2.97	2.82	2.77	2.67	2.59	2.54
16	3.63	3.24	3.01	2.85	2.74	2.59	2.49	2.33	2.28	2.16	2.07	2.01
20	3.49	3.10	2.87	2.71	2.60	2.45	2.35	2.18	2.12	1.99	1.90	1.84
40	3.23	2.84	2.61	2.45	2.34	2.18	2.07	1.90	1.84	1.69	1.59	1.51
100	3.09	2.70	2.46	2.30	2.19	2.03	1.92	1.75	1.68	1.51	1.39	1.28
∞	2.99	2.60	2.37	2.21	2.09	1.94	1.83	1.64	1.57	1.40	1.24	1.00

| EXAMPLE 10.12 | In Example 10.4 the means of row mean values were calculated and found to be 29.2 and 30.6, for rows with 10 and 9 data points, respectively. The respective standard deviations were 1.6 and 3.5. These standard deviations appear to be quite different, but is the difference statistically significant at the 95% confidence level? |

The ratio of interest is $3.5^2/1.6^2$ or 4.785. There were 4 values used to determine the highest standard deviation and 10 values for the lowest one, so the degrees of freedom are 3 and 9, respectively. From Table 10.2, F equals about 3.89, which is lower than 4.78. Thus the null hypothesis is rejected and we conclude at the 95% confidence level that the standard deviations are significantly different from one another.

This appears to be a strange conclusion for this case since the data rows were arbitrarily produced from a large pool of random data. On this basis, one is forced to question the conclusion. We must further understand that there is 1 chance in 20 that we will reach a wrong conclusion using this statistical approach. Is this one of those times? If the conclusion is important and we have reason to question the results, we may wish to study further.

It is important for the analyst to understand the limitations involved in analytical measurements and data analyses. If we had used an F table giving values for 99% confidence (not included here), we would have found an F value of 6.99. At this level of confidence, we would have reached a different conclusion and would find at the 99% probability level that the standard deviations were not significantly different.

10.7 | DETECTION LIMITS

The continuous discovery of new contaminants that can have adverse impacts on human health and the environment, and the very low concentrations at which standards for contaminants are being set by regulators in order to reduce risks from exposure, have necessitated an increased ability to precisely and accurately measure low concentrations of a variety of materials. Maximum contaminant levels for many constituents in drinking waters are in the low microgram per liter range (see Chap. 34), and at times are set in the nanogram per liter range or lower. Such low concentrations can severely test the ability of analytical instruments and the analyst. At times, exposure standards are set at higher levels than regulators wish simply because available analytical procedures are incapable of reliable measurement of the low concentrations. In several cases, a standard is set close to the analytical limit for its analysis. It is obvious that the method of analysis selected for measuring a substance must have a detection limit that is consistent with the lower limit of analysis desired. How does the analyst determine the detection limit for a substance by a given analytical procedure? There are standard protocols for this. The analyst needs to be familiar with them.

Analytical instruments, such as pH meters and other methods using electronic probes, gas chromatographs, and ion chromatographs, are being used increasingly for the determination of contaminant concentrations in environmental samples. Such instruments produce an electronic response to sample introduction that is stored in a database or used to draw a graph of some type from which the substance concentration is deduced. In the calibration of such instruments, blanks or samples containing none of the substance being measured are subject to the procedure, and the electronic response produced is recorded. The responses (i.e., peak height or area under a peak on a graph) resulting from analysis of a series of blanks are commonly different from each other, which is an indication of instrumental errors and limitations. The responses of a series of blanks can be analyzed statistically to determine their standard deviation s_b. This gives a measure of the uncertainty in the analytical measurement and is said to represent the *instrument noise* involved in the procedure. In order to calibrate the instrument, standard samples with different concentrations are subject to the procedure and the response from each concentration is recorded. If a certain low concentration of standard can be found that provides a response lying close to s_b, then it should be obvious that the uncertainty in the analysis itself is too great to allow precise measurements at that concentration.

In most analyses, a sample is extracted and processed by some standard procedure before it is subjected to instrumental analysis. When blank samples are extracted and processed in this manner, the standard deviation s_s will be larger than the standard deviation s_b from blank injection into the instrument alone. With such a value of s_s, we can estimate the method detection limit (MDL) using Eq. (10.22), which tells us whether two measured means are significantly different or not. Thus,

$$\text{MDL} = \bar{x}_1 - \bar{x}_2 = ts_s \sqrt{\frac{n_1 + n_2}{n_1 n_2}} \qquad (10.24)$$

An overall extraction and analysis method for lindane gave the following results when applied to lindane-free soil samples: $\mu g/L$ lindane = 0.3, 1.5, −0.5, 0.7, −1.0, 1.2, 0.1. Calculate the lindane detection limit for the analysis (at the 99% confidence level) for (*a*) a single analysis, and (*b*) the mean of four analyses.

EXAMPLE 10.13

We calculate s_s as follows:

$$\sum x_i = 0.3 + 1.5 - 0.5 + 0.7 - 1.0 + 1.2 + 0.1 = 2.3$$

$$\sum x_i^2 = 0.09 + 2.25 + 0.25 + 0.49 + 1.0 + 1.44 + 0.01 = 5.53$$

$$s_s = \sqrt{\frac{5.53 - (2.3)^2/7}{7 - 1}} = 0.89 \ \mu g/L$$

(*a*) For a single analysis, $n_1 = 1$ and the number of degrees of freedom is $1 + 7 - 2 = 6$. From Table 10.1, we see $t = 3.707$. Thus,

$$\text{MDL} = 3.707(0.89)\sqrt{\frac{1 + 7}{1(7)}} = 3.96(0.89) = 3.5 \ \mu g/L$$

(*b*) For four analyses, $n_1 = 4$, the number of degrees of freedom is 9 and $t = 3.25$. For this case,

$$\text{MDL} = 3.25(0.89)\sqrt{\frac{4 + 7}{4(7)}} = 2.04(0.89) = 1.8 \ \mu g/L$$

"Standard Methods" defines other detection limits as listed in Table 10.3. The different limits serve different purposes. They are related to one another approximately as follows: IDL:LLD:MDL:LOQ:POQ = 1:2:4:10:20. Thus when someone indicates the detection limit for a given analysis, it is necessary to understand what value they are using in order to judge how much confidence one might place in the values they are reporting.

In Example 10.13, the coefficient by which s_s was multiplied to obtain the MDL at the 99% confidence level varied between 2.04 and 3.96, depending upon the number of samples used to determine s_s and the number of samples used to determine the average concentration. We see that Table 10.3 uses a coefficient of 3.14 for determining MDL. This is based upon some slightly different assumptions. The 99% confidence level t value used to arrive at the Table 10.3 coefficient is called a one-sided value, which is equivalent to the 98% probability level of 3.143 in Table 10.1 for 7 samples and 6 degrees of freedom.

The statistical bases for the other particular coefficients by which s_b or s_s is multiplied in Table 10.3 are described in detail in "Standard Methods" and will not be repeated here. In general, the IDL is the initial starting point and is used to determine the contaminant concentration to be used for making a standard solution for use in more precise determination of the LLD. The concentration in this standard solution should be near, but no more than five times, s_b. This solution is analyzed numerous times to obtain a good measure of the standard deviation, which is then taken to equal the better s_b for use in determining LLD. The MDL differs from the LLD in that the errors in the complete analysis are included, such as those resulting from extractions and other sample preparation procedures. The LOQ has a higher coefficient still for laboratories where many different individuals use the same procedure and instruments, while the highest coefficient, which is for POQ, provides a limit for the case where many different laboratories with many different individuals and instruments are involved. This greatly increases the uncertainty in the analysis, and so a higher coefficient is justified. It is for this reason that when the U.S. EPA specifies a detection limit to be used for a given analysis, it is generally much higher than might seem reasonable to a single individual using great care on an analysis

Table 10.3 | Comparison between different analytical detection limits

Detection limit	Definition
Instrument detection limit (IDL)	$1.645(s_b)$
Lower limit of detection (LLD)	$2(1.645)(s_b)$
Method detection limit (MDL)	$3.14(s_s)$
Limit of quantification (LOQ)	$10(s_b)$
Practical quantification limit (POQ)	$5(3.14)(s_s)$

Source: L. S. Clesceri, A. E. Greenberg, and A. D. Eaton (eds.), "Standard Methods for the Examination of Water and Wastewater, 20th ed., American Public Health Association, Washington D.C., 1998.

conducted frequently with a single instrument in a single laboratory. Thus, a good researcher may feel for his or her work that lower detection limits than specified nationally may be justified.

10.8 | LOGNORMAL DISTRIBUTION

A normal distribution is just one of several different kinds of distributions that data may take. It is the distribution normally assumed to occur in analytical measurements. A property of normal distribution is that the distribution of data is symmetric around the median, and the median and mode are essentially the same. When they are not, the data are called *skewed*. If the median is larger than the mode, the data are said to be skewed to the right, while if it is smaller, the data are skewed to the left. When data are skewed, the distribution may not be best described by a normal distribution. Other distribution types may be involved. The discussion of the many other distributions that data at times follow is beyond the scope of this book, except for one distribution that is quite common among environmental analytical data sets. That is the *lognormal* distribution. A brief discussion of the lognormal distribution is given.

Nature of Lognormal Distribution

A lognormal distribution often occurs for hazardous contaminants that result from periodic discharges from discrete sources. Thus, at a wastewater treatment plant, the concentration of a hazardous chemical may commonly be in the low microgram or nanogram per liter range, but occasionally, much higher concentrations result from discharges, say from an industrial plant, causing concentrations at the treatment plant to increase by an order of magnitude or more. Such large concentration ranges often follow a lognormal distribution.[3] Daily stream flow, flood peak discharges, annual floods, and rainfall data also tend to follow a lognormal distribution.

Illustration data obtained for the concentration of dichloromethane (DCM) in the effluent of Water Factory 21, an advanced wastewater treatment plant used for reclamation of biologically treated municipal wastewater, were analyzed to determine whether they best fit a normal or lognormal distribution. Table 10.4 contains a summary of these data arranged in order of increasing concentration. Figure 10.3 represents a histogram of the data. The data are greatly skewed to the left with maximum frequency in the range below 10 μg/L. Natural logarithms of the data were then taken, and a histogram of the results is shown in Fig. 10.4. The logarithms tend to follow a normal distribution pattern much better.

Arranging data to draw bar diagrams takes time. Perhaps a better general method to determine what distribution is followed is to use probability paper. Normal probability and lognormal probability paper can be obtained from stores

---

[3]R. B. Dean: Estimating the Reliability of Advanced Waste Treatment, *Water and Sewage Works*, (p. 87, June 1976, and p. 57, July 1976).

Table 10.4 | Analysis of effluent dichloromethane data from Water Factory 21

Datum number	DCM, μg/L	ln DCM	$100[F^*(X_i)]$ % less than	Datum number	DCM, μg/L	ln DCM	$100[F^*(X_i)]$ % less than
1	1.8	0.59	1.1	32	14	2.64	57.2
2	2.8	1.03	2.9	33	14	2.64	59.0
3	3.6	1.28	4.8	34	14	2.64	60.9
4	3.7	1.31	6.6	35	15	2.71	62.7
5	3.8	1.34	8.4	36	16	2.77	64.5
6	4.0	1.39	10.2	37	17	2.83	66.3
7	4.0	1.39	12.0	38	18	2.89	68.1
8	4.3	1.46	13.8	39	18	2.89	69.9
9	4.3	1.46	15.6	40	19	2.94	71.7
10	4.8	1.57	17.4	41	21	3.04	73.5
11	5.0	1.61	19.2	42	23	3.14	75.3
12	5.0	1.61	21.0	43	23	3.14	77.1
13	5.2	1.65	22.9	44	24	3.18	79.0
14	5.2	1.65	24.7	45	25	3.22	80.8
15	5.6	1.72	26.5	46	25	3.22	82.6
16	5.7	1.74	28.3	47	28	3.33	84.4
17	5.8	1.76	30.1	48	32	3.47	86.2
18	5.8	1.76	31.9	49	33	3.50	88.0
19	6.0	1.79	33.7	50	38	3.64	89.8
20	6.2	1.82	35.5	51	50	3.91	91.6
21	8.8	2.17	37.3	52	60	4.09	93.4
22	9.0	2.20	39.1	53	60	4.09	95.2
23	9.2	2.22	41.0	54	65	4.17	97.1
24	9.3	2.23	42.8	55	72	4.28	98.9
25	9.9	2.29	44.6				
26	10.0	2.30	46.4	Mean	17.0	2.44	
27	10.5	2.35	48.2	Std. dev.	16.7	0.89	
28	11	2.40	50.0	$\bar{x}_g$		11.5	
29	12	2.48	51.8	s_g		2.4	
30	13	2.56	53.6	CL	Lower	9.0	
31	14	2.64	55.4		Upper	14.6	

providing graph paper. The procedure followed here is to arrange the data in ascending order as in Table 10.4. The fraction of data less than values are obtained using a formula such as

$$F^*(x_i) = \frac{i - 3/8}{n + 1/4} \qquad (10.25)$$

where i is the number of ith data point in the series of n data points. Thus, $i = 1$ for the lowest value. If there are 40 data points, then the fraction less than the lowest data point is $(1 - 3/8)/(40 + 1/4)$, or 0.0155. Thus, only 1.55 percent of the data would be less than this number. For the highest data point, the fraction would be $(40 - 3/8)/(40 + 1/4)$ or 0.984, or 98.4 percent of the data is less than this value. Equation (10.25) is used instead of a simpler formula, such as i/n in order to avoid end effects.

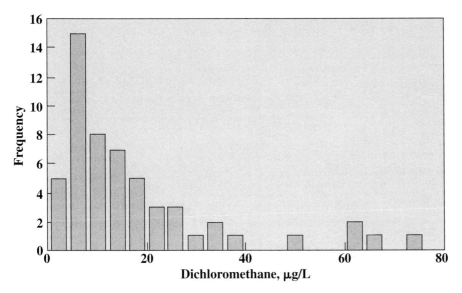

Figure 10.3
Frequency distribution of influent dichloromethane concentration at Water Factory 21, Fountain Valley, California.

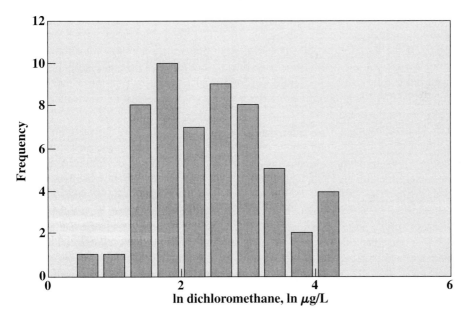

Figure 10.4
Frequency distribution of the natural log of influent dichloromethane concentrations at Water Factory 21.

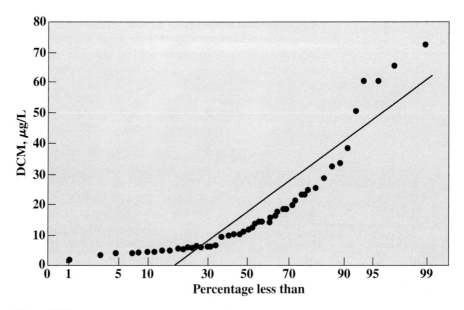

Figure 10.5
Normal probability distribution of influent dichloromethane concentrations at Water Factory 21.

The dichloromethane data from Water Factory 21 are shown as they would look on normal probability paper (Fig. 10.5) and lognormal probability paper (Fig. 10.6). If the data follow the selected distribution well, they will fall on a straight line. We see that the dichloromethane data do not follow a normal distribution very well but can be represented quite well by the lognormal distribution. Other procedures that provide a more quantitative measure of goodness of fit are available, but their discussion is beyond the scope of this chapter.

Data following a lognormal distribution can be analyzed by tools similar to that of the normal distribution as indicated by Dean.[4] In order to do this, the natural logarithm of each datum is taken, and the average and standard deviation of the logs are determined by the statistical techniques for a normal distribution as illustrated in Table 10.4. The antilog of the mean so taken, $\bar{x}_g$, represents the *geometric mean,* and the antilog of the standard deviation, s_g, represents the *geometric standard deviation* or *spread factor.* For a lognormal distribution, 68.3 percent of the data (equivalent to one standard deviation) will lie between concentrations represented by $\bar{x}_g/s_g$ and $\bar{x}_g s_g$, and 95.5 percent of the data (equivalent to two standard deviations) will lie between $\bar{x}_g/s_g^2$ and $\bar{x}_g s_g^2$. If we wish to determine the uncertainty in calculated $\bar{x}_g$ values, we can determine the confidence limits as follows:

$$\mathrm{CL} = \bar{x}_g s_g^{\pm t/\sqrt{n}} \tag{10.26}$$

[4]Dean, Estimating the Reliability of Advanced Waste Treatment.

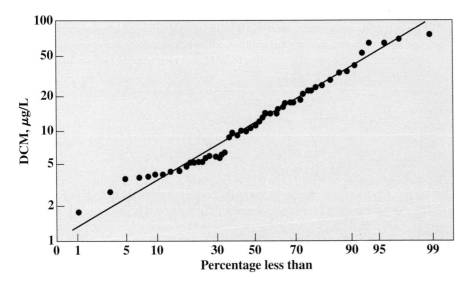

Figure 10.6
Lognormal probability distribution of influent dichloromethane concentrations at Water Factory 21.

where t is an appropriate value from Table 10.1 for $n - 1$ degrees of freedom. Results of such calculations for the dichloromethane data are given at the end of Table 10.4. The geometric mean for the data is 11.5, which can be seen as the 50% point in Fig. 10.6. With a geometric standard deviation of 2.4, and a 95% t of 2.004 from Table 10.1, the 95% confidence limits for the mean are found to be 9.0 and 14.6 $\mu g/L$.

10.9 | REGRESSION ANALYSIS

Linear Regression

A common analytical procedure for calibration of an instrument is to determine the instrument response to each of a series of different standard solution concentrations. The instrument response for each concentration is then plotted on graph paper to develop a standard curve for the analysis. Figure 10.7 represents such a curve for gas chromatographic analysis of benzene concentration in water. A response curve that follows a straight line is generally desired although this is not absolutely necessary and may not always result. For data following a reasonable straight line, such as that in Fig. 10.7, the standard curve can be analyzed statistically in a straightforward manner to provide a measure of uncertainty for the analysis. The procedure followed is termed *regression analysis.*

A straight line is characterized by the following equation,

$$y_i = a + bx_i \qquad (10.27)$$

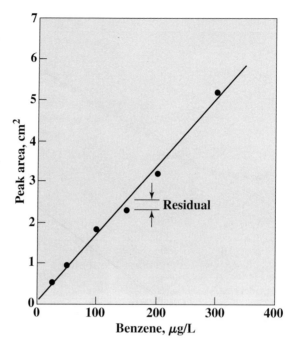

Figure 10.7
Calibration curve for gas chromatographic analysis of
benzene concentration in water.

where a equals the y intercept at x equals 0, and b represents the slope of the line.
The objective of regression analysis here is to determine the values of a and b and
their uncertainties. This assumes that the data follow the straight-line model chosen.
The common approach used is called *least-squares analysis.* For the straight-line
case, we assume first that the values x_i are known exactly. The indeterminate errors
are assumed to be related to the y_i measurements. When x_i represents careful dilutions of known standards, or in some cases, time, this assumption is reasonable. The
line generated by the least-squares method is the one that minimizes the squares of
the vertical displacements or *residuals* from that line. Such a residual is indicated in
Fig. 10.7. In essence the procedure finds values for a and b such that the sum of the
squares of residuals for the y_i measurements is the smallest possible value. The values of a and b so derived are as follows:

$$b = \frac{\sum[(x_i - \bar{x})(y_i - \bar{y})]}{\sum(x_i - \bar{x})^2} = \frac{n\sum x_i y_i - \sum x_i \sum y_i}{n\sum x_i^2 - (\sum x_i)^2} \tag{10.28}$$

$$a = \bar{y} - b\bar{x} = \frac{\sum x_i^2 \sum y_i - \sum x_i \sum x_i y_i}{n\sum x_i^2 - (\sum x_i)^2} \tag{10.29}$$

Two different equations are provided for solving each of b and a. They are exactly equivalent. The first is more illustrative of the concept involved, while the second is generally easier to solve when using a calculator or spreadsheet.

We can write four different standard deviations for this case, the standard deviation of the regression s_y, the standard deviation of the intercept s_a, the standard deviation of the slope s_b, and the standard deviation of the results obtained from the calibration curve s_c. Generally of interest is the latter value, s_c. However, for completeness, formulas for calculation of all four are given. Again, two different but equivalent equations are given for each.

$$s_y = \sqrt{\frac{\sum[y_i - (a - bx_i)]^2}{n - 2}} = \sqrt{\frac{[\sum y_i^2 - (\sum y_i)^2/n] - b^2[\sum x_i^2 - (\sum x_i)^2/n]}{n - 2}} \quad (10.30)$$

$$s_a = s_y\sqrt{\frac{\sum x_i^2}{n\sum(x_i - \bar{x})^2}} = s_y\sqrt{\frac{\sum x_i^2}{n\sum x_i^2 - (\sum x_i)^2}} \quad (10.31)$$

$$s_b = \frac{s_y}{\sqrt{\sum(x_i - \bar{x})^2}} = s_y\sqrt{\frac{n}{n\sum x_i^2 - (\sum x_i)^2}} \quad (10.32)$$

$$s_c = \frac{s_y}{b}\sqrt{\frac{1}{m} + \frac{1}{n} + \frac{(\bar{y}_c - \bar{y})^2}{b^2[\sum x_i^2 - (\sum x_i)^2/n]}} \quad (10.33)$$

In Eq. (10.33), m represents the number of analyses conducted on a given sample, and $\bar{y}_c$ represents the average of the m values. Equation (10.33) can be used to determine confidence limits for the value using $x_i \pm ts_c$, with the degrees of freedom for t equal to $n - 2$.

| | | | | | EXAMPLE 10.14 |

Determine the slope and intercept for the regression line in Fig. 10.7, and determine the standard deviation for each.

The data upon which Fig. 10.7 is based and relevant calculations are contained in the following table.

	Benzene x_i, μg/L	Peak area y_i, cm^2	x_i^2	y_i^2	$x_i y_i$
	25	0.54	625	0.2916	13.5
	50	0.96	2,500	0.9216	48
	100	1.82	10,000	3.3124	182
	150	2.30	22,500	5.2900	345
	200	3.20	40,000	10.2400	640
	300	5.17	90,000	26.7289	1,551
$\Sigma =$	825	13.99	165,625	46.7845	2,780
Average $=$	137.5	2.331			

Relevant calculations then are

$$a = \frac{165{,}625(13.99) - 825(2780)}{6(165{,}625) - (825)^2} = 0.077$$

$$b = \frac{6(2780) - 825(13.99)}{6(165{,}625) - (825)^2} = 0.0164$$

$$s_y = \sqrt{\frac{[46.7845 - (13.99)^2/6] - (0.0164)^2[165{,}625 - (825)^2/6]}{6 - 2}} = 0.179$$

$$s_a = \sqrt{\frac{0.179^2(165{,}625)}{6(165{,}625) - (825)^2}} = 0.130$$

$$s_b = \sqrt{\frac{6(0.179)^2}{6(165{,}625) - (825)^2}} = 0.00078$$

EXAMPLE 10.15

A sample is analyzed by gas chromatography for benzene, giving a peak area of 1.97 cm^2. Determine the benzene concentration and the 95% CL assuming (*a*) one sample only is analyzed, and (*b*) four replicates of the sample are analyzed.

For both cases, the benzene concentration is given by Eq. (10.27). Using the results of Example 10.14,

$$x_i = \frac{y_i - a}{b} = \frac{1.97 - 0.075}{0.0164} = 115 \ \mu g/L$$

(*a*) For the first case, $m = 1$. There is only one sample, and thus $\bar{y}_c = 1.97$ cm^2. The standard deviation is found with Eq. (10.33):

$$s_c = \frac{0.179}{0.0164} \sqrt{\frac{1}{1} + \frac{1}{6} + \frac{6(1.97 - 2.331)^2}{(0.0164)^2[6(165{,}625) - (825)^2]}} = 11.8$$

For the 95% confidence interval we have $6 - 2$ or 4 degrees of freedom, as we lose one degree of freedom each for the determination of a and b. From Table 10.1, t equals 2.776, and

$$95\% \ CL = \bar{x} \pm t s_c = 115 \pm 2.776(11.8) = 115 \pm 33 \ \mu g/L$$

(*b*) For the second case, $m = 4$, for which,

$$s_c = \frac{0.179}{0.0164} \sqrt{\frac{1}{4} + \frac{1}{6} + \frac{6(1.97 - 2.331)^2}{(0.0164)^2[6(165{,}625) - (825)^2]}} = 7.12$$

and, $95\% \ CL = \bar{x} \pm t s_c = 115 \pm 2.776(7.12) = 115 \pm 20 \ \mu g/L$

First-Order Reactions

First-order reactions are common in biological and chemical systems. A good example is the BOD test, which as described in Chap. 23 is often assumed to follow the first-order relationship,

$$\frac{dy}{dt} = k'L_t \quad \text{or} \quad \frac{dy}{dt} = k'(L_0 - y) = k'L_0 - k'y \quad (10.34)$$

The integrated form of Eq. (10.34) is

$$y = L_0(1 - e^{-k't}) \quad (10.35)$$

Here, y represents oxygen uptake values that are measured over time t, and L_0 represents the ultimate BOD of the sample. The BOD remaining at any time is $L_0 - y$. Equation (10.35) is shown plotted in Fig. 10.8a. It is represented by a curve that cannot be analyzed directly by linear regression. However Eq. (10.34) is shown in Fig. 10.8b to plot as a straight line with intercept at $k'L_0$ and slope $-k'$. It is thus subject to linear regression analysis. The trick here is to determine the reaction rate dy/dt from measurements only of y and t. This is generally done by taking the differences between the y and t values just before and after the one being considered and dividing the two to obtain dy/dt.

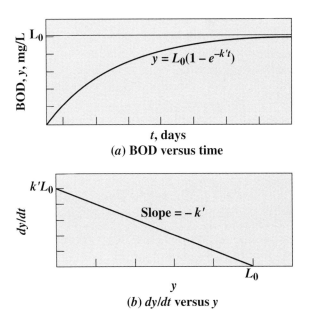

Figure 10.8
BOD first-order reaction rate curves.

EXAMPLE 10.16

Determine the rate constant k' and ultimate BOD from the following set of BOD measurements. Determine the standard deviation of each value as well.

Time t, days	1	2	3	5	7	9	12
BOD y, mg/L	27	36	53	65	80	89	98

In order to use linear regression for the BOD analysis, we use Eq. (10.34). For this, we need to calculate values for dy/dt from the BOD versus time data. The procedure for doing this is indicated in the following table.

t	y (x)	dt	dy	dy/dt (y)	$(dy/dt)^2$ (y^2)	y^2 (x^2)	$y(dy/dt)$ (xy)
0	0						
1	27	2	36	18	324.0	729	486.0
2	36	2	26	13	169.0	1,296	468.0
3	53	3	29	9.67	93.4	2,809	512.3
5	65	4	27	6.75	45.6	4,225	438.8
7	80	4	24	6	36.0	6,400	480.0
9	89	5	18	3.6	13.0	7,921	320.4
12	98						
Sums	350*			57.0	681	23,380	2,705
Averages	58.3*			9.5			

*The sum and average for the y column does not include values for days 0 and 12 since the dy/dt column does not include values for these days because of the computation method.

For example, for day 1, dt and dy are obtained from the difference between day 2 and day 0. That is dt is the difference between 2 and 0, or 2 days, and dy is the difference between 36 and 0, or 36 mg/L. The ratio 36/2 or 18 mg/L-day is thus taken as the value of dy/dt, which corresponds with the day 1 y-value of 27 mg/L. Using this procedure, we see that we can obtain only six values of dy/dt for the eight values of t and y listed, and thus only the six corresponding values of y can be used in the subsequent analysis.

Next, using the procedure for linear regression analysis described at the beginning of this section, we need to identify coefficients that correspond to the x and y values in Eqs. (10.28) to (10.33). From Fig. 10.8b it is apparent that y is the abscissa (x axis) and dy/dt is the ordinate (y axis) as indicated within parentheses at the top of the table. Thus, by analogy, we obtain the y-axis intercept a to be $k'L_0$ and the slope b to equal $-k'$ (see Fig. 10.8). Using Eqs. (10.28) and (10.29) and appropriate values from the table we obtain:

$$k'L_0 = \frac{23,380(57.0) - (350)(2,705)}{6(23,380) - (350)^2} = 21.7 \text{ mg/L-day}$$

$$k' = -\frac{6(2,705) - 350(57.0)}{6(23,380) - (350)^2} = 0.21 \text{ day}^{-1}$$

From which,

$$L_0 = 21.7/0.21 = 103 \text{ mg/L}$$

The standard deviations are obtained through use of Eqs. (10.30) to (10.32).

$$s_y = \sqrt{\frac{[(681) - (57.0)^2/6] - (-0.21)^2[23,380 - (350)^2/6]}{6 - 2}} = 1.52$$

$$s_{k'L} =$$

$$\sqrt{\frac{(1.52)^2(23,380)}{6(23,380) - (350)^2}} = 1.74 \text{ mg/L} - \text{d}$$

$$s_{k'} = \sqrt{\frac{6(1.52)^2}{6(23,380) - (350)^2}} = 0.0279 \text{ d}^{-1}$$

In order to obtain the standard deviation for L_0, we note that L_0 was determined from the quotient of $k'L_0$ and k'. Thus, Eq. (10.14) is appropriate for use, allowing us to obtain:

$$\frac{s_L}{L_0} = \sqrt{\left(\frac{s_{k'L}}{k'L_0}\right)^2 + \left(\frac{s_{k'}}{k'}\right)^2} = \sqrt{\left(\frac{1.74}{21.7}\right)^2 + \left(\frac{0.0279}{0.21}\right)^2} = 0.155$$

from which, $s_L = 0.155(103) = 16 \text{ mg/L}$

In summary, $k' = 0.21 \pm 0.028 \text{ day}^{-1}$ and $L_0 = 103 \pm 16 \text{ mg/L}$.

As a further summary, Fig. 10.9 illustrates the goodness of fit between the raw data and the curves drawn from the respective values found for k' and L_0.

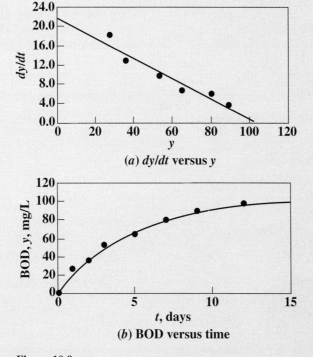

(a) dy/dt versus y

(b) BOD versus time

Figure 10.9
Results of regression analysis for BOD data.

Coefficient of Determination

A statistical parameter often used to judge subjectively how one variable is correlated with another is the *coefficient of determination,* or r^2, which is applied to the results of a linear regression analysis. The coefficient of determination is defined as

$$r^2 = \frac{\sum[(a - bx_i) - \bar{y}]^2}{\sum(y_i - \bar{y})^2} \qquad (10.36)$$

The coefficient of determination thus equals the ratio of the sum of squares of the calculated $y - \bar{y}$ values and the measured $y - \bar{y}$ values. If a perfect correlation exists between x and $y,$ that is, if the straight-line fit is perfect, then r^2 will equal 1. If there is no correlation between the two, that is, no reasonable fit exists between the data, r^2 will equal 0. Goodness of fit then is judged by how closely r^2 approaches 1.

EXAMPLE 10.17

Find the coefficient of determination for the results of the correlation analysis between analytical peak area and benzene concentration in Example 10.14.

Borrowing relevant information from Example 10.14, the calculated values of peak area, y_{ic}, are found by solving the regression line:

$$y_{ic} = a + bx_i = 0.077 + 0.0164x_i$$

The value for $\bar{y}$ is given in Example 10.14 as 2.331. The appropriate calculations then are

x_i	y_i	y_{ic}	$(y_{ic} - \bar{y})^2$	$(y_i - \bar{y})^2$
25	0.54	0.4867	3.404	3.210
50	0.96	0.8967	2.059	1.881
100	1.82	1.7167	0.378	0.262
150	2.3	2.5367	0.042	0.001
200	3.2	3.3567	1.051	0.754
300	5.17	4.9967	7.102	8.056
$\Sigma =$	13.99		14.036	14.164

From the summary data in the table,

$$r^2 = 14.036/14.164 = 0.991$$

Since the coefficient of determination is very close to 1, the correlation between peak area and benzene concentration is judged to be excellent, as is readily obvious from Fig. 10.7.

Many calculators have built-in programs for least-squares linear regression analysis. Spreadsheet software such as Excel can also be used to readily determine the statistical coefficients described herein.

Nonlinear Regression

Many equations we may wish to use in environmental engineering and science do not follow a linear relationship and thus cannot be analyzed directly by the linear regression procedures. Nevertheless, the least-squares concept can be applied here as well. A detailed presentation of nonlinear regression is beyond the scope of this chapter. Those interested can find good descriptions in the statistics textbooks listed in the references at the end of this chapter.

This one brief example will be used to illustrate the approach for nonlinear regression analysis. It will be applied to the BOD data given in Example 10.16. Here, rather than transforming the first-order reaction into its linear form, we use Eq. (10.35) directly to calculate values for BOD (y_c) as a function of t. We do this for each t_i for which we have a measured BOD value y_i. We then square the differences between y_i and y_c and sum these values. The trick then is to find values for k' and L that cause the sum of the squared values to be as small as possible. In order to do this, we first have to make reasonable guesses for the values of k' and L and use these to obtain an initial sum of squares to begin the process. Then, by changing the values of k' and L individually by small amounts and seeing the effect on the sum of squares, we can by trial and error eventually find the combination of values that minimize the sum of squares. This obviously would be tedious work if done with a hand calculator, but it can be quite readily accomplished through use of a computer spreadsheet. The Excel® spreadsheet has a solver routine that takes the burden out of the search and provides a quick and easy solution to the problem. Here, a table illustrates a possible spreadsheet setup for this approach and indicates the best k' and L values found using it:

EXAMPLE 10.18

$k' = 0.217$

$L = 104$

t_i	y_i	y_c	$y_i - y_c$	$(y_i - y_c)^2$
0	0	0.00	0.00	0.00
1	27	20.33	6.67	44.54
2	36	36.69	−0.69	0.47
3	53	49.86	3.14	9.88
5	65	68.99	−3.99	15.93
7	80	81.39	−1.39	1.92
9	89	89.42	−0.42	0.18
12	98	96.49	1.51	2.28
			Σ squares =	75.19

We see that the resulting values for k' and L in this case are quite close to the respective values of 0.21 day^{-1} and 103 mg/L obtained through linear regression analysis in Example 10.16.

10.10 | QUALITY ASSURANCE AND QUALITY CONTROL

Quality assurance (QA) is the following of a set of operating principles for collection and analysis of environmental samples in order to produce data of known and defensible quality. If one properly follows QA principles, then the accuracy and precision of results can be stated with a high level of confidence. Quality control (QC) represents the procedures followed by the analyst to produce credible results as part of an overall QA program. Details of QA and QC procedures are provided in "Standard Methods" and are only briefly discussed here.

The set of operating principles for QA are outlined in a QA plan. This plan includes plan approval signatures; staff organization and responsibilities; sample collection, control, and documentation procedures; training requirements for analysts; standard operating procedures (SOP) for each analytical method used; equipment calibration and maintenance procedures; internal QC activities, data assessment procedures for bias and precision; data reduction, validation, and reporting; and corrective actions. Development and following of a good QA plan is essential when many different individuals are involved from sample collection and documentation, transport of samples, and analysis as the potential for error along the way is large. Often large financial stakes are involved in the outcome of analysis of field samples. Legal requirements demand that samples be carefully taken and correctly labeled and that a chain of custody be carefully established to ensure that everyone is carrying out assigned duties.

Quality control is the part of the QA plan that becomes the responsibility of the analyst, who must have a good understanding of statistical analysis of the methods used and of the data produced. A QC program consists of the certification of operator competence and the following of a set of procedures for determining and then assuring the quality of the data produced. This includes the analysis of externally supplied standards and the calibration of the analytical procedure with standards, analysis of reagent blanks, determining the recovery of known standard additions to samples, analysis of duplicates to provide an indicator of possible analytical error and for determining the precision of the analysis, and the maintenance of control charts to help check upon the quality of the data being obtained. It is essential that the analysts understand the limits of the analyses being performed and be on constant vigil to ensure that data produced are always of high quality. More about QA and QC can be found in "Standard Methods."

PROBLEMS

10.1   State the difference between each of the two terms.

(*a*)   Mean and median

(*b*)   Accuracy and precision

(*c*)   Standard deviation and variance

(d)   Confidence interval and confidence limits

(e)   Instrument detection limit and method detection limit

10.2   (a)   For what is the t test used?

(b)   For what is the F test used?

10.3   The following set of replicate measurements for sulfate (mg/L) were obtained from the analysis of different groundwater samples.

A	B	C	D	E	F
1333	1494	1330	2096	1612	1556
1496	1399	1333	1974	1644	1519
1525	1487	1314	2087	1654	1414
1547	1493	1215	1949	1881	1489
	1402	1218	2314	1793	1449
	1488			1529	

Calculate the following for each set of data: (a) mean, (b) standard deviation, (c) coefficient of variance, and (d) standard error of the mean.

10.4   For each set of data in Prob. 10.3, calculate the 95% confidence limits.

10.5   For the data in Prob. 10.3 and at the 95% confidence level, is there a difference between the averages for (a) columns A and B? (b) columns C and D? (c) columns E and F?

10.6   For each set of data in Prob. 10.3 and at the 95% confidence level, is there a difference in the precision between columns B and E?

10.7   The following set of replicate measurements for nitrate (mg/L) were obtained from the analysis of different groundwater samples:

A	B	C	D	E	F
9	30	24	41	35	35
10	28	20	36	34	57
7	28	22	28	33	44
10	39	26	31	38	52
10		19	28	26	33
10		20	39		

Calculate the following for each set of data: (a) mean, (b) standard deviation, (c) coefficient of variance, and (d) standard error of the mean.

10.8   For each set of data in Prob. 10.7, calculate the 95% confidence limits.

10.9   For the data in Prob. 10.7, and at the 95% confidence level, is there a difference between the averages for (a) columns A and B? (b) columns C and D? (c) columns E and F?

10.10   For each set of data in Prob. 10.7 and at the 95% confidence level, is there a significant difference in the precision between columns B and E?

10.11   Pool the data sets in Prob. 10.3 to determine the standard deviation for the analytical method.

10.12   Pool the data sets in Prob. 10.7 to determine the standard deviation for the analytical method.

10.13 From the following set of BOD data, determine (a) the ultimate BOD, L_0, (b) the BOD rate constant, k', (c) the standard deviation for the rate constant, and (d) the standard deviation for L_0.

t, days	y, mg/L
0	0
1	55
2	128
4	170
6	220
8	260
10	262

10.14 From the following set of BOD data, determine (a) the ultimate BOD L_0, (b) the BOD rate constant k', (c) the standard deviation for the rate constant, and (d) the standard deviation for L_0.

t, days	y, mg/L
0	0
2	8.1
4	11
7	15.1
9	19.2
11	19.9

10.15 A standard solution of copper sulfate was analyzed by atomic-absorption spectrometry to prepare a standard curve for copper, with the following results.

Concentration, μg/L	0	100	200	300	400	500
Absorbance	0.003	0.140	0.276	0.429	0.548	0.688

Determine the slope and intercept of the calibration curve, as well as the 95% confidence limits for each.

10.16 An ion-selective electrode method was compared with the standard iodometric method for determining the concentration of dissolved sulfide in groundwater with the following results.

Sample	A	B	C	D	E	F	G	H	I	J
Electrode	130	14	182	4	127	13	154	14	192	154
Iodemetric	126	13	136	1	130	13	169	13	218	142

Comment upon the suitability of the electrode method for this analysis.

10.17 The concentration of ammonia in water is to be determined spectrophotometrically. In order to do this, a series of standard concentrations was analyzed with the following results, from which a calibration curve is to be prepared.

Concentration, mg/L	1	2	4	6
Absorbance	0.08	0.16	0.27	0.50

Determine the slope and intercept of the calibration curve, as well as the 95% confidence limits for each.

10.18 A water sample is analyzed for copper by atomic-absorption spectrometry using the calibration curve prepared in Prob. 10.15, giving an absorbance reading of 0.35.

Determine the copper concentration and the 95% confidence interval assuming (a) one sample is analyzed, and (b) six samples are analyzed.

10.19 A water sample is analyzed for ammonia spectroscopically using the calibration curve prepared in Prob. 10.17, giving an absorbance reading of 0.42. Determine the ammonia concentration and the 95% confidence interval assuming (a) one sample is analyzed, and (b) three samples are analyzed.

10.20 Two students are given a sample to analyze for chemical oxygen demand. Student A makes four determinations with a standard deviation of 20 mg/L, and student B makes seven determinations with a standard deviation of 9 mg/L. Does the difference in standard deviations imply that there is a significant difference in the analytical techniques of the two students?

10.21 Two students are given a sample to analyze for dissolved oxygen. Student A makes five determinations with a standard deviation of 0.4 mg/L, and student B makes eight determinations with a standard deviation of 0.3 mg/L. Does the difference in standard deviations imply that there is a significant difference in the analytical techniques of the two students?

10.22 The standard deviation of a method for determining benzene in groundwater is found to be 5 μg/L. How many analyses of the same sample are required if the 95% confidence interval of the mean is to be (a) ± 4 μg/L, and (b) ± 3 μg/L?

10.23 The standard deviation of a method for determining arsenic in water is 2 μg/L. How many analyses of the same sample are required if the 95% confidence interval of the mean is to be (a) ± 2 μg/L, and (b) ± 1.5 μg/L?

10.24 Express the results of the following calculations using only significant figures.

(a) $\dfrac{0.275 + 12.20 - 1.6532}{12.71 \times 0.6752}$

(b) $\dfrac{0.76 - 0.123 + 12.1}{0.0123 \times 8.6}$

(c) $7.651 \times 0.1223 - \dfrac{6.245}{12.47}$

(d) $3.1 \times 5.2 - 4.3 + 2.2$

10.25 Express the results of the following calculations using only significant figures.

(a) $\dfrac{0.25 \times 3.1 - 5.3 + 7.2}{4.5 \times 7.3}$

(b) $\dfrac{0.753 \times 67.25 - 22.68 + 13.44}{0.1234 \times 8.762}$

(c) $0.0023 \times 0.045 - \dfrac{0.088}{78}$

(d) $352. \times 0.111 - 11.22 + 0.115$

10.26 Estimate the absolute standard deviation for y for each of the following calculations, rounding results to significant figures. The numbers in parentheses are absolute standard deviations.

(a) $y = 7.23(\pm 0.04) + 12.76(\pm 0.21) - 5.11(\pm 0.11) - 0.11(\pm 0.01)$

(b) $y = 20.1(\pm 1.2) \times 123(\pm 9.7) - 132(\pm 1.6)$

(c) $y = \dfrac{0.123(\pm 0.07) - [0.102(\pm 0.021) \times 1.11(\pm 0.09)]}{0.0165(\pm 0.0021)}$

(d) $y = 27(\pm 3.2)(1 - e^{-5 \times 0.31(\pm 0.032)})$

10.27 Estimate the absolute standard deviation for y for each of the following calculations, rounding results to significant figures. The numbers in parentheses are absolute standard deviations.

(a) $\quad y = 8.65(\pm0.55) - 1.01(\pm0.02) - 12.7(\pm1.0) + 22.1(\pm0.87)$

(b) $\quad y = \dfrac{1.25(\pm0.08)}{0.26(\pm0.042)} + 3.72(\pm0.93)$

(c) $\quad y = 236(\pm11) \times 2.43(\pm0.35)/1.62(\pm0.063)$

(d) $\quad y = 3.45(\pm0.88)^{2.3(\pm0.56)}$

10.28 Determine whether the following data appear to have been drawn from a normal population by plotting on normal probability paper.

29.55, 28.13, 22.57, 22.31, 26.08, 30.06, 24.89, 29.58, 24.11, 26.06, 23.20, 25.40, 26.94, 27.60, 26.11, 22.04, 24.87, 22.52, 26.31

10.29 An atomic adsorption spectroscopic method for antimony in an urban atmosphere was compared with a colorimetric method with the following results. Do the results obtained by the different methods differ significantly? Use the 95% confidence level for judgment.

Atomic Adsorption: $n = 12$, $\bar{x} = 10.5\ \mu g/m^3$, $s = 2.4\ \mu g/m^3$

Colorimetric: $n = 10$, $\bar{x} = 8.9\ \mu g/m^3$, $s = 1.6\ \mu g/m^3$

10.30 The following results in mg/L were obtained from repeated analysis of nitrate in water. Determine the mean and median for the data set. Also from these data, prepare a frequency distribution table and a histogram.

10.7, 10.7, 10.5, 10.7, 10.3, 10.7, 10.9, 11.1, 10.5, 9.9, 10.7, 10.9, 11.1, 10.1, 10.3, 10.5, 10.9, 10.3, 10.3, 10.5 10.3, 10.1, 9.7, 10.3, 10.3, 10.1, 10.3, 10.3, 10.7, 9.9, 10.7, 10.7, 10.7, 10.1, 10.5, 9.9, 10.5, 10.7, 10.3, 10.1, 10.7, 10.5, 10.5, 11.1, 10.9, 10.9, 10.5, 10.5, 10.7, 10.7

10.31 Twelve analyses for sodium concentration in a given water sample produced a mean value of 76.0 mg/L and a standard deviation of 4.3 mg/L. Calculate the 95% confidence limits for the mean.

10.32 Seven analyses for calcium concentration in a given water sample produced a mean value of 122 mg/L and a standard deviation of 6 mg/L. Calculate the 95% confidence limits for the mean.

10.33 Thirty effluent samples were collected from a wastewater treatment plant and analyzed for BOD, resulting in an average concentration of 20.1 mg/L with a standard deviation of 8.2 mg/L. Assume a normal distribution of data.

(a) Calculate the 95% confidence limits for the mean.

(b) How many samples would one need to collect to reduce the confidence interval to one-half of the value obtained in part (a)?

10.34 Seventeen effluent samples were collected from a drinking water treatment plant and analyzed for chloroform, resulting in an average concentration of 57 μg/L with a standard deviation of 35 μg/L. Assume a normal distribution of data.

(a) Calculate the 95% confidence limits for the mean.

(b) How many samples would one need to collect to reduce the confidence interval to one-half of the value obtained in part (a)?

10.35 Four effluent samples were collected from a wastewater treatment plant effluent and analyzed for dichlorobenzene, resulting in a geometric mean concentration of 1.2

μg/L with a geometric standard deviation of 2.92. Assume a lognormal distribution of data.

(a) Calculate the 95% confidence limits for the mean.

(b) How many samples would one need to collect so that the upper 95% confidence limit would be no more than twice the geometric mean?

10.36 Thirteen effluent samples were collected from a wastewater treatment plant effluent and analyzed for ethylbenzene, resulting in a geometric mean concentration of 1.1 μg/L with a geometric standard deviation of 2.7. Assume a lognormal distribution of data.

(a) Calculate the 95% confidence limits for the mean.

(b) How many samples would one need to collect so that the upper 95% confidence limit would be no more than 1.5 times the geometric mean?

10.37 The following concentrations (μg/L) for dichlorobenzene were measured in a wastewater treatment plant effluent:

      0.12, 0.10, 0.20, 0.32, 0.50, 0.52, 0.56, 0.80, .89, 1.31, 1.41, 1.62, 4.1

(a) Assuming lognormal distribution, what is the geometric mean concentration and geometric standard deviation?

(b) What are the 95% confidence limits for the geometric mean?

10.38 The following concentrations (μg/L) for dichloromethane were measured in a wastewater treatment plant effluent:

      2.8, 4.0, 4.8, 5.2, 6.0, 6.2, 9.0, 9.8, 10.5, 12.0, 12.5, 17.0, 19.0, 22, 27, 35, 63

(a) Assuming lognormal distribution, what is the geometric mean concentration and geometric standard deviation?

(b) What are the 95% confidence limits for the geometric mean?

REFERENCES

Anderson, R. L. "Practical Statistics for Analytical Chemists," Van Nostrand Reinhold, New York, 1987.

Caria, M.: "Measurement Analysis: An Introduction to the Statistical Analysis of Laboratory Data in Physics, Chemistry and the Life Sciences," Imperial College Press, London, 2001.

Clesceri, L. S., A. E. Greenberg, and A. D. Eaton (eds.), "Standard Methods for the Examination of Water and Wastewater," 20th ed., American Public Health Association, Washington DC, 1998.

de Levie, R.: "How to use Excel® in Analytical Chemistry and in General Scientific Data Analysis," Cambridge University Press, Cambridge, MA, 2001.

Graham, R. C.: "Data Analysis for the Chemical Sciences: A Guide to Statistical Techniques," John Wiley & Sons, New York. 1993.

Meier, P. C., and R. E. Zund: Statistical Methods in Analytical Chemistry, 2nd ed., John Wiley & Sons, New York, 2000.

Miller, J. N., and J. C. Miller: "Statistics and Chemometrics for Analytical Chemistry," 4th ed., Prentice Hall, Upper Saddle River, NJ, 2000.

Otto, M.: "Chemometrics: Statistics and Computer Application in Analytical Chemistry," John Wiley & Sons, London. 1999.

Basic Concepts from Quantitative Chemistry

11.1 | GENERAL OPERATIONS

Quantitative chemistry may be considered a keystone in the training of environmental engineers and scientists. It serves as the basis for most research work and many field investigations. Unfortunately, most college courses in quantitative analysis, although they furnish excellent fundamental information, do not provide laboratory instruction and practice that are particularly valuable to environmental engineers. For this reason, specialized courses in quantitative analysis, often described as courses in "environmental chemistry," "aquatic chemistry," or "water and wastewater analysis," have been developed to meet the need.

Most schools concerned with the training of environmental engineers and scientists use as a reference textbook "Standard Methods for the Examination of Water and Wastewater."[1] This book is designed to be used by trained analysts and therefore presumes a certain background of information. Since many environmental engineers and scientists do not have such a background, the purpose of this chapter is to fill that need. It is recommended that a copy of a standard, up-to-date text on quantitative analysis be added to their library for reference purposes. Possible choices are listed at the end of this chapter. It is impossible and unnecessary in this chapter to cover in detail the many aspects of analytical chemistry that are adequately dealt with in all standard textbooks.

Sampling

It is an axiom that the analytical results obtained in the laboratory can never be more reliable than the sample upon which the tests are performed. It may be safely stated that more results are in error because of inadequate sampling than because of

[1]L. S. Clesceri, A. E. Greenberg, and A. D. Eaton, (eds.), "Standard Methods for the Examination of Water and Wastewater," 20th ed., American Public Health Association, New York, 1998.

faulty laboratory techniques. The subject of sample collection and care of samples is treated quite adequately in "Standard Methods," but additional emphasis is felt to be useful on the subject of *grab* versus *composited* samples.

Grab samples are those taken more or less instantaneously and analyzed separately. In general, most sampling in environmental engineering practice is of the grab variety. The major problem confronting the engineer is the decision of frequency of sampling. In this decision the engineer must always balance the issues of a sufficient number of samples for reliability versus costs. The number of samples may vary from one to over a hundred per day, depending upon the nature of the material to be sampled. At this point a good deal of engineering judgment is involved. A few examples will serve to illustrate.

1. Consider a deep well that has been in service for some time. The water quality will be uniform, and a single grab sample will give a true picture of conditions.

2. Oftentimes changes occur slowly, as in large rivers, and daily grab samples are adequate.

3. Where the character of a material changes considerably within a 24-h period, the use of grab samples at frequent intervals is dictated. In sampling industrial wastes, it may be necessary to obtain such samples every 10 or 15 min. The individual samples may be analyzed for certain characteristics such as pH, acidity, and alkalinity, and then pooled into 2-,4-, 8-, 12-, or 24-h composites for more complete analysis.

Composited samples are used mainly in evaluating the efficiency of wastewater treatment facilities, where average results are adequate. Such samples are collected at regular intervals, usually every hour or two, and pooled into one large sample over a 24-h period. Under such conditions, detention times can be considered self-canceling, and the only requirement that must be met is that the amount of each individual sample be taken in proportion to the flow existing at the time.

When composited samples are taken for periods shorter than 24 h (8-h composited samples have been quite common in waste treatment practice), serious errors may be introduced by ignoring detention time. In conventional activated sludge plants, the theoretical detention time from influent to effluent is about 8 h; thus, samples of influent collected from 8 A.M. to 4 P.M. measure the strength of daytime waste, but samples of effluent collected over the same period of time measure the treatment given to weak waste that entered the plant prior to 8 A.M. In taking short-term composited samples, sampling at downstream locations should be adjusted to detention times in the treatment units involved. In the average conventional activated sludge treatment plant with a total of 8 h of detention, sampling of the final effluent should not begin until 8 h after sampling of the influent is started; thus, 16 h are needed to collect a set of 8-h composited samples.

Laboratory Apparatus and Reagents

A good grade of laboratory apparatus is adequate for all practical uses. Pyrex or a similar glassware of low solubility and low coefficient of expansion is highly recommended. Reagents should be of analytical-reagent grade or known to meet the specifications of purity established by the American Chemical Society. Lower grades of chemicals, even technical grade, may be used for some purposes, but the analyst must make sure that such grades do not contain undesirable amounts of certain impurities. It is not practical to buy a reagent grade of sodium hydroxide for purposes that require a high-purity sodium hydroxide. The best grades obtainable are too impure. It is just as easy to purify a cheaper grade in the laboratory, and it serves the purpose just as well, at a much lower cost. As an alternative, purified sodium hydroxide solutions may be purchased.

Precipitation

Some analytical methods depend upon precipitation of an ion to allow its separation and measurement by actual weighing of the precipitated material. There are several requirements that must be met to make such procedures reliable. The major ones are as follows:

1. The precipitated material must have a very low solubility in water; i.e., its K_{sp} must be a very small number.
2. It must precipitate in a high state of purity or be capable of reprecipitation for further purification.
3. It must be capable of drying or of ignition at temperatures above 100°C to a definite compound of fixed composition.
4. It should not be hydroscopic at room temperatures.

Precipitation may also be used as a method of purification. Under such conditions, the desired ion is precipitated from solution, filtered, placed into solution again by the use of special reagents, and measured by other means than weight analysis.

Filtration

Filtration of precipitates or solids is accomplished by means of paper, membrane, or glass fiber filters. Many materials that the environmental engineer or scientist wishes to measure cannot be subjected to temperatures above 103°C without losing chemically bound water. Filtration through either filter paper (analytical, ashless grade) or glass fiber filters is commonly used where ignition at temperatures of around 550°C is required or permissible. At such temperatures filter paper is destroyed, and the desired residue remains. An ashless paper or one whose ash weight is known should be used. Today, however, glass fiber filters are commonly used at the higher temperatures as they show little weight loss at 550°C. However, they can melt at a temperature of 600°C, the former temperature used for organic combustion, and so the lower temperature of 550°C is now generally used, regardless of the type of filter.

Filters vary greatly in their porosity. It is essential to use a filter that is fine enough to retain all the precipitate but as coarse as possible to obtain rapid filtration rates. Barium sulfate requires the use of a very fine paper. Ferric hydroxide is retained on very coarse fast-filtering paper.

Separation of a precipitate can be accomplished by the use of a proper filter. Quantitative results, however, depend upon two other important factors:

1. All the precipitate or solid material must be transferred from the original vessel to the filter.

2. The precipitate and filter must be washed with water or a suitable solvent to remove dissolved solids that remain in the precipitate and wetted filter. Three washings with complete drainage between washings are considered sufficient.

Drying or Ignition

There are now two standard temperatures used for drying residues or solids; one is 103°C and the other is 180°C. A temperature of 103 to 105°C was the only temperature stipulated in the past because many of the residues or solids involved in environmental engineering practice are organic in nature and release water of composition in significant amounts at higher temperatures. Drying at 103°C ensures the removal of all free water, provided that the drying period is long enough, and minimizes the loss of water of crystallization. However, sometimes it is desirable with water supplies not containing significant concentrations of organic material to have a total solids analysis that more closely approximates the value obtained through a summation of individually determined mineral species. For this purpose a drying temperature of 180°C may be specified. Obviously, the drying temperature should be specified when reporting results because of the differences that can be obtained.

Ignitions are commonly conducted at 550°C, unless specified otherwise, to ensure the destruction of all organic matter by oxidation to carbon dioxide and water while minimizing the loss of inorganic salts by volatilization or decomposition. Calcium carbonate is a major component of many residues and is stable at 550°C.

Desiccation

Following drying or ignition operations, the residues and their containers (crucibles, evaporating dishes, or filters) *must be cooled to room temperature* before weighing on the analytical balance. If such cooling were allowed to take place in the open air, moisture would be picked up from the air by the residue and container. The amount of moisture pickup would depend upon the time of exposure and the relative humidity. To overcome this difficulty, residues and their containers are cooled in *desiccators* where the relative humidity is kept near zero percent by means of a *desiccant* such as anhydrous calcium chloride.

11.2 | THE ANALYTICAL BALANCE

The construction of analytical balances and the theory of weighing are adequately covered in standard texts on quantitative analysis. The care of such instruments cannot be stressed too much. It is important to recognize that analytical balances fall into the realm of *delicate* instruments and that great care and scrupulous cleanliness must be maintained with respect to both the balance and the weights used.

Automatic analytical balances, like the one shown in Fig. 11.1, are by all odds the most practical devices for modern laboratories. Such balances have weighing accuracies in the range of 0.01 to 1 mg, providing correct procedures are followed.

The importance of having the objects to be weighed at room temperature cannot be overemphasized. The average student in quantitative analysis should not have difficulty in translating his or her knowledge of physics to explain how air cur-

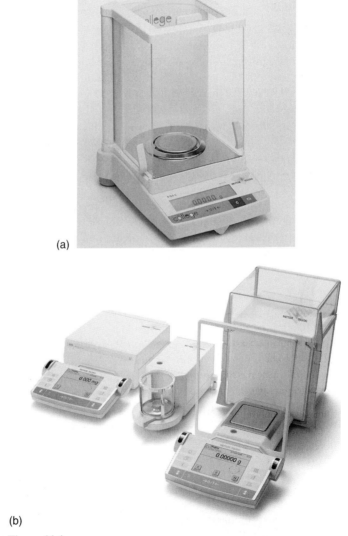

(a)

(b)

Figure 11.1
Automated balances. (*a*) Analytical balance. (*b*) Pan Balance.
(Courtesy of Mettler-Toledo, Inc.)

rents cause an object having a temperature above room temperature to weigh less than it should and an object that is colder than room temperature to weigh more than it should.

11.3 | GRAVIMETRIC ANALYSIS

Gravimetric analysis means analysis by weight and pertains to all determinations wherein the final results are obtained by means of the analytical balance. Determination of total solids, suspended solids, fixed solids, or volatile solids is made by gravimetric procedures because there is no better way of gaining the desired information. In general, however, gravimetric procedures are avoided as much as possible because they are time-consuming. In other than solids determinations, there is little occasion to use the analytical balance except to make up standard solutions, and the like.

Because of the importance of solids determinations that must be done by gravimetric procedures, it is essential that the fundamentals of gravimetric analysis be known. Many of these have been discussed in Sec. 11.1; a few more follow.

Constant Weight

All gravimetric measurements require some sort of crucible or dish to hold the residue or precipitate. The weight of this container (tare) must be known and deducted from the gross weight to obtain the net weight of the material being measured. Weighing of such objects with accuracies of 0.1 to 0.01 g can now be performed on an automated pan balance (Fig. 11.1).

Porcelain or high-silica crucibles or dishes are commonly used for gravimetric work. Platinum ware has many advantages but is not ordinarily used, except for special determinations, because of its cost. All objects exposed to the air for any length of time have a film of dust and adsorbed moisture. Porcelain ware is manufactured with some of its surface unglazed; consequently, it will also adsorb moisture into its interior. To use such containers for gravimetric work without proper conditioning would lead to relatively large errors in the final results. Oftentimes negative results are obtained when small quantities are involved.

The conditioning of crucibles and dishes for gravimetric work involves pretreatment to eliminate the dirt and moisture. All containers should be thoroughly cleaned with water and then heat-treated under exactly the same conditions as those to which the container will be subjected in the actual determination. If the container is to be used to measure total solids at 103 or 180°C, it should be conditioned at that temperature. To heat the dish, particularly porcelain ware, at higher temperatures, such as at 550°C, will drive out more moisture than desired. The dishes should be heated or fired at the desired temperature, cooled in the desiccator and weighed, heated and fired, cooled and weighed, repeatedly, until the container reaches what is known as constant weight. For small dishes with a weight of less than 25 g, this is assumed to be ±0.0002 g. For heavier dishes the allowance is greater but never over ±0.0005 g. With a little experience, the beginner will learn that a preliminary heating or firing of less than 2 h is seldom worthwhile. This is particularly true of new dishes and of old dishes that have not been used for some time. After dishes have been brought to constant weight, they should be kept in a desiccator to avoid collection of dust or adsorption of moisture prior to use.

Theoretically, the principles outlined for getting the tare weight of containers should be applied to the dish plus the residue to be measured. This is true when inorganic residues are involved; however, it has been demonstrated that when organic substances are concerned, such materials will usually continue to yield moisture in small amounts for long periods of time. Since free water is given off rather rapidly, barring physical interference to its release, the drying time for such solids is usually specified, and the weight obtained at the end of that time is accepted as being constant. The drying time for total solids in water (following evaporation to practical dryness on a steam bath or in a drying oven) and for suspended solids is 1 h. The drying time for total solids in sludges, many industrial wastes, and other semisolid samples (following evaporation to dryness) is considered to be 10 h. Usually, drying overnight is more practical and is recommended for sludges and other semisolid materials.

11.4 | VOLUMETRIC ANALYSIS

Volumetric analysis is a phase of quantitative analysis that depends upon the measurement of liquid reagent volumes of *standard solutions* needed to complete particular reactions in samples submitted to test. A standard solution is defined as follows: *a solution whose strength or reacting value per unit volume is known.* The facilities needed for conducting a simple volumetric analysis are (1) equipment to measure the sample accurately, either an analytical balance or volumetric glassware such a pipets; (2) a standard solution of suitable strength; (3) an indicator to show when the stoichiometric end point has been reached; and (4) a carefully calibrated buret for measuring the volume of standard solution needed to reach the stoichiometric end point as shown by the indicator

Analysis by volumetric methods is very popular, as compared with gravimetric methods, because of the time that usually can be saved by such procedures. Volumetric methods are used for many determinations such as dissolved oxygen, BOD, COD, and chlorides. There are a number of concepts and techniques involved in volumetric measurements that must be understood in order to obtain accurate results.

Calibrated Glassware

Calibrated glassware is of two types: (1) that which is calibrated to contain a definite volume, e.g., volumetric flasks and graduated cylinders; and (2) that which is calibrated to deliver prescribed volumes, e.g., pipets and burets. The former should never be used as a substitute for the latter when accuracy is required because some liquid will always remain in the container, and the delivered amount will be less than the calibrated amount. On the other hand, pipets are calibrated to deliver definite amounts under specified conditions and must always retain a small amount of the liquid after delivery.

Pipets are of two types, transfer and Mohr. Transfer, or volumetric, pipets have an enlarged section and only one graduation mark. They will deliver the specified amount provided that the pipet is clean, it is held in a near vertical position during delivery, contact is made between the tip of the pipet and the wall of the receiving vessel, and drainage is allowed to occur for 5 seconds after the level of the liquid in the tip appears to have reached a static condition. Transfer pipets should be used where volumes must be measured with a high degree of accuracy. Mohr-style pipets are made from glass tubing of uniform bore and have multiple graduation marks. The small sizes of 5-mL capacity or less may be used to deliver small quantities of liquids quite accurately, but with 10-mL and larger sizes the bore and calibrations are such that delivery of accurate amounts cannot be expected. The Mohr pipet is used largely for measuring fractional volumes of 1 mL or for measurement of volumes when accuracy is of little importance.

In general, it may be stated that transfer or volumetric pipets should be used for measurement of samples and of standard solutions. Mohr pipets may be used for the addition of nonstandardized reagents.

Calibrated glassware should be kept thoroughly clean. This is particularly true of transfer pipets and burets. Cleaning may be accomplished by means of some of the modern detergents designed for such purposes or by the use of commercial preparations.

Equivalent or Normal Solutions

By definition, *a standard solution is one whose strength or reacting value per unit volume is known.* Volumetric analysis could be practiced solely with such solutions, but it is much more convenient to prepare and use solutions that are equivalent to one another in strength, so that 1.0 mL of reagent A will react with exactly 1.0 mL of reagent B, and so on.

We can establish the basis for a system of equivalent solutions for the measurement of acids and bases (acidity and alkalinity) from the fundamental equation involved:

$$\underset{1.008}{H^+} + \underset{17.007}{OH^-} \rightarrow H_2O \tag{11.1}$$

From this equation it is seen that 1 g atomic weight or 1.008 g of hydrogen ion reacts with 1 g ionic weight or 17.007 g of hydroxyl ion. The number 17 serves for all practical purposes in the latter case. Solutions containing these amounts of hydrogen ion or hydroxyl ion can be considered equivalent.

The *equivalent weight* of a compound is defined as *that weight of the compound which contains one gram atom of available hydrogen or its chemical equivalent.* The equivalent weight can be determined as follows:

$$\text{Equivalent weight} = \frac{FW}{Z} \tag{11.2}$$

where FW is the formula weight of the compound and Z is a positive integer whose value depends upon the chemical context.

For acids, the value of Z is equal to the number of moles of H^+ obtainable from 1 mol of the acid. For HCl, $Z = 1$; and for H_2SO_4, $Z = 2$. For acetic acid (CH_3COOH), $Z = 1$, since only one of the hydrogen atoms in the acetic acid molecule will ionize to yield available H^+ ions in solution.

$$CH_3COOH \rightarrow CH_3COO^- + H^+$$

For bases, the value of Z is equal to the number of moles of H^+ with which 1 mol of the base will react. For NaOH, $Z = 1$; for $Ca(OH)_2$, $Z = 2$.

A *normal solution* is defined as *one that contains one equivalent weight of a substance per liter of solution.* Thus, for the preparation of 1 liter of a normal solution of an acid or a base, all that is needed is enough of the compound that furnishes the ion to yield 1.008 g of H^+ or 17 g of OH^- and enough distilled water to make a solution having a volume of 1 liter. The *normality* of a solution is its relation to the normal solution, and the symbol N is used as the abbreviation for "normal." Thus, a half-normal solution is expressed either as 0.5 N or as N/2. The preparation and standardization of normal solutions will be discussed in Chap. 15.

Normal solutions are also used for volumetric measurements that do not involve acidimetry or alkalimetry. For example, chloride can be measured by titration with a reagent such as silver nitrate. The reaction involves precipitation of chloride ion as silver chloride in the following manner:

$$\underset{35.5}{Cl^-} + \underset{108}{Ag^+} \rightarrow AgCl(s) \tag{11.3}$$

In this reaction 1 mol of Ag^+ is equivalent to 1 mol of Cl^-. From the equation

$$HCl \rightarrow H^+ + Cl^- \tag{11.4}$$

it can be reasoned that 1 mol of Cl^- is equivalent to 1 mol of H^+. Since things equal to the same thing are equal to each other, it may be stated safely that 1 mol of Ag^+ is equivalent to 1 mol of H^+. Thus, a normal solution of Ag^+ is one that contains 1 mol of Ag^+ per liter.

For all practical purposes, the equivalent weight of a compound involved in a precipitation reaction can be found from Eq. (11.2) by letting Z equal the oxidation number of the ion. The equivalent weight of $BaCl_2$ is FW/2. The equivalent weight of $Al_2(SO_4)_3$, considered either as an aluminum or as a sulfate salt, is FW/6. In certain cases, a salt may have more than one equivalent weight, depending on how it is used. Thus, the equivalent weight of K_2HPO_4 is FW/2 when potassium is the reacting ion, and FW/3 when the reacting ion is phosphate.

A third form of normal solutions involves those used for their oxidizing or reducing values. Examples are solutions of potassium dichromate, ferrous ammonium sulfate, and sodium thiosulfate. The equivalent weight or that weight of a compound needed to prepare 1 liter of a normal oxidizing or reducing solution is derived from the following equation:

$$\underset{23}{Na^0} + \underset{1.008}{H^+} \rightarrow Na^+ + H^0$$

Table 11.1 | Equivalent weights of some common oxidizing and reducing agents

Agent	Nature	Conditions	Oxidation number change	Equivalent weight
$KMnO_4$	Oxidizing	Acid	5	FW/5
$KMnO_4$	Oxidizing	Alkaline	3	FW/3
$K_2Cr_2O_7$	Oxidizing	Acid	2×3	FW/6
I	Oxidizing	Acid	1	AW/1
$KH(IO_3)_2$	Oxidizing	Acid	2×5	?*
$Na_2C_2O_4$	Reducing	Acid	2×1	FW/2
KI	Reducing	Acid	1	FW/1
As_2O_3	Reducing	Acid	2×2	FW/4
$Na_2S_2O_3$	Reducing	Acid	?	FW/1†
$Fe(NH_4)_2(SO_4)_2$	Reducing	Acid	1	FW/1

*In the case of $KH(IO_3)_2$, the reaction must be known in order to calculate the equivalent weight. For ordinary reactions it would be FW/10, but for the usual reaction with KI it is FW/12.
†The equivalent weight of $Na_2S_2O_3$ must be calculated indirectly because the oxidation number change of sulfur is not known. In the reaction with iodine, one $Na_2S_2O_3$ is equivalent to one atom of iodine. Therefore, the equivalent weight is FW/1.

In this reaction one H^+ takes one electron from one atom of sodium, or 1 mol of hydrogen ions can be considered as equivalent to "1 mol of electrons." Sodium undergoes a change in oxidation number of 1; therefore, the equivalent weight of a compound to be used as an oxidizing or reducing agent can be determined by letting Z equal the oxidation number change that the compound undergoes in the reaction involved. It is very important that this oxidation number change be known. For example, potassium permanganate undergoes an oxidation number change of 5 under acid conditions, and its equivalent weight for such use would be FW/5. Under alkaline conditions, it undergoes an oxidation number change of 3, and its equivalent weight for such use is FW/3. Table 11.1 shows the equivalent weights of some common oxidizing and reducing agents.

Primary Standards

The standardization or measurement of the exact strength of a normal or other solution depends upon the use of some standard material whose purity is known. Primary standards are usually salts or acid salts of high purity that can be dried at some convenient temperature without decomposing and that can be weighed with a high degree of accuracy. Examples are sodium carbonate and potassium acid phthalate, which are used to standardize acid and base solutions, respectively; potassium biiodate and potassium dichromate for reducing solutions; potassium oxalate for oxidizing solutions; and sodium chloride for solutions of silver ion. Analytical-reagent-grade chemicals are usually satisfactory for most purposes. For some research purposes and possibly for referee work, analyzed primary standards may be obtained from the U.S. National Institute of Standards and Technology.

Secondary Standards

Any solution that has been standardized against a primary standard is considered a secondary standard and may be used as such. In the preparation and use of solutions of a base, this practice is often followed and is a valuable time saver with all solutions that are not stable and therefore have to be restandardized frequently.

Choice of Indicators

Analysis by volumetric procedures requires that some method be employed of indicating the stoichiometric end point, that is, the *equivalence point,* of the titration. Internal indicators are greatly preferred. In any event, they should signal the reaching of the completed reaction as closely as possible. Improper choice of indicators can introduce serious errors into volumetric work. Indicators in common use today are of several major types, viz., electrometric, acid-base, precipitation, adsorption, and oxidation-reduction. All types are used regularly for environmental analysis. Because of the variety, it is ordinarily best to discuss indicators according to type or type reactions. The major type reactions are discussed in the following.

Acid and Base Titrations

Before proceeding with a discussion of the choice of indicators for measurements involved in acid and base titrations, it is best to review some of the theoretical aspects developed in Sec. 4.5. In the first place, the titrating agent used is always a strong acid or a strong base, meaning, of course, highly ionized; therefore, the combinations involved in titrations are strong plus strong, or weak plus strong. It is important that the pH changes occurring during acid and base titrations be understood. Several examples are presented in Sec. 4.5.

The pH of equivalence points involved in acid and base titrations vary, depending upon the ionization constants and concentrations of the materials used. For this reason, the use of a potentiometric indicator (pH meter) is superior for measuring end points in acid or base titrations, and is standard practice in most laboratories. Such titrations are commonly referred to as *potentiometric titrations* and require the use of a standard pH meter.

Color Indicators   A number of naturally occurring or synthetically prepared organic compounds undergo definite color changes in well-defined pH ranges. In general these compounds are weak acids or bases that change color when changed from the neutral to the ionized form. The pH at which the color change takes place depends upon the ionization constant for the particular indicator. A number of indicators that are useful for various pH ranges are listed in Table 11.2.

From the titration curves described in Sec. 4.5, it will be noted that the concept of neutrality at pH 7 has little application in acid or base titrations. Therefore, there is little common need for an indicator for this range. Only the curves for titration of strong versus strong can be said to have inflections at pH 7. It will

Table 11.2 | Properties of various acid-base indicators

Indicator	Acid color	Base color	pH Range
Methyl violet	Yellow	Violet	0–2
Malachite green (acidic)	Yellow	Blue-green	0–1.8
Thymol blue (acidic)	Red	Yellow	1.2–2.8
Bromphenol blue	Yellow	Blue	3.0–4.6
Methyl orange	Red	Yellow-orange	3.1–4.6
Bromcresol green	Yellow	Blue	3.8–5.4
Methyl red	Red	Yellow	4.4–6.2
Litmus	Red	Blue	4.5–8.3
Bromthymol blue	Yellow	Blue	6.0–7.6
Phenol red	Yellow	Red	6.8–8.4
Metacresol purple	Yellow	Purple	7.6–9.2
Thymol blue (alkaline)	Yellow	Blue	8.0–9.6
Phenolphthalein	Colorless	Red	8.2–9.8
Thymolphthalein	Colorless	Blue	9.3–10.5
Alizarin yellow	Yellow	Lilac	10.1–11.1
Malachite green (alkaline)	Green	Colorless	11.4–13.0

be noted that the titration curves for all acids with ionization constants greater than 10^{-7} show inflection points at or below pH 8.3. Thus, the stoichiometric end point for all such measurements can be said to have been reached at pH 8.3, and any indicator that gives a well-defined color change at such a pH is satisfactory for measuring acids. Phenolphthalein changes from colorless to pink in the pH range 8.2 to 8.4 and has been the color indicator commonly used in environmental analysis. Phenolphthalein is also of great value in that it may be used to indicate the end point when strong bases (caustic alkalinity) are being measured, because it changes from pink to colorless at a pH of about 8.3. It is also used to measure carbonate ion (carbonate alkalinity) by indicating when the carbonate ion has been converted to bicarbonate ion, and for carbon dioxide, showing when all the carbonic acid (CO_2) has been converted to bicarbonate. An alternative sometimes preferred to phenolphthalein as a pH 8.3 indicator because of sharper color change is metacresol purple.

The measurement of weakly basic substances requires the use of an indicator that will change color at pH levels of 3.7 to 4.5. The indicator commonly used in the past for such purposes was methyl orange. It does not yield well-defined color changes that are easily detected by all people and is readily bleached by chlorine. Other indicators have been proposed as a substitute but most suffer from interference by carbon dioxide. This is particularly true of methyl red. "Standard Methods" now suggests as alternatives to methyl orange that bromphenol blue be used as a pH 3.7 indicator for acidity measurements and bromcresol green or mixed bromcresol green-methyl red be used as a pH 4.5 indicator for total alkalinity measurements. The best substitute, however, is potentiometric measurement.

Precipitation Methods

The best example of a volumetric method involving formation of a precipitate in environmental analysis is the determination of chloride (chloride ion) by titration with silver nitrate. The indicator commonly used is potassium chromate (K_2CrO_4). Like Cl^-, chromate ion (CrO_4^{2-}) also forms a precipitate with Ag^+.

$$2Ag^+ + CrO_4^{2-} \rightarrow Ag_2CrO_4(s) \tag{11.5}$$

Silver chromate is red in color, and its appearance is used to show completion of the precipitation of Cl^-.

In order for CrO_4^{2-} to serve in this capacity, the solubility of $Ag_2CrO_4(s)$ must be sufficiently greater than that of AgCl so that essentially all chloride ions are precipitated before detectable amounts of $Ag_2CrO_4(s)$ are formed. This means that the *effective* solubility product (K_{sp}) of $Ag_2CrO_4(s)$ must be slightly greater than that of AgCl(s). Since indicators of this kind require an excess of reagent to form enough colored precipitate for visual detection, determination of this excess, commonly referred to as *indicator error* or blank, must be made and applied to all titrations.

Oxidation-Reduction Methods

Two types of indicators are commonly used: adsorption and those that change with oxidation-reduction potential. The former is illustrated by the use of starch solution to indicate the end point when solutions of iodine are titrated with sodium thiosulfate. Titration is "by eye" until the iodine concentration is near extinction, as shown by a pale yellow color. Upon the addition of a good starch indicator a blue solution results. This is due to adsorption of iodine upon the surface of the colloidal starch particles. As the titration proceeds, iodine is released from the starch, and disappearance of the blue color is taken as the end point. Other adsorption indicators show when an excess of titrant has been added. Starch acts in this manner when iodine solutions are used as the titrant.

Internal indicators that will change color with a change of oxidation-reduction potential (ORP) are commonly used. Such indicators are usually soluble organic substances that exist in two states of oxidation, the two forms being in equilibrium and of markedly different color. An example is Ferroin (ferrous 1,10-phenanthroline sulfate), which is used to indicate when sufficient ferrous ammonium sulfate titrant has been added to measure excess dichromate ion in the chemical oxygen demand (COD) test. It should be mentioned that selection of an ORP indicator depends upon the ORP at the stoichiometric end point for the particular reaction involved, just as the selection of an indicator for acidimetry or alkalimetry depends upon the pH of the solution resulting at the equivalence point.

The end point of an oxidation-reduction reaction can be determined electrometrically, most frequently using the amperiometric technique (Sec. 12.3). Different electrode systems may be used, such as a rotating platinum electrode as employed in the Wallace & Tiernan amperometric titrator (Fig. 11.2) or a system employing twin platinum microelectrodes.

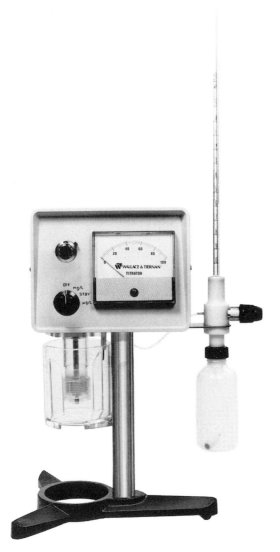

Figure 11.2
An amperometric titrator. *(Courtesy of US Filter Wallace and Tiernan Products.)*

Calculations

The data obtained during a titration must be translated into terms of mass to be of practical value. Since the unit of volume commonly used for measuring the amount of titrant added to a sample is the milliliter, the weight of active material per milimeter is important. For the normal system of standard solutions, one equivalent weight of a substance, in grams, in a liter of solution is a 1.0 normal solution. Each milliliter of such a solution contains one one-thousandth of an equivalent weight or what is more commonly referred to as 1 milliequivalent (meq). Thus, when working with solutions of normality N, we find that

$$\text{mL titrant} \times N = \text{meq of active material in titrant used}$$

The advantage of using equivalent solutions is that the meq of active material in the titrant used is equal to the meq of active material in the sample being titrated. In order to determine the concentration of active material in a sample, one must know the sample volume:

$$\text{meq/liter active material in sample} = \frac{\text{mL titrant} \times N \times 1000}{\text{sample volume in mL}} \quad (11.6)$$

In environmental chemistry, it is common practice to report the concentration of a material in mass units rather than in equivalent units. Also, the concentration of materials measured is usually so small that it is inconvenient to think in terms of grams. Therefore, most thinking is done in terms of milligrams, and results are expressed in terms of milligrams per liter (mg/L). The number of milligrams of active material in a meq is equal to the equivalent weight (EW) of the active material, so

$$\text{mg/L of active material in sample} = \frac{\text{mL titrant} \times N \times EW \times 1000}{\text{sample volume in mL}} \quad (11.7)$$

Samples are commonly measured by volume, and to avoid unnecessary and repeated calculations, the sample size and the normality of titrant are often chosen so that the buret reading in milliliters times some whole number, such as 1, 10, 20, 50, or 100, gives the number of milligrams of material per liter. This requires the use of solutions whose normality is some exact value. The preparation, standardization, and use of such solutions will be discussed in Chap. 15.

11.5 | COLORIMETRY

Analytical chemists as well as others who use analytical procedures are constantly striving to find faster, more economical, and convenient ways of obtaining quantitative data. To this end, colorimetric methods of analysis have been developed for many determinations of interest to environmental engineering and science. Colorimetric methods are most applicable in the realm of dilute solutions. This is fortunate, because a large majority of environmental samples fall in this classification.

In order for a colorimetric method to be quantitative, it must form a compound with definite color characteristics and in amounts directly proportional to the concentration of the substance being measured. Solutions of the colored compound or complex must have properties that conform to Beer's law and to Lambert's law.

Lambert's Law

Lambert's law, sometimes referred to as Bouguer's law, relates the absorption of light to the depth or thickness of the colored liquid. This law states that each layer of equal thickness absorbs an equal fraction of the light that traverses it. Thus, when a ray of monochromatic light passes through an absorbing medium, its intensity decreases exponentially as the length of the medium increases,

$$T = \frac{I}{I_0} = 10^{-a_1 l} \qquad (11.8)$$

or

$$A = \log \frac{I_0}{I} = a_1 l \qquad (11.9)$$

where
I_0 = intensity of light entering solution
I = intensity of light leaving solution
l = length of absorbing layer
a_1 = constant for particular solution
T = transmittance of solution
$100T$ = percentage transmission of solution
A = absorbance, or optical density of solution

If light intensity decreases exponentially with increase in depth or thickness, the colored solution behaves in conformity with this law. There are no known exceptions to this law as long as homogeneous materials are involved. This law is discussed solely because a knowledge of it is germane when the depth or thickness of colored samples is varied to decrease a color to a level in the range of a series of prepared standards or of spectrophotometric equipment. Knowledge and application of this law can save much time and expense.

Beer's Law

Beer's law is concerned with light absorption in relation to solution concentration. It states that the intensity of a ray of monochromatic light decreases exponentially as the concentration of the absorbing medium increases,

$$T = \frac{I}{I_0} = 10^{-a_2 c} \qquad (11.10)$$

or

$$A = \log \frac{I_0}{I} = a_2 c \qquad (11.11)$$

where a_2 is a constant for the particular solution and c is the concentration of the solution. If light is absorbed exponentially with concentration over a reasonable and practical range of concentration, the colored material is said to conform to Beer's law. The best way to determine whether a colored compound or complex obeys Beer's law is to prepare a series of samples in the desired range of concentration and submit them to test on a photoelectric colorimeter or spectrophotometer. If the observations of percent light transmission plot along a straight line on a semilog graph, the material can be considered to obey Beer's law. Many colored systems do not conform to Beer's law, and therefore development of any new colorimetric method should involve such a test procedure.

By combining the two absorption laws, the Lambert-Beer or Bouguer-Beer law is obtained:

$$T = \frac{I}{I_0} = 10^{-a'cl} \tag{11.12}$$

or
$$A = \log \frac{I_0}{I} = a'cl \tag{11.13}$$

where a' is the *absorptivity* constant.

It follows from the Lambert-Beer law that if light of the same intensity enters two different solutions and adjustments of depths are made so that the emerging beams are of the same intensity, then the transmissions are the same, and

$$c_1 l_1 = c_2 l_2 \tag{11.14}$$

Equation (11.14) indicates that if the depth of a sample is varied so that the intensity of color matches that of a standard, then the sample concentration is related to the standard concentration by the ratio of their depths.

Color-Comparison Tubes

Colorimetric measurements may be made in a wide range of equipment. Examples are standard color-comparison tubes, photoelectric colorimeters, or spectrophotometers. Each has its place and particular application in water and wastewater analysis.

Color-comparison tubes, sometimes referred to as Nessler tubes, have been the standard equipment for making colorimetric measurements for many years. Tubes of this type are shown in Fig. 14.1. Their use has largely been replaced, however, because of the convenience of photoelectric and spectrophotometric methods. Precise work with color-comparison tubes requires that tubes of matched size or bore be used in order to comply with Lambert's law. The chief difficulty with their use is that standard color solutions are seldom stable, and every time a determination has to be made it becomes necessary to prepare a series of fresh standards. This adds greatly to the labor and time required. Another objection is that all comparisons are made by eye, and the "human error" involved is often considerable because sensitivity to different colors varies. Furthermore, the analyst is required to interpolate values between standards.

Photoelectric Colorimeters

Photoelectric colorimeters have been used quite extensively in colorimetric work and are very satisfactory within their limitations. They make use of an electrometric device employing a photoelectric cell as the sensing element. The current developed by the photoelectric cell is translated into percent transmission or absorbance through a suitable galvanometer. The light source is an ordinary light bulb, and monochromatic light is obtained by allowing a beam of light to pass through a color filter. The monochromatic light is directed through a cell containing the sample, and

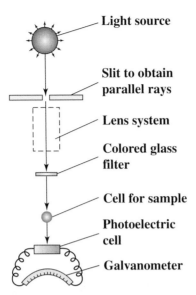

Light source

Slit to obtain parallel rays

Lens system

Colored glass filter

Cell for sample

Photoelectric cell

Galvanometer

Figure 11.3
Schematic diagram of a photoelectric colorimeter.

Figure 11.4
A photoelectric colorimeter. *(Courtesy of Hach Company, USA.)*

the light that penetrates hits the photoelectric cell. The instrument is adjusted to yield a light transmission corresponding to 100 percent with the cell containing a "blank sample." The "blank sample" is a portion of distilled or deionized water that has been treated in the same manner as regular samples. A schematic diagram of the essentials of a photoelectric colorimeter is shown in Fig. 11.3, and a typical instrument is shown in Fig. 11.4.

Photoelectric colorimeters require a separate color filter for each different chemical determination on which they are to be used; thus, the investment may become considerable, and the range of application is somewhat limited. They are not suitable for research purposes and should be considered for use principally where a very few well-established chemical determinations are involved.

Spectrophotometers

The modern spectrophotometer employing either a glass prism or a diffraction grating to produce monochromatic light is an extremely valuable instrument for colorimetric analyses. It has a wide range of adaptability that allows selection of monochromatic light of any wavelength in the visible spectrum. In addition, some instruments provide light in the ultraviolet and near-infrared regions. One filter usually suffices for the entire visible range of wavelengths. Separate filters are needed for the ultraviolet and for the infrared regions. The principle upon which spectrophotometers are based is the same as that for photoelectric colorimeters, except for the manner in which the monochromatic light is obtained. Figure 11.5 shows a schematic diagram of a spectrophotometer, and Fig. 11.6 shows a typical instrument.

A spectrophotometer is particularly recommended where a wide variety of determinations is made. Its versatility allows the best wavelength of light to be used at all times. The optimum wavelength can be determined at any time by establishing a spectral transmission curve. This is an essential part of research aimed at developing new methods of colorimetric analysis. The curve is established by making a series of observations of light transmission at several different wavelengths of light while using a typical colored solution in the cell. The results when plotted on linear coordinate paper yield a curve like that shown in Fig. 11.7 for nitrite. The wavelength that is absorbed to the greatest extent, in the case of nitrite 525 nanometers (nm), is the optimum wavelength to use.

Calibration and Use

Photoelectric colorimeters and spectrophotometers are calibrated for use in any particular determination by preparing a series of standards *in the same manner as regular tests are to be run* and by making observations on light transmission, using the wavelength specified for the determination. When such data are plotted on semilog paper, the curve should be essentially a straight line. The calibration curve, when carefully prepared, should serve for years and eliminate the need for the preparation of a series of standards. A typical calibration curve is shown in

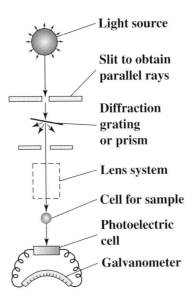

Light source

Slit to obtain
parallel rays

Diffraction
grating
or prism

Lens system

Cell for sample

Photoelectric
cell

Galvanometer

Figure 11.5
Schematic diagram of a spectrophotometer.

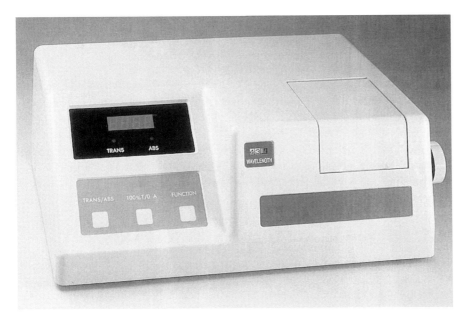

Figure 11.6
A spectrophotometer. *(Courtesy of Fisher Scientific Company, Pittsburgh, PA.)*

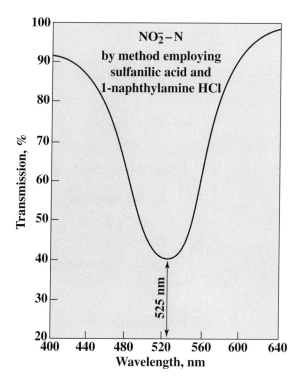

Figure 11.7
Spectral transmission curve for nitrite nitrogen, showing
optimum wavelength of light for photometric determination.

Fig. 11.8. It is usually good practice to include a standard in each set of samples, however, to make sure that unknown errors in reagents and the like do not lead to faulty results.

It should be emphasized that instrumental analysis does not necessarily ensure accurate results. Analysts must be constantly on guard to be sure that their instruments are in good working order, cells are kept scrupulously clean, and turbidity of samples is eliminated.

11.6 | PHYSICAL METHODS OF ANALYSIS

Environmental engineers and scientists often employ other methods of analysis that are based upon physical measurements. Sensitive instruments are frequently employed to make these measurements. Such instruments not only allow rapid measurements to be made, but permit identification of both organic and inorganic materials occurring in extremely small concentration. Instrumental methods are also adaptable for the continuous monitoring of rivers and waste streams so that significant fluctuations in concentration can be recorded for evaluation. Because of their

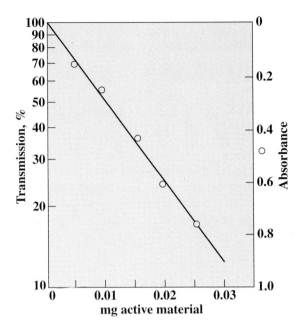

Figure 11.8
A typical calibration curve for photoelectric or spectrophotometric analysis.

importance, a separate chapter (Chap. 12) is included in this book to describe the various instrumental methods of particular value.

Some physical measurements commonly made in water and wastewater analysis do not require the use of expensive instruments. Inexpensive methods used to evaluate turbidity are discussed next. Sophisticated instruments relying on these principles are also used for more detailed characterization of suspended particles in water.

Turbidimetry

The method in which ordinary white light transmitted through a finely divided suspension is compared with that transmitted by a standard suspension is known as turbidimetry. While the original methods used to measure water turbidity were based upon this principle, the alternative approach (nephelometry) is now considered the accepted standard method for evaluating turbidity in water. However, sulfate concentration may still be determined in one approach through use of the normal turbidity measurement.

Nephelometry

Nephelometric methods are also employed to measure turbidity. In this method, light is allowed to strike a suspension at right angles to the eye of the observer or photoelectric cell of the instrument. The light reflected by the dispersed particles

(Tyndall effect) is recorded. When this principle is employed, very low turbidities can be determined in filtered water, and this is one of the major advantages of nephelometry. Bacteriologists often use nephelometry in following bacterial growth rates.

11.7 | PRECISION, ACCURACY, AND STATISTICAL TREATMENT OF DATA

As a reminder to reinforce concepts presented in Chap. 10, every analyst should have a clear understanding of the differences between precision and accuracy insofar as analytical results are concerned. A knowledge of statistics is of considerable importance as an aid to establishing sampling programs so that data obtained may be subjected to statistical treatment when necessary.

PROBLEMS

11.1 Under what circumstances may composite samples be preferred over grab samples, and for what reasons?

11.2 Briefly describe the differences between volumetric, colorimetric, and gravimetric analyses.

11.3 What are the requirements to make precipitation methods reliable?

11.4 What precautions must be taken in the use of crucibles or dishes in gravimetric work?

11.5 To what temperature should dishes be pretreated in order to condition them prior to gravimetric analysis?

11.6 Explain the significance of the drying and ignition temperatures of 103 and 600°C, respectively, in analysis of environmental samples.

11.7 Explain why it is necessary (*a*) to cool a sample to room temperature before weighing, and (*b*) to allow the sample to cool in a desiccator.

11.8 What facilities are needed for volumetric analysis?

11.9 What two different types of calibrated glassware are available for volumetric analysis?

11.10 What precautions must be taken with the use of a volumetric pipet?

11.11 What is a standard solution? What is an equivalent solution?

11.12 What is the difference between a primary standard and a secondary standard?

11.13 Determine the equivalent weights of the following materials in grams: (*a*) H_2SO_4; (*b*) HCl; (*c*) $Ca(OH)_2$; (*d*) CH_3COOH.
   Answer: (*a*) 49; (*b*) 36.5; (*c*) 37; (*d*) 60

11.14 Determine the equivalent weights of the following materials in grams:

 (*a*) $KMnO_4$ as an oxidizing agent in acid solutions

 (*b*) Ag_2SO_4 in a precipitation reaction involving silver

 (*c*) $K_2Cr_2O_7$ as an oxidizing agent

 (*d*) $BaCl_2$ in a precipitation reaction involving barium

11.15 How many grams of $AgNO_3$ are required to prepare 500 mL of a 0.1 N solution to be used in a precipitation reaction?

Answer: 18.5 g

11.16 How many grams of $Fe(NH_4)_2(SO_4)_2 \cdot 6H_2O$ are required to prepare 1 liter of a 0.25 N solution to be used in a reduction reaction?

11.17 Name five different types of indicators for volumetric analyses and give an example of each.

11.18 Are strong or weak acids and bases used in acid and base titrations, and why?

11.19 (*a*) Give an example of a potentiometric indicator. (*b*) Are potentiometric or color indicators generally preferred for pH measurements, and why?

11.20 What different types of indicators are available for oxidation-reduction methods?

11.21 A 50-mL sample containing $Ca(OH)_2$ requires 10 mL of 0.02 N H_2SO_4 to reach the equivalence point in an acid-base titration. Determine the concentration in mg/L of $Ca(OH)_2$ in the sample.

Answer: 148 mg/L

11.22 A 100-mL sample containing chloride ion is titrated with $AgNO_3$ in a precipitation reaction. Calculate the concentration of chloride ion in mg/L if 10 mL of 0.01 N $AgNO_3$ is required to reach the equivalence point.

11.23 (*a*) What is the difference between Lambert's law and Beer's law? (*b*) What is required of a solution for it to follow Beer's law?

11.24 If the color in a sample of drinking water viewed through a depth of 10 cm corresponds in intensity to a standard having a color of 40 units and viewed through a depth of 30 cm, how many color units does the water sample contain?

11.25 How are colorimetric methods calibrated?

11.26 Many spectrophotometers have two adjacent scales, one registering percent transmission and the other registering absorbance. What numerical value on the latter scale corresponds to 70 on the former?

11.27 A sample of drinking water is analyzed colorimetrically for ammonia nitrogen concentration. If the sample color when viewed through a depth of 5.5 cm corresponds in intensity to a 1 mg/L standard sample when viewed through a depth of 40 cm, what is the concentration of ammonia nitrogen in the sample?

Answer: 7.3 mg/L

11.28 If a sample containing 0.1 mg/L of nitrite-nitrogen gives 85 percent transmission when submitted to colorimetric analysis, what percent transmission should a sample containing 0.2 mg/L produce if the analysis follows Beer's law?

11.29 What are two different methods for making turbidity measurements in water, and how do they differ?

REFERENCES

Clesceri, L. S., A. E. Greenberg, and A. D. Eaton (eds.): "Standard Methods for the Examination of Water and Wastewater," 20th ed., American Public Health Association, Washington DC, 1998.

Day, Jr., R. A., and A. L. Underwood: "Quantitative Analysis," 6th ed., Prentice Hall, Englewood Cliffs, NJ, 1991.

Enke, C. G.: "The Art and Science of Chemical Analysis," John Wiley & Sons, New York, 2001.

Fifield, F. W., and P. J. Haines (eds.): "Environmental Analytical Chemistry," 2nd ed., Blackwell Science, Malden, MA, 2000.

Rubinson, J. F., and K. A. Rubinson: "Centemporary Chemical Analysis," Prentice Hall, Englewood Cliffs, NJ, 1998.

Skoog, D. A., D. M. West, and F. J. Holler: "Fundamentals of Analytical Chemistry," 7th ed., College Pub., Fort Worth, TX, 1996.

CHAPTER 12

Instrumental Methods of Analysis

12.1 | INTRODUCTION

A variety of sensitive instruments have been developed in recent years that have increased considerably the analyst's ability to measure and to cope with pollutional materials of increasing complexity. Instrumental methods of analysis are finding wide usage for routine monitoring of air quality, groundwater and surface water quality, and soil contamination, as well as during the course of water or waste treatment. Such methods have allowed analytical measurements to be made immediately at the source and have permitted the recording of such measurements to be made at some distance from the place of actual measurement. In addition, they have extended considerably the range of inorganic and organic chemicals that can be monitored and the concentrations that can be detected and quantified. A variety of instrumental methods are now routinely used both for investigating the extent of contamination and for routine monitoring of treatment effectiveness.

The pH meter and spectrophotometer are analytical instruments that were developed several decades ago and have become indispensable in environmental control. They were described in Chap. 11. Some of the instruments of more recent origin are rapidly becoming of equal value.

Nearly any physical property of an element or compound can be used as the basis for an instrumental measurement. The ability of a colored solution to absorb light, the capacity of a solution to carry a current, or the ability of a gas to conduct heat can all be used as the basis for an analytical method to measure the quantity of a material as well as to detect its presence. Some of the instrumental methods of analysis come under the heading either of optical methods or of electrical methods. Other highly useful methods are based on the measurement of other properties of materials, such as their tendency to partition between solid, liquid, and gaseous phases. Some of the more common instrumental methods of analysis, their methods of operation, and their major uses are briefly discussed in the following sections.

12.2 | OPTICAL METHODS OF ANALYSIS

Optical methods of analysis measure the results of interactions between radiant energy and matter. The first instruments here were developed for use in the visible region and were thus called *optical instruments*. Now this term is used for instruments measuring a much broader range of radiant energy, although such broad usage is not strictly correct. The radiant energy used in such measurements may vary from X rays, through visible light, to radio waves. The parameter most frequently used to characterize radiant energy is the wavelength, which is the distance between adjacent crests of the wave in a beam of radiation. Wavelength is normally measured in angstroms (Å, $1 \text{ Å} = 10^{-8}$ cm), or in nanometers (nm, $1 \text{ nm} = 10^{-7} \text{ cm} = 10 \text{ Å}$). Figure 12.1 indicates the normal range of wavelength for various types of radiant energy.

Radiant energy may also be considered to consist of *photons* or packets of energy that can interact with matter. The energy of a photon is related to its wavelength λ as follows:

$$E = \frac{hc}{\lambda} \tag{12.1}$$

where h is Planck's constant and is equal to 6.63×10^{-27} erg $\cdot$ s, and c is the velocity of light, which is equal to 3×10^{10} cm/s. This equation indicates that X rays, with short wavelength, are relatively high in energy and for this reason they can cause marked changes in matter. Microwaves and radio waves, on the other hand, have long wavelengths and are relatively low in energy. The changes that they cause upon interaction with matter are quite small and difficult to detect. Use of radiation of different energy content permits the determination of different properties of materials.

Optical methods of analysis may be designed to measure the ability of a material or solution to *absorb* radiant energy, to *emit* radiation when excited by an energy source, or to *disperse* or *scatter* radiation. Methods based upon these three principles will be described separately.

Absorption Methods

When a source of radiant energy, such as a beam of white light, is passed through a solution, the emergent beam will be lower in intensity than that entering. If the solu-

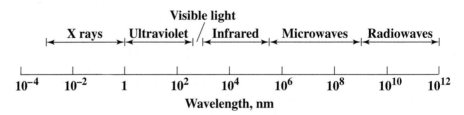

Figure 12.1
Range of wavelengths for different types of radiant energy.

tion does not contain suspended particles that scatter light, then the reduction in intensity is due primarily to *absorption* by the solution. The extent of absorption of white light is generally greater for some colors than for others, with the result that the emerging beam is colored.

The use of a spectrophotometer to determine the extent of absorption of various wavelengths of visible light by a given solution is discussed in Sec. 11.5, and the result of such a measurement is illustrated in Fig. 11.7. Colorimetry, of course, refers to the visible region of the spectrum, but the same principles apply to other regions. Thus, analytical methods based on absorption of ultraviolet or infrared radiation are also in common usage.

All instruments designed to measure the absorption of radiant energy have the basic components indicated in Fig. 12.2, which include an *energy source* to provide radiation of the desired wavelength, an *energy spreader* which permits separation of radiation of the desired wavelength from other radiation, and an *energy detector* which measures the portion of radiation that passes through the sample. Instruments for such measurements may vary from the simple and relatively inexpensive colorimeters discussed in Sec. 11.5 to highly complicated and expensive spectrophotometers that automatically scan the ability of a solution to absorb radiation over a wide range of wavelengths and automatically record the results of these measurements. One instrument cannot be used to measure absorbance at all wavelengths because a given energy source, energy spreader, and energy detector is suitable for use over only a limited range of wavelengths. Instruments that measure the absorption of visible light are widely used in water analysis. However, instruments measuring the absorption of ultraviolet or infrared radiation are finding increasing usage and will be briefly discussed.

Ultraviolet Spectroscopy   When a molecule absorbs radiant energy in either the ultraviolet or the visible region, valence or bonding electrons in the molecule are

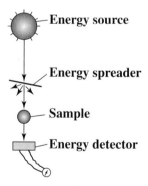

Figure 12.2
Basic components of
instruments for measurement
of absorption.

raised to higher-energy orbits. Some smaller molecular changes can also take place, but these are usually masked by the above electronic excitations. The result is that fairly broad absorption bands are usually observed in both the ultraviolet and the visible regions. Many instruments are designed to make measurements in both regions. A typical instrument of this type is shown in Fig. 12.3. The ultraviolet region is of more limited general usage, although it is particularly suitable for the selective measurement of low concentrations of organic compounds such as benzene-ring-containing compounds or unsaturated straight-chain compounds containing a series of double bonds.

Infrared Spectroscopy   Nearly all organic chemical compounds show marked selective absorption in the infrared region. However, infrared spectra are exceedingly complex compared to ultraviolet or visible spectra. Infrared radiation is of low energy, and its absorption by a molecule causes all sorts of subtle changes in the vibrational or rotational energy of the molecule. An understanding of these changes requires some knowledge of quantum mechanics, which is beyond the scope of this book. A person who is particularly knowledgeable in this area can use infrared spectra to identify particular atomic groupings that are present in an unknown molecule. To illustrate, the absorption of short-wavelength infrared radiation causes vibrations of hydrogen atoms in molecules, longer wavelengths cause vibrations of triple bonds, while still longer wavelengths cause vibrations of double bonds. Because of the complexity of infrared spectra, it is highly unlikely that any two different compounds will have identical curves. This fact has made infrared spectrophotometry a valuable aid in the identification of pesticides and other complex organic chemicals that have been extracted from environmental samples.

While infrared analysis gives much more information about a molecule than either ultraviolet or visible analysis, it is less sensitive, and thus a relatively high concentration (10^{-4} to 10^{-5} molar) of the substance to be analyzed is needed. Another problem stemming from the complex infrared spectra produced by even simple

Figure 12.3
An ultraviolet and visible-range spectrophotometer with recorder and computer and graphics interface. *(Courtesy of the Perkin-Elmer Corp.)*

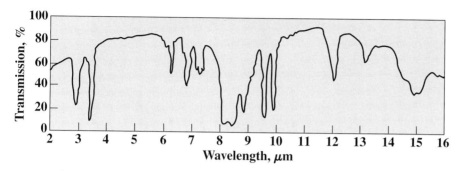

Figure 12.4
Infrared spectrogram of sodium tetrapropene benzene sulfonate.

molecules is that when many materials are present in a solution, interferences between their individual spectra make the use of infrared analysis for identification or quantification almost impossible. For this reason a lengthy separation procedure is often required to isolate compounds of interest from interferences prior to analysis. A typical infrared spectrum of an anionic detergent, tetrapropene benzene sulfonate, is shown in Fig. 12.4.

In principle, quantitative infrared analysis is the same as in the ultraviolet or visible regions. By examining the spectra of a pure substance, a wavelength may be found at which absorption is considerably greater than for other compounds present in the mixture. For analysis it is simply necessary to measure absorbance at the selected wavelength. Use is made of this principle in some instruments designed for the rapid measurement of total organic carbon (TOC), when only small quantities of organic carbon are present in water. Here a liquid sample is injected into a furnace, where the water is evaporated and the organic carbon catalytically burned to carbon dioxide. The carbon dioxide is carried by a stream of oxygen through an infrared analyzer that is specifically designed to measure and record the concentration of carbon dioxide present. Actually, total carbon is measured by this procedure, but if the sample is first prepared by acidification and aeration to remove inorganic carbon, then a selective measurement of the organic carbon present can be obtained. An instrument used to measure TOC is illustrated in Fig. 12.5.

Emission Methods

Several heavy metals are of environmental concern because of adverse effects they cause to human health and the environment. Thus, analytical methods for their detection and quantification are of great interest. It has long been known that many metallic elements, when subjected to suitable excitation, will emit radiations of characteristic wavelengths. This is the basis for the familiar flame test for sodium (which emits a yellow light) and for other alkali and alkaline-earth metals. When a much more powerful method of excitation is used in place of the flame, most metallic and some nonmetallic elements will emit characteristic radiations. Under prop-

Figure 12.5
A total organic carbon (TOC) analyzer. *(Courtesy of Shimadzu Scientific Instruments, Inc.)*

erly controlled conditions, the intensity of the emitted radiation at some particular wavelength can be correlated with a quantity of the element present. Thus, both a quantitative and a qualitative determination can be made. The various analytical procedures that make use of emission spectra are characterized by the excitation method used, the nature of the sample (whether solid or liquid), and the method of detecting and recording the spectra produced. The more widely used methods are discussed here.

Atomic Emission Spectroscopy This method in its simplest form as a flame photometer is used in water analysis for determining the concentration of alkali and alkaline-earth metals such as sodium, potassium, and calcium. A diagram showing the basic elements of a flame photometer is given in Fig. 12.6, and a typical instrument is shown in Fig. 12.7. A liquid sample to be analyzed is sprayed under controlled conditions into a flame where the water evaporates, leaving the inorganic salts behind as minute particles. The salts decompose into constituent atoms or radi-

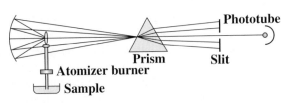

Figure 12.6
Schematic diagram of a flame photometer.

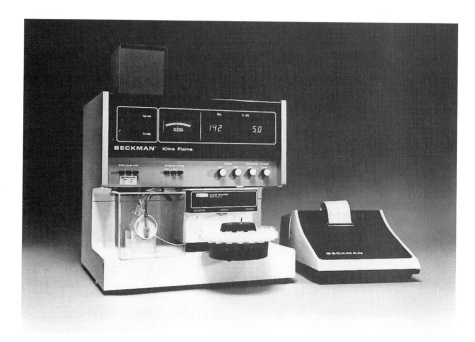

Figure 12.7

A flame photometer. *(Courtesy of Beckman Instruments, Inc.)*

cals and may become vaporized. The vapors containing the metal atoms are excited by the thermal energy of the flame, and this causes electrons of the metallic atoms to be raised to higher energy levels. When the electrons fall back to their original positions or to a lower level, they give off discrete amounts of radiant energy. The emitted radiation is passed through a prism, which separates the various wavelengths so that the desired region can be isolated. A photocell and some type of amplifier are then used to measure the intensity of the isolated radiation.

A number of fuel gases, such as acetylene, hydrogen, or the normal natural or manufactured gas used for heating in most laboratories can be used for the flame. The oxidant used is usually tank oxygen rather than air. Flames yielding higher temperatures are capable of exciting more elements and so are more versatile.

The emission spectrum for each metal is different, and its intensity depends upon the concentration of atoms in the flame, the method of excitation used, and the after-history of the excited atoms. Sodium produces a characteristic yellow emission at 589 nm, lithium a red emission at 671 nm, and calcium a blue emission at 423 nm. Each also gives a less intense emission at shorter wavelengths. Concentrations of these elements can be measured down to 1 mg/L or less with some degree of accuracy, depending upon the sensitivity of the instrument used. Detection limits by flame photometry for some elements are listed in Table 12.1 as indicated in "Standard Methods."

Table 12.1 | Partial list of emission methods of analysis listed in "Standard Methods" for metals and related elements and their detection limits (μg/L)

Element	Flame	Atomic absorption Direct aspiration	Electrothermal	ICP*	ICP/MS[†]
Aluminum		100	3	40	0.03
Antimony		70	3	30	0.07
Arsenic			1	50	0.025
Barium		30	2	2	0.008
Beryllium		5	0.2	0.3	0.025
Bismuth		6			
Boron				5	
Cadmium		2	0.1	4	0.006
Calcium		3		10	
Cesium		20			
Chromium		20	2	7	0.04
Cobalt		30	1	7	0.002
Copper		10	1	6	0.004
Germanium		10			
Iridium		600			
Iron		20	1	7	
Lead		50	1	40	0.005
Lithium	0.1	2		4	
Magnesium		0.5		30	
Manganese		10	0.2	2	0.002
Mercury		1[a]			
Molybdenum		100	1	8	0.003
Nickel		20	1	15	0.004
Osmium		80			
Platinum		100			
Potassium	100	5		100	
Rhodium		500			
Ruthenium		70			
Selenium			2	75	0.093
Silicon		300		20	
Silver		10	0.2	7	0.002
Sodium	5	2		30	
Strontium	200	30		0.5	0.001
Thallium				40	0.03
Tin		800	5		
Titanium		300			
Uranium					0.032
Vanadium		200	8		0.02
Zinc		5	2		0.02

*ICP = inductively coupled plasma (spectroscopy).
[†]ICP/MS = ICP instruments connected with mass spectrometers.
[a]Mercury when analyzed by cold-vapor atomic absorption spectroscopy; the detection limit for this method is not listed in "Standard Methods," but it is about 1 μg/L.

Atomic Absorption Spectroscopy   Although this is really an absorption method, it is included under emission spectroscopy because of its similarity to flame photometry. Atomic absorption spectroscopy has gained wide application for environmental analysis because of its versatility for the measurement of trace quantities of most elements in water. Elements such as copper, iron, magnesium, nickel, and zinc can be measured accurately to a small fraction of a milligram per liter as indicated in Table 12.1.

The similarity between atomic absorption spectroscopy and flame photometry is indicated by a comparison of the illustrations in Figs. 12.6 and 12.8. Like the flame photometer, an atomic absorption system consists of a flame unit into which a sample is directly aspirated, a prism to disperse and isolate the resulting emission lines, and a detector with appropriate amplifiers. In addition, the absorption system has a light source that emits a stable and intense light of a particular wavelength. Each element has characteristic wavelengths that it will readily absorb. A light source with wavelength readily absorbed by the element to be determined is directed through the flame, and a measure of its intensity is made without the sample, and then with the sample introduced into the flame. The decrease in intensity observed with the sample is a measure of the concentration of the element. A disadvantage of this method is that a different light source must be used for each element. Fortunately, sources such as hollow cathode lamps are available for most naturally occurring elements. A typical atomic absorption spectrophotometer is shown in Fig. 12.9.

The advantage of atomic absorption spectroscopy is that it is quite specific for many elements. Absorption depends upon the presence of free unexcited atoms in the flame, and these are in much greater abundance than excited atoms. Thus, elements, such as zinc and magnesium, that are not easily excited in flames and so give poor results with the flame photometer, can be readily measured by the atomic absorption method.

A modification to this, termed electrothermal atomic absorption spectroscopy, employs a small graphite furnace, which allows analysis for many heavy metals in the lower microgram per liter range as indicated in Table 12.1. Here, the atomizer burner is replaced with a small cylindrical graphite tube that can be programmed through a series of different temperatures. The radiation from the cathode lamp source passes through the open ends of the horizontal cylinder and selected wavelengths are measured by the phototube as with the atomizer-burner method. A small

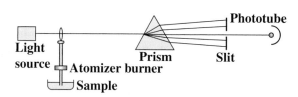

Figure 12.8
Schematic diagram of an atomic absorption spectrophotometer.

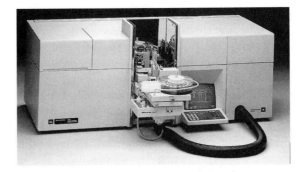

Figure 12.9
An atomic absorption spectrophotometer with graphite
furnace. *(Courtesy of the Perkin-Elmer Corp.)*

quantity (5 to 100 μL) of a water sample is inserted into the cylinder through a hole
in the side and the temperature program is initiated. The temperature first rises to
just over 100°C to allow the sample water to evaporate, leaving the metal containing salts behind. The temperature then increases to several hundred degrees Celsius,
which volatilizes the cations so they fill the cylindrical space, and the particular
cation to be determined absorbs the characteristic radiation from the cathode tube.
The graphite furnace, as it is called, allows the development of a greater density of
atoms and thus effects greater sensitivity to the atomic absorption procedure. Use of
the graphite furnace is now routine for water quality analyses.

Inductively Coupled Plasma Spectroscopy   While the preceding methods until
recently have been the most widely used for water analyses, there are many other
emission methods that have come into wider use and employ more powerful excitation in place of the flame. This can extend analysis to all metallic and many nonmetallic elements. Emission instruments are particularly adapted to the analysis of
solid as well as aqueous samples. Thus, they are frequently employed for analysis
of metals in sludges and other complex wastes. The basic components of most
emission methods are the same as those of the flame photometer. The older emission methods here made use of higher-energy excitation sources such as an alternating current (ac) arc, a direct current (dc) arc, or a high-voltage spark to excite a
greater variety of elements. However, the use of plasma sources, such as in the inductively coupled plasma (ICP) method, is a relatively new addition, having been
developed in the 1970s, and now is represented as a standard method for many metals of interest in water as summarized in Table 12.1. The advantage of these more
energetic atomization sources is there is lower interelement interference, good spectra can be obtained for many elements under the same excitation conditions, and
hence spectra for many elements can be recorded simultaneously. The disadvantages are higher instrument and operating costs, the need for more skilled operators,
and often less precision than with atomic absorption. A typical ICP spectrometer is
shown in Fig. 12.10.

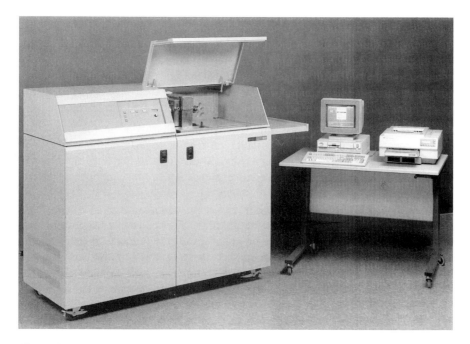

Figure 12.10
An inductively coupled plasma spectrometer. *(Courtesy of the Perkin-Elmer Corp.)*

In the ICP method, a stream of argon gas flows through three concentric quartz tubes, which are surrounded by a water-cooled induction coil that is powered by a radio-frequency generator to form a strong magnetic field. When a spark initiates ionization of the argon, the ions with their associated electrons are caused to follow a spiral flow pattern within the tubes as a result of the magnetic field and heating is the result of their collisions and resistance to this movement. The resulting temperature is 6000 to 8000 K, which is two to three times hotter than obtained with the hottest of the combustion flame temperatures. This temperature is sufficient to almost completely dissociate molecules so that little interference between them results, and atomic emission becomes highly efficient. The sample to be analyzed is introduced at the head of the argon flow and into the central tube. The emissions produced by the elements are focused through an entrance slit for either a monochromator or polychromator, and a portion of the spectrum is isolated for intensity measurements. The wavelength ranges analyzed by instruments may encompass the entire ultraviolet/visible spectrum from 180 to 900 nm. ICP instruments connected with mass spectrometers for detection (ICP/MS) is a newer addition that increases sensitivity of ICP down to the low nanogram per liter range (Table 12.1). Laboratories equipped with ICP are becoming quite common because of their great versatility for handling many samples and providing a broad range of analytical data.

Dispersion and Scattering

The turbidity of a sample may be measured either by its effect on the transmission of light, which is termed *turbidimetry,* or by its effect on the scattering of light, which is termed *nephelometry.* As indicated in Chap. 13, these properties are used in the "Standard Methods" procedures for turbidity measurement. While these procedures use the human eye to detect the light emitted, methods employing common electric photometers can also be employed and have the advantage that continuous measurements of turbidity can be made and recorded. They can also overcome some of the human error in making turbidity observations. Nephelometric measurement is most sensitive for very dilute suspensions, but for moderately heavy turbidity, either this or turbidimetric measurements can be made.

In turbidimetry, as indicated in Fig. 12.11, the amount of light passing through a solution is measured. The higher the turbidity, the smaller the quantity of light transmitted. In nephelometry, on the other hand, the detecting cell is placed at right angles to the light source to measure light scattered by the turbidity particles. Any spectrophotometer or photometer is satisfactory as a turbidimeter without modification. However, a special attachment is required for nephelometry.

Although turbidimetric analysis can be conducted at any wavelength of light, the "Standard Methods" procedure for determining sulfates by turbidimetric analysis recommends a wavelength of 420 nm. This results in a more sensitive analysis, because the blue light of this wavelength is scattered more strongly than red light at longer wavelengths.

Fluorimetry

Many organic and some inorganic compounds have the ability to absorb radiant energy of one wavelength and then to emit the energy as radiation at some longer

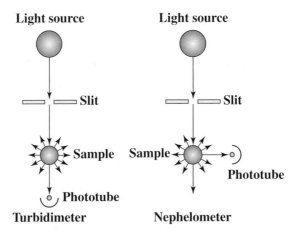

Figure 12.11
Schematic diagram of a turbidimeter and a nephelometer.

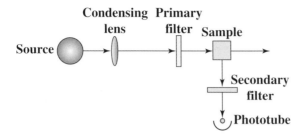

Figure 12.12
Schematic diagram of a spectrofluorometer.

wavelength. This phenomenon is known as *fluorescence,* and it provides the basis for a very sensitive analytical tool. A schematic diagram of a *fluorometer* that uses this principle is shown in Fig. 12.12, and a typical instrument is shown in Fig. 12.13. Fluorescence may be measured with a simple instrument termed a fluorometer, which employs filters for wavelength selection. A spectrofluorometer, however, is a more complex instrument similar to a spectrophotometer, and has a monochromator for obtaining the fluorescence spectra from a sample. In many ways fluorometers and spectrofluorometers are similar to a nephelometer. Ultraviolet or visible radiation is passed first through a filter to exclude unwanted wavelengths, and then through the sample. Fluorescent materials in the sample absorb the radiation and emit visible light, which passes through a second filter placed at right angles to the original radiation source. This filter excludes any scattered radiation and permits the photocell to

Figure 12.13
A spectrofluorometer. *(Courtesy of Hitachi Instruments, Inc.)*

measure only the wavelengths emitted by the sample. This method is extremely sensitive for fluorescent materials. Fluorimetric methods are commonly used to measure chlorophyll content of surface waters as an indicator of algae content and to detect important biochemical compounds such as ATP and F_{430} (Chap. 6).

Another use of fluorimetry in water quality studies has been to follow the movement of water and pollution. This is done by adding highly fluorescent dyes to the water and detecting their movement by fluoroscopic measurements. Rhodamine B and Pontacyl Pink B are dyes that have been used for this purpose and are detectable to concentrations of less than 1 μg/L. Their use is not always possible or practical with wastewaters, however, because of the high dosages needed to overcome background levels of fluorescent whitening agents from some detergents.

12.3 | ELECTRICAL METHODS OF ANALYSIS

Electrical methods of analysis make use of the relationships between electrical and chemical phenomena as outlined in Sec. 3.9. Such methods are particularly useful in water chemistry, as they lend themselves to continuous monitoring and recording. The pH meter is probably the most widely used electrical method of analysis. In this case a glass electrode and a reference electrode are inserted in a solution, and the electrical potential or voltage across these electrodes is a measure of the concentration of hydrogen ions in the solution. Methods based upon this principle are said to be *potentiometric.*

In other electrical methods, suitable electrodes are introduced into a solution and a small measured voltage is applied. The current that flows is dependent upon the composition of the solution and so may be used to make analytical measurements. Methods based upon this principle are said to be *polarographic.* There are many modifications of the different electrical methods of analysis, and the distinction between them is not always readily apparent. Other electroanalytical methods that have been adequately discussed in Sec. 3.9 are *conductimetry,* which measures the ability of a solution to carry a current, and *coulometry,* which is a measure of the equivalence relationship between the quantity of electricity required to effect a given quantity of chemical change.

Potentiometric Analysis

The principles of the electrochemical cell were discussed in Sec. 3.9. Equation (3.42) gives the relationship between the relative potential of an electrode and the concentration of a corresponding ionic species in solution. If use is made of this equation, the measurement of the potential of an electrode can permit the calculation of the activity or concentration of a component of the solution. A potentiometric analysis requires a special electrode designed for measurement of the specific ion of interest, a reference electrode, and a potential measuring device.

As an example, the chloride concentration in a solution can be measured using the cell assembly diagrammed in Fig. 12.14. A reference electrode shown here as a saturated calomel electrode is used in conjunction with the electrode used to mea-

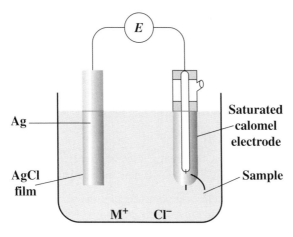

Figure 12.14
Cell assembly for potentiometric analysis for chlorides.

sure the constituent of interest. This particular reference electrode with potential E_{SCE} is a metal contacting a slightly soluble salt described later in this section. The chloride measuring electrode is actually a silver–silver chloride electrode, at which the following reaction occurs:

$$AgCl(s) + e^- \rightarrow Ag(s) + Cl^- \qquad (12.2)$$

The potential of this half-cell, on the basis of Eq. (3.42), is given by the following:

$$E_{Cl} = E_{Cl}^0 + \frac{RT}{zF} \ln[Cl^-] \qquad (12.3)$$

The measured potential of the complete cell thus becomes

$$E_{cell} = E_{Cl} - E_{SCE} = E_{Cl}^0 - E_{SCE} + \frac{RT}{zF} \ln[Cl^-] \qquad (12.4)$$

This equation indicates that the potential of the whole cell varies with the log of the chloride activity or approximately with the log of the chloride concentration. Once this cell is calibrated with standard chloride solutions, it can be used to measure the chloride concentration in unknown solutions. This is the basis for all potentiometric methods of analysis. The major requirement is for an electrode that is specific for the ion of interest and that is also relatively free from interactions with other constituents of the solution. Various electrodes of interest are described in the following.

Gas Electrode A gas electrode consists of a strip of nonreactive metal, such as platinum or gold, in contact with both the solution and a gas stream. Both electrodes in Fig. 3.8 are of this type, and the reactions indicated by Eqs. (3.39) and (3.40) are typical of gas electrodes. One gas electrode of particular importance is the *hydrogen electrode,* the standard electrode to which the potentials of other electrodes are related. It consists of a strip of sheet platinum coated with platinum black, immersed in a solution that is 1.0 N with respect to hydrogen ions and bathed with a stream of hydrogen gas under 1 atm of pressure. The reaction occurring at this electrode is that given by Eq. (3.39). The hydrogen electrode is assigned a value of zero, and potentials related to it are designated by the prefix E_H. The hydrogen electrode is cumbersome to use, and in practice other electrodes are commonly used for reference.

In shorthand notation the hydrogen electrode can be described as follows:

$$Pt \,|\, H_2(P \text{ atm}) \,|\, HCl(C \text{ M})$$

The vertical lines separate the different phases of the electrode, and the pressure of the gas and concentration of the solution can be included as indicated.

Metal Electrode   A metal electrode consists of a metal in contact with its ions in solution as indicated in Fig.12.15. An example would be an iron wire immersed in water. The notation for this electrode would be $Fe \,|\, Fe^{2+}(C)$, and the half-cell reaction at the electrode would be $\frac{1}{2}Fe \rightleftharpoons \frac{1}{2}Fe^{2+} + e^-$. Generally, this type of electrode might be used for analysis of metals such as silver, copper, mercury, lead, and cadmium that show reversible electron transfer between the metal and its ions in solution. Other metals cannot be used that do not develop reproducible potentials, generally due to strain or crystal deformations in their structures or because of oxide coatings on their surfaces. Metals in this category include iron, nickel, cobalt, and chromium.

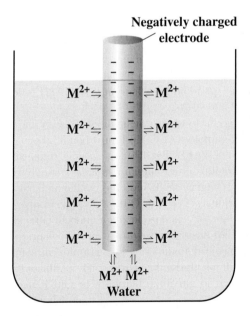

Figure 12.15
Development of a single-electrode potential on a metal electrode immersed in water.

Oxidation-Reduction Electrode   This electrode consists of a nonreactive electrode immersed in a solution of ions in both reduced and oxidized form. An example would be a platinum wire immersed in a solution containing both ferrous and ferric chloride. An oxidation-reduction electrode used in conjunction with a reference electrode is illustrated in Fig. 12.16. An oxidation-reduction electrode could be designated as $Pt\,|\,FeCl_2(C)$, $FeCl_3(C)$. The reaction occurring at this electrode would be $Fe^{2+} \rightleftharpoons Fe^{3+} + e^-$. Sometimes oxidation-reduction electrodes are used as internal indicators to show the stoichiometric end point during the oxidation-reduction type of titrations, as discussed in Sec. 11.4. They are also of interest as a means of showing the conditions that exist in biological systems. The ability to measure the relative oxidation-reduction (redox) potential of such a system would be highly desirable. Unfortunately, in complex wastewater systems that are generally not in equilibrium and that contain many oxidants and reductants, meaningful measurement of the redox potential is often not possible. However, oxidation-reduction measurement with the electrode system has sometimes been useful in an empirical way to indicate whether a system is aerobic and oxidizing or anaerobic and reducing.

Electrode with Metal Contacting Slightly Soluble Salt   This type of electrode consists of a metal in contact with one of its slightly soluble salts, while the salt, in turn, is in contact with a solution containing a common anion. The silver–silver chloride electrode illustrated in Fig. 12.14 is of this type. The example of greatest interest is the calomel electrode, which is used extensively as a reference electrode. The elements of a calomel electrode are shown in Fig. 12.17. The electrode contains mercury in contact with the slightly soluble Hg_2Cl_2, which, in turn, is in contact with a solution of KCl; $Hg\,|\,Hg_2Cl_2\,|\,KCl(C)$. The reaction that occurs at this electrode is $Hg + Cl^- \rightleftharpoons$

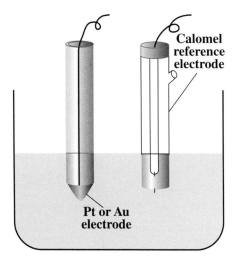

Figure 12.16
Electrode assembly for measurement of oxidation-reduction potentials (ORP).

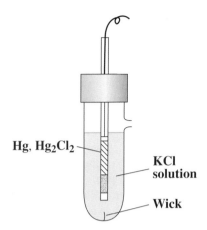

Figure 12.17
Calomel reference electrode.

$\frac{1}{2}Hg_2Cl_2 + e^-$. The calomel electrode is the reference electrode used in electrometric pH determinations, in oxidation-reduction measurements, and in most other electrochemical analyses for which a stable easy-to-use reference electrode is desired.

Calomel electrodes are of three types, normal, tenth normal, and saturated, depending on the concentration of KCl solution used in preparing them. The potential that each develops is a function of the concentration of the potassium chloride solution used. The potential of each with respect to the hydrogen electrode is shown in Table 12.2. From the data listed, it becomes apparent that knowledge of the concentration of KCl used in preparing a calomel reference electrode is necessary for proper interpretation of the information gathered with its aid. In general, the saturated type of calomel electrode is employed, and its use is so common that results are often reported in terms of E_{SCE} instead of E_H.

Membrane Electrode   Another general class of electrodes that is important in water quality measurements is the membrane electrode. The development of membrane electrodes began around the turn of the century after an empirical observation by Cremer that a potential developed across a thin glass membrane when it was placed between two solutions with different hydrogen-ion concentrations. This led to the development of the glass membrane electrode that is now so widely used for potentiometric determination of pH. Since then, many membranes have been devel-

Table 12.2 | Standard potential of calomel reference electrodes at 25°C

Concentration of KCl	E_H, volts
0.1 N	−0.334
1.0 N	−0.281
Saturated	−0.242

oped that are selectively sensitive to other ions. These ion-selective electrodes are available for direct potentiometric determination of K^+, Na^+, F^-, Ca^{2+}, NO_3^-, NH_4^+, S^{2-} and a number of others. Membrane electrodes are generally divided into four classes (1) glass electrodes, (2) liquid membrane electrodes, (3) gas-permeable membranes, and (4) crystalline membrane electrodes. The mechanism of potential development appears to be the same for the four classes. The glass electrode is the most widely used electrode and is discussed in more detail in the following, along with a general description of the other membrane electrode systems.

Glass Electrode The glass electrode is used universally for measurement of pH. For this purpose the electrode functions in highly colored solutions for which colorimetric methods are useless and in oxidizing media, reducing media, and colloidal systems for which other electrodes have failed almost completely. Earlier glass electrodes also developed a potential with Na^+ in alkaline solution, but electrodes are now available that can minimize this sodium interference. Advantage is taken of this interference, however, in the development of a sodium-specific glass membrane electrode.

The action of the glass electrode is now fairly well understood. The glass used in the sensitive part of the electrode must have special characteristics with respect to composition and thickness. Construction of the electrode is essentially the same as that of the calomel reference electrode except that it does not have an opening or a wick to make direct electrical connection with the surrounding fluid. For pH measurements, the electrolyte within the glass electrode is an acid solution of definite strength rather than a KCl solution. The glass electrode for pH determination is used in conjunction with a standard calomel reference electrode, and the system may be described as shown in Fig. 12.18.

The single-electrode potential established on the pH electrode is determined by the concentration, or more specifically the activity, of hydrogen ions in the solution, $\{H_c^+\}$, in relation to the activity in the electrolyte within the electrode, $\{H_e^+\}$. It is

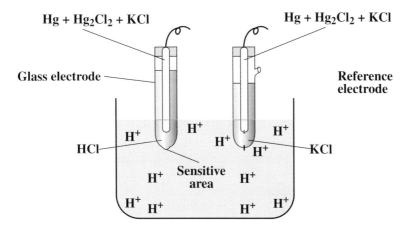

Figure 12.18
Electrode system employing the glass electrode for measurement of pH.

generally considered that the potential is actually determined by the relative activities of adsorbed H^+ on the two sides of the glass membrane as follows:

$$E_{obs} = k + 0.059 \log \frac{\{H_c^+\}}{\{H_e^+\}} \tag{12.5}$$

where k is a constant. Since $\{H_e^+\}$ remains constant the equation can be rewritten as follows:

$$E_{obs} = K + 0.059 \log \{H_c^+\}$$

or $\qquad\qquad\qquad E_{obs} = K - 0.059 \text{ pH} \tag{12.6}$

An electronic voltmeter or potentiometer (pH meter) is commonly employed for pH measurements and is calibrated to read pH directly.

Advantage has been taken of the observation that glasses with different compositions selectively develop potentials with different ions. Glass electrodes are commercially available for measuring concentration of the following ions: Na^+, K^+, NH_4^+, and Ag^+. "Standard Methods" procedures using these ion-selective electrodes are now available. The selectivity of the membrane to the various ions varies, however, as does the sensitivity. Before using glass electrodes for measuring cations other than H^+, the presence in the water and possible effects of interfering ions must be assessed.

Liquid Membrane Electrode Liquid membrane electrodes are of more recent development than glass electrodes. They provide a means for direct potentiometric determination of the activities of several polyvalent cations such as Ca^{2+} and Mg^{2+}. They can be used for certain anions, as well; a liquid membrane electrode forms the basis for one of the "Standard Methods" procedures for nitrate. In place of the glass membrane in the glass electrode, a liquid supported between porous glass or plastic disks is used. The liquid is an organic compound that is immiscible in water and contains functional groups that can exchange with ions from solution. By selecting functional groups that have strong affinity for particular cations or anions, the liquid membrane electrode can become quite specific. When placed in a solution, the ion for which the electrode is developed exchanges with ions within the water-liquid exchanger interface and establishes a potential that is measured. These electrodes must be used with the same precautions as other membrane electrodes.

Gas-Permeable Membrane Probes Certain electrochemical devices that are sometimes called electrodes are actually better called probes since in effect they are electrochemical cells. They consist of a pH electrode and a reference electrode that are immersed together in an internal solution that is retained by a gas-permeable membrane. The membrane separates the solution to be analyzed from the internal solution. Gases that may be analyzed include ammonia and carbon dioxide. For example, with the carbon dioxide probe, the internal solution consists of sodium bicarbonate and sodium chloride. When carbon dioxide diffuses through the membrane, it mixes with the internal solution and affects the pH through the carbon dioxide/bicarbonate equilibrium discussed in Chap. 4. The output pH thus provides a quantitative measure of the carbon dioxide content of the sample. A "Standard Method" procedure for ammonia also makes use of a gas-permeable membrane probe. Here the internal solution is ammonium chloride. The sample pH is first elevated to above 11 to convert sample ammonium to the gaseous ammonia. The ammonia diffuses through the membrane to mix with the ammonium chloride solution and affects pH according to the NH_3/NH_4^+ equilibrium, also discussed in Chap. 4. Any gas that forms a weak acid or base in water can be analyzed by gas-permeable membrane probes, including HCN, HF, H_2S, SO_2, and NO_2.

Crystalline Membrane Electrode   These electrodes contain a solid membrane (other than glass) between the sample and a reference solution that is selective for anions. The membrane generally consists of a sparingly soluble salt of the anion of interest. Such salts can bind that anion selectively, and, as with liquid membrane electrodes, a potential that can be measured is established. A crystalline membrane electrode for fluoride is presently used in one "Standard Methods" procedure for water. The membrane consists of a crystal of lanthanum fluoride that has been treated with a rare earth to increase its electrical conductivity. It shows theoretical response to changes in fluoride ion activity, and shows a thousandfold selectivity for fluoride over other common anions.

Precautions in Use of Electrodes   The fluoride-specific electrode illustrates some of the limitations of all electrodes when used for direct measurement of concentration. Electrodes measure activities and not concentrations. For a given concentration, the activity varies with ionic strength as discussed in Secs. 2.12 and 4.3. Thus, to use an electrode for concentration measurements, the electrode must be calibrated in a standard solution of similar ionic strength. Also, electrodes measure specific ionic species and not complexes of the species. Fluoride forms complexes with Al(III), Fe(III), and Si(IV), and that portion of the fluoride associated with these cations will not be measured by the fluoride electrode. In order to overcome this interference, a method must be found for breaking up the complex or else appropriate corrections must be made. Change in solution pH or addition of materials that preferentially complex the interfering ions may be appropriate.

Ion-specific electrodes offer great promise for rapid, routine, and automatic monitoring of water quality. They have limitations, however, and these must be known and dealt with by the analyst for each particular ion. New electrode systems are rapidly evolving. Those concerned with water and wastewater analysis should keep abreast of these developments because of their potential for solving many difficult analytical problems.

Polarographic Analysis

Analytical procedures where the relationship between voltage applied across two electrodes and current flow that results are used to analyze a solution are termed *voltammetry. Polarography* is a specialized form of voltammetry that differs from other forms in that a special inert electrode such as a *dropping mercury electrode* is used. While polarographic analysis is not represented as a recommended procedure itself for any analyses in "Standard Methods," it does form the basis for some important procedures that are standard or proposed standard, including *anodic stripping voltammetry* for heavy metals, *amperometric titration* for chlorine, and *membrane probes* for dissolved oxygen. The environmental engineer or scientist should understand the underlying principles for these important analyses.

Polarographic Analysis   A schematic diagram of an electric circuit for polarographic analysis is given in Fig. 12.19, and a typical instrument is shown in Fig. 12.20. A voltage varying from 0 to 3 volts is impressed across the electrodes. The exact voltage at any setting of the control is recorded with a voltmeter. This voltage results in a flow of current through the solution, the amount being read with a microammeter. When the electrodes are immersed in a given solution and the applied voltage is gradually increased, the current will remain near zero until the potential reaches a point that is sufficient to bring about the reduction of some ion present in the solution. At this point, electrolytic reduction starts at the inert electrode, and an increased voltage causes a sharp increase in current in accordance with Ohm's law, $E = IR$. As reduction occurs, there is a depletion of ions next to the inert electrode, and fresh ions reach the electrode by diffusion. As the voltage is continually increased, a point is reached at which the rate of diffusion of ions to the electrode limits the current. At this voltage, the rate of diffusion and, hence, the current are proportional to the concentration of the ions in solution that are being reduced.

An illustration of a typical current-voltage curve for a particular ion is shown in Fig. 12.21. Point *A* on the curve is called the half-wave potential and is equal to the potential at which the current is one-half the diffusion current. This potential is related to the standard reduction potential of the ion involved and serves as a means of determining the nature of the substance being reduced. Thus, the half-wave potential identifies the ion and the diffusion current determines the quantity present. When several reducible ions are present in the solution, an increase in current will take place when the half-wave potential of each is reached, as illustrated in Fig. 12.22. If the respective individual curves are sufficiently separated so they do not interfere with one another, then the half-wave potential will allow identifi-

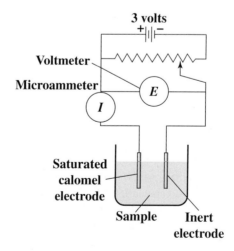

Figure 12.19
Electric circuit for polarographic analysis.

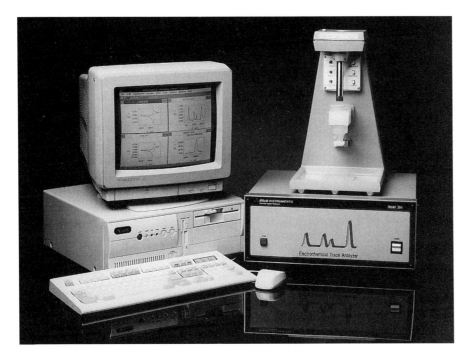

Figure 12.20
A polarographic analyzer. *(Courtesy of Advanced Measurement Technology, Inc.)*

cation of the particular ion, and the increase in diffusion current over that produced by the preceding ion will allow an estimate of the ion concentration.

The inert electrode used for polarographic analysis is usually a dropping-mercury electrode, which consists of metallic mercury dropping at a steady rate from a capillary tube. As the mercury drop grows, it continuously exposes fresh mercury

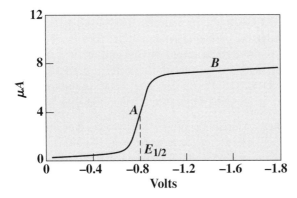

Figure 12.21
Typical current-voltage curve for polarographic analysis
of a single ion.

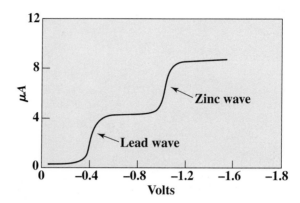

Figure 12.22
Current-voltage curve for lead and zinc in a single solution.

to the solution so that unwanted poisoning by the ions being reduced can be avoided. Other electrodes such as streaming-mercury electrodes, rotating electrodes, or noble-metal electrodes have also been used, and under certain circumstances may have advantages over the dropping-mercury electrode.

Anodic Stripping Voltammetry   Direct polarographic analysis of water or wastewater solutions is often not sensitive enough to measure the low concentrations of metals in water that are of environmental and health concern. However, *anodic stripping voltammetry* is a modification of polarography that provides sensitivity in the nanogram per liter range for certain metals such as lead, zinc, and cadmium, and is a listed procedure for these metals in "Standard Methods." Here, a cell similar to that used in the polarograph is used together with either a hanging-drop mercury electrode or a thin mercury film electrode. Analysis is carried out as a two-step procedure. First, a fresh hanging drop of mercury is formed, and a potential a few tenths of a volt more negative than the half-wave potential for the metal ion of interest is applied. Deposition of the metal on the drop then is allowed to occur for a measured period of time. In the second step, the voltage is then decreased at a linear fixed rate, and under the oxidizing potential that then exists, the metal is stripped rapidly from the electrode. The accompanying current flow is proportional to metal concentration. The much higher sensitivity of this procedure results because of the concentration on the mercury drop that is achieved in the first step, the sensitivity being a function of the length of time employed in this step.

Membrane Probes   Membrane probes are based upon the principle of polarographic analysis, and are useful for measurement of gaseous or non-ionized molecules. They are used in a "Standard Methods" procedures for dissolved oxygen. Oxygen is one of the most significant interfering substances in polarographic analysis, as it is readily reduced at the inert electrode. This interference is normally removed by flushing the sample with a nonreducible gas such as nitrogen. However, advantage is taken of the reducibility of oxygen in the polarographic determination of dissolved oxygen. A typical oxygen polarogram is indicated in Fig. 12.23. Two

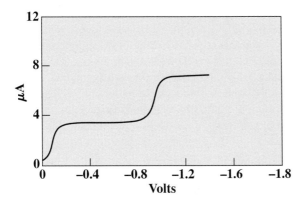

Figure 12.23
Typical current-voltage curve for oxygen.

waves are observed. The first, with a half-voltage of about -0.05 volt, is caused by the reduction of O_2 to H_2O_2, while the second, at about -0.9 volt, corresponds to the reduction of O_2 to H_2O. Normally the center of the current plateau between the two oxygen waves is used to make quantitative measurements.

Instruments specifically designed for dissolved oxygen determinations and using the dropping-mercury electrode have been available for some time. However, the development of gas-permeable membrane probes, which permit more specific analysis for dissolved oxygen, have proved to be very useful, not only for water quality monitoring, but also for control of the rate of aeration in biological treatment processes. A schematic diagram of such an electrode system is shown in Fig. 12.24. An inert metal such as gold or platinum serves as the cathode, and silver is

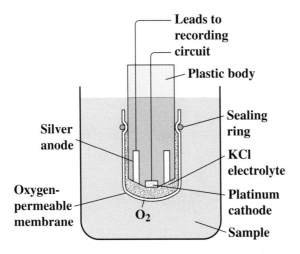

Figure 12.24
Schematic diagram of an oxygen electrode system.

used for the anode. These are electrically connected with a potassium chloride or other electrolytic solution. The complete cell is separated from the sample by means of a gas-permeable membrane, usually made of polyethylene or Teflon. The membrane shields the cathode and anode from contamination by interfering liquids and solids. When a potential of about 0.5 to 0.8 volt is applied across the anode and cathode, any oxygen that passes through the membrane will be reduced at the cathode, causing a current to flow. The magnitude of the current produced is proportional to the amount of oxygen in the sample. In another modification of this method, a lead cathode is used in place of the inert electrode, and a basic electrolyte such as potassium hydroxide is used. This, in combination with the silver anode, produces a galvanic cell having a sufficient potential to reduce the oxygen without the need for an externally applied voltage. An instrument equipped with this type of probe is shown in Fig. 12.25. Instruments employing such probes require that the sample be agitated to yield reliable results. Often dissolved oxygen probes are termed electrodes, but because they represent a complete electrochemical cell rather than a single electrode, they are more properly called probes.

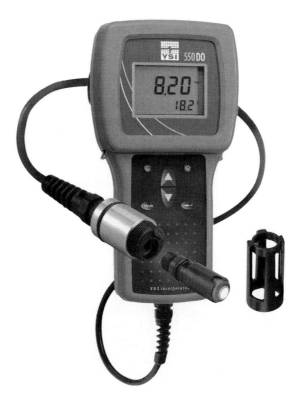

Figure 12.25
A dissolved oxygen analyzer. *(Courtesy of YSI, Inc.)*

12.4 | CHROMATOGRAPHIC METHODS OF ANALYSIS

Chromatography is the general term used to describe the set of different procedures used to separate components in a mixture based upon their relative affinity for partitioning between different phases. For example, carbon dioxide is more soluble in water than methane, so if both gases were present in a sample of air that was brought into contact with water, carbon dioxide would partition more strongly into the water than would methane. This property, which is different for different molecules, can be used to bring about their separation. The first paper to provide a modern description of chromatography was presented in 1906 by Michael Tswett, a biologist who separated the chlorophylls and other pigments from plant extracts by passing a petroleum ether solution in which they were contained through a glass column containing calcium carbonate particles. The pigments moved down through the column at different speeds because of their relative affinity for petroleum ether in the mobile phase, and calcium carbonate in the immobile phase. This separation could be readily seen because of the color of the pigments, and the intensity of the color noted could be used to quantify the amount of each pigment present.

As in Tswett's description, two different phases are generally used in modern chromatography to effect separation of a mixture; one is the stationary phase and the other the moving phase. The stationary phase may be either a liquid or a solid, and the moving phase may be either a liquid or a gas. When the mobile phase is a gas, the procedure is termed *gas chromatography,* and when it is a liquid, it is called *liquid chromatography.* Other descriptive terms are used depending upon the nature of the stationary phase. For example, when the mobile phase is a gas and the immobile phase is a liquid, the procedure is termed *gas-liquid chromatography.* When a liquid is the mobile phase, then we may have *paper chromatography* with paper as the stationary phase. Liquid chromatography has advanced considerably in recent years, leading to advanced instruments to which the name *high-performance liquid chromatography* (HPLC) is now applied. Some instruments for HPLC use ion exchange resins as the stationary phase and are called *ion chromatographs.*

Chromatographic separation is one of the two major features of chromatographic instruments, the other is compound detection and quantification. Numerous detectors are now available, each having its own particular sensitivity to a given set of compounds in a mixture. In the following brief discussion, gas chromatography and liquid chromatography are described along with the particular detectors that have been found useful for each in environmental analysis. Chromatographic instruments are now widely used in the environmental engineering and science fields, and permit rapid quantitative measurements to be made of chemicals present in complex mixtures and at concentrations in the ppb or μg/L range and lower. The development of these versatile and sensitive instruments has been one of the most significant factors making it possible to address the multitude of chemical hazards created by the ever-increasing industrialization of society.

Gas Chromatography

Gas chromatography is a highly versatile instrumental method of analysis that has been developed since 1951. It is currently used by environmental engineers and scientists for a variety of routine environmental analyses that prior to its development were difficult and highly time-consuming. Gas chromatography entails the vaporization of a liquid sample followed by the separation of the various gaseous components formed so that they can be individually identified and quantitatively measured. A schematic diagram of the components of a typical gas chromatograph is shown in Fig. 12.26, and an instrument is illustrated in Fig. 12.27. The basic components are a gas cylinder with reducing valve, a constant-pressure regulator, a port for the injection of the sample, a chromatographic column, a detector, an exit line, and a recorder.

The gas cylinder contains a carrier gas such as hydrogen, helium, or nitrogen, which is continuously swept through the chromatographic column at a predetermined temperature and flow rate. A small sample for analysis is injected, usually with a syringe, into the sample port where it is flash-evaporated to convert its components into a gaseous state. The constantly flowing stream of carrier gas carries the gaseous constituents through the chromatographic column. The gases travel through at different rates, so they emerge from the column at different times. Their presence in the emerging carrier gas is detected by chemical or physical means, and the response of the detector is fed into a recorder, perhaps after electronic integration. A typical chromatogram produced by this process is shown in Fig. 12.28. Each peak represents a specific chemical compound or a mixture of compounds with the same rate of movement through the column. The time for each compound to emerge from a given column is a characteristic of the compound and is known as its *retention time.* The area under the peak is proportional to the concentration of the compound in the sample.

Chromatographic columns are generally glass or metal tubes varying from about 1 to 10 m in length. With some columns the tube is 3 to 6 mm in diameter and is packed with an inert solid impregnated with any one of a large number of possible

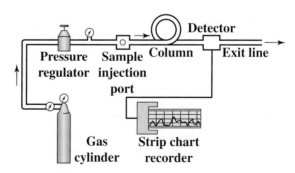

Figure 12.26
Schematic diagram of a gas chromatograph.

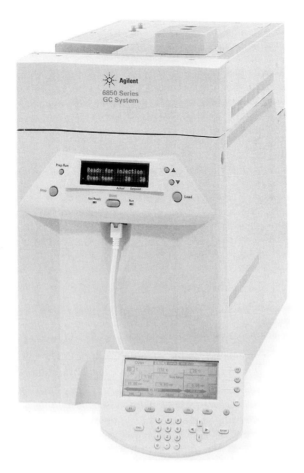

Figure 12.27
A gas chromatograph with automatic sample injection.
(Courtesy of Agilent Technologies.)

nonvolatile liquids, such as silicone oil or polyethylene glycol. As a gaseous sample passes through the column, its components are partitioned between the stationary liquid phase of the column and the moving gas phase. Components that are relatively soluble in the liquid phase move through the column at slower rates than components that are not so soluble. Through the election of suitable stationary phases and column lengths, most gaseous materials can be separated by this procedure. Much better resolution of peaks can be obtained with capillary columns having diameter of only 0.2 to 0.4 mm and length of 20 to 30 m. They are the columns of choice today for environmental analysis. They contain the solid phase on the inside wall of the capillary tube. The gas containing the organic components passes through the center of the capillary tube, and the organics partition themselves between the gas and the station-

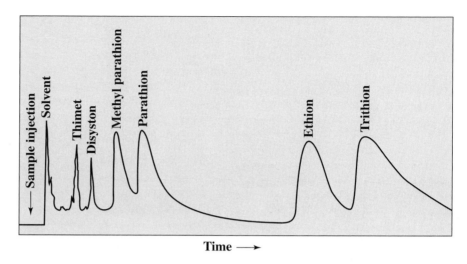

Figure 12.28
Gas chromatogram produced from pesticide analysis, using an electron capture detector and a packed column.

ary phase, emerging from the column at different times. The absence of packing minimizes the dispersion of the contaminants, so excellent resolution of the different compounds is obtained as illustrated in Fig. 12.29, even when there are 20 compounds or more present. These columns have made possible the identification of a much greater number of compounds in highly complex mixtures.

The temperature of the sample and column must be held sufficiently high during movement through the column and subsequent detection to maintain them in the gaseous state. Only materials that can be volatilized or that can be converted to volatile compounds can be detected by gas chromatography. More versatility is provided by instruments in which the temperature can be programmed to increase in a defined manner over predetermined time periods. This allows separation within a single column of materials covering a wide range of solubility in the liquid phase.

There are several different detectors in common usage. Each responds to some chemical or physical property of the eluting gases and converts this response to an electrical signal that can be recorded. One detector responds to the difference in *thermal conductivity* of the eluting gases. In this detector, a heated wire is placed in the stream of gases emerging from the chromatographic column. Each gas has a different thermal conductivity or different ability to carry away heat from the wire, and thus, as they pass, the temperature of the wire changes. This results in a change in resistance of the wire and hence the electrical response, which can be integrated and/or recorded. Thermal-conductivity detectors are not as sensitive as some detectors, but are particularly useful for the analysis of methane, carbon dioxide, and other gases resulting from anaerobic digestion.

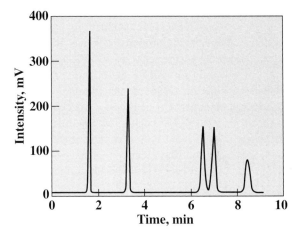

Figure 12.29
Gas chromatograph illustrating the higher peak resolution obtained when using a capillary column and a photoionization detector. The peaks in order from the left represent benzene, toluene, ethylbenzene, *m*-xylene, and *o*-xylene.

Other detectors are available that are highly sensitive for specific organic or inorganic materials. A *flame ionization* detector is perhaps the most widely used. It is sensitive to organic compounds, but will not respond to water vapor. This permits the direct injection of water samples for analysis without prior separation. Organic compounds yield ions and electrons when burned in a flame. The flame ionization detector makes use of this principle by measuring the current carried by these charged particles when a potential of a few hundred volts is applied across the burner and a collector electrode as gases are eluted from the chromatographic column and burned. Organic acids and petroleum hydrocarbons in concentrations of only a few milligrams per liter can be separated and measured, using instruments equipped with this detector. The *photoionization* detector operates on a similar principle, only an intense beam of ultraviolet radiation in the 105- to 150-nm range, rather than a flame, is used to ionize molecules. This works particularly well with small organic molecules having a double bond such as chlorinated ethenes.

Pesticides, trihalomethanes, and chlorinated solvents in concentrations of a microgram per liter or less can be measured after extraction from water by use of *electron capture, coulometric,* or *electrical conductivity* detectors. The electron capture detector measures the ability of compounds with halogen atoms or polar functional groups to capture beta particles emitted by a radioactive source such as ^{63}Ni or tritium. Coulometric detectors are not as sensitive, but are specific for organic compounds containing specific elements, such as chlorine. Here, the effluent from the chromatographic column is burned, the products are absorbed in a titration cell, and the chloride ions released are titrated continuously with silver ions generated electrically. In another variation, the HCl resulting from combustion is absorbed into a solution and the resulting increase in electrical conductivity is measured.

Several other detectors measuring properties such as gas density, change in potential at electrodes, and mass spectrometry are being used today. The mass spectrometer has the ability not only to indicate the quantity of a given organic compound as it emerges from a gas chromatographic column, but also to make positive identification of each organic material. The use of a gas chromatograph in conjunction with a mass spectrometer (GC/MS) can allow one to identify hundreds of organic compounds in air or water, and is in common usage to determine the compounds present at sites throughout the country that have been contaminated with hazardous wastes. The possibilities for gas chromatography are extensive, and new developments are occurring rapidly.

Gas chromatography in its numerous variations is now the "Standard Methods" procedure for detecting and quantifying a variety of individual volatile organic compounds. GC is used for measuring the Environmental Protection Agency–designated priority pollutants, taste- and odor-producing organic compounds, and a broad range of compounds designated as *volatile organic compounds* (VOCs). In order to increase sensitivity, organic contaminants at trace concentrations are often first concentrated by the *purge-and-trap* procedure or a variation of it. Here, volatile organic compounds in a water sample varying from a few mL up to 1 liter in size are transferred from the liquid phase by bubbling an inert gas such as nitrogen through the sample and passing the vapor through a column or trap containing a packing to which the volatile organics are sorbed. The trap column is then connected to a gas chromatograph, carrier gas is passed through the trap, and it is heated in a programmed fashion to elute the sorbed compounds into the GC. This procedure not only efficiently concentrates the volatile organics for increased sensitivity, but also separates them from interferences in the water sample, including water itself. In this manner, compounds can be determined with good precision down to concentrations in the low μg/L range, and sometimes even lower.

High-Performance Liquid Chromatography

HPLC is especially useful for separating nonvolatile species and those that are thermally unstable; thus, it extends the range of compounds that can be separated for analysis beyond that possible with gas chromatography. Among the compounds that can be separated by this procedure are amino acids, proteins, nucleic acids, hydrocarbons, fatty acids, carbohydrates, phenols, pesticides, antibiotics, metal-organic species, and a great variety of inorganic substances. In this section, HPLC methods based upon either adsorption or partitioning are discussed. In the subsequent section, Ion Chromatography, the method based upon ion exchange is presented.

The elements of a typical HPLC system are illustrated in Fig. 12.30. One or more solvent reservoirs hold degassed and prefiltered solvent, which is pumped through the column system at pressures up to 400 atm. A pulse damper is used to even the flow rate, and a precolumn or guard column is used to increase the life of the analytical column by removing particulates and contaminants from the solvent.

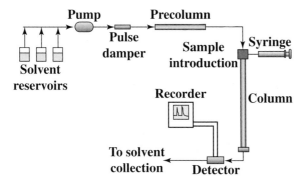

Figure 12.30
Schematic diagram of a high-performance liquid
chromatographic system.

Sometimes the guard column is used after sample injection to remove undesirable
contaminants that it may contain. The sample is introduced into the system either
with a syringe or more generally with a sampling loop, and as it is pushed through
the analytical column by the solvent, separation of components occurs. A suitable
detector for the species of interest is used at the column exit, and the response is
stored, integrated, and recorded. Waste solvent is collected for safe disposal. The
basic components of the system are similar to those of a gas chromatograph, and the
recorded output is similar to that in Figs. 12.28 and 12.29. A typical HPLC is illustrated in Fig. 12.31.

The solvents, column packing, and detector are selected based upon suitability
for the compounds of interest and the sensitivity of detection required. Columns
may or may not be heated, but it is best to have temperature control to maintain a
constant temperature in order to better control the retention time for the species of
interest. Solvents may range from organic compounds with low polarity such as
hexane to those of high polarity such as methanol and water. They may also be mix-

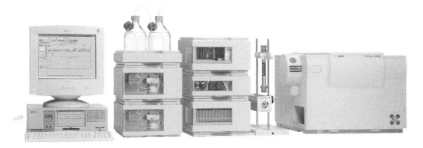

Figure 12.31
A high-performance liquid chromatograph (HPLC). *(Courtesy of Agilent
Technologies.)*

tures of compounds. Separations may be made using a single solvent *(isocratic elution)* or two or more solvents *(gradient elution)*. In the latter case, the solvents used generally differ significantly in polarity, and the ratio of the different solvents fed is programmed to vary in a defined way with time. This greatly enhances separation efficiency.

Analytical columns are generally 10 to 30 cm in length, much shorter than used for gas chromatography, and are kept straight. Typical inside diameters are 4 to 10 mm. The packing is uniform and fine, with diameters of 3 to 10 μm, in order to reduce dispersion. The packing is normally made of inert materials such as silica that is coated with an organic phase of desired characteristics to promote the desired degree of partitioning for the compounds of interest between the mobile and stationary phases. A recent tendency is to use shorter columns with smaller diameter in order to reduce the time for analysis and the quantity of solvent for disposal.

HPLC is sometimes distinguished by the relative polarity of the mobile and stationary phases. *Normal-phase chromatography* is when the mobile phase is a nonpolar solvent such as hexane and the stationary phase is a highly polar material such as water. With *reverse-phase chromatography* the mobile phase is a relatively polar solvent such as methanol or water, and the stationary phase is nonpolar, such as a hydrocarbon. In normal-phase chromatography, the least polar components in a mixture elute from the column first, while in reverse-phase chromatography, they elute last.

In addition to the many possible variations for solvents, columns, and methods of operation, there are many different HPLC detectors available. These may provide response based upon the bulk properties of the mobile phase, such as density, dielectric constant, or refractive index, or the response may be based upon the properties of the solute or compounds of interest, such as fluorescence, ultraviolet absorbance, or conductivity. Obviously, HPLC affords a great variety of possibilities to the analyst, the proper choice for a system depending upon the compounds of interest and the analytical sensitivity required. One of the current major deficiencies is that positive identification of compounds often cannot be made. If two or more compounds elute from a column at the same time, then the analyst cannot be sure which compound is present. Use of different columns, solvents, or detectors may help, but this can be time-consuming and may lead to interferences from other compounds. In gas chromatography, this problem was solved to a large extent by the use of mass spectrometers for detection. Mass spectrometers are now being used with HPLC as well, so that this limitation is disappearing.

Ion Chromatography

Ion chromatography is a variation of HPLC in which the analytical column consists of ion exchange resin. A typical instrument is illustrated in Fig.12.32. Natural ion exchangers such as clays and zeolites have long been used in the environmental engineering field to remove multivalent cations such as Ca^{2+} and Mg^{2+} that cause water hardness (Chap. 19). Synthetic ion exchange resins that are more efficient than the natural materials have long been available. Here, by passing a solution of NaCl through a

Figure 12.32
An ion chromatograph. *(Courtesy of Dionex Corporation.)*

cation exchanger, the sodium ions are retained by the resins to neutralize the negative charge on the resin. When water containing hardness is passed through the resin, the multivalent cations, which have a greater affinity for the resin, displace the sodium ions and they become part of the hardness-free effluent. Advantage is taken of resins that exchange both anions and cations in ion chromatography. Virtually any compound that exists in an ionized form is susceptible to measurement by ion chromatography, including both inorganic and organic species and both cations and anions.

As with normal HPLC, there are many variations of ion chromatography. In principle, a sample is first introduced at the head of an analytical column as with HPLC (Fig. 12.30), but here the column is packed with a suitable ion-exchange resin. The ions within the sample are taken up by the resin and concentrated there. Next, a solvent containing an ion that competes with the ions of interest is passed through the column, displacing those originally contained in the water sample. Since different ions have different relative affinities for the resin, they emerge from the column at different times, and thus can be detected and quantified. Conductivity detectors (see Sec. 3.9) are most frequently used, and measure the relative difference in conductivity between the eluting ions and those in the solvent. However, if the difference in conductivity is small, then an alternative is needed. The most common approach has been to use a suppressor column following the analytical column.

The suppressor column contains a second ion-exchange resin that converts the ions from the solvent into a species with limited ionization without affecting the conductivity of those of interest in the water sample. In this way, the ions in the water sample can be readily detected by the increase in sample conductivity that they cause. In other variations, detectors based upon absorption of ultraviolet or visible radiation can be used.

Ion chromatography has found its greatest application for analysis of the common anions found in water, including chloride, bromide, nitrite, nitrate, phosphate, and sulfate. These anions were measured separately in the past by time-consuming and often inaccurate analytical procedures. Now they can all be determined in a single analysis by ion chromatography, and generally with both speed and accuracy. Ion chromatography is now one of the "Standard Methods" procedures for these important water constituents, and undoubtedly is the method of choice by those having the instrument available.

Capillary Electrophoresis

Electrophoresis is a separation process that takes advantage of the fact that charged species migrate at different rates in an electrical field. This principle was discussed in Sec. 3.9 under conductivity. This phenomena is used in capillary electrophoresis to separate a wide variety of materials such as drugs, foods, natural products, biological materials, and pesticides. In "Standard Methods" capillary electrophoresis is a proposed method for analysis of common inorganic ions, such as fluoride, chloride, sulfate, nitrate, nitrite, bromide, and acetate. It is thus a competitive analytical method to ion chromatography. The capillary electrophoresis instrument consists of a buffer-filled capillary tube 50 to 100 cm long with an internal diameter of 25 to 100 μm that is placed in two containers filled with the same buffer. Platinum foil electrodes in the two buffer vessels are connected to direct current power supply that can develop a potential of 10 to 30 kV. For anion analysis, a small sample is placed at the cathode end of the capillary and the anions migrate toward the anode, but at different rates. The arrival of the anions at a determined point along the capillary is measured by a detector and recorded. For anion analysis, an ultraviolet detector is used along with a buffer that contains an ultraviolet-absorbing anion salt (sodium chromate). As sample anions move through the capillary, they displace the chromate anions, one for one, and thus decrease the amount of ultraviolet absorbance recorded as they pass through the detector. The method is fast, being complete in less than 5 min.

12.5 | OTHER INSTRUMENTAL METHODS

There are many other instrumental methods of analysis available that are of growing interest to environmental engineering and science because of the increasing complexity of the problems that must be addressed, and the expanding concern over health and environmental impacts of very low concentrations of organic and inorganic contaminants in the environment. Only a brief review of some of these methods is given in the following.

Mass Spectrometry

As indicated in Sec. 12.4, mass spectrometry when used in conjunction with gas chromatography (GC/MS), HPLC (LC/MS), or IPC (IPC/MS) can give positive identification and quantification for a large number of individual organic and inorganic compounds present in water and wastewater, soils, or air. A typical GC/MS system is illustrated in Fig.12.33. A mass spectrometer is an instrument that will sort out charged gas molecules or ions according to their masses. The substance to be analyzed is vaporized and converted to positive ions by bombardment with rapidly moving electrons. The ions formed are pulled from the gas stream by an electrical field. The ions are accelerated in some fashion depending upon the type of instrument and are separated by their mass-to-charge ratio. A suitable detector can then record the particles of different mass either qualitatively, quantitatively, or

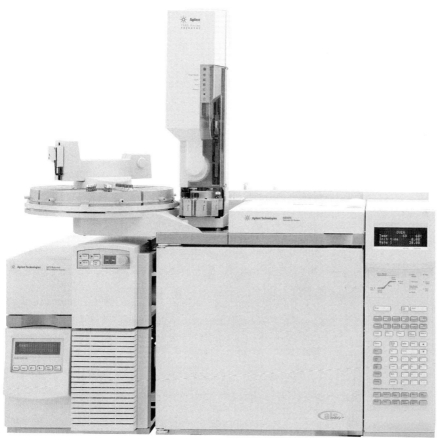

Figure 12.33
A gas chromatograph/mass spectrometer (GC/MS) system. *(Courtesy of Agilent Technologies.)*

both. Mass spectrometry is useful for tracer experiments using stable isotopes such as ^{15}N, or for determining isotope ratios such as for oxygen when determining the age of water from different sources.

When used for the positive identification of organic materials, such as those emerging from a gas chromatograph, the bombardment of the organic molecule by the rapidly moving electrons breaks the organic molecules into a number of charged fragments. The mass-to-charge (m/z) ratio of each fragment is measured as is the relative quantity of each fragment. Every organic molecule has its own pattern of fragmentation when bombarded under a given set of conditions as illustrated in Fig. 12.34. By comparing the mass-to-charge ratio and density of fragments of an unknown with that of known materials, positive identification can be made. MS is a powerful analytical method that is now widely used in environmental analysis. In combination with GC, HPLC, and IPC, it is helping to solve many difficult analytical problems and is greatly adding to our knowledge of the nature and fate of organic materials in the environment, and in developing technical measures for their control.

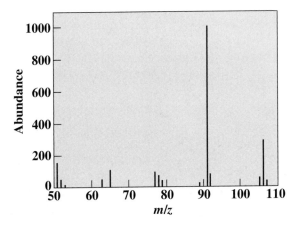

Figure 12.34

Mass spectrum for ethylbenzene as obtained with a GC/MS system. The m/z value refers to the mass-to-charge ratio for molecular ions resulting from electron impact on the molecule. The base peak at 91 corresponds to the ion formed from loss of a methyl group ($^{12}C_6H_5 = {}^{12}CH_2^+$), while the smaller peak at 106 represents $^{12}C_6H_5{}^{12}C_2H_5^+$. The smaller peaks adjacent to the larger peaks are generally attributable to ions having the same chemical formula, but different isotopic compositions. For example, a ^{13}C atom in the ion would increase the m/z by one unit.

X-Ray Analysis

X rays are high-energy electromagnetic radiations with short wavelength. They are used for analytical purposes much in the same way as radiations of longer wavelength, such as visible light. *X-ray absorption* follows the same absorption laws as for other radiation, except that the phenomenon is on an atomic, rather than a molecular, level. X-ray absorption is used for the measurement of the presence of heavy elements in substances composed primarily of low-atomic-weight materials. An example is the determination of the quantity of uranium in solution.

X-ray emission or *X-ray fluorescence* is used for the study of metals and other massive samples. A sample is bombarded with high-energy X rays that are absorbed by certain elements and re-emitted as X rays of lower energy. The bombarding X rays can dislodge an electron from an atom, leaving an unstable atom with a "hole" in one of the electron shells. Stability is regained by single or multiple transitions of electrons from outer shells to fill the hole. Each time an electron is transferred, the atom moves to a less energetic state and radiation is emitted at a wavelength corresponding to the energy difference between the initial and final states. The energy of the emitted radiation is characteristic for each emitting element and so can be used for analysis. This is similar to fluorimetry as previously discussed, except for the different energies of the respective radiations. The usefulness of X-ray fluorescence for rapid analysis for many heavy metals in water has been demonstrated. The heavy metals are first concentrated by passing a water sample through filter paper containing ion-exchange resins; the filter paper is then subjected to X-ray fluorescence. This technique is likely to see wider usage for water and wastewater analysis in the future.

X-ray diffraction is primarily of value for the study of crystalline material. X rays are reflected off the surfaces of crystals, and by studying the patterns of reflection as the crystalline material is rotated in the path of the X rays, much information about the structure of the material can be obtained.

Nuclear Magnetic Resonance Spectroscopy

NMR is used to detect and distinguish between the nuclei of atoms in a molecule. It is based upon absorption by the nuclei of radiation in the radio-frequency range. In order for absorption to be measured, the compound of interest must first be placed in a fixed magnetic field that causes nuclei to develop the energy states required for absorption. This method can be used to carry out specific chemical analyses or for ascertaining the structure of both inorganic and organic species. It is a highly specialized instrumental method that is finding increasing applications in the environmental field.

Radioactivity Measurements

Instrumental methods of analysis are routinely used for the measurement of radioactivity in the environment or in research studies that make use of radioactive tracers. Various instruments are available to measure particular types of radiations, the frequency of emissions, or both. A discussion of the many different instruments available is beyond the scope of this book.

PROBLEMS

12.1 List the physical properties of elements or compounds that can be used as the basis for an instrumental measurement.

12.2 What different kinds of interactions between radiant energy and materials or solutions are optical methods designed to measure?

12.3 Why cannot a single instrument be used to measure absorbance at all wavelengths?

12.4 What wavelength range is characteristic of (*a*) visible radiation, (*b*) ultraviolet radiation, and (*c*) infrared radiation? Which range is characterized by (*d*) the highest radiation energy and (*e*) the lowest radiation energy?

12.5 What are the basic components of instruments designed to measure absorption of radiant energy?

12.6 What different changes take place in molecules when absorbing ultraviolet or visible radiation in comparison with those absorbing infrared radiation?

12.7 What similarities and differences exist between the following: (*a*) emission methods, (*b*) methods based upon dispersion and scattering, and (*c*) fluorimetry?

12.8 What different instrumental methods are available for analysis of metals, and what are the advantages and disadvantages of each?

12.9 What is the basic difference between a fluorometer and a fluorescence spectrometer?

12.10 What is the basic difference between optical methods of analysis and emission methods?

12.11 How does flame photometry differ from atomic absorption spectroscopy?

12.12 What advantages does ICP have over atomic absorption spectroscopy for metal analysis?

12.13 What is the difference between turbidimetry and nephelometry?

12.14 What general instrumental methods of analysis might be used for the following substances: (*a*) mercury, (*b*) trichloroethene, (*c*) ammonia, (*d*) sodium, (*e*) benzene, (*f*) cadmium, (*g*) acetate?

12.15 What general instrumental methods of analysis might be used for the following substances: (*a*) chloroform, (*b*) potassium, (*c*) lead, (*d*) nitrate, (*e*) selenium, (*f*) toluene, (*g*) zinc?

12.16 What is the basic difference between potentiometric analysis and polarographic analysis?

12.17 List at least six different substances in water that can be determined by potentiometric analysis.

12.18 List at least four different substances in water that can be determined by polarographic analysis.

12.19 Give the basic concept behind polarographic analysis.

12.20 What difference in metal analysis is achieved by using electrothermal as opposed to direct aspiration atomic absorption spectroscopy?

12.21 Discuss the advantages and disadvantages of ion-specific electrodes versus ion chromatography for analysis of common inorganic cations and anions in water.

12.22 What are four different kinds of membrane electrodes and what is the principle behind the operation of each?

12.23 Give the basic concept behind chromatographic analysis.

12.24 What is the difference between gas chromatography, liquid chromatography, and gas-liquid chromatography?

12.25 What are the basic components of a chromatographic system?

12.26 What differences are there in the manner of preparation and manner of operation of packed columns versus capillary columns for gas chromatography?

12.27 What different detectors are used in gas chromatography and for what general classes of compounds is each particularly useful?

12.28 What is a necessary property of a compound for it to be analyzed by gas chromatography?

12.29 What advantage does high-performance liquid chromatography have over gas chromatography? What is one of HPLC's major limitations, and how has this limitation been overcome in gas chromatography?

12.30 What is the difference between normal-phase and reverse-phase HPLC?

12.31 How does ion chromatography differ from other liquid chromatographic systems? For what compounds is ion chromatography particularly suited?

12.32 What different instrumental methods of analysis are useful in general for analyses of each of the following: (*a*) metals, (*b*) inorganic ions, (*c*) volatile organic compounds, and (*d*) nonvolatile organic compounds?

12.33 What use is often made of mass spectrometry in environmental analysis?

12.34 What is the basic principle behind mass spectrometry?

REFERENCES

Clesceri, L. S., A. E. Greenberg, and A. D. Eaton (eds.): "Standard Methods for the Examination of Water and Wastewater," 20th ed., American Public Health Association, Washington, DC, 1998.

Robinson, J. W.: "Undergraduate Instrumental Analysis," 5th ed., Marcel Dekker, New York, 1995.

Rouessac, F., and A. Rouessac: "Chemical Analysis, Modern Instrumentation Methods and Techniques," John Wiley & Sons, Chichester, England, 2000.

Settle, F. A. (ed.): "Handbook of Instrumental Techniques for Analytical Chemist," Prentice Hall, Englewood Cliffs, NJ, 1997.

Skoog, D. A., F. J. Holler, and T. A. Nieman: "Principles of Instrumental Analysis," 5th ed., Saunders College Pub., Philadelphia, 1998.

Willard, H., J. Merritt, J. Dean, R. Settle (eds.): "Instrumental Methods of Analysis," 7th ed., Wadsworth Publishing, Belmont, CA, 1988.

13

Turbidity

13.1 | GENERAL CONSIDERATIONS

The term *turbid* is applied to waters containing suspended matter that interferes with the passage of light through the water or in which visual depth is restricted. The *turbidity* may be caused by a wide variety of suspended materials that range in size from colloidal to coarse dispersions, depending upon the degree of turbulence. In lake or other waters existing under relatively quiescent conditions, most of the turbidity will be due to colloidal and extremely fine dispersions. In rivers under flood conditions, most of the turbidity will be due to relatively coarse dispersion.

Turbidity may be caused by a wide variety of materials. In glacier-fed rivers and lakes most of the turbidity is due to colloidal rock particles produced by the grinding action of the glacier. The beautiful blues and greens of the lakes and rivers in Glacier National Park are typical examples. As rivers descend from mountain areas onto the plains, they receive contributions of turbidity from farming and other operations that disturb the soil. Under flood conditions, great amounts of topsoil are washed to receiving streams. Much of this material is inorganic in nature and includes clay and silt, but considerable amounts of organic matter are included. As the rivers progress toward the ocean, they pass through urban areas where domestic and industrial wastewaters, treated or untreated, may be added. The domestic waste may add great quantities of organic and some inorganic materials that contribute turbidity. Certain industrial wastes may add large amounts of organic substances and others inorganic substances that produce turbidity. Street washings contribute much inorganic and some organic turbidity. Organic materials reaching rivers serve as food for bacteria, and the resulting bacterial growth and other microorganisms that feed upon the bacteria produce additional turbidity. Inorganic nutrients such as nitrogen and phosphorus present in wastewater discharges and agricultural runoff stimulate the growth of algae, which also contribute turbidity.

From these considerations, it is safe to say that the materials causing turbidity may range from purely inorganic substances to those that are largely organic in nature. This disparity in the nature of the materials causing turbidity makes it impossible to establish hard-and-fast rules for its removal.

13.2 | ENVIRONMENTAL SIGNIFICANCE

Turbidity is an important consideration in public water supplies for three major reasons.

Aesthetics

Consumers of public water supplies expect and have a right to demand turbidity-free water. Most people are aware that domestic wastewater is highly turbid. Any turbidity in the drinking water is automatically associated with possible wastewater pollution and the health hazards occasioned by it. This fear has a sound basis historically, as anyone knows who is familiar with the waterborne epidemics that formerly plagued the water works industry.

Filterability

Filtration of water is rendered more difficult and costly when turbidity increases. The use of slow sand filters has become impractical in most areas because high turbidity shortens filter runs and increases cleaning costs. Satisfactory operation of rapid sand filters generally depends upon effective removal of turbidity by chemical coagulation before the water is admitted to the filters. Failure to do so can result in short filter runs and production of an inferior-quality water, unless filters with special construction and operation are used.

Disinfection

Disinfection of public water supplies is usually accomplished by means of chlorine, ozone, chlorine dioxide, or untraviolet radiation. To be effective, there must be contact between the agent and the organisms that the disinfectant is to kill.

In turbid waters, most of the harmful organisms are exposed to the action of the disinfectant. However, in cases in which turbidity is caused by municipal wastewater suspended solids or runoff from animal feed lots, many of the pathogenic organisms may be encased in the particles and protected from the disinfectant. As an example, a major outbreak of waterborne disease caused by contaminated drinking water in Milwaukee, Wisconsin, in 1993, brought diarrheal illness to about 370,000 people and caused about 100 deaths and 4000 people to be hospitalized. The causative agent was the protozoan *Cryptosporidium parvum,* a common intestinal pathogen in dairy cattle, and was believed to have reached the water supply intake following heavy runoff from farmlands. *Giardia lamblia* is another protozoan that has caused numerous waterborne disease outbreaks in recent years. These organisms resist disinfection by chlorination, making good turbidity removal by filtration one of the important lines of defense. Because of these more recent problems, the U.S. Environmental Protection Agency has set more stringent turbidity standards for drinking waters beginning in January 2002. Turbidity in treated drinking water must never exceed 1 nephelometric turbidity unit (NTU) and must not exceed 0.3 NTU in 95 percent of daily samples in any one month. The World Health Organization guidelines for drinking water suggests 5 NTU turbidity to prevent consumer complaints from appearance, but recommends a median turbidity of only 1 NTU in order to achieve adequate terminal disinfection.

13.3 | STANDARD UNIT OF TURBIDITY

Because of the wide variety of materials that cause turbidity in natural waters, it has been necessary to use an arbitrary standard. The original standard chosen was

$$1 \text{ mg SiO}_2/L = 1 \text{ unit of turbidity}$$

and the silica used had to meet certain specifications as to particle size.

Standard suspensions of pure silica are not now used for measuring turbidity. They were used originally to calibrate the Jackson candle turbidimeter, the former standard instrument for turbidity measurement. This was a rather crude instrument in which the turbidity of a suspension was measured by the depth of suspension through which the outline of a flame from a standard candle just disappeared. The Jackson candle turbidimeter was removed as a standard procedure from the 17th edition of "Standard Methods" as it has generally been replaced in practice by more reliable, sensitive, and easier to use instruments that depend upon the principle of nephelometry. Also, silica as a standard reference material has been replaced by standardized preparations of formazin polymer. The formazin suspensions were first calibrated against the Jackson candle turbidimeter, and thus there is some relationship between turbidity measurements by the Jackson candle turbidimeter and nephelometry. However, the Jackson candle turbidimeter measures the interference to light passage in a straight line while nephelometry measures the scattering of light from particles. Because of the basic difference in the phenomena measured, results from the two different procedures on different suspensions can vary widely. In order to avoid any confusion this may cause, turbidity measurements by the standard nephelometry procedure are now reported in NTU.

13.4 | METHOD OF DETERMINATION

The current standard method for measurement of turbidity depends upon instruments that employ the principles of nephelometry. In the instrument, a light source illuminates the sample and one or more photoelectric detectors are used with a readout device to indicate the intensity of scattered light at right angles to the path of the incident light. It is customary to use a particular formazin polymer suspension as a standard, the stock solutions for which should be prepared monthly. Commercially available preparations such as styrene divinylbenzene beads may also be used. Details are given in "Standard Methods." When using the formazin standard, 40 NTU are about equivalent to 40 Jackson candle turbidity units (JTU). Turbidities as low as 0.02 NTU can be determined by this procedure providing that water with sufficiently low turbidity can be obtained for use in instrument calibration. Samples with turbidities greater than 40 NTU are diluted with turbidity-free water until values within the range of 30 to 40 NTU are obtained. The turbidity is then determined by multiplying the measured turbidity by the dilution factor. In this manner, a single instrument can be used to measure a very broad range of turbidity values. In addition, the instruments can be automated to provide a continuous reading of water turbidity for process control.

13.5 | APPLICATION OF TURBIDITY DATA

Turbidity measurements are of particular importance in the field of water supply. They have limited use in the field of domestic and industrial waste treatment.

Water Supply

Knowledge of the turbidity variation in raw-water supplies is of prime importance to water treatment plant operation. Such measurements are used in conjunction with other information to determine whether a supply requires special treatment by chemical coagulation and filtration before it may be used for a public water supply. Many large cities, such as New York, Boston, and San Francisco, have upland or mountain supplies whose turbidities are so low that treatment other than chlorination is not required, although this may change because of growing concerns with protozoan pathogens.

Water supplies obtained from rivers usually require chemical flocculation because of high turbidity. Turbidity measurements are used to determine the effectiveness of the treatment produced with different chemicals and the dosages needed. Thus, they aid in selection of the most effective and economical chemical to use. Such information is necessary to design facilities for feeding the chemicals and for their storage.

Turbidity measurements help to gauge the amount of chemicals needed from day to day in the operation of treatment works. This is particularly important on "flashy" rivers where no impoundment is provided. Measurement of turbidity in settled water prior to filtration is useful in controlling chemical dosages so as to prevent excessive loading of rapid sand filters. Finally, turbidity measurements of the filtered water are needed to check on faulty filter operation, and to conform with regulatory requirements.

Domestic and Industrial Waste Treatment

The suspended-solids determination is usually employed in waste treatment plants to determine the effectiveness of suspended-solids removal. The determination is slow and time-consuming, and in plants employing chemical treatment, changes in chemical dosages have to be made rather frequently. Turbidity measurements can be used to advantage, because of the speed with which they can be made, to gain the necessary information. By their use, chemical dosages can be adjusted to use the minimum amount of chemical while producing a high-quality effluent.

PROBLEMS

13.1 Discuss the nature of materials causing turbidity in
    (*a*) River water during a flash flood
    (*b*) Polluted river water
    (*c*) Domestic wastewater

13.2 Discuss why turbidity in general cannot be correlated with the weight concentration of suspended matter in water samples (see "Standard Methods").

13.3 What limit is placed on turbidity in water supplies by the present U.S. Environmental Protection Agency standards and why has such a limit been set?

13.4 Why are nephelometric methods now preferred over the Jackson candle turbidimeter?

13.5 Why is silica no longer the preferred material for turbidity standards?

13.6 What is a NTU?

13.7 What is the general detection limit for turbidity measurements?

13.8 What uses are made of turbidity measurements in water treatment plant operation?

13.9 Under what conditions are turbidity measurements useful in wastewater treatment plant operation?

REFERENCE

Clesceri, L. S., A. E. Greenberg, and A. D. Eaton (eds.): "Standard Methods for the Examination of Water and Wastewater," 20th ed., American Public Health Association, Washington, DC, 1998.

C H A P T E R

14

Color

14.1 | GENERAL CONSIDERATIONS

Many surface waters, particularly those emanating from swampy areas, are often colored to the extent that they are not acceptable for domestic or some industrial uses without treatment to remove the color. The coloring materials,[1] many of which are humic substances, result from contact of the water with organic debris, such as leaves, needles of conifers, and wood, all in various stages of decomposition. It consists of vegetable extracts of a considerable variety. Iron is sometimes present as ferric humate and produces a color of high potency.

Natural color exists in water primarily as negatively charged colloidal particles.[2] Because of this fact, its removal can usually be readily accomplished by coagulation with the aid of a salt having a trivalent metallic ion, such as aluminum or iron.

Surface waters may appear highly colored, because of colored suspended matter, when in reality they are not. Rivers that drain areas of red clay soils, such as those in the Piedmont area of the South Atlantic states, become highly colored during times of flood. Color caused by suspended matter is referred to as *apparent color* and is differentiated from color due to vegetable or organic extracts that are colloidal and which is called *true color*. In water analysis it is important to differentiate between apparent and true color. Color intensity generally increases with an increase in pH. For this reason recording pH along with color is advised.

Surface waters may become colored by pollution with highly colored wastewaters. Notable among these are wastes from dyeing operations in the textile industry and from pulping operations in the paper industry. Dye wastes may impart colors of wide variety that are readily recognized and traced. The pulping of wood produces considerable amounts of waste liquors containing lignin derivatives and other materials in dissolved form. The lignin derivatives are highly colored and

---

[1] T. R. J. McCrea, *J. Amer. Water Works Assoc.,* **25:** 931 (1933).
[2] T. Saville, *J. New Engl. Water Works Assoc.,* **31:** 78 (1917).

quite resistant to biological attack. When such material is disposed of into natural watercourses, color is added that persists for great distances. For this reason, disposal of such colored wastes is closely regulated.

14.2 | PUBLIC HEALTH SIGNIFICANCE

Waters containing coloring matter derived from natural substances undergoing decay in swamps and forests are not generally considered to possess harmful or toxic properties. The natural coloring materials, however, give a yellow-brownish appearance to the water, somewhat like that of urine, and there is a natural reluctance on the part of water consumers to drink such waters because of the associations involved. Also, disinfection by chlorination of waters containing natural organics results in the formation of chloroform, other trihalomethanes, and a range of other chlorinated organics, leading to problems of much current concern.

It is the responsibility of any water purveyor, public or private, to produce a product that is hygienically safe. Public health officials are aware of the fact that consumers will seek other sources of drinking water if the public water supply is not aesthetically acceptable, no matter how safe it may be from the hygienic viewpoint. Where waters are not aesthetically acceptable, consumers often shun safe domestic supplies and use waters from uncontrolled springs or private wells which may serve as foci for dissemination of pathogenic organisms. For this reason, waters intended for human use should not exceed 15 color units, the recommended or secondary standard set by the U.S. EPA and the World Health Organization (WHO) guidelines.

14.3 | METHODS OF DETERMINATION

Natural color, like turbidity, is due to a wide variety of substances, and it has been necessary to adopt an arbitrary standard for its measurement. This standard is employed directly and indirectly in the measurement of color. Many samples require pretreatment to remove suspended matter before true color can be determined. The method of pretreatment must be carefully selected to avoid introduction of errors.

Standard Color Solutions

Waters containing natural color are yellow-brownish in appearance. Through experience, it has been found that solutions of potassium chloroplatinate (K_2PtCl_6) tinted with small amounts of cobalt chloride yield colors that are very much like the natural colors. The shading of the color can be varied to match natural hues very closely by increasing or decreasing the amount of cobalt chloride.

The color produced by 1 mg/L of platinum (in the form of K_2PtCl_6) is taken as the standard unit of color. The usual procedure is to prepare a stock solution of K_2PtCl_6 that contains 500 mg/L of platinum. Cobalt chloride is added to provide the proper tint. The stock solution has a color of 500 units, and a series of working standards may be prepared from it by dilution. A matched set of color-comparison tubes, commonly called Nessler tubes, as shown in Fig. 14.1, are usually used to

Figure 14.1
Color-comparison tubes, commonly called Nessler tubes.

contain the standards. A series ranging from 0 to 70 color units is employed and will serve for several months, provided that it is protected from dust and evaporation. The color-comparison tubes should be a matched set conforming to American Public Health Association standards as described in the introductory chapter of "Standard Methods."

Samples subjected to analysis may contain suspended matter that will interfere with the measurement of true color. Apparent color is determined on the sample "as is." Suspended matter must be removed to enable determination of true color. This can usually be accomplished by centrifuging the sample to separate the suspended solids. Analysis is performed on the clarified liquor. Filtration is not recommended because of possible adsorption of color on the filtering medium.

Samples with color less than 70 units are tested by direct comparison with the prepared standards. For samples with a color greater than 70 units, a dilution is made with demineralized water to bring the resulting color within the range of the standards, and calculation of color is made, using a correction factor for the dilution employed.

Methods Employing Proprietary Devices

A number of instruments have been developed for the measurement of color to eliminate the need for renewing standard color solutions from time to time. Most of these instruments employ colored glass disks that simulate the various color standards when used in the particular instrument.

The proprietary devices find their greatest use in water works laboratories where trained chemists are not employed, or for field measurements where the use of standard color solutions is not practical. They are not accepted as a standard procedure for measuring color because of variations in the color of the glass disks and their tendency to change characteristics, owing to fingerprints, dust, and so on. They should always be standardized against standard solutions of K_2PtCl_6 for highly important work.

Spectrophotometric Methods Applicable to Domestic and Industrial Wastewaters

Many industrial wastes are highly colored, and some contain colored substances that are quite resistant to biological destruction. Regulations concerning the color of effluents that may be discharged to streams are common. Evaluation of the color of yellow-brownish-hue wastes can be made by the standard procedures previously described. Other systems of measurement have to be used to measure and describe colors that do not fall into this classification. "Standard Methods" lists three possible spectrophotometric methods for color determination. One involves use of a normal spectrophotometer with an operating range from 400 to 700 nm and the collection of percent transmissive values on a sample at several different wavelengths. A calculation procedure together with a color table is then used to express sample color in terms of dominant wavelength, luminance, and purity. The other two procedures make use of filter photometers and three different color filters to achieve similar characterization as provided by the spectrophotometer approach. These procedures obviously provide much more detail about color than given by the simple platinum standard approach. They may be appropriate for use when looking for the source of color in a water supply or in evaluating the effectiveness of color removal procedures. The results, however, do not translate directly into the U.S. EPA secondary standard or WHO guideline of 15 color units for drinking water. Thus they may not be appropriate for routine monitoring to ensure compliance with drinking water requirements.

14.4 | INTERPRETATION AND APPLICATION OF COLOR DATA

The color of surface waters utilized for domestic supplies is of major concern for reasons mentioned previously. Many industrial processes also require the use of color-free water. Removal of color is an expensive matter when capital investment and operating costs are considered. In addition, color in natural waters is an indirect indicator of the potential for trihalomethane formation during disinfection with chlorine. Therefore, a water supply is generally desired with a color low enough so that chemical treatment will not be required and trihalomethane formation will not constitute a burdensome treatment problem. This "prospecting" may or may not be successful. If it is, color data can be used as one of the parameters to satisfy the client that expensive chemical treatment or alternative means of disinfection are not necessary. If it is not successful, color data can be used along with other information to support the case that more costly forms of treatment are needed to produce an acceptable supply.

Before a chemical treatment plant is designed, research should be conducted to ascertain the best chemicals to use and amounts required. In dealing with colored waters, color determinations serve as the basis of the decisions. Such data must be obtained for proper selection of chemical feeding machinery and the design of storage space.

Once operation of the treatment facilities has begun, color determinations on the raw and finished wastes serve to govern the dosages of chemicals used, to ensure economical operation, and to produce a low-color water that is well within accepted limits.

PROBLEMS

14.1 Discuss briefly the causes of color in water.

14.2 Differentiate between "apparent" and "true" color.

14.3 What limit is generally placed on color and why are such standards set?

14.4 What is used as the standard unit of color?

14.5 What is the purpose of adding cobalt chloride to color standards?

14.6 What is the difference between the spectrophotometric methods for measuring color and the platinum standard procedures?

14.7 What uses are generally made of color information in practice?

14.8 What problems might be caused by color-causing materials in water other than the color itself?

14.9 How is color generally removed from water?

REFERENCE

Clesceri, L. S., A. E. Greenberg, and A. D. Eaton (eds.): "Standard Methods for the Examination of Water and Wastewater," 20th ed., American Public Health Association, Washington, DC, 1998.

15

Standard Solutions

15.1 | GENERAL CONSIDERATIONS

Chemists familiar with the analysis of water and wastewater involving volumetric procedures have learned that the use of standard reagents of definite normality saves a great deal of time in calculating results in terms of milligrams per liter. The preparation of solutions of a definite normality is not a tedious procedure if a logical system is used. Two or three important steps are involved, depending upon the nature of the materials.

Selection of the Proper Normality

In water and wastewater analysis it is usually desirable to report results in terms of milligrams per liter of some particular ion, element, or compound. As a rule, it is most convenient to have the standard titrating agent of such strength that 1 mL is equivalent to 1 mg of the material being measured. Thus, 1 liter of a standard solution is ordinarily equivalent to 1000 mg or 1 g of the measured substance. The desired normality of the titrant is obtained by the relationship of 1 to the equivalent weight (EW) of the measured material (see Secs. 2.2 and 11.4). For example, the normality of acid solutions used to measure ammonia, ammonia nitrogen, and alkalinity (as $CaCO_3$) might be as follows:

	Ammonia	*Ammonia nitrogen*	*Alkalinity (as $CaCO_3$)*
$\dfrac{1}{EW} =$	$1/17 = N/17$	$1/14 = N/14$	$1/50 = N/50$
$=$	$0.0588\ N$	$0.0714\ N$	$0.02\ N$

The normalities of basic solutions for measuring carbon dioxide (as CO_2) and mineral acidity (as $CaCO_3$) are

$$
\begin{array}{cc}
& Acidity \\
CO_2 & (as\ CaCO_3) \\
\dfrac{1}{EW} = 1/44 = N/44 & 1/50 = N/50 \\
= 0.0227\ N & 0.02\ N
\end{array}
$$

The normalities of silver nitrate solutions for measuring chloride or sodium chloride are

$$
\begin{array}{cc}
Chloride & Sodium\ chloride \\
\dfrac{1}{EW} = 1/35.45 = N/35.45 & 1/58.44 = N/58.44 \\
= 0.0282\ N & = 0.0171\ N
\end{array}
$$

The normality of reducing agents for measuring oxygen[1] is

$$
\begin{array}{c}
Oxygen \\
\dfrac{1}{EW} = 1/8 = N/8 \\
= 0.125\ N
\end{array}
$$

The normality of oxidizing agents is obtained in a similar manner.

From the fact that standard solutions of titrating agents are of such strength that 1 mL is equivalent to 1 mg of the measured material, it will be readily apparent that when 1-liter samples are titrated the buret reading will give milligrams per liter directly. Usually, it is inconvenient to use 1-liter samples, and calculations are easily made by the simple formula

$$
\text{mL titrant used} \times \frac{1000}{\text{mL sample}} = \text{mg/L}
$$

In some instances, as in the determination of dissolved oxygen where a fixed sample size is used, the strength of the titrant is adjusted so that each milliliter of titrant is equivalent to 1 mg/L.

Preparation of a Solution of Proper Normality

In cases in which a standard solution can be prepared from materials of known purity that can be accurately weighed on the analytical balance, the desired amount can be weighed, transferred to a volumetric flask, and diluted to the proper volume. Such solutions may be used without standardization against primary standards.

Many materials from which standard solutions are prepared are of such a character that their purity is not accurately known, or it may be impossible to weigh them ex-

---

[1]Actually a N/40 or 0.025 N solution is used in practice, for reasons given in Sec. 22.4.

actly. In these cases, a solution is prepared that is known to be slightly stronger than desired, and it can be kept most conveniently in a graduated cylinder until standardization is completed. Standardization is accomplished by using a suitable primary standard.

Standardization of Solutions with Primary Standards

The procedure for standardizing solutions to an exact normality is somewhat peculiar to water and wastewater analysis and is not usually described in quantitative chemistry textbooks. Six fundamental steps are involved, as follows:

1. Calculate the weight of the primary standard that is exactly equivalent to 1 liter of the solution to be standardized.
2. Weigh three or four samples of the heat-dried primary standard that are sufficient to use about 20 mL of the solution. Weighings must be exact, and corrections must be made for percent purity if it differs significantly from 100 percent.
3. Calculate the volume of titrant of the desired normality needed to react with the corrected weight of each sample.
4. Add sufficient demineralized water, and other reagents as needed, to each sample of primary standard to accomplish solution, and titrate two of them with the reagent to be standardized to the required end point. The titrations should be less than the calculated amounts obtained in step 3. If so, the solution is stronger than desired, and the difference between the actual and calculated titrations represents the deficiency of water.
5. Calculate the amount of water to be added to the remaining solution. Drain the contents of the buret into the stock supply in the graduated cylinder and measure the total volume remaining. The amount of water to be added may be calculated from the following expression:

$$\frac{\text{Volume remaining}}{\text{Actual titration}} \text{ (calculated titration} - \text{actual titration)} = \text{mL}$$

Make separate calculations for each of the titrations and add the least amount of water indicated.[2] *Mix thoroughly after adding the water, rinse, and fill the buret with the new solution.*

6. Titrate additional samples of primary standard and repeat steps 4 and 5 until the proper strength has been reached as shown by a correlation of actual titrations with calculated values. The solution may then be considered to have the desired normality.

15.2 | PREPARATION OF 1.00 N AND 0.020 N H₂SO₄ SOLUTIONS

Standard solutions of sulfuric acid are used for the determination of alkalinity which is normally expressed in terms of $CaCO_3$, with an equivalent weight of 50;

---

[2]Beginners should add slightly less water to avoid overdilution.

therefore, N/50 or 0.020 N solutions are required. Because large amounts of this reagent are used, it is most convenient to prepare a stock solution of 1.00 N acid and prepare the 0.020 N solution from it by simple dilution.

The purity of sulfuric acid as purchased usually varies from about 96 to 98 percent. In addition, it is very difficult to weigh accurately because of its hygroscopic properties. Solutions prepared from it must be standardized by means of some primary standard. Sodium carbonate is the primary standard usually used. The analytical grade is satisfactory, provided that it has been dried for 1 h at 140°C and kept in a desiccator prior to use.

Calculation of Concentrated H_2SO_4 Needed

In order to simplify calculations, they will be made on the basis of 1-liter amounts. By definition, 1 liter of 1.00 N acid contains 1.008 g of available hydrogen ion. Calculations are as follows:

$$1 \text{ FW or 98 g pure } H_2SO_4 = 2.016 \text{ g H}^+$$

$$\frac{FW}{2} \text{ or 49 g pure } H_2SO_4 = 1.008 \text{ g H}^+$$

Assume that concentrated acid is 96 percent pure.

Then $\qquad \dfrac{49}{0.96} = 51 \text{ g concentrated acid} = 1.008 \text{ g H}^+$

It is desirable that the solution be slightly stronger than 1.00 N. To be sure of this, take 5 percent excess:

$$51 \times 1.05 = 53.5 \text{ g}$$

Preparation of 1.00 N Acid Solution

Weigh approximately 53 ± 1 g of concentrated acid into a small beaker on a trip balance. Place about 500 mL of demineralized water in a 1-liter graduated cylinder and add the acid to it. Rinse the contents of the beaker into the cylinder with demineralized water, and add water to the 1-liter mark. Mix thoroughly by stirring or pouring back and forth from the cylinder into a large beaker. Cool to room temperature before use.

Calculation of Primary Standard Needed

Sodium carbonate is a convenient primary standard. It has a molecular weight of 106 and an equivalent weight of 53 when reacting with H_2SO_4 to a pH of 3.7. Thus, using the symbol "≎" to mean *equivalent to*:

$$53 \text{ g Na}_2CO_3 ≎ 1000 \text{ mL } 1.00 \text{ N } H_2SO_4$$

or $\qquad 1.06 \text{ g Na}_2CO_3 ≎ 20 \text{ mL } 1.00 \text{ N } H_2SO_4$

Weigh four samples of Na_2CO_3 ranging from 1.00 to 1.10 g and proceed as described in the third part of Sec. 15.1.

Preparation of 0.020 N Acid Solution

When a 1.00 N acid solution is available, solutions of any normality less than 1.00 N can be prepared from it by dilution, provided that proper care is used in measuring the amount of 1.00 N acid needed and dilutions are made in volumetric flasks. The amount of acid of any normality needed to make a definite volume of an acid of another normality may be calculated from the relationship

$$mL \times N = mL \times N$$

If it is desired to make 1 liter of 0.020 N acid from a stock supply of 1.00 N acid, the calculation is as follows:

$$mL \times 1.0 = 1000 \times 0.02$$
$$mL = 20$$

Twenty milliliters of 1.00 N acid when diluted to 1000 mL with demineralized water and *thoroughly mixed* yields a 0.020 N solution that is satisfactory for most purposes. For referee work, it would be advisable to check the normality of the 0.020 N acid against weighed samples of a primary standard.

15.3 | PREPARATION OF 1.00 N AND 0.020 N NaOH SOLUTIONS

Standard solutions of sodium hydroxide are used to measure carbon dioxide and acidity. The equivalent weight of carbon dioxide when reacting with sodium hydroxide, to pH 8.3 or the phenolphthalein end point, is 44, as may be calculated from the equation

$$\underset{44}{CO_2} + Na^+ + \underset{17}{OH^-} \rightarrow Na^+ + HCO_3^- \tag{15.1}$$

Therefore, N/44 or 0.0227 N solutions of NaOH are best suited for determination of carbon dioxide. Mineral acidity is always expressed in terms of calcium carbonate, which has an equivalent weight of 50, and 0.020 N solutions of bases are used for its determination.

In practice, it is most convenient to prepare a 1.00 N solution of NaOH as a stock supply and make solutions of lower normality by dilution, as with sulfuric acid. In some laboratories, so few determinations of carbon dioxide are made that it is impractical to keep a supply of both 0.0227 N and 0.020 N solutions on hand. In such cases, the 0.020 N solution is used for both determinations, and a factor of 0.88 is applied to correct buret readings when carbon dioxide is measured.

Crystal or pellet forms of sodium hydroxide cannot be purchased in a pure form. They are always contaminated with sodium carbonate as a result of reaction with carbon dioxide of the air during manufacture. Even the so-called analytical reagent grade contains several percent of sodium carbonate and is unfit for the preparation of standard solutions without purification. Several methods are used, but only two are commonly used in water quality analytical laboratories. Purified sodium hydroxide can be purchased in 50 percent solution form.

A primary standard is required in the preparation of standard solutions of sodium hydroxide. Potassium acid phthalate ($KHC_8H_4O_4$) is excellent because of its high equivalent weight and other desirable properties. It should be dried at 100 to 105°C for 1 h and kept in a desiccator prior to use.

Purification of NaOH

All sodium hydroxide must be subjected to a purification process to free it of sodium carbonate which also has basic properties. Two methods will be described in order of their preference.

1. Sodium carbonate is relatively insoluble in concentrated (approximately 50 percent or 18 M) solutions of sodium hydroxide. If 500 g of stick- or pellet-form sodium hydroxide is added to 500 mL of distilled water, the sodium hydroxide will dissolve, leaving the sodium carbonate undissolved. The resulting solution will be rather turbid because of the suspended sodium carbonate. On standing several days, the carbonate will float or settle, and a clear solution will result that is sufficiently free of carbonate for the preparation of standard solutions. Siphon the purified 50 percent solution into a Pyrex bottle, and use a rubber stopper to exclude the air. It is good practice to keep a considerable supply of purified 50 percent sodium hydroxide on hand because of the time involved in its preparation.

2. Dilute solutions of sodium hydroxide may be freed of carbonate by precipitation with barium hydroxide.

$$2Na^+ + CO_3^{2-} + Ba(OH)_2 \rightarrow BaCO_3(s) + 2NaOH \qquad (15.2)$$

The resulting barium carbonate precipitate may be removed by filtration (if protected from carbon dioxide of air), or it may be allowed to settle and the clarified solution siphoned to another bottle. The purified solution must be standardized after this treatment, as barium hydroxide is used in excess and an amount of sodium hydroxide equivalent to the sodium carbonate originally present remains.

Calculation of NaOH Needed

For purposes of these calculations, it is assumed that a purified solution of sodium hydroxide, 50 percent, is available and that approximately 1 liter of a 1.00 N solution is to be prepared. By definition, 1 liter of 1.00 N base contains the equivalent of 1.008 g of H^+ or 17 g of OH^- per liter. Calculations are as follows:

$$1 \text{ FW or } 40 \text{ g pure NaOH} = 17 \text{ g OH}^-$$

The stock solution of purified NaOH is 50 percent strength. Therefore,

$$\frac{40}{0.5} = 80 \text{ g } 50\% \text{ NaOH} = 17 \text{ g OH}^-$$

Since the stock solution of NaOH may not contain exactly 50 percent and it is desirable to prepare a solution that is slightly stronger than 1.00 N, take 10 percent excess:

$$80 \times 1.1 = 88 \text{ g}$$

Preparation of 1.00 N NaOH Solution

As rapidly as possible, so as to minimize absorption of carbon dioxide from the air, weigh 88 ± 1 g of the purified 50 percent sodium hydroxide solution into a small Erlenmeyer flask on a trip balance. Place 500 mL of carbon-dioxide-free demineralized water in a 1-liter graduated cylinder, and add the sodium hydroxide solution. Rinse the Erlenmeyer flask with carbon-dioxide-free water, and add rinsings to the cylinder. Dilute to approximately 1 liter with carbon-dioxide-free water and mix thoroughly with a plunger-type stirrer to minimize contact with the air. Protect the solution from the air by keeping an inverted beaker or some other form of cap over the top of the cylinder until standardization is completed.

Calculation of Primary Standard Needed

Potassium acid phthalate is an excellent primary standard. It may be obtained in essentially 100 percent pure form. Its EW is equal to its FW, which is 204:

$$204 \text{ g KHC}_8\text{H}_4\text{O}_4 \backsimeq 1000 \text{ mL } 1.00 \text{ N NaOH}$$

or

$$2.04 \text{ g KHC}_8\text{H}_4\text{O}_4 \backsimeq 10 \text{ mL } 1.00 \text{ N NaOH}$$

Weigh four samples of $KHC_8H_4O_4$ ranging from 2.0 to 2.2 g, and proceed as described in the third part of Sec. 15.1. Titration should be to pH 8.3 or the phenolphthalein end point. When standardization is completed, the 1.00 N solution should be stored in a Pyrex bottle fitted with a rubber stopper to exclude air.

Preparation of 0.020 N NaOH Solution

Proceed in a manner similar to the instructions for preparing 0.020 N acid. The 1.00 N hydroxide should be diluted with carbon-dioxide-free water, stored in a Pyrex bottle, and protected from the carbon dioxide of the atmosphere.

Standardization with Secondary Standards

The 1.00 N and 0.020 N solutions of sodium hydroxide may be standardized against corresponding solutions of sulfuric acid. The acid solutions serve as secondary standards, and any error made in their preparation will be reflected in the hydroxide solutions. A pH of 3.7 or an indicator such as methyl orange or bromphenol blue can be used for the titration end point, and whether the acid or the hydroxide is used as the titrant is optional. It is usually most convenient to use the acid.

PROBLEMS

15.1 Define: molar solution, normal solution, standard solution, mole, equivalent weight, and milliequivalent.

15.2 How may carbonate-free sodium hydroxide solutions be prepared?

15.3 (a) What is the normality of a solution of H_2SO_4 if 16.2 mL were required to neutralize 1.22 g of Na_2CO_3 to the methyl orange end point?

   (b) How many mL of water must be added to the solution in part (a) to make it exactly 1.00 N? Assume you have 927 mL of the solution.

15.4 How many mL of 1.00 N NaOH are required to neutralize 0.20 g of HCl?

15.5 A sample of potassium acid phthalate weighing 3.75 g required 15.0 mL of a NaOH solution for titration to the phenolphthalein end point. If 460 mL of NaOH remain, how much water should be added to it to make it exactly 1.00 N?

15.6 A sample of Na_2CO_3 weighing 1.50 g required 25.0 mL of a H_2SO_4 solution for titration to the methyl orange end point. If 980 mL of H_2SO_4 remain, how much water should be added to it to make it exactly 1.00 N?

15.7 How many mL of 1.00 N NaOH are required to prepare 500 mL of 0.0227 N NaOH?

15.8 How many mL of 1.00 N H_2SO_4 are required to prepare 2 liters of 0.050 N H_2SO_4?

15.9 In the preparation of two liters of 1.00 N acid from 35 percent hydrochloric acid, what weight of the impure acid should be taken, assuming standardization in the recommended manner?

15.10 A potassium dichromate ($K_2Cr_2O_7$) solution is commonly used under acid conditions to measure the chemical oxygen demand (COD) of wastewater. The results are expressed as mg/L oxygen. How many grams of potassium dichromate must be added to 1 liter of solution so that 1.00 mL is equivalent to 10.0 mg of oxygen? What is the normality of this solution?

15.11 A ferrous ammonium sulfate $[Fe(NH_4)_2(SO_4)_2]$ solution is used under acid conditions for titration in the COD test. How many grams of ferrous ammonium sulfate must be added to 1 liter of solution so that 1.00 mL is equivalent to 1.00 mg of oxygen? What is the normality of this solution?

REFERENCE

Clesceri, L. S., A. E. Greenberg, and A. D. Eaton (eds.): "Standard Methods for the Examination of Water and Wastewater," 20th ed., American Public Health Association, Washington, DC, 1998.

pH

16.1 | GENERAL CONSIDERATIONS

pH is a term used rather universally to express the intensity of the acid or alkaline condition of a solution. It is a way of expressing the hydrogen-ion concentration, or more precisely, the hydrogen-ion activity. It is important in almost every phase of environmental engineering and science. In the field of water supplies, it is a factor that must be considered in chemical coagulation, disinfection, water softening, and corrosion control. In wastewater treatment employing biological processes, pH must be controlled within a range favorable to the particular organisms involved. Chemical processes used to coagulate wastewaters, dewater sludges, or oxidize certain substances, such as cyanide ion, require that the pH be controlled within rather narrow limits. For these reasons and because of the fundamental relationships that exist between pH, acidity, alkalinity, and chemical speciation in general, it is very important to understand the theoretical as well as the practical aspects of pH.

16.2 | THEORETICAL CONSIDERATIONS

The concept of pH evolved from a series of developments that led to a fuller understanding of acids and bases. Acids and bases were originally distinguished by their difference in taste and later by the manner in which they affected certain materials that came to be known as indicators. With the discovery of hydrogen by Cavendish in 1766, it soon became apparent that all acids contained the element hydrogen. Chemists soon found that neutralization reactions between acids and bases always produced water. From this and other related information, it was concluded that bases contained hydroxyl groups.

In 1887 Arrhenius announced his theory of ionization. Since that time acids have been considered to be substances that dissociate to yield hydrogen ions or protons, and bases have been considered to be substances that dissociate to yield hydroxide ions (also called hydroxyl ions). According to the concepts of Arrhenius, strong acids and bases are highly ionized and weak acids and bases are poorly ionized in aqueous solution. Proof of these claims had to await the development of suitable devices for the measurement of hydrogen-ion concentration or activity.

Measurement of Hydrogen-Ion Activity

The hydrogen electrode (Sec. 12.3) was found to be a suitable device for measuring hydrogen-ion activity. With its use, it was found that pure water dissociates to yield a concentration of hydrogen ions equal to about 10^{-7} mol/L.

$$H_2O \rightleftharpoons H^+ + OH^- \tag{16.1}$$

Since water dissociates to produce one hydroxide ion for each hydrogen ion, it is obvious that about 10^{-7} mol/L of hydroxide ion is produced simultaneously. By substitution into the equilibrium equation, we obtain

$$\frac{\{H^+\}\{OH^-\}}{\{H_2O\}} = K \tag{16.2}$$

but, since the concentration of water is so extremely large and is diminished so very little by the slight degree of ionization, it may be considered as constant (its activity equals 1.0), and Eq. (16.2) may be written as

$$\{H^+\}\{OH^-\} = K_W \tag{16.3}$$

and for pure water at about 25°C,

$$\{H^+\}\{OH^-\} = 10^{-7} \times 10^{-7} = 10^{-14} \tag{16.4}$$

This is known as the ion product or ionization constant for water.

When an acid is added to water, it ionizes in the water and the hydrogen-ion activity increases; consequently, the hydroxide-ion activity must decrease in conformity with the ionization constant. For example, if acid is added to increase the $\{H^+\}$ to 10^{-1}, the $\{OH^-\}$ must decrease to 10^{-13}:

$$10^{-1} \times 10^{-13} = 10^{-14}$$

Likewise, if a base is added to water to increase the $\{OH^-\}$ to 10^{-3}, the $\{H^+\}$ decreases to 10^{-11}. It is important to remember that the $\{OH^-\}$ or the $\{H^+\}$ can never be reduced to zero, no matter how acidic or basic the solution may be.

The pH Concept

Expression of hydrogen-ion activity in terms of molar concentrations is rather cumbersome. In order to overcome this difficulty, Sorenson (in 1909) proposed to express such values in terms of their negative logarithms and designated such values as p_H^+. His symbol has been superseded by the simple designation pH. The term may be represented by

$$pH = -\log \{H^+\} \quad \text{or} \quad pH = \log \frac{1}{\{H^+\}} \tag{16.5}$$

and the pH scale is usually represented as ranging from 0 to 14, with pH 7 at 25°C representing absolute neutrality.

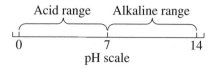

Because K_W changes with change in temperature, the pH of neutrality changes with temperature as well, being 7.5 at 0°C and 6.5 at 60°C. Acid conditions increase as pH values decrease, and alkaline conditions increase as the pH values increase.

16.3 | MEASUREMENT OF pH

The hydrogen electrode is the absolute standard for the measurement of pH. It is rather cumbersome and not well adapted for universal use, particularly in field studies or in solutions containing materials that are adsorbed on platinum black. A wide variety of indicators were calibrated with the hydrogen electrode to determine their color characteristics at various pH levels. From these studies it became possible to determine pH values fairly accurately by choosing an indicator that exhibited significant color changes in the particular range involved. With the use of about six to eight indicators, it is possible to determine pH values in the range of interest (see Sec. 11.4). Their use has been superseded by development of the glass electrode.

About 1925 it was discovered that an electrode could be constructed of glass (Sec. 12.3) that would develop a potential related to the hydrogen-ion activity without interference from most other ions. Its use has become the standard method of measuring pH.

Measurement with the Glass Electrode

pH meters employing the glass electrode are manufactured by many companies. They range from portable battery-operated units selling for a few hundred dollars to highly precise instruments selling for over a thousand dollars. Units that could be operated on 110-volt alternating current were developed about 1940 and are highly satisfactory for most routine laboratory purposes, being capable of measuring pH within a ±0.1 pH unit. The small portable battery-operated units are most suitable for field work. Figure 16.1 shows two examples of available instruments.

pH measurements can be made in a wide variety of materials and under extreme conditions, provided that attention is paid to the type of electrode used. Measurement of pH values above 10 and at high temperatures is best made with special glass electrodes designed for such service. The pH of semisolid substances can be made with a spear-type electrode. The instruments are normally standardized with buffer solutions of known pH values. Preferably, a buffer solution having a pH within 1 to 2 units of the sample pH should be used.

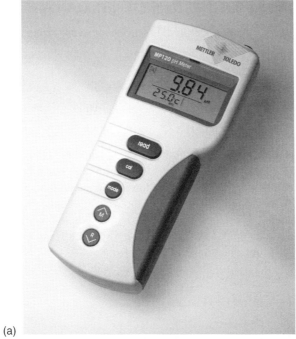

(a)

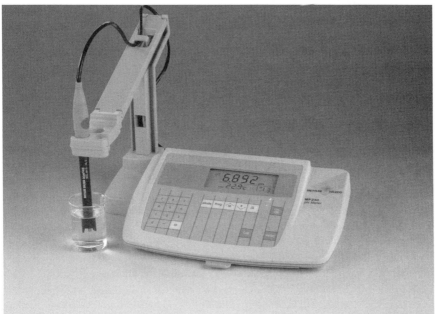

(b)

Figure 16.1

Commercial models of pH meters. (*a*) Small battery-operated model for use in the field or laboratory. (*b*) Line-operated laboratory benchtop model. (*Courtesy of Mettler-Toledo, Inc.*)

16.4 | INTERPRETATION OF pH DATA

pH data should always be interpreted in terms of hydrogen-ion activity, which, of course, is a measure of the intensity of acidic or basic conditions. For approximation, however, we can assume $[H^+] = \{H^+\}$. Thus,

$$\text{at pH 2} \qquad [H^+] = 10^{-2}$$
$$\text{at pH 10} \qquad [H^+] = 10^{-10}$$
$$\text{at pH 4.5} \qquad [H^+] = 10^{-4.5}$$

pH does not measure total acidity or total alkalinity. This can be illustrated by comparing the pH of 0.10 N solutions of sulfuric acid and acetic acid, which have the same neutralizing value. The pH of the former is approximately 1 because of its high degree of ionization, and the pH of the latter is about 3 because of its low degree of ionization.

In some instances the pOH, or hydroxide-ion activity, of a solution is of major interest. It is customary to calculate pOH from pH values, using the relationship given in Eq. (16.3). Approximations are often made from the relationship

$$\text{pH} + \text{pOH} = 14 \qquad (16.6)$$
or
$$\text{pOH} = 14 - \text{pH} \qquad (16.7)$$

The $[OH^-]$ of a solution can never be reduced to zero, no matter how acidic the solution is, nor can the $[H^+]$ ever be reduced to zero, no matter how alkaline a solution becomes. However, pH itself can reach zero or even less in highly acidic solutions ($\{H^+\} \geq 1.0$).

Concepts of pOH, or hydroxide-ion activity, are of particular importance in precipitation reactions involving formation of hydroxides. Examples are the precipitation of Mg^{2+} in the softening of water with lime and in chemical coagulation processes employing iron and aluminum salts.

PROBLEMS

16.1   What is the relationship (a) between pH and hydrogen-ion activity and (b) between pH and hydroxide-ion activity?

16.2   What would be the pH of a solution containing (a) 1.008 g of hydrogen ion per liter, (b) 0.1008 g of hydrogen ion per liter, and (c) 1.7×10^{-8} g of OH^- per liter? Assume ion concentration equals ion activity.

16.3   One solution has a pH of 4.0 and another a pH of 6.0. What is (a) the hydrogen-ion activity and (b) the hydroxide-ion activity in each of the solutions?

16.4   A decrease in pH of one unit represents how much of an increase in hydrogen-ion activity?

16.5   A 50 percent decrease in hydrogen-ion activity represents how much of an increase in pH units?

16.6   Approximately what is the pH of 2.00 N HCl solution?

16.7   Approximately what is the pH of 0.020 N NaOH solution?

16.8 What is the hydroxide-ion concentration if the hydrogen-ion concentration is 3.0×10^{-2} mol/L? Assume ion concentration equals ion activity.

16.9 What type of electrode is used for pH measurements, and how is it calibrated?

REFERENCES

Clesceri, L. S., A. E. Greenberg, and A. D. Eaton (eds.): "Standard Methods for the Examination of Water and Wastewater," 20th ed., American Public Health Association, Washington, DC, 1998.

Day, Jr., R. A., and A. L. Underwood: "Quantitative Analysis," 6th ed., Prentice Hall, Englewood Cliffs, NJ, 1991.

Enke, C. G.: "The Art and Science of Chemical Analysis," John Wiley & Sons, New York, 2001.

Skoog, D. A., D. M. West, and F. J. Holler: "Fundamentals of Analytical Chemistry," 7th ed., College Pub., Fort Worth, TX, 1996.

CHAPTER 17

Acidity

17.1 | GENERAL CONSIDERATIONS

Most natural waters, domestic wastewaters, and many industrial wastes are buffered principally by a carbon dioxide–bicarbonate system. By reference to Fig. 4.8, which shows titration curves for several weak acids, it will be noted from the curve for carbonic acid that the stoichiometric end point is not reached until the pH has been raised to about 8.5. On the basis of this information, it is customary to consider that all waters having a pH lower than 8.5 contain acidity. Usually the phenolphthalein end point at pH 8.2 to 8.4 is taken as the reference point. Inspection of the curve for carbonic acid in Fig. 4.8 shows that at pH 7.0 considerable carbon dioxide remains to be neutralized. It also shows that carbon dioxide (carbonic acid) alone will not depress the pH below a value of about 4.

Figure 4.7 shows a titration curve for a strong acid, and from the nature of the curve, it may be concluded that neutralization of the acid is essentially complete at pH 4. Thus, from the nature of the titration curves for carbonic acid and for strong acids, it becomes obvious that the acidity of natural waters is caused by carbon dioxide or by strong mineral acids, the former being the effective agent in waters having pH values greater than 4 and the latter the effective agent in waters with pH values less than 4, as shown in Fig. 17.1.

17.2 | SOURCES AND NATURE OF ACIDITY

Carbon dioxide is a normal component of all natural waters. It may enter surface waters by absorption from the atmosphere, but only when its concentration in water is less than that in equilibrium with carbon dioxide in the atmosphere, in accordance with Henry's law. Carbon dioxide may also be produced in waters through biological oxidation of organic matter, particularly in polluted water. In such cases, if photosynthetic activity is limited, the concentration of carbon dioxide in the water may exceed equilibrium with that of the atmosphere and carbon dioxide will escape from the liquid. Thus, it may be concluded that surface waters are constantly ab-

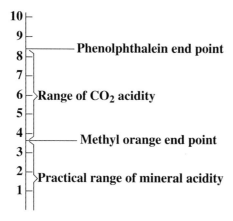

Figure 17.1
Types of acidity of importance in ordinary
water and wastewater analysis, and the pH
ranges in which they are significant.

sorbing or giving up carbon dioxide to maintain an equilibrium with the atmosphere. The amount that can exist at equilibrium is very small because of the low partial pressure of carbon dioxide in the atmosphere.

Groundwaters and waters from the hypolimnion of stratified lakes and reservoirs often contain considerable amounts of carbon dioxide. This concentration results from bacterial oxidation of organic matter with which the water has been in contact, and under these conditions, the carbon dioxide is not free to escape to the atmosphere. Carbon dioxide is an end product of both aerobic and anaerobic bacterial oxidation; therefore, its concentration is not limited by the amount of dissolved oxygen originally present. It is not uncommon to encounter groundwaters with 30 to 50 mg/L of carbon dioxide. This is particularly true of waters that have percolated through soils that do not contain enough calcium or magnesium carbonate to neutralize the carbon dioxide through formation of bicarbonate.

$$CO_2 + CaCO_3(s) + H_2O \rightarrow Ca^{2+} + 2HCO_3^- \tag{17.1}$$

Mineral acidity is present in many industrial wastes, particularly those of the metallurgical industry and some from the production of synthetic organic materials. Certain natural waters may also contain mineral acidity. The drainage from abandoned mines and lean ore dumps will contain significant amounts of sulfuric acid or salts of sulfuric acid if sulfur, sulfide, or iron pyrite are present. Conversion of these materials to sulfuric acid and sulfate is brought about by sulfur-oxidizing bacteria under aerobic conditions.

$$2S(s) + 3O_2 + 2H_2O \xrightarrow{\text{bact.}} 4H^+ + 2SO_4^{2-} \tag{17.2}$$

$$FeS_2(s) + 3\tfrac{1}{2}O_2 + H_2O \xrightarrow{\text{bact.}} Fe^{2+} + 2H^+ + 2SO_4^{2-} \tag{17.3}$$

Salts of heavy metals, particularly those with trivalent metal ions, such as Fe(III) and Al(III), hydrolyze in water to release mineral acidity.

$$FeCl_3 + 3H_2O \rightleftharpoons Fe(OH)_3(s) + 3H^+ + 3Cl^- \qquad (17.4)$$

Their presence is indicated by the formation of a precipitate as the pH of solutions containing them is increased during neutralization.

Many industrial wastes contain organic acids. Their presence and nature can be determined by use of electrometric titration curves or gas chromatography.

17.3 | SIGNIFICANCE OF CARBON DIOXIDE AND MINERAL ACIDITY

Acidity is of little concern from a sanitary or public health viewpoint. Carbon dioxide is present in malt and carbonated beverages in concentrations greatly in excess of any concentrations known in natural waters, and no deleterious effects due to the carbon dioxide have been recognized. Waters that contain mineral acidity are so unpalatable that problems related to human consumption are nonexistent.

Acid waters are of concern because of their corrosive characteristics and the expense involved in removing or controlling the corrosion-producing substances. The corrosive factor in most waters is carbon dioxide, but in many industrial wastes it is mineral acidity. Carbon dioxide must be reckoned with in water-softening problems where the lime or lime–soda ash method is employed.

Where biological processes of treatment are used, the pH must ordinarily be maintained within the range of 6 to 9.5. This criterion often requires adjustment of pH to favorable levels, and calculation of the amount of chemicals needed is based upon acidity values in most cases.

Combustion of fossil fuels in power plants and automobiles leads to the formation of oxides of nitrogen and sulfur, which when mixed with rain, hydrolyze to form sulfuric and nitric acids. The resulting acid rain can lower the pH in poorly buffered lakes, adversely affecting aquatic life, and can increase the amount of chemicals, such as aluminum, leached from soil into surface runoff. For these reasons, control has been placed on the amount of sulfur and nitrogen oxides that can be discharged to the atmosphere through combustion.

17.4 | METHODS OF MEASUREMENT

Both carbon dioxide and mineral acidity can be measured by means of standard solutions of alkaline reagents. Mineral acids are measured by titration to a pH of about 3.7, the methyl orange end point. For this reason mineral acidity is also called *methyl orange acidity*. Titration of a sample to the phenolphthalein end point of pH 8.3 measures both mineral acidity plus acidity due to weak acids. This total acidity is also termed *phenolphthalein acidity*.

Carbon Dioxide

If reliable results are to be obtained, special precautions must be taken during the collection, handling, and analysis of samples for carbon dioxide, regardless of the method used. In waters where carbon dioxide is an important consideration, its partial pressure is greatly in excess of that in the atmosphere; therefore, exposure to the air must be avoided or kept at a minimum. For this reason analysis can be accomplished to best advantage at the point of collection where exposure to the air and temperature change can be avoided.

The sample should be collected in the same manner that is used to obtain a sample for dissolved oxygen, i.e., by using a submerged tube or pipe inlet, to exclude air bubbles, and allowing the container to overflow in order to displace any water that has come in contact with the air. If the sample must be transported to the laboratory for analysis, the bottle should be filled completely and capped or stoppered so as to leave no air pocket. The temperature should be kept as near that at which it was collected as possible.

Titration Method   In order to minimize contact with the air, it is best to collect and titrate the sample in a graduated cylinder or a color-comparison tube. The tube or cylinder should be filled to overflowing and the excess siphoned off or removed with a pipet to get the proper sample size. After addition of the proper amount of phenolphthalein indicator, the titration is conducted in a manner to minimize loss of carbon dioxide. Ordinarily, appreciable amounts of carbon dioxide will be lost in the first titration because of the excessive stirring needed. Reliable results may be obtained by taking a second sample and adding the indicated amount of titrant to it before stirring. The titration may then be completed without significant loss of carbon dioxide. The final end point is reached somewhat slowly, and so it is recommended that the titration not be considered complete until a pinkish color persists for 30 s.

When sodium hydroxide is used as the standard reagent, it is important that it be free of sodium carbonate. The reaction involved in the neutralization may be considered to occur in two steps,

$$2NaOH + CO_2 \rightarrow Na_2CO_3 + H_2O \qquad (17.5)$$

$$Na_2CO_3 + CO_2 + H_2O \rightarrow 2NaHCO_3 \qquad (17.6)$$

and, from Eq. (17.6), it should be obvious that if sodium carbonate is originally present in the sodium hydroxide, it will cause erroneous results. In order to overcome this problem, a sodium carbonate solution is one of the standard titrants recommended for carbon dioxide measurements. Sodium carbonate can be used in this capacity because it reacts quantitatively with carbon dioxide, as shown in Eq. (17.6). It has a definite advantage in that it may be purchased in analytical-grade form.

Calculation from pH and Alkalinity   It is possible to calculate the amount of carbon dioxide in a water sample from ionization expressions for carbonic acid. When the pH is less than about 8.5, the primary ionization constant for carbonic acid can be used, provided that the hydrogen-ion and bicarbonate-ion concentrations (ignoring activity corrections) and the value for K_{A1} are known:

$$\frac{[H^+][HCO_3^-]}{[H_2CO_3^*]} = K_{A1} \qquad (17.7)$$

In practice, $[H_2CO_3^*]$ in this expression is set equal to the sum of the molar concentrations of carbonic acid and free carbon dioxide because of the difficulty in distinguishing between these two forms. Since the free carbon dioxide represents about 99 percent of this total, the expression is only an approximation of a true equilibrium expression, although a fairly good one.

The use of Eq. (17.7) is illustrated in the following example. If $K_{A1} = 4.3 \times 10^{-7}$, $[H^+] = 10^{-7}$, and $[HCO_3^-] = 4.3 \times 10^{-3}$, then the CO_2 concentration would equal $(10^{-7})(4.3 \times 10^{-3})/(4.3 \times 10^{-7}) = 10^{-3}$ mol/L or 44 mg/L. However, in order for such a calculation to be accurate, the effect of other ions, as discussed in Sec. 4.3, must be considered, as must the effect of temperature on K_{A1} and K_W. Since these considerations can make the calculation of free carbon dioxide a complicated process, a nomographic chart is included in "Standard Methods" to facilitate the determination of free carbon dioxide from sample pH, alkalinity, dissolved solids, and temperature.

Determinations of carbon dioxide from pH and alkalinity measurements can result in highly accurate results, but not necessarily so. The method suffers from the fact that the dissolved solids concentration must be known. This usually requires a separate determination by gravimetric or conductivity methods. Also, the pH must be measured very accurately, as small variations can introduce serious errors. For example, an inaccuracy of 0.1 in the pH determination causes a carbon dioxide error of about 25 percent. Therefore, it is questionable whether results obtained by this method under ordinary laboratory or field conditions are more reliable than results obtained by the titration procedure, if proper attention is paid to details described for the titration method. Considering the difficulties with each procedure, it would appear that the titration procedure would normally be the method of choice for carbon dioxide concentrations greater than 2 mg/L, while for smaller concentrations the titration errors could be excessive, and thus the calculation procedure would be preferred.

Field Method   The titration procedure has many advantages and is sufficiently accurate for all practical purposes.

Methyl Orange Acidity

All natural waters and most industrial wastes that have a pH below 4 contain mineral or methyl orange acidity. Mineral acids are essentially neutralized by the time the pH has been raised to about 3.7 (see Fig. 4.7), and a color pH indicator is normally used where a pH meter is not available. While methyl orange was formerly used for this purpose, bromphenol blue is now recommended as it has a sharper color change at pH 3.7. Results are reported in terms of methyl orange acidity expressed as $CaCO_3$. Since $CaCO_3$ has an equivalent weight of 50, N/50 or 0.020 N NaOH is used as the titrating agent so that 1 mL is equivalent to 1 mg of acidity.

Phenolphthalein Acidity

Occasionally, it is desirable to measure the total acidity resulting both from mineral acids and from weak acids in the sample. Since most weak acids are essentially neutralized by titration to pH 8.3, either phenolphthalein or metacresol purple indicators can be used for this titration. When heavy-metal salts are present, it is usually desirable to heat the sample to boiling and then carry out the titration. The heat speeds the hydrolysis of the metal salts, allowing the titration to be completed more rapidly. "Standard Methods" suggests adding hydrogen peroxide as well to further speed hydrolysis. Again, 0.020 N NaOH is used as the titrating agent, and results are reported in terms of phenolphthalein acidity expressed as $CaCO_3$.

17.5 | APPLICATION OF ACIDITY DATA

Carbon dioxide determinations are particularly important in the field of public water supplies. In the development of new supplies, it is an important factor that must be considered in the treatment method and the facilities needed. Many underground supplies require treatment to overcome corrosive characteristics resulting from carbon dioxide. The amount present is an important factor in determining whether removal by aeration or simple neutralization with lime or sodium hydroxide will be chosen as the treatment method. The size of equipment, chemical requirements, storage space, and cost of treatment all depend upon the amounts of carbon dioxide present. Carbon dioxide is an important consideration in estimating the chemical requirements for lime or lime–soda ash softening.

Most industrial wastes containing mineral acidity must be neutralized before they may be discharged to rivers or sewers or subjected to treatment of any kind. Quantities of chemicals, size of chemical feeders, storage space, and costs are determined from laboratory data on acidity.

PROBLEMS

17.1 What causes acidity in natural waters?

17.2 (*a*) What pH range is used to measure mineral acidity in water?

  (*b*) What pH range is used to measure total acidity in water?

  (*c*) How are methyl orange and phenolphthalein acidity related to mineral and total acidity?

17.3 Why may we be concerned with acidity in water?

17.4 Can the pH of a water sample be calculated from a knowledge of its acidity? Why?

17.5 Can the carbon dioxide content of a waste sample known to contain a significant concentration of acetic acid be determined by the titration procedure? Why?

17.6 What methods can be used to determine carbon dioxide concentration in water, and what are the relative advantages and disadvantages of each?

17.7 A sample of water collected in the field had a pH of 6.8. By the time the water sample reached the laboratory for analysis, the pH had increased to 7.5. Give a possible explanation for this change.

17.8 Estimate the carbon dioxide content of a natural water sample having a pH of 7.3 and a bicarbonate-ion concentration (as HCO_3^-) of 30 mg/L. Assume that the effect of the dissolved solids on ion activity is negligible and the water temperature is 25°C.

17.9 (a) A water supply was found to have a bicarbonate-ion concentration of 50 mg/L and a CO_2 content of 30 mg/L. Estimate the approximate pH of the water (temperature equals 25°C).

    (b) If the CO_2 content of the water in part (a) were reduced to 3 mg/L by aeration, what would the pH then be?

17.10 Air contains an average of about 0.032 percent by volume of carbon dioxide, and Henry's law constant K_H for carbon dioxide in water at 20°C is 26 atm/M.

    (a) Using Eq. (2.15), calculate the concentration of carbon dioxide in a water sample when in equilibrium with atmospheric carbon dioxide at sea level.

    (b) On the basis of the calculation in part (a), what will occur when a sample containing 10 mg/L of carbon dioxide is vigorously exposed to the air prior to titration for carbon dioxide?

17.11 A water sample has a methyl orange acidity of 60 mg/L. Calculate the quantity of lime in mg/L of $Ca(OH)_2$ required to raise the pH to 4.3.

REFERENCE

Clesceri, L. S., A. E. Greenberg, and A. D. Eaton (eds.): "Standard Methods for the Examination of Water and Wastewater," 20th ed., American Public Health Association, Washington, DC, 1998.

18

Alkalinity

18.1 | GENERAL CONSIDERATIONS

The *alkalinity* of a water is a measure of its capacity to neutralize acids. The term *acid neutralization capacity* (ANC) is also sometimes used. The alkalinity of natural waters is due primarily to the salts of weak acids, although weak or strong bases may also contribute. Bicarbonate represents the major form of alkalinity, since it is formed in considerable amounts from the action of carbon dioxide upon basic materials in the soil, as shown in Eq. (17.1). Other salts of weak acids, such as borate, silicate, and phosphate, may be present in small amounts. A few organic acids that are quite resistant to biological oxidation—for example, humic acid—form salts that add to the alkalinity of natural waters. In polluted or anaerobic waters, salts of weak acids such as acetic, propionic, and hydrogen sulfide may be produced and would also contribute to alkalinity. In other cases, ammonia or hydroxide may make a contribution to the total alkalinity of a water.

Under certain conditions, natural waters may contain appreciable amounts of carbonate and hydroxide alkalinity. This condition is particularly true in surface waters where algae are flourishing. The algae remove carbon dioxide, free and combined, from the water to such an extent that pH values of 9 to 10 are often obtained. The chemistry involved is discussed in Sec. 18.7. Boiler waters always contain carbonate and hydroxide alkalinity. Chemically treated waters, particularly those produced in lime or lime–soda ash softening of water, contain carbonate and excess hydroxide.

Although many materials may contribute to the alkalinity of a water, the major portion of the alkalinity in natural waters is caused by three major classes of materials which may be ranked in order of their association with high pH values as follows: (1) hydroxide, (2) carbonate, and (3) bicarbonate. For most practical purposes, alkalinity due to other materials in natural waters is insignificant and may be ignored.

The alkalinity of waters is due principally to salts of weak acids and strong bases, and such substances act as buffers to resist a drop in pH resulting from acid addition, as discussed in Sec. 4.6. Alkalinity is thus a measure of the buffer capacity and in this sense is used to a great extent in wastewater treatment practice.

18.2 | PUBLIC HEALTH SIGNIFICANCE

As far as is known, the alkalinity of a water has little public health significance. Highly alkaline waters are usually unpalatable, and consumers tend to seek other supplies. Chemically treated waters sometimes have rather high pH values which have met with some objection on the part of consumers. For these reasons, the U.S. EPA has set an upper bound pH secondary standard of 8.5.

18.3 | METHOD OF DETERMINING ALKALINITY

Alkalinity is measured volumetrically by titration with N/50 or 0.020 N H_2SO_4 and is reported in terms of equivalent $CaCO_3$. For samples whose initial pH is above 8.3, the titration is made in two steps. In the first step the titration is conducted until the pH is lowered to 8.3, the point at which phenolphthalein indicator turns from pink to colorless. The second phase of the titration is conducted until the pH is lowered to about 4.5, corresponding to the bromcresol green end point. When the pH of a sample is less than 8.3, a single titration is made to a pH of 4.5.

The choice of pH 8.3 as the end point for the first step in the titration is in accord with the fundamentals of base titration developed in Sec. 4.5. This value corresponds to the equivalence point for the conversion of carbonate ion to bicarbonate ion:

$$CO_3^{2-} + H^+ \rightarrow HCO_3^- \tag{18.1}$$

The use of a pH of about 4.5 for the end point for the second step of the titration corresponds approximately to the equivalence point for the conversion of bicarbonate ion to carbonic acid:

$$HCO_3^- + H^+ \rightarrow H_2CO_3 \tag{18.2}$$

On the basis of Eq. (4.66), the exact end point for this titration would be dependent upon the initial bicarbonate-ion concentration in the sample. Using pK_B from Table 4.2 for the bicarbonate salt of carbonate we see that Eq. (4.66) becomes

$$\text{pH (bicarbonate equivalence point)} = 3.19 - \tfrac{1}{2} \log [HCO_3^-] \tag{18.3}$$

A $[HCO_3^-]$ of 0.01 M corresponds to an alkalinity of 500 mg/L as $CaCO_3$, for which the equivalence point would be 4.19. These considerations require that the carbonic acid or carbon dioxide formed from bicarbonate during the titration not be lost from solution.

Phenolphthalein or metacresol purple color indicators may be used for the pH 8.3 end point, and bromcresol green or a mixed bromcresol green–methyl red indicator may be used for pH 4.5. However, the actual pH of the stoichiometric end point in alkalinity determinations can be best determined by potentiometric titration. This fact is particularly important in natural waters where the total alkalinity is a summation of the effects resulting from salts of weak acids of which bicarbonates are only one. The pH at which the inflection in the titration curve occurs (see Fig.

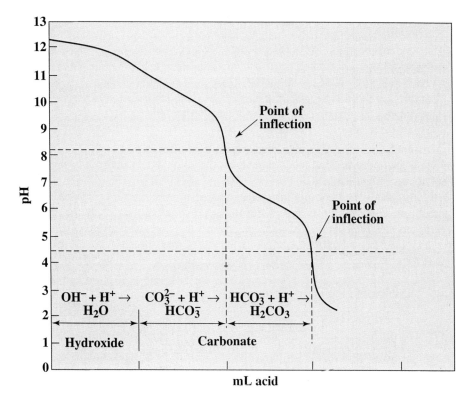

Figure 18.1
Titration curve for a hydroxide-carbonate mixture.

18.1) is taken as the true end point. The pH values given for the equivalence points for various alkalinities from Eq. (18.3) or in "Standard Methods" hold only for essentially pure bicarbonate solutions and should not be applied indiscriminately to domestic or industrial wastewaters, or even natural waters.

18.4 | METHODS OF EXPRESSING ALKALINITY

Alkalinity measurements are made on a wide variety of materials. These range from relatively pure waters through polluted waters, such as municipal and industrial wastewaters, to digesting sludges. The methods of expressing alkalinity values vary considerably; therefore, it is necessary to explain the methods in some detail and to indicate the areas where the various methods are employed.

Phenolphthalein and Total Alkalinity

Inspection of the titration curves for a strong base (hydroxide alkalinity), shown in Fig. 4.7, and for sodium carbonate, in Fig. 4.9, shows that essentially all the hydroxide has been neutralized by the time the pH has been decreased to 10 and that the carbonate has been converted to bicarbonate by the time the pH has been lowered to about 8.3. In a mixture containing both hydroxide and carbonate, the carbonate modifies the titration curve to the extent that only the inflection at pH 8.3 occurs, as shown in Fig. 18.1. Because of this, it has become common practice to express the alkalinity measured to the phenolphthalein end point as *phenolphthalein alkalinity*. This term is quite widely used at present in the field of waste treatment and is still used to some extent in the area of water analysis.

If the titration of a sample that originally contained both carbonate and hydroxide alkalinity is continued beyond the phenolphthalein end point, the bicarbonate reacts with the acid and is converted to carbonic acid. The reaction is essentially complete when the pH has been lowered to about 4.5 (see Fig.18.1). The amount of acid required to react with the hydroxide, carbonate, and bicarbonate represents the *total alkalinity*. It is customary to express alkalinity in terms of $CaCO_3$; therefore, 0.020 N H_2SO_4 is used in its measurement. Calculations are made as follows:

$$\text{Phenol. alk.} = (\text{mL } 0.020 \text{ N } H_2SO_4 \text{ to pH } 8.3) \ \frac{1000}{\text{mL sample}} \qquad (18.4)$$

$$\text{Total alk.} = \text{total mL } 0.020 \text{ N } H_2SO_4 \text{ to pH} \begin{cases} 5.0 \\ 4.8 \\ 4.6 \\ 4.0 \end{cases} \times \frac{1000}{\text{mL sample}} \qquad (18.5)$$

In the determination of total alkalinity, the pH at the stoichiometric end point is directly related to the amount of carbonate alkalinity originally present in the sample, as discussed in Sec. 18.3.

Hydroxide, Carbonate, and Bicarbonate Alkalinity

In water analysis it is often desirable to know the kinds and amounts of the various forms of alkalinity present. This information is especially needed in water-softening processes and in boiler-water analysis. It is customary to calculate hydroxide, carbonate, and bicarbonate alkalinities from the fundamental information given by the titration curves for strong bases and sodium carbonate. (See Figs. 4.7 and 4.9.) There are three procedures commonly used to make these calculations: (1) calculation from alkalinity measurements alone, (2) calculation from alkalinity plus pH measurements, and (3) calculation from equilibrium equations. The first is the classical method and is based on empirical relationships for the calculation of the various forms of alkalinity from phenolphthalein and total alkalinities. This is intended for use by technicians and others who do not have a knowledge of the fundamental chemistry involved. The results from this method are only approximate for samples with pH greater than 9. However, water chemists and engineers concerned with water softening, corrosion control, and prevention of scaling at elevated pH levels are vitally concerned with ionic species and concentrations. For these reasons it has become necessary to be able to calculate hydroxide-, carbonate-, and bicarbonate-ion concentrations at all pH levels with considerable accuracy. This can be done with either the second or the third procedure.

The second procedure gives sufficiently accurate estimates for most practical purposes, and also makes use of phenolphthalein and total alkalinity measurements. In addition, an accurate initial pH measurement is required for the direct calculation of hydroxide alkalinity. In the third procedure, the various equilibrium equations for carbonic acid are used to compute the concentrations of the various alkalinity forms. This method gives reasonably accurate results for constituents, even when present in the fractional mg/L range, provided that an accurate pH measurement is made. The concentration of constituents in low concentration is sometimes of importance. A total alkalinity, as well as a pH measurement, is required. In addition, a dissolved-solids measurement to correct for ion activity (see Sec. 4.3) and a temperature measurement for the selection of a proper equilibrium constant must be made. It is important to understand the basis for these procedures. This is presented in the following.

Calculation from Alkalinity Measurements Alone

In this procedure, phenolphthalein and total alkalinities are determined, and from these measurements the calculation of three types of alkalinity, hydroxide, carbonate, and bicarbonate, are made. This can be done by assuming (incorrectly) that hydroxide and bicarbonate alkalinity cannot exist together in the same sample. This permits only five possible situations to be present, which are as follows: (1) hydroxide only, (2) carbonate only, (3) hydroxide plus carbonate, (4) carbonate plus bicarbonate, and (5) bicarbonate only. Reference to Figs. 4.7 and 4.9 will demonstrate that neutralization of hydroxides is complete by the time enough acid has been added to decrease the pH to 8.3, and that carbonate is exactly one-half neutralized when the pH has been decreased to the same degree. Upon continuation of the titration to reach a pH of about 4.5, a negligible amount of acid is needed in the case of the hydroxide, and an amount exactly equal to that needed to reach pH 8.3 is required for the carbonate. This is the fundamental information needed to determine which forms of alkalinity are present and the amounts of each. A graphical representation of typical titrations obtained with the various combinations of alkalinity is shown in Fig. 18.2.

Hydroxide Only   Samples containing only hydroxide alkalinity have a high pH, usually well above 10. Titration is essentially complete at the phenolphthalein end point. In this case hydroxide alkalinity is equal to the phenolphthalein alkalinity.

Carbonate Only   Samples containing only carbonate alkalinity have a high pH of 8.5 or higher. The titration to the phenolphthalein end point is exactly equal to one-half of the total titration. In this case carbonate alkalinity is equal to the total alkalinity.

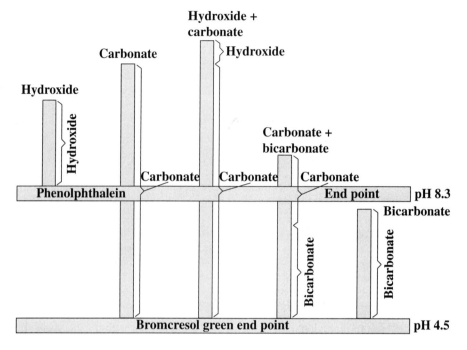

Figure 18.2
Graphical representation of titration of samples containing various forms of alkalinity.

Hydroxide-Carbonate   Samples containing hydroxide and carbonate alkalinity have a high pH, usually well above 10. The titration from the phenolphthalein to the bromcresol green end point represents one-half of the carbonate alkalinity. Therefore, carbonate alkalinity may be calculated as follows:

$$\text{Carbonate alk.} = 2 \text{ (titration from pH 8.3 to pH 4.5)} \times \frac{1000}{\text{mL sample}}$$

and   Hydroxide alk. = total alk. − carbonate alk.

Carbonate-Bicarbonate   Samples containing carbonate and bicarbonate alkalinity have a pH > 8.3 and usually less than 11. The titration to the phenolphthalein end point represents one-half of the carbonate. Carbonate alkalinity may be calculated as follows:

$$\text{Carbonate alk.} = 2 \text{ (titration to pH 8.3)} \times \frac{1000}{\text{mL sample}}$$

and   Bicarbonate alk. = total alk. − carbonate alk.

Bicarbonate Only Samples containing only bicarbonate alkalinity have a pH of 8.3 or less, usually less. In this case bicarbonate alkalinity is equal to the total alkalinity.

The foregoing methods of approximate calculation generally have been superseded by the more precise methods that will now be described.

Calculation from Alkalinity plus pH Measurements

In this procedure, measurements are made for pH, phenolphthalein, and total alkalinity. This will allow calculation of hydroxide, carbonate, and bicarbonate alkalinity.

Hydroxide First, the hydroxide alkalinity is calculated from the pH measurement, using the dissociation constant for water:

$$[OH^-] = \frac{K_W}{[H^+]} \tag{18.6}$$

This calculation requires a precise pH measurement for the determination of $[H^+]$. Since a hydroxide concentration of 1 mol/L is equivalent to 50,000 mg/L of alkalinity as $CaCO_3$, the relationship in Eq. (18.6) can be expressed more conveniently as

$$\text{Hydroxide alk.} = 50,000 \times 10^{(pH-pK_w)} \tag{18.7}$$

At 25°C, $pK_W = 14.00$. However, it varies from 14.94 at 0°C to 13.53 at 40°C. Therefore, it is important that a temperature measurement be made and the correct pK_W be used. The relationship between pH, temperature, and hydroxide alkalinity is shown graphically in Fig. 18.3. To be more precise, a dissolved-solids measurement should be made to correct for ion activity, although in this case the correction is fairly negligible and not necessary for most practical purposes. A nomograph is available in "Standard Methods" that allows rapid calculation of hydroxide alkalinity, using pH, temperature, and dissolved-solids measurements.

Carbonate Once the hydroxide alkalinity is determined, use can be made of the principles from the first procedure to calculate the carbonate and bicarbonate alkalinity. The phenolphthalein alkalinity represents all the hydroxide alkalinity plus one-half the carbonate alkalinity. Therefore, carbonate alkalinity may be calculated as follows:

$$\text{Carbonate alk.} = 2 \text{ (phenol. alk.} - \text{hydroxide alk.)} \tag{18.8}$$

Bicarbonate The titration from pH 8.3 to pH 4.5 measures the remaining one-half of the carbonate alkalinity plus all the bicarbonate alkalinity. It is also apparent that the bicarbonate alkalinity represents the remaining alkalinity after the hydroxide plus carbonate alkalinities are subtracted. From either standpoint, the bicarbonate alkalinity becomes

$$\text{Bicarbonate alk.} = \text{total alk.} - \text{(carbonate alk.} + \text{hydroxide alk.)} \tag{18.9}$$

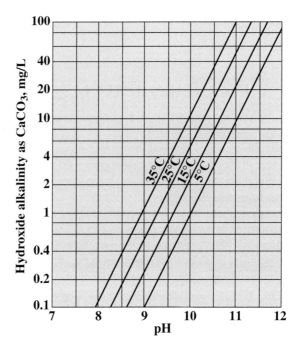

Figure 18.3
Relationship between hydroxide alkalinity and pH at
various temperatures.

Calculation from Equilibrium Equations[1]

The distribution of the various forms of alkalinity can be calculated from equilibrium equations plus a consideration of electroneutrality (charge balance) in solution. In order to preserve electroneutrality, the sum of the equivalent concentrations of the cations must equal that of the anions. Total alkalinity is a measure of the equivalent concentration of all cations associated with the alkalinity-producing anions, except the hydrogen ion. Thus, as shown in Example 4.17, the balance of equivalent concentrations of alkalinity-associated cations and anions is given by

$$[H^+] + \frac{\text{alkalinity}}{50,000} = [HCO_3^-] + 2[CO_3^{2-}] + [OH^-] \tag{18.10}$$

The equilibrium equations that must be considered are those for water [Eq. (18.6)] and for the second ionization of carbonic acid (ignoring activity corrections),

$$\frac{[H^+][CO_3^{2-}]}{[HCO_3^-]} \tag{18.11}$$

---

[1]E. W. Moore, *J. Amer. Water Works Assoc.*, **31:** 51 (1939).

From a pH measurement, $[H^+]$ and $[OH^-]$ can be determined, using Eq. (18.6). The only other unknowns are $[HCO_3^-]$ and $[CO_3^{2-}]$, and these can be determined from a simultaneous solution of Eqs. (18.10) and (18.11). The following equations result:

$$\text{Carbonate alkalinity} \atop \text{(mg/L as CaCO}_3\text{)}} = \frac{50,000[(\text{alkalinity}/50,000) + [H^+] - (K_W/[H^+])]}{1 + ([H^+]/2K_{A2})} \quad (18.12)$$

$$\text{Bicarbonate alkalinity} \atop \text{(mg/L as CaCO}_3\text{)}} = \frac{50,000[(\text{alkalinity}/50,000) + [H^+] - (K_W/[H^+])]}{1 + (2K_{A2}/[H^+])} \quad (18.13)$$

At 25°C, K_W is 10^{-14} and K_{A2} is 4.7×10^{-11}. However, these values vary quite radically with temperature. Also, the activities of the ions vary considerably with ionic concentration, as indicated in Sec. 4.3. These corrections are rather tedious; consequently, "Standard Methods" presents nomographs for the evaluation of carbonate and bicarbonate based on these considerations. The nomographs, as well as Eqs. (18.12) and (18.13), yield results in terms of alkalinity expressed as $CaCO_3$. At times the actual concentrations of carbonate or bicarbonate ion may be desired. Conversions to milligrams per liter of CO_3^{2-} or HCO_3^- are as follows:

$$\text{mg/L } CO_3^{2-} = \text{mg/L carbonate alk.} \times 0.6 \quad (18.14)$$

$$\text{mg/L } HCO_3^- = \text{mg/L bicarbonate alk.} \times 1.22 \quad (18.15)$$

Molar concentrations may be obtained by dividing milligrams per liter by the mole ionic weight in milligrams:

$$[CO_3^{2-}] = \frac{\text{mg/L } CO_3^{2-}}{60,000} \quad \text{and} \quad [HCO_3^-] = \frac{\text{mg/L } HCO_3^-}{61,000} \quad (18.16)$$

18.5 | CARBON DIOXIDE, ALKALINITY, AND pH RELATIONSHIPS IN NATURAL WATERS

From the equations

$$CO_2 + H_2O \rightleftharpoons H_2CO_3 \rightleftharpoons HCO_3^- + H^+ \quad (18.17)$$

$$M(HCO_3)_2 \rightleftharpoons M^{2+} + 2HCO_3^- \quad (18.18)$$

$$HCO_3^- \rightleftharpoons CO_3^{2-} + H^+ \quad (18.19)$$

$$CO_3^{2-} + H_2O \rightleftharpoons HCO_3^- + OH^- \quad (18.20)$$

it is obvious that carbon dioxide and the three forms of alkalinity are all part of one system that exists in equilibrium, since all equations involve HCO_3^-. A change in concentration of any one member of the system will, of course, cause a shift in the equilibrium, alter the concentration of the other ions, and result in a change of pH. Conversely, a change in pH will shift the relationships. Figure 18.4 shows the relationship between carbon dioxide and the three forms of alkalinity in a water with 100 mg/L of total alkalinity, over the pH range of importance in practice. Use was made of Eqs. (17.7), (18.7), (18.12), and (18.13) for its construction. The informa-

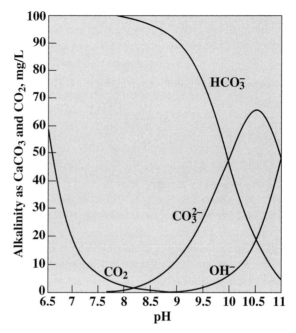

Figure 18.4
Relationship between carbon dioxide and the three forms
of alkalinity at various pH levels (values calculated for
water with a total alkalinity of 100 mg/L at 25°C).

tion given in Fig. 18.4 is for illustrative purposes only, as the relationships differ
with total alkalinity, temperature, and so on.

18.6 | APPLICATION OF ALKALINITY DATA

Information concerning alkalinity is used in a variety of ways in practice.

Chemical Coagulation

Chemicals used for coagulation of water and wastewater react with water to form
insoluble hydroxide precipitates. The hydrogen ions released react with the alkalinity of the water. Thus, the alkalinity acts to buffer the water in a pH range where
the coagulant can be effective. Alkalinity must be present in excess of that destroyed by the acid released by the coagulant for effective and complete coagulation to occur.

Water Softening

Alkalinity is a major item that must be considered in calculating the lime and soda-
ash requirements in softening of water by precipitation methods. The alkalinity of

softened water is a consideration in terms of whether such waters meet drinking water standards.

Corrosion Control

Alkalinity is an important parameter involved in corrosion control. It must be known in order to calculate the Langelier saturation index.[2]

Buffer Capacity

Alkalinity measurements are made as a means of evaluating the buffering capacity of wastewaters and sludges. They can also be used to assess a natural waters ability to resist the effects of acid rain.

Industrial Wastes

Many regulatory agencies prohibit the discharge of wastes containing caustic (hydroxide) alkalinity to receiving waters. Municipal authorities usually prohibit the discharge of wastes containing caustic alkalinity to sewers. Alkalinity as well as pH is an important factor in determining the amenability of wastewaters to biological treatment.

18.7 | OTHER CONSIDERATIONS

A number of situations are encountered in practice that involve carbon dioxide–alkalinity–pH relationships and deserve some explanation.

pH Changes during Aeration of Water

It is common practice to aerate water to remove carbon dioxide, ammonia, and volatile organic chemicals. Since carbon dioxide is an acidic gas, its removal tends to decrease $[H^+]$ and thus raise the pH of the water in accordance with Eq. (18.17). Normal air contains about 0.035 percent by volume of carbon dioxide, which is equivalent to $10^{-3.46}$ atm at sea level. The Henry's law constant [see Eq. (2.15)] for carbon dioxide at 25°C is about 31.6 atm/M; therefore, the equilibrium concentration of carbon dioxide with air is $10^{-3.46}$ (44,000)/31.6 or about 0.48 mg/L. From Eq. (17.7) it can then be calculated that a water with an alkalinity of 100 mg/L, aerated until in equilibrium with the carbon dioxide in air, would have a pH of about 8.6. A water with higher alkalinity would tend to have a higher pH upon aeration, and one with lower alkalinity would tend to have a lower pH.

pH Changes in the Presence of Algal Blooms

Many surface waters support extensive algal blooms. pH values as high as 10 have been observed in areas where algae are growing rapidly, particularly in shallow

---

[2]W. F. Langelier, *J. Amer. Water Works Assoc.,* **28:** 1500 (1936); T. E. Larson and A. M. Buswell, *J. Amer. Water Works Assoc.,* **34:** 1667 (1942).

water. Algae use carbon dioxide in their photosynthetic activity, and this removal is responsible for such high pH conditions. We have seen that aeration for removal of carbon dioxide tends to increase the pH to between 8 and 9 in waters with moderate alkalinity. Algae, however, can reduce the free carbon dioxide concentration below its equilibrium concentration with air and consequently can cause an even greater increase in pH. As the pH increases, the alkalinity forms change, with the result that carbon dioxide can also be extracted for algal growth both from bicarbonates and from carbonates in accordance with the following equilibrium equations:

$$2HCO_3^- \rightleftharpoons CO_3^{2-} + H_2O + CO_2 \qquad (18.21)$$

$$CO_3^{2-} + H_2O \rightleftharpoons 2OH^- + CO_2 \qquad (18.22)$$

Thus, the removal of carbon dioxide by algae tends to cause a shift in the forms of alkalinity present from bicarbonate to carbonate, and from carbonate to hydroxide. It should be noted that during these changes the total alkalinity remains constant unless removal results through precipitation of carbonate salts such as $CaCO_3(s)$. Algae can continue to extract carbon dioxide from water until an inhibitory pH is reached, which is usually in the range of pH 10 to 11.

During the dark hours of the day, algae produce rather than consume carbon dioxide. This is because their respiratory processes in darkness exceed their photosynthetic processes. This carbon dioxide production has the opposite effect and tends to reduce the pH. Diurnal variations in pH due to algal photosynthesis and respiration are common in surface waters.

In natural waters containing appreciable amounts of Ca^{2+}, calcium carbonate precipitates when the carbonate-ion concentration, according to Eq. (18.21), becomes great enough so that the $CaCO_3$ solubility product is exceeded:

$$Ca^{2+} + CO_3^{2-} \rightleftharpoons CaCO_3(s) \qquad (18.23)$$

This precipitation usually happens before pH levels have exceeded 10, and it places a ceiling over the pH values obtainable. The calcium carbonate precipitated as a result of removal of carbon dioxide through algal action produces the marl deposits in lakes. Marl deposits are the precursors of limestone.

Alkalinity of Boiler Waters

Boiler waters contain both carbonate and hydroxide alkalinity. Both are derived from bicarbonate in the feed water. Carbon dioxide is insoluble in boiling water and so is removed with the steam. This causes an increase in pH and a shift in alkalinity forms from bicarbonate to carbonate and from carbonate to hydroxide, as indicated in Eqs. (18.21) and (18.22). Under these extreme conditions, pH levels in excess of 11.0 are often obtained. If Ca^{2+} levels are high, precipitation of $CaCO_3(s)$ may occur.

PROBLEMS

18.1   What is alkalinity and what produces it in natural waters?

18.2   Of what benefit is alkalinity in natural waters?

18.3 What pH end points are used to characterize the alkalinity of water?

18.4 What end-point indicators are commonly used for alkalinity measurements?

18.5 What are the three major kinds of alkalinity found present in natural waters?

18.6 Alkalinity is normally measured in terms of $CaCO_3$. What is the formula weight of this compound? In what way might this formula weight make $CaCO_3$ a convenient reference material for alkalinity?

18.7 Calculate the alkalinity as $CaCO_3$ of a water that contains 85 mg/L of HCO_3^-, 120 mg/L of CO_3^{2-}, and 2 mg/L of OH^-.

18.8 What effect does the removal of carbon dioxide from water through aeration have on each of the three kinds of alkalinity found present in natural waters?

18.9 What effect does the addition of carbon dioxide have on the total alkalinity of water?

18.10 On analysis, a series of samples was found to have the following pH values: 5.5, 3.0, 11.2, 8.5, 7.4, and 9.0. What can you conclude regarding the possible presence of a significant bicarbonate, carbonate, or hydroxide alkalinity in each sample?

18.11 Calculate the phenolphthalein and total alkalinities of the following samples:

(a) A 50-mL sample required 5.3 mL 0.020 N H_2SO_4 to reach the phenolphthalein end point and a total of 15.2 mL to reach the methyl orange end point.

(b) A 100-mL sample required 20.2 mL of 0.020 N H_2SO_4 to reach the phenolphthalein end point and a total of 25.6 mL to reach the methyl orange end point.

18.12 For the following, calculate the hydroxide, carbonate, and bicarbonate alkalinity by the procedures (1) using alkalinity measurements alone and (2) using alkalinity plus pH measurements. The sample size is 100 mL, 0.020 N H_2SO_4 is used as the titrant, and the water temperature is 25°C (BG = bromcresol green).

Sample	Sample pH	Total mL titrant to reach end point	
		Phenol.	BG
A	11.9	10.0	15.5
B	10.0	14.4	38.6
C	11.2	8.2	8.4
D	7.0	0	12.7

18.13 For the following, calculate the hydroxide, carbonate, and bicarbonate alkalinity (1) using alkalinity measurements alone and (2) using alkalinity plus pH measurements. The sample size is 50 mL, 0.020 N H_2SO_4 is used as the titrant, and the water temperature is 25°C (BG = bromcresol green).

Sample	Sample pH	Total mL titrant to reach end point	
		Phenol.	BG
A	8.3	0	30
B	10.3	7.4	14.4
C	11.0	8.0	13.5
D	7.0	0	28.1

18.14 Assuming that the effect of dissolved salts on ion activity is negligible and that the temperature of the water sample is 25°C, do the following:

    (*a*)   Estimate the bicarbonate-ion concentration (in mg/L as HCO_3^-) if the pH of the sample is 10.3 and the carbonate concentration is 120 mg/L (in mg/L as CO_3^{2-}).

    (*b*)   Estimate the hydroxide, carbonate, bicarbonate, and total alkalinities for the solution in part (*a*).

REFERENCE

Clesceri, L. S., A. E. Greenberg, and A. D. Eaton (eds.): "Standard Methods for the Examination of Water and Wastewater," 20th ed., American Public Health Association, Washington, DC, 1998.

CHAPTER 19

Hardness

19.1 | GENERAL CONSIDERATIONS

Hard waters are generally considered to be those waters that require considerable amounts of soap to produce a foam or lather and that also produce scale in hot-water pipes, heaters, boilers, and other units in which the temperature of water is increased materially. To the layperson, the soap-consuming capacity is most important because of economic aspects and because of the difficulty encountered in obtaining suitable conditions for optimum cleansing; to the engineer, the scaling problem is the most challenging.

With the advent of synthetic detergents, many of the disadvantages of hard waters for household use have been diminished. However, soap is preferred for some types of laundering and for personal hygiene, and hard waters remain as objectionable as ever for these purposes. The scaling problem continues to be a consideration in spite of advances in our knowledge of water chemistry and the development of many proprietary devices that are claimed to prevent scaling through the application of principles not fully explainable.

Although there is less demand today from the general public for the removal of hardness through water-softening processes, the need is still great. The trend is toward private and industrial installations in preference to municipal softening plants, except where hardness is considered unreasonably high.

The hardness of waters varies considerably from place to place. In general, surface waters are softer than groundwaters. The hardness of water reflects the nature of the geological formations with which it has been in contact. Figure 19.1 shows the general character of the water supplies in the United States. The softest waters are found in the New England, South Atlantic, and Pacific Northwest states. Iowa, Illinois, Indiana, Arizona, New Mexico, and the Great Plains states have the hardest waters.

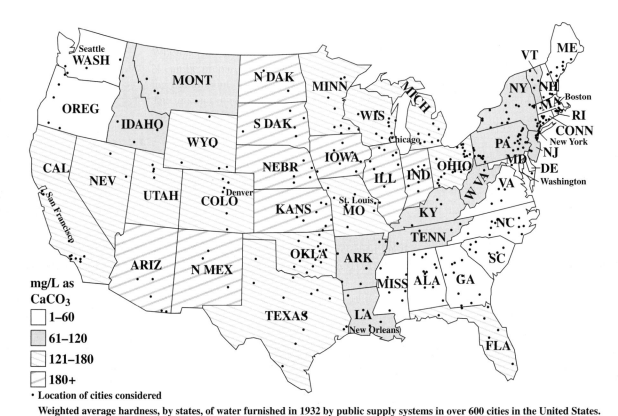

mg/L as CaCO$_3$

☐ 1–60

▨ 61–120

▨ 121–180

▨ 180+

• **Location of cities considered**

Weighted average hardness, by states, of water furnished in 1932 by public supply systems in over 600 cities in the United States.

Figure 19.1
Hardness characteristics of U.S. water supplies. *(From U.S. Geological Survey Paper 658.)*

Waters are commonly classified in terms of the degree of hardness as follows:

mg/L as CaCO$_3$	Degree of hardness
0–75	Soft
75–150	Moderately hard
150–300	Hard
300 up	Very hard

19.2 | CAUSE AND SOURCE OF HARDNESS

Hardness is caused by multivalent metallic cations. Such ions are capable of reacting with soap to form precipitates and with certain anions present in the water to form scale. The principal hardness-causing cations are the divalent calcium, magnesium, strontium, ferrous iron, and manganous ions. These cations, plus the most important anions with which they are associated, are shown in Table 19.1 in the order of their relative abundance in natural waters. Aluminum and ferric ions are sometimes con-

Table 19.1 | Principal cations causing hardness in water and the major anions associated with them

Cations causing hardness	Anions
Ca^{2+}	HCO_3^-
Mg^{2+}	SO_4^{2-}
Sr^{2+}	Cl^-
Fe^{2+}	NO_3^-
Mn^{2+}	SiO_3^{2-}

sidered as contributing to the hardness of water. However, their solubility is so limited at the pH values of natural waters that ionic concentrations are negligible.

The hardness in water is derived largely from contact with the soil and rock formations. Rain water as it falls upon the earth is incapable of dissolving the tremendous amounts of solids found in many natural waters. The ability to dissolve

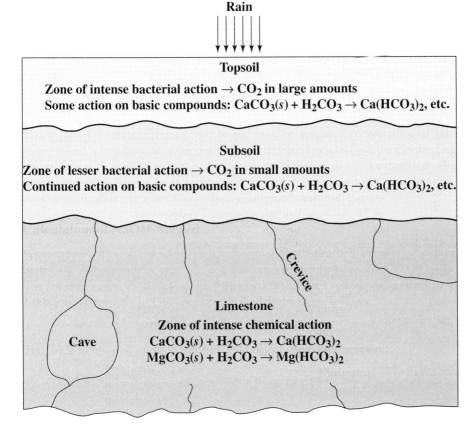

Rain

Topsoil
Zone of intense bacterial action → CO_2 in large amounts
Some action on basic compounds: $CaCO_3(s) + H_2CO_3 \rightarrow Ca(HCO_3)_2$, etc.

Subsoil
Zone of lesser bacterial action → CO_2 in small amounts
Continued action on basic compounds: $CaCO_3(s) + H_2CO_3 \rightarrow Ca(HCO_3)_2$, etc.

Crevice

Limestone
Zone of intense chemical action
$$CaCO_3(s) + H_2CO_3 \rightarrow Ca(HCO_3)_2$$
$$MgCO_3(s) + H_2CO_3 \rightarrow Mg(HCO_3)_2$$

Cave

Figure 19.2
Source of carbon dioxide and the solution of substances causing hardness.

is gained in the soil where carbon dioxide is released by bacterial action. The soil water becomes highly charged with carbon dioxide, which, of course, exists in equilibrium with carbonic acid. Under the low pH conditions that develop, basic materials, particularly limestone formations, are dissolved. Figure 19.2 shows where the carbon dioxide originates and how it attacks the insoluble carbonates in the soil and in limestone formations to convert them to soluble bicarbonates. Since limestone is not pure carbonate but includes impurities such as sulfate, chloride, and silicate, these materials become exposed to the solvent action of the water as the carbonate is dissolved, and they pass into solution, too.

In general, hard waters originate in areas where the topsoil is thick and limestone formations are present. Soft waters originate in areas where the topsoil is thin and limestone formations are sparse or absent.

19.3 | PUBLIC HEALTH SIGNIFICANCE

Hard waters are as satisfactory for human consumption as soft waters. There is some evidence that Ca^{2+} and Mg^{2+} are protective against heart disease. Because of their adverse action with soap, however, their use for cleansing purposes is quite unsatisfactory, unless soap costs are disregarded.

19.4 | METHODS OF DETERMINATION

Hardness is normally expressed in terms of $CaCO_3$ as is alkalinity. Many methods of determination have been proposed over the course of the years. Two are presently used as standard methods.

Calculation Method

Perhaps the most accurate method of determining hardness is by a calculation based upon the divalent ions found through a complete cation analysis. This method is to be preferred where complete analyses are available. The concentrations of the individual divalent cations of importance can also be determined separately by the standard atomic absorption or inductively coupled plasma procedures described in Chap. 12, or by ion chromatography or ion-specific electrodes, which are also described in Chap. 12.

It has been shown that some hard waters contain appreciable amounts of strontium. Thus, if strontium is not included in the calculations for total hardness, the calculated values may be in considerable error.

Calculation of the hardness caused by each ion is performed by use of the general formula

$$\text{Hardness (in mg/L) as } CaCO_3 = M^{2+} \text{ (in mg/L)} \times \frac{50}{\text{EW of } M^{2+}} \qquad (19.1)$$

where M^{2+} represents any divalent metallic ion.

Calculate the hardness of a water with the following analysis:

EXAMPLE 19.1

Cation	Concentration, mg/L	Anion	Concentration, mg/L
Na^+	20	Cl^-	40
Ca^{2+}	15	SO_4^{2-}	16
Mg^{2+}	10	NO_3^-	1
Sr^{2+}	2	Alkalinity	50

Only the divalent cations, Ca^{2+}, Mg^{2+}, and Sr^{2+} cause hardness:

Cation	EW	Hardness, mg/L as $CaCO_3$
Ca^{2+}	20.0	$(15)(50)/(20.0) = 37.5$
Mg^{2+}	12.2	$(10)(50)/(12.2) = 41.0$
Sr^{2+}	43.8	$(2)(50)/(43.8) = \underline{2.3}$
		Total hardness $= 80.8$

EDTA Titrimetric Method

This method involves the use of solutions of ethylenediaminetetraacetic acid (EDTA) or its sodium salt as the titrating agent.

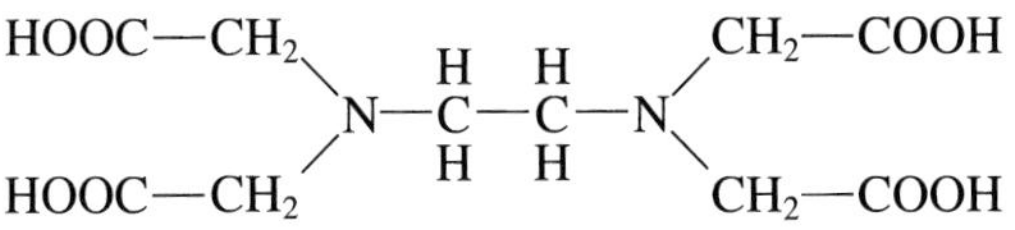

Acid

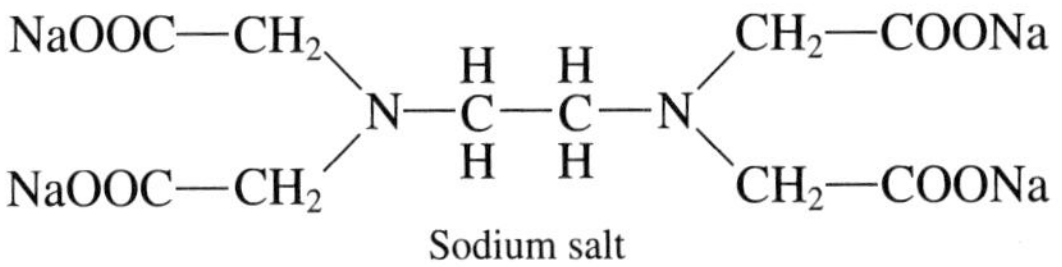

Sodium salt

These compounds, usually represented by EDTA, are chelating agents and form extremely stable complex ions with Ca^{2+}, Mg^{2+}, and other divalent ions causing hardness, as shown in the equation

$$M^{2+} + EDTA \rightarrow [M \cdot EDTA]_{complex} \qquad (19.2)$$

The successful use of EDTA for determining hardness depends upon having an indicator present to show when EDTA is present in excess, or when all the ions causing hardness have been complexed.

Dyes such as Eriochrome Black T or Calmagite serve as excellent indicators to show when all the hardness ions have been complexed. When a small amount of either, which alone have a blue color, is added to a hard water with a pH of about

10.0, it combines with a few of the Ca^{2+} and Mg^{2+} ions to form a weak complex ion that is wine red in color, as shown in the equation with Eriochrome Black T.

$$M^{2+} + \text{Eriochrome Black T} \rightarrow (M \cdot \text{Eriochrome Black T})_{\text{complex}} \qquad (19.3)$$
$$\quad\quad\quad\quad\quad \text{Blue} \quad\quad\quad\quad\quad\quad\quad\quad \text{Wine red}$$

During the titration with EDTA, all free hardness ions are complexed according to Eq. (19.2). Finally, the EDTA disrupts the wine red (M Eriochrome Black T) complex because it is capable of forming a more stable complex with the hardness ions. This action frees the Eriochrome Black T indicator, and the wine red color changes to a distinct blue color, heralding the end of the titration.

Although the EDTA method is subject to certain interferences, most of them can be overcome by proper modifications. The method yields very precise and accurate results and does not require analyses for individual cations. It is the method of choice in most laboratories.

19.5 | TYPES OF HARDNESS

In addition to total hardness, it is desirable and sometimes necessary to know the types of hardness present. Hardness is classified in two ways: (1) with respect to the metallic ion and (2) with respect to the anions associated with the metallic ions.

Calcium and Magnesium Hardness

Calcium and magnesium cause by far the greatest portion of the hardness occurring in natural waters. In some considerations it is important to know the amounts of calcium and magnesium hardness in water. For example, it is necessary to know the magnesium hardness or the amount of Mg^{2+} in order to calculate lime requirements in lime-soda ash softening. The calcium and magnesium hardness may be calculated from the complete chemical analysis, as discussed in Sec. 19.4. Such information is not always available, and recourse is made to some method of analysis that allows separate measurement of calcium or magnesium hardness. If calcium hardness is determined, magnesium hardness is obtained by subtracting calcium hardness from total hardness, as follows:

$$\text{Total hardness} - \text{calcium hardness} = \text{magnesium hardness} \qquad (19.4)$$

This procedure yields reasonably reliable results because most of the hardness in natural waters is due to these two cations. Most methods for measuring calcium hardness will also include strontium hardness.

Carbonate and Noncarbonate Hardness

The part of the total hardness that is chemically equivalent to the bicarbonate plus carbonate alkalinities present in a water is considered to be *carbonate hardness*. Since alkalinity and hardness are both expressed in terms of $CaCO_3$, the carbonate hardness can be found as follows:

When alkalinity < total hardness,

$$\text{Carbonate hardness (in mg/L)} = \text{alkalinity (in mg/L)} \qquad (19.5)$$

When alkalinity $\geq$ total hardness,

$$\text{Carbonate hardness (in mg/L)} = \text{total hardness (in mg/L)} \qquad (19.6)$$

Carbonate hardness is singled out for special recognition because the bicarbonate and carbonate ions with which it is associated tend to precipitate this portion of the hardness at elevated temperatures such as occur in boilers or during the softening process with lime.

$$Ca^{2+}(aq) + 2HCO_3^-(aq) \rightarrow CaCO_3(s) + CO_2(g) + H_2O(l) \qquad (19.7)$$

$$Ca^{2+}(aq) + 2HCO_3^-(aq) + Ca(OH)_2(s) \rightarrow 2CaCO_3(s) + 2H_2O(l) \qquad (19.8)$$

It may also be considered as that part of the total hardness that originates from the action of carbonic acid on limestone, as illustrated in Fig. 19.2. Carbonate hardness was formerly called *temporary hardness* because it can be caused to precipitate by prolonged boiling [see Eq. (19.7)].

The amount of hardness that is in excess of the carbonate hardness is called *noncarbonate hardness* and can be estimated as follows:

$$\text{Noncarbonate hardness (NCH)} = \text{total hardness} - \text{carbonate hardness} \qquad (19.9)$$

Since all forms of hardness as well as alkalinity are expressed in terms of $CaCO_3$, the calculations for Eqs. (19.5), (19.6), and (19.9) can be made directly. These are excellent examples of the reason why alkalinity and hardness are normally expressed in terms of $CaCO_3$. Noncarbonate hardness was formerly called *permanent hardness* because it cannot be removed or precipitated by boiling. Noncarbonate hardness cations are associated with sulfate, chloride, and nitrate anions.

Pseudo-Hardness

Sea, brackish, and other waters that contain appreciable amounts of Na^+ interfere with the normal behavior of soap because of the common ion effect (Sec. 2.13). Sodium is not a hardness-causing cation, and so this action which it exhibits when present in high concentration is termed *pseudo-hardness.*

19.6 | APPLICATION OF HARDNESS DATA

Hardness of a water is an important consideration in determining the suitability of a water for domestic and industrial uses. The environmental engineer uses it as a basis for recommending the need for softening processes. The relative amounts of calcium and magnesium hardness and of carbonate and noncarbonate hardness present in a water are factors in determining the most economical type of softening process to use, and become important considerations in design. Determinations of hardness serve as a basis for routine control of softening processes.

PROBLEMS

19.1 What is hardness in water and by what is it caused?

19.2 For what reasons is hardness in water undesirable?

19.3 What is a precipitate sometimes formed in water upon boiling?

19.4 What are the "Standard Methods" procedures for measuring hardness?

19.5 Discuss the principles involved in the EDTA titrimetric method of measuring hardness.

19.6 What is the difference between permanent and temporary hardness?

19.7 Why might we wish to know how much of the hardness in water is caused by magnesium and how much by calcium?

19.8 A water has the following analysis:

Cation	mg/L	Anion	mg/L
Na^+	20	Cl^-	40
K^+	30	HCO_3^-	67
Ca^{2+}	5	CO_3^{2-}	0
Mg^{2+}	10	SO_4^{2-}	5
Sr^{2+}	2	NO_3^-	10

What is the total hardness, carbonate hardness, and noncarbonate hardness in mg/L as $CaCO_3$?

19.9 A water has the following analysis:

Cation	mg/L	Anion	mg/L
Na^+	30	Cl^-	60
K^+	5	HCO_3^-	45
Ca^{2+}	36	CO_3^{2-}	1
Mg^{2+}	12	SO_4^{2-}	60
Sr^{2+}	1	NO_3^-	2

What is the total hardness, carbonate hardness, and noncarbonate hardness in mg/L as $CaCO_3$?

REFERENCE

Clesceri, L. S., A. E. Greenberg, and A. D. Eaton (eds.): "Standard Methods for the Examination of Water and Wastewater," 20th ed., American Public Health Association, Washington, DC, 1998.

20

Residual Chlorine and Chlorine Demand

20.1 | GENERAL CONSIDERATIONS

The prime purpose of disinfecting public water supplies and wastewater effluents is to prevent the spread of waterborne diseases. The practice of disinfection with chlorine has become so widespread and generally accepted that the real reason is frequently taken very much for granted. One needs to become familiar with the history of the great plagues that have afflicted humans and the developments that led to the proof that water is the major vehicle of transmission for some diseases. The historical aspects that led to the practice of disinfection cannot be dealt with here adequately; hence, supplementary reading is highly recommended.[1] In more recent years, chlorination has been found to produce trihalomethanes and other organics of health concern. Thus, use of alternative disinfectants, such as chloramines, chlorine dioxide, ultraviolet radiation, and ozone, which do not cause this particular problem, is increasing. The chemistry of these disinfectants as well as their limitations need to be understood. One important limitation is that chlorination alone is not sufficiently protective against some important disease-causing protozoa, such as *Giardia lamblia* and *Cryptosporidium parvum*. Good filtration is also required.

Early History of Diseases

Communicable diseases have been a curse of human beings since time immemorial. The intensity of the problem appears to have been magnified as the density of the population increased. During the fourteenth century a plague known as the "Black Death" swept over Europe, leaving about 25 percent of the people dead in its wake. An epidemic in London in the winter of 1664–1665 caused 70,000 deaths, equal to 14 percent of the population.

---

[1]M. J. Rosenau, "Preventive Medicine and Hygiene," Appleton-Century-Crofts, New York, 1935; S. C. Prescott and M. P. Horwood, "Sedgwick's Principles of Sanitary Science and Public Health," Macmillan, New York, 1946.

With the development of the industrial revolution, which attracted people to urban areas and caused them to live under more crowded conditions, the frequency of epidemics increased. Up until 1854, there had been a great deal of theorizing concerning the causes and modes of transmission of disease but there was no general consensus as to the cause. The science of bacteriology, upon which definite proof depended, was still unborn.

In 1854 a localized epidemic of Asiatic cholera broke out in London. Through the careful investigations of two men, John Snow and John York, it was demonstrated, as well as could be by the means available at that time, that the source of infection was water from the Broad Street Pump. It was further demonstrated that the well was contaminated by wastewater from a damaged sewer nearby and that the sewer carried wastewater from a home housing one suffering from the disease. The Broad Street Pump epidemic is a milestone in public health engineering practice, for it established without doubt that water was a major vehicle for the spread of Asiatic cholera, one of the greatest of human plagues. This discovery stimulated and gave real purpose to the practice of slow sand filtration which had been initiated about 1830.

The science of bacteriology is considered to have originated about 1870. Robert Koch, in 1875, was successful in growing a pure culture of the bacterium causing anthrax. This was another event of great significance, for within a few years the causative organisms of typhoid (1884), Asiatic cholera (1883), and many other diseases were grown in pure culture. These developments provided the means for absolute proof that water can serve as a major vehicle for disease transmission.

The cholera epidemic in Hamburg, Germany, in 1892 served as another milestone in the knowledge concerning waterborne disease. During the epidemic, cholera organisms were actually found in the river waters used for the water supply. In addition, the efficacy of slow sand filters for removing disease organisms was demonstrated.

The typhoid epidemic at Lausen, Switzerland, in 1872 was caused by contamination of a spring water supply. It is noteworthy because of the remoteness of the point of contamination and the considerable distance the water traveled underground without freeing itself of the disease organisms. This characteristic has been shown to be particularly true in limestone areas where cracks and crevices occur.

History of Disinfection Practice

Chlorination of water supplies on an emergency basis has been practiced since about 1850. With definite evidence at hand that certain diseases were transmitted by water, emergency treatment with hypochlorite became quite common during periods of epidemics.

It was not until 1904 that continuous chlorination of a public water supply was attempted in England. Shortly thereafter, in 1908, George A. Johnson initiated treatment with calcium hypochlorite of the water at the Bubbly Creek filter plant of the Union Stock Yards in Chicago. In 1909, Jersey City, New Jersey, started hypochlorite treatment of its Boonton supply. This was the first attempt to chlorinate a public water supply in the United States and led to a celebrated court case in which a wise judge upheld the right of the city to chlorinate the water supply in the best interests of public health. From that day, disinfection of public water supplies has spread so that today it is almost routine practice.

The practice of chlorinating public water supplies did not spread rapidly at first because of the instability during storage of the calcium hypochlorite then available. The development of facilities for feeding gaseous chlorine occurred about 1912, and from that time chlorination practice has grown rapidly. With the increased use of chlorine for disinfecting purposes, there has been a corresponding decrease in the incidence of waterborne disease. The gross effect that modern sanitation practices, of which chlorination of water supplies and pasteurization of milk are major ones, have had upon the typhoid death rate in the United States is shown in Fig. 20.1. The continual decline in the importance of typhoid fever is illustrated in Fig. 20.2 by the number of waterborne outbreaks between 1946 and 1970. Of more current concern is the increase in cases of waterborne diseases such as infectious hepatitis, a viral infection, and even more recently, giardiasis and cryptosporidiosis, both of which are caused by protozoa.

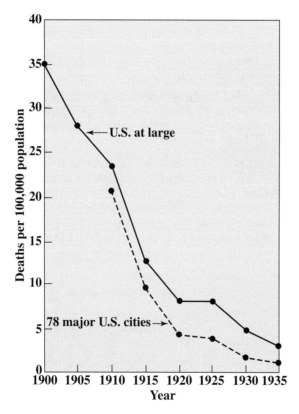

Figure 20.1
Typhoid and paratyphoid death rate in the United States
and in 78 major U.S. cities. *(A. E. Gorman and A. Wolman, J.*
Amer. Water Works Assoc., February 1939.)

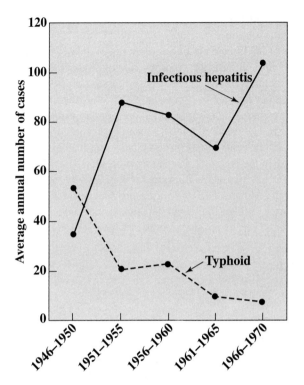

Figure 20.2
Average annual number of typhoid and hepatitis cases
occurring in waterborne outbreaks—1946 to 1970. *(After
G. F. Craun and L. J. McCabe, J. Amer. Water Works Assoc., 65:
74, 1973.)*

20.2 | CHEMISTRY OF CHLORINATION

Chlorine is used in the form of free chlorine or as hypochlorite. In either form it acts
as a potent oxidizing agent and often dissipates itself in side reactions so rapidly
that little disinfection is accomplished until amounts in excess of the chlorine demand have been added.

Reactions with Water

Chlorine combines with water to form hypochlorous and hydrochloric acids.

$$Cl_2 + H_2O \rightleftharpoons HOCl + H^+ + Cl^- \tag{20.1}$$

$$\frac{[H^+][Cl^-][HOCl]}{[Cl_2]} = 4 \times 10^{-4} \qquad \text{(at 25°C)} \tag{20.2}$$

This equilibrium is the dominant one in chlorine-containing water from vacuum chlorinators with a pH of 2 to 3, and at the instant of contact when chlorine

water is added to any water for disinfection (Fig. 20.3). The nature of the reactions is dominated by the free Cl_2. These often result in the development of obnoxious compounds such as trichloramine, NCl_3. To minimize these effects, a high-quality water is often used as chlorinator feed water and flash mixing is required at the point where the chlorine water is applied to prevent the development of localized low pH conditions. In dilute solution and at pH levels above about 4, the equilibrium given in Eq. (20.2) is displaced greatly to the right, and very little Cl_2 exists as such in solution. The hypochlorous acid formed is a weak acid and is very poorly dissociated at pH levels below 6.

$$HOCl \rightleftharpoons H^+ + OCl^- \qquad (20.3)$$

$$\frac{[H^+][OCl^-]}{[HOCl]} = 2.7 \times 10^{-8} \qquad \text{(at 20°C)} \qquad (20.4)$$

The relative amounts of OCl^- and $HOCl$ in solution as a function of pH and in accordance with the equilibrium given in Eq. (20.4) are shown in Fig. 20.4. This equilibrium is established very rapidly.

Hypochlorite is used in the form of solutions of sodium hypochlorite and high-test Ca hypochlorite in the dry form. The former is popular where large amounts are necessary such as in wastewater disinfection where local supplies are available or on-site generation from salt solutions is feasible. Ca hypochlorite is popular for situations where limited amounts are required or intermittent usage is dictated. Both

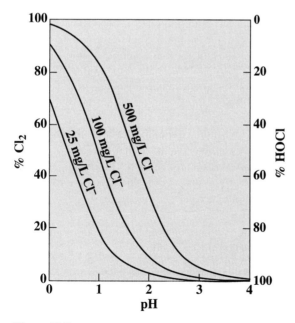

Figure 20.3
Effect of pH and chloride concentration on the distribution of chlorine and hypochlorous acid in water at 25°C.

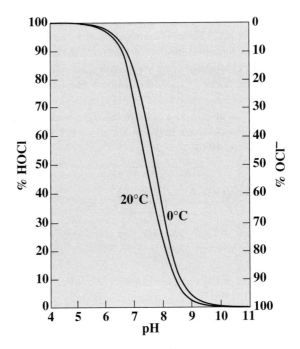

Figure 20.4
Effect of pH on the distribution of hypochlorous acid and hypochlorite ion in water.

compounds ionize in water to yield hypochlorite ion as illustrated here:

$$Ca(OCl)_2 \rightarrow Ca^{2+} + 2OCl^-$$
$$NaOCl \rightarrow Na^+ + OCl^- \tag{20.5}$$

This ion, of course, establishes an equilibrium with hydrogen ions in accordance with Eq. (20.3). Thus, it may be concluded that the same equilibria are established in water regardless of whether chlorine or hypochlorite is added. The significant difference would be in pH effects and their influence on the relative amounts of OCl^- and HOCl at equilibrium. Chlorine tends to decrease the pH, whereas hypochlorite tends to increase the pH.

Reactions with Impurities in Water

Chlorine and hypochlorous acid react with a wide variety of substances, including ammonia and naturally occurring humic materials.

Reactions with Ammonia   Ammonium ion exists in equilibrium with ammonia and hydrogen ion [Eq. (2.48)]. The ammonia reacts with chlorine or hypochlorous acid to form monochloramine, dichloramine, and trichloramine, depending upon the relative amounts of each and to some extent upon the pH, as follows:

$$NH_3 + HOCl \rightarrow NH_2Cl + H_2O \qquad \text{(monochloramine)} \qquad (20.6)$$

$$NH_2Cl + HOCl \rightarrow NHCl_2 + H_2O \qquad \text{(dichloramine)} \qquad (20.7)$$

$$NHCl_2 + HOCl \rightarrow NCl_3 + H_2O \qquad \text{(trichloramine)} \qquad (20.8)$$

The mono- and dichloramines have significant disinfecting power and are therefore of interest in the measurement of chlorine residuals. The chemistry of these reactions is discussed in more detail in Sec. 20.3.

Extraneous Reactions Chlorine combines with a wide variety of materials, particularly reducing agents. Many of the reactions are very rapid, while others are much slower. These side reactions complicate the use of chlorine for disinfecting purposes. Their demand for chlorine must be satisfied before chlorine becomes available to accomplish disinfection. They result in the production of numerous disinfection by-products which are now regulated by the U.S. EPA and listed in the WHO Guidelines for drinking water because of their potential health significance (see Chap. 34).

The reaction between hydrogen sulfide and chlorine will serve to illustrate the type of reaction that occurs with reducing agents:

$$H_2S + Cl_2 \rightarrow 2HCl + S \qquad (20.9)$$

Fe^{2+}, Mn^{2+}, and NO_2^- are examples of other inorganic reducing agents present in water supplies that will react with chlorine.

Organic compounds that possess unsaturated linkages will also add hypochlorous acid and increase the chlorine demand:

$$\begin{array}{ccc} & & Cl \quad OH \\ -C{=}C- + HOCl \rightarrow & -C-C- \\ H \quad H & & H \quad H \end{array} \qquad (20.10)$$

Chlorine also reacts with other halogens in water. For example, hypochlorous acid reacts with bromide to form hypobromous acid:

$$Br^- + HOCl \rightarrow HOBr + Cl^- \qquad (20.11)$$

HOBr is also a disinfectant, but reacts more rapidly than chlorine. When bromide is present in water, the chlorine appears to be more reactive for this reason. HOBr also reacts with organics.

Chlorine can react with phenols to produce mono-, di-, or trichlorophenols, which can impart tastes and odors to waters. Chlorine, as well as hypobromous acid formed as noted previously, also react with humic substances present in most water supplies, forming a variety of halogenated products including trihalomethanes (THMs) and haloacetic acids. The THMs were the first discovered,[2] and are suspected human carcinogens which are regulated in drinking water with a sum total maximum contaminant level of 80 μg/L. The major THMs formed are chloroform,

---

[2]J. J. Rook, *Water Treat. Exam.,* **23:** 234 (1974).

bromodichloromethane, dibromochloromethane, and bromoform. The brominated species result when trace amounts of bromide are present in the water being chlorinated. An early survey of 80 water supplies in the United States indicated the average concentration of chloroform alone resulting from disinfection with chlorine was 21 μg/L, and the maximum was over 300 μg/L.[3] Because of the interest in reducing the concentrations of THMs, haloacetic acids, and other currently unregulated halogenated organic products of chlorination in drinking waters, other possible disinfectants, such as chlorine dioxide, ozone, or ultraviolet radiation, are being considered or used by many utilities. However, the benefits from chlorine disinfection are immense, as already discussed. Further, it is the only recognized method of disinfection that is capable of providing protective residuals within the distribution system to guard against inadvertent bacterial contamination, from back siphonage or cross connections. Other disinfectants have their own advantages and disadvantages. While exploration of other disinfectants is worthwhile, continued efforts directed toward minimizing the production of chlorinated organics from chlorination is needed as well. Many possibilities for improvement exist. For example, ozone may be used as the primary disinfectant at the water treatment plant, followed by addition of ammonia and chlorination to form a chloramine residual in the distribution system that is protective, but much less reactive with natural organics than free chlorine. Disinfection with combinations of chemicals can make determination of disinfectant residuals more challenging for the analyst.

20.3 | PUBLIC HEALTH SIGNIFICANCE OF DISINFECTION RESIDUALS

Disinfection is a process designed to kill harmful organisms, and it does not ordinarily produce a sterile water. These generalizations hold for disinfection with chlorine, chlorine dioxide, ultraviolet radiation, and ozone. Two factors are extremely important in disinfection: time of contact and concentration of the disinfecting agent. Where other factors are constant, the disinfecting action may be represented by

$$\text{Kill} \propto C^n \times t \qquad (n > 0) \tag{20.12}$$

The important point is that with long contact times a low concentration of disinfectant suffices, whereas short contact times require a high concentration to accomplish equivalent kills.

It has become common practice to refer to chlorine, hypochlorous acid, and hypochlorite ion as *free chlorine residuals,* and the chloramines are called *combined chlorine residuals.* Research has shown that with free chlorine residuals, a lower pH, which favors the formation of HOCl over OCl$^-$, is more effective for disinfection. Research has also shown that a greater concentration of combined chlorine residual than of free chlorine residual is required to accomplish a given kill in a

---

[3]A. A. Stevens et al., *J. Amer. Water Works Assoc.* **68** (11): 615 (1976).

specified time. For these reasons it is necessary to know both the concentration and the kind of residual chlorine acting.

The rate of the reaction between ammonia and hypochlorous acid varies considerably, depending upon the pH and temperature. The reaction rate is most rapid at pH 8.3 and decreases rapidly as the pH is decreased or increased. For this reason, it is common to find free chlorine and combined chlorine residuals coexisting after contact periods of 10, 15, or even 60 min.

The action of excess chlorine on waters containing ammonia merits special consideration. With mole ratios of chlorine to ammonia up to 1:1, both monochloramine and dichloramine are formed. The relative amounts of each are a function of the pH, which affects the relative rate of formation of the two species as well as the thermodynamic equilibrium between them. Greater proportions of dichloramine appear at lower pH values.

Further increases in the mole ratio of chlorine to ammonia result in formation of some trichloramine and oxidation of part of the ammonia to N_2 or NO_3^-. With chlorination within the pH range of 6 to 7, these reactions are essentially complete when 1.5 mol of chlorine has been added for each mole of ammonia nitrogen originally present in the water. Chloramine residuals usually reach a maximum when about 1 mol of chlorine has been added for each mole of ammonia, and then decline to a minimum value at a chlorine-to-ammonia ratio of 1.5. Further additions of chlorine produce free chlorine residuals. Chlorination of a water to the extent that all the ammonia is converted to N_2 or a higher oxidation state is referred to as *breakpoint chlorination* because of the peculiar character of the chlorine residual curve, as illustrated in Fig. 20.5.

Theoretically, it would require 3 mol of chlorine for the complete conversion of ammonia to nitrogen trichloride (trichloramine) and 4 mol for the complete oxidation to nitrate. The fact that in the pH range of 6 to 7, only 1.5 mol is required can be accounted for if the following overall reaction were predominant:

$$2NH_3 + 3Cl_2 \rightarrow N_2 + 6H^+ + 6Cl^- \tag{20.13}$$

At lower pH, a higher ratio of chlorine to ammonia is required, presumably because some trichloramine is formed, and at higher pH, a higher ratio is required, presumably because some nitrate is formed.[4] The nature and rates of the intermediate reactions involved have been investigated.[5] Equation (20.13) indicates that 7.6 g of chlorine per gram of ammonia nitrogen would be required to reach the breakpoint.

Breakpoint chlorination is required to obtain a free chlorine residual for better disinfection if ammonia is present in a water supply. Breakpoint chlorination has also been proposed as one method for removal of ammonia when required in wastewater treatment for reasons discussed in Sec. 25.4.

---

[4]Pressly, et al., "Nitrogen Removal by Breakpoint Chlorination," U.S. Dept. of Interior Report (Sept. 1970).

[5]I. W. Wei. "Chlorine-Ammonia Breakpoint Reactions; Kinetics and Mechanisms." Ph.D. thesis, Harvard University, Cambridge, MA, May 1972; B. Savnier, and R. E. Selleck, "Kinetics of Breakpoint Chlorination and Disinfection," SERL Rept. 76–2. University of California, Berkeley. May, 1976.

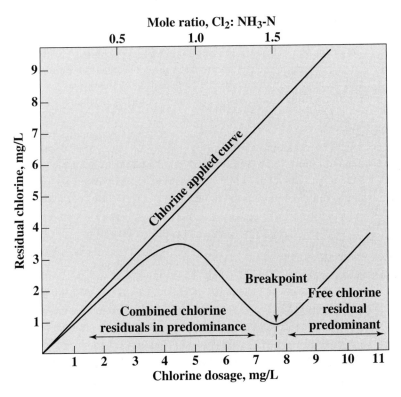

Figure 20.5
A residual-chlorine curve showing a typical breakpoint. Ammonia-nitrogen content of water, 1.0 mg/L.

While free chlorine residuals have good disinfecting powers, they are usually dissipated quickly in the distribution system. For this reason final treatment with ammonia is often practiced to convert free chlorine residuals to longer-lasting combined chlorine residuals.

20.4 | METHODS OF CHLORINE RESIDUAL DETERMINATION

It was not until about 1940 that the difference in disinfecting power of chloramine and free chlorine residuals was demonstrated. Prior to that time no attempt had been made to develop analytical procedures for differentiation. Therefore, methods for measuring chlorine residuals may be classed into old methods that measure total chlorine and new methods that allow measurement of free and combined forms.

Total Chlorine Residuals

All common methods of measuring chlorine residuals depend upon their oxidizing power; consequently, any other oxidizing agents present may interfere with the test. Manganese in valences above 2 and nitrite are the most common interferences.

Orthotolidine Method   This method is described primarily for historical reasons. In 1909 Phelps proposed the use of orthotolidine as a colorimetric indicator for chlorine residuals. Ellms and Hauser in 1913 incorporated the use of color standards and thereby made the test quantitative. Shortly thereafter, proprietary devices employing colored glass disks or sealed colored liquid standards, suitable for field as well as laboratory use, were developed. The introduction of a simple testing procedure, which allowed close control of chlorination, was an important factor in the widespread acceptance of chlorination of public water supplies. It made chlorination a practical method of disinfection on even the smallest supplies because of the simplicity of the test. However, the orthotolidine test gives relatively poor accuracy and precision compared with other procedures now available, and also orthotolidine is now known to be a toxic compound with carcinogenic potential. Thus, its use as a standard procedure has been eliminated from "Standard Methods."

Iodometric Method   The iodometric method is of interest, historically, because it served as the basis of controlling chlorination until about 1913 when alternate tests were first developed. The method is still used and depends upon the oxidizing power of free and combined chlorine residuals to convert iodide ion to free iodine, the reactions being represented as

$$Cl_2 + 2I^- \rightarrow I_2 + 2Cl^- \tag{20.14}$$

$$I_2 + \text{starch} \rightarrow \text{blue color} \qquad \text{(qualitative test)} \tag{20.15}$$

In the presence of starch, the iodine produced a blue color, which was accepted as evidence of the presence of residual chlorine but, of course, did not indicate the amount of residual present, except as people were able to judge the intensity of the blue color.

The iodometric method provides a means of quantitative measurement of total residual if the iodine released is titrated with a standard solution of a reducing agent. The usual reagent is sodium thiosulfate, and the end point is indicated by the disappearance of the blue color.

$$I_2 + 2Na_2S_2O_3 \rightarrow Na_2S_4O_6 + 2NaI \tag{20.16}$$

or

$$I_2 + 2S_2O_3^{2-} \rightarrow S_4O_6^{2-} + 2I^- \tag{20.17}$$

The iodometric procedure is appropriate for total chlorine concentrations greater than 1 mg/L. In an alternative iodometric method used for measuring chlorine residual in wastewater the procedure is reversed. Here, a standard solution of the reducing agent (thiosulfate or phenylarsine oxide) is added to the sample to react with the chlorine, and the reducing agent remaining is determined by titration with a standard iodine or iodate solution. The end point can be signaled by a change of starch from colorless initially to blue following the addition of excess iodine or iodate. A unit used for ampherometric titration discussed next can also be used here. The amount of chlorine present is determined from the difference between the amount of reducing agent initially added and that remaining after reacting with residual chlorine.

Free and Combined Chlorine Residuals

With the development of knowledge concerning the relative disinfecting powers of free and combined chlorine residuals, it became important to have ways of differentiating and measuring them. There are several methods that are now available, the most widely used being amperometric titration and the DPD method.

Amperometric Titration Method   Oxidation-reduction titration procedures were among the first methods for measuring free and combined chlorine residuals. One device used for such titrations is the amperometric titrator, which is based upon principles of polarography. It contains an electrode to show when the reactions for the oxidation-reduction titration involved are completed. Such an instrument with a rotating platinum electrode is shown in Fig. 11.2.

Phenylarsine oxide (C_6H_5AsO) is the reducing agent normally used as the titrating agent. It reacts with free chlorine residuals at pH 6.5 to 7.5 in a quantitative manner. The reaction becomes sluggish at pH greater than 7.5. Also, chloramines are reduced at pH levels below 6.0, provided that iodide ion is present. The chloramines oxidize iodide to free iodine, and the phenylarsine oxide reduces the free iodine, thereby measuring the amount of chloramines present. By conducting a two-stage titration, with the pH adjusted at about 7 and then at about 4, it is possible to measure separately free chlorine residuals and combined chlorine residuals. Interferences from nitrite and oxidized forms of manganese are eliminated by conducting the titrations at pH levels above 3.5. A separate estimation of monochloramine and dichloramine can also be made through recognition of the fact that both free chlorine and monochloramine can be measured together at neutral pH if a small amount of iodide is present. Thus, a third titration permits estimates to be made of the three major forms of chlorine residual. The amperometric titration procedure is not subject to interference from color or turbidity, which is a particular advantage when measuring chlorine residuals in some wastewaters.

DPD Method   With the DPD method, the principle is similar to that for the amperometric method. When *N,N*-diethyl-*p*-phenylenediamine (DPD) is added to a sample containing free chlorine residual, an instantaneous reaction occurs, producing a red color. If a small amount of iodide is then added to the sample, monochloramine reacts to produce iodine, which in turn oxidizes more DPD to form additional red color. If a large quantity of iodide is then added, dichloramine will react to form still more red color. By measuring the intensity of the red color produced in the neutral pH range of 6.2 to 6.5 after each of the three steps outlined, then free chlorine, monochloramine, and dichloramine residuals can be determined. The intensity of the red color produced can be determined either by titration with ferrous ion (ferrous ammonium sulfate) until the red color disappears, or directly by colorimetric analysis. This is a relatively easy procedure to use.

20.5 | MEASUREMENT OF CHLORINE DEMAND

The *chlorine demand* of a water is the difference between the amount of chlorine applied and the amount of free, combined, or total available chlorine remaining at the end of the contact period. The chlorine demand is different with different waters, and even with a given water will vary with the amount of chlorine applied, the desired residual, time of contact, pH, and temperature. The test should be conducted with chlorine or with hypochlorite, depending upon the form that will be used in practice.

Measurement of chlorine demand can be readily made by treating a series of samples of the water in question with known but varying dosages of chlorine or hypochlorite. The water samples should be at a temperature within the range of interest, and after the desired contact period, determination of residual chlorine in the samples will demonstrate which dosage satisfied the requirements of the chlorine demand, in terms of the desired residual.

20.6 | DISINFECTION WITH CHLORINE DIOXIDE

A discussion of chlorination should also include chlorine dioxide, as it has certain advantages over free chlorine and hypochlorite. It is about as effective an oxidizing agent as hypochlorous acid and is a particularly good disinfectant at high pH values. A new application in late 2001 by the U.S. EPA was the use of about 800 ppm chlorine dioxide in air in the Hart Senate Office Building in Washington, DC, to decontaminate rooms of anthrax spores contained in a letter sent by an unknown assailant.

Chlorine dioxide does not combine with ammonia to produce chloramines nor with natural organics to form chloroform or other trihalomethanes. Another advantage of chlorine dioxide is that it is very effective in the destruction of phenolic compounds that combine with other types of chlorine to create the undesirable taste-producing chlorinated phenols.

Chlorine dioxide is an unstable gas and so is usually produced at the plant site by mixing a solution of sodium chlorite ($NaClO_2$) with a strong chlorine solution:

$$2NaClO_2 + Cl_2 \rightarrow 2ClO_2 + 2NaCl \qquad (20.18)$$

The yield of chlorine dioxide is increased if the pH is depressed to less than 4 by the addition of a small excess of chlorine. The more extensive use of chlorine dioxide in water treatment has been hampered by its relatively high cost compared to other forms of chlorine, and the lack of experience with its use. Chlorite can be formed in water disinfected with chlorine dioxide, and there is some concern over its potential health effects.

Either the iodometric approach, the amperometric titration procedure, or the DPD method can be used to determine chlorine dioxide residuals in water. In order to differentiate chlorine dioxide from the various forms of chlorine residual, some aspects of the chemistry of chlorine dioxide need to be understood. At a high pH of about 12, chlorine dioxide dissociates into chlorite and chlorate, neither of which is measured in the test at neutral pH or above:

$$2ClO_2 + H_2O \rightarrow 2H^+ + ClO_2^- + ClO_3^- \qquad (20.19)$$

At neutral pH, the oxidizing power of chlorine dioxide is only partially measured through titration with reducing agents such as DPD or phenylarsine oxide, forming chlorite in the process:

$$ClO_2 + e^- \rightarrow ClO_2^- \qquad (20.20)$$

However, at a low pH of 2, the chlorite formed is reduced further to chloride, oxidizing an equivalent amount of iodine for measurement in the process:

$$ClO_2^- + 4H^+ + 4e^- \rightarrow Cl^- + 2H_2O \qquad (20.21)$$

By a series of titrations at different pH levels, and with other chemical additions to suppress or enhance the reduction of the different chlorine-containing species, the concentrations of free chlorine, combined chlorine, chlorine dioxide, chlorite, and chlorate can be determined by taking advantage of this chemistry. If it is known in advance that only certain chlorine-containing species are present, then the number of titrations required can generally be reduced through an appropriate selection of titration conditions based upon knowledge of the reactions involved.

20.7 | DISINFECTION WITH OZONE

Ozone (O_3) is another strong disinfectant that is being used as an alternative to chlorine. The advantages of ozone are that it is an especially effective disinfectant at low concentrations and against protozoa such as *Giardia* and *Cryptospiridium*. The disadvantages include its relatively high cost, especially in capital cost for the ozone-producing equipment, which must be maintained on site. Ozone also does not produce a long-lasting residual, and so an alternative disinfectant, such as chloramine, is often added to provide protection throughout the distribution system. Another major advantage of ozone is that no halogenated organic compounds are produced through its usage. However, ozone does react with the natural humic materials in water to produce organic compounds that are more susceptible to biological degradation than the natural humic materials. This can result in the growth of bacteria in the distribution line, which can be detrimental to water quality and the flow of water in the pipe lines. Research to better understand this potential problem and to reduce its effects is under way. Because of its strong interactions with organic materials in waters, ozone is sometimes used for destruction of organic taste- and odor-causing materials. It is also used at times to oxidize reduced iron and manganese salts to insoluble oxides that can be removed from the water before distribution.

The proposed analytical procedure for measurement of ozone residuals in water depends upon its ability to oxidize organic materials. Here, indigo, a blue dye, is used in a colorimetric procedure. Under acid conditions, ozone rapidly decolorizes indigo through oxidation. The reduction in color of standard indigo solutions by the ozone-containing water of interest is measured quantitatively with a spectrophotometer.

20.8 | APPLICATION OF DISINFECTANT DEMAND AND DISINFECTANT RESIDUAL DATA

Determination of the disinfectant demand of a water or wastewater is an important consideration in design. It serves as the basis for determining the capacity of disinfection units required, the amount of disinfectant needed, the type of shipping containers, and all appurtenances required for handling and storage.

Disinfectant residuals are used universally in disinfection practice to control addition of disinfectant so as to ensure effective disinfection without waste of chemicals. A measurable disinfectant residual at the consumer's tap is required in many localities. They are also used to control disinfection of domestic and industrial wastes and usually are the sole criteria immediately available to determine whether or not desired objectives are being attained.

PROBLEMS

20.1 Why is it important to disinfect water supplies?

20.2 Why is it important to determine chlorine residual in water treatment practice?

20.3 What important chlorinated species are formed upon water chlorination due to the presence of ammonia?

20.4 What are the relative advantages and disadvantages of free chlorine versus combined chlorine?

20.5 Discuss the significance of contact time, chlorine residual, and pH as factors influencing the degree of disinfection obtained by chlorination.

20.6 Why is the orthotolidine test for chlorine residual no longer recommended for use?

20.7 What is the principle behind the iodometric procedure from chlorine residual?

20.8 Outline one procedure for determining residual chlorine to allow differentiation between free chlorine, chloramines, and interferences.

20.9 By use of appropriate equilibrium equations, show why the addition of chlorine tends to decrease the pH of water, while hypochlorite tends to increase the pH.

20.10 Compute the relative proportions of free chlorine occurring as HOCl and as OCl$^-$ at a pH of 6.8 and a temperature of 20°C.

20.11 What is breakpoint chlorination and how does it change the different chlorinated species in water?

20.12 According to Chick's law, the rate of kill of bacteria by chlorination follows first-order reaction kinetics. Assuming this to be true, how much contact time is required to kill 99 percent of the bacteria with a chlorine residual of 0.1 mg/L, if 80 percent are killed in 2 min with this residual? (See Sec. 3.10.)

20.13 Discuss applications of the chlorine demand test.

20.14 Write three equations to illustrate type reactions that cause chlorine demand.

20.15 What major disadvantage with chlorine has led to seeking other disinfectants for drinking water supplies?

20.16 Discuss the advantages and disadvantages of different possible disinfectants for pathogens in water.

20.17 (*a*) Under what conditions are trihalomethanes formed in drinking water, (*b*) what are the different trihalomethanes that may be formed, and (*c*) what leads to the formation of more brominated than chlorinated species?

REFERENCE

Clesceri, L. S., A. E. Greenberg, and A. D. Eaton (eds.): "Standard Methods for the Examination of Water and Wastewater," 20th ed., American Public Health Association, Washington, DC, 1998.

CHAPTER 21

Chloride

21.1 | GENERAL CONSIDERATIONS

Chloride occurs in all natural waters in widely varying concentration. The chloride content normally increases as the mineral content increases. Upland and mountain supplies usually are quite low in chloride, whereas river and groundwaters usually have a considerable amount. Sea and ocean waters represent the residues resulting from partial evaporation of natural waters that flow into them, and chloride levels are very high.

Chloride salts gain access to natural waters in many ways. The solvent power of water dissolves chloride from topsoil and deeper formations. Spray from the ocean is carried inland as droplets or as minute salt crystals, which result from evaporation of the water in the droplets. These sources constantly replenish the chloride in inland areas where they fall. Ocean and sea waters invade the rivers that drain into them, particularly the deeper rivers. The salt water, being more dense, flows upstream under the fresh water that is flowing downstream. There is a constant intermixing of the salt water with the fresh water above. In the case of the Hudson River, which has a deep channel and rather slight gradient, seawater invades for a distance of about 80 km upstream. This invasion has been a major factor in preventing New York City from developing the Hudson River as a source of water supply. Groundwaters in areas adjacent to the ocean are in hydrostatic balance with seawater. Overpumping of groundwaters produces a difference in hydrostatic head in favor of the seawater, and it intrudes into the fresh-water area. Such intrusion has occurred at many locations in Florida and California.

Most of the water used for irrigation in the Southwest is lost to the atmosphere through evapotransporation, leaving the salts originally present in the irrigation water behind in the soil. These salts must be removed to prevent destruction of the soil's crop-growing potential, and the resulting highly saline irrigation return waters tend to increase the chloride content of surface waters into which they are discharged. This and normal evaporation cause significant increases in the chloride content of the Colorado River as it flows from the mountains in the north into Mexico in the south.

Human excreta, particularly the urine, contain chloride in an amount about equal to the chloride consumed with food and water. This amount averages about 6 g of chloride per person per day and increases the amount of Cl^- in municipal wastewater about 15 mg/L above that of the carriage water. Thus, wastewater effluents add considerable chloride to receiving streams. Many industrial wastes contain appreciable amounts of chloride. Control of contamination of surface waters by chloride contained in industrial wastes is a major consideration in the Ohio River valley and in all areas where oil-field brines and other salt brines are allowed to reach receiving streams.

21.2 | SIGNIFICANCE OF CHLORIDE

Chloride in reasonable concentrations is not harmful to humans. At concentrations above 250 mg/L it gives a salty taste to water, which is objectionable to many people. For this reason a secondary standard of 250 mg/L for chloride in drinking water has been set by the U.S. EPA, the same value contained in the WHO guidelines. In many areas of the world where water supplies are scarce, sources containing as much as 2000 mg/L are used for domestic purposes without the development of adverse effects, once the human system becomes adapted to the water.

The chloride content of waters used for irrigation of agricultural crops is generally controlled along with the total salinity of the water. Evapotransportation tends to increase the chloride and salinity at the root zone of irrigated plants, making it difficult for crops to take up water due to osmotic pressure differences between the water outside the plants and within the plant cells. For this reason, chloride and total salinity concentrations at or below the drinking water standards are normally specified for waters used to irrigate salt-sensitive crops.

Before the development of bacteriological testing procedures, chemical tests for chloride and for nitrogen, in its various forms, served as the basis of detecting contamination of groundwaters by wastewater. Chloride is used to some extent as a tracer in environmental engineering practice. It is inconvenient to use in many instances because of the quantities required to produce significant increases in chloride level and because of their tendency to produce density currents. Their use as tracers has been superseded to a great extent by organic dyes which can be measured accurately in trace amounts. In groundwater studies, however, where the ratio of soil to water is very high, organic dyes tend to sorb to a sufficient degree such that they move slower than the water itself. Here, chloride, or perhaps a less prevalent halogen such as bromide, may still be useful as a tracer.

21.3 | METHODS OF DETERMINATION

There are several "Standard Methods" procedures for measurement of chloride. Among these are a potentiometric procedure (Sec. 12.3) using silver nitrate as a titrant to complex chloride with silver and employing a silver-silver chloride electrode system to detect the end point, and ion chromatography (Sec. 12.4), which is useful for determining concentrations of several common anions in one analysis. These instrumental methods require somewhat expensive equipment. Two other procedures that are much less expensive involve titrations and can be carried out readily in the field. These are the Agentometric or Mohr method and the mercuric nitrate procedure. Another alternative is the ferricyanide method, which is colorimetric, and lends itself readily to automatic analysis of large numbers of samples. The principles behind the two titration procedures and the colorimetric procedure are discussed in the following.

Mohr Method (Argentometric)

The Mohr method employs a solution of silver nitrate for titration, and "Standard Methods" recommends the use of a 0.0141 N solution. This corresponds to a N/71 solution or one in which each milliliter is equivalent to 0.5 mg of chloride ion. The silver nitrate solution can be standardized against standard chloride solutions prepared from pure sodium chloride. In the titration the chloride ion is precipitated as white silver chloride.

$$Ag^+ + Cl^- \rightleftharpoons AgCl(s) \qquad (K_{sp} = 3 \times 10^{-10}) \qquad (21.1)$$

The end point cannot be detected visually unless an indicator capable of demonstrating the presence of excess Ag^+ is present. The indicator normally used is potassium chromate, which supplies chromate ions. As the concentration of chloride ions approaches extinction, the silver-ion concentration increases to a level at which the solubility product of silver chromate is exceeded and it begins to form a reddish-brown precipitate.

$$2Ag^+ + CrO_4^{2-} \rightleftharpoons Ag_2CrO_4(s) \qquad (K_{sp} = 5 \times 10^{-12}) \qquad (21.2)$$

This is taken as evidence that all the chloride has been precipitated. Since an excess of Ag^+ is needed to produce a visible amount of $Ag_2CrO_4(s)$, the indicator error or blank must be determined and subtracted from all titrations.

Several precautions must be observed in this determination if accurate results are to be obtained:

1. A uniform sample size must be used, preferably 100 mL, so that ionic concentrations needed to indicate the end point will be constant.
2. The pH must be in the range of 7 to 8 because Ag^+ is precipitated as $AgOH(s)$ at high pH levels and the CrO_4^{2-} is converted to $Cr_2O_7^{2-}$ at low pH levels.
3. A definite amount of indicator must be used to provide a certain concentration of CrO_4^{2-}; otherwise $Ag_2CrO_4(s)$ may form too soon or not soon enough.

The indicator error or blank varies somewhat with the ability of individuals to detect a noticeable color change. The usual range is 0.2 to 0.4 mL of titrant.

If the silver nitrate solution used for titration is exactly 0.0141 N, the calculation for chlorides given in "Standard Methods" may be simplified as

$$Cl^- \text{ (in mg/L)} = \frac{(mL\ AgNO_3 - blank) \times 0.5 \times 1000}{mL\ sample} \qquad (21.3)$$

since $0.0141 \times 35.45 = 0.5$.

Mercuric Nitrate Method

The mercuric nitrate method of determining chloride is much less subject to interferences than the Mohr method because the titration is performed in a sample whose pH is adjusted to a value of about 2.5. Under these conditions, Hg^{2+} ion combines with Cl^- to form the $HgCl_2$ complex which is soluble,

$$Hg^{2+} + 2Cl^- \rightleftharpoons HgCl_2(aq) \qquad (\beta_2 = 1.7 \times 10^{13}) \qquad (21.4)$$

therefore making end-point detection easier than with the Mohr procedure. As the Cl^- concentration approaches zero, the Hg^{2+} concentration increases to a level where it becomes significant as the mercuric nitrate is added.

Diphenylcarbazone is the indicator used to show the presence of excess Hg^{2+} ions. It combines with them to form a distinct purple color. A blank correction is needed as with the Mohr procedure, although the value is usually less. Nitric acid is added to the indicator to reduce the sample pH to 2.5, a value that must be maintained uniformly in unknown samples, standards, and blanks. A pH indicator, xylene cyanol FF, which is blue-green at pH 2.5 is also included and improves the end point by masking the pale color developed by diphenylcarbazone during the titration.

"Standard Methods" recommends that a 0.0141 N solution of mercuric nitrate be used as the titrant in determining chloride. Each milliliter of such a solution is equivalent to 0.5 mg of Cl^-, and therefore the comments made concerning calculations with similar strength solutions of $AgNO_3$ also apply in this case.

Ferricyanide Method

An instrumental procedure is the automated ferricyanide method, which is a colorimetric procedure. In this case, mercuric ion contained in the mercuric thiocyanate titrant forms a soluble complex with chloride. This releases the thiocyanate to react with ferric ion, which is also added, to form intensely red ferric thiocyanate, the intensity of which is proportional to the chloride concentration.

21.4 | APPLICATION OF CHLORIDE DATA

In many areas the level of chloride in natural waters is an important consideration in the selection of supplies for human, industrial, and agricultural uses. Where brackish waters must be used for domestic purposes, the amount of chloride present is an important factor in determining the type of desalting apparatus to be used. The chloride determination is used to control pumping of groundwater from locations where intrusion of seawater is a problem.

In areas where the discharge of salt-water brines and industrial wastes containing high concentrations of chloride must be controlled to safeguard receiving waters, the chloride determination serves to excellent advantage for regulatory purposes.

Chloride interferes in the determination of chemical oxygen demand (COD). A correction must be made on the basis of the amounts present or else a complexing agent such as $HgSO_4$ can be added.

Sodium chloride has a considerable history as a tracer. One of its principal applications has been in tracing pollution of wells. It is admirably suited for such purposes for five reasons:

1. Its presence is not visually detectable.
2. It is a normal constituent of water and has no toxic effects.
3. The chloride ion is not adsorbed by soil formations.
4. It is not altered or changed in amount by biological processes.
5. The chloride ion is easily measured.

It is to be expected that chloride will continue in limited use as a tracer where other methods are not applicable.

PROBLEMS

21.1 Discuss the significance of the presence of high chloride concentrations in water supplies.

21.2 Why has a secondary standard for chloride in drinking water been set by the U.S. EPA and the WHO, and what is the recommended value?

21.3 What different "Standard Methods" procedures are available for determining chloride concentration, and what are the relative advantages and disadvantages of each?

21.4 Explain why a blank correction must be applied to the titration values in both the Mohr and mercuric nitrate methods in the calculation of chloride content.

21.5 Would the analytical results by the Mohr method for chloride be higher, lower, or the same as the true value if an excess of indicator were accidentally added to the sample? Why?

21.6 Why must the sample pH be neither high nor low in the Mohr method for chloride?

21.7 (a) In the determination of chloride by the Mohr method, what will be the equilibrium concentration of silver ions in mg/L, on the basis of the solubility-product principle, when the chloride concentration has been reduced to 0.2 mg/L?

   (b) If the concentration of chromate indicator used is 5×10^{-3} M, how much excess silver ion in mg/L must be present before the formation of a red precipitate will begin?

21.8 What purpose is served by the nitric acid added to the indicator in the mercuric nitrate method for chloride?

21.9 Estimate the concentration of mercuric ions in the mercuric nitrate method of determining chloride under the same conditions as given in Prob. 21.7(a).

21.10 How is color formed in the ferricyanide method for chloride?

REFERENCES

Clesceri, L. S., A. E. Greenberg, and A. D. Eaton (eds.): "Standard Methods for the Examination of Water and Wastewater," 20th ed., American Public Health Association, Washington, DC, 1998.

Day, Jr., R. A., and A. L. Underwood: "Quantitative Analysis," 6th ed., Prentice Hall, Englewood Cliffs, NJ, 1991.

Skoog, D. A., D. M. West, and F. J. Holler: "Fundamentals of Analytical Chemistry," 7th ed., Saunders College Pubs., Fort Worth, TX, 1996.

Dissolved Oxygen

22.1 | GENERAL CONSIDERATIONS

All living organisms are dependent upon oxygen in one form or another to maintain the metabolic processes that produce energy for growth and reproduction. Aerobic processes are the subject of greatest interest because of their need for free oxygen. Humans are vitally concerned with the oxygen content of the air that they breathe, since they know from experience that an appreciable reduction in oxygen content will lead to discomfort and possibly death. For this reason, the number of occupants within enclosures must be carefully restricted to the ventilating capacity.

Environmental engineers and scientists are, of course, interested in atmospheric conditions in relation to humans, but, in addition, they are vitally concerned with the "atmospheric conditions" that exist in liquids, water being the liquid in greatest abundance and importance.

All the gases of the atmosphere are soluble in water to some degree. Both nitrogen and oxygen are classified as poorly soluble, and since they do not react with water chemically, their solubility is directly proportional to their partial pressures. Hence, Henry's law may be used to calculate the amounts present at saturation at any given temperature. The solubility of both nitrogen and oxygen varies greatly with the temperature over the range of interest for natural waters. Figure 22.1 shows solubility curves for the two gases in distilled or low-solids-content water in equilibrium with air at 760-mm Hg (1 atm) pressure. The solubility is less in saline waters. It will be noted that under the partial-pressure conditions that exist in the atmosphere, more nitrogen than oxygen dissolves in water. At saturation, the dissolved gases contain about 38 percent oxygen on a molar basis, or nearly twice as much oxygen as in the normal atmosphere.

The solubility of atmospheric oxygen in fresh waters ranges from 14.6 mg/L at 0°C to about 7 mg/L at 35°C under 1 atm of pressure. Since it is a poorly soluble gas, its solubility varies directly with the atmospheric pressure at any given temperature. This is an important consideration at high altitudes. Because rates of biological oxidation increase with temperature, and oxygen demand increases accordingly, high-temperature conditions, where dissolved oxygen is least soluble, can be

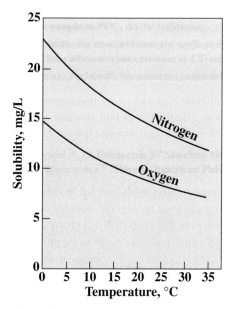

Figure 22.1
Solubility of oxygen and nitrogen in distilled
water saturated with air at 1 atm.

of concern. Most of the critical conditions related to dissolved-oxygen deficiency
in natural waters occur during the summer months when temperatures are high and
solubility of oxygen is at a minimum. For this reason it is customary to think of
dissolved-oxygen levels of about 8 mg/L as being the maximum available under
critical conditions.

The low solubility of oxygen is the major factor that limits the purification capacity of natural waters and necessitates treatment of wastes to remove pollutional matter
before discharge to receiving streams. In aerobic biological treatment processes, the
limited solubility of oxygen is of great importance because it governs the rate at
which oxygen will be absorbed by the medium and therefore the cost of aeration.

The solubility of oxygen is less in salt-containing water than it is in clean water.
For this reason the solubility for a given temperature decreases as one progresses from
fresh water to estuary water to the ocean. The extent of this effect is indicated in Table
22.1, which contains a listing of oxygen solubility as a function of temperature and
chloride content. Chloride concentration is used as a measure of the seawater–fresh
water mix in a sample. The chloride content of seawater is about 19,000 mg/L.

In polluted waters the saturation value is also less than that of clean water. The
ratio of the value in polluted water to that in clean water is referred to as the β
value. The rate of absorption of oxygen in polluted waters is normally less than in
clean water and the ratio is referred to as the α value. They may range as low as 0.8
for β and 0.4 for α in some wastewaters. Both α and β values are important design
factors in selection of aeration equipment.

Table 22.1 | Solubility of dissolved oxygen in water in equilibrium with dry air at 1 atm and containing 20.9 percent oxygen*

Temperature, °C	Chloride concentration, mg/L				
	0	5000	10,000	15,000	20,000
0	14.6	13.8	13.0	12.1	11.3
1	14.2	13.4	12.6	11.8	11.0
2	13.8	13.1	12.3	11.5	10.8
3	13.5	12.7	12.0	11.2	10.5
4	13.1	12.4	11.7	11.0	10.3
5	12.8	12.1	11.4	10.7	10.0
6	12.5	11.8	11.1	10.5	9.8
7	12.2	11.5	10.9	10.2	9.6
8	11.9	11.2	10.6	10.0	9.4
9	11.6	11.0	10.4	9.8	9.2
10	11.3	10.7	10.1	9.6	9.0
11	11.1	10.5	9.9	9.4	8.8
12	10.8	10.3	9.7	9.2	8.6
13	10.6	10.1	9.5	9.0	8.5
14	10.4	9.9	9.3	8.8	8.3
15	10.2	9.7	9.1	8.6	8.1
16	10.0	9.5	9.0	8.5	8.0
17	9.7	9.3	8.8	8.3	7.8
18	9.5	9.1	8.6	8.2	7.7
19	9.4	8.9	8.5	8.0	7.6
20	9.2	8.7	8.3	7.9	7.4
21	9.0	8.6	8.1	7.7	7.3
22	8.8	8.4	8.0	7.6	7.1
23	8.7	8.3	7.9	7.4	7.0
24	8.5	8.1	7.7	7.3	6.9
25	8.4	8.0	7.6	7.2	6.7
26	8.2	7.8	7.4	7.0	6.6
27	8.1	7.7	7.3	6.9	6.5
28	7.9	7.5	7.1	6.8	6.4
29	7.8	7.4	7.0	6.6	6.3
30	7.6	7.3	6.9	6.5	6.1

*After G. C. Whipple and M. C. Whipple, Solubility of Oxygen in Sea Water, *J. Amer. Chem. Soc.,* **33:** 362 (1911).

22.2 | ENVIRONMENTAL SIGNIFICANCE OF DISSOLVED OXYGEN

In liquid wastes, dissolved oxygen is the factor that determines whether the biological changes are brought about by aerobic or by anaerobic organisms. The former use free oxygen for oxidation of organic and inorganic matter and produce innocuous end products, whereas the latter bring about such oxidations through the reduction of certain inorganic salts such as sulfate, and the end products are often very obnoxious. Since both types of organisms are ubiquitous in nature, it is highly important that conditions favorable to the aerobic organisms (aerobic conditions) be maintained; otherwise the anaerobic organisms will take over, and development of nuisance conditions will result. Thus, dissolved-oxygen measurements are vital for maintaining aerobic conditions in natural waters that receive pollutional matter and in aerobic treatment processes intended to purify domestic and industrial wastewaters.

Dissolved-oxygen determinations are used for a wide variety of other purposes. It is one of the most important single tests that environmental engineers and scientists use. In most instances involving the control of stream pollution, it is desirable to maintain conditions favorable for the growth and reproduction of a normal population of fish and other aquatic organisms. This condition requires the maintenance of dissolved-oxygen levels that will support the desired aquatic life in a healthy condition at all times.

Determinations of dissolved oxygen serve as the basis of the BOD test; thus, they are the foundation of the most important determination used to evaluate the pollutional strength of domestic and industrial wastes. The rate of biochemical oxidation can be measured by determining residual dissolved oxygen in a system at various intervals of time.

All aerobic treatment processes depend upon the presence of dissolved oxygen, and tests for it are indispensable as a means of controlling the rate of aeration to make sure that adequate amounts of air are supplied to maintain aerobic conditions and also to prevent excessive use of air and energy.

Oxygen is a significant factor in the corrosion of iron and steel, particularly in water distribution systems and in steam boilers. Removal of oxygen from boiler-feed waters by physical and chemical means is common practice in the power industry. The dissolved-oxygen test serves as the means of control.

22.3 | COLLECTION OF SAMPLES FOR DETERMINATION OF DISSOLVED OXYGEN

A certain amount of care must be exercised in the collection of samples to be used for dissolved-oxygen determinations. In most cases of interest, the dissolved-oxygen level will be below saturation, and exposure to the air will lead to erroneous results. For this reason, a special sampling device similar to the one described in "Standard Methods" is needed. All such samplers are designed on the principle that contact with air cannot be avoided during the time the sample bottles are being filled. However, if space is available to allow the bottles to overflow, a sample of water that is representative of the mixture being sampled can be obtained. Most samplers are designed to provide an overflow of two or three times the bottle volume to ensure collection of representative samples. Also, as oxygen readily penetrates most rubber and plastic tubing, great caution is required in their use for sampling.

Most samples for dissolved oxygen are collected in the field, where it is not convenient to perform the entire determination. Since oxygen values may change radically with time because of biological activity, it is customary to "fix" the samples immediately after collection. The usual procedure is to treat the samples with the conventional reagents used in the dissolved-oxygen test and then perform the titration when the samples are brought to the laboratory. This procedure will give low results for samples with a high iodine demand, and in this case it is better to preserve the sample by addition of 0.7 mL concentrated sulfuric acid and 0.02 g sodium azide. When this is done, it is necessary to add 3 mL of alkali-iodide reagent rather than the usual 2 mL because of the extra acid the sample contains. Better results are also obtained if the "fixed" samples are stored in the dark and on ice until the analyses can be completed.[1] The chemical treatment employed in "fixing" is radical enough to arrest all biological action, and the final titration may be delayed up to 6 h.

There are two standard procedures for measuring dissolved oxygen. The older and still primary procedure is through oxidation-reduction titration using iodometric procedures. The second is a specialized adaptation of polarography in which a membrane-covered probe is employed after calibration using the iodometric procedure.

22.4 | STANDARD REAGENT FOR MEASURING DISSOLVED OXYGEN

Most volumetric methods of determining dissolved oxygen depend upon reactions that release an amount of iodine equivalent to the amount of oxygen originally present, with subsequent measurement of the amount of iodine released by means of a standard solution of a reducing agent. Sodium thiosulfate is the reducing agent normally used, and starch solution is used to determine the end point. All reactions in the determination of oxygen involve oxidation and reduction. However, starch is used as the end-point indicator. It forms a starch-iodine complex with iodine from dilute solutions to produce a brilliant blue color and returns to a colorless form when the iodine is all reduced to iodide ion.

Selection of 0.025 N Thiosulfate Solution

The equivalent weight of oxygen is 8 g. Since the normality of most titrating agents used in water and wastewater analysis is adjusted so that each milliliter is equivalent to 1.0 mg of the measured material, it would follow that a N/8 or 0.125 N solution of thiosulfate should be used. However, such a solution is too concentrated to allow accurate determinations of dissolved oxygen unless unreasonably large samples are titrated. It has become standard practice to use 200-mL samples for titration. This is one-fifth of a liter. By using a titrating agent which is one-fifth as strong as conventionally used, the results obtained on 200-mL samples, in terms of milliliters of titrant used, are the same as if 1-liter samples had been treated with the 0.125 N reagent. Thus, when a N/40 (N/8 $\times$ $\frac{1}{5}$) or 0.025 N solu-

---

[1]R. Porges, Dissolved Oxygen Determinations for Field Surveys, *J. Water Pollution Control Federation,* **36:** 1247 (1964).

tion of thiosulfate is used to titrate 200-mL samples, dissolved-oxygen values in milligrams per liter are equal to the titration volume in milliliters. This eliminates the need for calculations.

Preparation and Standardization of 0.025 N Thiosulfate

Sodium thiosulfate ($Na_2S_2O_3 \cdot 5H_2O$) can be obtained in relatively pure form. However, because of its water of hydration, it cannot be dried to a compound of definite composition, and even loses water at room temperature under conditions of low humidity. It is necessary, therefore, to prepare solutions that are slightly stronger than desired and to standardize them against a primary standard.

The equivalent weight of sodium thiosulfate cannot be calculated from its formula and anticipated valence change, as is the case with most reducing agents (see Table 11.1). Rather, it must be calculated from its reaction with an oxidizing agent, in this case iodine.

$$2Na_2S_2O_3 \cdot 5H_2O + I_2 \rightarrow Na_2S_4O_6 + 2NaI + 5H_2O \qquad (22.1)$$

From Eq. (22.1) it may be concluded that each molecule of sodium thiosulfate is equivalent to one atom of iodine. Since each atom of iodine gains one electron in the conversion to iodine ion, it is obvious that each molecule of thiosulfate supplies one electron when oxidized to tetrathionate, or

$$2S_2O_3^{2-} + I_2 \rightarrow S_4O_6^{2-} + 2I^- \qquad (22.2)$$

From these considerations, it may be concluded that the equivalent weight of sodium thiosulfate is equal to the formula weight; and something in excess of 1/40 of the formula weight, 6.205 g, should be taken to prepare 1 liter of a solution that is slightly stronger than 0.025 N. Usually 6.5 g is sufficient.

The thiosulfate solution is standardized with potassium bi-iodate. It can be obtained in essentially 100 percent pure form. It is customary to prepare a N/40 or 0.025 N solution by weighing out an exact amount on the analytical balance and diluting to the proper volume in a volumetric flask. The primary standard reacts with iodide ion in acid solution to release iodine:

$$2IO_3^- + 10I^- + 12H^+ \rightarrow 6I_2 + 6H_2O \qquad (22.3)$$

The iodine released is chemically equivalent to the oxidizing agent used. Thus, if 20 mL of 0.025 N $KIO_3 \cdot HIO_3$ (see Table 11.1) is used in the standardization procedure, exactly 20 mL of 0.025 N thiosulfate solution should be used in the titration. If the solution of thiosulfate is stronger than desired, it may be brought to proper strength by diluting with water in accordance with the principles set forth in Sec. 15.1.

Thiosulfate solutions are subject to attack by bacterial action and by carbon dioxide. Sulfur bacteria oxidize thiosulfate to sulfate under aerobic conditions, and carbon dioxide can depress the pH sufficiently to cause the thiosulfate ion to decompose into SO_3^{2-} and S. The SO_3^{2-} is converted to SO_4^{2-} by dissolved oxygen. Thiosulfate solutions may be protected from bacterial and CO_2 attack by adding 0.4

g of NaOH per liter. The resulting high pH prevents bacterial growth and keeps the pH from falling as a result of small amounts of carbon dioxide that may gain access to the solution. Excess NaOH must be avoided, as this too can make the solution less stable.

22.5 | METHODS OF DETERMINING DISSOLVED OXYGEN

Originally, measurement of dissolved oxygen was made by heating samples to drive out the dissolved gases and analyzing the collected gases for oxygen by methods applied in gas analysis. Such methods require large samples and are very cumbersome and time-consuming.

The Winkler or iodometric method and its modifications are the standard volumetric procedures for determining dissolved oxygen at the present time. The test depends upon the fact that oxygen oxidizes Mn^{2+} to a higher state of valence under alkaline conditions and that manganese in higher states of valence is capable of oxidizing I^- to I_2 under acid conditions. Thus, the amount of I_2 released is equivalent to the dissolved oxygen originally present. The iodine is measured with standard sodium thiosulfate solution and interpreted in terms of dissolved oxygen.

The Winkler Method

The unmodified Winkler method is subject to interference from a great many substances. Certain oxidizing agents such as nitrite and Fe^{3+} are capable of oxidizing I^- to I_2 and produce results that are too high. Reducing agents such as Fe^{2+}, SO_3^{2-}, S^{2-}, and polythionate, reduce I_2 to I^- and produce results that are too low. The unmodified Winkler method is applicable only to relatively pure waters.

The reactions involved in the Winkler procedure are as follows:

$$Mn^{2+} + 2OH^- \rightarrow Mn(OH)_2(s) \qquad \text{(white precipitate)} \qquad (22.4)$$

If no oxygen is present, a pure white precipitate of $Mn(OH)_2(s)$ forms when $MnSO_4$ and the alkali-iodide reagent (NaOH + KI) are added to the sample. If oxygen is present in the sample, then some of the Mn(II) is oxidized to Mn(IV) and precipitates as a brown hydrated oxide. The reaction is usually represented as follows:

$$Mn^{2+} + 2OH^- + \tfrac{1}{2}O_2 \rightarrow MnO_2(s) + H_2O \qquad (22.5)$$

or
$$Mn(OH)_2 + \tfrac{1}{2}O_2 \rightarrow MnO_2(s) + H_2O \qquad (22.6)$$

The oxidation of Mn(II) to $MnO_2(s)$, sometimes called fixation of the oxygen, occurs slowly, particularly at low temperatures. Furthermore, it is necessary to move the flocculated material throughout the solution to enable all the oxygen to react. Vigorous shaking of the samples for at least 20 seconds is needed. In the case of brackish or seawaters, much longer contact times are required.

After shaking the samples for a time sufficient to allow all oxygen to react, the floc is allowed to settle so as to leave at least 5 cm of clear liquid below the stopper;

then sulfuric acid is added. Under the low pH conditions that result, the MnO_2 oxidizes I^- to produce I_2.

$$MnO_2(s) + 2I^- + 4H^+ \rightarrow Mn^{2+} + I_2 + 2H_2O \qquad (22.7)$$

I_2 is rather insoluble in water, but forms a complex with the excess iodide present to reversibly form the more soluble tri-iodate, thus preventing escape of I_2 from solution:

$$I_2 + I^- \rightleftharpoons I_3^- \qquad (22.8)$$

The sample should be stoppered and shaken for at least 10 seconds to allow the reaction to go to completion and to distribute the iodine uniformly throughout the sample.

The sample is now ready for titration with 0.025 N thiosulfate. The use of 0.025 N thiosulfate is based upon the premise that a 200-mL sample will be used for titration. In adding the reagents used for the Winkler test, a certain amount of dilution of the sample occurs; therefore, it is necessary to take a sample somewhat greater than 200 mL for the titration. When 300-mL bottles are used in the test, 1 mL of $MnSO_4$ and 1 mL of alkali-KI solutions are used. These are added in such a manner as to displace approximately 2 mL of sample from the bottle, and a correction should be made. When the 2 mL of acid is added, none of the oxidized floc is displaced; thus, no correction need be made for its addition. To correct for the addition of the first two reagents, 201 mL of the treated sample is taken for titration.

Titration of a sample of a size equivalent to 200 mL of the original sample with 0.025 N thiosulfate solution yields results in milliliters, which can be interpreted directly in terms of milligrams per liter of dissolved oxygen.

The Azide Modification of the Winkler Method

The nitrite ion is one of the most frequent interferences encountered in the dissolved-oxygen determination. It occurs principally in effluents from wastewater treatment plants employing biological processes, in river waters, and in incubated BOD samples. It does not oxidize Mn^{2+} but does oxidize I^- to I_2 under acid conditions. It is particularly troublesome because its reduced form, N_2O_2, is oxidized by oxygen, which enters the sample during the titration procedure and is converted to NO_2^- again, establishing a cyclic reaction that can lead to erroneously high results, far in excess of amounts that would be expected. The reactions involved may be represented as follows:

$$2NO_2^- + 2I^- + 4H^+ \rightarrow I_2 + N_2O_2 + 2H_2O \qquad (22.9)$$

and $$N_2O_2 + \tfrac{1}{2}O_2 + H_2O \rightarrow 2NO_2^- + 2H^+ \qquad (22.10)$$

When interference from nitrite is present, it is impossible to obtain a permanent end point. As soon as the blue color of the starch indicator has been discharged, the nitrite formed by the reaction in Eq. (22.10) will react with more I^- to produce I_2 and the blue color of the starch indicator will return.

Nitrite interference may be easily overcome by the use of sodium azide (NaN_3). It is most convenient to incorporate the azide in the alkali-KI reagent. When sulfuric acid is added, the following reactions occur and the NO_2^- is destroyed.

$$NaN_3 + H^+ \rightarrow HN_3 + Na^+ \qquad (22.11)$$

$$HN_3 + NO_2^- + H^+ \rightarrow N_2 + N_2O + H_2O \qquad (22.12)$$

By this procedure, nitrite interference is eliminated and the method of determination retains the simplicity of the original Winkler procedure.

Rideal-Stewart Modification of the Winkler Method

The Rideal-Stewart or permanganate modification is designed to overcome the effects of a wide variety of interferences caused by reducing substances, including nitrite. It involves pretreatment of the sample with potassium permanganate under acid conditions. The permanganate is added in excess and oxidizes the reducing agents present. The excess is destroyed by adding a reducing agent, potassium oxalate, which in slight excess does not react with iodine. Some of the reactions involved are as follows:

$$5NO_2^- + 2MnO_4^- + 6H^+ \rightarrow 5NO_3^- + 2Mn^{2+} + 3H_2O \qquad (22.13)$$

$$5Fe^{2+} + MnO_4^- + 8H^+ \rightarrow 5Fe^{3+} + Mn^{2+} + 4H_2O \qquad (22.14)$$

$$\text{Aldehydes} + MnO_4^- + H^+ \rightarrow \text{acids} + Mn^{2+} + H_2O \qquad (22.15)$$

$$\begin{matrix} \text{H} & \text{H} \\ | & | \\ -\text{C}=\text{C}- \end{matrix} + MnO_4^- + H^+ \rightarrow \text{acids} + Mn^{2+} + H_2O \qquad (22.16)$$

The NO_3^- formed in Eq. (22.13) does not oxidize I^- under the conditions of the test. Fe^{3+}, in concentration below 10 mg/L, does not interfere. At levels above 10 mg/L it must be treated to lower its ionic concentration to avoid interference. Potassium fluoride is usually added for this purpose, since it supplies F^-, which combines with Fe^{3+} to form poorly ionized FeF_3.

$$Fe^{3+} + 3F^- \rightarrow FeF_3 \qquad (22.17)$$

Excess $KMnO_4$ is destroyed by adding potassium oxalate.

$$5(COO^-)_2 + 2MnO_4^- + 16H^+ \rightarrow 10CO_2 + 2Mn^{2+} + 8H_2O \qquad (22.18)$$

After the excess permanganate has been destroyed, the regular Winkler procedure is followed, except that additional amounts of alkali-KI are needed to overcome the effects of the acid added originally to facilitate the action of the permanganate. Proper corrections must be made for the volumes of reagents added, in order to calculate the volume of sample required for the titration with thiosulfate.

22.6 | DISSOLVED-OXYGEN MEMBRANE PROBES

The use of membrane probes, which allow *in situ* measurements of dissolved oxygen to be made, has increased significantly since their development. These probes, described under polarographic analysis in Sec. 12.3, are especially useful for taking dissolved-oxygen profiles of reservoirs and streams. They can be lowered to various depths, and the dissolved-oxygen concentration can be read from a connecting microammeter located at the surface. They can also be suspended in biological waste treatment tanks to monitor the dissolved-oxygen level at any point. The rate of biological oxygen utilization in such a tank can be determined by placing a sample of the mixed liquor in a BOD bottle and then inserting a dissolved-oxygen probe to observe the rate at which oxygen is depleted. They can also be used to obtain rapid dissolved-oxygen measurements as part of the BOD test, a particular advantage when large numbers of samples must be analyzed. The portability makes the membrane-probe procedure excellent for field use.

Membrane probes are usually calibrated by making measurements in water samples that have been analyzed for dissolved oxygen by the Winkler procedure. Thus, any errors in the Winkler analysis will be carried over to the probe calibration. During dissolved-oxygen measurements, it is important that sufficient movement of the sample by the probe be maintained to prevent low readings which result if oxygen is depleted at the membrane as it is reduced at the cathode. Membrane electrodes measure oxygen fugacity, which correlates with partial pressure. In order to translate this into concentration in solution, measurement of temperature and salt concentration or conductivity is essential as suggested in Table 22.1. It is thus obvious that either accurate temperature and conductivity measurements must be made along with dissolved-oxygen measurements so that a correction can be applied, or else instruments that are equipped with a thermistor and conductance meter or other devices to compensate automatically for temperature changes must be used.

22.7 | APPLICATION OF DISSOLVED-OXYGEN DATA

Dissolved-oxygen data are used in a wide variety of applications. Many of these have been discussed under general considerations in Sec. 22.1.

PROBLEMS

22.1 Discuss why it is desirable to maintain a significant dissolved-oxygen concentration in rivers and streams.

22.2 What factors affect the solubility of oxygen in water?

22.3 What precautions must be followed in the collection of samples for dissolved oxygen analyses?

22.4 Give two reasons why fixation of dissolved oxygen should be performed in the field if at all feasible.

22.5 Two samples were collected simultaneously at the same spot in a river for dissolved-oxygen analysis. One sample was "fixed" immediately after collection, and the other was treated later in the laboratory. Indicate two possible factors that could cause lower results to be obtained in the second sample.

22.6 What is the solubility of oxygen in water in contact with air when the atmospheric pressure is 0.88 atm, the temperature is 16°C, and the chloride concentration is 1000 mg/L?

22.7 Calculate the percent saturation of dissolved oxygen in a water sample with a temperature of 22°C and a dissolved-oxygen concentration of 5.3 mg/L when the atmospheric pressure is 1 atm. Assume the sample salinity is less than 100 mg/L.

22.8 Write chemical equations summarizing all the essential reactions involved in the unmodified Winkler method for dissolved oxygen.

22.9 Prepare a table showing five substances that interfere with the Winkler method, and indicate which modification would be used to overcome each interference.

22.10 How does azide eliminate the nitrite interference in the iodometric method for determining dissolved oxygen?

22.11 What is the purpose of the Rideal-Stewart modification of the iodometric dissolved oxygen method?

22.12 What is the function of the NaOH sometimes used in preparing the thiosulfate solution used for dissolved-oxygen determinations?

22.13 What advantages do dissolved-oxygen probes have over the Winkler test for dissolved-oxygen measurements? What are some disadvantages?

22.14 What water characteristics other than dissolved oxygen must be estimated in order to determine dissolved-oxygen concentration in water when using the dissolved-oxygen probe?

REFERENCES

Clesceri, L. S., A. E. Greenberg, and A. D. Eaton (eds.): "Standard Methods for the Examination of Water and Wastewater," 20th ed., American Public Health Association, Washington, DC, 1998.

Day, Jr., R. A., and A. L. Underwood: "Quantitative Analysis," 6th ed., Prentice Hall, Englewood Cliffs, NJ, 1991.

Skoog, D. A., D. M. West, and F. J. Holler: "Fundamentals of Analytical Chemistry," 7th ed., Saunders College Pubs., Fort Worth, TX, 1996.

23

Biochemical Oxygen Demand

23.1 | GENERAL CONSIDERATIONS

Biochemical oxygen demand (BOD) is usually defined as the amount of oxygen required by bacteria while stabilizing decomposable organic matter under aerobic conditions. The term "decomposable" may be interpreted as meaning that the organic matter can serve as food for the bacteria, and energy is derived from its oxidation.

The BOD test is widely used to determine the pollutional strength of domestic and industrial wastes in terms of the oxygen that they will require if discharged into natural watercourses in which aerobic conditions exist. The test is one of the most important in stream-pollution-control activities. This test is of prime importance in regulatory work and in studies designed to evaluate the purification capacity of receiving bodies of water.

The BOD test is essentially a bioassay procedure involving the measurement of oxygen consumed by living organisms (mainly bacteria) while utilizing the organic matter present in a waste, under conditions as similar as possible to those that occur in nature. In order to make the test quantitative, the samples must be protected from the air to prevent reaeration as the dissolved-oxygen level diminishes. In addition, because of the limited solubility of oxygen in water, about 9 mg/L at 20°C, strong wastes must be diluted to levels of demand in keeping with this value to ensure that dissolved oxygen will be present throughout the period of the test. Since this is a bioassay procedure, it is extremely important that environmental conditions be suitable for the living organisms to function in an unhindered manner at all times. This condition means that toxic substances must be absent and that all accessory nutrients needed for bacterial growth, such as nitrogen, phosphorus, and certain trace elements, must be present. Biological degradation of organic matter under natural conditions is brought about by a diverse group of organisms that carry the oxidation essentially to completion, i.e., almost entirely to carbon dioxide and water. Therefore, it is important that a mixed group of organisms, commonly called "seed," be present in the test.

The BOD test may be considered as a wet oxidation procedure in which the living organisms serve as the medium for oxidation of the organic matter to carbon dioxide and water. A quantitative relationship exists between the amount of oxygen required to convert a definite amount of any given organic compound to carbon dioxide, water, and ammonia, and this can be represented by the following generalized equation:

$$C_nH_aO_bN_c + \left(n + \frac{a}{4} - \frac{b}{2} - \frac{3}{4}c\right)O_2 \rightarrow nCO_2 + \left(\frac{a}{2} - \frac{3}{2}c\right)H_2O + cNH_3 \quad (23.1)$$

On the basis of this relationship, it is possible to interpret BOD data in terms of organic matter, as well as the amount of oxygen used during its oxidation. This concept is fundamental to an understanding of the rate at which BOD is exerted.

The oxidative reactions involved in the BOD test are a result of biological activity, and the rate at which the reactions proceed is governed to a major extent by population numbers and temperature. Temperature effects are held constant by performing the test at 20°C, which is, more or less, a median value as far as natural bodies of water are concerned. The predominant organisms responsible for the stabilization of organic matter in natural waters are forms native to the soil. The rate of their metabolic processes at 20°C and under the conditions of the test is such that time must be reckoned in days. Theoretically, an infinite time is required for complete biological oxidation of organic matter, but for all practical purposes, the reaction may be considered complete in 20 days. However, a 20-day period is too long to wait for results in most instances. It has been found by experience that a reasonably large percentage of the total BOD is exerted in 5 days; consequently, the test has been developed on the basis of a 5-day incubation period. It should be remembered, therefore, that 5-day BOD values represent only a portion of the total BOD. The exact percentage depends upon the character of the "seed" and the nature of the organic matter, and can be determined only by experiment. In the case of domestic and many industrial wastewaters, it has been found that the 5-day BOD value is about 70 to 80 percent of the total BOD. This is a large enough percentage of the total so that 5-day values are used for many considerations. The 5-day incubation period was selected also to minimize interference from oxidation of ammonia as discussed in Section 23.2.

23.2 | THE NATURE OF THE BOD REACTION

Studies of the kinetics of BOD reactions have established that they are for most practical purposes "first-order" in character (see Sec. 3.10), or the rate of the reaction is proportional to the amount of biodegradable organic matter remaining at any time, as modified by the population of active organisms. Once the population of organisms has reached a level at which only minor variations occur, the reaction rate is controlled by the amount of food available to the organisms and may be expressed as follows:

$$-\frac{dC}{dt} = k'C \tag{23.2}$$

where C represents the concentration of biodegradable organic matter (pollutants) at time t, and k' is the rate constant for the reaction. This means that the rate of the reaction gradually decreases as the concentration C of biodegradable organic matter or food for the bacteria decreases.

The major concern with biodegradation of organic matter in natural waters is the resulting oxygen consumption as given in Eq. (23.1), and the impact this has on oxygen content of the water. Thus it became customary to describe biodegradable organic matter in terms of its equivalent oxygen consumption potential. Toward this end for BOD considerations it is customary to use L_t in place of C, where L_t represents the ultimate BOD (BOD_L) or oxygen equivalent of the biodegradable organic matter remaining at any time t.

EXAMPLE 23.1

The concentration C_0 of biodegradable organic matter in a water sample is 37 mg/L and can be represented by the empirical molecular formula $C_6H_{11}ON_2$. Estimate the oxygen equivalent concentration L_0 of this organic matter.

From Eq. (23.1),

$$C_6H_{11}ON_2 + \left[6 + \frac{11}{4} - \frac{1}{2} - \frac{3}{4}(2)\right]O_2 \rightarrow 6CO_2 + \left[\frac{11}{2} - \frac{3}{2}(2)\right]H_2O + 2NH_3$$

or, $\qquad\qquad C_6H_{11}ON_2 + 6.75O_2 \rightarrow 6CO_2 + 2.5H_2O + 2NH_3$

The formula weight (FW) of the organic matter is $6(12) + 11 + 16 + 2(14) = 127$. The quantity of oxygen required to oxidize one FW of the organic matter is $6.75(32) = 216$. From this information we can determine the oxygen equivalents of the organic matter to be

$$L_0 = \frac{37}{127}(216) = 63 \text{ mg/L}$$

Thus, the oxygen equivalents or ultimate BOD (L_0) for the water sample is 63 mg/L.

Using oxygen equivalence of BOD in place of biodegradable organic matter concentration, Eq. (23.2) can be rewritten in the form

$$-\frac{dL_t}{dt} = k'L_t \tag{23.3}$$

Upon integration of Eq. (23.3), the expression

$$\frac{L_t}{L_0} = e^{-k't} = 10^{-kt} \tag{23.4}$$

is obtained. Some prefer to use base 10 rather than e, and for this case, $k = k'/2.303$. Equation (23.4) states that the fraction of the original biodegradable organic matter that remains after any time t has elapsed is equal to $e^{-k't}$.

In many cases the interest is in how much oxygen has been consumed by biodegradation over a given time interval, rather than in how much BOD_L remains. That is, the interest is in how much of the original BOD has been exerted. The BOD exerted is represented by y, where

$$y = (L_0 - L_t) \qquad (23.5)$$

Combining Eqs. (23.4) and (23.5), we obtain

$$y = L_0(1 - e^{-k't}) = L_0(1 - 10^{-kt}) \qquad (23.6)$$

In the BOD test, it is y that is measured rather than L_t. This value, the oxygen uptake through biooxidation of the degradable organic matter, is readily measured through oxygen concentration decrease in a sample in a closed bottle with time. If t is long enough, say 20 days, then the value $e^{-k't}$ approaches zero, so then $y = L_0$. Thus, measurements of y can and are used to estimate L_0.

We have no good procedure for measuring L_0 more directly, primarily because we have no analytical test, other than a bioassay procedure such as the BOD test, that is capable of distinguishing between biodegradable and nonbiodegradable organic material. Natural waters as well as industrial and municipal waters often contain a large fraction of nonbiodegradable organic materials, such as humic acids, and so total organic content measurements do not provide a good alternative to the bioassay procedure for measuring BOD.

The BOD_L of a sample is 120 mg/L. If k' for this sample is 0.28 day^{-1}, how much of the BOD is exerted, and how much remains after the following time intervals: (*a*) 3 days, (*b*) 5 days, and (*c*) 10 days?

EXAMPLE 23.2

Since $L_0 = BOD_L$, we can make this substitution into Eq. (23.6) and solve for y, the BOD exerted, as a function of time. Then, we use Eq. (23.5) and solve for L_t to determine the amount of BOD_L that remains:

(*a*) $\quad y_3 = 120(1 - e^{-0.28 \times 3}) = 68$ mg/L $\qquad L_3 = 120 - 68 = 52$ mg/L
(*b*) $\quad y_5 = 120(1 - e^{-0.28 \times 5}) = 90$ mg/L $\qquad L_5 = 120 - 90 = 30$ mg/L
(*c*) $\quad y_{10} = 120(1 - e^{-0.28 \times 10}) = 113$ mg/L $\qquad L_{10} = 120 - 113 = 7$ mg/L

We see here that after 10 days, about 94 percent of the ultimate BOD has been exerted. Thus, oxygen consumption over the first 10 days would provide a close approximation to the ultimate value of 120 mg/L.

Figure 23.1 illustrates more completely for an ideal case the relationship between oxygen consumption y and BOD_L remaining (L_t) as a function of time for Example 23.2. In practice, however, the ideal is not necessarily followed, and an

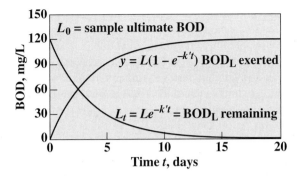

Figure 23.1 Changes in biodegradable organic matter, measured in oxygen equivalents or BOD_L, as a function of time.

oxygen uptake curve similar to that shown in Fig. 23.2 might result. Normal BOD first-order kinetics and the BOD test itself are based upon oxidation of biodegradable organic matter only. However, reduced forms of nitrogen such as ammonia, which are formed from organic oxidation as indicated in Eq. (23.1) or otherwise present in the sample, may also be oxidized and will exhibit an oxygen uptake. The effect is illustrated in Fig. 23.2. The total oxygen uptake is the summation of that due to carbonaceous oxidation plus nitrogen oxidation or nitrification. Bacteria that oxidize reduced forms of nitrogen often are small in number in samples, so the ef-

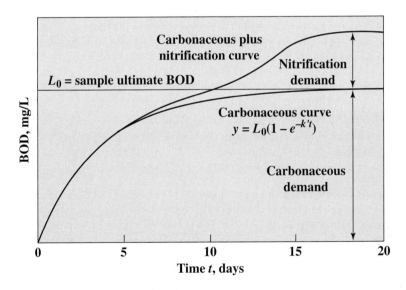

Figure 23.2 BOD curve resulting from sample exhibiting both carbonaceous and nitrogenous oxygen demand.

fects of nitrification may not show themselves until after 5 to 10 days as illustrated in the figure. Thus BOD readings before that time are then truly representative of carbonaceous demand. However, nitrification may take place earlier, and the analyst must be able to recognize when this might occur and how to deal with it. This is addressed in the following.

The importance of having a mixed culture of organisms corresponding to those in the soil, for proper measurement of BOD, has been mentioned. Such cultures, when derived from the soil or domestic wastewater, contain large numbers of heterotrophic bacteria that oxidize the carbonaceous matter present in BOD samples. In addition, they normally contain certain autotrophic bacteria, particularly nitrifying bacteria, that oxidize noncarbonaceous matter for energy. The nitrifying bacteria are usually present in relatively small numbers in untreated domestic wastewater, and fortunately, their reproductive rate at 20°C is such that their populations do not become sufficiently large to exert an appreciable demand for oxygen until about 8 to 10 days have elapsed in the regular BOD test as previously mentioned. Once the organisms become established, they oxidize nitrogen in the form of ammonia to nitrous and nitric acids in amounts that introduce serious error into BOD work.

$$2NH_3 + 3O_2 \xrightarrow{\text{nitrite-forming bacteria}} 2NO_2^- + 2H^+ + 2H_2O \tag{23.7}$$

$$2NO_2^- + O_2 \xrightarrow{\text{nitrite-forming bacteria}} 2NO_3^- \tag{23.8}$$

It is true that the oxidation of inorganic nitrogen can deplete the dissolved oxygen in streams, and this effect must be taken into account in stream analysis. However, it is not desirable to use normal BOD measurements for such estimates, because ammonia nitrogen is added to BOD dilution water as a required nutrient and its oxidation could lead to erroneous conclusions about the water sample itself. The potential dissolved oxygen utilization by nitrification is best evaluated by an analysis of the waste for the different forms of nitrogen present and use of the stoichiometric relationships between oxygen and nitrogen given by Eqs. (23.7) and (23.8).

For the BOD results in Example 23.1, estimate the nitrogenous oxygen demand that would result from complete oxidation of the ammonia nitrogen released during the carbonaceous oxidation of the biodegradable organic matter, and the summation of the total carbonaceous plus nitrogenous oxygen demand of the organics.

EXAMPLE 23.3

Based upon the stoichiometric equation developed in Example 23.1, 2 mol of ammonia nitrogen are produced per empirical mole of organics oxidized. From the sum of Eqs. (23.7) and (23.8), 4 mol or $4 \times 32 = 128$ g of oxygen would be consumed by the oxidation of these 2 mol. Thus, using proportions,

$$\text{Nitrogenous oxygen demand} = \frac{37}{127}(128) = 37 \text{ mg/L}$$

Total carbonaceous plus nitrogenous oxygen demand $= 65 + 37 = 102$ mg/L

The interference caused by nitrifying organisms makes the actual measurement of total carbonaceous BOD difficult unless provision is made to eliminate it. The interference caused by the nitrifying bacteria was a major reason for selecting a 5-day incubation period for the regular BOD test. However, in cases in which the effluent from biological treatment units such as trickling filters and activated sludge are to be analyzed for BOD, the effluents often contain populations of nitrifying organisms sufficient to utilize significant amounts of oxygen for nitrification during the regular 5-day incubation period. It is important to know the amount of residual carbonaceous BOD in such cases in order to measure plant efficiency.[1] Fortunately, the action of the nitrifying bacteria can be arrested by the use of specific inhibiting agents such as 2-chloro-6-(trichloro methyl) pyridine (TCMP), thus allowing measurement of residual carbonaceous BOD without interference from nitrification. Furthermore, algae when present, can introduce another variable that makes BOD data on rivers and estuaries difficult to interpret because of their ability to both produce and consume oxygen.[2]

23.3 | METHOD OF MEASURING BOD

The BOD test is based upon determinations of dissolved oxygen; consequently, the accuracy of the results is influenced greatly by the care given to its measurement. BOD may be measured directly in a few samples, but in general, a dilution procedure is required.

Direct Method

For samples with 5-day BOD less than 7 mg/L, it is not necessary to dilute them, provided that they are aerated to bring the dissolved-oxygen level nearly to saturation at the start of the test. Many river waters fall into this category.

The usual procedure is to adjust the sample to about 20°C and aerate with diffused air to increase or decrease the dissolved gas content of the sample to near saturation. Two or more BOD bottles are then filled with the sample; at least one is analyzed for dissolved oxygen immediately, and the others are incubated for 5 days at 20°C. After 5 days, the amount of dissolved oxygen remaining in the incubated samples is determined, and the 5-day BOD is calculated by subtraction of the 5-day results from those obtained on day 0.

The direct method of measuring BOD involves no modification of the sample, and therefore produces results under conditions as nearly similar as possible to the

---

[1]C. N. Sawyer and L. Bradney, *Sewage Works J.*, **18:** 1113 (1946).

[2]T. F. Wisniewski, Oxygen Relationships in Streams, U.S. Public Health Service Seminar, *R. A. Taft Sanitary Eng. Center Tech. Rept.* W58-2. Cincinnati, (March 1958); G. P. Fitzgerald, The Effect of Algae on BOD Measurements, *J. Water Pollution Control Federation,* **36:** 1524 (1964).

natural environment. Unfortunately, the BOD of very few samples falls within the range of dissolved oxygen available in this test.

Dilution Method

The dilution method of measuring BOD is based upon the fundamental concept that the rate of biochemical degradation of organic matter is directly proportional to the amount of unoxidized material existing at the time, as discussed in Sec. 23.2. According to this concept, the rate at which oxygen is used in dilutions of the waste is in direct ratio to the percent of waste in the dilution, provided that all other factors are equal. For instance, a 10 percent dilution uses oxygen at one-tenth the rate of a 100 percent sample. Experience has served as the basis for the mathematical development of the BOD reaction; therefore, it is safe to assume the validity of the concept.

In any bioassay work, it is important to control all environmental and nutritional factors in a manner that will not interfere with the desired action. In the BOD test, this means that everything influencing the rate at which organic matter is biologically stabilized must be kept under close control and highly reproducible from test to test. The major items of importance are (1) freedom from toxic materials, (2) favorable pH and osmotic conditions, (3) presence of available accessory nutrient elements, (4) standard temperature, and (5) presence of a significant population of mixed organisms of soil origin.

A wide variety of waste materials are subject to the BOD test. These range from industrial wastes that may be free of microorganisms to domestic wastewater with an abundance of organisms. Many industrial wastes have extremely high BOD values, and very high dilutions must be made to meet the requirements imposed by the limited solubility of oxygen. Domestic wastewater has an ample supply of accessory nutrient elements, such as nitrogen and phosphorus, but many industrial wastes are deficient in one and sometimes both of these elements. Because of these limitations, the dilution water used in BOD work must compensate for the limitations imposed by any sample subjected to analysis. Since these limitations are not always known, it is safe practice to use a dilution water that will provide for all contingencies. This is not necessary and may be undesirable, however, when domestic wastewater is the sole consideration.

The Dilution Water A wide variety of waters have been used for BOD work. Natural surface waters would appear to be ideal, but they have a number of disadvantages, including variable BOD, variable microorganism population (often including algae and significant populations of nitrifying bacteria), and variable mineral content. Tap water has been used, but it suffers from most of the limitations found in surface waters plus the possibility of toxicity from chlorine residuals. Through long experience, it has developed that a synthetic dilution water prepared from distilled or demineralized water is best for BOD testing because most of the variables mentioned can be kept under control.

Whether distilled or demineralized water is used, it must be free from toxic substances. Chlorine, or chloramines and copper are the two most commonly found.

In many cases it is necessary to dechlorinate the water used. Copper contamination with distilled water is normally due to exposed copper in the condenser. The BOD of distilled waters prepared from potable supplies is usually sufficiently low to allow use of the water without storage other than that needed to bring its temperature into a favorable range.

The pH of the dilution water may range anywhere from 6.5 to 8.5 without affecting the action of the heterotrophic bacteria. It is customary to buffer the solution by means of a phosphate system at about pH 7.0. The buffer is essential to maintain favorable pH conditions at all times.

The proper osmotic conditions are maintained by the potassium and sodium phosphates added to provide buffering capacity. In addition, calcium and magnesium salts are added which contribute to the total salt content.

The potassium, sodium, calcium, and magnesium salts added to give buffering capacity and proper osmotic conditions also serve to provide the microorganisms with any of these elements that are needed in growth and metabolism. Ferric chloride, magnesium sulfate, and ammonium chloride supply the requirements for iron, sulfur, and nitrogen. The phosphate buffer furnishes any phosphorus that may be needed. The nitrogen should be eliminated in cases where nitrogenous oxygen demand is being measured.

The dilution water now contains all the essential materials for the measurement of BOD except the necessary microorganisms. A wide variety of materials have been used for "seeding" purposes. Experience has shown that domestic wastewater, particularly from combined sewer systems, provides about as well balanced a population of mixed organisms as anything, and usually 2 mL of wastewater per liter of dilution water is sufficient. Some river waters are satisfactory, but care must be taken to avoid using waters that contain algae or nitrifying bacteria in significant amounts.

The dilution water should always be "seeded" with wastewater or other material to ensure a uniform population of organisms in various dilutions and to provide an opportunity for any organic matter present in the dilution water blanks to be exposed to the same type of organisms as those involved in the stabilization of the waste. The latter is a point that is often ignored; this has led to erroneously high results in many cases.

Finally, the dilution water should be aerated to saturate it with oxygen before use.

The Need for Blanks and Seed Controls   In the determination of BOD by the dilution technique, it is safe to assume that the dilution water containing the "seeding" material will contain organic matter and that addition of the diluting water to the sample will increase the amount of oxidizable organic matter; therefore, a correction must be applied. "Standard Methods" recommends conducting a separate BOD determination on the seed itself to obtain a good measure of its contribution to oxygen uptake in the diluted sample. However, with normal seed, this correction is rather small and the need for this extra step must be questioned. As an alternative, a BOD analysis of the seeded dilution water itself, or blanks as they are termed here, should serve adequately to provide the needed seed correction. With this approach no separate seed study is required; analysis of the dissolved oxygen concentration in the blanks initially and at the end of the test provide the information that is needed with adequate accuracy for the BOD analysis.

At least three blanks should be included with each set of BOD samples. In any bioassay test there is a certain amount of biological variation. Since the blanks provide the reference value from which all calculations of BOD are made, it is important that it have some statistical reliability. Usually three blanks provide such reliability, but each analyst should satisfy his or her particular requirements.

Dilutions of Waste   The analyst has a real responsibility in deciding what dilutions should be set for determination of BOD. Usually it is best to set three different dilutions. When the strength of a sample is known with some assurance, two dilutions may suffice. Where samples of unknown strength are involved, the dilutions should cover a considerable range, and in some instances it may be necessary to set as many as four or five dilutions. In any case there should be an overlapping of the BOD measurable by successive dilutions.

It has been demonstrated that BOD is not influenced by oxygen concentrations as low as 0.5 mg/L. It has also been learned that it is not statistically reliable to base BOD values upon dilutions that produce a depletion of oxygen less than 2 mg/L. Therefore, it has become customary to base calculations of BOD on samples that produce a depletion of at least 2 mg/L and have at least 0.5 mg/L of dissolved oxygen remaining at the end of the incubation period. This restriction usually means a range of 2 to 7 mg/L. With this information at hand it is possible to construct a table showing the range of BOD measurable by various dilutions. Table 23.1 presents

Table 23.1 | BOD measurable with various dilutions of samples

Using percent mixtures		By direct pipeting into 300-mL bottles	
% mixture	Range of BOD	mL	Range of BOD
0.01	20,000–70,000	0.02	30,000–105,000
0.02	10,000–35,000	0.05	12,000– 42,000
0.05	4,000–14,000	0.10	6,000– 21,000
0.1	2,000– 7,000	0.20	3,000– 10,500
0.2	1,000– 3,500	0.50	1,200– 4,200
0.5	400– 1,400	1.0	600– 2,100
1.0	200– 700	2.0	300– 1,050
2.0	100– 350	5.0	120– 420
5.0	40– 140	10.0	60– 210
10.0	20– 70	20.0	30– 105
20.0	10– 35	50.0	12– 42
50.0	4– 14	100	6– 21
100	0– 7	300	0– 7

such information for dilutions prepared on a percentage basis and also for dilutions prepared by direct pipeting into bottles of about 300-mL capacity. It is customary to estimate the BOD of a sample and set one dilution based upon the estimate. Two other dilutions, one higher and one lower, are also set up. For example, a sample is estimated to have a BOD of 1000 mg/L. Reference to Table 23.1 will show that a 0.5 percent mixture should be used. If a 0.2 and a 1.0 percent mixture are included, the range of measurable BOD is extended from 200 to 3500 mg/L and should compensate for any errors in the original estimate.

In the direct-pipeting technique, preliminary dilutions should be made of all samples that require less than 0.5 mL of the sample so that amounts added to the bottles can be measured without serious error. The volumes of all bottles must be known in order to allow calculation of the BOD when this method is used.

Incubation Bottles   The bottles used for BOD analysis should be equipped with glass stoppers that are ground to a point to prevent trapping of air when the stopper is inserted. The bottles should be equipped with some form of water seal to prevent air from entering the bottle, and this seal should remain in place during the entire incubation period.

It is extremely important that bottles used for BOD work be free of organic matter. Cleaning can be best accomplished by use of a good grade of detergent. Bottles should be rinsed with hot water to kill nitrifying organisms which tend to develop on the walls of the bottles. Care must be exercised to make sure that all the cleaning agent is removed from the bottle before use. This assurance can usually be accomplished by four rinses with tap water and a final rinse with distilled or demineralized water.

Initial Dissolved Oxygen   For samples with BOD below 200 mg/L, it is necessary to use amounts of sample in excess of 1.0 percent. Serious errors may be introduced in results if the dissolved oxygen of the sample differs materially from that of the dilution water and corrections are not made. With sample dilutions of less than 20 percent, it is usually sufficient to adjust the samples to 20°C, aerate to saturation, and then assume the dissolved-oxygen concentration is the same as in the dilution water. This eliminates the need for measuring dissolved oxygen on such samples, and also satisfies any immediate oxygen demand. With sample dilutions greater than 20 percent, the dissolved oxygen of the sample should be determined separately.

Calculation of BOD   In the direct method of BOD analysis, no seed or dilution corrections are required and so the following simple formula results:

$$\text{BOD}_5 \text{ (mg/L)} = D_1 - D_2 \qquad \text{(when no seed or dilution is used)} \qquad (23.9)$$

where D_1 and D_2 represent dissolved-oxygen concentrations (in mg/L) of the sample bottle contents initially and at the end of the 5-day incubation period, respectively.

If dilution is used, but no seed has been added to the dilution water, then Eq. (23.9) must include a correction for dilution:

$$\text{BOD}_5 \text{ (mg/L)} = \frac{D_1 - D_2}{P} \qquad \text{(when unseeded dilution water is used)} \quad (23.10)$$

where P represents that fraction of the BOD bottle that is represented by the sample. Referring to Table 23.1, if the percent mixture approach is used, P equals the percent mixture divided by 100. If direct pipeting is used, P equals mL pipeted divided by 300 mL when a 300-mL bottle is used.

Finally, if seeded dilution water is used, then corrections must be applied both for the seed and for the dilution water. In this case, the following equation results:

$$\text{BOD}_5 \text{ (mg/L)} = \frac{(D_1 - D_2) - f(B_1 - B_2)}{P} \qquad \begin{array}{l}\text{(when seeded dilution} \\ \text{water is used)}\end{array} \quad (23.11)$$

where B_1 and B_2 are the seed dissolved-oxygen concentrations (in mg/L) before and after incubation of the seed, respectively, and f represents the fraction of seed in the incubated sample. If the approach as suggested here is used and seed corrections are determined directly from blank dissolved-oxygen measurements of seeded dilution water, then B_1 and B_2 represent before and after dissolved-oxygen measurements of the blanks themselves. In this case, f is determined from the fraction of dilution water in the sample, which is simply $1 - P$. Making this substitution into Eq. (23.11) results in the following:

$$\text{BOD}_5 \text{ (mg/L)} = \frac{(D_1 - D_2) - (1 - P)(B_1 - B_2)}{P} \qquad (23.12)$$

or

$$\text{BOD}_5 \text{ (mg/L)} = \frac{(B_2 - D_2) - (B_1 - D_1)}{P} - (B_1 - B_2) \qquad (23.13)$$

With a good seed in the correct amount, $B_1 - B_2$ should only be 0.1 to 0.2 mg/L, and thus the last term in Eq. (23.13) can generally be ignored. Further, when the sample is adjusted to 20°C and aerated initially to increase its dissolved-oxygen level to near saturation like the dilution water, $B_1 - D_1$ will be very close to zero, and likewise can be ignored, in which case a highly simplified equation results

$$\text{BOD}_5 \text{ (mg/L)} = \frac{B_2 - D_2}{P} \qquad \text{(simplified equation for most uses)} \quad (23.14)$$

The use of simplified Eq. (23.14) is often justified since the coefficient of variation for the test as indicated in "Standard Methods" is quite high, about 15 percent, and thus the additional analytical effort required to make use of the more exact Eq. (23.11) must be questioned.

EXAMPLE 23.4

In determining the BOD$_5$ of a sample, an analyst added 2, 5, and 10 mL of sample to three different 300-mL BOD bottles and filled them with seeded dilution water. The analyst also prepared three blank bottles with the same dilution water and incubated the set at 20°C for 5 days. Dissolved-oxygen (DO) measurements were made on the samples before and after with the following results.

Sample size in bottle, mL	Initial DO, mg/L	Final DO, mg/L
2	8.1	5.6
5	8.0	1.7
10	8.1	0.0
Blank average	8.2	8.0

What is BOD$_5$ for the sample?

Only the 2- and 5-mL sample bottles are in the acceptable range with final DO values of at least 2 mg/L below initial values and with more than 0.5-mg/L DO remaining at the end of the 5-day period. Using Eq. (23.13) for these two bottles gives

For 2-mL sample bottle, $\text{BOD}_5 = \dfrac{(8.0 - 5.6) - (8.2 - 8.1)}{2/300} - (8.2 - 8.0) = 345$ mg/L

For 5-mL sample bottle, $\text{BOD}_5 = \dfrac{(8.0 - 1.7) - (8.2 - 8.0)}{5/300} - (8.2 - 8.0) = 366$ mg/L

The BOD$_5$ can be taken as the average of the two values, or 356 mg/L. However, the error in the $B_2 - D_2$ calculation is greater on a relative basis for the 2-mL sample bottle than for the 5-mL sample bottle, and so it might be questioned whether the average is better to use than 366 mg/L. One could conduct a statistical analysis to decide, although "Standard Methods" simply recommends taking the average.

We might also compare results using simplified Eq. (23.14), for which less data is required. Here,

For 2-mL sample bottle, $\text{BOD}_5 = \dfrac{(8.0 - 5.6)}{2/300} = 360$ mg/L

For 5-mL sample bottle, $\text{BOD}_5 = \dfrac{(8.0 - 1.7)}{5/300} = 378$ mg/L

and the average BOD$_5$ is 369 mg/L, or within 4 percent of the value obtained using Eq. (23.13), which requires more analytical data and expense.

23.4 | RATE OF BIOCHEMICAL OXIDATIONS

For a great many years the BOD reaction was considered to have a rate constant k' equal to 0.23 per day at 20°C. This value was established by extensive studies on polluted river waters and domestic wastes in the United States and England. As application of the BOD test spread to the analysis of industrial wastes, and the use of synthetic

Table 23.2 | Significance of reaction rate constant k' upon BOD

Time, days	Percent of total BOD exerted				
	$k' = 0.10$	0.20	0.30	0.40	0.50
1	10	18	26	33	39
2	18	33	45	55	63
3	26	45	59	70	78
4	33	55	70	80	86
5	39	63	78	86	92
6	45	70	83	91	95
7	50	75	88	94	97
10	63	86	95	98	99
15	78	95	99	99+	99+
20	86	98	99+	99+	99+

dilution waters became established, it was soon noted that k' values considerably in excess of 0.23 per day were involved and that an appreciable variation occurred for different waste materials. In addition, it was found that k' values for domestic wastes varied considerably from day to day and averaged about 0.40 per day, rather than 0.23 per day as originally determined. Also, the k' values for effluents from biological waste treatment plants were found to be significantly lower than those for the raw wastes. Another factor affecting BOD rates is temperature. The magnitude of this effect and a method for correction when projecting laboratory-determined rates to field conditions were discussed in Sec. 3.10. The importance of the reaction rate k' with respect to the BOD developed at any time is shown in Table 23.2.

The significance of k' in determining the course of the BOD reaction is illustrated in Fig. 23.3. For a waste having a given L_0 value, the BOD values on any given day will vary widely until about 15 days have elapsed. In the past it was common practice to interpret 5-day BOD in terms of L_0 values by assuming a k' value of

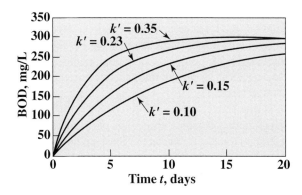

Figure 23.3 Effect of rate constant k' on BOD for L_0 value of 300 mg/L.

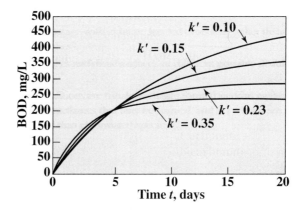

Figure 23.4 Effect of rate constant k' on the course of the BOD reaction with an assumed BOD_5 of 200 mg/L.

0.23 per day. Figure 23.4 shows how the L_0 value of a sample with a 5-day BOD of 200 varies with the value of k'.

The variation in k' values leaves considerable room for speculation as to why such differences in rates of reaction occur. Two factors of major importance are involved: (1) the nature of the organic matter and (2) the ability of the organisms present to utilize the organic matter.

Organic matter occurring in domestic and industrial wastes varies greatly in chemical character and availability to microorganisms. That part which exists in true solution is readily available, but that part which occurs in colloidal and coarse suspension must await hydrolytic action before it can diffuse into the bacterial cells where oxidation can occur. The rate of hydrolysis and diffusion are probably the most important factors in controlling the rate of the reaction. It is well known that simple substrates, such as glucose, are removed from solution at very rapid rates, and k' values are correspondingly high. More complex materials are removed much more slowly, and k' values are lower. In a complex material such as domestic waste, reaction rates are modified greatly by the more complex substances, whereas in an industrial waste containing soluble compounds of simple character, the reaction rate is usually very rapid. Certain organic compounds, such as lignin, are very slowly attacked by bacteria. Some of the synthetic detergents also fall into this category.

A lag period is often noted in the BOD reaction with some industrial wastes, particularly those containing organic compounds of synthetic origin or with chemical structure not found in natural materials. The "seeding" organisms used in the BOD test may or may not have specific bacteria that can utilize the material as food. If not, the substance will not exert a BOD. Oftentimes only a few bacteria are present that can oxidize the substance, and the rate of oxidation is so slow for a period, possibly several days, that a measurable BOD cannot be detected. With sufficient time, however, the population of the specific bacteria will increase to levels at

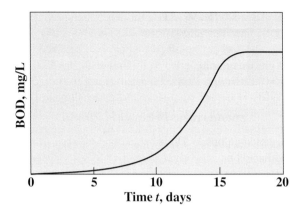

Figure 23.5 Possible BOD change with time resulting when organics requiring acclimation by organisms are present.

which the oxidation progresses at normal rates. In cases of this kind, as illustrated in Fig. 23.5, the lag period can usually be overcome by using for "seeding" purposes water from a river into which such wastes are discharged. The water should be taken well downstream from the point of discharge. Growths attached to rocks downstream from the point of waste discharge sometimes will furnish an adequate seed. Adapted seed can sometimes be found in soils that have been exposed to the waste materials for a long period, perhaps through accidental spills. A properly adapted seed may also sometimes be developed in the laboratory by aerating for several days a mixture composed of the neutralized waste and a small portion of domestic wastewater. However, with some xenobiotic compounds, microorganisms with the ability to adapt to them may not be ubiquitous in the environment, and will need to be sought more broadly. Indeed, they may not exist!

The phenomenon of the lag period is sometimes explained on the basis that the "seeding" organisms do not have the proper enzyme systems to utilize the organic matter. With time, however, and if the right organisms are present, they adapt themselves to the new food supply and will then furnish the necessary enzymes. Another factor at times is the presence of toxic organic or inorganic compounds, which may cause delay in the onset of oxidation while bacteria adapt to the inhibition, or may even prevent the bacteria from growing and oxidizing the organic material present. The analyst needs to understand these limitations of the BOD test and how to deal with them.

The rate k' and extent L_0 of biochemical reactions can be evaluated in a number of ways. Procedures for evaluation of k' and L_0, together with their uncertainties, through a series of BOD measurements with time were discussed in Sec. 10.9 on regression analysis. In order to determine carbonaceous BOD itself in such cases, measurements need to be made of changes in nitrogen species with time in order to correct for oxygen consumption from nitrification. "Standard Methods" now has two proposed procedures specifically designed for determining oxygen uptake with time and for estimating L_0. One uses a series of bottles, or alternately, a large reservoir for the diluted sample, and oxygen uptake measurements are made over time together with analyses for changes in nitrogen forms. Carbonaceous BOD at any time is determined by subtracting oxygen usage for nitrification from total oxygen consumption. Regression analysis is then used to estimate the ultimate BOD. The other approach uses a respirometer for continuous oxygen uptake measurements. Here, the objective is not necessarily to determine L_0, but to obtain a better understanding of the characteristics of a wastewater. Observing oxygen uptake with waste dilution can help evaluate whether toxicants to biodegradation are present. Observations of oxygen uptake with time can help evaluate whether biological acclimation to organics is required. This is evident when resulting BOD curves are similar to that illustrated in Fig. 23.5. Corrections for nitrification need to be made here as this can cause similar late oxygen uptake as noted in Fig. 23.2. In any event, oxygen uptake measurements with time are essential when first characterizing a wastewater for biological treatment in order to determine waste characteristics that are crucial for process design.

23.5 | DISCREPANCY BETWEEN L_0 VALUES AND THEORETICAL OXYGEN DEMAND VALUES

The ultimate BOD, or L_0, value of organic substances is often considered to be equal to the theoretical oxygen demand as calculated from the chemical equation involved [see Eq. (23.1)]. For example, oxidation of glucose to carbon dioxide and water requires 192 g of oxygen per mole or 1.065 mg of oxygen per milligram of glucose.

$$\underset{180}{C_6H_{12}O_6} + \underset{192}{6O_2} \rightarrow 6CO_2 + 6H_2O \tag{23.15}$$

A great deal of BOD work has been done with glucose solutions in a concentration of 300 mg/L. Such a solution has a theoretical oxygen demand of 320 mg/L. Actual BOD measurements made upon such solutions have yielded 20-day[3] and calculated L_0 values in the range of 250 to 285, or about 85 percent of the theoretical amount. Thus, it is evident that not all the glucose is completely converted to carbon dioxide and water. The explanation of the discrepancy involves an understanding of the transformations that organic matter undergoes when subjected to biological attack.

In order for organic matter to be oxidized by bacteria, it must serve as food material from which the organisms can derive energy for growth and reproduction. This means that part of the organic matter is converted to cell tissue. The part that is converted to cell tissue will remain unoxidized until such time as the organisms must draw upon cell tissue to derive energy to maintain life (endogenous respiration). When bacteria die, they become food material for other bacteria, and a further transformation to carbon dioxide, water, and cell tissue occurs. Living bacteria, as well as dead ones,

---

[3]A. Q. Y. Tom, "Investigations on Improving the BOD Test," Sc.D. Thesis, Massachusetts Institute of Technology, 1951.

serve as food material for higher organisms such as protozoans. In each transformation further oxidation occurs, but in the final analysis there remains a certain amount of cellular organic matter that is quite resistant to further biological attack. This is commonly referred to as humus and represents an amount of organic matter corresponding to the discrepancy between the total BOD and the theoretical oxygen demand.

23.6 | DISCREPANCY BETWEEN OBSERVED RATES AND FIRST-ORDER RATES

BOD studies[4] using soluble organic materials have shown certain limitations of the assumption that the reaction follows "first-order" kinetics. In many cases the exertion of carbonaceous BOD has been observed to occur in two phases similar to that pictured in Fig. 23.2. The second phase, however, is not due to nitrification as in the illustration, but is speculated to result from the secondary action of protozoa. The sequence of biochemical action can be explained as follows. During the first one or two days of incubation, the soluble organic material is rapidly consumed by the bacteria, about 30 to 50 percent is oxidized, and the remainder is converted to bacterial cells as discussed in Sec. 23.5. When this conversion is completed, a "plateau" representing a reduced rate of oxidation occurs and is attributed to the endogenous respiration phase of bacterial metabolism. Within a day or two a secondary rise in the rate of oxidation occurs and is attributed to a rise in the population of protozoa that are predators and consume the bacteria for food. The occurrence and duration of a noticeable plateau between these phases depend upon the interval of time between the peak of the bacterial population and that of the protozoan population. These observations indicate protozoa can play a significant role in the BOD test.

While these observations indicate caution should be used in the interpretation of BOD data, they do not invalidate the use of "first-order" kinetics for solving practical stream pollution problems involving complex wastes. The BOD exertion curves for most such wastes, which contain both soluble and particulate organic matter, tend to be "composite" curves representing the summation of the oxidations for each individual compound. Since the rates of breakdown of different compounds vary so widely, the bumps and valleys in the curve tend to be leveled out, with the result that a relatively smooth "first-order" type of curve is obtained.

23.7 | APPLICATION OF BOD DATA

BOD data have wide application in practice. It is the principal test applied to domestic and industrial wastes to determine strength in terms of oxygen required for stabilization. It is the only test applied that gives a measure of the amount of biologically oxidizable organic matter present that can be used to determine the rates at

---

[4]A. W. Busch, BOD Progression in Soluble Substrates, *Sewage and Ind. Wastes,* **30:** 1336 (1958); M. N. Bhatla and A. F. Gaudy, Role of Protozoa in the Diphasic Exertion of BOD. *J. Sanit. Eng. Div., Am. Soc. Civil Engrs.,* **91:** SA3, 63 (1965).

which oxidation will occur, or BOD will be exerted, in receiving bodies of water. BOD is therefore the major criterion used in stream pollution control where organic loading must be restricted to maintain desired dissolved-oxygen levels. The determination is used in studies to measure the purification capacity of streams and serves regulatory authorities as a means of checking on the quality of effluents discharged to such waters.

Information concerning the BOD of wastes is an important consideration in the design of treatment facilities. It is a factor in the choice of treatment method and is used to determine the size of certain units, particularly trickling filters and activated-sludge units. After treatment plants are placed in operation, the test is used to evaluate the efficiency of various processes.

Many municipalities and sewer authorities finance wastewater treatment operations through sewer rental charges. Industries contributing wastes to municipal systems are often required to contribute a fair share of the operation and maintenance costs. BOD is one of the factors normally used in calculating such charges, particularly where secondary treatment employing biological processes is employed.

PROBLEMS

23.1   What use is made of the BOD test in water pollution control?

23.2   List five requirements that must be complied with in order to obtain reliable BOD data.

23.3   List five requirements of a satisfactory dilution water for BOD work.

23.4   What purpose or purposes are served by each of the following in BOD dilution water: (*a*) $FeCl_3$, (*b*) $MgSO_4$, (*c*) K_2HPO_4, (*d*) NH_4Cl, and (*e*) $CaCl_2$?

23.5   Explain how a sample of river water having a temperature below 20°C should be pretreated in preparation for BOD analysis.

23.6   Explain how a proper seed might be obtained in order to determine the BOD of an industrial waste that is not readily oxidized biologically.

23.7   Why is a seed control needed in the BOD test?

23.8   Why is it best not to use the normal BOD test to estimate the possible nitrogenous oxygen demand potential of a waste?

23.9   A wastewater has an estimated 5-day BOD of 160 mg/L. Assuming you were going to use a three-bottle dilution series and 310-mL bottles were used, how many mL of the wastewater would you put in each bottle?

23.10   A wastewater has an estimated 5-day BOD of 300 mg/L. Assuming you were going to use a three-bottle dilution series, what percent mixture of sample would you prepare for adding to each bottle?

23.11   What is the major factor affecting the time for onset of nitrification in the BOD test?

23.12   What method can be used to control nitrification in the 5-day BOD test?

23.13   What justification does the engineer have for using first-order reaction kinetics to describe the complex biochemical processes occurring in the BOD test?

23.14   What factors affect the rate of biochemical oxidation in the BOD test?

23.15 What significant part do protozoa play in the BOD test?

23.16 How can one tell that the normal biological seed used in a given BOD test is not well acclimated to biological degradation of organic matter in an industrial wastewater?

23.17 Describe a procedure you might follow to determine k' and L_0 for an industrial wastewater.

23.18 The following data were obtained in the analysis of an industrial waste: After 5 days of incubation at 20°C, the residual dissolved oxygen in blanks was 7.80 mg/L, and in a 0.1 percent dilution of the waste was 2.80 mg/L.

(*a*) What is the 5-day BOD of the waste?

(*b*) How many pounds of 5-day BOD are contained in 10,000 gallons of the waste?

23.19 Determine the 10-day carbonaceous BOD of a river sample from the following data (assume no dilution used).

Analysis, mg/L	Day 0	Day 10
DO	8.3	1.4
NH_3-N	2.1	0.8
NO_2^--N	0.0	0.1
NO_3^--N	0.5	2.1

Note: The formula weight for each nitrogen species is taken as 14 here since the values are reported as nitrogen, not as the respective ammonia, nitrite, or nitrate.

23.20 Determine the 12-day carbonaceous BOD of a river sample from the following data in which the diluted sample mixture contained 10 percent of sample.

Analysis, mg/L	Day 0	Day 12
Blank	8.2	8.1
Diluted sample		
DO	8.0	1.4
NH_3-N	1.1	0.6
NO_2^--N	0.0	0.0
NO_3^--N	0.2	1.1

Note: The formula weight for each nitrogen species is taken as 14 here since the values are reported as nitrogen, not as the respective ammonia, nitrite, or nitrate.

23.21 In Probs. 23.19 and 23.20, the summation of the inorganic nitrogen species shown at the end of incubation are higher than at day 0. Explain why this might occur.

23.22 The following dissolved-oxygen values were found after 5 days of incubation in 310-mL BOD bottles: 7.7, 7.9, and 7.9 in three blank samples; 6.5, 4.0, and 0.5 mg/L in bottles containing 2, 5, and 10 mL of sample, respectively. The 0-day dissolved oxygen of the sample was 0.0 mg/L. What is the 5-day BOD of the sample?

23.23 The following dissolved-oxygen values were found after 5 days of incubation in 305-mL BOD bottles: 8.2, 8.0, and 8.1 in three blank samples and 7.4, 6.5, 4.9, 1.5, and 0.0 mg/L in bottles containing 0.5, 1, 2, 4, and 8 mL of sample, respectively. The 0-day dissolved oxygen of the sample was 6.5 mg/L. What is the 5-day BOD of the sample?

23.24 A sample of wastewater was incubated for 7 days at 20°C and showed a BOD of 208 mg/L. (Assume k' equals 0.35/day.)

(*a*) Calculate its 5-day BOD.

(*b*) Calculate its 10-day BOD.

(*c*) Calculate the ultimate BOD.

23.25 Approximately what would be the ultimate and 5-day carbonaceous BOD values for samples containing 200 mg/L of the following materials, assuming that they were readily oxidized in the BOD test? (Assume that k' equals 0.20/day.)

(*a*) Acetic acid

(*b*) Butanol

(*c*) Glucose

(*d*) Benzoic acid

(*e*) Alanine

REFERENCE

Clesceri, L. S., A. E. Greenberg, and A. D. Eaton (eds.): "Standard Methods for the Examination of Water and Wastewater," 20th ed., American Public Health Association, Washington, DC, 1998.

CHAPTER 24

Chemical Oxygen Demand

24.1 | GENERAL CONSIDERATIONS

The chemical oxygen demand (COD) test is widely used as a means of measuring the organic strength of domestic and industrial wastes. This test allows measurement of a waste in terms of the total quantity of oxygen required for oxidation to carbon dioxide and water in accordance with Eq. (23.1). It is based upon the fact that all organic compounds, with a few exceptions, can be oxidized by the action of strong oxidizing agents under acid conditions. The amino nitrogen (with an oxidation number of -3) will be converted to ammonia nitrogen as indicated in Eq. (23.1). However, organic nitrogen in higher oxidation states will be converted to nitrate.

During the determination of COD, organic matter is converted to carbon dioxide and water regardless of the biological assimilability of the substances. For example, glucose and lignin are both oxidized completely. As a result, COD values are greater than BOD values and may be much greater when significant amounts of biologically resistant organic matter is present. Wood-pulping wastes are excellent examples because of their high lignin content. For reasons presented in Sec. 23.5, the COD of materials such as glucose is always greater than the L_0 value.

One of the chief limitations of the COD test is its inability to differentiate between biologically oxidizable and biologically inert organic matter. In addition, it does not provide any evidence of the rate at which the biologically active material would be stabilized under conditions that exist in nature.

The major advantage of the COD test is the short time required for evaluation. The determination can be made in about 3 h rather than the 5 days required for the measurement of BOD. For this reason it is used as a substitute for the BOD test in many instances. COD data can often be interpreted in terms of BOD values after sufficient experience has been accumulated to establish reliable correlation between COD and BOD.

24.2 | HISTORY OF THE COD TEST

Chemical oxidizing agents have long been used for measuring the oxygen demand of polluted waters. Potassium permanganate solutions were used for many years, and the results were referred to as *oxygen consumed* from permanganate. The oxidation caused by permanganate was highly variable with respect to various types of compounds, and the degree of oxidation varied considerably with the strength of reagent used. Oxygen-consumed values were always considerably less than 5-day BOD values. This fact demonstrated the inability of permanganate to carry the oxidation to any particular end point.

Ceric sulfate, potassium iodate, and potassium dichromate are other oxidizing agents that have been studied extensively for the determination of chemical oxygen demand. Potassium dichromate has been found to be the most practical of all, since it is capable of oxidizing a wide variety of organic substances almost completely to carbon dioxide and water. Because all oxidizing agents must be used in excess, it is necessary to measure the amount of excess remaining at the end of the reaction period in order to calculate the amount actually used in the oxidation of the organic matter. It is relatively easy to measure any excess of potassium dichromate, an important point in its favor.

In order for potassium dichromate to oxidize organic matter completely, the solution must be strongly acidic and at an elevated temperature. As a result, volatile materials originally present and those formed during the digestion period are lost unless provision is made to prevent their escape. Reflux condensers are ordinarily used for this purpose and allow the sample to be boiled without significant loss of volatile organic compounds.

Certain organic compounds, particularly low-molecular-weight fatty acids, are not oxidized by dichromate unless a catalyst is present. It has been found that silver ion acts effectively in this capacity. Aromatic hydrocarbons and pyridine are not oxidized under any circumstances.

24.3 | CHEMICAL OXYGEN DEMAND BY DICHROMATE

Potassium dichromate is a relatively cheap compound that can be obtained in a high state of purity. The analytical-reagent grade, after drying at 103°C, can be used to prepare solutions of an exact normality by direct weighing and dilution to the proper volume. The dichromate ion is a very potent oxidizing agent in solutions that are strongly acidic. The reaction involved in the usual case, where organic nitrogen is all in a reduced state (oxidation number of -3), may be represented in a general way as follows:

$$C_nH_aO_bN_c + dCr_2O_7^{2-} + (8d + c)H^+ \xrightarrow{\Delta} nCO_2$$
$$+ \frac{a + 8d - 3c}{2} H_2O + cNH_4^+ + 2dCr^{3+} \qquad (24.1)$$

where $d = 2n/3 + a/6 - b/3 - c/2$. For these and other reasons mentioned previously, dichromate approaches an ideal reagent for the measurement of COD.

Selection of Normality

COD results are reported in terms of milligrams of oxygen. Since the equivalent weight of oxygen is 8 g, it would seem logical to use a N/8 or 0.125 N solution of oxidizing agent in the determination so that results can be calculated in accordance with the general procedure described in Sec. 15.1. Experience with the test has shown that it has sufficient sensitivity to allow the use of a stronger solution of dichromate, and a N/4 or 0.25 N solution is recommended. This allows the use of larger samples by doubling the range of COD that can be measured in the test procedure, since each milliliter of a 0.25 N solution of dichromate is equivalent to 2 mg of oxygen.

Measurement of Excess Oxidizing Agent

In any method of measuring COD, an excess of oxidizing agent must be present to ensure that all organic matter in oxidized as completely as is within the power of the reagent. This requires that a reasonable excess be present in all samples. In it necessary, of course, to measure the excess in some manner so that the actual amount reduced can be determined. A solution of a reducing agent is ordinarily used.

Nearly all solutions of reducing agents are gradually oxidized by oxygen dissolved from the air unless special care is taken to protect them from oxygen. Ferrous ion is an excellent reducing agent for dichromate. Solutions of it can be best prepared from ferrous ammonium sulfate which is obtainable in rather pure and stable form. In solution, however, it is slowly oxidized by oxygen, and standardization is required each time the reagent is to be used. The standardization is made with the 0.25 N solution of dichromate. The reaction between ferrous ammonium sulfate and dichromate may be represented as follows:

$$6Fe^{2+} + Cr_2O_7^{2-} + 14H^+ \rightarrow 6Fe^{3+} + 2Cr^{3+} + 7H_2O \qquad (24.2)$$

Blanks

Both the COD and BOD tests are designed to measure oxygen requirements by oxidation of organic matter present in the samples. It is important, therefore, that no organic matter from outside sources be present if a true measure of the amount present in the sample is to be obtained. Since it is impossible to exclude extraneous organic matter in the BOD test and impractical to do so in the COD test, blank samples are required in both determinations.

Indicator

A very marked change in oxidation-reduction potential (ORP) occurs at the end point of all oxidation-reduction reactions. Such changes may be readily detected by electrometric means if the necessary equipment is available. Oxidation-reduction indicators may also be used; Ferroin (ferrous 1,10-phenanthroline sulfate) is an excellent one to indicate when all dichromate has been reduced by ferrous ion. It gives a very sharp brown color change that is easily detected in spite of the blue color produced by the Cr^{3+} formed on reduction of the dichromate.

Calculations

Although an oxidizing agent is used in the measurement of COD, it does not figure directly in the calculation of COD. This is because a solution of a reducing agent must be used to determine how much of the oxidizing agent was used, and it is simpler to relate everything to the reducing agent in this case, because its strength varies from day to day and its normality is seldom, if ever, exactly equal to 0.25 N.

Calculation of COD is made using the following formula:

$$COD \ (mg/L) = \frac{8000 \ (\text{blank titr.} - \text{sample titr.})[\text{norm. } Fe(NH_4)_2(SO_4)_2]}{mL \ \text{sample}} \qquad (24.3)$$

Alternate Procedure for Low COD Samples

The COD test is precise and accurate for samples with a COD of 50 mg/L or greater. For more dilute samples it is preferred that a more dilute dichromate solution be used so that a significant relative difference between the quantity of dichromate added and that remaining after refluxing results. With dilute samples, care must be exercised to avoid sample contamination, and good analytical techniques must be used if reasonably accurate results are to be obtained. It is also important in any modification that the volume of concentrated sulfuric acid to volume of sample plus dichromate solution be maintained at a 1:1 ratio. If it is smaller, the oxidizing power of the solution will decrease significantly, while if it is larger, the blank consumption of dichromate becomes excessive.

Methods to Reduce Hazardous Waste Generation

The COD test can generate a large volume of liquid hazardous waste. In the past, common practice was to dilute completed samples with tap water and discharge them down the drain with a good flushing of water. This meant that considerable quantities of acid, chromium, silver, and also mercury (added for chloride complexation as noted in Sec. 24.4) could reach a treatment plant and perhaps surface waters. For this reason drain disposal is now discouraged and sometimes prohibited, and so spent solutions must be stored, packaged, and disposed into approved hazardous waste storage sites. It is possible to reduce this problem by recovering silver and mercury from the samples, but this requires proper permitting. In order to reduce this problem alternate procedures can be used. "Standard Methods" now offers two closed-reflux methods in which smaller sample and reagent volumes are used. Refluxing here is conducted in sealed containers. However, the principles are essentially the same as in the more historical open-reflux method. In order to maintain sufficient sensitivity with the reduced volumes, the concentration of the ferrous ammonium sulfate titrant is reduced. In one variation, a colorimetric rather than a volumetric procedure is used. This takes advantage of the change during organic oxidation from orange color of Cr(VI), which absorbs at a wavelength of 600 nm, to the blue color of Cr(III) solutions, which absorb at 420 nm. Measurements of color change from sample oxidation at either wavelength can be used for quantification. Although costs of prepared reagents for the closed-flux COD procedures from commercial companies tend to be high, many analysts prefer them in comparison with reagents for the conventional reflux procedure because of their ease in use and reduction in quantities of resulting waste chemicals requiring disposal.

While these considerations tend to support the use of the closed-reflux procedures, a similar variation can be made with the open-reflux procedure as well. Here, for example, a 10-mL rather than a 50-mL sample can be used. In this case, only 5.00 mL of dichromate solution is added together with only 15 mL of Ag^+-amended concentrated sulfuric acid. By reducing the concentration of the ferrous ammonium sulfate titrant from 0.25 N to 0.025 N, suitable sensitivity can still be maintained. The same reflux apparatus as used with larger samples works satisfactorily here. With this modification, only one-fifth of the volume of waste solutions are generated, and little sacrifice in analytical precision is made. As noted under the procedure for low COD samples, the essential aspect of varying sample size is that the ratio of concentrated sulfuric acid volume to total water volume must be maintained constant and equal to 1:1.

24.4 | INORGANIC INTERFERENCES

Certain reduced inorganic ions can be oxidized under the conditions of the COD test and thus can cause erroneously high results to be obtained. Chloride causes the most serious problem because of its normally high concentration in most wastewaters,

$$6Cl^- + Cr_2O_7^{2-} + 14H^+ \rightarrow 3Cl_2 + 2Cr^{3+} + 7H_2O \tag{24.4}$$

Fortunately, this interference can be eliminated by the addition of mercuric sulfate to the sample prior to the addition of the other reagents. The mercuric ion combines with the chloride ions to form a poorly ionized mercuric chloride complex (see Secs. 2.13 and 4.8).

$$Hg^{2+} + 2Cl^- \rightleftharpoons HgCl_2(aq) \qquad (\beta_2 = 1.7 \times 10^{13}) \tag{24.5}$$

In the presence of excess mercuric ions the chloride-ion concentration is so small that it is not oxidized to any extent by dichromate.

Nitrite is oxidized to nitrate and this interference can be overcome by the addition of sulfamic acid to the dichromate solution. However, significant amounts of nitrite seldom occur in wastes or in natural waters. This also holds true for other possible interferences such as ferrous iron and sulfide.

24.5 | APPLICATION OF COD DATA

The COD test is used extensively in the analysis of industrial wastes. It is particularly valuable in surveys designed to determine and control losses to sewer systems. Results may be obtained within a relatively short time and measures taken to correct errors on the day they occur. In conjunction with the BOD test, the COD test is helpful in indicating toxic conditions and the presence of biologically resistant organic substances. The test is widely used in the operation of treatment facilities because of the speed with which results can be obtained.

PROBLEMS

24.1 Give four different applications for the COD analysis in environmental engineering practice.

24.2 What general groups of organic compounds are not oxidized in the COD test?

24.3 (*a*) What chemical changes take place during refluxing in the COD test?

  (*b*) Does the COD analysis result in oxidation of reduced nitrogen ($-$III) species?

24.4 What is the most prevalent inorganic species that interferes with the COD analysis, and how is this interference dealt with?

24.5 Saturated organic aliphatic acids are difficult to oxidize with dichromate. How is this difficulty circumvented in the standard COD procedure?

24.6 Indicate whether COD results would probably be higher, lower, or the same as the true value under the following conditions, and briefly explain why: (*a*) mercuric sulfate was not added; (*b*) silver sulfate was not added; (*c*) the ferrous ammonium sulfate was assumed to have the same normality as it did 2 weeks prior to the current analysis.

24.7 Why do the COD and BOD analyses usually give different results for the same waste?

24.8 Compare the relative advantages and disadvantages of the open-reflux and closed-reflux COD procedures.

24.9 What wavelengths might be used for colorimetric analysis of COD and why?

24.10 What change in COD analysis is used in order to measure waters with low COD?

24.11 (*a*) What chemicals in the waste from COD analyses are of most environmental concern?

  (*b*) What changes in COD analytical procedures can be used to reduce the quantity of hazardous waste resulting from the analysis?

24.12 What is the theoretical COD of samples containing 300 mg/L of (*a*) ethanol, (*b*) phenol, and (*c*) leucine?

24.13 (*a*) Estimate the COD of a solution containing 500 mg/L of butanol.

  (*b*) If the compound were readily degradable biologically, about what would you expect the 5-day BOD to be?

24.14 What could be inferred from the following analytical results concerning the relative ease of biodegradability of each waste?

Waste	5-day BOD, mg/L	COD, mg/L
A	240	300
B	100	500
C	120	240

REFERENCES

Clesceri, L. S., A. E. Greenberg, and A. D. Eaton (eds.): "Standard Methods for the Examination of Water and Wastewater," 20th ed., American Public Health Association, Washington, DC, 1998.

Skoog, D. A., D. M. West, and F. J. Holler: "Fundamentals of Analytical Chemistry," 7th ed., Saunders College Pubs., Fort Worth, TX, 1996.

C H A P T E R

Nitrogen

25.1 | GENERAL CONSIDERATIONS

The compounds of nitrogen are of great importance in water resources, in the atmosphere, and in the life processes of all plants and animals. The chemistry of nitrogen is complex because of the several oxidation states that nitrogen can assume and the fact that changes in oxidation state can be brought about by living organisms. To add even more interest, the oxidation state changes wrought by bacteria can be either positive or negative, depending upon whether aerobic or anaerobic conditions prevail.

Nitrogen can exist in seven oxidation states, and essentially all are of environmental interest.

$$\begin{array}{ccccccc}
-\text{III} & \text{O} & \text{I} & \text{II} & \text{III} & \text{IV} & \text{V} \\
NH_3 & N_2 & N_2O & NO & N_2O_3 & NO_2 & N_2O_5
\end{array}$$

Three forms combine with water to form inorganic ionized species that can reach high concentrations,

$$NH_3 + H_2O \rightarrow NH_4^+ + OH^- \tag{25.1}$$

$$N_2O_3 + H_2O \rightarrow 2H^+ + 2NO_2^- \tag{25.2}$$

$$N_2O_5 + H_2O \rightarrow 2H^+ + 2NO_3^- \tag{25.3}$$

The respective water-soluble species formed, ammonium, nitrite, and nitrate, are of historical environmental concern in water, and their concentrations in drinking water supplies and surface waters have been regulated for decades.

The other inorganic oxidation states, N_2, N_2O (nitrous oxide), NO (nitric oxide), and NO_2 (nitrogen dioxide) exist as gases that have somewhat limited solubility in water. Additionally, nitrogen is an important element in many organic chemicals. The reduced form N(III) is a major element in proteins and the amino acids of which they are composed, and in nucleic acids as well. Thus, nitrogen is essential to all life.

However, except for N_2, the major component of the earth's atmosphere, nitrogen compounds in all oxidation states can result in environmental problems of concern as will be discussed. Adding to the complexity, nitrogen is readily changed in oxidation state and in chemical form through natural biological, chemical, and photochemical processes. In order to help capture the beneficial uses of nitrogen while avoiding the harmful changes some of its forms can cause, environmental engineers and scientists need to have a good understanding of nitrogen chemistry and the nitrogen cycle. The nitrogen cycle is illustrated in Fig. 25.1.

From the nitrogen cycle, it is seen that the atmosphere serves as a very large reservoir from which nitrogen is constantly removed by the action of electrical discharge, nitrogen-fixing bacteria and algae, and combustion processes. During electrical storms nitrogen is oxidized to NO, which is oxidized by ozone in the atmosphere to form NO_2. NO_2 in turn is reduced back to NO by photolysis. This forward and backward reaction establishes a steady-state concentration between NO and NO_2, with NO generally being by far the dominant species. The two species together are generally referred to as NO_x. Combustion processes, such as in the internal combustion engine of automobiles, also lead to conversion of N_2 to NO and NO_2. Other oxidative reactions with NO_2 in the atmosphere lead to the conversion of NO_2 to N_2O_5, which according to Eq. (25.3) can combine with water in the atmosphere to produce the nitrate of nitric acid, which reaches the earth's surface with falling rain. Nitrate is also produced by direct oxidation of nitrogen or of ammonia in the production of commercial fertilizers. The nitrate serves to fertilize plant life and is converted to proteins (organic nitrogen).

$$NO_3^- + CO_2 + \text{green plants} + \text{sunlight} \rightarrow \text{protein} \qquad (25.4)$$

The major means by which atmospheric nitrogen enters the nitrogen cycle is through conversion to protein by nitrogen-fixing bacteria. This includes the photosynthetic *cyanobacteria,* which have many similarities to algae.

$$N_2 + \text{nitrogen-fixing bacteria} \rightarrow \text{protein} \qquad (25.5)$$

In addition, ammonia and ammonium compounds are applied to soils to supply plants with ammonia for further production of proteins. Urea is one of the popular ammonium compounds because it releases ammonia gradually.

$$NH_3 + CO_2 + \text{green plants} + \text{sunlight} \rightarrow \text{protein} \qquad (25.6)$$

Animals and human beings are incapable of utilizing nitrogen from the atmosphere or from inorganic compounds to produce proteins. They are dependent upon plants, or other animals that feed upon plants, to provide protein, with the exception

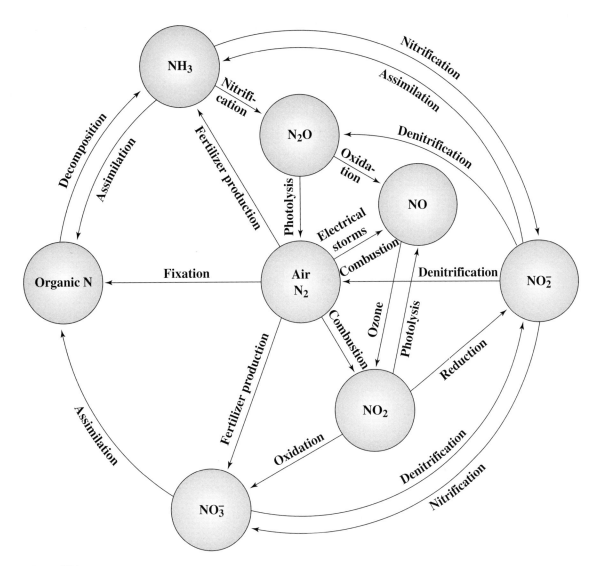

Figure 25.1
The nitrogen cycle.

of ruminants. The multiple-stomached animals are capable of producing part of their protein requirement from carbohydrate matter and urea through bacterial action. Within the animal body, protein matter is used largely for growth and repair of muscle tissue. Some may be used for energy purposes. In any event, nitrogen compounds are released in the waste products of the body during life. At death the proteins stored in the body become waste matter for disposal. The urine contains the nitrogen resulting from the metabolic breakdown of proteins. The nitrogen exists in urine principally as urea which is hydrolyzed rather rapidly by the enzyme urease to ammonium carbonate:

$$\underset{NH_2}{\overset{NH_2}{C=O}} + 2H_2O \xrightarrow{\text{urease enzyme}} (NH_4)_2CO_3 \qquad (25.7)$$

The feces of animals contain appreciable amounts of unassimilated protein matter (organic nitrogen). It and the protein matter remaining in the bodies of dead animals and plants are converted in large measure to ammonia by the action of heterotrophic bacteria, under aerobic or anaerobic conditions:

$$\text{Protein (organic-N)} + \text{bacteria} \rightarrow NH_3 \qquad (25.8)$$

Some nitrogen always remains in nondigestible matter and becomes part of the nondigestible residue sink. As such it becomes part of the *detritus* in water or sediments, or the *humus* in soils.

The ammonia released by bacterial action on urea and proteins may be used by plants directly to produce plant protein. If it is released in excess of plant requirements, the excess is oxidized by the autotrophic nitrifying bacteria. One group (*Nitrosomonas*) convert ammonia under aerobic conditions to nitrite and derive energy from the oxidation:

$$2NH_3 + 3O_2 \xrightarrow{\text{\textit{Nitrosomonas}}} 2NO_2^- + 2H^+ + 2H_2O \qquad (25.9)$$

The nitrite is oxidized by another group of nitrifying bacteria (*Nitrobacter*) to nitrate:

$$2NO_2^- + O_2 \xrightarrow{\text{\textit{Nitrobacter}}} 2NO_3^- \qquad (25.10)$$

The nitrate formed may serve as fertilizer for plants. Nitrate produced in excess of the needs of plant life is carried away in water percolating through the soil because the soil does not have the ability to hold nitrate. This frequently results in relatively high concentrations of nitrate in groundwaters and is an extensive problem in Illinois, Iowa, and other midwestern states. Nitrification also can result in some production of N_2O.

Under anaerobic conditions nitrate and nitrite are both reduced by a process called *denitrification*. Nitrate is reduced to nitrite, and then reduction of nitrite occurs. Reduction of nitrite is carried all the way to ammonia by a few bacteria for protein formation, but mostly the nitrate is reduced to nitrogen gas, which escapes to the atmosphere. This constitutes a serious loss of fertilizing matter in soils when anaerobic conditions develop. Also, some denitrifying bacteria produce N_2O from nitrate reduction, and this too is a gas that leaves soil or water to enter the atmosphere. N_2O in the atmosphere can be reduced through photolysis to produce N_2 and an excited state of oxygen, which oxidizes another portion of the N_2O to NO. We thus see that transformations of nitrogen occur in water, in soil, and in the atmosphere through a variety of chemical, photochemical, and biological processes.

25.2 | ENVIRONMENTAL SIGNIFICANCE OF NITROGEN SPECIES

While nitrogen is seen to be an essential component of all living things, excessive concentrations of certain nitrogen species in some compartments of the environment can lead to significant environmental problems. This is true of some nitrogen species in the atmosphere as well as in terrestrial and aquatic environments.

Atmospheric Concerns with Nitrogen Species

The three major environmental problems associated with nitrogen species in the atmosphere are photochemical smog, global warming, and stratospheric ozone depletion. Photochemical smog results when partially oxidized organic matter, NO_x, and sunlight come together under certain meteorological conditions, resulting in a series of complex chemical and photochemical reactions that lead to the production of high ozone concentrations and organic chemicals that together produce eye irritation, reduced air visibility, crop damage, and severe adverse health impacts in humans. The automobile has been a primary producer of two of the ingredients of photochemical smog, partially oxidized organic matter and NO_x. The most serious problems occur in dense urban areas where there are many automobiles.

Wide use of fossil fuels over the past century has resulted in an increase in atmospheric carbon dioxide, which acts as a blanket to prevent heat from radiating from the earth, a phenomenon, come to be known as the greenhouse effect, that is increasing the earth's temperature. However, the gaseous oxides of nitrogen also exhibit a greenhouse effect. While carbon dioxide is believed responsible for about 55 percent of the increased changes to the radiative temperature between 1980 and 1990, increased NO_x production from fuel and biomass combustion and particularly from denitrification (N_2O) as a result of increased commercial fertilizer usage is estimated to be responsible for 6 percent of the increase.[1] While the amount of NO_x in the atmosphere would appear to be small compared with CO_2, one molecule of N_2O has a heat-trapping ability equivalent to 200 molecules of CO_2.

While NO_x is partially responsible for increased ozone production as part of photochemical smog production in urban areas near the earth's surface, it is somewhat surprising that it also plays a role in the destruction of ozone in the stratosphere. Stratospheric ozone plays a key role in protecting life on earth from the harmful effects of excessive ultraviolet radiation. The widespread use of chlorofluorocarbons (CFCs) is known to have resulted in significant destruction of the protective stratospheric ozone, but NO_x is playing a role as well. N_2O and NO_2 are both converted to NO in the atmosphere, and NO reaching the stratospheres reacts with ozone to result in its depletion.

---

[1]R. Rosswall, "Greenhouse Gases and Global Change: International Collaboration," *Env. Sci. Tech.,* **25**(4): 567, 1991.

Aquatic Concerns with Nitrogen Species

An Indicator of Sanitary Quality   It has long been known that polluted waters will purify themselves, provided that they are allowed to age for sufficient periods of time. The hazard to health or the possibility of contracting disease by drinking such waters decreases markedly with time and temperature increase, as shown in Fig. 25.2.

Prior to the development of bacteriological tests for determining the sanitary quality of water (about 1893), those concerned with the public health were largely dependent upon chemical tests to provide circumstantial evidence of the presence of contamination. The chloride test was one of these (see Sec. 21.2), but it gave no evidence of how recently the contamination had occurred. Chemists working with wastes and freshly polluted waters learned that most of the nitrogen is originally present in the form of organic (protein) nitrogen and ammonia. As time progresses, the organic nitrogen is gradually converted to ammonia nitrogen, and later on, if aerobic conditions are present, oxidation of ammonia to nitrite and nitrate occurs. The progression of events was found to occur somewhat as shown in Fig. 25.3, and more refined interpretations of the sanitary quality of water were based upon this knowledge. For example, waters that contained mostly organic and ammonia nitrogen were considered to have been recently polluted and therefore of great potential danger. Waters in which most of the nitrogen was in the form of nitrate were considered to have been polluted a long time previously and therefore offered little threat to the public health. Waters with appreciable amounts of nitrite were of highly questionable character. The bacteriological test for coliform organisms provides circumstantial evidence of much greater reliability concerning the hygienic safety of water, and it has eliminated the need for extended nitrogen analysis in most water supplies.

In 1940 it was found that drinking waters with high nitrate content often caused methemoglobinemia in infants. From extended investigations in Iowa, Minnesota,

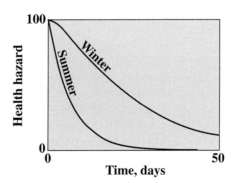

Figure 25.2
Surface waters; health hazard in relation to age of pollution.

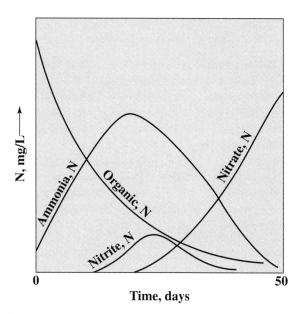

Figure 25.3
Changes occurring in forms of nitrogen present in polluted
water under aerobic conditions.

and Ohio, where the problem has been most acute, it has been concluded that the nitrate content should be limited.[2] For this reason, the U.S. EPA has set a maximum contaminant level requiring that the nitrate-nitrogen concentration not exceed 10 mg/L and the nitrite-nitrogen concentration not exceed 1 mg/L in public water supplies.

Methemoglobinemia is actually a result of interaction of nitrite with hemoglobin, the nitrite being formed from nitrate reduction in the digestive system. For this reason, a maximum contaminant level for drinking water has now been set for nitrite as well as nitrate. Nitrite can also interact with amines chemically (especially when chlorinating for disinfection[3]) or enzymatically[4] to form nitrosamines, which are strong carcinogens. The formation of N-nitrosodimethylamine (NDMA) by these processes has been found to result during wastewater treatment and has become an issue recently in wastewater reuse projects and contaminated groundwater supplies. Ammonia reacts with chlorine to form chloramines, which are slower-acting disinfectants than free chlorine as discussed in Sec. 20.2. Ammonia is sometimes added

[2]K. F. Maxcy. Report on Relation of Nitrate Nitrogen Concentration in Well Waters to the Occurrence of Methemoglobinemia in Infants. *Natl. Acad. Sci.-Research Council Sanit. Eng. and Environment Bull.,* 264, 1950.

[3]J. Choi and R. L. Valentine, "Formation of *N*-nitrosodimethylamine (NDMA) from reaction of monochloramine: a new disinfection by-product," *Water Research,* **36:** 817–824 (2002).

[4]M. Alexander, "Biodegradation and Bioremediation," Academic Press, San Diego, 1994.

to drinking water supplies when a disinfection residual in water mains is desired as chloramines do not decompose as rapidly as chlorine.

Nutritional and Related Problems   All biological processes employed for wastewater treatment are dependent upon reproduction of the organisms employed, as discussed in Sec. 6.6. In planning waste treatment facilities it becomes important to know whether the waste contains sufficient nitrogen for the organisms. If not, any deficiency must be supplied from outside sources. Determinations of ammonia and organic nitrogen are normally made to obtain such data.

Nitrogen is one of the fertilizing elements essential to the growth of algae. Such growth is often stimulated to an undesirable extent in bodies of water that receive either treated or untreated effluents, because of the nitrogen and other fertilizing matter contributed by them. Nitrogen analyses are an important means of gaining information on this problem.

Oxidation in Rivers and Estuaries   The autotrophic conversion of ammonia to nitrite and nitrate requires oxygen, indicated in Eqs. (25.9) and (25.10), and so the discharge of ammonia nitrogen and its subsequent oxidation can seriously reduce the dissolved-oxygen levels in rivers and estuaries, especially where long residence times required for the growth of the slow-growing nitrifying bacteria are available. Also, these organisms are produced in large numbers by highly efficient aerobic biological waste treatment systems, and their discharge with the treated effluent can cause rapid nitrification to occur in waterways. Disinfection of effluents (e.g., with chlorine or ultraviolet light) minimized this problem. Nitrogen analyses are important in assessing the possible significance of the problem, and in the operation of treatment processes designed to reduce ammonia discharge.

Control of Biological Treatment Processes   Determinations of nitrogen are often made to control the degree of purification produced in biological treatment. With the use of the BOD test, it has been learned that effective stabilization of organic matter can be accomplished without carrying the oxidation into the nitrification stage. This results in reducing time of treatment and air requirements where ammonia removal is not otherwise mandated.

Advantage is taken of denitrification for removing nitrogen from wastes where this is required to prevent undesirable growths of algae and other aquatic plants in receiving waters. Ammonia and organic nitrogen are first biologically converted to nitrite and nitrate by aerobic treatment. The waste is then placed under anoxic conditions, where denitrification converts the nitrite and nitrate to nitrogen gas, which escapes to the atmosphere. For denitrification to occur, organic matter, which is oxidized for energy while the nitrogen is being reduced, must be present. Methanol was a favorite form of organic matter, but, due to its increased cost, other materials including untreated wastewater itself are commonly used. When nitrification is required to protect oxygen resources in streams, advantage can be taken of the oxidizing potential of nitrate present in the plant effluent to reduce oxygen requirements for treatment. Here the nitrate-containing effluent is recycled back to be mixed with settled wastewater under anoxic conditions to effect some organic oxidation through denitrification. The reduced quantity of organics remaining are then subject to the usual aerobic step, but now less oxygen is required for treatment. An additional advantage here is that nitrate is simultaneously removed in the denitrification step, which now occurs as a first step, rather than as the last step in the treatment train. The formation of N_2 by denitrification is sometimes a problem in the activated sludge process of wastewater treatment. Prolonged detention of activated sludge in final settling tanks allows formation of sufficient nitrogen gas to buoy the sludge, if nitrate is present in adequate amounts. This is often referred to as the "rising" sludge problem. As this discussion suggests, analyses for nitrogen forms in wastewater treatment can be of great importance in process control and in achieving overall treatment objectives.

In some states, ammonia-nitrogen limitations have been imposed because of suspected toxic effects upon fish life. It is well known that un-ionized ammonia is toxic but that the ammonium ion is not. Since the relationship between the two is pH-dependent,

$$NH_3 + H^+ \rightleftharpoons NH_4^+$$

a discussion is in order. Figure 25.4 shows the relationship between free ammonia and ammonium ion that exists for several concentrations of ammonia nitrogen over the pH range of interest in most natural waters. Free ammonia in concentrations above about 0.2 mg/L can cause fatalities in several species of fish. Applying the usual safety factor, a National Research Council Committee has recommended that no more than 0.02 mg/L free ammonia be permitted in receiving waters.[5] From this

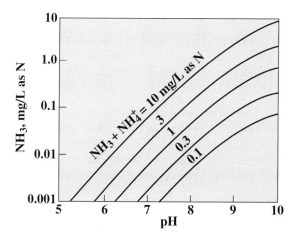

Figure 25.4
The effect of pH and ammonia nitrogen concentration
($NH_3 + NH_4^+$) on the concentration of free ammonia in
water.

[5]Committee on Water Quality Criteria, "Water Quality Criteria, 1972," Superintendent of Documents, Washington, DC, 1972.

and the data presented in Fig. 25.4, it appears that ammonia toxicity will not be a problem in receiving waters with pH below 8 and ammonia-nitrogen concentrations less than about 1 mg/L.

Control of ammonia for the given reasons is generally accomplished by nitrification. In some cases, limitations are placed upon the amount of total nitrogen that can be present in an effluent. This condition can be met by removal of ammonia in some instances, but more often it requires nitrification and denitrification. Because of these new requirements, methods of measuring all forms of nitrogen have become important.

25.3 | METHODS OF ANALYSIS

Nitrogen exists in four forms that are of interest in water resources. These are the forms located in the outer ring of the nitrogen cycle (Fig. 5.1). These are ammonia, nitrite, nitrate, and organic nitrogen. A discussion of the determination of each form is required. Because of differences in the amounts present in potable and polluted waters, methods applied to water may differ somewhat from those applied to wastewaters. It is customary to report all results in terms of nitrogen so that values may be interpreted from one form to another without use of a factor.

Ammonia Nitrogen

All nitrogen that exists as ammonium ion or in the equilibrium

$$NH_4^+ \rightleftharpoons NH_3 + H^+ \qquad (25.11)$$

is considered to be ammonia nitrogen. Essentially four different methods for determining ammonia nitrogen are included in "Standard Methods": two colorimetric procedures, a volumetric procedure, and an instrumental method using an ammonia-selective membrane probe.

By Direct Nesslerization   This historical colorimetric method, which yields a yellow color from the reaction of Nessler's reagent with ammonia, has now been dropped from "Standard Methods" because Nessler's reagent contains mercury. Use of analytical methods that result in waste analytical solutions that contain the hazardous chemical mercury are no longer being favored if satisfactory alternatives are available.

By Direct Phenate Addition   The phenate method is now the standard colorimetric procedure for ammonia. It involves the addition of an alkaline phenol solution together with hypochlorite and a manganous salt. The manganese catalyzes the reaction involving phenol, hypochlorite, and ammonia to produce indophenol, which has an intense blue color. This procedure is subject to similar interferences as with Nesslerization. Ammonia nitrogen can be determined over a range of 0.02 to 2 mg/L by an automated phenate procedure.

By Distillation   The phenate procedure is subject to serious error from extraneous color and turbidity. Thus, direct application is impractical to use on many samples. The distillation procedure is used to separate the ammonia from interfering substances, and measurement of ammonia nitrogen can then be made in a number of ways, including the phenate procedure.

Ammonium ion exists in equilibrium with ammonia and hydrogen ion, as shown in Eq. (25.11). At pH levels above 8, the equilibrium is displaced far enough to the right so that ammonia is liberated as a gas along with the steam produced when a sample is boiled, as follows:

$$NH_4^+ \xrightarrow{\Delta} NH_3 \uparrow + H^+ \tag{25.12}$$

Removal of ammonia allows the hydrogen ions released in the decomposition of ammonium ion to accumulate in the residue, and a decrease in pH will result unless a buffer is present to combine with the hydrogen ions. A borate buffer is added to maintain the pH in a range of 9.5 in order to displace the equilibrium far to the right. Higher pH levels are not recommended because of the danger of some ammonia being released from organic sources at the temperature of boiling water. Experience has shown that essentially all the free ammonia will be expelled from solutions whose pH is maintained at 9.5 by the time 200 mL of water has been distilled when samples of 500 to 1000 mL are used. The distillate is collected by passage of the vapor through a condenser and then into an acid solution contained in a flask. The acid solution converts the free ammonia in the vapor to ammonium ion [the reverse of Eq. (25.12)], which cannot volatilize from the condensate solution. The acid solution used depends upon the analysis that follows. A boric acid solution is used for analyses by titration, and a dilute sulfuric acid solution for analyses by the phenate or membrane probe procedures. The titration and membrane probe procedures are described in the following sections on volumetric analysis and the ammonia-selective electrode.

For water samples that contain small amounts of ammonia nitrogen, the usual procedure is to measure the amount of nitrogen in the distillate by the phenate method. The calculation of ammonia nitrogen in terms of milligrams per liter must take into consideration the volume of distillate as well as the sample size. Many analysts find it convenient to distill an amount that exceeds 200 mL to avoid close attention at the end of the distillation. The calculation can be made, provided that the volume of distillate is known, by the following method:

$$\text{mg/L } NH_3-N = \frac{V_D}{V_{DA}} \times A \times \frac{1000}{s} \tag{25.13}$$

where
V_D = mL of distillate
V_{DA} = mL of distillate actually analyzed
A = mg of NH_3-N found in the analyzed portion of the distillate
s = mL of sample used for distillation

By Volumetric Analysis When samples contain more than 2 mg/L of ammonia nitrogen, as is the case with domestic and many industrial wastes, the concentration can be determined by titration with a standard solution of sulfuric acid after distillation and absorption of ammonia in boric acid as previously described. The chemistry involved is as follows: First, boric acid is an excellent buffer. It combines with ammonia in the distillate to form ammonium and borate ions, as shown in the equation:

$$NH_3 + H_3BO_3 \rightarrow NH_4^+ + H_2BO_3^- \qquad (25.14)$$

This causes the pH to increase somewhat, as was discussed under buffer action in Sec. 4.6, but the pH is held in a favorable range for absorption of ammonia by the use of an excess of boric acid. The ammonia may then be measured by back titration with a strong acid such as sulfuric acid. Actually, the acid measures the amount of borate ion present in the solution as follows:

$$H_2BO_3^- + H^+ \rightarrow H_3BO_3 \qquad (25.15)$$

When the pH of the boric acid solution has been decreased to its original value, an amount of strong acid equivalent to the ammonia has been added. The titration is most easily conducted by potentiometric methods which eliminates the need for internal indicators. The proper pH for the end point is best determined by diluting the specified volume of boric acid solution with ammonia-free distilled water in an amount equal to the volume of distillate desired and measuring the pH of the mixture.

Many water chemists prefer to use a N/14 or 0.0714 N, solution of sulfuric acid for the measurement of ammonia nitrogen. Since each milliliter of 0.0714 N acid is equivalent to 1.0 mg of nitrogen, its use eliminates the need for the 0.28 factor required when 0.020 N acid is used. Furthermore, much smaller volumes of 0.0714 N acid are needed for titration, an important factor in conserving reagent and preventing undue dilution of the sample during titration.

By Ammonia-Selective Electrode   This approach makes use of a gas-permeable membrane probe as discussed in Sec. 12.3. The sample pH is raised above 11 to convert ammonium to the ammonia form, which diffuses through the membrane and changes the pH of the internal solution. This change is measured by an internal pH glass electrode and is interpreted in terms of sample ammonia nitrogen concentration using a calibration curve. The calibration curve must be prepared with a standardized ammonia solution. In an alternate approach, the change in pH following addition of measured amounts of a standardized ammonia solution to the sample can be used, rather than a calibration curve, to determine the ammonia nitrogen concentration of a sample.

Organic Nitrogen

All nitrogen present in organic compounds may be considered organic nitrogen. This includes the nitrogen in amino acids, nucleic acids, amines, amides, imides, nitro derivatives, and a number of other compounds. Most of these have very little significance in water analysis unless specific industrial wastes are involved.

Most of the organic nitrogen that occurs in domestic wastes is in the form of proteins or their degradation products: polypeptides and amino acids. Therefore, the methods employed in water analysis have been designed to ensure measurement of these forms without particular regard to other organic forms. Actually, most forms except the nitrogen in nitro compounds are measured by the methods used.

Most organic compounds containing nitrogen are derivatives of ammonia, and thus the nitrogen has an oxidation state of -3. Destruction of the organic portion of the molecule by oxidation frees the nitrogen as ammonia. The Kjeldahl method employing sulfuric acid as the oxidizing agent is standard procedure. A concentrated salt-copper mixture is used to hasten the oxidation of some of the more resistant organic materials. The reaction that occurs may be illustrated by the oxidation of alanine (α-aminopropionic acid). In the reaction, carbon is oxidized to carbon dioxide, while the sulfate ion is reduced to sulfur dioxide. The amino group is released as ammonia but, of course, cannot escape from the acid environment and is held as an ammonium salt.

$$\underset{\text{Alanine}}{CH_3CHNH_2COOH} + 7H_2SO_4 \xrightarrow[\text{cat}]{\Delta} 3CO_2 + 6SO_2 + 8H_2O + NH_4HSO_4 \quad (25.16)$$

The oxidation proceeds rapidly at temperatures slightly above the boiling point of sulfuric acid (340°C). The boiling point of the acid is increased to about 360 to 370°C to enhance the rate of oxidation by addition of sodium or potassium sulfate.

The complete digestion of organic matter is essential if all organic nitrogen is to be released as ammonia. A misinterpretation often occurs as to what conditions exist when digestion is complete, and some explanation seems in order. This can best be done by listing the changes that samples undergo during digestion.

1. Excess water is expelled, leaving concentrated sulfuric acid to attack the organic matter.
2. Copious white fumes form in the flask at the time sulfuric acid reaches its boiling point. Digestion is just beginning at this stage.
3. The mixture turns black, owing to the dehydrating action of the sulfuric acid on the organic matter.
4. Oxidation of carbon occurs. Boiling during this period is characterized by extremely small bubble formation due to the release of carbon dioxide and sulfur dioxide.
5. Complete destruction of organic matter is indicated by a clearing of the sample to a "water-clear" solution.
6. Digestion should be continued for at least 20 min after the samples appear clear to ensure complete destruction of all organic matter.

Once the organic nitrogen has been released as ammonia nitrogen, it may be measured in a manner similar to that described in the discussion of ammonia nitrogen. The excess sulfuric acid must be neutralized and the pH of the sample adjusted to about 11. Under such conditions, the equilibrium shown in Eq. (25.12) is displaced greatly to the right, and ammonia can be distilled with ease. The ammonia nitrogen in the vapor is collected as in the distillation procedure described for ammonia nitrogen, and further analysis is conducted by any of the procedures there described. Calculation of organic nitrogen is made in the same manner as for ammonia nitrogen.

Nitrite Nitrogen

Nitrite nitrogen seldom appears in concentrations greater than 1 mg/L, even in waste-treatment-plant effluents. Its concentration in surface waters and groundwaters is normally much below 0.1 mg/L. For this reason sensitive methods are needed for its measurement. The colorimetric procedure provides this sensitivity. However, another standard procedure is an instrumental approach using ion chromatography as described in Sec. 12.4. Several anions can be determined simultaneously on a single sample by ion chromatography, but the colorimetric procedure is cheaper and preferred when analysis for nitrite alone is desired or nitrite concentrations are especially low.

In the colorimetric procedure a modification of the Griess-Ilosvay diazotization method is used. This employs the use of two organic reagents: sulfanilamide and *N*-(1 naphthyl)-ethylenediamine dihydrochloride. The reactions involved may be represented as follows:

$$\text{Sulfanilamide} + HNO_2 + HCl \rightarrow \text{A diazonium salt} + 2H_2O \tag{25.17}$$

$$\xrightarrow{HCl} \text{A red-colored azo dye} + HCl \tag{25.18}$$

Under acid conditions nitrite ion as nitrous acid reacts with an amino group of sulfanilamide to form a diazonium salt that combines with *N*-(1-naphthyl)-ethylenediamine dihydrochloride to a bright-colored, pinkish-red azo dye. The color produced is directly proportional to the amount of nitrite nitrogen present in the sample, and determination of the amount can be made by comparison with color standards or by means of photometric measurement. Photometric measurement is preferred because standards for visual comparison are not permanent and must be prepared each time analyses are performed.

Nitrate Nitrogen

Obtaining reliable nitrate nitrogen analysis is difficult. In recent years several procedures have been developed. All have limitations with which the analysts must be familiar. The ultraviolet spectrophotometric procedure can be used for initial screening to help decide on initial sample dilutions if needed and what particular method of analysis is most appropriate. This screening procedure and four standard methods are described in the following.

Screening by Ultraviolet Spectrophotometry   Nitrate ions in water absorb ultraviolet radiation with a wavelength of 220 nm. For this reason optical methods of analysis as described in Sec. 12.2 can be used to measure nitrate. The method is quite sensitive but requires a spectrophotometer that can be operated in the ultraviolet range. Any materials that absorb 220 nm radiation will interfere in this analysis. This includes nitrite, hexavalent chromium, and many different organic compounds. Thus, this method is generally used only for screening; caution in the use of this method for other than simple screening is required.

Ion Chromatographic and Capillary Ion Electrophoresis Methods   The ion chromatographic procedure described in Sec. 12.4 is highly useful for analysis with nitrate nitrogen concentrations greater than about 0.2 mg/L. In this way many of the normal interferences in nitrate analysis can be avoided. The procedure also permits the simultaneous analysis for other anions in the sample. The disadvantage of this procedure is the cost for purchase and maintenance of the instrument. The capillary ion electrophoresis method, also described in Sec. 12.4, is similar to the ion chromatographic method in permitting analysis for other ions, but its operating costs are significantly less. Ultraviolet absorption is used for detection.

Nitrate Electrode Method   The nitrate electrode is a liquid membrane electrode as described in Sec. 12.3 and can detect the presence of nitrate nitrogen down to concentrations of about 1 mg/L. The advantage of the electrode method is that once calibrated, analysis for nitrate is rapid. Also, the electrode is readily adapted to continuous monitoring and to automatic process control. Disadvantages are that several common ions such as chloride and bicarbonate cause interference, and the electrode is not sensitive with the lower nitrate concentrations frequently encountered.

Cadmium Reduction Method   The cadmium reduction method offers a highly sensitive procedure for nitrate analysis. A filtered sample with added NH_4Cl–EDTA solution is passed through a specially prepared column containing amalgamated cadmium granules. During passage, nitrate is quantitatively reduced by the cadmium to nitrite. The nitrite produced is determined by the diazotization method previously described. Since nitrite originally present in the sample is also measured, this procedure in effect measures the sum of nitrate plus nitrate nitrogen. In order to determine the nitrate concentration, a separate analysis of nitrite alone is required, and this value is subtracted from the results of the cadmium reduction procedure. Because of the sensitivity of the diazotization method for nitrite, nitrate concentrations as low as 0.01 mg/L can be detected. Samples with nitrate nitrogen concentrations greater than 1 mg/L can also be measured if the samples are diluted prior to passage through the cadmium column. A standard nitrate solution should be passed through the column on occasion to establish that quantitative conversion of nitrate to nitrite is being obtained. If not, then the column should be reactivated or the sample flow rate should be adjusted to obtain quantitative conversion. This procedure has successfully been used in an automated method of analysis.

Total Nitrogen

Total nitrogen in water is taken to mean the sum total of the concentrations of ammonia, nitrite, nitrate, and organic nitrogen. The gaseous forms of nitrogen, including N_2, are not included in the total. The concentration of N_2 in water is generally quite high, roughly 15 to 20 mg/L, due to equilibrium with N_2 in the atmosphere as indicated in Fig. 22.1. However, because it is so nonreactive, it plays little role in water quality considerations associated with nitrogen. Knowledge of the concentration of the individual four forms of nitrogen present in water is generally more important than their "total value," and thus the analytical procedures for individual species will always be of importance. The total can be readily obtained from their summation. Nevertheless, "Standard Methods" now contains analytical procedures that provide measurement of the "total nitrogen" content. This may be useful for obtaining a quick overview of the total nitrogen present, or as a method of determining the concentration of a particular species by difference when concentrations of the other three are known. For example, the normal Kjeldahl digestion procedure for determining organic nitrogen can be avoided if one knows the total nitrogen concentration and the concentration of the three inorganic forms. Organic nitrogen concentration is then equal to the total minus the sum of the three inorganic forms. Of course, the error in the organic nitrogen determination then reflects the errors inherent in all the other analyses combined as discussed in Chap. 10, and whether this error is acceptable needs to be evaluated.

The "Standard Methods" procedure for total nitrogen analysis involves the oxidation of ammonia, nitrite, and organic nitrogen to nitrate, and subsequent reduction of the nitrate to nitrite using the cadmium reduction procedure followed by the sensitive colorimetric diazotization method for nitrite already discussed. For the original oxidation of reduced nitrogen forms to nitrate, an alkaline digestion solution containing persulfate $(S_2O_8^{2-})$ is mixed with the sample and heated under pressure at 100 to 110°C. This results in release of organic nitrogen as ammonia and oxidation of the reduced inorganic nitrogen forms to nitrate, while reducing the persulfate sulfur (VII) to sulfate sulfur (VI). The procedure offers speed and simplicity over the Kjeldahl method and lends itself to automated methods that are available commercially.

25.4 | APPLICATION OF NITROGEN DATA

At the present time, data concerning the nitrogen compounds that exist in drinking water supplies are used largely in connection with disinfection practice. The amount of ammonia nitrogen present in a water determines to a great extent the chlorine needed to obtain free chlorine residuals in breakpoint chlorination and determines to some extent the ratio of monochloramine to dichloramine when combined chlorine residuals are involved. Nitrate and nitrite determinations are important in determining whether water supplies meet U.S. EPA maximum contaminant levels for the control of methemoglobinemia in infants.

Nitrogen data are important in connection with wastewater treatment. By controlling nitrification when not required, aerobic-treatment costs can be kept at a minimum. When nitrogen oxidation is required, then understanding nitrogen changes through nitrification and denitrification is essential for process control. Ammonia and organic nitrogen analyses are important in determining whether sufficient available nitrogen is present for biological treatment. If not, they are needed to calculate the amounts that must be supplied from outside sources, an important economic consideration in many instances.

Where wastewater sludges are sold for their fertilizing value, the nitrogen content of the sludges is a major factor in determining their value for such purposes.

The productivity of natural waters in terms of algal growths is related to the fertilizing matter that gains entrance to them. Nitrogen in its various forms is a major consideration. Also, reduced forms of nitrogen are oxidized in natural waters, thereby affecting the dissolved-oxygen resources. For these reasons nitrogen data are often part of the information needed in stream-pollution-control programs.

PROBLEMS

25.1 In what forms does nitrogen normally occur in natural waters?

25.2 Discuss the significance of nitrogen analysis in water pollution control.

25.3 Analyses for various forms of nitrogen were made at three points in a stream as follows:

Point	Location	DO, mg/L	Nitrogen, mg/L			
			Org-N	NH_3	NO_2^-	NO_3^-
1	Point of waste discharge	7	3	4	0	0
2	5 km downstream	2	1	2	1	3
3	10 km downstream	0	1	0	0	2

On the basis of your knowledge of the nitrogen cycle, explain the processes involved in the relative changes in each nitrogen form, as well as the decrease in total nitrogen in moving downstream from point 1 to point 3.

25.4 Would you expect to find the highest concentration of each of the following in raw domestic wastewater or in the effluent from an aerobic biological waste treatment plant? Why? (a) Organic-N; (b) NH_3-N; (c) NO_2^--N; (d) NO_3^--N.

25.5 What biological processes might be used to reduce the total nitrogen content of a wastewater?

25.6 What limit is placed on the nitrogen content of drinking water and why?

25.7 What is NO_x and what environmental problems are associated with it?

25.8 What is the general source of N_2O in the environment, and what environmental problems can it cause?

25.9 What are the oxidation states of nitrogen, and what environmental problem or problems may be associated with each?

25.10 What environmental problems may be associated with ammonia discharge to a river?

25.11 Why might one wish to suppress nitrification in a biological wastewater treatment plant? Is it always a good idea to do this? Why?

25.12 How can advantage be taken of the need to nitrify a wastewater in order to reduce the oxygen demand for treatment?

25.13 Of what significance is ammonia nitrogen in water disinfection?

25.14 What four analytical methods can be used for ammonia nitrogen measurements?

25.15 How may organic nitrogen concentration in a wastewater be determined?

25.16 Describe the chemistry of the colorimetric method for nitrite determination.

25.17 What methods are available for analysis of nitrate nitrogen?

25.18 Describe two procedures for determining total nitrogen in a water sample.

25.19 Determine the total nitrogen concentration in mg/L of a water sample containing 0.25 mM ammonia, 0.02 mM nitrite, 0.75 mM nitrate, and 1.3 mg/L organic nitrogen.

25.20 Determine the total nitrogen concentration in mg/L of a water sample containing 1.3 mg/L ammonia, 0.25 mg/L nitrite, 15.1 mg/L nitrate, and 0.9 mg/L organic nitrogen.

25.21 The total nitrogen concentration of a water sample is found to be 11.2 mg/L. How might the organic nitrogen content be determined without chemical analysis for the organic nitrogen content of the sample? What is the advantage and disadvantage of your method over direct chemical analysis for organic nitrogen?

REFERENCES

Clesceri, L. S., A. E. Greenberg, and A. D. Eaton (eds.): "Standard Methods for the Examination of Water and Wastewater," 20th ed., American Public Health Association, Washington, DC, 1998.

Skoog, D. A., F. J. Holler, and T. A. Nieman: "Principles of Instrumental Analysis," 5th ed., Saunders College Pubs., Philadelphia, 1998.

Skoog, D. A., D. M. West, and F. J. Holler, "Fundamentals of Analytical Chemistry," 7th ed., Saunders College Pubs., Fort Worth, TX, 1996.

26

Solids

26.1 | GENERAL CONSIDERATIONS

Environmental engineers and scientists are concerned with the measurement of solid matter in a wide variety of liquid and semiliquid materials ranging from potable waters through polluted waters, domestic and industrial wastes, and sludges produced in treatment processes. Strictly speaking, all matter except the water contained in liquid materials is classified as solid matter. The usual definition of *solids,* however, refers to the matter that remains as *residue upon evaporation* and drying at 103 to 105°C. All materials that exert significant vapor pressure at such temperatures are, of course, lost during the evaporation and drying procedures. The residue, or solids, remaining represent only those materials present in a sample that have a negligible vapor pressure at 105°C.

Because of the wide variety of inorganic and organic materials encountered in the analyses for solids, the tests are empirical in character and relatively simple to perform. Gravimetric methods are used in almost all cases, and reference should be made to Sec. 11.3 in preparation for such determinations. Exceptions are the measurement of settleable solids and the estimation of dissolved solids by specific conductance measurements. The major problems in the analyses for solids are concerned with specific tests designed to gain information on the amounts of various kinds of solids present, e.g., dissolved, suspended, volatile, and fixed.

Dissolved and Suspended Solids

The *total solids* in a liquid sample consist of *total dissolved solids* and *total suspended solids.* Total dissolved solids are materials in the water that will pass through a filter with a 2.0-μm or smaller nominal average pore size. The material retained by the filter is the total suspended solids. The amount and nature of dissolved and suspended matter occurring in liquid materials vary greatly. In potable waters, most of the matter is in dissolved form and consists mainly of inorganic salts, small amounts of organic matter, and dissolved gases. The total-dissolved-solids content of potable waters usually ranges from 20 to 1000 mg/L, and, as a rule, hardness increases with total dissolved solids. Unlike the measurement of total solids where sample drying is conducted at 103 to 105°C, total-*dissolved*-solids analysis for water supplies is conducted at 180°C. As discussed in Sec. 11.1, the reason for the higher temperature used in the latter is to remove all mechanically occluded water. Here, organic material is generally very low in concentration, so losses due to the higher drying temperature will be negligible. In all other liquid materials, the amounts of suspended colloidal and larger matter increase with the degree of pollution. Sludges represent an extreme case in which most of the solid matter is suspended, and the dissolved fraction is of minor importance. Determination of the amounts of dissolved and suspended matter is accomplished by making tests upon filtered and unfiltered portions of samples.

Volatile and Fixed Solids

One of the major objectives of performing solids determinations upon domestic wastes, industrial wastes, and sludge samples is to obtain a measure of the amount of organic matter present. This test is accomplished by a combustion procedure in which organic matter is converted to gaseous carbon dioxide and water, and thus volatilized, while the temperature is controlled to prevent decomposition and volatilization of inorganic substances as much as is consistent with complete oxidation of the organic matter. *The loss in weight through such high-temperature oxidation and volatilization is interpreted in terms of organic matter.* One must be careful in this context not to confuse the term *volatile solids* as used here for total organics in a sample with the newer term, *volatile organics,* as discussed in Chap. 34. The latter is a specific set of individual trace organic compounds that can be readily purged from solution by simple aeration at room temperature. Such volatile organics are not measured by the solids test as they are lost from solution during the initial evaporation and drying phase.

The standard procedure for volatile-solids analysis is to conduct ignitions at 550°C. It is about the lowest temperature at which organic matter, particularly carbon residues resulting from pyrolysis of carbohydrates and other organic matter, as shown in Eq. (26.1), can be oxidized at reasonable speed.

$$C_x(H_2O)_y \xrightarrow{\Delta\Delta} xC + yH_2O \uparrow \qquad (26.1)$$

$$C + O_2 \xrightarrow{\Delta} CO_2 \qquad (26.2)$$

Also, at 550°C decomposition of inorganic salts is minimized. Any ammonium compounds not released during drying are volatilized, but most other inorganic salts are relatively stable, with the exception of magnesium carbonate, as shown in the equation

$$MgCO_3 \xrightarrow[350°C]{\Delta} MgO + CO_2 \uparrow \qquad (26.3)$$

In the determination of the volatile content of suspended solids, dissolved inorganic salts are not a consideration because they are removed during the filtration procedure. In sludge analysis, ammonium exits mainly as ammonium bicarbonate, which is completely volatilized during the evaporation and drying procedures, and is not present to interfere in a volatile-solids determination.

$$NH_4HCO_3 \xrightarrow[105°C]{\Delta} NH_3 \uparrow \; + H_2O \uparrow \; + CO_2 \uparrow \tag{26.4}$$

Other unstable inorganic salts in sludge are normally present in such small amounts in relation to the amount of total solids that their influence is usually ignored.

Serious errors can be introduced in volatile-solids determinations by conducting ignitions at uncontrolled temperatures. For this reason it is standard practice to conduct combustions in a muffle furnace where the temperature can be accurately controlled. Calcium carbonate is decomposed at temperatures above 825°C, and since it is a major component of the inorganic salts normally present in samples subjected to volatile-solids analysis, its decomposition could introduce significant errors.

Unless care is exercised in the initial stages, the determination of volatile solids in sludges is often subject to serious error because of physical or mechanical losses due to decrepitation during the procedure. Decrepitation may be eliminated by a preliminary controlled firing of samples with a bunsen burner to destroy all flammable materials before placing the samples in the muffle furnace. If ignitions are properly performed, the weight loss incurred is a reasonably accurate measure of organic matter for most municipal sludges, and the residue remaining as ash or fixed solids is a good representation of inorganic materials present. Further caution is required when using the volatile-solids determination with industrial wastewaters, the organic portion of which may consist of organics such as short-chain fatty acids, alcohols, ketones, aldehydes, and hydrocarbons, all of which are highly volatile and lost during the evaporation and drying process. With such wastewaters, COD and TOC measurements are likely to provide a better measure of total organic content.

Settleable Solids

The term *settleable solids* is applied to solids in suspension that will settle, under quiescent conditions, because of the influence of gravity. Only the coarser suspended solids with a specific gravity sufficiently greater than that of water will settle. Sludges are accumulations of settleable solids. Their measurement is important in practice to determine the need for sedimentation units and the physical behavior of waste streams entering natural bodies of water.

26.2 | ENVIRONMENTAL SIGNIFICANCE OF SOLIDS DETERMINATIONS

The amount of dissolved solids present in water is a consideration in the water's suitability for domestic use. In general, water with a total-solids content of less than 500 mg/L is most desirable for such purposes. A higher total-solids content imparts taste to the water and often has a laxative and sometimes the reverse effect upon people whose bodies are not used to the higher levels. This is important to people who travel and to the transportation companies who are interested in the welfare of their passengers. Water with a high dissolved-solids content tends to stain glassware and has adverse impacts on irrigated crops, plants, and grasses. In many areas, it is impossible to find natural waters with a solids content under 500 mg/L; consequently it is impossible to meet this desired level without some form of treatment. In many instances, treatment to reduce the solids content is not practiced, and residents who regularly use such waters appear to suffer no ill effects. Standards generally recommend an upper limit of 1000 mg/L on potable waters.

The suspended-solids content of wastewaters discharged before or after treatment to natural waters is generally regulated. Suspended solids can float and form unsightly scum layers or sink and cause sediment buildup. For obvious aesthetic reasons, those using surface waters for recreational swimming, boating, or viewing, object to the sight of suspended material from wastewater discharge. The suspended-solids analysis is used to ensure that an important wastewater discharge requirement is being met.

26.3 | DETERMINATION OF SOLIDS IN WATER SUPPLIES

Because of the wide variety of materials subjected to solids determinations, the tests applied vary somewhat, and it is best to discuss them in terms of water, polluted waters, and sludges. Dissolved solids are the major concern in water supplies; therefore, the total-solids determination and the specific conductance measurement are of greatest interest. Suspended-solids tests are seldom made because of the small amounts present. They are more easily evaluated by measurement of turbidity.

Total Solids

The determination of total solids is easily made by evaporation and drying of a measured sample in a tared container. The use of platinum dishes is highly recommended because of the ease with which they can be brought to constant weight before use. Vycor ware is a good substitute. The use of porcelain dishes is to be avoided because of their tendency to change weight.

The determination of volatile solids (organic content) is ordinarily not a consideration in waters intended for domestic use. The results obtained cannot be interpreted in terms of organic matter with any degree of reliability. In cases in which the organic content of a water is important, it is usually best to obtain such information by means of a COD, BOD, or TOC determination.

Specific Conductance

A rapid estimation of the dissolved-solids content of a water supply can be obtained by specific-conductance measurements. Such measurements indicate the capacity of a sample to carry an electric current, which in turn is related to the concentration of ionized substances in the water. Most dissolved inorganic substances in water supplies are in the ionized form and so contribute to the specific conductance. Although this measurement is affected by the nature of the various ions, their relative concentrations, and the ionic strength of the water, such measurements can give a practical estimate of the variations in dissolved mineral content of a given water supply. Also, by the use of an empirical factor, specific conductance can allow a rough estimate to be made of the dissolved mineral content of water samples. The use of specific conductance and the factors affecting this measurement are discussed in Sec. 3.9.

Dissolved and Suspended Matter

In cases in which turbidity measurements are not deemed adequate to provide the necessary information, the suspended solids may be determined by filtration through a glass-fiber filter. Because of the small amounts of suspended solids usually present in water supplies, the determination is subject to considerable error unless an abnormally large sample is filtered. Another technique is to filter a sample of water through filter paper and determine total solids in the filtrate. The difference between total solids in unfiltered and filtered samples is a measure of the suspended solids present.

26.4 | DETERMINATIONS APPLICABLE TO POLLUTED WATERS AND DOMESTIC WASTEWATERS

The settleable- and suspended-solids determinations are of greatest value in assessing the strength of domestic wastes and lightly polluted waters.

Settleable Solids

The determination of settleable solids can be important in the analysis of wastewaters. The test is ordinarily conducted in an Imhoff cone (see Fig. 26.1), allowing 1 h settling time under quiescent conditions. Samples should be adjusted to near room temperature and the test conducted in a location where direct sunlight does not interfere with normal settlement of the solids. Results are measured and reported in terms of milliliters per liter.

Total Solids

Total-solids determinations are ordinarily of little value in the analysis of polluted waters and domestic wastewaters because they are difficult to interpret with any degree of accuracy. In most instances the dissolved solids present in the original water represent such a large and variable percentage of the total amount found in the polluted wastewaters that it is impossible to evaluate the data, unless the amount of solids originally present in the carriage water is known. The test was originally devised as a means of evaluating the amount of polluting matter present in wastewaters. Since the BOD and COD tests are capable of evaluating the strength of such materials much more exactly, there is little justification for running total-solids tests for such purposes.

Figure 26.1
Imhoff cones used to measure settleable solids.

Some waste treatment processes, particularly those involving sedimentation, are adversely affected by radical changes in the density of wastewater. In coastal cities, seawater often gains access to sewer systems at times of high tide, and in industrial cities, intermittent discharge of highly mineralized waste may occur. Both can cause significant changes in density. The total-solids test can be used to good advantage to detect such changes, although such contamination can be detected more easily by the chloride determination or by specific-conductance measurements.

Suspended Solids

The suspended-solids determination is important in the analysis of polluted waters. It is one of the major parameters used to evaluate the strength of domestic wastewaters and to determine the efficiency of treatment units. In stream-pollution-control work, all suspended solids are considered to be settleable solids, as time is not a limiting factor. Deposition is expected to occur through biological and chemical flocculation; therefore, measurement of suspended solids is considered fully as significant as BOD.

The suspended-solids determination is subject to considerable error if proper precautions are not taken, as discussed in Secs. 11.1 and 11.3. Usually, the sample size is limited to 50 mL or less because of difficulties encountered in filtration of larger samples. The weight of solids removed seldom exceeds 20 mg and is often less than 10 mg. Small errors in weighing can be quite significant. Sufficient sample should be filtered, if possible, to yield an increase in weight of about 10 mg. This often requires the filtration of 500 mL or more of samples of biologically treated wastewaters or lightly polluted waters.

The volatile content of suspended solids can be determined by direct ignition of the glass-fiber filter in the muffle furnace because the amount of solids involved is too small for decrepitation to occur. Suspended solids often contain as much as 80 percent of volatile matter. The fixed solids remaining frequently weigh less than 2 mg. The filter is not destroyed by this procedure, provided that certain precautions are taken. A temperature of 600°C is near the melting point of the filter, and for this reason combustion no higher than the recommended temperature of 550°C should be used. A slight weight loss occurs in the filter during drying and combustion, and for this reason a blank filter should be subjected to the same treatment as the sample filter so that an appropriate correction can be made.

Suspended solids are reported in terms of milligrams per liter, and volatile suspended solids are normally reported in terms of percent of the suspended solids.

26.5 | DETERMINATIONS APPLICABLE TO INDUSTRIAL WASTEWATERS

Industrial wastewaters include such a wide variety of materials that analyses for exploratory purposes should include all determinations that can possibly provide significant information. For this reason all the solids tests commonly applied to domestic wastewaters are important. The settleable-solids test serves as the principal basis of determining whether primary sedimentation facilities are required for treatment. In addition, the total-solids determination has special significance. Many industrial wastes contain unusual amounts of dissolved inorganic salts, and their presence is easily detected by the total-solids test. Their concentration and nature are factors in determining the susceptibility of wastes to anaerobic treatment. Before the development of the COD and TOC tests, the volatile content of total solids was used extensively to measure the amount of organic matter present. It was very helpful in assessing the amount of biologically inert organic matter, such as lignin in the case of wood-pulping waste liquors. However, because of the high concentrations of organics that are lost upon sample evaporation with many industrial wastewaters, as discussed in Sec. 26.1, the TOC and COD tests tend to be more reliable indicators of the total organic content of industrial wastewaters.

26.6 | DETERMINATION OF SOLIDS IN SLUDGES

The total- and volatile-solids determinations are of value in the analysis of raw and digested sludges. Both are subject to some error because of the loss of volatile organic compounds during the drying process. This factor is not particularly significant in samples of raw or well-digested sludges, but it is important when partially digested sludges containing appreciable amounts of volatile acids are being analyzed. This loss often leads to faulty interpretation of volatile-solids destruction.

Because it is impossible to pipet samples of raw and digested sludges, it is common practice to weigh the samples in previously tared dishes. It is customary to use small porcelain evaporating dishes about 7 cm in diameter. It is important that the dishes be previously ignited to constant weight in order to provide reliable results for volatile solids. Because of the rather nonuniform character of sludges, it is necessary to use relatively large samples of 25 to 50 g, unless some method of homogenization has been employed. As a result, considerable amounts of residue are obtained upon evaporation of the samples. It is usually necessary to dry the samples at 103°C for several hours to be sure that all the moisture has had a chance to escape. Most analysts prefer to dry the samples overnight.

The volatile-solids content of sludges is an important determination. Accurate results can be obtained, provided that care is taken to control decrepitation, as described in Sec. 26.1.

The steps involved in the collection of data for total- and volatile-solids determinations are not always understood. A typical set of data and calculations are as follows:

Wt of dish on analytical balance		30.160 g
Wt of dish on trip balance	30.5 g	
Wt of dish + sample	70.8	
Wt of sample	40.3	
Wt of dish + dry sludge solids		32.780
Wt of sludge solids		2.620

Percent solids in sludge $\dfrac{2.62}{40.3} \times 100 = 6.50\%$

Wt of dish + ash		30.720
Wt of volatile matter		2.06

Percent volatile matter $\dfrac{2.06}{2.62} \times 100 = 78.7\%$

It is unnecessary to report the percent of fixed solids, as this can be readily determined from the percent of volatile solids.

The measurement of solids in activated sludges is a special case. Because of the relatively low concentrations usually involved, serious errors can be introduced by measuring total solids, which would also include dissolved solids. It is customary to measure activated sludge by procedures used to determine suspended solids. This allows the dissolved solids to pass into the filtrate. A number of modifications are used to measure activated-sludge solids, but if volatile solids are desired, as is often the case, the test is best performed with the glass-fiber-filter technique. The sample should be limited to 5 or possibly 10 mL because of potential filtration problems.

26.7 | APPLICATIONS OF SOLIDS DATA IN ENVIRONMENTAL ENGINEERING PRACTICE

In the realm of public and industrial water supplies, the total-dissolved-solids determination is the only one of importance. It is used to determine the suitability of potential supplies for development. In cases in which water softening is needed, the type of softening procedure used may be dictated by the dissolved-solids content, since precipitation methods decrease, and exchange methods increase, the solids. Corrosion control is frequently accomplished by the production of stabilized waters through pH adjustment. The pH at stabilization depends to some extent upon the total solids present as well as the alkalinity and temperature.

The settleable-solids determination has two significant applications. First, it is used extensively in the analysis of industrial wastes to determine the need for and design of primary settling tanks in plants employing biological treatment processes. The test is also widely used in waste-treatment-plant operation to determine the efficiency of sedimentation units. It is fully as important in the operation of large as well as small treatment plants.

The suspended- and volatile-suspended-solids determinations are used to evaluate the strength of domestic and industrial wastes. The tests are particularly valuable in determining the amount of suspended solids remaining after settleable solids have been removed in primary settling units, for the purpose of determining the loading of remaining materials on secondary biological treatment units. In the larger treatment plants, suspended-solids determinations are used routinely as a measure of the effectiveness of treatment units. From the viewpoint of stream pollution control, the removal of suspended solids is usually as important as BOD removal. Both suspended- and volatile-suspended-solids determinations are used to control the biological solids in the activated-sludge process.

The total- and volatile-solids tests are the only solids determinations normally applied to sludges. They are indispensable in the design and operation of sludge-digestion, vacuum-filter, and incineration units.

PROBLEMS

26.1   What is it that differentiates between total solids, total dissolved solids, and total suspended solids in liquid samples?

26.2   Why are each of the following solids analyses of interest in water quality control?

    (*a*)   Total dissolved solids in municipal water supplies

    (*b*)   Total and volatile suspended solids in domestic wastewater

    (*c*)   Total and volatile solids in sludge

    (*d*)   Settleable solids in domestic wastewater

26.3   What significant information is furnished by the determination of volatile solids?

26.4 (*a*) Why is 103 to 105°C the drying temperature generally used for total-solids analyses?

(*b*) Under what conditions and why is 180°C sometimes used as the drying temperature?

26.5 (*a*) Why was 550°C chosen as the combustion temperature in the volatile-solids analysis?

(*b*) What possible problems would result if either a lower or a higher temperature were used?

26.6 What problems result when attempting to determine organic content using the total-volatile-solids analysis with industrial wastewaters containing short-chain organic acids, alcohols, or alkanes?

26.7 What limitation does the volatile-solids analysis have for analysis of the organic content of partially digested sludges?

26.8 What precaution must be taken when using glass-fiber filers for volatile-solids analyses?

26.9 What information of value is provided by the Imhoff cone test?

26.10 What caution is needed in determining the volatile-solids content of sludge prior to combustion in a muffle furnace, and why?

26.11 What precautions must be taken in the determination of volatile solids? Why?

26.12 Would you expect the analytical results to be higher than, lower than, or the same as the true value under the following conditions, and why?

(*a*) Weighing a warm crucible

(*b*) Estimating organic content by combustion at 550°C of a sludge sample with a high magnesium carbonate content

(*c*) Estimating the organic content by volatile-solids analysis of a sample containing a large quantity of organic materials having a high vapor pressure

(*d*) Estimating the organic content of a sample by combustion at 800°C rather than at 550°C

26.13 A domestic wastewater contains 350 mg/L of suspended solids. Primary sedimentation facilities remove 65 percent. Approximately how many cubic meters of sludge containing 5.0 percent solids will be produced per million cubic meters of wastewater settled?

REFERENCE

Clesceri, L. S., A. E. Greenberg, and A. D. Eaton (eds.): "Standard Methods for the Examination of Water and Wastewater," 20th ed., American Public Health Association, Washington, DC, 1998.

Iron and Manganese

27.1 | GENERAL CONSIDERATIONS

Both iron and manganese create serious problems in public water supplies. The problems are most extensive and critical with underground waters, but difficulties are encountered at certain seasons of the year in waters drawn from some rivers and some impounded surface supplies. Why some underground supplies are relatively free of iron and manganese and others contain so much has always been something of an enigma that defied explanations when viewed solely from the viewpoint of inorganic chemistry. Biochemical changes, or more exactly, changes in environmental conditions brought about by biological reactions, are major considerations. Since both manganese and iron are present in insoluble forms in significant amounts in nearly all soils, any explanation of how appreciable amounts can gain entrance to water flowing through or coming in contact with the soil must consider how the iron and manganese are converted to soluble forms.

Iron exists in soils and minerals mainly as insoluble ferric oxides and iron sulfide (pyrite). It occurs in some areas also as ferrous carbonate (siderite), which is very slightly soluble. Since groundwaters usually contain significant amounts of carbon dioxide (see Sec. 17.2), appreciable amounts of ferrous carbonate may be dissolved by the reaction shown in the equation

$$FeCO_3(s) + CO_2 + H_2O \rightarrow Fe^{2+} + 2HCO_3^- \qquad (27.1)$$

in the same manner that calcium and magnesium carbonates are dissolved. However, iron problems are prevalent where it is present in the soil as insoluble ferric compounds. Dissolution of measurable amounts of iron from such solids does not occur, even in the presence of appreciable amounts of carbon dioxide, as long as dissolved oxygen is present. Under reducing (anaerobic) conditions, however, the ferric iron is reduced to ferrous iron, and solution occurs without difficulty.

Manganese exists in the soil principally as manganese dioxide, which is very insoluble in water containing carbon dioxide. Under reducing (anaerobic) conditions, the manganese in the dioxide form is reduced from an oxidation state of IV to II, and solution occurs, as with ferric oxides.

Confirmation that iron and manganese gain entrance to water supplies through changes produced in environmental conditions as a result of biological reactions has stemmed from five lines of evidence:

1. Groundwaters that contain appreciable amounts of iron or manganese or both are always devoid of dissolved oxygen and are high in carbon dioxide content. The iron and manganese are present as Fe(II) and Mn(II). The high carbon dioxide content indicates that bacterial oxidation of organic matter has been extensive, and the absence of dissolved oxygen shows that anaerobic conditions were developed.

2. Wells producing good-quality water, low in iron and manganese, for many years have been known to produce poor-quality water when organic wastes have been discharged on the soil around or near the well, thereby creating anaerobic conditions in the soil.

3. The iron and manganese problem in impounded surface supplies has been correlated with reservoirs that stratify, but occurs only in those in which anaerobic conditions develop in the hypolimnion. The soluble iron and manganese released from the bottom muds are contained in the waters of the hypolimnion until the fall overturn occurs. At that time they are distributed throughout the reservoir and cause trouble in the water supply until sufficient time has elapsed for oxidation and sedimentation to occur under natural conditions.

4. It has been shown, on the basis of thermodynamic considerations, that Mn(IV) and Fe(III) are the only stable oxidation states for iron and manganese in oxygen-containing waters.[1] Thus, these forms can be reduced to the soluble Mn(II) and Fe(II) only under highly anaerobic reducing conditions.

5. Studies have clearly demonstrated that dissimilatory iron-reducing bacteria can use Fe(III) and Mn(IV) as electron acceptors for energy metabolism under anaerobic conditions, leading to formation of the reduced forms, Fe(II) and Mn(II).[2] Thus, bacteria may not only create the reducing conditions required for iron and manganese reductions, but may also benefit from direct involvement in the reductions.

When oxygen-bearing water is injected into the ground for recharge of the groundwater aquifer, it is sometimes noted that the soluble iron content of the water increases, an observation which seems to contradict the previously stated need for anaerobic conditions. The explanation is that the oxygen is consumed through the

[1]J. J. Morgan and W. Stumm, The Role of Multivalent Metal Oxides in Limnological Transformations, as Exemplified by Iron and Manganese. *Proc. Intern. Conf. on Water Pollution Research,* Pergamon Press, New York, 1964.

[2]D. R. Lovely, E. J. P. Phillips, and D. J. Lonergan, Hydrogen and Formate Oxidation Coupled to Dissimilatory Reduction of Iron or Manganese by *Alteromonas putrefaciens, Appl. Environ. Microbiol.,* **55:** 700–706 (1989).

oxidation of insoluble pyrite (FeS_2), leading to anaerobic conditions and the formation of soluble iron sulfate.

$$2FeS_2 + 7O_2 + 2H_2O \rightarrow 2Fe^{2+} + 4SO_4^{2-} + 4H^+ \qquad (27.2)$$

In summary, the evidence is clear that the development of anaerobic conditions is essential for appreciable amounts of iron and manganese to gain entrance to a water supply. Only under anaerobic conditions are the soluble forms of iron, Fe(II), and manganese, Mn(II), thermodynamically stable.

27.2 | ENVIRONMENTAL SIGNIFICANCE OF IRON AND MANGANESE

As far as is known, humans suffer no harmful effects from drinking waters containing iron and manganese. Such waters, when exposed to the air so that oxygen can enter, become turbid and highly unacceptable from an aesthetic viewpoint, owing to the oxidation of iron and manganese to Fe(III) and Mn(IV) states, which can form colloidal precipitates. The rates of oxidation are not rapid, and thus reduced forms can persist for some time in aerated waters. This is especially true when the pH is below 6 with iron oxidation and below 9 with manganese oxidation. In addition, iron and manganese can form stable complexes with humic substances in water that can be even more resistant to oxidation than the inorganic species alone. Oxidation rates may be increased by the presence of certain inorganic catalysts or through the action of microorganisms. Both iron and manganese interfere with laundering operations, impart objectionable stains to plumbing fixtures, and cause difficulties in distribution systems by supporting growths of iron bacteria. Iron also imparts a taste to water which is detectable at very low concentrations. For these reasons public water supplies ought not to contain more than 0.3 mg/L of iron or 0.05mg/L of manganese, the U.S. EPA secondary standards.

27.3 | METHODS OF DETERMINING IRON

A great many methods of determining iron have been developed. Precipitation methods are commonly used where quantities are relatively large, as in some industrial wastes. However, in water supplies the amounts present are normally so small that colorimetric procedures are more satisfactory. The colorimetric procedures have a major advantage in that they are usually highly specific for the ion involved, and a minimum of pretreatment is required. Iron may also be determined by atomic absorption spectroscopy or inductively coupled plasma spectroscopy (Sec. 12.2).

Phenanthroline Method

The phenanthroline method is the preferred standard procedure for the measurement of iron in water at the present time, except when phosphate or heavy-metal interferences are present. The method depends upon the fact that 1,10-phenanthroline combines with Fe^{2+} to form a complex ion that is orange-red in color. The color produced conforms to Beer's law and is readily measured by visual or photometric comparison.

Water samples subjected to analysis have usually been exposed to the atmosphere; consequently, some oxidation of Fe(II) to Fe(III) and precipitation of ferric hydroxide may have occurred. It is necessary to make sure that all the iron is in a soluble condition. This is done by treating a portion of the sample with hydrochloric acid to dissolve the ferric hydroxide:

$$Fe(OH)_3(s) + 3H^+ \rightarrow Fe^{3+} + 3H_2O \qquad (27.3)$$

Since the reagent 1,10-phenanthroline is specific for measuring Fe(II), all iron in the form of Fe(III) must be reduced to the ferrous condition. This is most readily accomplished by using hydroxylamine as the reducing agent. The reaction involved may be represented as follows:

$$4Fe(III) + 2NH_2OH \rightarrow 4Fe(II) + N_2O + H_2O + 4H^+ \qquad (27.4)$$

Three molecules of 1,10-phenanthroline are required to sequester or form a complex ion with each Fe^{2+}. The reaction may be represented as shown in the equation:

1,10-Phenanthroline   Orange-red complex

By proper modifications of the test procedure, measurements of total, dissolved, and suspended iron can be made. These considerations are not normal, however, and, whenever they are, special precautions must be taken in sampling and transportation of samples to ensure that no changes occur before analyses are performed. Because of the possible errors that may result, it is best that the analyst assume full responsibility for sampling as well as for analysis. When interfering materials such as phosphates and heavy metals are present, satisfactory results can be obtained by use of a procedure that involves acidification of the sample with HCl and extraction of the iron into diisopropyl-ether prior to the addition of the phenanthroline solution.

27.4 | METHODS OF DETERMINING MANGANESE

Manganese is principally of concern in water supplies. Its concentration seldom exceeds a few milligrams per liter; therefore, colorimetric methods are most applicable. The colorimetric method recommended in "Standard Methods" depends upon oxidation of the manganese from its lower oxidation state to VII, where it forms the highly colored permanganate ion. The color produced is directly proportional to the concentration of manganese present over a considerable range of concentration in accordance with Beer's law, and it is easily measured visually or by photometric means. Chloride interferes because of its reducing action in an acid medium, and so provisions must be made to overcome chloride influence. Most other reducing agents are rendered inactive by the strong oxidizing agents used to form the permanganate ion. Manganese may also be determined by atomic absorption spectroscopy or inductively coupled plasma spectroscopy (Sec. 12.2).

Persulfate Method

The persulfate method is suited for routine determinations of manganese because pretreatment of samples is not needed to overcome chloride interference. Ammonium persulfate is commonly used as the oxidizing agent. It is subject to deterioration during prolonged storage; for this reason, it is always good practice where samples are not run routinely to include a standard sample with each set of samples to verify the potency of the persulfate used.

Chloride interference is overcome by adding Hg^{2+} to form the neutral $HgCl_2$ complex. Since the stability constant of $HgCl_2$ is about 1.7×10^{13}, the concentration of chloride ion is decreased to such a low level that it cannot reduce the permanganate ions formed.

The oxidation of manganese in lower valences to permanganate by persulfate requires the presence of Ag^+ as a catalyst. The reaction involved in the oxidation may be represented as follows:

$$2Mn^{2+} + 5S_2O_8^{2-} + 8H_2O \xrightarrow{Ag^+} 2MnO_4^- + 10SO_4^{2-} + 16H^+ \qquad (27.6)$$

The color produced by the permanganate ion is stable for several hours, provided that a good-quality distilled water is used for dilution purposes and reasonable care is taken to protect the sample from contamination by dust of the atmosphere.

27.5 | APPLICATIONS OF IRON AND MANGANESE DATA

In explorations for new water supplies, particularly from underground sources, iron and manganese determinations are an important consideration. Supplies may be rejected on this basis alone. When supplies containing amounts in excess of 0.3 mg/L iron or 0.05 mg/L manganese are developed, the engineer must decide whether treatment is justified and, if so, the best method of treatment. The ratio of iron to manganese is a factor that determines the type of treatment used, as well as the amount of organic matter present in the water. The efficiency of treatment units is determined by routine tests for iron and manganese. They are also used to aid in the solution of problems in distribution systems where iron-oxidizing bacteria are troublesome.

Corrosion of cast-iron and steel pipelines often produces "red-water" troubles in distribution systems. The iron determination is helpful in assessing the extent of corrosion and aiding in the solution of these problems. Research on corrosion and methods of corrosion control require the use of many types of tests to evaluate the extent of metal loss. The iron determination is one of them.

PROBLEMS

27.1   What is the environmental significance of iron and manganese in water supplies?

27.2   Discuss the different oxidation states of iron and manganese occurring in natural water supplies, and discuss the conditions under which each form is stable.

27.3   Discuss briefly how iron and manganese get into underground water supplies.

27.4   What analytical methods are generally used for measuring the concentration of iron in water supplies?

27.5   What analytical methods are generally used for measuring the concentration of manganese in water supplies?

27.6   In what form must iron exist in order to be measured by the phenanthroline method, and how is iron converted to this form?

27.7   In the colorimetric determination of manganese:

(*a*)   What is the function of ammonium persulfate?

(*b*)   What is the function of Ag^+?

(*c*)   What is the function of $HgSO_4$?

(*d*)   In what oxidation state must the manganese be for colorimetric measurement?

Fluoride

28.1 | GENERAL CONSIDERATIONS

Water engineers have at least a dual interest in the determination of fluoride. They are responsible for designing and operating units for the removal of fluoride from water supplies that contain excessive amounts and, on the other hand, it becomes their responsibility to supervise the addition of fluoride to optimum levels in water supplies that are deemed to be deficient in fluoride by local health agencies. In some areas, industrial contamination of the atmosphere and vegetation by fluoride has been a serious problem. Control methods have had to be employed to protect cattle and other herbivorous animals from damage to bones and teeth.

Significance of High Fluoride Levels in Water Supplies

A disfigurement in the teeth of humans known as *mottled enamel* (dental fluorosis) has been recognized for many years. United States immigration authorities, at an early date, noticed that people arriving from certain areas of Europe were severely afflicted, whereas people from other areas showed little or no evidence of mottling. This led dental authorities to believe that the diseased condition was due to a local factor. Shortly after this information became known, reports of mottled enamel among people native to the United States began to appear. These cases came largely from cities in the Great Plains and Rocky Mountain states, but no real clue was offered to explain the cause of the defective teeth until about 20 years later.

Substantial evidence that fluoride is the cause of mottled enamel was obtained by Churchill of the Aluminum Co. of America in 1930.[1] The city of Bauxite, Arkansas, was one of the communities that had reported a high incidence of mottled enamel among its people. Churchill, through spectrographic analysis, found appreciable amounts of fluoride ion in the Bauxite water supply. In collaboration with McKay, a dentist of Colorado Springs, Colorado, he studied waters from five areas

---

[1] H. V. Churchill, *Ind. Eng. Chem.,* **23**: 996 (1931).

where mottling was endemic and from 40 areas where it was not a problem. From these studies it was concluded that excessive fluoride levels in drinking water are the cause of mottled enamel. Their data showed that mottling did not appear unless the fluoride-ion concentration was in excess of 1.0 mg/L and that the degree and severity of mottling increased as the fluoride level rose.

As soon as excessive amounts of fluoride in water supplies had been established as the cause of dental fluorosis, research on methods of removal were initiated. The passing of water through various types of defluoridation media such as tricalcium phosphate, bone char, bone meal, and activated alumina was found to accomplish fluoride removal by a combination of ion exchange and sorption. Fluoride can also be removed during lime softening through coprecipitation with magnesium hydroxide, or by alum coagulation, but at quite high alum doses. Such methods are used in practice, but the number of plants providing facilities for fluoride removal is small.

Significance of Low Fluoride Levels in Water Supplies

As a result of the great interest focused on the fluoride content of public water supplies in relation to the dental fluorosis problem, much information became available on fluoride. It was natural that the dental profession would use this information to determine whether fluoride was correlated with other dental disease.

In 1938, Dean[2] presented information that demonstrated that dental caries is less prevalent when mottled enamel occurs. This led to extensive correlation studies on dental caries versus fluoride levels in public waters at many places in the United States. The results obtained, as summarized by Dean,[3] are presented in Fig. 28.1. From this information, a dental caries–fluoride hypothesis evolved: Approximately 1 mg/L of fluoride ion is desirable in public waters for optimal dental health. At decreasing levels, dental caries becomes a serious problem, and at increasing levels, dental fluorosis becomes a problem.

The dental caries–fluoride hypothesis has served as the basis for programs of supplementing public water supplies having low fluoride levels with fluoride to bring the concentration up to about 1 mg/L. Because there was some question about the physiological effects of supplemental fluoride, as well as its efficacy as a substitute for "natural" fluoride for controlling dental caries, extensive 10-year pilot studies were conducted at Newburgh, New York; Grand Rapids, Michigan; and several other cities. The results of the investigations have been unanimous in demonstrating the safety of supplemental fluoridation and showing that added fluoride is as effective in controlling dental caries as so-called "natural" fluoride. The program for fluoridation of public water supplies deficient in natural fluoride has been sponsored by many organizations interested in the public health, including the American Dental and Medical Associations. There has been considerable opposition to the program from several quarters, but the program has advanced steadily. By 1970 over 4500 community water supplies in the United States serving over 84 million people

---

[2]H. T. Dean, *Public Health Repts. (U.S.)*, **53:** 1443 (1938).

[3]H. T. Dean, *J. Amer. Water Works Assoc.*, **35:** 1161 (1943).

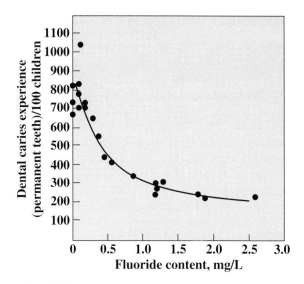

Figure 28.1
Relationship between dental caries and fluoride level in drinking water. *(After Dean.)*

were being supplemented with fluoride.[4] Control of additions is based upon the fluoride determination.

28.2 | CHEMISTRY OF FLUORINE AND ITS COMPOUNDS

Fluorine is among the most active elements known and is not used in the elemental form in water treatment. It forms simple fluoride compounds and many complex ions. The principal forms in which fluoride is added to public water supplies are

NaF	Na_2SiF_6 (sodium silicofluoride)
CaF_2	H_2SiF_6 (hydrofluosilicic acid)
HF	$(NH_4)_2SiF_6$ (ammonium silicofluoride)

All the compounds and complex ions containing fluorine dissociate to yield fluoride ion. At the concentrations of about 1 mg/L involved in water treatment practice, it is generally considered that hydrolysis of the fluosilicate ion is essentially complete, as shown in the equation

$$SiF_6^{2-} + 3H_2O \rightleftharpoons 6F^- + 6H^+ + SiO_3^{2-} \tag{28.1}$$

On this basis, the fluoride in silicofluorides can be determined by any method that is sensitive to fluoride ion.

---

[4]F. J. Maier, 25 Years of Fluoridation, *J. Amer. Water Works Assoc.,* **62:** 3 (Jan. 1970).

28.3 | METHODS OF DETERMINING FLUORIDE

There are four standard methods for determination of fluoride at the present time: the electrode method, two colorimetric procedures, and ion chromatography. The electrode method is the simplest but does require use of an expanded-scale pH meter and a special electrode. The principles behind and limitations of the electrode method are described under Crystalline Membrane Electrode in Sec. 12.3. The ion chromatographic technique, described in Sec. 12.4, permits analysis for fluoride and several other anions simultaneously. However, when fluoride is to be determined by this method, the analysts must be certain that the column and eluants used can effectively separate fluoride from possible interfering anions. All four methods are subject to interferences from other ions. It is often necessary to separate the fluorides from the interfering ions before making the colorimetric test. The separation is accomplished by a distillation procedure such as that used in the determination of ammonia nitrogen. In the case of fluoride, however, the distillation is performed from acidified solutions. Under these conditions, all fluoride ions are converted to poorly ionized hydrogen fluoride, according to the equation

$$F^- + H^+ \xrightarrow{\Delta} HF \uparrow \tag{28.2}$$

The hydrogen fluoride is evolved with the steam and held in the condensate.

Samples of water that do not contain significant amounts of interfering ions, and distillates obtained from the purification procedure may be analyzed by any one of the four methods. The colorimetric methods involve the bleaching of a preformed color by the fluoride ion. The preformed color is the result of the action between zirconium ion and either alizarin dye or SPADNS dye, depending upon the method used. The color produced is commonly referred to as a "lake," and the intensity of the color produced is reduced if the amount of zirconium present is decreased. Fluoride ion combines with zirconium ion to form a stable complex ion, ZrF_6^{2-}, and the intensity of the color lake decreases accordingly. The reaction involved may be represented as follows when the alizarin dye is used:

$$\underset{\text{Reddish color}}{(Zr-\text{alizarin lake})} + 6F^- \rightarrow \text{alizarin} + \underset{\text{Yellow}}{ZrF_6^{2-}} \tag{28.3}$$

The bleaching action is a function of the fluoride-ion concentration and is directly proportional to it. Thus, Beer's law is satisfied in an inverse manner. Comparisons may be made visually or photometrically. The bleaching action of fluoride is slow in the alizarin dye procedure, and a 1-h contact period is recommended before comparisons are made. Because temperature and time are important variables, it is necessary that standards be prepared for making comparisons each time analyses are to be made. When photometric measurements are used, care must be exercised to keep contact time and temperature the same as employed in developing the calibration curve. Good practice requires that at least one standard be included with samples each time photometric measurements are made.

The SPADNS method is presently the preferred colorimetric procedure. It overcomes the strict time limitations of the alizarin approach and so is particularly useful when photometric methods of analysis are preferred. Here, use is made of the almost instantaneous reaction rate between fluoride and zirconium ions which result when more acid is added to the reaction mixture. By this procedure, readings can be made immediately after the sample and reagents are mixed. Under the somewhat different conditions of this modification, less interference is encountered when SPADNS dye rather than alizarin dye is used. The alizarin dye procedure is sometimes used for automated colorimetric methods of analysis when large numbers of samples are to be analyzed.

28.4 | APPLICATION OF FLUORIDE DATA

Because of the public health significance of fluoride in water supplies intended for human use, determination of fluoride has become important. In situations where fluoride is added to provide an optimum level for the control of dental caries, it is necessary to know the amount of natural fluoride present so that proper amounts of supplemental fluoride can be added. Wherever supplementation is practiced, it is necessary to maintain surveillance on the finished water to be sure that proper amounts of chemicals are being fed. The usual practice is to collect samples from the distribution system as well as at the treatment plant.

In areas where natural fluoride exceeds the U.S. EPA maximum contaminant levels, the fluoride content of a water may determine the suitability of a supply for development. The current MCL for fluoride is 4 mg/L to protect against crippling skeletal fluorosis. Also, a secondary standard (not enforceable) of 2 mg/L has been set to protect against objectionable dental fluorosis. The size and design of treatment units will depend upon the level of fluoride present in the water. Fluoride determinations, of course, serve as the basis for determining when removal units require regeneration.

PROBLEMS

28.1 Describe three situations in water works practice which would require fluoride determinations.

28.2 (*a*) What problems result with excessive fluoride concentration in water supplies?

(*b*) What problem can result when fluoride is absent?

28.3 What analytical methods are available for fluoride analysis?

28.4 How can excessive fluoride be removed from drinking water supplies?

28.5 What is the purpose of sample acidification prior to distillation to separate fluoride from interfering ions?

28.6 What precautions must be exercised in determining fluoride by the electrode method and by ion chromatography?

28.7 In what way does the color formation in the colorimetric procedures for fluoride differ from that with most other colorimetric analyses?

CHAPTER 29

Sulfate

29.1 | GENERAL CONSIDERATIONS

The sulfate ion is one of the major anions occurring in natural waters. It is of importance in public water supplies because of its cathartic effect upon humans when it is present in excessive amounts. For this reason the U.S. EPA's secondary standard for sulfate is 250 mg/L in waters intended for human consumption. Sulfate is important in both public and industrial water supplies because of the tendency of waters containing appreciable amounts to form hard scales in boilers and heat exchangers.

Sulfate is of considerable concern because it is indirectly responsible for two serious problems often associated with the handling and treatment of wastewaters. These are odor and sewer-corrosion problems resulting from the reduction of sulfate to hydrogen sulfide under anaerobic conditions, as shown in the following equations:

$$SO_4^{2-} + \text{organic matter} \xrightarrow[\text{bacteria}]{\text{anaerobic}} S^{2-} + H_2O + CO_2 \tag{29.1}$$

$$S^{2-} + H^+ \rightleftharpoons HS^- \tag{29.2}$$

$$HS^- + H^+ \rightleftharpoons H_2S \tag{29.3}$$

A knowledge of the sulfur cycle, as represented in Fig. 29.1, is essential to an understanding of the transformations that occur.

Odor Problems

In the absence of dissolved oxygen and nitrate, sulfate serves as an electron acceptor for biochemical oxidations catalyzed by anaerobic bacteria. Under anaerobic conditions, the sulfate ion is reduced to sulfide ion, which establishes an equilibrium with hydrogen ion to form hydrogen sulfide in accordance with its primary ionization constants $K_{A1} = 9.1 \times 10^{-8}$ and $K_{A2} = 1.3 \times 10^{-13}$. The relationships existing between H_2S, HS^-, and S^{2-} at various pH levels in a 10^{-3} molar solution are shown in Fig. 29.2. At pH values of 9 and above, most of the reduced sulfur ex-

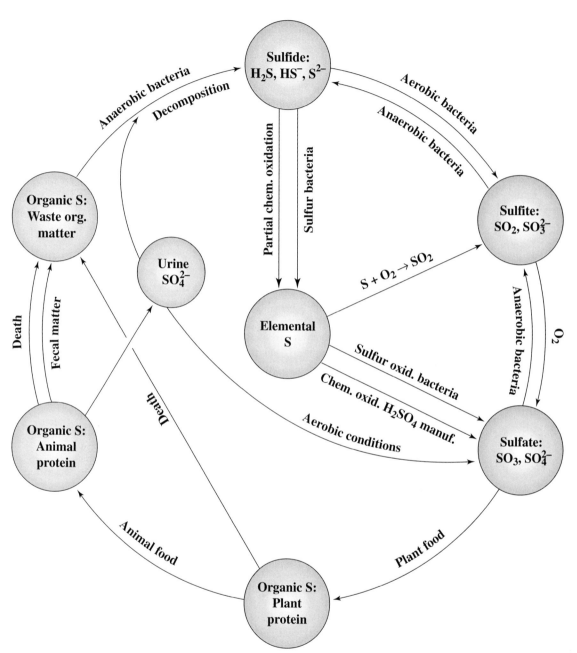

Figure 29.1
The sulfur cycle.

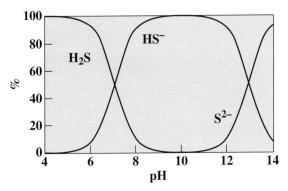

Figure 29.2
Effect of pH on relative distribution of sulfide species in
water.

ists in solution as HS⁻ an S²⁻ ions, and the amount of free H_2S is so small that its
partial pressure is very low and odor problems are minimal. At pH levels below 9,
the equilibrium shifts toward the formation of un-ionized H_2S and is about 50 percent complete at pH 7. Under such conditions the partial pressure of hydrogen sulfide becomes great enough to cause serious odor problems whenever sulfate reduction yields a significant amount of sulfide ion. Concentrations in air above 20 ppm
should be avoided because of toxicity.

Corrosion of Sewers

In many areas of the United States—particularly in the southern part of the country
where domestic wastewater temperatures are high, detention times in the sewers are
long, and sulfate concentrations are appreciable—"crown" corrosion of concrete
sewers can be an important problem. The difficulty is always associated with reduction of sulfate to hydrogen sulfide, and the hydrogen sulfide is often blamed for the
corrosion. Actually, H_2S, or hydrosulfuric acid as its aqueous solutions are called, is
a weaker acid than carbonic acid and has little effect on good concrete. Nevertheless, "crown" corrosion of gravity-type sewers does occur, and hydrogen sulfide is
indirectly responsible.

Gravity-type sewers provide an unusual environment for biological changes in
the sulfur compounds present in wastewaters. Sewers are really part of a treatment
system, for biological changes are constantly occurring during transportation. These
changes require oxygen, and if sufficient amounts are not supplied through natural
reaeration from air in the sewer, reduction of sulfate occurs, and sulfide is formed.
At the usual pH level of domestic wastewaters, most of the sulfide is converted to
hydrogen sulfide and some of it escapes into the atmosphere above the wastewater.
Here it does no damage if the sewer is well ventilated and the walls and crown are
dry. In poorly ventilated sewers, however, moisture collects on the walls and crown.
Hydrogen sulfide dissolves in this water in accordance with its partial pressure in
the sewer atmosphere. As such it does no harm.

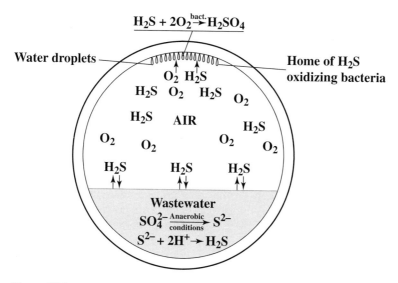

Figure 29.3
Formation of hydrogen sulfide in sewers and "crown" corrosion resulting from oxidation of hydrogen sulfide to sulfuric acid.

Bacteria capable of oxidizing hydrogen sulfide to sulfuric acid are ubiquitous in nature and are always present in domestic wastewater. It is natural that some of these organisms should infect the walls and crown of sewers at times of high flows or in some other manner. Because of the aerobic conditions normally prevailing in sewers above the wastewater, these bacteria oxidize the hydrogen sulfide to sulfuric acid,

$$H_2S + 2O_2 \xrightarrow{\text{bacteria}} H_2SO_4 \tag{29.4}$$

and the latter, being a strong acid, attacks the concrete. Bacteria of the genus *Thiobacillus* are capable of sulfide oxidation to sulfuric acid at a pH as low as 2 and are thought to be responsible for this problem. Sulfuric acid formation from sulfide is particularly serious in the crown, where drainage is at a minimum. Figure 29.3 summarizes the important aspects of odor and corrosion problems in sewer systems.

Other Concerns

High sulfate concentrations as well as low pH conditions can result in streams that are fed by drainage from abandoned coal mines and other exploited mineral-bearing deposits. The sulfide minerals present are oxidized through a combination of bacterial and chemical action to produce sulfuric acid,

$$2FeS_2(\text{pyrite}) + 7O_2 + 2H_2O \xrightarrow{\text{bacteria}} 2Fe(II) + 4SO_4^{2-} + 4H^+ \tag{29.5}$$

Not only does the sulfate content increase in streams to which mine drainage discharges, but the lowered pH and high iron content produce added harm to water quality. For these reasons, care is exercised today to replace the cover or to seal mines to prevent the introduction of oxygen and water that lead to the Eq. (29.5) reaction.

Combustion of fossil fuels leads to formation of gaseous oxides of sulfur, which hydrolyze when dissolved in rainwater to form sulfuric acid. The resulting "acid rain" is of concern as discussed in Sec. 17.3.

29.2 | METHODS OF ANALYSIS

Five methods of determining sulfate are currently considered standard. Ion chromatography and capillary ion electrophoresis as discussed in Sec. 12.4 are among the best procedures for sulfate measurement and can determine concentrations as low as 0.1 mg/L. The other three procedures depend upon the formation of insoluble barium sulfate from addition of excess barium chloride to the sample.

The main difference between the three procedures is in the method used to quantify the results of barium sulfate formation. In the gravimetric method, the precipitate formed is weighed. In the turbidimetric procedure, quantification is based upon the interference the precipitate causes to the passage of light. In the methylthymol blue method excess barium is determined colorimetrically to provide a measure of the amount of added barium that combines with sulfate. The choice of method depends to a considerable extent upon the purpose for which the determination is being made and the concentration of sulfate in the sample. The three procedures that depend upon barium sulfate formation are described in the following.

Gravimetric

The gravimetric method is considered to yield accurate results and of the three barium addition procedures is the recommended standard approach for sulfate concentrations above 10 mg/L. The quantitative aspects of this method depend upon the fact that barium ion combines with sulfate ion to form poorly soluble barium sulfate as follows:

$$Ba^{2+} + SO_4^{2-} \rightarrow BaSO_4(s) \tag{29.6}$$

The precipitation is normally accomplished by adding barium chloride in slight excess to samples of water acidified with hydrochloric acid and kept near the boiling point. The samples are acidified to eliminate the possibility of precipitation of $BaCO_3(s)$, which might occur in highly alkaline waters maintained near the boiling temperature. Excess barium chloride is used to produce sufficient common ion to precipitate sulfate ion as completely as possible.

Because of the great insolubility of barium sulfate ($K_{sp} = 1 \times 10^{-10}$), there is a considerable tendency for much of the precipitate to form in a colloidal condition that cannot be removed by ordinary filtration procedures. Digestion of the samples at temperatures near the boiling point for a few hours usually results in a transfer of the colloidal to crystalline forms, in accordance with the principle discussed in Sec. 3.6, and filtration can then be accomplished. The crystals of barium sulfate are usually quite small; for this reason, a special grade of filter paper (suitable for sulfate determinations) should be used. With reasonable care to make sure that all crystals have been transferred to the filter and with sufficient washing to remove all excess barium chloride and other salts, this method is capable of measuring sulfate with a high order of accuracy. Its major limitation is the time required. The barium sulfate precipitate formed may be weighed after filtration either following combustion to destroy the filter paper or by weighing the precipitate and filter together and then subtracting the weight of the previously tared filter from the results to obtain the weight of the precipitate alone.

Turbidimetric

The turbidimetric method of measuring sulfate is based upon the fact that barium sulfate formed following barium chloride addition to a sample tends to precipitate in a colloidal form and that this tendency is enhanced in the presence of an acidic buffer solution containing magnesium chloride, potassium nitrate, sodium acetate, and acetic acid. By standardizing the procedure used to produce the colloidal suspension of barium sulfate, it is possible to obtain results that are quantitative and acceptable for many purposes. The method is rapid and has wide application, since samples with sulfate concentrations greater than 10 mg/L can be analyzed by taking smaller portions and diluting them to the recommended 50-mL sample size. Because of the variables that enter into a determination of this sort, it is recommended that at least one standard sample of sulfate ion be included in each set of samples to verify that conditions used in the test are comparable to those used in establishing the calibration curve.

Automated Methylthymol Blue

An automated procedure is advantageous when many analyses for sulfate are required. Here, a continuous-flow analytical instrument is used in which chemicals are automatically added to and mixed with samples in a flowing stream. After a standard time passes to allow for chemical reaction to occur, the sample enters a cell where measurement of color, or in this case turbidity, is made for quantification. In the automated procedure for sulfate, barium chloride is first automatically added to the samples of low pH to form a barium sulfate precipitate; the sample pH is then adjusted to about 10. Methylthymol blue reagent is then added and combines with the excess barium added to form a blue chelate. The uncomplexed methylthymol blue remaining forms a grey color which is automatically measured. The amount of sulfate in the original sample is based upon the instrument response that is obtained. Obviously, the instrument must be calibrated with standard sulfate solutions, the additions of chemicals must be precise, and interferences must be absent. This can be accomplished by an automated approach as applied here.

29.3 | APPLICATIONS OF SULFATE DATA

The sulfate content of natural waters is an important consideration in determining their suitability for public and industrial water supplies. The amount of sulfate in wastewaters is a factor of concern in determining the magnitude of problems that can arise from reduction of sulfate to hydrogen sulfide. In anaerobic digestion of sludges and industrial wastes, the sulfate is reduced to hydrogen sulfide, which is evolved with methane and carbon dioxide. If the gas is to be used in gas engines, the hydrogen sulfide content should not exceed 750 ppm by volume. A knowledge of the sulfate content of the sludge or waste fed to digestion units provides a means of estimating the hydrogen sulfide content of the gas produced. From this information, the designing engineer can determine whether scrubbing facilities will be needed to remove hydrogen sulfide, and the size of the units required.

In the engineering and operation of treatment processes, especially anaerobic ones, knowledge of sulfate content can be of great importance. Sulfate-reducing bacteria generally outcompete methanogens kinetically for organic carbon in anaerobic treatment of high-sulfate organic wastewaters. Additionally, the sulfide produced can be toxic to methanogens. Thus, sulfate can have a highly adverse impact on the methanogenic process. In a similar manner, high sulfate concentrations in groundwater can hinder the natural anaerobic biodegradation of chlorinated solvents such as trichloroethene and tetrachloroethene. Many organic compounds contain sulfur as sulfate, sulfonate, or sulfide. During aerobic treatment of such wastes, complete utilization or dissimilation results in release of the organically bound sulfur as sulfate ion, but under anaerobic treatment, the sulfur is generally released as sulfide.

PROBLEMS

29.1   What is the significance of a high-sulfate concentration in water supplies and in wastewater disposal?

29.2   List three conditions that must occur more or less simultaneously for "crown" corrosion of a sewer to take place.

29.3   What analytical methods are available for the analysis of sulfate?

29.4   List four precautions that must be observed to ensure an accurate gravimetric determination of sulfate concentration.

29.5   What is the purpose of digestion of the sample in the gravimetric analysis for sulfate?

29.6   In the determination of sulfate concentration by the gravimetric procedure, a 400-mL sample yielded 0.0360 g of $BaSO_4(s)$. How many mg/L of sulfate was in the sample?

29.7   Why must exactly the same procedure be followed each time an analysis is made for sulfate concentration by the turbidimetric method?

29.8   What are two purposes for the conditioning reagent used in the turbidimetric determination of sulfate concentration?

29.9   In the methylthymol blue method of sulfate analysis, the color formed in the sample decreases with increase in sulfate concentration. What causes this?

29.10   From equilibrium considerations, calculate the relative proportions of sulfide in the H_2S, HS^-, and S^{2-} forms at (a) a pH of 6.0; (b) a pH of 7.5; and (c) a pH of 10.0.

29.11   Why would an engineer be interested in the sulfate content of a wastewater to be subjected to anaerobic treatment?

REFERENCE

Clesceri, L. S., A. E. Greenberg, and A. D. Eaton (eds.): "Standard Methods for the Examination of Water and Wastewater," 20th ed., American Public Health Association, Washington, DC, 1998.

30

Phosphorus and Phosphate

30.1 | GENERAL CONSIDERATIONS

The phosphate determination has grown rapidly in importance in environmental engineering and science practice because of the many ways in which phosphorus compounds affect environmental phenomena of interest. The inorganic compounds of phosphorus of significance are the phosphates or their molecularly dehydrated forms, usually referred to as polyphosphates or condensed phosphates. Organically bound phosphorus is usually a minor consideration.

Water Supplies

Polyphosphates are used in some public water supplies as a means of controlling corrosion. They are also used in some softened waters for stabilization of calcium carbonate to eliminate the need for recarbonation.

All surface water supplies support growth of minute aquatic organisms. The free swimming and floating organisms are called *plankton* and are of great interest because of effects on water quality. The plankton are composed of animals, *zooplankton,* and plants, *phytoplankton.* The latter are predominantly algae and cyanobacteria, and since they are chlorophyll-bearing organisms, their growth is influenced greatly by the amount of the fertilizing elements nitrogen and phosphorus in the water. Research has shown that nitrogen and phosphorus are both essential for the growth of algae and cyanobacteria and that limitation in amounts of these elements is usually the factor that controls their rate of growth. Where both nitrogen and phosphorus are plentiful, algal blooms occur, which may produce a variety of nuisance conditions. Experience has shown that such blooms do not occur when nitrogen or phosphorus or both are present in very limited amounts. The critical level for inorganic phosphorus has been established as somewhere near 0.005 mg/L or 5 μg/L under summer growing conditions.

Wastewater Treatment

Domestic wastewater is relatively rich in phosphorus compounds. Prior to the development of synthetic detergents, the content of inorganic phosphorus usually ranged from 2 to 3 mg/L and organic forms varied from 0.5 to 1.0 mg/L. Most of the inorganic phosphorus was contributed by human wastes as a result of the metabolic breakdown of proteins and nucleic acids and elimination of the liberated phosphate in the urine. The amount of phosphorus released is a function of protein intake and, for the average person in the United States, this release is considered to be about 1.5 g/day.[1]

Most heavy-duty synthetic detergent formulations designed for the household market contain large amounts of polyphosphates as "builders." Many of them contain from 12 to 13 percent phosphorus or over 50 percent of polyphosphates. The use of these materials as a substitute for soap has greatly increased the phosphorus content of domestic wastewater. It has been estimated from sales of polyphosphates to the detergent industry that domestic wastewater probably contains from two to three times as much inorganic phosphorus at the present time as it did before synthetic detergents became widely used, unless local ordinances limit the use of phosphate-based detergents.

The organisms involved in the biological processes of wastewater treatment all require phosphorus for reproduction and synthesis of new cellular material. Domestic wastewater contains amounts of phosphorus far in excess of the amount needed to stabilize the limited quantity of organic matter present. This fact is demonstrated by the presence of appreciable amounts in effluents from biological wastewater treatment plants. Many industrial wastes, however, do not contain sufficient quantities of phosphorus for optimum growth of the organisms used in treatment. In such cases, deficiencies may be overcome by the addition of inorganic phosphates.

Fertilizing Value of Sludges

A major problem in wastewater treatment practice is the disposal of sludges remaining from aerobic and anaerobic treatment processes. All sludges contain nitrogen and phosphorus in significant amounts and have value for fertilizing purposes. The total phosphorus content of digested sludges is ordinarily about 1 percent and that of heat-dried activated sludge about 1.5 percent.

Boiler Waters

Phosphate compounds are widely used in steam power plants to control scaling in boilers. If polyphosphates are used, they are rapidly hydrolyzed to orthophosphate at the high temperatures involved. Control of phosphate levels is accomplished through determinations of orthophosphate.

---

[1] C. N. Sawyer, Factors Involved in Disposal of Sewage Effluents to Lakes, *Sewage and Ind. Wastes,* **26:** 317 (1954).

30.2 | PHOSPHORUS COMPOUNDS OF IMPORTANCE

Phosphorus compounds of wide variety are encountered in environmental engineering and science practice. A list of the more important ones is given in Table 30.1.

All the polyphosphates (molecularly dehydrated phosphates) gradually hydrolyze in aqueous solution and revert to the ortho form from which they were derived:

$$Na_4P_2O_7 + H_2O \rightarrow 2Na_2HPO_4 \qquad (30.1)$$

The rate of reversion is a function of temperature and increases rapidly as the temperature approaches the boiling point. The rate is also increased by lowering the pH, and advantage is taken of this fact in the preparation of samples for the determination of polyphosphates. The hydrolysis of polyphosphates is also influenced by bacterial enzymes. The rate of reversion is very slow in pure waters but is more rapid in wastewaters. Experiments have shown that pyrophosphates are hydrolyzed more rapidly than tripolyphosphates in some waters, while in others the reverse is true. A matter of several hours and possibly days is required for the complete reversion of polyphosphate to orthophosphate, particularly at low temperatures or at high pH values.

From these considerations, it should be obvious that determinations for phosphorus or phosphate must employ procedures to measure polyphosphates if a true measure of the total inorganic forms present is to be obtained.

30.3 | METHODS OF DETERMINING PHOSPHORUS OR PHOSPHATE

The amounts of ortho, poly, and organic phosphorous present are of interest. Fortunately, it is possible to measure orthophosphate with very little interference from polyphosphates because of their stability under the conditions of pH, time, and temperature used in the test. Both poly and organic forms of phosphorus must be converted to orthophosphate for measurement.

Table 30.1 | Phosphorus compounds commonly encountered in environmental engineering and science

Name	Formula
Orthophosphates	
Trisodium phosphate	Na_3PO_4
Disodium phosphate	Na_2HPO_4
Monosodium phosphate	NaH_2PO_4
Diammonium phosphate	$(NH_4)_2HPO_4$
Polyphosphates	
Sodium hexametaphosphate	$Na_3(PO_3)_6$
Sodium tripolyphosphate	$Na_5P_3O_{10}$
Tetrasodium pyrophosphate	$Na_4P_2O_7$

Orthophosphate

Phosphorus occurring as orthophosphate (H_3PO_4, $H_2PO_4^-$, HPO_4^{2-}, PO_4^{3-}) can be measured quantitatively by gravimetric, volumetric, or colorimetric methods. The gravimetric method is applicable where large amounts of phosphate are present, but such situations do not occur in ordinary practice. The volumetric method is applicable when phosphate concentrations exceed 50 mg/L, but such concentrations are seldom encountered except in boiler waters and anaerobic digester supernatant liquors. The method involves formation of a precipitate, filtration, careful washing of the precipitate, and titration. The procedure is time-consuming. The standard procedures for water and wastewater, however, use colorimetric methods, possibly at some sacrifice of accuracy.

Three colorimetric methods are used for measuring orthophosphate. They are essentially the same in principle but differ in the nature of the agent added for final color development. The chemistry involved is essentially as follows: Phosphate ion combines with ammonium molybdate under acid conditions to form a molybdophosphate complex,

$$PO_4^{3-} + 12(NH_4)_2MoO_4 + 24H^+ \rightarrow$$
$$(NH_4)_3PO_4 \cdot 12MoO_3 + 21NH_4^+ + 12H_2O \quad (30.2)$$

When large amounts of phosphate are present, the molybdophosphate forms a yellow precipitate that can be filtered and used for volumetric determination. At lower concentrations of phosphate, a yellow colloidal sol is formed, which has been proposed as a basis of colorimetric measurement of intermediate concentrations. With concentrations of phosphates under 30 mg/L, the usual range in water analysis, the yellow color of the colloidal sol is not discernible, and other means of color development are necessary. In one modification, vanadium is added and forms a vanado-molybdophosphoric acid complex that yields a much more intense yellow color, permitting analysis for phosphorus down to the mg/L or lower range.

The molybdenum contained in ammonium phosphomolybdate is also readily reduced to produce a blue-colored sol that is proportional to the amount of phosphate present. Excess ammonium molybdate is not reduced and therefore does not interfere. Either ascorbic acid or stannous chloride may be used as the reducing agent. The colored compound formed is referred to as molybdenum blue or heteropoly blue. The chemistry involved with stannous chloride as the reducing agent may be represented in a qualitative manner as follows:

$$(NH_4)_3PO_4 \cdot 12MoO_3 + Sn^{2+} \rightarrow (molybdenum\ blue) + Sn^{4+} \quad (30.3)$$

When using stannous chloride, an extraction procedure can be applied both for increased sensitivity and to obtain accurate results when excessive interferences are present in the sample. In this case, the phosphomolybdate is extracted from the sample into a benzene-isobutanol solution prior to addition of the stannous chloride.

Polyphosphates

Polyphosphates may be converted to orthophosphate by boiling samples that have been acidified with sulfuric acid for at least 90 min. The hydrolysis may be hastened by heating in an autoclave at 20 psi. The excess acid added to speed the hydrolysis must first be neutralized before proceeding with the addition of the ammonium molybdate solution. The orthophosphate formed from the polyphosphates is measured in the presence of orthophosphate originally present in the sample by one of the methods applicable to orthophosphate. The amount of polyphosphates is obtained by difference as follows:

$$\text{Total inorganic phosphate} - \text{orthophosphate} = \text{polyphosphate} \qquad (30.4)$$

Organic Phosphorus

The amount of organic phosphorus present in industrial wastes or in sludges is sometimes of interest. This analysis requires that the organic matter be destroyed so that the phosphorus is released as phosphate ion. The organic matter may be destroyed by any of the three wet oxidation or digestion procedures listed in "Standard Methods." The oxidant used differs with the procedure and may be perchloric acid, nitric acid–sulfuric acid, or persulfate. Perchloric acid is the most rigorous oxidant, but is also the most hazardous to use. In order to avoid and prevent damage from explosions, special hoods for the digestion must be used, and care in the order of adding chemicals is essential. For these reasons, perchloric acid digestion should be carried out only by an experienced chemist and only if the added rigor of this procedure is necessary. Persulfate digestion is recommended if experience proves the results obtained are suitable for the need.

Once digestion has been accomplished, measurement of the phosphorus released can be made by any of the methods applied to orthophosphate. All forms of phosphorus (total) are measured in an organic phosphorus determination. Therefore, the organic phosphorus is obtained as follows:

$$\text{Total phosphorus} - \text{inorganic phosphorus} = \text{org-P} \qquad (30.5)$$

30.4 | APPLICATIONS OF PHOSPHORUS DATA

Phosphorus data are important because of the significance of phosphorus as a vital factor in life processes. In the past, the data have been used principally to control phosphate dosages in water systems for corrosion prevention and in boilers for control of scale. Phosphorus determinations are also important in assessing the potential biological productivity of surface waters, and in many areas limits have been established on the amounts of phosphorus that may be discharged to receiving bodies of water, particularly lakes and reservoirs. Phosphorus determinations are routine in the operation of wastewater treatment plants and in stream pollution studies in many areas. Because of the importance of phosphorus as a nutrient in biological methods of wastewater treatment, its determination is essential with many industrial wastes and in the operation of waste treatment plants.

PROBLEMS

30.1   What practical uses are made of polyphosphates?

30.2   Why are limits sometimes placed on the discharge of phosphates to receiving waters?

30.3 Why is phosphate sometimes added in the biological treatment of industrial wastewaters?

30.4 What analytical procedures are available for the analysis of orthophosphate?

30.5 What is the difference between orthophosphate, polyphosphates, and organic phosphorus? In which form must the phosphorus be for colorimetric analysis?

30.6 How is the analysis for phosphorus conducted to differentiate between the three forms of phosphorus?

30.7 What are the different commonly used polyphosphate forms?

30.8 Would you expect the analytical results for orthophosphate to be higher than, lower than, or the same as the original value in a sample of domestic wastewater that had been acidified to prevent bacterial action and stored for several days prior to analysis?

REFERENCE

Clesceri, L. S., A. E. Greenberg, and A. D. Eaton (eds.): "Standard Methods for the Examination of Water and Wastewater," 20th ed., American Public Health Association, Washington, DC, 1998.

Oil and Grease

31.1 | GENERAL CONSIDERATIONS

The oil and grease content of domestic and certain industrial wastes, and of sludges, is an important consideration in the handling and treatment of these materials for ultimate disposal. Oil and grease are singled out for special attention because of their poor solubility in water and their tendency to separate from the aqueous phase. Although this characteristic is advantageous in facilitating the separation of oil and grease by use of flotation devices, it does complicate the transportation of wastes through pipelines, their destruction in biological treatment units, and their disposal into receiving waters.

Wastes from the meat-packing industry, particularly where hard fats from the slaughtering of sheep and cattle are involved, and from restaurants have resulted in serious decreases in the carrying capacity of sewers. Such experiences, and other factors related to treatment or ultimate disposal, have served as the basis for ordinances and regulations governing the discharge of greasy materials to sewer systems or receiving waters and have forced the installation of preliminary treatment facilities by many industries and restaurants for the recovery of grease or oil before discharge is permitted.

A number of problems are caused by oil and grease in waste treatment practice. Very few plants have provisions for the separate disposal of these materials to scavengers or by incineration; consequently, that which separates as scum in primary settling tanks is normally transferred with the settled solids to disposal units. In sludge digestion tanks, oil and grease tend to separate and float to the surface to form dense scum layers, because of their poor solubility in water and their low specific gravity. Scum problems have been particularly severe where high-grease-content wastes, such as those from the meat-packing and oil and fat industries, have been admitted to public sewer systems. The vacuum filtration of sludge is also complicated by high grease content.

Not all the oil and grease is removed from sewage by primary settling units. Appreciable amounts remain in the clarified wastewater in a finely divided emulsified form. During subsequent biological attack in secondary treatment units or in the receiving stream, the emulsifying agents are usually destroyed, and the finely divided oil and grease particles become free to coalesce into larger particles that separate from the water. In activated-sludge plants, the grease often accumulates into "grease balls" that give an unsightly appearance to the surface of final settling tanks. Both trickling filters and the activated-sludge process are adversely affected by unreasonable amounts of grease that seems to coat the biological solids sufficiently to interfere with oxygen transfer from the liquid to the interior of the living cells. This is sometimes described as a "smothering" action.

Separation of floating grease in final settling tanks has been a problem in some treatment plants employing high-rate processes. This has been attributed to short-term contact of the waste with limited amounts of biological growths that destroy the emulsifying agents present but do not have sufficient adsorptive powers to hold the grease that is released, nor time to oxidize it. As a result, the grease is free to separate under quiescent conditions such as occur in final settling tanks or receiving waters.

All too frequent spills of crude and refined petroleum from ships used for their transport have resulted in loss of fish, mammals, and waterfowl, and the fouling of beaches. Such spills may result from accidents at sea or from the discharge of oil-laden bilge water as ships approach port to pick up new cargo. Oil and grease leaking from automobiles result in high concentrations in storm runoff from streets, contaminating waterways into which storm water drains. Thus, it is not only the oil and grease in wastewaters that are of environmental concern. In oceanic spills, detailed analyses of oil and grease components are commonly used to determine the source of spills, which many times are not readily apparent.

31.2 | OIL AND GREASE AND THEIR MEASUREMENT

The term *grease* applies to a wide variety of organic substances that are extracted from aqueous solution or suspensions by hexane. Hydrocarbons, esters, oils, fats, waxes, and high-molecular-weight fatty acids are the major materials dissolved by this solvent. All these materials have a "greasy feel" and are associated with the problems in waste treatment related to grease.

Hexane is currently the "Standard Methods" solvent of choice for oil and grease determinations because it is a good solvent for all the materials normally associated with the term grease and has a minimum solvent power for other organic compounds. While hexane has been the standard for this purpose for many years, it was replaced in recent times by 1,1,2-trichloro-1,2,2-trifluoroethane, commonly known as CFC-113, because it exhibited less of an explosion hazard. However, all chlorofluorocarbons (CFCs) including CFC-113, which are excellent solvents and refrigerants and were in the past widely used industrially and commercially, are responsible for the depletion of the beneficial ozone in the stratosphere. For this reason they are being phased out of commercial use. As a result, "Standard Methods" has returned to hexane, except in the partition-infrared procedure. Those using hexane must be aware of the potential explosion risk it presents. Extraction procedures that recover the solvent upon evaporation are of importance for environmental and safety reasons.

Hexane extraction does not measure low-molecular-weight hydrocarbons such as gasoline as they do not partition well into the solvent. In addition, one of the procedures for water samples requires sample drying at 103°C prior to extraction. As a result, all materials with boiling points below this temperature are lost, as well as significant amounts of all other materials that have appreciable vapor pressures at 103°C. Such compounds, except in unusual cases, are normally present in relatively small amounts in domestic wastewaters and are of little concern except in wastes from the petroleum industry. Most of the materials generally classified as "grease" have very low vapor pressures at 103°C and can be recovered essentially 100 percent by hexane extraction. In cases in which drying oils are present, some oxidation occurs at the unsaturated linkages during the drying procedure and may render them insoluble. Such oils, however, do not normally occur in domestic waste to any great extent.

Although the methods employed for the determination of grease may seem highly unrefined and inaccurate, they are the results of years of effort to obtain a reasonable measure of those things in water, domestic and industrial wastes, and sludges that tend to separate from the aqueous phase and create special problems.

31.3 | METHODS OF ANALYSIS

All the standard methods of determining oil and grease depend upon a preferential solution of the greasy materials using extractions with organic solvents and are subject in some degree to the limitations already discussed. The methods employed for water samples and sludges differ somewhat, and separate discussions are needed.

Water and Wastewater

The oil or grease content of relatively clean waters is not a routine determination and is seldom performed except in special cases in which accidental contamination has occurred. The choice of method of analysis depends upon the volatility of the contaminants. All the standard procedures measure high-boiling-point materials, but those with appreciable vapor pressures at 70°C should be measured by the use of partition-infrared or hydrocarbon procedures.

With domestic wastewaters, oils, fats, waxes, and fatty acids are the principal substances classed as oil and grease. Industrial wastewaters may contain simple esters and, possibly, a few other compounds in the same category. The term *oil* represents a wide variety of substances ranging from low- to high-molecular-weight hydrocarbons from petroleum, spanning the range from gasoline through heavy fuel and lubricating oils. Grease represents higher-molecular-weight hydrocarbons and all glycerides of animal and vegetable origin. The fatty acids occur principally in a precipitated form as calcium and magnesium soaps. As such, they are insoluble in the solvents. Samples are acidified with hydrochloric acid to a pH of about

1.0 to release the free fatty acids for analysis. The reaction involved may be represented by the following equation:

$$Ca(C_{17}H_{35}COO)_2 + 2H^+ \rightarrow 2C_{17}H_{35}COOH + Ca^{2+} \tag{31.1}$$

Four different procedures are available for measuring oil and grease in water and wastewater samples. They all involve an initial extraction into hexane or CFC-113 (infrared procedure only). In the partition-gravimetric method, the hexane is then separated from water and evaporated, the residue remaining being used as a measure of the oil and grease content. In the partition-infrared method, the CFC-113 extracted materials are measured with infrared scanning. The accuracy of this procedure depends upon the use of oil and grease standards for calibration that are similar in composition to the oil and grease in the samples being analyzed. The advantage over the gravimetric procedure is speed of analysis. The third, or Soxhlet extraction, procedure involves an initial step of acidification and filtration to remove oil and grease from the aqueous phase, and then hexane extraction. This procedure tends to retain more of the volatile hydrocarbons than the gravimetric procedure, but is more time intensive.

Filtration is considered acceptable practice for the third procedure, since it effectively separates those materials normally referred to as oil and grease and allows low-molecular-weight and soluble materials, which are of no consequence, to escape in the filtrate. Drying of the filtered material removes water so that the solvent can penetrate the sample readily and accomplish separation of the grease in the 4-h extraction period normally provided. It also eliminates the possibility of appreciable amounts of water being carried into the extract and thereby simplifies the drying procedure. A Soxhlet type of extractor that provides intermittent batchwise extraction is used (Fig. 31.1).

A fourth standard method for water and wastewater is designed to more selectively determine hydrocarbons associated with petroleum products and to exclude fatty acids and other fatty materials that are more associated with animal and vegetable materials. Thus, this is a *hydrocarbon* analysis, rather than an oil and grease analysis. Here, silica gel is added to the hexane extract and selectively removes the fatty materials, the remaining materials in the hexane solvent are then analyzed by any of the first three procedures.

Sludges

Sludges are frequently of such consistency and character that they are difficult to filter, and prolonged periods are often required to dry them sufficiently for solvent extraction. The current standard procedure involves the use of a chemical dehydration technique that eliminates the need for filtration and drying. The method consists of weighing a definite amount of sample, acidifying it to release fatty acids as shown in Eq. (31.1), and then adding a quantity of $MgSO_4 \cdot H_2O$ sufficient to combine with all free water by forming higher hydrated forms, $MgSO_4 \cdot 7H_2O$ being the ultimate. With the water in a chemically bound form, the sample is pulverized to facilitate extraction of grease. A Soxhlet extractor, as shown in Fig. 31.1, is used with hexane to separate the grease from the $MgSO_4 \cdot 7H_2O$ and the organic matter that is not grease.

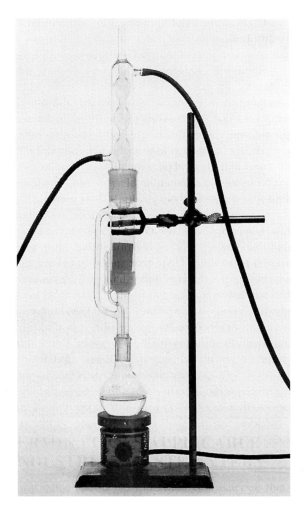

Figure 31.1
Soxhlet apparatus for determination of grease.

31.4 | APPLICATIONS OF OIL AND GREASE DATA

Oil and grease determinations are made more or less routinely for a number of purposes. Many municipalities and other local authorities have ordinances for the regulation of the discharge of oil and grease-bearing industrial wastes to sewer systems or to receiving waters and, of course, use oil and grease determinations for regulatory purposes. Industries that must treat their wastes to remove such materials use the test or some modification of it to determine the effectiveness of treatment units and to keep a record of the oil and grease content of their discharges.

 One of the major purposes of waste treatment facilities is to remove unsightly and obnoxious floating matter, of which oil and grease are major constituents. Oil and grease determinations on raw and settled wastewaters give a measure of the effectiveness of primary settling tanks, and determinations on final effluents provide a record of the efficacy of secondary treatment units, as well as the amounts actually discharged to receiving waters. The latter is particularly important where disposal is into recreational areas.

Oil and grease determinations are used extensively in sludge disposal practice. Determinations on the raw and digested sludges, when properly adjusted to volumes involved in a given plant, allow calculation of oil and grease destruction during anaerobic digestion. When scum problems occur in digestion units, determinations of oil and grease often yield information of considerable value. The oil and grease content of sludge is a factor in determining its suitability for use as a fertilizer, and purchasers often require a guarantee that the percentage shall not exceed a certain value.

PROBLEMS

31.1 What are sources of oil and grease in receiving waters, and what adverse environmental effects can result?

31.2 List four important sources that contribute oil and grease to municipal wastewaters.

31.3 What are the organic components of grease and oil?

31.4 Name five operating difficulties that are caused or intensified by grease at waste treatment plants.

31.5 What analytical procedures are available for analysis of oil and grease in municipal and industrial wastewaters?

31.6 What analytical procedure is used for analysis of sludges?

31.7 What solvent is used for extraction of grease for analysis, and what are its advantages and disadvantages?

31.8 Why are samples of wastewater and sludge acidified prior to the separation of grease?

31.9 Define the following: (*a*) fats; (*b*) waxes; (*c*) soaps; (*d*) animal and vegetable oils; (*e*) mineral oils.

REFERENCE

Clesceri, L. S., A. E. Greenberg, and A. D. Eaton (eds.): "Standard Methods for the Examination of Water and Wastewater," 20th ed., American Public Health Association, Washington, DC, 1998.

Volatile Acids

32.1 | GENERAL CONSIDERATIONS

The volatile-acids determination is widely used in the control of anaerobic waste treatment processes. In the biochemical decomposition of organic matter that occurs, facultative and anaerobic bacteria of wide variety hydrolyze and convert the complex materials to low-molecular-weight compounds, as discussed in Secs. 6.7 to 6.11. Among the low-molecular-weight compounds formed, the short-chain fatty acids, such as acetic, propionic, butyric, and to a lesser extent isobutyric, valeric, isovaleric, and caproic, are important components. These low-molecular-weight fatty acids are termed *volatile acids* because they can be distilled at atmospheric pressure. An accumulation of volatile acids can have a disastrous effect upon anaerobic treatment if the buffering capacity of the system is exceeded and the pH falls to unfavorable levels.

In anaerobic digestion units that are operating in a stabilized condition, three groups of bacteria work in harmony to accomplish the destruction of organic matter. Following hydrolysis and fermentation to complex acids the acidogenic and dehydrogenating organisms carry the degradation to acetic acid and hydrogen and then the methane-forming bacteria complete the conversion into methane and carbon dioxide (Fig. 32.1). When a sufficient population of methane-forming bacteria is present and environmental conditions are favorable, they utilize the end products produced by the acidogenic bacteria as fast as they are formed. As a result, acids do not accumulate beyond the neutralizing ability of the natural buffers present, and the pH remains in a favorable range for the methane bacteria. Under such conditions the volatile-acid content of digesting sludges or anaerobically treated wastewaters usually runs in the range of 50 to 250 mg/L, expressed as acetic acid.

Methane-forming bacteria are ubiquitous in nature, and some are always present in domestic wastewater and sludge derived therefrom. Their population, however, is very small compared with that of many of the fermentative and acidogenic bacteria. This disparity in numbers is the reason for the troubles encountered in starting anaerobic treatment units without benefit of "seeding." Untreated municipal wastewater sludges and many industrial wastewaters have a relatively low buffering capacity, and

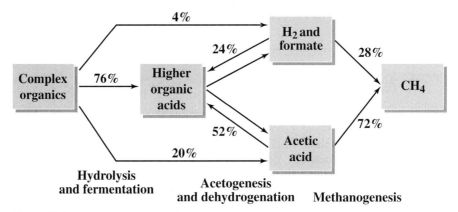

Figure 32.1
Stages in the methane fermentation of complex wastes. Percentages represent conversion of waste COD by various routes. *(After P. L. McCarty, One Hundred Years of Anaerobic Treatment, in "Anaerobic Digestion 1981," Hughes et al., eds., Elsevier Biomedical Press, Inc., B.V., Amsterdam, pp. 3–22, 1981.)*

when they are allowed to ferment anaerobically, volatile acids are produced so much faster than the few methane bacteria present can consume them that the buffers are soon spent and free acids exist to depress the pH. At pH values below 6.5, methane bacteria are seriously inhibited, but many fermentative and acidogenic bacteria are not until pH levels fall to about 5. Under such unbalanced conditions, the volatile-acids concentration continues to increase to levels of 2000 to 6000 mg/L or more, depending upon the solids content of the sludge. Active methane treatment may never develop in such mixtures unless the sludge or wastewater is diluted or neutralizing agents, such as lime, are added to produce a favorable pH for the methane bacteria. The volatile-acids determination, in conjunction with pH measurements, is valuable in control of environmental conditions during the initiation of anaerobic treatment.

Successful operation of anaerobic treatment units depends upon maintaining a satisfactory balance between the methane and acidogenic bacteria. The methane bacteria appear to be the most susceptible to changes in environmental conditions and food load. They are affected much more radically by changes in pH and temperature than the acidogenic bacteria. Inhibitions caused by changes in either of these factors result in a decreased rate of destruction of volatile acids; consequently, volatile acids begin to accumulate in the system. Many of the acidogenic bacteria are known to reproduce more rapidly than the methane bacteria. Under increased food loads, volatile acids may be formed faster than the slow-growing methane organisms can take care of them. This discrepancy results in an accumulation of volatile acids in the system. Sludge must be removed or transferred from anaerobic treatment units on occasion; however, removal of too large an amount will deplete the methane organism population to levels where volatile acids cannot be destroyed as fast as they are formed and accumulations will develop. Volatile-acids determinations are important in detecting the presence of unbalanced conditions in anaerobic treatment units caused by any of these factors. The onset of unfavorable conditions can be detected almost immediately, and usually several days in advance of other methods, such as through pH measurements.

32.2 | THEORETICAL CONSIDERATIONS

Because of the importance of volatile-acid formation in anaerobic treatment, it is of value to provide more details of the somewhat complicated biological processes involved. Volatile acids are formed as intermediates during the anaerobic degradation of carbohydrates, proteins, and fats, as discussed in Secs. 6.7 to 6.11. Figure 32.1 illustrates the four major stages through which a complex waste such as domestic sludge must pass during its conversion to methane gas. The first two stages of hydrolysis and fermentation may be carried out by the same organisms. Here, complex organic materials such as carbohydrates, proteins, and fats and oils are hydrolyzed into basic components, which are then fermented to fatty acids, alcohols, carbon dioxide, ammonia, and some hydrogen. In the third stage of acetogenesis and dehydrogenation, a more specialized group of anaerobic bacteria ferment the higher organic acids to form acetic acid, hydrogen, and perhaps formic acid. In the fourth stage, hydrogen, formate, and acetic acid are converted to methane by the anaerobic methane-forming bacteria. The third and fourth stages are closely coupled as here the third-stage acetic-acid-forming bacteria depend upon the methanogens to reduce the concentrations of hydrogen, formate, and acetic acid to sufficiently low levels for the third-stage conversion to be energetically favorable. Because of this close coupling, it was formerly thought that methanogens carried out both the third and fourth stages. Through painstaking studies to isolate the organisms involved, the two separate third and fourth stages are now known to exist.

Acetic acid is the most abundantly volatile acid produced, and is formed as an intermediate during the anaerobic treatment of almost all organic material. With a complex waste such as domestic sludge, about 72 percent of the organic matter based upon COD or electron equivalents is converted to acetic acid before finally being changed into methane gas. A similar high percentage is found with most industrial wastewaters as well.

The anaerobic biological degradation of wastes thus results in the production of large quantities of organic acids. If these acids are not converted to methane gas as rapidly as they are formed, their concentration will increase and will lower the pH. The major buffering material in anaerobic treatment that tends to prevent a drop in pH is bicarbonate, which, in equilibrium with carbonic acid, tends to regulate the hydrogen-ion concentration (Sec. 4.5):

$$H_2CO_3^* \rightleftharpoons H^+ + HCO_3^- \tag{32.1}$$

$$\frac{[H^+][HCO_3^-]}{[H_2CO_3^*]} = K_{A1} \tag{32.2}$$

or
$$[H^+] = K_{A1}\frac{[H_2CO_3^*]}{[HCO_3^-]} \tag{32.3}$$

K_{A1} at 35°C, a common temperature of operation of anaerobic digestion, is 4.8×10^{-7}. On the basis of Henry's law, an equilibrium also exists between the partial pressure of carbon dioxide in the gas phase and the carbon dioxide and carbonic acid concentrations in solution.

$$CO_2(aq) = CO_2(g) \tag{32.4}$$

$$CO_2(aq) + H_2O = H_2CO_3 \tag{32.5}$$

and, by definition, $\qquad [H_2CO_3^*] = [CO_2(aq)] + [H_2CO_3] \tag{32.6}$

The relationship between the concentration of carbon dioxide in the gas and aqueous phases is given by Henry's law,

$$\frac{[CO_2(g)]}{[H_2CO_3^*]} = K_H \tag{32.7}$$

At 35°C, the temperature commonly used in anaerobic treatment, K_H is 38 atm/M. Combining Eqs. (32.3) and (32.7), and solving for pH, we obtain (ignoring activity corrections),

$$pH = \log \frac{K_H}{K_{A1}} + \log \frac{[HCO_3^-]}{[CO_2(g)]} = 7.9 + \log \frac{[HCO_3^-]}{[CO_2(g)]} \tag{32.8}$$

This relationship between pH, bicarbonate concentration, and carbon dioxide content in the gas is shown graphically in Fig. 32.2.

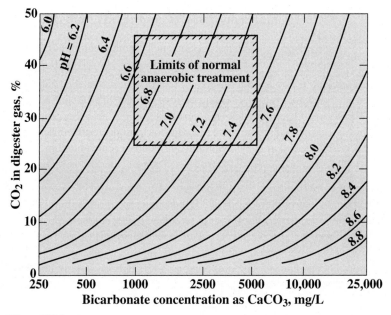

Figure 32.2
Relationship between pH, bicarbonate concentration, and carbon dioxide concentration at 35°C and 1 atm. *(P. L. McCarty, Anaerobic Waste Treatment Fundamentals, Public Works, 123: October 1964.)*

EXAMPLE 32.1

The bicarbonate concentration measured as alkalinity in a digester is found to equal 2500 mg/L as $CaCO_3$, and an analysis of the digester gas indicated the carbon dioxide content to be 30 percent. If atmospheric pressure of the total gas inside the digester is 1 atm, what should be the digester pH?

Since the bicarbonate concentration is reported as $CaCO_3$, we use 50,000 as the equivalent weight. Thus, $[HCO_3^-]$ = 2500/50,000, or 0.05. The carbon dioxide gaseous concentration equals its partial pressure in atmospheres. Since the digester gas is at 1 atm total pressure, and the gaseous partial pressure is equal to its volume fraction, $[CO_2(g)]$ = 0.3. Thus from Eq. (32.8),

$$pH = 7.9 + \log\frac{0.05}{0.3} = 7.1$$

We find we obtain the same result if we use Fig. 32.2.

We see that the higher the bicarbonate concentration and the lower the percent carbon dioxide in the digester gas, the higher the pH.

As volatile acids accumulate in an anaerobic reactor, they destroy the bicarbonate buffer and increase the carbon dioxide concentration:

$$\text{R-COOH} + HCO_3^- \rightarrow \text{R-COO}^- + H_2O + CO_2 \qquad (32.9)$$

These changes, as illustrated in Fig. 32.2, result in a gradual drop in pH. When the bicarbonate concentration decreases below 1000 mg/L as $CaCO_3$, further accumulation of acid decreases the pH very rapidly. For these reasons it is necessary to maintain buffer capacity by adding basic materials from outside sources or to exercise some control over the rate of volatile-acids formation if pH conditions favorable to methane-forming bacteria are to be maintained. Impending pH problems are indicated well before they occur by an increase in volatile-acids concentration.

EXAMPLE 32.2

What will the pH be for the case of Example 32.1 if the volatile-acid concentration suddenly increases from near zero to 2000 mg/L as acetic acid and the carbon dioxide content of the gas increases to 35 percent?

According to Eq. (32.9), volatile acids will destroy bicarbonate alkalinity. Since the equivalent weight of acetic acid is 60,000 mg, the bicarbonate alkalinity will decrease to 0.05 − (2000/60,000), or 0.0167 M or 835 mg/L as $CaCO_3$. The carbon dioxide partial pressure will increase to 0.35 atm. Thus, using Eq. (32.8),

$$pH = 7.9 + \log\frac{0.0167}{0.35} = 6.6$$

This represents a serious drop in pH that could adversely affect digester operation. Some pH control would be advisable.

Volatile-acid concentration is traditionally expressed in mg/L as acetic acid, which has an equivalent weight of 60 g. Thus, if a chromatographic analysis is used to determine the concentrations of the individual volatile acids, then one would need to sum up their respective millimolar concentrations and multiply by 60 to obtain the volatile-acid concentration in mg/L as acetic acid.

| EXAMPLE 32.3 | Determine the volatile acid concentration in mg/L if the following analyses were obtained for individual volatile acids in digester sludge. |

Volatile acid	Concentration, mg/L	Equivalent weight, g	Concentration, mM	Concentration, mg/L as acetic acid
Formic	5	46	0.11	7
Acetic	760	60	12.67	760
Propionic	310	74	4.19	251
Butyric	110	88	1.25	75
Isobutyric	40	88	0.45	27
Pentanoic	10	102	0.10	6
Hexanoic	20	116	0.17	10
Total volatile acids				1140

The weight concentration in mg/L of each volatile acid is divided by its equivalent weight (which equals the formula weight) to obtain the millimolar concentration of each acid. This is then multiplied by the equivalent weight of acetic acid to obtain the concentration as acetic acid, as indicated in the Example 32.3 table. A summation of the acetic-acid equivalent concentrations of each acid is then made, and this equals the total-volatile-acid concentration of 1140 mg/L as acetic acid. As an alternative, the mM concentrations could be totaled, and this total could be multiplied by 60 to obtain the same result.

32.3 | METHODS OF DETERMINING VOLATILE ACIDS

Two methods of determining volatile acids are currently described in "Standard Methods." One uses column-partition chromatography, and the other involves distillation. Several laboratories today also use gas chromatography or ion chromatography (Sec. 12.4) for the measurement of the concentration of individual volatile acids, although these procedures are not yet considered as standard. They may be in the future as we learn more about the significance of individual volatile acids in the control of the overall process.

Chromatographic Separation

The column-partition chromatographic method allows measurement of nearly 100 percent of all volatile acids present and is fairly rapid. Other organic acids such as pyruvic, phthalic, fumaric, lactic, succinic, and oxalic acids are also measured to some degree by this procedure. However, these acids generally do not occur in high concentrations in anaerobic treatment, so results from the column-partition method are usually about the same as from the distillation procedure. Because acids other than the volatile acids are included in the procedure, "Standard Methods" indicates this method is for "organic acids" rather than for volatile acids.

The chromatographic method is a modification of a procedure originally designed to allow separation and measurement of the individual volatile acids.[1] In partition chromatography two solvents are involved. One is fixed on or in a solid adsorbent, such as silicic acid, and placed in a column. A sample containing the materials to be separated is placed on top of the column. The second or mobile solvent is then added and carries the materials to be separated through the column, where they are continuously partitioned between the immobile solvent and the moving solvent, thus producing a differential migration of the various materials. Materials that are more soluble in the moving phase than in the stationary phase will emerge from the column first. This is the same general principle involved in gas, liquid, and ion chromatography as described in Sec. 12.4.

In routine operation of an anaerobic waste treatment system, the volatile-acids analysis is used mainly for control to determine whether or not the system is in balance; thus, the total-volatile-acids concentration rather than the concentration of each individual acid is of most interest. For this purpose, the column-partition chromatography procedure has been simplified, all volatile acids being separated from inorganic acids and other salts in the sample so that a total-volatile-acids analysis might be made. The suspended solids are first removed from the sample either by filtration or centrifugation, and the sample is then acidified with a strong inorganic acid, such as sulfuric, to convert the volatile acids from the ionized form to the un-ionized form:

$$R\text{-}COO^- + H^+ \rightarrow R\text{-}COOH \tag{32.10}$$

which is highly soluble in the mobile chloroform-butanol solvent used.

The acidified sample is distributed uniformly over a short column containing oven-dried silicic acid, which readily absorbs the sample water, thus forming the stationary solvent phase. A given quantity of the chloroform-butanol mixture is then passed through the column and selectively carries with it the un-ionized volatile acids, which are much more soluble in this mixture than in the stationary water phase. The sulfuric acid and other ionized salts, however, are more soluble in the water and so are left behind. It is important that sufficient mobile solvent be added to extract all the volatile acids present, but not so much that the inorganic acids are carried through.

The extracted volatile acids are measured by titration with sodium hydroxide to the phenolphthalein end point:

$$R\text{-}COOH + NaOH \rightarrow R\text{-}COO^- + Na^+ + H_2O \tag{32.11}$$

[1]H. F. Mueller, A. M. Buswell, and T. E. Larson, Chromatographic Determination of Volatile Acids, *Sewage and Ind. Wastes,* **28:** 255 (1956).

A sodium hydroxide in methanol solution is used for this titration because water is insoluble in the organic solvent. This solution, however, is unstable and must either be standardized frequently or prepared fresh daily. It can be made readily by diluting a proper quantity of 1.00 N aqueous sodium hydroxide with methanol. The separation and titration using the chloroform-butanol mixture should be done within a fume hood because of the carcinogenic potential of chloroform.

Distillation

The direct-distillation method is commonly used for the routine determination of volatile acids. The method is rapid and sufficiently accurate for most practical purposes, since it is not usually necessary to know volatile-acids concentrations more accurately than ± 50 mg/L. In this procedure use is made of the fact that all the low-molecular-weight fatty acids up to octanoic acid have significant vapor pressures at 100°C, the boiling point of water, so that sample distillates will contain appreciable amounts of these acids. The vapor pressures of the acids are shown in Table 32.1.

Anaerobic treatment reactors normally have a pH maintained in the range of 6.5 to 7.5, and under such conditions organic acids exist largely in the ionic form and cannot be distilled. By the addition of a strong nonvolatile acid, such as sulfuric, the organic acids are converted to the un-ionized form, as indicated in Eq. (32.10), which can be distilled. Usually, sufficient acid is added to reduce the pH to below 1.0.

It would be expected from the vapor-pressure data in Table 32.1 that the low-molecular-weight acids would distill most readily because of their greater vapor pressures. This is not the case, however. Formic, acetic, and propionic acids are difficult to separate by distillation in dilute aqueous solution.[2] This is because aqueous solutions are not ideal in character. Molecular association occurs between the acid and water molecules and varies with the particular acid. As a result, binary mixtures of all three classes described in Sec. 3.5 are formed.

Table 32.1 | Vapor pressures of volatile acids at 100°C

Acid	Vapor pressure at 100°C, atm*
Formic	0.99
Acetic	0.55
Propionic	0.24 (calc.)
Butyric	0.092 (calc.)
Valeric	0.037
Hexanoic	0.014
Heptanoic	0.005
Octanoic	0.002

*Data from "International Critical Tables," McGraw-Hill, New York.

---

[2]A. M. Buswell and S. L. Neave, Laboratory Studies on Sludge Digestion, *Illinois State Water Survey Bull.,* **30:** 76 (1930).

Formic acid and water form a Class II mixture, and water is the major component of the vapor phase when dilute solutions are distilled. Acetic acid and water form a Class I mixture, and the vapor phase is predominantly water vapor. Fractionation of the vapor in each case produces a distillate that is essentially pure water and the acids remain in the undistilled liquor. It is for this reason that fractionation of the vapors must be kept at a minimum in the volatile-acids determination.

Propionic, butyric, and other fatty acids that are soluble in water form Class III mixtures. The distillate contains the acid and water in a constant ratio until all the acid is distilled.

It is unnecessary to separate suspended solids from samples prior to distillation if care is used to prevent excessive heating of the walls of the distillation flask above the level of the mixture in the flask. If this is allowed to happen, some decomposition of suspended solids may occur in the strong acid environment, which may give rise to the production of organic acids. Also, reduction of sulfuric acid may occur with the production of sulfur dioxide. The latter, being an acid anhydride, will dissolve in the distillate to produce sulfurous acid and give abnormally high results. There is little chance of this happening when electric heaters are used, but it frequently happens if uncontrolled gas flames are used for heat.

The distillation rate is an important consideration in this determination and should be closely controlled. At low rates of distillation considerable fractionation may occur in the neck of the flask and the volatile acids will not carry over in the distillate as desired. Care must be taken to stop the distillation when the appropriate amount of distillate has been formed. The residue remaining in the flask is in contact with the excess sulfuric acid used, and the acid becomes more concentrated as the distillation progresses. If the distillation is carried too far, the sulfuric acid will cause decomposition of other organic materials present and be reduced as previously described.

The distilled acids are measured quantitatively by titration with a standardized NaOH solution to the phenolphthalein end point [Eq. (32.11)]. Approximately 70 percent of the volatile acids is distilled from samples treated by the recommended procedure. This matter is taken into account in calculations of the volatile acids present.

32.4 | APPLICATIONS OF VOLATILE-ACIDS DATA

Volatile-acids determinations have been valuable in providing information concerning the anaerobic degradation of organic matter and the environmental conditions best suited for optimum activity of methane-producing bacteria. An understanding of the complex chemical and biological processes involved in anaerobic treatment is essential to the intelligent application of anaerobic treatment to the wide variety of industrial wastes as well as domestic sludges. Toward this end, knowledge of the role volatile acids play in the process, and the importance of their detection as indicators of when methane organisms fail to keep pace with the fermentative and acidogenic organisms, is vital.

The value of volatile-acids data in routine control of anaerobic treatment units at wastewater treatment plants and other installations has been amply demonstrated.

These data are needed especially in the control of units that are operating at or near design capacity. With the development of high-rate treatment processes, the test has become even more important, as corrective measures have to be made more promptly. The volatile-acids test offers the most promise of providing pertinent information as quickly as possible and at reasonable expense.

PROBLEMS

32.1 Discuss the significance of the volatile-acids determination in anaerobic sludge digestion.

32.2 (*a*) Define what is meant by volatile acids.

(*b*) Indicate what are the most prevalent volatile acids formed during anaerobic treatment.

(*c*) Indicate the general classes of organic compounds from which each of the most prevalent volatile acids result.

32.3 What procedures are available for determining volatile-acid concentration?

32.4 What is the purpose of acidification of samples for the volatile-acid analysis?

32.5 What is the principle involved in the determination of volatile acids by column-partition chromatography?

32.6 What is the principle involved in the distillation procedure for volatile acids?

32.7 Explain why the lower-molecular-weight volatile acids that have the higher vapor pressure at 100°C have lower fractional recoveries in distillates than the higher-molecular-weight acids.

32.8 What precautions must be exercised in the direct distillation procedure for volatile acids?

32.9 The concentration of individual volatile acids was found by gas chromatography to be as follows: formic acid, 5 mg/L; acetic acid, 160 mg/L; propionic acid, 220 mg/L; butyric acid, 90 mg/L; isobutyric acid, 70 mg/L; pentanoic acid, 50 mg/L; and hexanoic acid, 60 mg/L. What is the total-volatile-acid concentration in mg/L as acetic acid?

32.10 Calculate the pH in a digester if the atmospheric pressure is 0.9 atm, the carbon dioxide content of the gas is 28 percent, and the bicarbonate concentration is 4000 mg/L as $CaCO_3$.

32.11 For Example 32.1, calculate what the new pH would be if an increase in volatile-acid concentration of 3000 mg/L was experienced and this was accompanied by an increase in the carbon dioxide content of the digester gas to 40 percent.

REFERENCE

Clesceri, L. S., A. E. Greenberg, and A. D. Eaton (eds.): "Standard Methods for the Examination of Water and Wastewater," 20th ed., American Public Health Association, Washington, DC, 1998.

C H A P T E R

Gas Analysis

33.1 | GENERAL CONSIDERATIONS

Anaerobic decomposition of sludges and of some liquid wastes, particularly those with BOD values exceeding 1500 mg/L, is considered the most economical method of treatment when the organic materials involved are amenable to methane fermentation. Anaerobic processes are popular because operating costs are low and energy is produced rather than consumed by the process.

The gas produced in anaerobic digestion of sludges usually contains 33 to 38 percent carbon dioxide, 55 to 65 percent methane, small amounts of hydrogen, some nitrogen, and traces of hydrogen sulfide; and it has a heating value in the neighborhood of 22,000 kJ/m^3 (600 Btu/ft^3 or 5,300 kcal/m^3). The caloric value of the gas produced is normally in excess of the heat requirements for maintaining proper temperatures in municipal sludge digestion units. It may be in excess for industrial wastewaters as well, depending upon waste strength and the heating needs for treatment. The excess gas has value for producing energy for heating or for useful work.

The composition of gas produced during anaerobic treatment varies somewhat with the environmental conditions in the reactor. It changes rapidly during the period that anaerobic treatment is being initiated, such as when reactors are first started, and when normal treatment is inhibited. For anaerobic reactors operating in a routine manner and being fed a given substrate regularly, the composition of the gas produced is fairly uniform; however, the ratio of carbon dioxide to methane varies radically with the character of the substrate undergoing decomposition. Buswell and Boruff[1] showed that methane fermentation of the three com-

---

[1]A. M. Buswell and C. S. Boruff, The Relation between the Chemical Composition of Organic Matter and the Quality and Quantity of Gas Produced during Sludge Digestion, *Sewage Works J.,* **4:** 454 (1932).

mon classes of organic materials produces carbon dioxide and methane in the following ratios:

	Carbon dioxide	Methane
Carbohydrates	50	50
Low-MW fatty acids	38	62
High-MW fatty acids	28	72
Proteins	$\begin{cases} 24 \\ 31 \end{cases}$	76 \\ 69

Because of the wide variety of organic substances (usually of unknown composition) that are subjected to methane fermentation, it has become common practice to analyze the gases produced to determine their fuel value and, in some cases, to maintain a check on the behavior of the treatment units. The idea has been advanced that the onset of anaerobic treatment troubles is accompanied by an increase in the carbon dioxide content of the gas produced and that this test can be used in place of the volatile-acids determination to detect such conditions. Experience will demonstrate whether carbon dioxide measurements can substitute for volatile-acids determinations in this capacity. Because of the influence of the substrate on the ratio of carbon dioxide to methane in the gas produced, it is reasonable to assume that interpretations based upon carbon dioxide content alone may lead to faulty conclusions.

Interest has been shown in recent years in the use of hydrogen concentration as an additional indicator of the performance of anaerobic treatment. Hydrogen plays a key role in anaerobic treatment as discussed in Sec. 32.2. In order for the anaerobic process to operate properly, the hydrogen concentration must be maintained quite low by the hydrogen-using methanogens, such that the concentration in the gas produced is normally below 0.01 percent. During digester upset or unbalance, the hydrogen concentration may exceed this value by one to two orders of magnitude, an increase that could signal the need for closer observation and control. Unfortunately, the standard methods for gas analysis given in the following will not measure hydrogen at such low values, and a different instrumental approach is required, such as use of a reduction gas detector.[2]

33.2 | METHODS OF ANALYSIS

Gas analysis can usually be accomplished either by the volumetric procedure or by gas chromatography, except for the measurement of hydrogen sulfide, which usually occurs in amounts too small to be measured by either of these procedures. In general, volumetric analyses are quite accurate and are suitable for the determination of oxygen, methane, hydrogen, and carbon dioxide. Nitrogen is usually determined in such analyses by an indirect procedure. Volumetric analyses are quite time-consuming. However, the equipment required is relatively simple, needs no

---

[2]D. P. Smith and P. L. McCarty, Energetic and Rate Effects on Methanogenesis of Ethanol and Propionate in Perturbed CSTRs. *Biotechnology and Bioengineering,* **34:** 39 (1989).

calibration before use, and therefore is particularly suitable when analyses are conducted on an infrequent basis.

Gas chromatographic analysis has the distinct advantage of speed; a gas analysis can be completed in only a few minutes. For this advantage to be utilized, however, the instrument must have been previously calibrated for each gas of interest, the oven must have reached a constant temperature, and the detector must be giving a stable response. Because of these requirements, this method of analysis is more suited to routine work where gas analyses are conducted several times each week, making it permissible to keep the instrument always in operating condition. This method is sufficiently accurate for most practical purposes, although not quite as good as volumetric analysis. Although it is possible to separate hydrogen, oxygen, nitrogen, methane, carbon dioxide, and hydrogen sulfide by gas chromatography, this is not presently practical on a routine basis, using one instrument with a single column and detector. In addition, the quantities of hydrogen or hydrogen sulfide occurring in digester gas are usually too small in concentration to be measured with any accuracy. Gas chromatography is mainly suited to routine analysis to determine the relative proportions of methane, carbon dioxide, and "air" (nitrogen plus oxygen) occurring in a sample of anaerobic reactor gas. Normally, these are the main gaseous components of interest.

33.3 | VOLUMETRIC ANALYSIS

Early methods of volumetric analysis employed separate measurement of carbon dioxide and oxygen, followed by a slow simultaneous combustion of hydrogen and methane. The analysis was completed by measuring the amount of carbon dioxide produced during the combustion of the methane and then employing a knowledge of Gay-Lussac's law of combining volumes (Sec. 2.9) to determine the amounts of methane and hydrogen present in the mixture.

In the past an apparatus was sometimes used that employed a slow-combustion unit in which hydrogen and methane were burned together in the same unit. However, its operation in the determination of hydrogen and methane was somewhat hazardous because of the possibility of explosions; and it is therefore not recommended. A preferred device, such as illustrated in Fig. 33.1, provides for separate oxidation of hydrogen and methane. Hydrogen is oxidized by passing the gas through a heated unit charged with cupric oxide, and methane is oxidized in a separate unit by bringing a mixture of it and oxygen in contact with a catalyst at relatively low temperatures. Explosion hazards are completely eliminated.

Carbon Dioxide

Carbon dioxide is measured by bringing a sample of known volume, usually 100 mL, into contact with a solution of potassium hydroxide. The carbon dioxide reacts with the hydroxide to form potassium carbonate, as shown in the following equation:

$$\underset{\text{1 vol.}}{CO_2} + 2KOH \rightarrow \underset{\text{0 vol.}}{K_2CO_3} + H_2O \qquad (33.1)$$

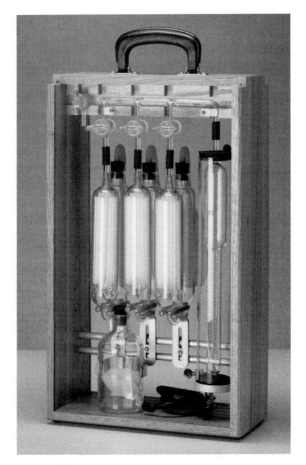

Figure 33.1
Gas-analysis apparatus equipped for separate low-temperature oxidation of hydrogen and catalytic oxidation of methane. *(Courtesy of Burrell Scientific, Inc.)*

In the reaction, the carbon dioxide disappears from the gaseous phase, and the potassium carbonate formed remains in the liquid phase; therefore, the loss in volume of the gas is equal to the carbon dioxide content. Additionally, any hydrogen sulfide present in the gas also combines with the potassium hydroxide,

$$H_2S + 2KOH \rightarrow K_2S + 2H_2O \qquad (33.2)$$
$$\text{1 vol.} \qquad\qquad \text{0 vol.}$$

but the volume of hydrogen sulfide is usually so small that its effect may be ignored.

Potassium hydroxide is used instead of sodium hydroxide to absorb carbon dioxide because of the greater solubility of potassium carbonate. If sodium hydroxide is used, sodium carbonate tends to precipitate. Some of it usually floats and clogs the capillary passages.

Oxygen

Theoretically, there is very little possibility of oxygen being present in the gas produced by anaerobic treatment. However, small amounts may gain entrance to the sample during the sampling procedure and during charging of the gas-analysis apparatus. It is therefore good practice to analyze for oxygen. The presence of more than 0.1 to 0.2 percent usually indicates poor technique in sampling and in transferring the sample to the gas-analysis apparatus.

Oxygen in gas from anaerobic treatment is generally measured by the use of alkaline pyrogallol. Under alkaline conditions, pyrogallol (1,2,3-trihydroxybenzene) is oxidized by oxygen. The end products of the reactions, such as carbon dioxide and organic acids, are held as potassium salts in the absorbing solution. If all carbon dioxide has been removed from the sample before bringing it into contact with the alkaline pyrogallol, any decrease in sample volume will be due to removal of oxygen.

Hydrogen

Hydrogen may be determined separately in the presence of methane by passing the gas mixture over cupric oxide maintained at a temperature in the range of 290 to 300°C. Under such conditions hydrogen is oxidized to water, but methane is not oxidized. The water vapor formed condenses at the temperatures to which the sample must be reduced for subsequent volume measurements, and therefore loss in volume after contact with heated cupric oxide is a measure of the hydrogen present.

Methane

After removal of hydrogen, methane may be determined in the residual gas by slow combustion or by catalytic oxidation. In either case oxygen is required, but the technique for each is very different.

Catalytic Oxidation   A mixture of the gas containing methane and oxygen is used in this method, and the oxidation is performed catalytically at temperatures below the kindling point; consequently, an explosion does not occur. The volume of oxygen required is determined from the equation

$$\underset{\text{1 vol.}}{CH_4} + \underset{\text{2 vol.}}{2O_2} \rightarrow \underset{\text{1 vol.}}{CO_2} + \underset{\text{0 vol.}}{2H_2O} \tag{33.3}$$

It will be noted that 2 volumes of oxygen are required to oxidize 1 volume of methane and that 1 volume of carbon dioxide is produced. The 2 volumes of water vapor produced are reduced to zero volume at the temperature at which final gas volumes are measured. Since 2 volumes of oxygen are required for each volume of methane, at least $2\frac{1}{2}$ volumes should be used to provide a sufficient excess to carry the reaction to completion.

After the analyses for carbon dioxide, oxygen, and hydrogen are finished, the volume of residual gas containing methane is about 60 to 70 mL if a 100-mL sample is used initially. It is impossible to mix 60 to 70 mL of gas with $2\frac{1}{2}$ times its volume of oxygen in the equipment provided. The maximum-size sample of methane that may be used in a 100-mL buret and be mixed with $2\frac{1}{2}$ times its volume of oxygen is about 28 mL. The usual procedure is to waste sufficient residual gas to leave a sample ranging in size from 20 to 25 mL. Some analysts prefer not to waste the gas but store the excess in the oxygen absorption pipet as a reserve supply in case it is needed. Oxygen is next admitted to the sample in proper amount, and then the mixture is brought into contact with the catalyst.

The methane content of the gas used for the combustion may be determined in two ways. Inspection of Eq. (33.3) will show that 2 volumes of oxygen are used for each volume of methane oxidized and 1 volume of carbon dioxide results. A contraction in gas volume occurs that is equal to the amount of oxygen used. Since this is equal to two times the methane.

$$\text{Volume of methane} = \tfrac{1}{2} \text{ total contraction}$$

The residual gas, after measurement to determine total contraction, still contains carbon dioxide in a volume equal to the methane originally present. It may be measured by bringing it into contact with potassium hydroxide and measuring the residual gas to get the volume absorbed.

Final calculations of methane content must be based upon an adjustment of the values obtained on the test portion of the methane-bearing sample to the total volume. For example, if 65 mL remained after carbon dioxide, oxygen, and hydrogen were removed and 25 mL was used for the methane determination, the values should be multiplied by a factor of 65/25 to obtain the percentage of methane.

Slow Combustion   The separate determination of methane by the slow-combustion method is not common. The procedure is essentially the same as for a combination of hydrogen and methane. Analysis of the data is exactly as described under catalytic oxidation.

Nitrogen

Nitrogen is a relatively inert gas and remains unchanged at the end of the usual gas-analysis procedure. It is assumed that it is the only component of any significance that behaves in this manner. It is customary to total the carbon dioxide, oxygen, hydrogen, and methane percentages and subtract them from 100. The difference represents inert gases and is reported as nitrogen.

Sources of Error

There are four major sources of error in gas analysis, as follows:

Collection, Storage, and Handling of Samples   Unless special care is taken in the collection of samples, contamination by air occurs. Samples should always be collected in glass or metal tubes, particularly if an appreciable time elapses before analysis can be made. Gum-rubber balloons are not suitable because they are permeable to hydrogen and methane. Transfer of gas from the sample tube to the gas-analysis apparatus requires the use of a displacing fluid, and some modification of the sample is apt to occur. Also, some air may gain entrance.

Confining Fluid   Mercury is the ideal confining fluid because of the insolubility of all gases in it but, because of its great density, cost, and potential for human health hazard, it is not used today. For ordinary purposes, a high degree of accuracy is not needed and less ideal confining fluids can be used. Water has much too great a solvent power for all the gases involved to serve satisfactorily. However, it has been found that an aqueous solution containing 20 percent sodium sulfate and 5 percent sulfuric acid has markedly reduced solvent powers. This is the mixture normally used in portable equipment but does introduce some error in analysis.

Incomplete Combustion of Methane   During the combustion of methane, a high concentration of oxygen is present at the start of the combustion, but as the combustion proceeds, the oxygen concentration decreases markedly, owing to use and to dilution by the carbon dioxide formed. Unless a volume of oxygen at least $2\frac{1}{2}$ times the size of the gas sample is used, an oxygen deficiency may occur. Incomplete combustion is a common cause of high nitrogen values and of low methane values.

Temperature Changes   In gas analysis, small changes in temperature can cause serious errors. This is a special problem during the measurement of hydrogen and methane, where the reactions are conducted at high temperatures. There is always a tendency to measure the volume remaining after combustion before the temperature of the gas has returned to the original value. This may lead to positive errors in some instances and negative errors in others.

33.4 | GAS CHROMATOGRAPHIC ANALYSIS

Gas chromatography affords a rapid and simple method of gas analysis when used on a routine basis. The principles of this method were outlined in Sec. 12.4. An instrument equipped with a thermal conductivity detector is usually used for analysis of gas from anaerobic treatment, and helium is normally used as a carrier gas to sweep the gas sample through the instrument. Only about 1 or 2 mL of a gas sample are required for the analysis, and this is usually introduced into the instrument with a syringe, although some instruments are equipped with a gas sampling port, which simplifies sample measurement and introduction.

One of the most important items in the gas chromatograph is the packing material used in the column for separation of the gaseous components. Several commercially available column packings are listed in "Standard Methods," together with the gases for which they are particularly suitable for separating. Columns that separate air (oxygen plus nitrogen), methane, and carbon dioxide are necessary for routine analysis of gas from anaerobic treatment. Columns that also give separation between the two components of air are more desirable, however. The ability to determine the presence of oxygen is desirable since this is a good check on the adequacy of sampling technique as discussed under the volumetric method.

Most instruments have a sufficiently stable operation, so at a given temperature of operation and gas flow rate a linear relationship will exist between the height of a peak on the gas chromatogram and the concentration of a gaseous component. Thus, by using a pure gas standard and injecting different volumes into the gas chromatograph, a curve of peak height versus gas volume may be prepared. The percentage by volume of a given gas in a sample is then equal to its measured volume divided by the total of the sample injected. Peak area is also widely used and can be determined automatically with an integrater.

Thermal-conductivity detectors tend to lose sensitivity with time, with the result that the peak height for a given volume of gas tends to decrease. This can be corrected by frequent standardization. However, it has been found that the decrease in peak height is proportionally the same for methane, carbon dioxide, and air. Since gas from anaerobic treatment is composed almost entirely of these gases, the summation of their respective concentrations should total almost 100 percent. Any significant decrease below this total will indicate a decrease in detector sensitivity, provided that other analytical errors have not been made. If this is true, then the correct percentage for each component can be determined by proportioning each up to give a total of 100 percent. For example, if the measured percentages of the three components total 90 and the measured methane percentage is 54 percent, its actual percentage would be 54(100/90) or 60 percent.

Sources of Error

Errors in collection, storage, and handling of samples will be the same as with the volumetric analysis method. In addition, changes in instrument temperature or carrier gas flow rate, changes in detector sensitivity, degeneration of column packing, and inaccuracies in measurement of volume of injected gas samples will all result in errors. With gas from anaerobic treatment the summation of the methane, carbon dioxide, and air percentages should total close to 100 percent. If it does not, then one of the discussed sources of errors might be the cause.

33.5 | HYDROGEN SULFIDE

The measurement of hydrogen sulfide is particularly important where gas is to be used for fuel in gas engines. Most engine manufacturers specify that the gas used should not have more than 50 grains of H_2S/100 ft^3 (1.14 mg/L), in order to prevent harm from corrosion.

Hydrogen sulfide is commonly measured in gas by means of the Tutweiler apparatus shown in Fig. 33.2. The procedure is essentially as follows: A sample of gas is introduced into the apparatus, which contains a small amount of starch indicator. Small amounts of a standard iodine solution are added intermittently, with vigorous shaking between additions of iodine. The iodine reacts with hydrogen sulfide as shown in the following equation:

$$H_2S + I_2 \rightarrow 2HI + S \tag{33.4}$$

When sufficient iodine has been added to oxidize all the hydrogen sulfide, excess iodine is indicated by the typical blue color produced by starch indicator. The strength of the iodine solution is selected to facilitate calculation of hydrogen sulfide in the particular units desired.

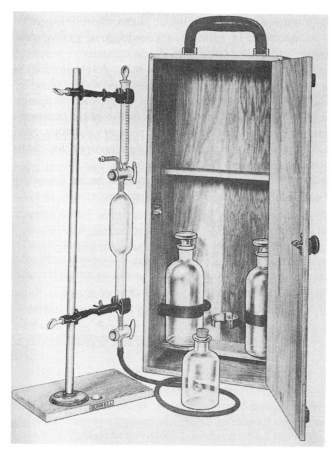

Figure 33.2
Tutweiler apparatus used to determine hydrogen sulfide in sludge
gas. *(Courtesy of Burrell Scientific, Inc.)*

33.6 | APPLICATIONS OF GAS-ANALYSIS DATA

Gas analyses are used largely at wastewater treatment plants where information on
the fuel value of gas is important. In addition, a knowledge of gas composition can
be of considerable help in the control of digestion units as a supplement to information provided by volatile-acids, pH, and other determinations or as a replacement
for some of them. Under normal conditions of operation, the percentages of
methane and carbon dioxide do not vary greatly. Sudden changes in gas composition can signal a change either in the operation of the treatment unit or in the
amount or composition of incoming wastewater or sludge. Such changes can thus
be used as a warning sign to suggest the need for closer observation and control of
the treatment unit.

The determination of hydrogen sulfide will continue to be an important consideration wherever gas is used for fuel in gas engines, particularly in areas where the sulfate content of wastewater is high.

PROBLEMS

33.1   How is gas analysis used in the control of the anaerobic waste treatment process?

33.2   What principle is involved in the separation of gases by chromatography?

33.3   (*a*)   How is the fraction of nitrogen determined in a gas sample using volumetric analysis?

     (*b*)   What sources of error would be of particular concern in determining nitrogen by this method?

REFERENCE

Clesceri, L. S., A. E. Greenberg, and A. D. Eaton (eds.): "Standard Methods for the Examination of Water and Wastewater," 20th ed., American Public Health Association, Washington, DC, 1998.

CHAPTER 34

Trace Contaminants

34.1 | GENERAL CONSIDERATIONS

The subject of trace organic and inorganic contaminants in foods has been under investigation for over a century[1] and, since 1960, has become an item of increasing concern in the field of water supply.[2] In the context used here, the term *trace* means low concentration, perhaps 1 mg/L or less. Although the primary interest here is in water, air, and their impurities, the effects of chemical impurities in water on human populations ordinarily cannot be easily isolated for epidemiological or statistical study, because both food and water involve the digestive tract and vital organs of the body, particularly the kidneys and liver. In general the "food factor" is so great as to overwhelm the "water factor." However, much recent epidemiological research has indicated that some adverse health outcomes (e.g., low birth weight, intrauterine growth retardation, cancer) can be correlated to drinking water contamination. Synthetic organic chemicals are of most concern in this regard, and drinking water standards for many of these compounds have been or are being developed.

Trace inorganic and organic substances in water supplies originate from natural sources, both plant and animal, and from the synthetic organic chemical industry. Surface water supplies are prone to carry significant amounts of trace materials, but groundwaters are not immune, particularly shallow well supplies which in some locations carry heavy burdens of industrial chemicals. In surface waters, natural plant exudates from growing algae and fungi carry materials of wide variety that cause taste and odors. Several are sulfur-containing compounds such as dimethyl sulfide and various mercaptans, including methyl mercaptan, isopropyl mercaptan, and

---

[1]National Research Council, Committee on Food Protection, "Toxicants Occurring Naturally in Foods," National Academy of Sciences, Washington, DC, 1973.

[2]AWWA Committee Report, Organic Contaminants in Water, *J. Amer. Water Works Assoc.,* **66:** 682 (1974).

butyl mercaptan. Geosmin contained in exudates from actinomycetes is another such compound.[3]

Geosmin

Decomposition of plant materials releases colored substances of a highly refractive nature. Except for tannins, the exact nature of these substances is not known, but they are polymeric, have high molecular weights, and are phenolic in character. A symbolic description according to Christman and Minear[4] is

These substances are commonly referred to as humic materials because of their similarity to humus in soil-organic matter. The excreta of humans and land animals contain soluble refractory organic matter that survives in the effluents of waste treatment plants, even those employing advanced treatment methods. It is reasonable to assume that the excreta of aquatic animals—fish, frogs, turtles, snakes, etc.—contain similar refractory materials. In human excreta, the soluble refractory organics originate largely from the unabsorbable fraction of foods and from wastes generated by kidney and liver function. Excluding aesthetic considerations, there is little evidence to indicate that water supplies containing trace organics derived from the excreta of warm- or cold-blooded animals have any public health significance. A possible exception is the finding that chlorination of humic substances in water results in the formation of chloroform and other halogenated organic compounds.

[3]D. Jenkins, Effects of Organic Compounds—Taste, Odor, Color, and Chelation, Organic Matter in Water Supplies; Occurrence, Significance, and Control, in V. L. Snoeyink and M. F. Whelan (eds.), *Univ. Illinois Bull.,* **70:** 15 (June 4, 1973).

[4]R. F. Christman and R. A. Minear, Organic Compounds in Aquatic Environments, in S. D. Faust and J. V. Hunter (eds.), p. 119, Marcel Dekker, New York, 1971.

The major concern related to trace contaminants in water supplies, however, stems from the wide variety of synthetic chemicals that gain access to our surface waters and some ground supplies through the discharge of wastewaters, from industries and municipal industrial complexes, from spills that result from accidents that occur during transportation by land and water, and from uncontrolled use or discharge of these materials to the environment.

The literature shows that lead was the first metal to be brought under scrutiny due to the prevalent use of lead service pipes. The problem was particularly severe in soft water areas, such as New England, because protective coatings did not develop. Zinc and copper were questioned in 1923 and 1926, respectively, because of the use of galvanized services and the growing use of copper salts for algae control. About the same time, iodide was found in significant amount in some supplies, notably in Pennsylvania. In 1931, Churchill showed the correlation between excessive fluoride and mottling of tooth enamel. The literature is replete with references to iron and manganese, largely because of the nuisances they produce rather than threats to the public health. It was not until the 1962 U.S. Public Health Drinking Water Standards were advanced that limits were set on the levels of inorganics of public health significance other than copper and zinc.

The attention of public health and water supply engineers was then refocused upon trace inorganics by four important factors: (1) the outbreak of "Itai-Itai" (ouch-ouch) disease that occurred among farmers who drank water containing cadmium from the Jintsu river in Japan: (2) the discovery that metallic mercury escaping from laboratories and industry, mainly chlorine manufacture, was capable of being converted to methyl mercury, concentrated by aquatic life, and passed along through natural food chains to humans, largely through fish: (3) the evidence that exists to indicate that certain forms of the inorganics may be carcinogenic; and (4) the realization that purposeful recycling of wastewaters for supplementation of drinking water supplies is imminent in many areas of the country.

The particular concern following publication of Rachel Carson's book "Silent Spring"[5] in 1962 shifted to insecticides and herbicides, widely used in agricultural and forestry operations, and then to polychlorinated biphenyls (PCB), and a variety of other halogenated organic compounds. Much of the past concern about these compounds was related to the ability of aquatic plants and animals to store and concentrate these materials in their tissues and, thereby, interfere with their reproductive processes. Of greater concern, however, was the potential hazard to humans using the animals for food. The Federal Drug Administration ban on the sale of Coho salmon taken from Lake Michigan because of their unusual PCB content is one example.

The chlorinated organic compounds—DDT; 2,4-D; 2,4,5-T; PCB; chlordane, and others—were singled out for special consideration because of their resistance to biological degradation, i.e., their refractory nature. Then in the 1970s, a suspected new source of chlorinated compounds was exposed. This was related to the use of chlorine for disinfection purposes. This fear was enhanced by reports of serious fish kills in streams below sewer outfalls where disinfection with chlorine was prac-

---

[5]R. Carson, *Silent Spring,* Houghton Mifflin Co., Boston (1962).

ticed. These results have raised serious doubt in the minds of many regarding the safety of using chlorine for the disinfection of drinking water supplies, in spite of a 70-year history showing its great benefit for disease prevention.

Studies in 1974 of the drinking water supplies of New Orleans and Cincinnati isolated and identified 166 separate organic compounds or derivatives; many chloro and some bromo derivatives. Most of the organics found at New Orleans were present at concentrations of 1 μg/L or less, with a few in the 1- to 10-μg/L range, but chloroform was found at levels of 40 to 150 μg/L in six separate finished water supplies. Similar levels were then found in water supplies throughout the country, and subsequent evidence proved these levels result from reactions between chlorine and naturally occurring organic matter.

At one time it was felt that only surface waters were subject to contamination with trace organics. However, groundwater supplies were also revealed in the early 1980s to be contaminated. Some estimates indicated that as many as one-quarter to one-third of the U.S. municipal wells are contaminated with xenobiotic (synthetic) organic and inorganic compounds. Chlorinated solvents such as trichloroethene, tetrachloroethene, 1,1,1-trichloroethane, and dichloromethane, degradation products such as vinyl chloride, and the BTEX group of petroleum hydrocarbons are among the most commonly found organic contaminants. Chromium, mercury, arsenic, and many other hazardous inorganic elements are also found. Levels of contamination range from the microgram per liter range to the 10- to 100-mg/L range at contaminated locations resulting from hazardous waste disposal, accidental spills, and leaking storage tanks.

It is generally conceded that the organic and inorganic materials in surface water supplies, under normal conditions, cannot be considered to be present in the toxic range capable of producing violent illness or death. Rather, some must be viewed in terms of hazardous materials because of their possible subtle, long-term effects. Among the chemicals identified have been some with known effects upon the central nervous system and others with *carcinogenic* (oncogenic or tumor causing) properties. Also, the possibility exists that some with *teratogenic* (disfiguring) or *mutagenic* characteristics may be present. Some of these compounds may be involved in birth defects, retardation of intrauterine growth, central nervous system disorders, and other adverse health outcomes. The importance of these materials in public water supplies is difficult to ascertain with certainty for the reasons previously expressed. Nevertheless, we should expend every effort to minimize their presence by source control and proper water and wastewater treatment. Newer concerns about water focus on endocrine-disrupting chemicals and pharmaceuticals.

Also, an ever-increasing environmental concern is the issue of trace contaminants in the atmosphere. Millions of kilograms of organic and inorganic materials are emitted into the atmosphere from a wide variety of industrial operations, as well as from the more well-known source of combustion of fossil fuels. The passage of the Clean Air Act of 1991 in the United States has resulted in a focus on air toxics, that is, the presence of trace contaminants of concern in air. This will be an even more difficult issue to address than the health impacts of trace contaminants in water because of the difficulty in assessing the exposure concentration and duration.

34.2 | THE SAFE DRINKING WATER ACT

The critical nature of the water supply problem brought congressional action resulting in Public Law 93-523, known as the Safe Drinking Water Act (SDWA), which was signed into law in 1974. It then became incumbent upon the newly formed Environmental Protection Agency (EPA) to interpret the law and establish Drinking Water Standards that apply to all public water systems in the United States, not just those that supply common carriers, as did the U.S. Public Health Standards. The SDWA was further strengthened by passage in 1986 of the SDWA amendments. The SDWA resulted in a significant increase in the number of trace contaminants to be regulated, particularly organic contaminants.

The SDWA charged the EPA to establish both enforceable primary drinking water standards for health-related contaminants and nonenforceable secondary drinking water guidelines for contaminants that may adversely affect the aesthetic quality of drinking water. With primary standards, maximum contaminant levels (MCLs) are established, and with the secondary guidelines, secondary standards are established for the contaminants of concern. The federal government is charged only with enforcing the MCLs. The secondary standards serve only as guidelines to water purveyors and the public, but are often incorporated into drinking water standards enforced at the state level.

In addition to enforceable MCLs, the EPA establishes a nonenforceable maximum contaminant level goal (MCLG) for each health-related constituent. MCLGs are set at a level at which no known or anticipated adverse effect on human health occurs and that allows for an adequate margin of safety, without regard to the cost of reaching these goals. MCLs are to be set as close to MCLGs as feasible, taking into consideration analytical instrument limitations, and limitations of the best technology, treatment techniques, other means of achieving standards that are available, and cost. The EPA has adapted the policy that MCLGs for known or probable carcinogens be set at zero. This policy is based upon the assumption that there is no threshold concentration for carcinogens; it is possible that a single molecule may cause cancer, although the probability of it doing so may be extremely small. An MCLG of zero means none. Since all current analytical methods have finite detection limits, zero is just a concept, not something that can in fact be measured or enforced. For noncarcinogens, MCLGs are based upon "no-effect" levels for chronic-lifetime exposures to the contaminants and include a factor of safety. The assumption here is made that an organism can tolerate or detoxify a certain amount of toxic agent with no ill effect up to a certain threshold. MCLGs and MCLs for noncarcinogens are often the same.

The EPA has also established a drinking water contaminant candidate list as required by the SDWA to aid in priority setting for the Agency's drinking water program. A chemical may be placed on this list if it is known to adversely affect public health and is likely to occur in public water systems with a frequency and at levels posing a threat to public health. The agency then enters a process of evaluation to determine whether the contaminants on this list deserve to be regulated under the SDWA. Two chemicals on this list that have received wide publicity in recent years are methyl *tert*-butyl ether (MTBE), a gasoline component, and perchlorate (ClO_4^-), a rocket fuel, both of which have contaminated several groundwater resources in recent years. MTBE is discussed in Sec. 5.7. Perchlorate is addressed in Sec. 34.5.

Guidelines for drinking water quality are also set by the World Health Organization (WHO). These guidelines "are intended to be used as a basis for the development of national standards that, if properly implemented, will ensure the safety of drinking-water supplies through the elimination, or reduction to a minimum concentration, of constituents of water that are known to be hazardous to health." The WHO guidelines are not mandatory limits, but are for use by countries in the context of local or national environmental, social, economic, and cultural conditions. In general, more constituents are included within the WHO guidelines than within the EPA primary and secondary standards.

34.3 | DRINKING WATER STANDARDS

Current (January 1, 2002) U.S. EPA drinking water standards and WHO guidelines for trace organic contaminants are listed in Table 34.1 and for trace inorganic contaminants in Table 34.2. Values of EPA secondary standards or WHO guidelines for substances causing consumer complaints are listed in Table 34.3. For some contaminants, such as epichlorohydrine, acrylamide, copper, and lead, the MCL specifies that best available treatment technology be used rather than providing a numerical concentration limit. This is permitted when it is determined that monitoring for the contaminant is not economically or technically feasible. Epichlorohydrine and acrylamide are found as impurities in coagulant aids. These compounds are difficult to measure and quantify, so the requirement here is to certify that they do not exceed specified levels in the water treatment chemicals used. Copper and lead are particularly difficult chemicals to regulate with MCLs since they can result as products from corrosion of utility lines, frequently within the home. Thus, their absence in water leaving a treatment plant or, indeed, in water taken from the distribution system, may not guarantee their absence in tap water. The treatment technologies used to satisfy the MCLs here are complicated and involve monitoring, corrosion control, public information, and pipe replacement.

Compounds that are readily degraded in the environment, even though they may be highly toxic, may not be subject to a drinking water standard. An example of this is the phosphorus-based pesticide parathion, which is subject to water hydrolysis. This should not imply that the presence of such contaminants in water would not be of concern but only that there is a very low probability that they would actually be found in drinking waters at hazardous levels.

Trace contaminants are conveniently divided into organic and inorganic contaminants. A broad discussion of organic contaminants is provided in Chap. 5 and will not be repeated here. Only generalizations are provided. However, trace inorganic contaminants are discussed only briefly in preceding chapters, and so a more general discussion of these is provided in Sec. 34.5.

Table 34.1 | Trace organic chemicals regulated through the U.S. EPA drinking water standards or contained in the WHO guidelines

Contaminant	EPA MCLG, mg/L	EPA MCL, mg/L	WHO guidelines, mg/L
Pesticides and herbicides			
Alachlor	Zero	0.002	0.02
Atrazine	0.003	0.003	0.002
Carbofuran	0.04	0.04	0.007
Chlordane	Zero	0.002	0.0002
Dibromochloropropane (DBCP)	Zero	0.0002	0.001
2,4-D	0.07	0.07	0.03
1,2-Dichloropropane	Zero	0.005	0.04
Diquat	0.02	0.02	0.01
Endrin	0.002	0.002	
Ethylene dibromide (EDB)	Zero	0.00005	
Heptachlor	Zero	0.0004	0.00003
Heptachlor epoxide	Zero	0.0002	0.00003
Lindane	0.0002	0.0002	0.002
Methoxychlor	0.04	0.04	0.02
Pentachlorophenol	Zero	0.001	0.009
Picloram	0.5	0.5	
Simazene	0.004	0.004	0.002
Toxaphene	Zero	0.005	
2,4,5-TP (Silvex)	0.05	0.05	0.009
Disinfection byproducts			
Bromate	Zero	0.010	0.025
Chlorite	0.8	1.0	0.2
Haloacetic acids (total)		0.060	
Monochloroacetic acid			
Dichloroacetic acid	Zero		0.05
Trichloroacetic acid	0.3		0.1
Bromoacetic acid			
Dibromoacetic acid			
Trihalomethanes (total)		0.080	
Chloroform			0.2
Bromodichloromethane	Zero		0.06
Dibromochloromethane	0.06		0.1
Bromoform	Zero		0.1
Cyanogen chloride (as CN)			0.07
Chlorinated solvents and related chemicals			
Carbon tetrachloride	Zero	0.005	0.002
1,2-Dichloroethane	Zero	0.005	0.03
1,1-Dichloroethene	0.007	0.007	0.03
cis-1,2-Dichloroethene	0.07	0.07	0.05
trans-1,2-Dichloroethene	0.1	0.1	0.05
Dichloromethane	Zero	0.005	0.02
Tetrachloroethene	Zero	0.005	0.04
1,1,1-Trichloroethane	0.02	0.2	2
1,1,2-Trichloroethane	0.003	0.005	
Trichloroethene	Zero	0.005	0.07
Vinyl chloride	Zero	0.002	0.005

Table 34.1 | (continued)

Contaminant	EPA MCLG, mg/L	MCL, mg/L	WHO guidelines, mg/L
Aromatic hydrocarbons			
Benzene	Zero	0.005	0.01
Benzo(a)pyrene	Zero	0.0002	0.0007
Ethylbenzene	0.7	0.7	0.3
Styrene	0.1	0.1	0.02
Toluene	1	1	0.7
Xylenes (total)	10	10	0.5
Other chlorinated synthetic organic compounds			
Chlorobenzene	0.1	0.1	0.3
p-Dichlorobenzene	0.075	0.075	0.3
o-Dichlorobenzene	0.6	0.6	1.0
1,2,4-Trichlorobenzene	0.07	0.07	0.02
Hexachlorobenzene	Zero	0.001	
Hexachlorobutadiene			0.0006
PCBs	Zero	0.0005	
Dioxin (2,3,7,8-TCDD)	Zero	$3(10)^{-8}$	
Epichlorohydrin	Zero	TT*	0.0004
Hexachlorocyclopentadiene	0.05	0.05	
Other nonchlorinated synthetic organic compounds			
Acrylamide	Zero	TT*	0.0005
Dalapone	0.2	0.2	
Di(2-ethylhexyl)adipate	0.4	0.4	0.08
Di(2-ethylhexyl)phthalate	Zero	0.006	0.008
Dinoseb	0.007	0.007	
EDTA			0.6
Endothall	0.1	0.1	
Glyphosate	0.7	0.7	
Nitrilotriacetic acid			0.2
Tributyltin oxide			0.002
Oxamyl (vydate)	0.2	0.2	

*Standards specify the best available treatment technology.

34.4 | TRACE ORGANIC CONTAMINANTS

Most trace organic contaminants regulated for drinking water are chlorinated anthropogenic compounds. The largest category represents pesticides and herbicides, which is not surprising since these compounds find use because of their lethal properties. These were among the first trace organic contaminants to be regulated. Most are halogenated, with the exceptions of carbofuran and diquat. Chlorinated organic compounds also make up a large portion of the remaining trace organic contaminants that are regulated in drinking water. Halogenated compounds tend to be sin-

Table 34.2 | Trace inorganic chemicals regulated through the U.S. EPA drinking water standards or contained in the WHO guidelines

| | EPA | | WHO |
| | MCLG, | MCL, | Guidelines, |
Contaminant	mg/L	mg/L	mg/L
Metals			
Antimony	0.006	0.006	0.005
Barium	2	2	0.7
Beryllium	0.004	0.004	
Boron			0.5
Cadmium	0.005	0.005	0.003
Chromium (total)	0.1	0.1	0.05
Copper	1.3	TT*	2
Lead	Zero	TT*	0.01
Manganese			0.5
Mercury	0.002	0.002	0.001
Nickel	0.1	0.1	0.02
Selenium	0.05	0.05	0.01
Thallium	0.0005	0.002	
Uranium			0.002
Others			
Arsenic	Zero	0.01	0.01
Asbestos	7 MFL[†]	7 MFL[†]	
Cyanide	0.2	0.2	0.07

*Standards specify the best available treatment technology.
[†]Million fibers per liter.

Table 34.3 | U.S. EPA secondary standards and WHO guidelines for aesthetic quality for trace contaminants in drinking water

Contaminant	EPA secondary standards, mg/L	WHO guidelines, mg/L
Aluminum	0.05 to 0.2	0.2
Copper	1	1
Foaming Agents	0.5	
Hydrogen sulfide		0.05
Iron	0.3	0.3
Manganese	0.05	0.1
Silver	0.1	
Zinc	5	

gled out because of their general toxicity, relatively high hydrophobicity (which tends to cause them to concentrate in fatty materials), large volume usage, and relative resistance to biodegradation.

Disinfection by-products is a category of growing concern because of potential toxicity from many of the compounds formed from disinfection with chlorine and related compounds, as well as from the disinfectants themselves. Disinfection of water supplies is essential for disease prevention, but at the same time, unwanted by-products of disinfection need to be minimized to the extent possible. The number of regulated disinfection by-products has grown over the past 10 years, and now includes haloacetic acids as well as trihalomethanes. Both haloacetic acids and trihalomethanes have MCLs for the sum total of compounds in each group. Some individual compounds within each group also have MCLGs. This suggests that these compounds are the most hazardous within the group.

Chlorinated solvents and related chlorinated aliphatic hydrocarbons (CAHs) constitute another large group of drinking water regulated contaminants. These are among the group of compounds termed volatile organic compounds (VOCs) that have come under regulation since their wide presence in groundwater was discovered. Also included as VOCs and listed in Table 34.1 are the aromatic hydrocarbons [except benzo(a)pyrene] and a few of the other chlorinated synthetic organic compounds such as the chlorinated benzenes. VOCs are generally not a problem in surface waters because they are rapidly removed from water by aeration processes and transferred to the atmosphere. However, in groundwaters contact with a gas phase and turbulence are low, and so following contamination, concentrations of VOCs can remain quite high. These compounds also tend to be less hydrophobic than the pesticides and so move more readily through the soil and into groundwater. The two most common groundwater groups of contaminants are the chlorinated solvents and the aromatic hydrocarbons. The predominant CAHs found in groundwater are the five chlorinated solvents, carbon tetrachloride, tetrachloroethene, trichloroethene, 1,1,1-trichloroethane, and dichloromethane, and their chemical and biological degradation products, which include dichloroethane and dichloroethene isomers, vinyl chloride, and chloroform. The major aromatic compounds found in groundwater are the water-soluble components of gasoline, which include benzene, toluene, ethylbenzene, and xylenes (BTEX).

The drinking water regulated trace organic contaminants include a series of other miscellaneous compounds that are used by society for a variety of purposes and have sufficient toxicity, solubility, resistance to biodegradation, and presence in drinking water to be regulated. Compounds other than those listed in Table 34.1 are also being considered by the EPA for establishment of MCLs.

34.5 | TRACE INORGANIC CONTAMINANTS

The principal source of all the trace inorganic contaminants listed in Table 34.2 with the exception of fluoride and selenium is industrial wastes from manufacturing or metal-finishing operations. Because of this, municipalities that accept industrial wastes into their sewerage systems are legally responsible for reducing the concentrations to acceptable levels before discharge into surface water or groundwater. Usually, treatment at the source is the only practical way of ensuring that the prescribed level of trace inorganics can be met. This poses particular problems in cities where industrial sewer connections have been long established.

Metals

Antimony   Antimony has oxidation states of (III) and (V), forming compounds with halides, oxygen, and sulfur. It is used for flameproofing textiles, in vulcanizing of rubber, in the manufacture of paint, semiconductors, thermoelectric devices, and fireworks, and in parasiticides to treat schistosomiasis and leishmaniasis. Antimony resembles arsenic both chemically and biologically and exhibits similar toxic effects. The major toxic symptoms involve the gastrointestinal tract, heart, respiratory tract, skin, and liver. Some evidence implicates antimony in reproductive problems for women.

Barium   Barium salts are used mainly in the manufacture of paints, linoleum, paper, and drilling muds. Fortunately, the principal form is the sulfate, which is highly insoluble. A limit of 2.0 mg/L has been placed on barium because prolonged tests with experimental animals has shown muscular and cardiovascular disorders and kidney damage.

Beryllium   Beryllium is used in metallurgy to make special alloys, in the manufacture of X-ray diffraction tubes and electrodes for neon signs, in nuclear reactors, and in rocket fuels. Beryllium does not occur at high concentrations in natural waters because of the rarity of its occurrence, the insolubility of its oxides and hydroxides, and its tendency to sorb onto soils at near neutral pH. Beryllium has been found to cause cancer in animals, and for that reason, is considered a possible human carcinogen.

Cadmium   Cadmium is used extensively in the manufacture of batteries, paints, and plastics. In addition, it is used to plate iron products, such as nuts and bolts, for corrosion prevention. It is from plating operations that most of the cadmium reaches the water environment. At extreme levels, it causes an illness called "Itai-Itai" disease, characterized by brittle bones and intense pain. At low levels of exposure over prolonged periods, it causes high blood pressure, sterility among males, kidney damage, and flu-like disorders. It has recently been discovered that significant amounts are contained in cigarette smoke.

Chromium   In the water environment, soluble chromium exists primarily in the form of chromate [Cr(VI)]. Trivalent forms [Cr(III)] are hydrolyzed completely in natural waters, and the chromium precipitates as the hydroxide, leaving minor amounts in solution. Furthermore, there is no evidence to indicate that the trivalent form is detrimental to human health. Chromium is used extensively in industry to make alloys, refractories, catalysts, chromic oxide, and chromate salts. Chromic oxide is used extensively to produce chromic acid in the plating industry. Chromate salts are used in paints and to produce "cleaning solution" in laboratories. Most of the latter eventually reaches the sewer system. Chromate poisoning causes skin disorders and liver damage. There is some reason to believe that chromates are oncogenic (carcinogenic). For this reason, the permissible level in drinking waters has been restricted to 0.1 mg/L.

Copper Although copper is used commercially for many purposes, its major source in drinking water is corrosion of copper pipes used for water conveyance. Copper is a gastrointestinal tract irritant but is generally not harmful to humans at low milligram per liter concentrations. However, a few individuals with a copper metabolism disorder termed Wilson's disease can be affected by such concentrations. The restriction to 1.3 mg/L of copper in drinking water helps avoid such problems and also guards against the copper taste that occurs at higher levels. In surface waters, copper is toxic to aquatic plants at concentrations sometimes below 1.0 mg/L and has frequently been used as the sulfate salt to control growth of algae in water supply reservoirs. Concentrations near 1.0 mg/L can be toxic to some fish. It thus tends to be much more of an environmental hazard than a human hazard.

Lead Lead is highly toxic and is also considered a probable carcinogen. Lead poisoning has been recognized for many years. It occurs at concentrations in the blood of a large number of children at levels known to cause adverse health effects. Lead has been identified as being a cause of brain and kidney damage. In youngsters it may result in mental retardation and even convulsions in later life. This has brought about the abandonment of the use of lead service pipes and lead-based paints for interior decoration; the latter because of the tendency of some children to gnaw wood and eat the paint. However, lead can still enter drinking water from solder used for connecting copper pipes. Recognition of the fact that most of the tetraethyl lead contained in gasolines is expelled to the atmosphere as lead oxide has resulted in the elimination of lead in gasoline.

Mercury Mercury is widely used in amalgams, scientific instruments, batteries, arc lamps, and extraction of gold and silver, and the electrolytic production of chlorine. Its salts are used as fumigants in combating plant diseases and insect pests. It is also used as an antifoulant in ship paints and mildew proofing of canvas tarpaulins and tents. The most spectacular incident of mercury poisoning in humans resulted from the ingestion of sea food taken from Minamata Bay, Japan, during the late 1950s. A chemical plant, whose wastewaters discharged into the bay, was found to be the source of the mercury. Of a total of 111 cases reported, 43 people died. Babies born of afflicted mothers suffered congenital defects.

In the United States and Canada, a potential problem of mercury poisoning was discovered from an unusual amount of mercury found in fish taken from Lake St. Clair. The major source of the mercury was found to be from chlor-alkali plants employing mercury electrodes. The high concentration of mercury in the fish was found to be due to methylated mercury compounds, CH_3Hg^+ and $(CH_3)_2Hg$, that were shown to be produced by bacterial action in bottom muds under anaerobic conditions.

Most of the recent research concerning the toxicity of mercury has involved the methylated mercury compounds. Because of their extreme effects, the current standard for drinking water has been set at the very low level of 0.002 mg/L or 2 μg/L.

Nickel Nickel is used in electroplating, and the rinse waters from these operations constitute the major avenue by which salts of these metals gain access to the aquatic environment. It appears to be of low toxicity in humans, but there is some evidence that inhaled nickel may be carcinogenic. However, evidence that it is carcinogenic from oral exposure is small.

Selenium Selenium occurs in natural waters in limited regions in arid areas of the United States such as in California and South Dakota. It resembles sulfur in many of its chemical forms, and is found as selinite (SeO_3^{2-}), selenate (SeO_4^{2-}), and selenium hydride (H_2Se), as well as elemental selenium metal. Selenium is not widely used in industry. Its major use is in the manufacture of electrical components: photoelectric cells and rectifiers.

Selenium at low levels is essential in the diet of humans and many animals, but at higher concentrations can be quite toxic. This makes standard setting complicated. Selenium is an example of an element that may occur in soils in trace amounts but may be accumulated in certain cereals and pasture plants in quantities harmful to animals and humans when ingested. In animals it prevents proper bone formation and causes "alkali disease" and "blind staggers." The problem is critical with animals because they are dependent upon the local plants for food. A major selenium problem, leaving dead and deformed waterfowl in the fertile San Joaquin Valley of California, was discovered in 1983 after saline groundwater containing selenium was brought to the surface through irrigation and stored in a depression known as Kesterson Reservoir. The reservoir has been eliminated, but this simply transferred the problem to numerous small ponds throughout the valley used by farmers to evaporate the salty water. The problem yet has no adequate solution. More often, however, selenium is deficient in the diet. As a result the drinking water standard for selenium has been increased from the 1962 U.S. PHS drinking water standard of 0.01 mg/L to the current MCL of 0.05 mg/L.

Thallium Salts of thallium are used in rodenticides, in semiconductor research, and for making alloys with mercury for use in switches operating at subzero temperatures. Thallium salts are extremely toxic, and the effects are cumulative. Thus, chronic exposure limits for humans are justifiably set quite low. Major toxic effects of thallium are to the digestive tract, the nervous system, skin, and the cardiovascular tract.

Nonmetals

Arsenic Arsenic occurs in natural waters in oxidation states III and V, in the form of arsenous acid (H_3AsO_3) and its salts, and arsenic acid (H_3AsO_5) and its salts, respectively. Arsenic is quite widely distributed in natural waters, occurring at levels of 5 μg/L or more in approximately 5 percent of those tested, and in many instances throughout the world, groundwaters contain concentrations exceeding 1000 μg/L, including many regions in the western United States. Arsenic also gains access to the water environment through mining operations, the use of arsenical insecticides, and from the combustion of fossil fuels, where part of the fallout occurs on aquatic areas. Arsenic poisoning (arsenicosis) can range from pigmentation (white or dark spots on the skin), skin hardening and development of raised wartlike nodules (keratosis), and skin cancer. Other forms of cancer, such as bladder, lung, and kidney, may also result. Other resulting problems are peripheral vascular disease (blackfoot disease), resulting in gangrene, hypertension and ischemic heart disease, liver damage, anemia, and diabetes mellitus.

An increased prevalence of skin cancer was documented in villages in Taiwan following a switch from surface water to arsenic-contaminated well water for drinking in the 1920s. The worst arsenic-related drinking water problems have occurred in West Bengal and Bangladesh. Resulting health problems were first identified in West Bengal in the late 1980s. Because of dense populations and lack of access to safe drinking water, 4 million tubewells were installed in the 1970s in Bangladesh to tap better-quality groundwater sources. While this resulted in halving a high infant mortality rate, in 1993 the tubewell water was found to contain high levels of arsenic, at places reaching up to 3000 μg/L. Around 5000 patients have been identified with arsenic-related health problems in West Bengal, although some estimates put the number of patients at more than 200,000. The number in Bangladesh is not known, but must be many times greater than in West Bengal. Similar concerns have recently been reported for Vietnam. The growing awareness of arsenic-related health problems has led to a rethinking of the acceptable concentration in drinking water. Following a thorough review, the U.S. EPA in 2001 decided to reduce the drinking water MCL to 0.010 mg/L, which is now the same as the WHO guidelines.

Asbestos   Asbestos is the name given to a group of naturally occurring long, thin, and strong highly fibrous hydrated silicate minerals that have commercial utility largely because of their ability to be woven into fabrics and their resistance to heat and chemical attack. Asbestos has been most widely used in the construction industry in cement products, as floor tile, and in insulation. A common usage in the water works industry in the past has been in asbestos cement pipe used for transport of drinking water. It has also been used in the brake linings of automobiles. Occupational exposure to asbestos in air is well known to lead to asbestosis and a greatly increased risk of cancer, especially in the lungs. The risk of cancer from asbestos exposure is much greater in smokers, leading to an incidence of cancer that is much greater than from either asbestos exposure alone or smoking alone. Evidence of health effects from drinking water containing asbestos is less clear, but nevertheless, because it is a known human carcinogen, a limit has been set on the concentration of asbestos in drinking water.

Asbestos in water occurs in the form of fine particles that can be too small to impede the passage of light. Thus, asbestos particles may be present in the MCL range of several million particles per liter without giving the appearance of turbidity. Asbestos reaches drinking water supplies both through natural runoff from areas with exposed asbestos-containing minerals as well as from mining operations, and in both cases, can reach concentrations exceeding the MCL. There are two basic forms of asbestos: chrysotile, the major commercially mined form, and amphiboles. These two separate groups have different sizes, structures, and chemical properties.

The drinking water MCL applies only to fibers with lengths greater than 10 μm because of evidence that cancer results only from the longer fibers. Most fibers in water tend to be shorter than this.

Cyanide   Cyanides gain access to the water environment through the discharge of rinse waters from plating operations and from refinery and coal coking wastewaters. The cyanide ion has a relatively short half-life because it can serve as a source of energy for aerobic bacteria, provided the concentration is kept below its toxic threshold to them. For this reason, it should be of little concern where biological treatment systems are employed in the treatment of municipal wastes or where several days detention has occurred in natural waters. The standard of 0.2 mg/L undoubtedly is included to protect against industries with direct discharges to natural waters.

Perchlorate   Perchlorate (ClO_4^-) is not currently regulated by the SDWA nor is it on the WHO guidance list. However, it is on the EPA's candidate list and deserves some comment here. Perchlorate is both a naturally occurring and an industrially made chemical, although most perchlorate of concern comes from human sources. Perchlorate is a primary ingredient of solid rocket propellant. Most perchlorate problems have resulted from perchlorate manufacturing processes or improper disposal of rocket fuels. An increasing number of incidences of perchlorate contamination of drinking waters during the 1990s, with confirmed releases in 20 different states in the United States, has led to research on methods to remove perchlorate from drinking water and consideration of perchlorate as a chemical for regulation under the SDWA. Perchlorate interferes with iodide uptake into the thyroid gland, which may result in thyroid gland tumors. Based upon a preliminary assessment of potential human health risks of perchlorate exposure, the EPA has developed a draft reference dose for human intact that is intended to be protective. This is 0.00003 mg/kg-day. This translates into a drinking water equivalent dosage of 1 μg/L. Whether an MCL will be developed for perchlorate depends upon the outcome of reviews under way.

34.6 | SECONDARY STANDARDS AND GUIDELINES

Secondary standards have been set for only a few trace contaminants by the EPA as summarized in Table 34.3. Equivalent WHO guidelines are also listed there.

Aluminum in the form of alum is commonly used as a coagulant in water treatment. The secondary standard was set for aluminum to prevent its postprecipitation and discoloration of drinking water in the distribution system.

Copper and zinc impart a metallic taste to water, and this is the primary reason for the secondary standards for these compounds. Copper also has an MCL as already discussed. The toxicity of zinc is very low, and so only a secondary standard is set for it. Zinc gains access to the water environment from mining operations, wastewaters from electroplating, and the corrosion of galvanized piping.

Foaming in water is rightfully objectionable to the consumer. The secondary standard of 0.5 mg/L is based upon the foam produced by that concentration of a standard synthetic detergent in water.

Problems with iron and manganese are discussed in Chap. 27 and will not be discussed further here.

Silver is not a significant pollutant of natural waters, but soluble silver salts are excellent disinfectants. Even small ingestion of soluble silver can cause a darkening of the skin and eyes (argyria). The secondary standard of 0.1 mg/L has been established primarily to discourage use of silver salts for disinfection purposes.

34.7 | TRACE CHEMICAL ANALYSES

Trace contaminants always present particular difficulties in analytical measurement because of their very low concentrations in water. It was only through the development of highly sophisticated and modern analytical instruments that measurement of many contaminants at concentrations in the microgram per liter level and below became possible. Not only do trace contaminants present problems in detection, but also the analysis may be subject to interferences from all the other constituents in water, many of which are in the much higher milligram per liter range. Chemicals that may be used as part of the analytical procedure, for example, solvents used for extraction of the contaminant from water, must also be free of interfering substances. For trace analyses, the analytical procedure used must be very specific for the contaminant of interest, or procedures need to be used that permit separation of the contaminant from most of the other constituents in solution, or both. Essentially all trace analyses make use of instrumental procedures such as described in Chap. 12.

Trace organic contaminants are generally measured with gas chromatography (GC). However, there is a growing number of applications that can use high-performance liquid chromatography (HPLC) as well. Both procedures are based upon the principles of separation first and then detection and quantification. For most organic contaminants the separation is generally carried out in two stages. First, the contaminant is separated by partitioning from the water sample into a gas phase (gas stripping) or a solvent phase (liquid/liquid extraction). These procedures permit separation of the contaminant or contaminants of interest from the majority of other constituents in water. Additional steps may be used to concentrate or separate the contaminants removed prior to injection into the mobile phase of a GC or HPLC. Further separation of contaminants takes place within the GC or HPLC as the contaminants pass through the packed or capillary chromatographic columns and partition to different extents between the mobile and immobile phases present.

Ideally, this series of steps allows separation of the contaminants of interest from all other contaminants so that they can be detected and quantified individually as they pass through a detector. In many cases, separation is not complete. However, if a detector is used that is particularly sensitive to the contaminant of interest but not to interfering contaminants, satisfactory quantification may still be possible. For example, with GC there are detectors (electron capture and coulometric detectors) that are especially sensitive to halogenated compounds. Thus, a nonhalogenated compound passing the detector at the same time would not cause interference. Perhaps the most specific detectors are mass spectrometers (MS), which can single out for measurement a single, defined contaminant from many other contaminants. MS often offers the only procedure for positive analysis of some contaminants in drinking water. Initial analysis of a water sample using combined GC/MS often is required to confirm that a peak being measured by less-expensive GC procedures is indeed the contaminant of interest. Once confirmation is obtained, the less-expensive instruments may be used for routine analyses. Instruments that combine HPLC and MS have also been developed, but as yet they have not attained the general level of usage now enjoyed by GC/MS.

Other, less-capital-intensive methods of confirmation than GC/MS may also be used. For example, two different contaminants may emerge at the same time and be detected with a given GC, but it is unlikely that they would emerge and be detected at the same time with a different GC employing a different chromatographic column and detector. Thus, confirmation may be obtained by comparing the response of the unknown contaminant in two different GC systems with the response of a known standard. If the responses are identical, there is a high probability that the contaminant is indeed the same compound as that of the standard.

VOCs are those organic trace contaminants that can be readily separated from water into a gas phase by gas stripping. This requires that the contaminants have a reasonably high Henry's constant, on the order of 0.2 atm/M or greater (see Sec. 5.34). Contaminants with Henry's constants greater than 10 atm/M readily partition into the air in the head space of a bottle that is only partially filled with a water sample. Advantage is taken of this fact in head-space analysis where a water sample is first given time and sufficient mixing to equilibrate the contaminant between the liquid and gas phases, and then a sample of the gas phase is removed with a syringe and analyzed directly by GC. Knowledge of the Henry's constant and the respective air and sample volumes is then needed to translate measured gas concentrations into an original water sample concentration. However, the more common procedure used for such highly volatile contaminants is purge and trap. Here, volatile organic contaminants are transferred from a liquid sample by bubbling an inert gas such as nitrogen or helium through the sample while in a specially designed purging chamber. The vapor is passed through a sorbent trap, where the organics are removed from the vapor. After purging is complete, the trap is heated and backflushed with inert gas to desorb the contaminants onto a gas chromatographic column.

The VOCs with less volatility can also be separated from water by gas stripping. This procedure called closed-loop stripping analysis (CLSA) generally requires that the sample be actively gased for a much longer period of time than in the purge and trap procedure to remove the contaminant from water. Again, the contaminant is concentrated by adsorbing it from the gas phase onto a solid adsorbing phase, such as activated carbon. The gas is generally recirculated continuously between the liquid sample and the adsorbing phase forming a closed loop to minimize the amount of stripping gas used and the impact of any interfering contaminants it might contain. The concentrated contaminant is generally removed from the sorbing phase by liquid extraction for injection into the GC system.

Pesticides in general do not have high Henry's constants and cannot be gas stripped from water. They are generally removed by liquid/liquid extraction using solvents such as a mixture of methylene chloride (dichloromethane) and hexane, which are immiscible in water. After extraction, the contaminant concentration is increased by evaporating a large portion of the solvent. Additional steps are sometimes necessary to remove interfering contaminants that may have also been extracted from the water prior to injection into the GC system. Some contaminants of interest may not be volatile, may not be subject to solvent extraction because they are highly hydrophilic, or may not be sufficiently volatile to form a gas phase in the GC. Here, derivitization to change their physical properties may be necessary. For example, carboxylic acids can become subject to extraction and GC analysis if they are converted to the respective methyl ester. Another procedure that is sometimes applicable is to carry out acid extraction, which maintains the acid in its nonionized and more hydrophobic and volatile form.

These procedures appear relatively complex, time consuming, and expensive. However, many different contaminants can generally be measured by a single overall analytical procedure. Thus, only a few different analyses may be sufficient to obtain quantification for most of the trace organic contaminants listed in Table 34.1. For example, the disinfection by-products, chlorinated solvents and related compounds, and the chlorobenzenes may be analyzed together by a single purge and trap GC procedure that uses either an electrolytic conductivity or microcoulometric detector. The aromatic hydrocarbons [with the exception of benzo(a)pyrene] may be included as well if a photoionization detector is added in series to either of the detectors. Many of the halogenated pesticides listed in Table 34.1 can be analyzed by solvent extraction and GC analysis using an electron capture detector. It should be obvious, however, that a good deal of calibration is required, as well as much skill, knowledge, and experience on the part of the analyst, before reliable analyses can be obtained.

The standard procedure for determining trace levels of the metals listed in Tables 34.2 and 34.3 is by atomic absorption as discussed in Sec. 12.2. Generally, the graphite furnace is required with atomic absorption in order to measure the low levels of trace inorganics associated with the drinking water standards. The method for mercury represents another modification termed flameless atomic absorption. Here, mercury is reduced to the elemental state and is swept as a vapor from the sample by an air stream into the path of radiation from a mercury cathode-ray tube. The mercury vapor absorbs the radiation in proportion to its concentration, and the resulting reduced radiation reaching a detector is recorded as with the usual atomic absorption procedure. As an alternative, the inductively coupled plasma method described in Sec. 12.2 may be used for heavy metals. It has the advantage that the concentration of several metals can be determined in one analysis but has the disadvantage not only of high capital cost, but also that the detection limit for many metals is well above the drinking water MCLs. Coupling an ICP with a mass spectrometer, however, can lower the detection limit greatly.

Other trace inorganics listed in Tables 34.2 and 34.3 are determined generally by specifically developed analytical procedures. Arsenic can be determined either by atomic absorption with a graphite furnace or by the inductively coupled plasma method. Asbestos particles can be identified and counted through use of electron microscopy with electron diffraction analysis. Cyanide can be determined after separation from interferences by aeration of an acidified sample and collection of HCN in a basic solution. The transferred cyanide can then be determined by titration, colorimetric procedures, or a cyanide-specific electrode, depending upon the concentration present.

The U.S. EPA is currently considering several additional trace contaminants for inclusion on the list of drinking water standards. In addition, there are many other hazardous chemicals that are regulated or required to be monitored in air, water, and at sites of hazardous waste disposal. A greater awareness of the broad range of chemicals that may have adverse effects on human health and the environment, coupled with the ability of modern analytical methods to detect and quantify contaminant concentrations at trace levels, have had major impacts on the environmental engineering and science field. It is apparent that solutions to the complex problems engineers and scientists are being asked to address require teamwork among chemists, biologists, physical scientists, social scientists, and engineers. Each must understand both the abilities and the limitations of analytical methods that are used to quantify concentrations of contaminants in trace amounts, since this will have a major impact on their ability to deal effectively with environmental problems.

PROBLEMS

34.1   What are the differences in public concerns and nature of possible federal enforcement between primary and secondary drinking water regulations?

34.2   What is the difference between an MCL and an MCLG?

34.3   What value is assigned to the MCLG for carcinogens, and what is the basis for this assignment?

34.4   Why are enforceable limits for drinking water standards (MCLs) never assigned a value of zero?

34.5   Why are numerical limits not set for drinking water MCLs for some trace contaminants, such as epichlorohydrine, acrylamide, copper, and lead?

34.6   What general class of organic compounds make up the majority of the organics that are regulated by MCLs, and what properties do they possess that cause them to be so regulated?

34.7   What is the general source of trihalomethanes in drinking water?

34.8   List three general classes of halogenated organic compounds within which are contained many of the compounds that have been assigned MCLs.

34.9   What components of gasoline are of particular concern as groundwater contaminants and have been assigned MCLs?

34.10 (*a*)   What heavy metals are regulated through drinking water MCLs?

  (*b*)   Which ones are regulated through drinking water secondary standards?

  (*c*)   What metal has both an MCL and a secondary standard?

34.11 (*a*)   What oxidation state of chromium is most harmful to human health?

  (*b*)   What form of mercury is most harmful to human health?

34.12 Which of the inorganic elements with drinking water MCLs are known to be both beneficial to human beings and harmful to human beings?

34.13 What two major analytical instruments would you select if you wanted to be able to measure a majority of the trace inorganic and organic contaminants with drinking water MCLs?

34.14 Why are trace contaminants difficult to measure accurately in water supplies?

34.15 What procedures are generally used for separating trace organic materials from interfering substances prior to detection and quantification?

34.16 What procedures may be used to confirm that a peak showing up on a chromatogram is actually the compound of interest?

REFERENCES

Pontius, F. W.: Regulatory Compliance Planning to Ensure Water Supply Safety, *J. AWWA,* **94**(3) 52-64 (2002).

Safe Drinking Water Committee: "Drinking Water and Health," Vol 1 (1977), Vol 2 (1980), Vol 3 (1980), Vol 4 (1982), Vol 5 (1983), Vol 6 (1986), Vol 7 (1987), Vol 8 (1987) and Vol 9 (1989), National Academy Press, Washington, DC.

Thermodynamic Properties at 25°C

Substance	ΔG_f° kJ/mol	Reference	ΔH_f° kJ/mol	Reference
Inorganics				
$H^+(aq)$	0.00	k	0.00	k
$H^+(10^{-7})(aq)$	−39.87	g		
$H_2(g)$	0.00	l	0.00	l
$H_2(aq)$	17.57	j	−4.18	j
$H_2O(g)$	−228.6	l	−241.8	l
$H_2O(l)$	−237.18	l	−285.80	l
$H_2O_2(aq)$	−134.097	g	−191.13	h
$OH^-(aq)$	−157.2	l	−230.0	l
$O_2(g)$	0.00	k	0.00	k
$O_2(aq)$	16.32	j	−11.71	j
$O_3(g)$	165.7	l	142	l
$Ba^{2+}(aq)$	−560.8	k	−537.6	k
$BaSO_4(s)$	−1362.2	k	−1473.2	k
$Br^-(aq)$	−104.0	j	−121.5	j
$Ca^{2+}(aq)$	−553.6	k	−542.8	k
$CaCO_3(s)$	−1128.76	l	−1206.87	l
$CaF_2(s)$	−1175.6	k	−1228.0	k
$Ca(OH)_2(s)$	−896.76	l	−986.6	l
$CaSO_4 \cdot 2H_2O(s)$	−1795.7	l	−2021.1	l
$CaSO_4(s)$	−1322.0	k	−1434.5	k
$CO(g)$	−137.28	h	−110.54	h
$CO_2(g)$	−394.38	l	−393.51	l
$CO_2(aq)$	−386.23	l	−412.92	l
$C(s)$	0.00	h	0.00	h
$H_2CO_3(aq)$	−623	l	−699	l
$HCO_3^-(aq)$	−586.85	k	−692.00	k
$CO_3^{2-}(aq)$	−527.80	k	−677.10	k
$Cl_2(g)$	0.00	k	0.00	k
$Cl_2(aq)$	6.90	j	−23.40	j
$Cl^-(aq)$	−131.30	l	−167.20	l
$ClO^-(aq)$	−36.8	k	−107.1	k

Substance	ΔG_f° kJ/mol	Reference	ΔH_f° kJ/mol	Reference
Inorganics				
$ClO_2(aq)$	117.6	j	74.9	j
$ClO_2^-(aq)$	17.1	j	-66.5	j
$ClO_3^-(aq)$	-3.4	j	-99.2	j
$ClO_4^-(aq)$	-8.6	j	-129.3	j
$Cr^{3+}(aq)$	-215.5	j	-256	j
$CrO_4^{2-}(aq)$	-727.9	j	-881.1	j
$Cr_2O_7^{2-}(aq)$	-1301.0	j	-1490	j
$HCrO_4^-(aq)$	-764.8	j	-878.2	j
$Cu^{2+}(aq)$	65.5	j	64.8	j
$F^-(aq)$	-278.8	k	-332.6	k
$Fe^{2+}(aq)$	-78.87	j	-89	k
$Fe^{3+}(aq)$	-4.60	g	-48.5	k
$Fe(OH)_2(s)$	-483.54	l	-568.2	l
$FeS(s)$	-100.4	k	-100.0	k
$FeS_2(s)$	-166.9	l	-178.2	k
$Hg^{2+}(aq)$	153.6	j	172.4	j
$I_2(s)$	0.00	j	0.00	j
$I^-(aq)$	-51.59	j	-55.19	j
$IO_3^-(aq)$	-128.0	j	-221.3	j
$Mg^{2+}(aq)$	-454.8	k	-466.9	k
$Mg(OH)_2(s)$	-833.5	k	-924.5	k
$Mn^{2+}(aq)$	-228.0	j	-220.7	j
$MnO_2(s)$	-465.1	k	-520.0	k
$Na^+(aq)$	-261.9	k	-240.1	k
$NH_3(g)$	-16.4	k	-45.9	k
$NH_3(aq)$	-26.65	l	-80.83	l
$NH_4^+(aq)$	-79.37	l	-132.50	l
$HNO_2(aq)$	-42.97	j	-119.20	j
$NO_2^-(aq)$	-37.2	g	-104.6	k
$HNO_3(aq)$	-111.3	j	-207.3	j
$NO_3^-(aq)$	-111.34	g	-207.4	k
$PO_4^{3-}(aq)$	-1018.7	k	-1277.4	k
$HPO_4^{2-}(aq)$	-1089.3	j	-1292.1	j
$H_2PO_4^-(aq)$	-1130.4	j	-1296.3	j
$H_3PO_4(aq)$	-1142.6	j	-1288.3	j
$S(s)$	0.00	k	0.00	k
$H_2S(g)$	-33.4	k	-20.6	k
$H_2S(aq)$	-27.87	g	-39.3	l
$HS^-(aq)$	12.05	g	-17.6	l
$H_2SO_4(l)$	-690.0	k	-814.0	k
$S^{2-}(aq)$	79.5	k	30.1	k
$SO_3^{2-}(aq)$	-486.6	g	-635.5	k
$SO_4^{2-}(aq)$	-744.63	g	-909.3	k
$S_2O_3^{2-}(aq)$	-513.4	g	-652.3	k
$Se(s)$	0.00	l	0.00	l
$SeO_3^{2-}(aq)$	-373.8	l	-512.1	l
$SeO_4^{2-}(aq)$	-441.1	l	-607.9	l
$H_2Se(aq)$	77.0	l	75.7	l
$Zn^{2+}(aq)$	-147.1	l	-153.9	k
$ZnS(s)$	-201.3	k	-206.0	k

Substance	ΔG_f^o kJ/mol	Reference	ΔH_f^o kJ/mol	Reference
	Organics			

Acid-dicarboxylic

Substance	ΔG_f^o kJ/mol	Reference	ΔH_f^o kJ/mol	Reference
Fumarate$(^{2-})(aq)$	−604.21	g		
Fumaric acid(aq)	−647.14	g		
l-Malate$(^{2-})(aq)$	−845.08	g		
Oxalacetate$(^{2-})(aq)$	−797.18	g		
Oxalate$(^{1-})(aq)$	−699.1	h	−818.0	h
Oxalate$(^{2-})(aq)$	−674.9	h	−824	h
Succinate$(^{2-})(aq)$	−690.23	g		
Succinic acid(aq)	−746.38	g		

Acid-monocarboxylic

Substance	ΔG_f^o kJ/mol	Reference	ΔH_f^o kJ/mol	Reference
Acetate$(^-)(aq)$	−369.41	g	−486.0	k
Acetic acid(aq)	−399.6	e	−488.4	e
Acetic acid(l)	−389.9	k	−484.3	k
Acrylate$(^-)(aq)$	[−286.19]	g		
Benzoic$(^-)(aq)$	−245.6	d		
Butyrate$(^-)(aq)$	−352.63	g		
Caproate$(^-)(aq)$	[−335.96]	g		
Crotonate$(^-)(aq)$	−277.4	g		
Formate$(^-)(aq)$	−351.0	k	−425.6	k
Formic acid(aq)	−356.1	h	−410	h
β-Gluconate$(^-)(aq)$	[−1128.3]	g		
Glycerate$(^-)(aq)$	[−658.1]	g		
Glycolate$(^-)(aq)$	−530.95	g		
Glyoxylate$(^-)(aq)$	−468.6	g		
β-Hydroxybutyrate$(^-)(aq)$	[−506.3]	g		
β-Hydroxypropionate$(^-)(aq)$	−518.4	g		
β-Ketobutyrate$(^-)(aq)$	−493.7	g		
α-Ketoglutarate$(^-)(aq)$	−797.55	g		
Lactate$(^-)(aq)$	−517.81	g		
Palmitate$(^-)(aq)$	−309.5	e		
Palmitic acid(s)	−305	g	−891.5	k
Propionate$(^-)(aq)$	[−361.08]	g		
Pyruvate$(^-)(aq)$	−474.63	g		
Valerate$(^-)(aq)$	[−344.34]	g		

Acid-tricarboxylic

Substance	ΔG_f^o kJ/mol	Reference	ΔH_f^o kJ/mol	Reference
cis-Aconitate$(^{3-})(aq)$	−922.61	g		
Citrate$(^{3-})(aq)$	−1168.34	g		
Isocitrate$(^{3-})(aq)$	−1161.69	g		

Alcohol

Substance	ΔG_f^o kJ/mol	Reference	ΔH_f^o kJ/mol	Reference
n-Butanol(aq)	−171.84	g		
Ethanol(l)	−174.8	e	−277.7	e
Ethanol(aq)	−181.75	g		
Ethylene glycol(l)	−323.21	g	−460.0	k
Ethylene glycol(aq)	[−330.5]	g		
Glycerol(l)	−477.06	g	−669.6	k
Glycerol(aq)	−488.52	g		
Mannitol(aq)	−942.61	g		
Methanol(aq)	−175.39	g	−245.9	l
n-Propanol(aq)	−175.81	g		

Substance	ΔG_f^o kJ/mol	Reference	ΔH_f^o kJ/mol	Reference
Organics				
Isopropanol(aq)	-185.94	g		
Sorbitol(aq)	-942.7	g		
Aldehyde				
Acetaldehyde(aq)	-139.9	g		
Acetaldehyde(l)	-137.6	k	-192.2	k
Acetaldehyde(g)	-133.0	k	-166.2	k
Butyraldehyde(l)	-119.67	g	-239.3	k
Formaldehyde(aq)	-130.54	g		
Formaldehyde(g)	-102.5	k	-108.6	k
Amino acid				
L-Alanine(aq)	-371.54	g		
L-Arginine(aq)	-240.2	g		
L-Asparagine·H$_2$O(aq)	-763.998	g		
L-Aspartate($^-$)(aq)	-700.4	g		
L-Aspartic acid(aq)	-721.3	g		
L-Cysteine(aq)	-339.78	g		
L-Cystine(aq)	-666.93	g		
L-Glutamate($^-$)(aq)	-699.6	g		
L-Glutamic acid(aq)	-723.8	g		
L-Glutamine(aq)	-529.7	g		
Glycine(aq)	-370.788	g		
Glycine($^+$)(aq)	-384.19	g		
Glycine($^-$)(aq)	-314.96	g		
L-Leucine(aq)	-343.1	g		
Isoleucine	-343.9	g		
L-Methionine(aq)	-502.92	g		
L-Phenylalanine(aq)	-207.1	g		
L-Serine(aq)	-510.87	g		
L-Threonine(aq)	$[-514.63]$	g		
L-Tryptophane(aq)	-112.6	g		
L-Tyrosine(aq)	-370.7	g		
L-Valine(aq)	-356.9	g		
Aromatic				
Benzene(g)	129.700	f	82.93	h
Benzene(aq)	133.9	i		
Benzoic acid(s)	-245.6	g		
Benzyl alcohol(l)	-31.25	g		
Biphenyl(aq)	275.2	c		
o-Cresol(g)	-37.1	g		
m-Cresol(g)	-40.54	g		
p-Cresol(g)	-32.09	g		
Ethylbenzene(g)	130.574	h	29.9	k
Ethylbenzene(aq)	136	i		
Phenol(s)	-47.6	g		
Phenol(aq)	-47.5	i		
Toluene(g)	122.290	h	50.5	k
Toluene(aq)	127	i		
m-Xylene(g)	118.847	h	17.3	k
m-Xylene(aq)	122.9	i		

Substance	ΔG_f^o kJ/mol	Reference	ΔH_f^o kJ/mol	Reference
Organics				
o-Xylene(*g*)	122.077	h	19.1	k
o-Xylene(*aq*)	124.6	i		
p-Xylene(*g*)	121.135	h	18.0	k
p-Xylene(*aq*)	125.2	i		
Carbohydrate				
Dihydroxyacetone(*aq*)	[−445.18]	g		
D-Erythrose(*aq*)	[−598.3]	g		
D-Fructose(*aq*)	−915.38	g		
α-D-Galactose(*aq*)	−923.53	g		
α-D-Glucose(*aq*)	−917.22	g		
Glyceraldehyde(*aq*)	[−437.65]	g		
Glycogen(per glucose)(*aq*)	−662.33	g		
D-Heptose(*aq*)	[−1,077]	g		
α-Lactose(*aq*)	−1515.24	g		
β-Lactose(*aq*)	−1570.09	g		
β-Maltose(*aq*)	−1497.04	g		
D-Ribose(*aq*)	[−757.3]	g		
Sucrose(*aq*)	−1551.85	g		
Haloaliphatic				
Bromoethane(*g*)	−54.39	k		
Bromoethane(*l*)	−85.35	k		
Bromoethane(*aq*)	−49.88	i		
1-Chloropropane(*g*)	−23.53	b	−131.9	k
1-Chloropropane(*aq*)	−7.32	b		
Chloroethane(*g*)	−43.01	b	−112.1	k
Chloroethane(*aq*)	−36.83	b		
Chloroethene(*g*)	51.46	b	37.2	k
Chloroethene(*aq*)	59.65	b		
Chloromethane(*g*)	−58.4	b	−81.9	k
Chloromethane(*aq*)	−52.71	b		
1,2-Dibromoethane(*l*)	−20.67	l	−80.8	l
1,2-Dibromoethane(*aq*)	−9.03	i		
1,1-Dichloroethane(*g*)	−73.23	b	−127.7	k
1,1-Dichloroethane(*aq*)	−68.69	b		
1,2-Dichloroethane(*g*)	−73.78	b	−126.4	k
1,2-Dichloroethane(*aq*)	−72.93	b		
1,1-Dichloroethene(*g*)	24.16	b	2.8	k
1,1-Dichloroethene(*aq*)	32.23	b		
cis-1,2-Dichloroethene(*g*)	24.33	b	4.6	k
cis-1,2-Dichloroethene(*aq*)	27.8	b		
trans-1,2-Dichloroethene(*g*)	26.54	b		
trans-1,2-Dichloroethene(*aq*)	32.06	b		
Dichloromethane(*g*)	−68.8	b	−95.4	k
Dichloromethane(*aq*)	−66.11	b		
1,2-Dichloropropane(*g*)	−83.01	b	−162.8	k
1,2-Dichloropropane(*aq*)	−79.88	b		
Hexachloroethane(*g*)	−54.88	b	−143.6	k
Hexachloroethane(*aq*)	−49.62	b		
Pentachloroethane(*g*)	−70.18	b	−142.0	k

Substance	ΔG_f° kJ/mol	Reference	ΔH_f° kJ/mol	Reference
Organics				
Pentachloroethane(*aq*)	−68.26	b		
1,1,1,2-Tetrachloroethane(*g*)	−80.26	b		
1,1,1,2-Tetrachloroethane(*aq*)	−77.75	b		
1,1,2,2-Tetrachloroethane(*g*)	−85.48	b	−149.2	k
1,1,2,2-Tetrachloroethane(*aq*)	−88.92	b		
Tetrachloroethene(*g*)	20.48	b	−10.9	k
Tetrachloroethene(*aq*)	27.59	b		
Tetrachloromethane(*g*)	−53.5	b	−95.7	k
Tetrachloromethane(*aq*)	−45.1	b		
1,1,1-Trichloroethane(*g*)	−76.12	b	−144.4	k
1,1,1-Trichloroethane(*aq*)	−69.04	b		
1,1,2-Trichloroethane(*g*)	−77.41	b	−151.3	k
1,1,2-Trichloroethane(*aq*)	−77.64	b		
Trichloroethene(*g*)	19.86	b	−9.0	k
Trichloroethene(*aq*)	25.41	b		
Trichloromethane(*g*)	−70.06	b	−102.7	k
Trichloromethane(*aq*)	−66.5	b		
Haloaromatic				
Chlorobenzene(*g*)	99.2	a		
Chlorobenzene(*aq*)	102.3	a		
1,2-Dichlorobenzene(*g*)	82.7	a	30.2	k
1,2-Dichlorobenzene(*aq*)	84.3	a		
1,3-Dichlorobenzene(*g*)	78.6	a	25.7	k
1,3-Dichlorobenzene(*aq*)	81.8	a		
1,4-Dichlorobenzene(*g*)	77.2	a	22.5	k
1,4-Dichlorobenzene(*aq*)	78.3	a		
1,3,5-Trichlorobenzene(*g*)	52.6	a	−13.4	k
1,3,5-Trichlorobenzene(*aq*)	56.8	a		
Hexachlorobenzene(*g*)	33.3	a	−35.5	k
Hexachlorobenzene(*aq*)	46	a		
2-Chlorobenzoate($^-$)(*aq*)	−237.9	a		
3-Chlorobenzoate($^-$)(*aq*)	−246	a		
4-Chlorobenzoate($^-$)(*aq*)	−239.5	a		
2,4-Dichlorobenzoate(*aq*)	−276.4	a		
2-Chlorobiphenyl(*aq*)	259.6	c		
2,4′-Dichlorobiphenyl(*aq*)	235	c		
2,2′,4,4′,6-Pentachlorobiphenyl(*aq*)	162.7	c		
2,2′,4,4′-Tetrachlorobiphenyl(*aq*)	185.9	c		
2,4,4′-Trichlorobiphenyl(*aq*)	210.6	c		
2-Chlorophenol(*aq*)	−56.8	a		
2-Chlorophenol($^-$)(*aq*)	−8.1	a		
2,4-Dichlorophenol(*aq*)	−84.4	a		
2,4-Dichlorophenol($^-$)(*aq*)	−39.4	a		
Pentachlorophenol(*aq*)	−112.3	a		
Pentachlorophenol($^-$)(*aq*)	−86.6	a		
Hydrocarbon				
Acetylene(*g*)	209.2	g	227.4	k
Butane(*aq*)	−0.21	b		
Butane(*g*)	−15.707	h	−124.7	h

	ΔG_f^o kJ/mol	Reference	ΔH_f^o kJ/mol	Reference
Substance				
Organics				
Ethane(*aq*)	−17.43	b		
Ethane(*g*)	−32.89	h	−84.67	h
Ethene(*aq*)	81.43	b		
Ethene(*g*)	68.180	b	52.4	k
Methane(*aq*)	−34.74	b		
Methane(*g*)	−50.79	h	−74.6	k
Pentane(*aq*)	9.24	b		
Pentane(*g*)	−8.20	h	−146.9	k
Propane(*aq*)	−7.32	b		
Propane(*g*)	−21.606	h	−103.8	k
Ketone				
Acetone(*aq*)	−161.17	g		
N-containing, other				
Creatine(*aq*)	−264.3	g		
Urea(*s*)	−196.82	g	−333.1	k
Urea(*aq*)	−203.76	g	−319.2	l
Purine				
Hypoxanthine(*aq*)	89.5	g		
Quanine(*s*)	46.99	g		
Urate(⁻)(*aq*)	−325.9	g		
Uric acid(*aq*)	−356.9	g		
Xanthine(*s*)	−165.85	g		

[a]J. Dolfing and B. K. Harrison, Gibbs Free Energy of Formation of Halogenated Aromatic Compounds and Their Potential Role as Electron Acceptors in Anaerobic Environments, *Env. Sci.Tech,* **26**(11): 2213–2218 (1992).

[b]J. Dolfing and D. B. Janssen, Estimates of Gibbs Free Energies of Formation of Chlorinated Aliphatic Compounds. *Biodegradation,* **5**(1): 21–28 (1994).

[c]D. A. Holmes, B. K. Harrison and J. Dolfing, Estimation of Gibbs Free Energies of Formation for Polychlorinated Biphenyls. *Env. Sci. Tech,* **27**(4): 725–731 (1993).

[d]M. T. Madigan, J. M. Martinko and J. Parker. *Brock Biology of Microorganisms,* 8th ed., Prentice Hall, Upper Saddle River, NJ, 1997.

[e]P. L. McCarty, Energetics and Bacterial Growth. In S. D. Faust and J. V. Hunter eds., *Organic Compounds in Aquatic Environments,* Marcel Dekker, Inc, New York, 1971, pp. 495–531.

[f]R. H. Perry, D. W. Green and J. O. Maloney, *Perry's Chemical Engineering Handbook,* 6th ed., McGraw-Hill Book Company, New York, 1984.

[g]R. K. Thauer, K. Jungermann and K. Decker. Energy Conservation in Chemotrophic Anaerobic Bacteria, *Bacteriological Reviews,* **41**(1): 100–180 (1977).

[h]R. C. Weast and M. J. Astle, CRC *Handbook of Chemistry and Physics,* 61st ed., CRC Press, Inc., Boca Raton, FL, 1980.

[i]Estimated from ΔG_f^o for gas or liquid and from aqueous solubility concentrations, with ΔG_f^o corrected by $-RT \ln K$.

[j]W. Stumm and J. J. Morgan, *Aquatic Chemistry, Chemical Equilibria and Rates in Natural Waters*, 3rd ed., John Wiley & Sons, New York, 1996.

[k]D. R. Linde, ed., *Handbook of Chemistry and Physics*, 82nd ed., CTC Press, Inc., Boca Raton, FL, 2001.

[l]F. D. Rossini, D. D. Wagman, W. H. Evans, S. Levine, and I. Jaffe, *Selected Values of Thermodynamic Properties*, Circular of the National Bureau of Standards 500, U.S. Government Printing Office, Washington, D.C., 1952.

Note: Values in brackets are reported to have been obtained by approximate calculation.

The physical state of each substance is indicated after the substance name as crystalline or solid (*s*), gas (*g*), liquid (*l*), or in water solution (*aq*).

APPENDIX

B

Acronyms, Roman Symbols, and Greek Symbols

Symbol	Definition
ACRONYMS	
ABS	Alkylbenzene sulfonate (branched detergent)
ADP	Adenosine diphosphate
ANC	Acid neutralization capacity
APHA	American Public Health Association
ATP	Adenosine triphosphate
AW	Atomic weight
AWWA	American Water Works Association
BF	Bioconversion factor in structure activity relationship
BHC	Benzene hexachloride
BOD	Biochemical oxygen demand (mg/L)
BOD_5	5-Day biochemical oxygen demand at 20°C (mg/L)
BOD_L	Ultimate biochemical oxygen demand (mg/L)
BTEX	Benzene, toluene, ethylbenzene, and xylene
CAH	Chlorinated aliphatic hydrocarbon
CFC	Chlorofluorocarbon
CI	Confidence interval
CL	Confidence limits
CLSA	Closed-loop stripping analysis
COD	Chemical oxygen demand (mg/L)
2,4-D	2,4-Dichlorophenoxyacetic acid
DBCP	1,2-Dibromo-3-chloropropane
DDT	2,2-bis(*p*-chloro-phenyl)-1,1,1-trichloroethane
DES	Diethylstilbestrol
DGGE	Denaturing gradient gel electrophoresis
DLVO	Colloidal stability theory by Derjagin, Landau, Verwey, and Overbeek
DMA	Dimethylamine
DNA	Deoxyribonucleic acid
DNAPL	Dense non-aqueous phase liquid
DO	Dissolved oxygen
DPD	*N,N*-diethyl-*p*-phenylenediamine

Symbol	Definition
ACRONYMS	
EDB	Ethylene dibromide (1,2-Dibromoethane)
EDC	Endocrine disrupting chemical
EDTA	Ethylenediaminetetraacetic acid
ELISA	Enzyme linked immunosorbent assay
E_{LUMO}	Lowest unoccupied molecular orbital energy (eV)
emf	Electromotive force
EPA	U.S. Environmental Protection Agency
EW	Equivalent weight (g)
F_{430}	Electron carrier enzyme in methanogenic bacteria
FAD	Flavin adenine dinucleotide (oxidized form)
FADH	Flavin adenine dinucleotide (reduced form)
FISH	Fluorescence in-situ hybridization
FW	Formula weight (g)
GC	Gas chromatograph
GC/MS	Gas chromatograph/ mass spectrometer
HAA	Haloacetic acid or hormonally active agent
HMX	Octahydro-1,3,5,7-tetranitro-1,3,5,7-tetrazocine
HPLC	High performance liquid chromatograph
IC_{50}	Concentration at which biological activity is reduced 50 percent
ICP	Inductively coupled plasma
IDL	Instrument detection limit
IPC	Isopropyl N-phenylcarbamate
IR	Infrared
ISO	International Organization for Standardization
IUPAC	International Union of Pure and Applied Chemistry
LAS	Linear alkylbenzene sulfonate detergent
LC	Liquid chromatography
LFER	Linear free-energy relationship
LLD	Lower limit of detection
LNAPL	Light non-aqueous phase liquid
LOQ	Limit of quantification
LSER	Linear solvation energy relationship
MCL	Maximum contaminant limit
MCLG	Maximum contaminant limit goal
MDL	Method detection limit
meq	Milliequivalent
MS	Mass spectrometer
MTBE	Methyl *tert*-butyl ether
MW	Molecular weight
NAD	Nicotinamide adenine dinucleotide (oxidized form)
NADH	Nicotinamide adenine dinucleotide (reduced form)
NAPL	Non-aqueous phase liquid
NDMA	N-Nitrosodimethylamine
NMR	Nuclear magnetic resonance spectroscopy
NO_x	Oxides of nitrogen
NTU	Nephelometric turbidity unit
ORP	Oxidation-reduction potential
OSHA	U.S. Occupational Safety and Health Administration
PAH	Polycyclic aromatic hydrocarbon
PCB	Polychlorinated Biphenyl
PCE	Tetrachloroethene (perchloroethylene)

Symbol	Definition
ACRONYMS	
PCR	Polymerase chain reaction
PFOS	Perfluro-octane-sulfonate
PhAC	Pharmaceutically active chemical
pH_{pzc}	The pH at the point of zero charge on a surface
PM	Particulate matter
PM_{10}	Particulate matter less than 10 μm in size
$PM_{2.5}$	Particulate matter less than 2.5 μm in size
POQ	Practical quantification limit
ppb	Parts per billion
ppm	Parts per million
ppt	Parts per trillion
PZC	Point of zero charge for a surface
QA	Quality assurance
QC	Quality control
QSAR	Quantitative structure-activity relationship
rad	Roentgen-absorption-dose for radiation
RDX	Hexahydro-1,3,5-trinitro-1,3,5-triazine
rem	Roentgen-equivalent-man dose for radiation
RNA	Ribonucleic acid
rRNA	Ribosomal ribonucleic acid
RTQ-PCR	Real time quantitative polymerase chain reaction
SCE	Standard calomel electrode
SDWA	Safe drinking water act
SI	International System of Units
SPADNS	Sodium 2-(parasulfophenylazo)-1,8-dihydroxy-3,6-naphthalene disulfonate
2,4,5-T	2,4,5-Trichlorophenoxyacetic acid
TCDD	2,3,7,8-Tetrachlorodibenzo-p-dioxin
TCE	Trichloroethene
TCMP	2-Chloro-6-(trichloromethyl) pyridine
TDML	Total daily maximum load
TDO	Toluene dioxygenase
THM	Trihalomethane
TOC	Total organic carbon
TOTX	Total concentration of species X
U.S. EPA	United States Environmental Protection Agency
UV	Ultraviolet
VOC	Volatile organic compound
WEF	Water Environment Federation
WHO	World Health Organization
ROMAN SYMBOLS	
a	Interfacial area to volume of water (m^{-1})
a_1	Constant related to Lambert's law
a_2	Constant related to Beer's law
a'	Constant in Lambert-Beer law
A	Constant in equation describing effect of temperature on reaction rate
A	Constant in Rankine formula
A	Absorbance
b	Coefficient in BET Isotherm equation
B	Constant in Rankine formula
B	Dissolved oxygen concentration of biological seed in BOD analysis
c	Concentration of particles affecting osmotic pressure (mol/m^3)

Symbol	Definition

ROMAN SYMBOLS

c	Velocity of light (3×10^{10} cm/s)
C	Concentration (mg/L or mol/L)
C	Specific heat (J/g-°C)
C	Constant in Rankine formula
C_{actual}	Actual solution concentration (mol/L or mg/L)
C_{equil}	Equilibrium concentration given by Henry's Law (mol/L)
C_s	Concentration of substance in solvent (mol/L)
C_T	Sum total molar concentration in solution of a compound or element in all its forms
C_w	Concentration of substance in water (mol/L)
D	Dielectric constant of liquid
D	Dissolved oxygen content of sample (mg/L)
e	Void fraction or porosity of soil
E	Internal energy of system (J, ergs, or eV)
E	Enzyme concentration (mol/L)
E	Electromotive force (volt)
E^0	Electromotive force under standard conditions (volt)
E_a	Activation energy for reaction (L-atm/mol)
E_a	Absolute error between measured and true value
$E_c S$	Enzyme-substrate complex (mol/L)
E_f	Free enzyme concentration (mol/L)
E_r	Relative error between measured and true value
E_t	Total enzyme concentration (mol/L)
f_{oc}	Fraction of organic carbon in a soil
F	Faraday (C/eq)
F	Value used in the statistical F-test
F_{430}	Electron carrier enzyme in methanogenic bacteria
G	Free energy of system (J)
g	Gravity (cm/s^2)
ΔG^o	Standard free energy for reaction (J)
ΔG_f^o	Standard free energy of formation (J/mol)
h	Height (m)
h	Planck's constant (6.63×10^{-27} erg·s)
H	Enthalpy of system (J)
H	Dimensionless coefficient for Henry's law
ΔH^o	Standard enthalpy for reaction (J)
ΔH_f^o	Standard enthalpy of formation (J/mol)
i	Interval within which frequency of values is obtained for a histogram
I	Current flow (amp)
I	Intensity of light leaving a solution
I_0	Intensity of light entering a solution
k	Rate constant
k'	First-order rate constant for BOD (d^{-1})
k_a	Acid-catalyzed rate constant
k_b	Base-catalyzed rate constant
k_g	Gas-phase mass-transfer rate (m/s)
k_h	Hydrolysis rate constant (d^{-1})
k_n	Rate constant that is not dependent upon proton or hydroxide concentration
k_r	Reduction rate constant
k_{sa}	Surface-area normalized rate coefficient (L/m^2-h)
k_w	Water-phase mass-transfer rate (m/s)

Symbol	Definition
ROMAN SYMBOLS	
K	Equilibrium, formation, stability, or distribution coefficient or constant
K_A	Ionization constant for an acid
K_{ads}	Adsorption coefficient in Langmuir isotherm equation (L/g)
K_B	Ionization constant for a base
K_H	Henry's constant for gas (atm/M)
K_{inst}	Instability constant
$K_L a$	Overall mass gas-liquid transfer rate (d^{-1})
K_{oc}	Contaminant water organic carbon partition coefficient
K_{ow}	Octanol/water partition coefficient
K_p	Contaminant water soil partition coefficient (mg organic sorbed/kg solids)
K_s	Half-velocity coefficient in bacterial reaction (mg/L)
K_{sp}	Solubility product
K_W	Ionization constant of water
l	Length (m)
L_0	Ultimate BOD (mg/L)
L_t	BOD remaining (mg/L)
M	Molarity of solution (moles/L)
n	Number of moles or of values
n	Exponent in Freundlich isotherm equation
N	Normality of solution (eq/L)
N	Number of neutrons in an atom
p	Vapor pressure of substance (atm)
pE	Negative of logarithm of hypothetical electron concentration
pE^0	Negative of logarithm of hypothetical electron concentration under standard conditions
pH	Negative of logarithm of proton molar concentration
pK	Negative of logarithm of an equilibrium constant
pOH	Negative of logarithm of hydroxide molar concentration
P	Pressure (atm)
P	Fraction of BOD bottle represented by sample
P_{gas}	Partial pressure of gas (atm)
P_i	Vapor pressure of compound in mixture (atm)
P_i^0	Vapor pressure of pure compound (atm)
q	Heat added (J)
q	Mass contaminant adsorbed per unit of adsorbent (g/g)
q	Charge at the shear surface of colloid
q_m	Maximum contaminant adsorption that can occur per unit of adsorbent (g/g)
q_p	Heat absorbed at constant pressure (J)
q_{rev}	Heat absorbed with slow reversible process (J)
q_v	Heat absorbed at constant volume (J)
Q	Flow rate through a capillary tube (m/s)
Q	Flow rate into or out of a reactor (m^3/d)
Q_{10}	Fractional increase in biological reaction rate for 10°C rise
r	Radius (m)
r	Coefficient of determination
R	Universal gas constant (atm-L/mol-K)
R	Electrical resistance (ohm)
s	Standard deviation for a small or finite data set
s_g	Geometric standard deviation or spread factor
s_m	Standard error of the mean
S	Entropy of system (J/K)

Symbol	Definition
ROMAN SYMBOLS	
S	Concentration of substrate for bacteria (mg/L)
S	Water solubility of compound (mg/L or mole fraction)
t	Time (s, min, hr, d, or year)
t	Value for a student's t-test
$t_{1/2}$	Half life (d)
t_r	Retardation factor of contaminant in groundwater
T	Temperature (°C or K)
T	Transmittance of solution
V	Volume (L)
V	Overall rate of bacterial reaction (mg/L-d)
V_A	Volume per mol of solvent (m³)
V_F	Applied electric field in zeta potential measurement (V/cm)
V_i	Intrinsic molar volume
V_s	Volume solvent (m³)
V_w	Volume water (m³)
w	Work of expansion (J)
W_n	Weight of substance in water after n extractions (mol/L)
$\bar{x}$	Mean, average, or arithmetic mean for a small or finite data set
$\bar{x}_g$	Geometric mean
x_i	Numerical value of ith measurement
X	Concentration of bacteria (mg/L)
X_i	Mole fraction of compound in mixture
y	BOD exerted at any time t (mg/L)
Y	Frequency of occurrence for small or finite data set
z	Electron equivalents per mole
Z	Charge on ion
Z	Number of protons in atom
GREEK SYMBOLS	
α	Ionization fraction for mass balance equation
β	Buffer index or buffer intensity
β	Proportionality constant in generalized gas law
β_n	Overall constant for complex where n = 1, 2, 3, etc.
$^0\chi$	Coefficient in structure-property relationship
$^0\chi^v$	Coefficient in structure-property relationship
$\Delta\delta_{x-y}$	Modulus of atomic charge differences across common bonds
δ	Thickness of diffuse layer of colloid or surface (m)
ϕ	Coefficient in structure-property relationship
γ	Activity coefficient
γ	Surface tension (dyne/cm or N)
Γ	Sorbed concentration or adsorption density (g/g)
κ	Specific conductance (1/ohm-cm)
κ_s	Standard specific conductance (1/ohm-cm)
λ	Equivalent ionic conductance (S-cm³/eq)
λ	Wavelength of radiation (cm)
Λ	Equivalent conductance (S-cm³/eq)
μ	Ionic strength of solution
μ	Liquid viscosity
μ	Mean value for an infinite data set
π	Osmotic pressure (atm)
θ	Temperature coefficient in generalized gas law (atm-L/K)
ρ	Density of soil or liquid (g/cm³)
ρ	Specific resistance (ohm-cm)
σ	Standard deviation for an infinite data set
ζ	Zeta potential (mV)

INDEX